AF738640

Biorefinery

Juan-Rodrigo Bastidas-Oyanedel
Jens Ejbye Schmidt
Editors

Biorefinery

Integrated Sustainable Processes for Biomass Conversion to Biomaterials, Biofuels, and Fertilizers

Springer

Editors
Juan-Rodrigo Bastidas-Oyanedel
Chemical Engineering Department
Khalifa University, Masdar Campus
Abu Dhabi, United Arab Emirates

Jens Ejbye Schmidt
SDU-Department of Chemical
Engineering, Biotechnology, and
Environmental Technology
University of Southern Denmark
Odense M, Denmark

ISBN 978-3-030-10960-8 ISBN 978-3-030-10961-5 (eBook)
https://doi.org/10.1007/978-3-030-10961-5

Library of Congress Control Number: 2019934724

© Springer Nature Switzerland AG 2019
This work is subject to copyright. All rights are reserved by the Publisher, whether the whole or part of the material is concerned, specifically the rights of translation, reprinting, reuse of illustrations, recitation, broadcasting, reproduction on microfilms or in any other physical way, and transmission or information storage and retrieval, electronic adaptation, computer software, or by similar or dissimilar methodology now known or hereafter developed.
The use of general descriptive names, registered names, trademarks, service marks, etc. in this publication does not imply, even in the absence of a specific statement, that such names are exempt from the relevant protective laws and regulations and therefore free for general use.
The publisher, the authors, and the editors are safe to assume that the advice and information in this book are believed to be true and accurate at the date of publication. Neither the publisher nor the authors or the editors give a warranty, express or implied, with respect to the material contained herein or for any errors or omissions that may have been made. The publisher remains neutral with regard to jurisdictional claims in published maps and institutional affiliations.

This Springer imprint is published by the registered company Springer Nature Switzerland AG
The registered company address is: Gewerbestrasse 11, 6330 Cham, Switzerland

Contents

Contributors

Jesper Ahrenfeldt Department of Chemical and Biochemical Engineering, Technical University of Denmark, DTU Risø Campus, Roskilde, Denmark

Mariam Ali AlMahri Department of Applied Mathematics and Sciences, College of Science, Khalifa University of Science and Technology, Abu Dhabi, United Arab Emirates

Saleha Almardeai Chemical Engineering Department, Masdar Institute, Khalifa University of Science and Technology, Abu Dhabi, United Arab Emirates

I. M. AlNashef Department of Chemical Engineering, Masdar City Campus, Khalifa University of Science and Technology, Abu Dhabi, United Arab Emirates

Mashael Alshehhi Chemical Engineering Department, Petroleum Institute, Khalifa University of Science and Technology, Abu Dhabi, United Arab Emirates

J. Aravind School of Bioscience and Technology, Vellore Institute of Technology, Vellore, Tamil Nadu, India

Muhammad Tahir Ashraf Department of Chemical Engineering, Khalifa University of Science and Technology, Abu Dhabi, United Arab Emirates

Alexandre Babin BioEngine—Research Team on Green Process Engineering and Biorefineries, Chemical Engineering Department, Université Laval, Québec, QC, Canada

Robert M. Baldwin National Bioenergy Center, National Renewable Energy Laboratory, Golden, CO, USA

Fawzi Banat Department of Chemical Engineering, Khalifa University, Abu Dhabi, United Arab Emirates

Juan-Rodrigo Bastidas-Oyanedel Chemical Engineering Department, Masdar Institute, Khalifa University of Science and Technology, Abu Dhabi, United Arab Emirates

Chemical Engineering Department, Petroleum Institute, Khalifa University of Science and Technology, Abu Dhabi, United Arab Emirates

Niclas Scott Bentsen Department of Geosciences and Natural Resource Management, University of Copenhagen, Copenhagen, Denmark

Nicolas Bernet LBE, Univ Montpellier, INRA, Narbonne, France

Tomasz Bochenski Department of Chemistry, Khalifa University of Science and Technology, Masdar Campus, Abu Dhabi, United Arab Emirates

Fabian Bonk Chemical Engineering Department, Masdar Institute, Khalifa University of Science and Technology, Abu Dhabi, United Arab Emirates

Department of Environmental Microbiology, UFZ-Helmholtz Centre for Environmental Research, Leipzig, Germany

J. Jed Brown Center for Sustainable Development, College of Arts and Sciences, Qatar University, Doha, Qatar

Grzegorz Przemyslaw Brudecki Chemical Engineering Department, Masdar Institute, Khalifa University of Science and Technology, Masdar City, Abu Dhabi, United Arab Emirates

Eduardo Caballero Centro Regional de Estudios en Alimentos Saludables (CREAS), Valparaíso, Chile

Juan Castilla-Archilla National University of Ireland, Galway (NUI Galway), Galway, Ireland

Juan A. Cecilia Dpto. Química Inorgánica, Cristalografía y Mineralogía, Facultad de Ciencias, Universidad de Málaga, Málaga, Spain

Wui Yarn Chan Department of Chemical Engineering, Massachusetts Institute of Technology, Cambridge, MA, USA

Tanmay Chaturvedi Department of Energy Technology, Aalborg University, Esbjerg, Denmark

Lasse Røngaard Clausen Section of Thermal Energy, Department of Mechanical Engineering, Technical University of Denmark, Kgs. Lyngby, Denmark

Iwona Cybulska Chemical Engineering Department, Masdar Institute, Masdar City, Khalifa University of Science and Technology, Abu Dhabi, United Arab Emirates

Catholic University of Louvain, Louvain-la-Neuve, Belgium

Faculty of Bioscience Engineering, Earth and Life Institute, Université Catholique de Louvain, Louvain-la-Neuve, Belgium

Prosper Dzidzienyo Chemical Engineering Department, Khalifa University of Science and Technology, Masdar City Campus, Abu Dhabi, United Arab Emirates

Chuanji Fang Chemical Engineering Department, Masdar Institute, Khalifa University of Science and Technology, Masdar City, Abu Dhabi, United Arab Emirates

Thomas J. Farmer Department of Chemistry, Green Chemistry Centre of Excellence, University of York, York, UK

Rashed H. Farzanah Chemical Engineering Department, Masdar Institute, Masdar City, Khalifa University of Science and Technology, Abu Dhabi, United Arab Emirates

Xavier Flotats GIRO Joint Research Unit IRTA-UPC, Department of Agrifood Engineering and Biotechnology, Universitat Politècnica de Catalunya - UPC BarcelonaTECH, Parc Mediterrani de la Tecnologia, Castelldefels, Barcelona, Spain

César Fonseca Section for Sustainable Biotechnology, Department of Chemistry and Bioscience, Aalborg University, Copenhagen, Denmark

Christian Grundahl Frankær Department of Chemistry, University of Copenhagen, København Ø, Denmark

Daniel Gapes Scion, Te Papa Tipu Innovation Park, Rotorua, New Zealand

Cristina Garcia-Sancho Dpto. Química Inorgánica, Cristalografía y Mineralogía, Facultad de Ciencias, Universidad de Málaga, Málaga, Spain

G. M. Guebitz Institute of Environmental Biotechnology, University of Natural Resources and Life Sciences, Vienna, Tulln, Austria

M. K. Hadj-Kali Chemical Engineering Department, King Saud University, Riyadh, Saudi Arabia

Sabeera Haris Chemical Engineering Department, Masdar Institute, Khalifa University of Science and Technology, Abu Dhabi, United Arab Emirates

Ping-Heng Hsieh Department of Environmental Engineering, National Chung Hsing University, Taichung, Taiwan

L. Infelise ARTES_Innovation for Sustainable Growth, Bologna, Italy

Kichul Jung Department of Civil and Environmental Engineering, Konkuk University, Seoul, South Korea

Zsófia Kádár Section for Sustainable Biotechnology, Department of Chemistry and Bioscience, Aalborg University, Copenhagen, Denmark

J. Kazimierczak Institute of Biopolymers and Chemical Fibres, Łódź, Poland

E. Kopania Institute of Biopolymers and Chemical Fibres, Łódź, Poland

Chyi-How Lay Green Energy Development Center, Feng Chia University, Taichung, Taiwan

Bronwyn Laycock School of Chemical Engineering, The University of Queensland, St Lucia, QLD, Australia

Piet N. L. Lens National University of Ireland, Galway (NUI Galway), Galway, Ireland

Chiu-Yue Lin Green Energy Development Center, Feng Chia University, Taichung, Taiwan

Rafael Luque Departamento de Quimica Orgánica, Universidad de Cordoba, Cordoba, Spain

Pedro J. Maireles-Torres Dpto. Química Inorgánica, Cristalografía y Mineralogía, Facultad de Ciencias, Universidad de Málaga, Málaga, Spain

Roman Moscoviz LBE, Univ Montpellier, INRA, Narbonne, France

Gibson S. Nyanhongo Institute of Environmental Biotechnology, University of Natural Resources and Life Sciences, Vienna, Tulln, Austria

Adrian Oehmen School of Chemical Engineering, The University of Queensland, St Lucia, QLD, Australia

Vincent O'Flaherty National University of Ireland, Galway (NUI Galway), Galway, Ireland

Bradley D. Olsen Department of Chemical Engineering, Massachusetts Institute of Technology, Cambridge, MA, USA

Alessandro Pellis Department of Chemistry, Green Chemistry Centre of Excellence, University of York, York, UK

Steven Pratt School of Chemical Engineering, The University of Queensland, St Lucia, QLD, Australia

Maria Puig-Arnavat Department of Chemical and Biochemical Engineering, Technical University of Denmark, DTU Risø Campus, Roskilde, Denmark

Biomass Gasification Group, Chemical and Biochemical Engineering Department, Technical University of Denmark, Roskilde, Denmark

Elie Desmond-Le Quéméner LBE, Univ Montpellier, INRA, Narbonne, France

Giulia Ravenni Department of Chemical and Biochemical Engineering, Technical University of Denmark, DTU Risø Campus, Roskilde, Denmark

Zsuzsa Sárossy Department of Chemical and Biochemical Engineering, Technical University of Denmark, DTU Risø Campus, Roskilde, Denmark

Jens Ejbye Schmidt SDU-Department of Chemical Engineering, Biotechnology, and Environmental Technology, University of Southern Denmark, Odense M, Denmark

Chemical Engineering Department, Khalifa University of Science and Technology, Abu Dhabi, United Arab Emirates

Munirasu Selvaraj Department of Chemical Engineering, Khalifa University, Abu Dhabi, United Arab Emirates

Biswarup Sen School of Environmental Science and Engineering, Tianjin University, Tianjin, China

Carmen Soto Centro Regional de Estudios en Alimentos Saludables (CREAS), Valparaíso, Chile

George Stephanopoulos Department of Chemical Engineering, Massachusetts Institute of Technology, Cambridge, MA, USA

Hanifa Taher Chemical Engineering Department, Masdar Institute, A Part from Khalifa University of Science and Technology, Abu Dhabi, United Arab Emirates

Mette Hedegaard Thomsen Department of Energy Technology, Aalborg University, Esbjerg, Denmark

Tobias Pape Thomsen Department of Chemical and Biochemical Engineering, Technical University of Denmark, DTU Risø Campus, Roskilde, Denmark

Ana Inés Torres Instituto de Ingenieńa Qúımica, Facultad de Ingenieńa, Universidad de la República, Montevideo, Uruguay

Eric Trably LBE, Univ Montpellier, INRA, Narbonne, France

Luigi-Jules Vandi School of Chemical Engineering, The University of Queensland, St Lucia, QLD, Australia

Céline Vaneeckhaute BioEngine—Research Team on Green Process Engineering and Biorefineries, Chemical Engineering Department, Université Laval, Québec, QC, Canada

CentrEau, Centre de recherche sur l'eau, Université Laval, Québec, QC, Canada

Eric Walling BioEngine—Research Team on Green Process Engineering and Biorefineries, Chemical Engineering Department, Université Laval, Québec, QC, Canada

CentrEau, Centre de recherche sur l'eau, Université Laval, Québec, QC, Canada

I. Wazeer Chemical Engineering Department, King Saud University, Riyadh, Saudi Arabia

Alan Werker Promiko AB, Lomma, Sweden

J. Wietecha Institute of Biopolymers and Chemical Fibres, Łódź, Poland

Ahmed F. Yousef Chemistry Department, Masdar Institute, Khalifa University of Science and Technology, Abu Dhabi, United Arab Emirates

Ahasa Yousuf Chemical Engineering Department, Masdar Institute, Khalifa University of Science and Technology, Abu Dhabi, United Arab Emirates

Raymond Jianxiong Zeng Fujian Provincial Key Laboratory of Soil Environmental Health and Regulation, College of Resources and Environment, Fujian Agriculture and Forestry University, Fuzhou, Fujian, China

Fang Zhang Fujian Provincial Key Laboratory of Soil Environmental Health and Regulation, College of Resources and Environment, Fujian Agriculture and Forestry University, Fuzhou, Fujian, China

Part I
Biorefineries: General Description

Biorefineries: Industrial Innovation and Tendencies

Juan Castilla-Archilla, Vincent O'Flaherty, and Piet N. L. Lens

1 Introduction

1.1 Background

Over the last number of decades, sustainability has become a key consideration. This has largely been driven by increased environmental awareness and decreasing reservoirs of fossil fuels and natural resources, together with the social benefits of a lower polluted environment. The result of this focus on sustainability is the ongoing development of a vast range of technologies for the valorization (waste remediation coupled to the recovery of high-value products) of industrial waste streams, many of which are fundamentally based on microbial processes. Furthermore, the use of a waste as the source of material will reduce the overall costs of the plant.

Biological treatments have been central to the waste valorization strategies in a wide range of industries and applications. This approach has greatly advanced in recent years through extensive research and enhanced understanding of microbial processes. The increased use of these technologies led to a new and emerging sector—industrial biorefining. This development was supported together with the advance of other physical and chemical technologies (Rocha et al. 2015; Safari et al. 2017). These different combinations result in a greater processing system capacity. To date, a huge range of industrial wastes have been studied and can be exploited for the production of energy, green chemicals and material compounds (Xie et al. 2012; Fava et al. 2015) for biorefining purposes.

From the point of view of sustainability, the biorefining concept aims to achieve zero waste production and a circular economy, using residual waste as a raw material. Essential to this biorefinery concept is the constant availability of biomass

J. Castilla-Archilla (✉) · V. O'Flaherty · P. N. L. Lens
National University of Ireland, Galway (NUI Galway), University Road, Galway, Ireland
e-mail: j.castillaarchilla1@nuigalway.ie

© Springer Nature Switzerland AG 2019

J.-R. Bastidas-Oyanedel, J. E. Schmidt (eds.), *Biorefinery*,
https://doi.org/10.1007/978-3-030-10961-5_1

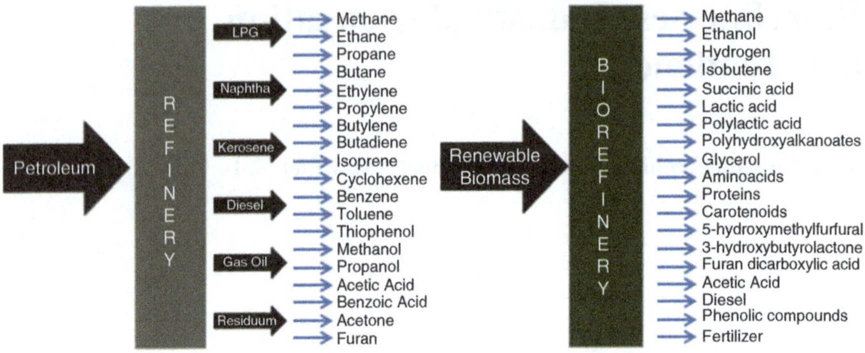

Fig. 1 Overview of products and chemical compounds that can be obtained in a traditional petroleum-based refinery and in a biorefinery plant

resources. To achieve this, it is vital that biomass is available throughout the year or can be stored whilst maintaining its characteristics. Otherwise, seasonal variations will result in inefficient processing. Currently, many laboratories and industrial companies are improving and applying the concept of biorefinery which can substitute petroleum derivative materials. Some of these studies are merging this biorefinery concept with the treatment of waste for the production of chemical compounds.

The biorefinery concept is analogous to a traditional petroleum refinery concept (Fig. 1), with the fractionation of the biomass into a range of products such as chemical building blocks, biomaterials or biofuels. The term "biorefinery" or "green biorefinery" is broadly used. It is sometimes difficult to draw a line between industries using a bioprocess and a biorefinery plant by itself.

The term "biorefinery" has become a trending topic in the last few years, and its use has started to be quite extended. A good example of this is the increasing number of publications using this term (Fig. 2). This chapter aims to analyse the various changes and evolution over time in the biorefinery industry, with a focus on collaborations, partnerships and joint ventures between industries and sectors. Some successful and nonsuccessful stories from the laboratory to the industrial production plant are presented as well.

1.2 What Is a Biorefinery?

The fuel production using a renewable source of biomass was developed during the latest of the nineteenth and beginning of the twentieth century, with the production of ethanol. In those days, the modern oil industry was in its infancy (Speight 2008). However, the use of ethanol as a fuel was quickly displaced by the use of gasoline as a cheaper fuel. It was until 1975 when the Brazil government launched the programme "Proalcool" to reduce the dependence on oil import using sugarcane for the production of ethanol (Amorim and Lopes 2005). However, this process was

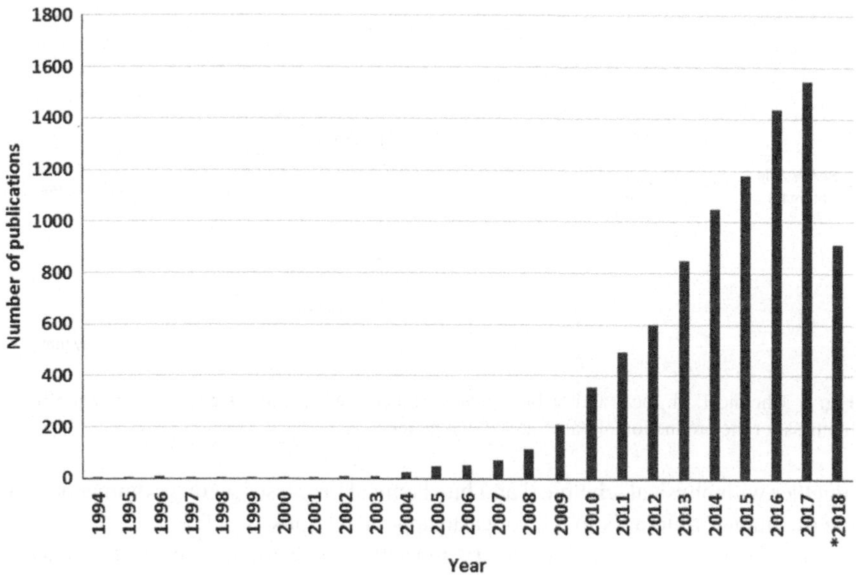

Fig. 2 Number of publications with "biorefinery" as a key word in ScienceDirect since the beginning of the 1990s until 2018 (*first quarter of 2018)

Table 1 Biorefinery and biorefining concept broadly used in the actual industry

Author	Year	Definition
Kamm et al.	2008	Biorefining involves the transfer of efficiency and logic from the fossil-based chemistry and substance conversion industry as well as energy production to the biomass industry
Biopol	2008	Biorefinery is the sustainable processing of biomass into a spectrum of marketable products
Biorefinery Euroview	2008	Biorefineries are integrated bio-based industries, using a variety of different technologies to produce chemicals, biofuels, food and feed ingredients, biomaterials and power from biomass raw materials
IEA Bioenergy Task 42	2009	Biorefining is the sustainable processing of biomass into a spectrum of bio-based products and energy
Morais and Bogel-Lukasik	2013	Green biorefinery is based on the ideal processes that use a non-polluting material as a feedstock, there is no production of any contaminant and solvents are avoided

focused on the fuel production and not yet on the biorefinery idea. This definition would be more developed early in the twenty-first century up to some of the current concepts Table 1.

The concept of a biomass-based refining process first appears at the beginning of the 1980s using biomass and suppressing the methane production to obtain different end products, organic chemicals and liquid fuels. In this way, a move was made from petroleum-based refinery to biorefining using biomass as a feedstock (Levy et al. 1981). In the mid-1980s, different applications of this biorefining process

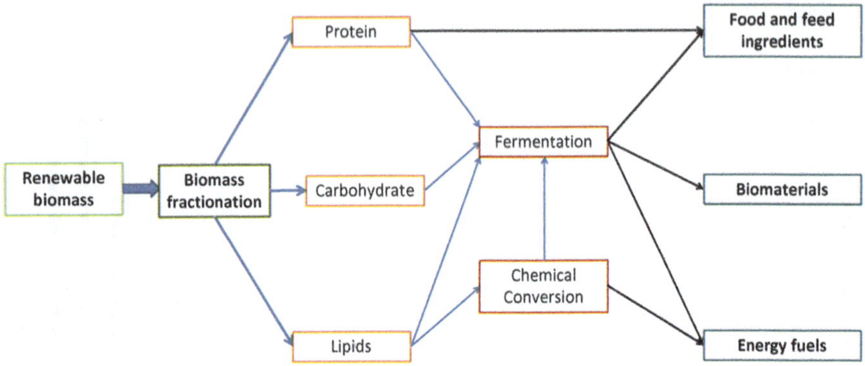

Fig. 3 Chemical, physical and/or biological treatment before and after the fermentation using biomass for the production of value-added compounds

together with physical, chemical and biochemical processes were recommended for the treatment of biomass to a zero waste approach (Koukios 1985).

This developed into the biorefinery concept at the beginning of the 1990s when the refining process exploited the entire crop as opposed to just the straw which was previously the case. This idea was first put forward by Galletti in 1991 (Galletti 1991). The concept of a green biorefinery was used for a factory producing fodder pellets. The plant juice obtained during the process was used as a raw material for the production of amino acids (Andersen and Kiel 2000). This concept of green biorefinery is used in the classification made by the International Energy Agency (IEA) Bioenergy Task 42. The biorefineries are classified by this office as platforms (such as sugar or syngas), products (energy or materials), feedstock (energy crops or residues) and conversion processes (biochemical, thermochemical, chemical or mechanical) (Cherubini et al. 2009).

A different meaning for the term "green biorefinery" was given by Morais and Bogel-Lukasik (2013) revising it into the biorefinery concept, based on the application of the 12 principles of green chemistry (Anastas and Warner 1998). The use of the reference "green" was based on the ideal processes that use non-polluting materials such as a feedstock, where there is no production of any contaminant during the processing steps and solvents are avoided.

A biorefinery plant combines different conversion processes and manufacturing technologies, as in Fig. 3, to give a balanced operation for the transformation of renewable biomass into a spectrum of intermediate and final products (fuels or materials) which aim to substitute the petrol-based products.

1.3 Classification of Biorefineries

Biorefineries have been classified in the literature by different conditions, platform technologies, final products, feedstock and conversion process (Sect. 1.2). However, the biorefinery plants have started to be more complex, for example, with different combinations of conversion processes and simultaneous production of material and energy (Cherubini 2010). Because of this, the biorefineries which are presented in this chapter are classified by the origin of the feedstock used as a raw material. Three different categories for biorefineries can be used, depending on whether the raw material used as a feedstock comes from a crop, waste or microalgae (first, second and third generation of biorefineries).

1.3.1 First-Generation Biorefineries

The first generation of biorefineries use crops as a raw material, such as maize, corn, sugar beet or sugarcane (Erwin and Steve 2015; Global-Bioenergies 2016a; Brinkman et al. 2017). The advantages of the first-generation biorefineries are the easy fermentation of the raw feedstock with a lower pretreatment required in comparison with the other generations, for example, for the production of ethanol, succinic acid or polyhydroxyalkanoates. However, the disadvantage of these biorefineries is that the raw material used as a feedstock can enter into competition with food production for human and/or animal consumption or with farmland dedicated to growing the plants used as raw material. Also, there are some concerns with the use of forest lands which could be destroyed by using them to grow the biorefinery crops (Fargione et al. 2008).

This feedstock can, on the other hand, come from a surplus of the current production through greater and more sophisticated agricultural techniques, which lead to a higher production yield (Mudombi et al. 2018). A sustainable way for this first-generation biorefinery would be by exploiting unused agricultural land, lands that used to be farmland or barren lands and have no impact on the environment (Simha 2015).

1.3.2 Second-Generation Biorefineries

The feedstock for second-generation biorefinery plants is residual waste from industries or abundant agro-industrial and forestry waste. This feedstock could vary from easily biodegradable raw materials, such as whey milk or bakery food waste, to more complex substrates like lignocellulosic materials which require a pretreatment step to break down the complex sugars into simple carbohydrates.

The main advantage of second-generation biorefineries is the noncompetition with food or farmland for human or animal feeding (Zhang 2013). Furthermore, the biorefinery process is an alternative method for some of the conventional strategies for solid waste management, such as landfilling or burning (Nizami et al. 2017). On the other hand, the factor of the availability of the feedstock has to be taken into account. This source could be seasonal, which negatively impacts the biorefinery plant if other feedstock cannot be used. To be as productive as possible, the plant must be producing all year round. Also, the amount of waste has to match to the volume of feedstock required.

1.3.3 Third-Generation Biorefineries

The third generation of biorefineries relies on the production of microalgae or cyanobacteria using nutrients, wastes, syngas and/or sunlight. The principal advantage of the third-generation biorefineries is the noncompetition with food or farmland. Furthermore, the same amount of biomass produced in this generation, compared with the first generation, can be obtained faster, due to the different growth periods for crops and microalgae. In addition, these third-generation biorefineries can be considered as carbon sinks.

The principal disadvantage of third-generation biorefineries is the complex technology involved, some of which are still under development and require a deeper knowledge and research to achieve an economically commercial production (Lam et al. 2018). However, the production of high valuable compounds could overcome any higher investment required in this generation, focused in sector such as fine chemicals, cosmetics or medicine (Zhu 2015; Khanra et al. 2018).

1.4 Industrial Cases

Table 2 shows some biorefineries applied around the world at industrial scale, classified by the generation type. Table 2 also includes the feedstock used, the final product and its production volume.

2 Bioethanol from First to Third Generation

2.1 Introduction

The bioethanol production had a surge in its production level at the end of the 1990s and beginning of 2000. Bioethanol is well established in the fuel market. The biggest market for ethanol production is the USA which represents 58.41% of the total market, followed by Brazil, EU, China and Canada (26.10%, 5.23%, 3.23% and 1.66%, respectively). The other 5.36% is spread across the rest of the world (Fig. 4).

Using the US market as a reference, the bioethanol production was quite constant during the twentieth century, having a huge increment from 2000 to 2010 (6.14–50.34 Mm^3) (Fig. 5). This was due to some policies introduced during the 1970s and 1980s that helped the production of ethanol to grow slightly, for example, the "Oxygenated Fuels Program and the Reformulated Gasoline" (Duffield et al. 2015), to introduce more oxygen compounds into the gasoline in order to decrease some contamination problems such as carbon monoxide and volatile organic compound emissions from cars (Vinuesa et al. 2003). The most used chemical for this was methyl tertiary butyl ether (MTBE), a petroleum derivate. To promote the use of

Table 2 Selection of some biorefineries applied at industrial scale across the world

Company	Country	Feedstock	Product	Volume produced (m^3)	Generation	Year
Pannonia Ethanol	Hungary	Corn	Ethanol	450,000	First	2012
CleanStar (CleanStar service, Novozymes and ndzilo)	Mozambique	Cassava	Ethanol	2,000	First	2012
Global Bioenergies	Germany	Sugar beet	Bio-isobutene	100	First	2017
Global Bioenergies	France	Sugar beet	Bio-isobutene	50,000	First	FUTURE
NatureWorks, Ingeo	USA	Corn/cassava/sugar cane	PLA	150,000	First	
Futerro	Belgium	–	PLA	1,500	First	2010
Reverdia (Royal DSM, global Life Science and Material Sciences and Roquette Frères)	Italy	Starch	Succinic acid	10,000	First	2012
Succinity GmbH (Corbion Puran and BASF)	Spain	–	Succinic acid	10,000	First	2014
BioAmber	Canada	Corn	Succinic acid	30,000	First	2015
St1	Finland	Potato flake industrial waste	Bioethanol 99.7%	850	Second	2008
BetaRenewables	Italy	–	Ethanol	40,000	Second	2013
GranBio	Brazil	Straw and bagasse	Ethanol	82,000	Second	2014
PDET-DSM	USA	Cellulosic	–	–	First/second	2015
St1	Sweden	Bakery food and industrial waste	Bioethanol	5,000	Second	2015
DuPont	USA	Corn stover	Ethanol	110,000	Second	2015
St1	Finland	Sawdust (wood mill residues)	Bioethanol	50,000	Second	2017
SEKAB	Sweden	Forest industry waste	Bioethanol	–	Second	–
St1	Sweden	–	Diesel	–	Second	FUTURE
Biorizon Biotech	Spain	Wastewater	Aminoacids, fertilizer	–	Third	–
Earthrise	USA	River water/CO_2/nutrients	Spirulina	500	Third	1981
Fitoplancton Marino	Spain	Sea water/CO_2/nutrients	Cosmetics	–	Third	2002

– not given

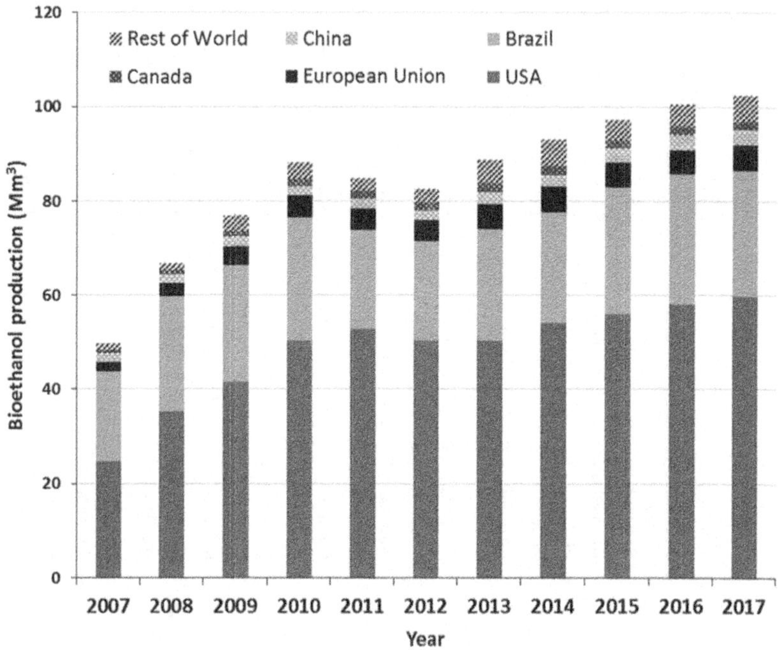

Fig. 4 Bioethanol production worldwide from 2007 to 2017 (Source: Renewable Fuels Association (RFA))

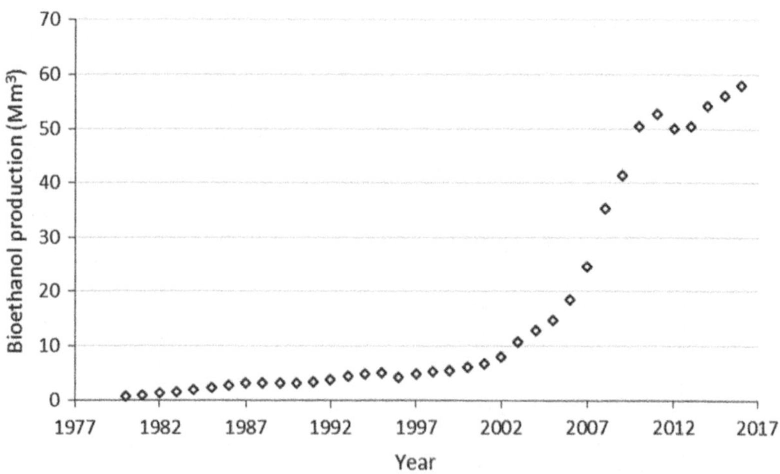

Fig. 5 Total bioethanol production in the USA since 1980 until 2017 (Source: Renewable Fuels Association (RFA))

ethanol rather than MTBE, a tax exemption was implemented (Duffield et al. 2015). Moreover, the ban on MTBE which was introduced by the state of California in 2003 was a major contributor to boost bioethanol use. Afterwards other states followed by also implementing the prohibition of MTBE, which led to a significant increment in ethanol production (Duffield et al. 2015).

Currently, in the USA there are approximately 214 plants dedicated to ethanol production. In 2017, the total production of bioethanol was 59.81 Mm^3 (RFA 2018). The majority of these plants are from first generation (199), using mainly corn as a feedstock; a few of them use sorghum. Only 15 of these plants use a residual waste as the feedstock, like cellulosic biomass, cheese whey, brewery waste, sugar waste and waste alcohol (RFA 2017).

Second-generation biorefinery plants have been gaining more attention during the last number of years. This is because of the development of the technology involved in the breaking down of complex lignocellulosic materials. The first plant was using solid waste for ethanol production with bakery waste (Wessberg and Eerola 2013). However, the most significant impact was the development of the technology to treat lignocellulosic material; the first plant using this source was built in Italy by Beta Renewables in 2013 (Nickel Institute 2014). These technologies allowed for the production of ethanol using non-food biomass. Also, some companies combined the use of first- and second-generation materials (Ward 2015; Dias et al. 2014).

2.2 *First-Generation Bioethanol Plant*

2.2.1 CleanStar

A good example of a sustainable first-generation plant was located in Mozambique, which used cassava as the feedstock. Local farmers produce a surplus of this crop by the application of more sophisticated and modern agricultural techniques, which lead to a higher yield using the same surface area (Novozymes 2012). This project was impulsed to substitute the use of charcoal by ethanol for cooking as a less polluting fuel in some areas in Maputo (Mozambique). However, this approach has some inconveniences related to the scale up of the process with discontinuous supply of ethanol, the design and the quality of the cooking stoves (Mudombi et al. 2018).

2.2.2 Pannonia Ethanol

The largest biorefinery in Europe for ethanol production is currently Pannonia (Hungary) owned by Ethanol Europe Renewables. This plant is using more than a million tonnes of maize and obtaining mainly ethanol with a production of 450,000 m^3

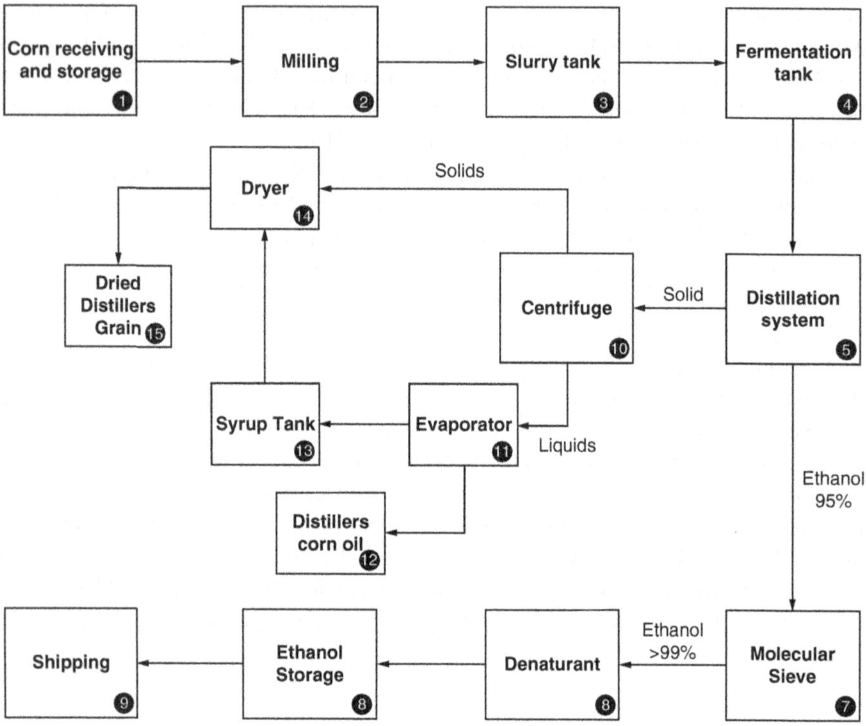

Fig. 6 Schematic industrial plant diagram for ethanol production from first-generation biorefinery using corn as a feedstock (adapted from http://www.pannoniaethanol.com/en/)

per year. At the same time, as co-products they obtain 325,000 tonnes of livestock feed and 10,000 m^3 of corn oil as by-product (Pannonia 2017). The commissioning of this plant started in 2012 and is currently the biggest ethanol producer in Europe. The plant uses corn in a spread area of 100 km around the plant.

Figure 6 shows a block diagram of the plant. The corn is delivered by trucks to the plant and discharged directly from the trucks by gravity (**1**). This storage terminal is operated by Cargill with a capacity to have stock stored for the plant for 2–3 weeks (Schill 2010). The grains are milled into flour (**2**). This powder is mixed with the enzymes and water in a previous step of the fermentation. Part of the grain is solubilized into the liquid fraction, and the mixture is introduced in the fermentation tank (**3**). In the fermentation tank, the saccharification using enzymes and fermentation by the yeast are carried out simultaneously (**4**).

The fermented mash is moved to a distillation system. This system consists of the separation of the solid and the liquid fraction and the concentration of the ethanol close to 95% (**5**). A molecular sieve is used to concentrate the ethanol to at least 99% (**6**). The ethanol is denatured (**7**) and stored prior transportation (**8**). The plant has a loading platform for trains and loading dock for ships (**9**).

The bagasse after the distillation process is centrifuged to separate the solid fraction in a wet cake and a liquid fraction **(10)**. An evaporator is used with the liquid fraction **(11)** to recover corn oil **(12),** and the rest of the fraction is collected in a syrup tank **(13)**. This fraction together with the solid fraction coming from the centrifuge is dried **(14)** to obtain dried distiller grains used as an animal feeding **(15)**.

2.3 Second-Generation Bioethanol Plants

2.3.1 St1 Company

The company St1 has developed different technologies for ethanol production using different kinds of solid waste. They have registered three different technologies depending on the nature of the solid waste: Etanolix®, Cellunolix® and Bionolix® (st1 2015a, 2018). There are currently five plants using the Etanolix technology (four in Finland and one in Sweden). The feedstock for these plants is industrial bakery waste and industrial process residues, as well as out-of-date waste bread from shops and markets (st1 2015b). The idea of this technology is the production of the ethanol close to the source of the feedstock, with small ethanol production facilities which produce 85% ethanol. After this wide range production, the ethanol is collected in one of the plants (Hamina) to be dehydrated up to 99.5% and blended with petrol to produce the final transportation fuel (st1 2008). Currently, the annual capacity of this plant in Hamina has a production of 88,000 m^3 of ethanol, with an expected capacity close to 300,000 m^3 by 2020 (Werner 2015).

The second technology developed by this company is Cellunolix®, which focusses heavily on lignocellulosic feedstock. There is currently an industrial plant in operation which works with this technology in Finland, using sawdust (wood mill residues) as a feedstock (Novozymes 2015b). Ethanol being the major product, they produce other by-products with high value like lignin, syrup, turpentine and furfural.

The latest technology (Bionolix®) developed on an industrial scale is for using domestic and commercial biowaste as feedstock (Wessberg and Eerola 2013). Through this new source, the company expects to diversify its biofuel production concept.

2.3.2 Proesa®

The first plant built for ethanol production from second-generation biofuels uses lignocellulosic materials and was built in Italy by Beta Renewables in 2013 with a production capacity of 75,000 m^3 per year (BetaRenewables 2013; Nickel Institute 2014). This company owns the Proesa® Technology to convert the lignocellulosic material into simple sugars. Prior to this industrial step, a pilot plant was built in Alessandria (Italy) with a capacity of 1 tonne per day of feedstock treated (BetaRenewables 2018).

The same technology (Proesa®) is applied by GranBio in its second-generation ethanol plant in Alagoas (Brazil). This plant has a production capacity of 82,000 m^3 per year using straw and sugarcane bagasse as a feedstock. In addition to the Proesa® Technology, this plant uses the enzymes from Novozymes (Denmark) and yeast from DSM (Holland) (GranBio 2014).

GranBio invested in 2013 190 million dollars in the plant and 75 million dollars on the steam and electricity generation system (GranBio 2014). The plant uses straw and sugarcane bagasse as a feedstock. The material is subjected to an explosion pretreatment prior to an enzymatic hydrolysis to obtain the simple sugars. These sugars are fermented into ethanol and distilled into lignin, with high calorific value, and vinasse (GranBio 2018).

2.3.3 SEKAB Technology

Another company that has been working on ethanol production using lignocellulosic material as a feedstock is SEKAB. The company was the result of a joint venture in 1985 between a pulp and paper company, MoDo, and a chemical company, Berol Kemi. After that, the company was favoured due to some of the actions taken by the Swedish government to improve the cellulosic ethanol production (Hellsmark et al. 2016a). A pilot plant was develop and owned by SEKAB and two Swedish universities, Luleå University of Technology and Umeå University (Mossberg et al. 2018). The plant was managed by SEKAB and EnergiCentrum Norr. In 2005, a new demo plant was inaugurated, and SEKAB merged with EnergiCentrum Norr; after that the company was named as SEKAB E-technology (Hellsmark et al. 2016a).

By 2008 the company had been investigating the use of diluted acid hydrolysis for cellulosic feedstock. They shifted this process to enzymatic hydrolysis treatment. However, the main problem for this was the price of the enzymes and finding pentose-fermenting organisms (Catalysts 2008b; Sekab 2015). The idea was to use forestry residues, branches and tree tops as a raw material for the ethanol production. This was delayed because the technology was not mature enough for a production plant using enzymatic hydrolysis, which has higher potential in terms of yield and energy efficiency compared with diluted acid hydrolysis (Catalysts 2008a). In that year, 2008, the company had a turnover of 250 million of dollars, mainly due to the gain of the production of ethanol with a first-generation process (Hellsmark et al. 2016a). The commercial plant was planned to be built by 2008–2009, but this was delayed with the start of the crisis and the drop of the gasoline price (Hellsmark et al. 2016b).

After that, in 2010, the pilot plant was fully operational for production of ethanol using lignocellulosic feedstock. The company, besides ethanol, targeted other compounds such as acetic acid, acetaldehyde and ethyl acetate (Catalysts 2010). By 2013, the CelluTech technology developed by SEKAB and some of the Swedish universities was marketed with Davy Process Technology and SEKAB.

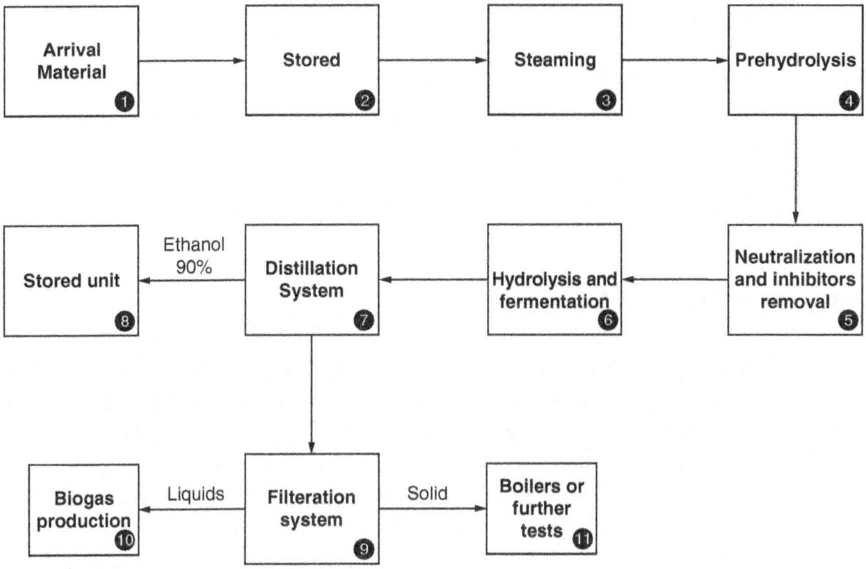

Fig. 7 Schematic demo plant for ethanol production using lignocellulosic material (forestry waste) for ethanol production (Source: website Sekab)

This technology can be adapted to a different range of products like wood, straw, corn residues or bagasse (Sekab 2013). At the end of 2013, the pilot plant has been operated as an open test centre (Mossberg et al. 2018).

The demo plant owned by SEKAB, Luleå University of Technology and Umeå University was using the CelluTech technology, where other technologies related with this process were developed. In this plant (see Fig. 7), the cellulose waste arrives to the plant **(1)**, the material is screened and a fan blows it to the silo on the roof **(2)**. Hot steam is added to preheat the raw material and expel the air; the diluted acid is added at this point as well **(3)**. Hemicellulose is dissolved in the acid at high temperature (170 °C); a second treatment for cellulose is used by increasing the temperature up to 200 °C **(4)**. The mash is neutralized, and also inhibitors produced in the previous step are removed in this step **(5)**. The hydrolysis and fermentation step can be carried out separately or simultaneously **(6)**. A distillation system is used to concentrate and recover the ethanol **(7)**; the stillage is sent to a process to separate the liquid and solid fractions. After the distillation, the ethanol concentration is around 90%, this is stored and used in boilers or for tests **(8)**.

A filtration system is used to separate the solids from the stillage **(9)**; the solids are used for energy production or as a substrate for an advanced recovery of chemical compounds. The process water is used in the biogas plant at the industrial site **(10)**. The solid material is removed and can be used for some different tests or in the boilers on site **(11)**.

2.3.4 Pelagonia Project

Ethanol Europe Renewables in collaboration with DuPont announced in 2015 the construction of the largest cellulosic ethanol plant in the world. The plant will be based in the Republic of Macedonia, using a site of around 20,000 hectares that was used in the past to grow tobacco and currently partially abandoned. The plant will have a total production capacity close to 114,000 m^3 (DuPont 2015).

2.3.5 POET-DSM Technology

Another second-generation biorefinery plant for ethanol production is POET-DSM, located in Emmetsburg (Iowa, USA), a joint venture between POET and DSM in collaboration with Novozymes and NREL as a part of the Liberty project. This plant for cellulosic biomass treatment is integrated with a first-generation biorefinery. The total production of ethanol from the cellulosic source is close to 95–75,000 m^3 per year. This technology is in operation since 2008 on pilot scale in Scotland. By running this pilot plant, the company solved some issues related to the management of feedstock and its storage, making it available throughout the year (Dishman 2017).

Figure 8 presents a scheme of the plant. The whole crop is used after collection (**1**), the lignocellulosic material is used in the second-generation biorefinery plant,

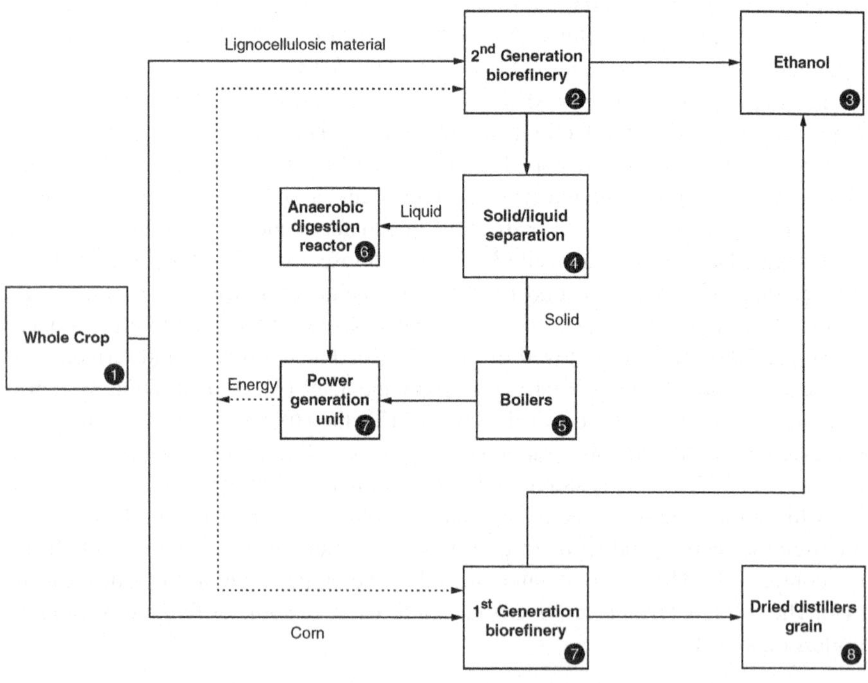

Fig. 8 Integrated plant with first- and second-generation biorefinery using the whole crop of corn as a feedstock (Adapted from Project Peer Review from DOE)

(2) and the corn is used in the first-generation biorefinery plant **(7)**. In the second-generation plant, the stillage obtained during the ethanol production is separated **(3)** into a solid fraction which is used as a fuel of the boilers **(4)** and a liquid fraction which is treated in an anaerobic reactor for biogas production **(5)**. This is part of the power generation unit to produce steam and energy **(6)** that are used in both plants. The ethanol is stored together with the ethanol coming from the first-generation plant **(7)**.

In the case of the first-generation biorefinery **(8)**, corn is used for the production of ethanol. Dried distillers grain is obtained as a by-product which is used as an animal feedstock **(9)** (Ward 2015; Dishman 2017).

2.4 Third-Generation Bioethanol: Algenol

Algenol built a pilot-scale Direct to Ethanol® integrated biorefinery. The technology of Algenol consists of the production of ethanol using carbon dioxide and seawater with a genetically engineered microorganism (Algenol 2010). By 2009, Algenol made a joint venture with Dow for the development of a pilot plant in Fort Myers (Florida, USA), with a capacity of approximately 38 m^3, where Dow would provide the materials and films for the system ('Dow continues specialities shift,' 2009; Elliott 2016). However, in 2011 when the construction of the plant started, Algenol announced the end of the joint venture with Dow (McMahon 2012; Chemicals-Technology 2012).

Algenol announced an investment of 190 million dollars to build a plant with a production capacity of approximately 1400 m^3 by 2015 (Khanra et al. 2018; Bailey 2015). In October of 2015, the main goal of the company changed and shifted its focus from fuel production to carbon capture and chemical compound production (Lane 2015).

3 Biopolymers

3.1 Polylactic Acid (PLA)

3.1.1 Chemtech

Sulzer Chemtech opened a pilot plant in Switzerland in 2011. The total production capacity of this plant was 1000 tonnes of PLA. This plant was equipped with their own technology which relied on a new cost-effective polymerization process (Corbion 2008). The aim of the plant was to show customers the polymerization process on a one-to-one basis before acquiring this technology.

Indeed, one of the developments of this company led to an improvement in the heat resistance of the current bioplastics and made them more competitive in price, employing a cost-efficient polymerization process using lactide monomers to produce high-quality PLA. This new PLA product endures temperatures of up to 180 °C, making it possible to use this material in the automotive, electronic and textile industries.

3.1.2 NatureWorks/Ingeo

The first pilot plant for PLA production was developed in Savage (USA) in 1994 by Cargill. They converted sugar via fermentation into lactic acid and produced PLA. For its scaling up and commercialization, Cargill joined with Dow Chemical to found Cargill-Dow as an equal partnership in 1997 to commercialize the PLA polymer, operating under the trade name "NatureWorks". They built an industrial manufacturing plant in Blair (Nebraska, USA) in 2001. Later, in 2005 Cargill bought out Dow's share and renamed the company "NatureWorks" and the PLA brand as "Ingeo" (Larson et al. 2010).

In 2009, NatureWorks doubled its production capability to 140,000 tonnes per year. The plant uses corn as feedstock. This feedstock is obtained in an area of 100 km around the production facility with a constant supply throughout the year. Furthermore, NatureWorks is currently involved in advanced research on the production of PLA from residues in second-generation biorefineries (Bopp 2012).

The Chemtech technology was installed in the NatureWorks production plant in 2013; this allowed to enlarge the production from 140,000 to 150,000 tonnes per year. Furthermore, with the application of this technology, NatureWorks increased the portfolio of its products (NatureWorks 2012).

By 2011, PTT Global Chemical Public Company Limited (PTT Chemical) made an investment of 150 million of dollars in NatureWorks. Together, they built a PLA production plant in Thailand designed by the Jacobs (NYSE:JEC) Engineering Group. This plant uses sugarcane and cassava roots as a feedstock. In 2012, NatureWorks and BioAmber reached an agreement to commercialize a new family of bio-based compounded polymer resin grades.

Figure 9 shows the Ingeo manufacturing plant (Erwin and Steve 2015; Vink et al. 2010). In this process, the first step is the transportation of the corn to a corn wet mill **(1)**, where the starch is separated from other components that can be recovered, such as proteins or fibres **(2)**. After that, the starch is hydrolysed to dextrose using enzymes **(3)**. The lactic acid is produced by microbial fermentation of the dextrose and some other media that can be added **(4)**. This fermentation is carried out at low pH; however, it also requires the addition of calcium hydroxide for pH correction. This mash obtained has to be neutralized in the next step using sulphuric acid **(5)**. The addition of the sulphuric acid leads to the precipitation of calcium sulphate (gypsum). The gypsum is recovered by filtration **(6)**, and the lactic acid is concentrated by evaporation **(7)**, where the condensate is recycled using it in the fermentation tank. The last step prior to obtaining the pure lactic acid is its purification **(8)**.

The pure lactic acid is then used in the lactide/Ingeo production **(9)**. The first step in this part of the process is to remove the water **(10)** prior to condensation to produce

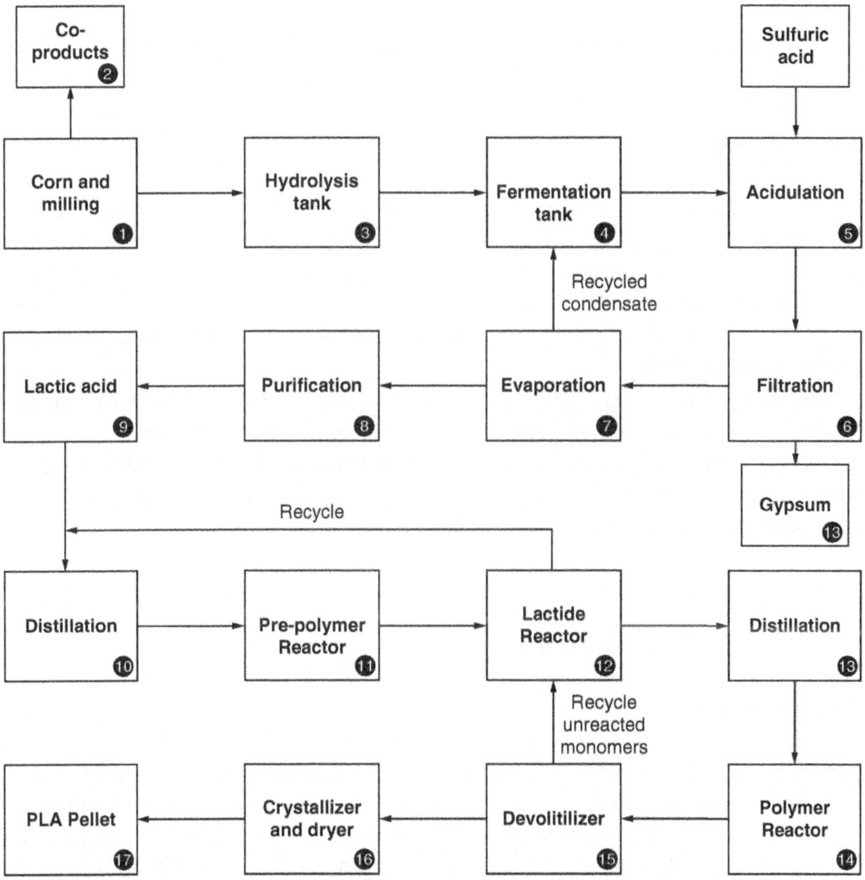

Fig. 9 Diagram of polylactic acid (PLA) production using the Ingeo process

low molecular weight PLA pre-polymer **(11)**. The pre-polymer is catalytically converted to a mixture of lactides **(12)** and purified by vacuum distillation and melt crystallization **(13)**. In the polymerization reactor, the Ingeo molecules are produced using ring-opening lactide polymerization **(14),** which avoids the use of any solvents. A devolatilizer is used to remove unreacted lactide and other molecules from the mixture containing the PLA **(15)**. This is crystallized and dried in the next steps **(16)** to obtain the final PLA pellet **(17)**.

3.1.3 Futerro

In September 2007, Galactic and Total Petrochemicals set up a partnership for the development of Futerro. They opened a production plant in Belgium in 2010 with a capacity of 2000 tonnes per year of polylactic acid. This investment was about 15 million euros, and Galactic was the supplier of lactic acid for the PLA production and Total Petrochemicals of the polymerization technology (Laird 2011).

3.1.4 PLAneo

In 2010, a smaller plant with a capacity of 500 tonnes per year was built in Guben (Germany). This plant was built by ThyssenKrupp in conjunction with Uhde Inventa-Fischer. They are currently building a plant with its patent product (PLAneo) in Changchun (China). The commissioning of the plant is scheduled for the first quarter of 2018. This plant will have a capacity of 10,000 tonnes per year (Thyssenkrupp 2016).

This process relies on a polymerization technology developed by Purac (current Corbion, owned by Dutch CSM group). Corbion built a production plant of D- and L-lactides with a production capacity of 75,000 tonnes in Rayong (Thailand). In addition to this, in collaboration with Total, Purac has recently begun building a PLA production with a capacity of 75,000 tonnes at the same location.

3.2 Polyhydroxyalkanoates

3.2.1 Metabolix

Metabolix, founded in 1992, was dedicated to the development of low-cost PHA production; this technology was developed at the Massachusetts Institute of Technology (McCarthy 2003). They developed different technologies and numerous patents related to the use of enzymes for PHA production. In 2006, Metabolix formed an alliance with the Archer Daniels Midland Company (ADM) under the name Telles. They announced that they will produce 50,000 tonnes of commercial grade Mirel annually commencing in the second quarter of 2009 (DiGregorio 2009), although the plant wasn't in operation until 2010. However, the closure of the plant and their withdrawal from the industry were announced in February 2012 by Mark Bemis (president of ADM) due to uncertainty around projected capital and production (Laird 2012; Metabolix 2012). From this point, Metabolix shifted its focus onto additive solutions based on PHA (Additives For Polymers 2014) developing other uses for the PHA such as copolymers with PVC or PLA resins (Kann 2016; Additives For Polymers 2012).

By 2016, Metabolix established a partnership with South Korea's CJ CheilJedang Corp to found, construct and operate a 10,000 tonnes PHA production per year. This plant would use the Metabolix's PHA technology and would be located in Iowa (USA) (Additives For Polymers 2016; Hazarika 2016). Nevertheless, Metabolix completely sold its intellectual property concerning PHA, together with certain related equipment and inventory to CJ CheilJedang Corp in August 2016 for ten million dollars (Laird 2018).

3.2.2 Bio-on: MINERV

Bio-on is an Italian Intellectual Property Company (IPC). This company works with bio-fermentation technologies involved in bioplastic and sustainable chemical production. Bio-on sells its PHA product under the title of MINERV; this includes over 100 different monomers with different properties, such as melting points ranging from 40 to 180 °C (BusinessWire 2011). This technology is focused on certain fields such as the toy, cosmetic and biomedical sector, as well as food packaging (Bio-on 2015c, 2017a).

As a feedstock for the fermentation process, beet or some industrial products, such as glycerol, fats and some oils can be used. This fermentation process is based on the ability of *Ralstonia eutropha* to metabolize carbon sources and convert them into PHA (de Paula et al. 2017). They recover and purify the PHA inside the cell using their own Bio-on's technology (Bio-on 2015b).

Bio-on signed some license agreements to build plants for PHA production around the world. With SECI (part of Gruppo Industriale Maccaferri holding), they made an agreement to build the first facility for PHA production using biodiesel by-products (mainly glycerol). SECI bought this technology for four million euro to build this production plant in San Quirico (Italy) (Bio-on 2015b). Moore Capital invested 5.5 million euro in acquiring the production licence to produce PHA from sugarcane by-products in Brazil with a capacity of 10,000 tonnes per year. Also, Bio-on signed an agreement with Cristal Union (France) to build the most modern PHA production plant from sugar beet with an initial capacity of 5000 tonnes per year expandable to 10,000 tonnes (Bio-on 2015a).

Bio-on is building its own PHA production plant in Bologna (Italy) by summer 2018. The investment for this plant is around 20 million euro and will have a capacity of 1000 tonnes per year, expandable to 2000 tonnes per year (Bio-on 2017b).

4 Biochemical Building Blocks

4.1 Introduction

The US Department of Energy analysed more than 300 compounds obtained from sugars to find the top ten opportunities for the production of high-value compounds from renewable biomass (Werpy et al. 2004). This report identified 12 chemical building blocks that can be produced via fermentation or chemical reaction (Fig. 10). Aside from those chemicals, other value compounds could be obtained in a biorefinery. For example, using lignocellulosic feedstock, some aliphatic and aromatic compounds could be recovered (Kawaguchi et al. 2016).

By 2010, Bozell and Petersen made a revision of the list published by the DOE. They added some extra criteria to pick the top ten compounds. Those were more focused from a point of view of the industrial case. For example, they took off glutamic acid because of the poor improvement of new derivatives. However, a

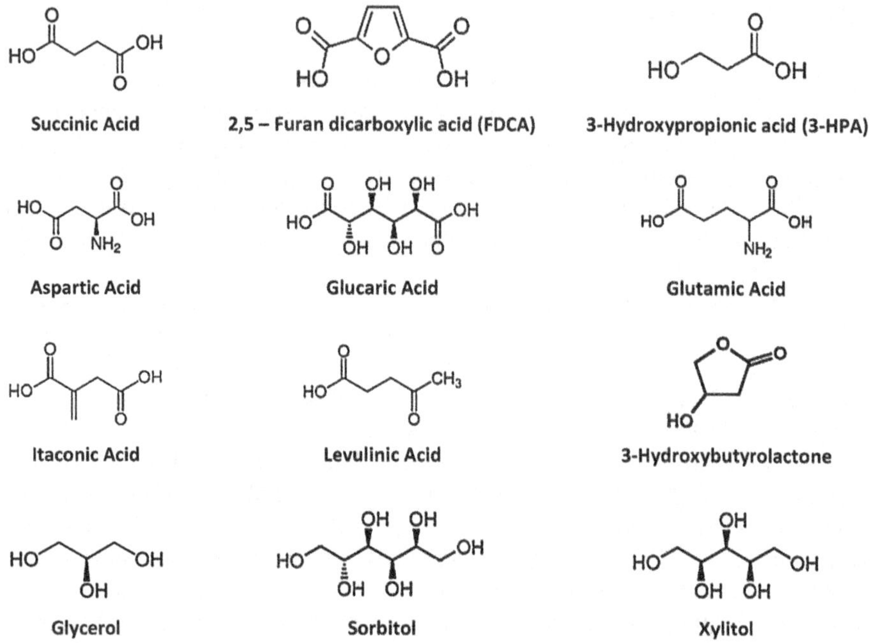

Fig. 10 Top 12 value-added chemicals from biomass selected by the National Renewable Energy Laboratory for the US Department of Energy

commercial success would return this compound in the list. The other compounds they took from the original list were aspartic, glucaric and itaconic acid. Those products were considered in the original list because of the researchers related to them, but Bozell and Petersen excluded them to avoid that limited research activities would give lower priority to other compounds that could be more interesting. They added into the new list ethanol as a biofuel and precursor of ethylene, lactic acid mainly as a precursor of PLA (see Sect. 3.1) and biohydrocarbons mostly isoprene, and they grouped the 2,5-furan dicarboxylic acid (FDCA) inside of furans together with furfural and hydroxymethylfurfural (Bozell and Petersen 2010).

Both lists have been developed to focus on the production of the denominated chemical building blocks as direct substitutes of the existing petrochemical blocks. The use of the products will lead to the same final current materials with the same characteristics as chemically synthesized ones. However, the biorefineries should develop new final products from new chemical building blocks and have completely different characteristics and properties, such as is the case of the bioplastics made entirely by PHA (see Sect. 3.2), which are made from renewable resources and in addition are completely biodegradable.

4.2 *Furandicarboxylic Acid (FDCA)*

4.2.1 Avantium: YXY Platform

Avantium developed a technology platform named YXY for the production of renewable chemicals, like furanic and levulinic for plastics and other applications. One of these materials is polyethylene furanoate (PEF), which is made by FDCA and aims to be a substitute for PET that can reduce the carbon dioxide and water footprint between 50 and 70% with the production of PEF instead of PET. The FDCA is obtained by dehydration of fructose to methoxymethylfurfural followed by an oxidation to produce the FDCA. Avantium opened a pilot plant in Geleen (The Netherlands) relying on the YXY technology in 2011 covered by 15 patents. Some of the industrial partners of Avantium by this time were NatureWorks and Teijin Aramid, for bio-based materials, and DAF Trucks for the development of biofuels (Chemicals-Techonology 2010; Sims 2011). Additionally by 2013, Avantium was collaborating on the production of PEF with the Coca-Cola Company, Danone and ALPLA Werke Alwin Lehner (Avantium 2013).

In 2016, BASF and Avantium established a joint venture that led to the creation of Synvina. This new company was focused on the construction and operation of a reference plant for 50,000 tonnes of FDCA per year (Basf 2016). In order to optimize its future commercial-scale production, they extended the work of the pilot plant from 24 to 36 months. This delayed the launching of the commercial facility by 2–3 years, until 2023–2024 (Avantium 2018).

4.2.2 MetGen: ENZINE Platform

MetGen, a Finnish company focussing on tailored enzymatic production by genetic engineering, developed a new technology platform named ENZINE (Metgen 2017). With the use of this technology, the company has developed a new chemo-enzymatic pathway for the production of FDCA (Metgen 2018).

4.3 *Levulinic Acid (LVA)*

4.3.1 GFBiochemicals

GFBiochemicals started the production of levulinic acid (LVA) in Caserta (Italy) in 2015. They have produced LVA at pilot scale via thermochemical conversion of renewable biomass since 2008 (GFBiochemicals 2015a). Corn was used as a feedstock with an initial production of 2000 tonnes per year and increasing the capacity to 10,000 tonnes per year by 2017 (GFBiochemicals 2015b; Capaldo 2015). Currently, GFBiochemicals is working in collaboration with Henkel and VITO to use lignocellulosic waste and other residues from agriculture and forestry for the production of the LVA. These products will have 70% less GHS compared to its fossil-based counterparts (GFBiochemicals 2018).

4.3.2 Bio-on

Also Bio-on, in addition to PHA research and production (see Sect. 3.2), started in 2017 to work in collaboration with the Sadam Group on the construction of a demo plant for LVA production using sugar industry by-products as a feedstock. The capacity of the demo plant in Parma (Italy) will be 5000 tonnes per year. This enzymatic process for the production of LVA is expected to be less polluting via thermochemical conversion and have a lower production cost (Bio-on 2017c).

4.4 3-Hydroxybutyrolactone (3-HP)

4.4.1 Joint Venture Cargill-Novozymes

Cargill and Novozymes started a collaboration in 2008 to develop a technology using microorganisms capable of converting renewable feedstock into 3-PH. By 2012, BASF joined the team, interested in the acrylic acid production as a final compound (Stephan 2012). In 2014, the group announced its intention to build a pilot facility for glacial acrylic acid production, after demonstrating the successful conversion of 3-HP into acrylic acid (Novozymes 2014; Bomgardner 2014). However, at the beginning of 2015, BASF left the group (Siripuram 2015). Nevertheless, Novozymes and Cargill announced that they would work on the project (Novozymes 2015a).

4.4.2 OPXBio

Dow and OPX Biotechnologies (OPXBio) announced in 2011 their agreement to develop acrylic acid from renewable feedstock. This technology consists of the fermentation of the sugars dextrose or sucrose by a genetically engineered *E. coli*; the biomass is killed and removed, followed by the decantation and recycling of the water. The acrylic acid is obtained by dehydration of the 3-HP (Tullo 2013). OPXBio announced its success in the production of acrylic acid in a 3 m^3 fermenter in 2012 (Sims 2012). Evonik, who was a previous partner of Dow in other projects, made a joint venture as well with OPXBio for the development of some bio-based specialty chemicals using OPXBio's technology EDGE (Efficiency Directed Genome Engineering). After that, OPXBio announced to scale up to 50 m^3 and industrial scale by 2017 (Tullo 2013). Nevertheless, by 2015 OPXBio sold all its fermentation-based processes and systems technology to Cargill (Cargill 2015).

4.5 *Succinic Acid*

Succinic acid, via biological production, can be more competitive than other technological routes in terms of minimum selling prices (US$/kg) (Santos et al. 2016). The first industrial production plant was set up by Reverdia in Cassano Spinola (Italy) in 2012, with a capacity of up to 10,000 tonnes per year. Another company Succinity (Corbion 2011) for succinic acid production was established in 2014 in Montmélo (Spain) GmbH with also a capacity of 10,000 tonnes per year.

BioAmber started the production of SA in 2015 in Sarnia (Ontario, Canada), with currently the highest production capacity up to 30,000 tonnes per year using corn mainly as feedstock for the production of succinic acid. BioAmber started as a spin-off from New York-based Diversified Natural Products (DNP). Originally, the company was named DNP Green Technology. The latter company worked in collaboration with France-based Agro-Industrie Recherches et Développements (ARD) for the development of a pilot-scale plant in 2010 in Pomacle (France) (Stadler and Chauvet 2018). However, in 2010 DNP Green Technology acquired the whole control of the company (shared previously with ARD) and changed the name of the corporation to BioAmber.

During the pilot-scale phase, *Escherichia coli* was used for succinic acid production; however the sensitivity of this microorganism to pH changes resulted in stability problems. It was therefore replaced by *Saccharomyces cerevisiae*, genetically modified for AS production with the company for yeast engineering production US-based Cargill (USA) (Nandy and Srivastava 2018). Because of the replacement of the bacteria with yeast, there was a delay of 1 year.

Currently, the company is working with DuPont, Evonik and Johnson Matthey Davy Technologies to develop a route from biosuccinic acid to 1,4-butanediol (BDO) and tetrahydrofuran. DNP has established a joint venture with France-based Agro-Industrie Recherches et Développements (ARD) to develop and commercialize bio-based succinic acid (ICIS 2016).

4.6 *Bio-Isobutene*

Global Bioenergies was founded in 2008 and based in France. This company is focused on the development of processes to obtain chemical building blocks from renewable sources via fermentation. In 2014 the company built its pilot plant for isobutene production with a capacity of 10 tonnes per year with a reactor of 500 L. This chemical building block was obtained using by-products from the sugar industry through fermentation. This technology was patented previously by the company (Marlière et al. 2015; Marc et al. 2015).

By the end of 2016, the company has completed the construction of its demo facility in Leuna (Germany) dedicated to the production of high-purity isobutene from renewable resources. The total capacity of this plant is 100 tonnes per year. It is the only plant dedicated to the direct fermentation of gaseous hydrocarbons. It combines two 5 m^3 fermenters and complete purification systems mimicking all aspects of a commercial-scale isobutene facility. The project was supported by a 5.7 million euros grant from the German Federal Ministry for Research, along with 4.4 million euros from a loan of a consortium of French banks (Global-Bioenergies 2016b).

Global Bioenergies is working now with Clariant (leader in specialty chemicals) as a supplier of isobutene to be used as a rheology modifier in the cosmetic industry. Thus Clariant can move into a more sustainable production, where bio-based ingredients relying on renewable resources are a key factor. Also, L'Oreal is testing this material to be used in their products (Global-Bioenergies 2017). In collaboration with Audi, the company has produced successfully e-fuels: isooctane produced from isobutene, which has been tested in their engines after more than 2 years of collaboration (Brindle 2016).

For the scaling up to industrial production level, Global Bioenergies formed a joint venture with Cristal Union to form IBN-One with the aim to build the first plant for bio-isobutene production (Global-Bioenergies 2015). This plant will have a capacity of 50,000 tonnes per year, and the engineering design will correspond to IPSB and Technip. The construction of the plant is expected by 2018 (Global-Bioenergies 2016a).

5 Microalgae Biorefineries: Third Generation

5.1 Microalgae Production

The third-generation biorefineries are gaining more attention lately, such as for fuel and energy production, where mainly just CO_2 and sunlight are required for their production. Many studies have analysed the necessity of a higher-value product than fuel for these biorefineries to be profitable (Roux et al. 2017; Moreno-Garcia et al. 2017).

However, there is a huge improvement required to achieve the industrial-scale production of this kind of biorefineries (Colling Klein et al. 2018). For example, the downstream processes in first- or second-generation biorefineries could represent around 30% of the total costs; for a microalgae biorefinery, this could be up to 60% (Lam et al. 2018).

One example is Algenol, who shifted from fuel production into higher-value compounds (Sect. 2.4). The company is selling products from food and cosmetic industry, such as natural colourants, protein and spirulina.

5.2 Spirulina Production

Spirulina is a cyanobacterium which was consumed in the past in some regions such as Africa and America. It was in the early 1960s when *Spirulina* started to gain more attention from the scientific community (Vonshak 1997). The first commercial production was by Dainippon Ink & Chemicals Inc. (DIC) in 1978 in Bangkok (Thailand). This group acquired in 1982 the *Spirulina* farms from Proteus Corporation founded in 1981 in California (USA) and renamed this company as Earthrise Nutritionals (Kumar et al. 2015).

Currently, there are more than 60 companies dedicated to *Spirulina* production focused in human food, dietary supplements, nutritionals products and colourants (Laurens 2017). The largest producer is Earthrise Nutritionals, with an outdoor facility using open ponds that has a total area of 180,000 m^2 (Maeda et al. 2018). This plant produces and refines completely the *Spirulina* until the finished products for human food with a production of 500 tonnes per year of *Spirulina* dry matter (Algae Biomass Organization 2015). As well, since 2015 Earthrise is using *Spirulina* for the production of natural blue food colouring under the commercial name Linablue® (Earthrise 2015). They made an inversion of ten million of dollars in 2013 and second investment in 2016 for a value of 13 million of dollars; the whole plant will be in operation in 2018 (Earthrise 2016).

5.3 Cosmetic

Microalgae as a rich source of active metabolites and a variety of enzymes have attracted the attention of the nutraceutical and cosmetic sector (Rizwan et al. 2018; Maeda et al. 2018). Laurens (2017) as part of the IEA Task 42 have included more than 20 industries focused on microalgae production for cosmetic products.

One example of this kind of industries is Fitoplancton Marino, which is one of the European leading companies in marine microalgae, with production of commodities for aquaculture, nutraceuticals, cosmetics and health industries. The company was established in 2002 in Cadiz (Spain) for the production of commodities. In 2011 the company was acquired by the Hisparroz Group, which is related with the Ebro Food Group, the highest company in the food processing industry in Spain and with high presence in the biotechnology sector (FitoplanctonMarino 2018). The company uses *Tetraselmis chuii* enriched in superoxide dismutase; they developed their own system to dehydrate, purify and extract this compound for application in functional foods and focused in cosmetics to reduce free radical damage to skin (Unamunzaga Escosura and Mantecon Galvez 2016). The plant has a total of 36 outdoor photobioreactors with 2 m^3 each of them, with the capacity to produce up to 6 different strains in parallel (MIRACLE 2017).

5.4 Biofertilizer Production

Microalgae compounds in crops have greater and faster effects than traditional systems used in agriculture, with the advantage that it does not generate any residue in the plant or the crop. For example, the actual use of a microalgae biorefinery is for the production of biofertilizer and aquafeed, which showed a positive effect as a fish meal (Vizcaíno et al. 2014).

Biorizon Biotech started as a spin-off of the University of Almeria (Spain). The main goal of this company is the development, production and commercialization of amino acids and fertilizer using microalgae technologies. The most important culture they were growing was *Spirulina*. In the case of the fertilizer, the company has developed its own technology, which allowed them to shift in a different profile of substances.

Through the project Regenera, Biorizon Biotech is studying the use of microalgae for wastewater treatment, whilst at the same time the biomass produced is used as a fertilizer and a source of different compounds of interest in the agro-food industry. The company is involved in other projects such as Bacagro, Valgest and Sabana as well (Biorizon-Biotech 2018).

The aim of the Sabana project, which started in December 2016, is to build a demo plant with a microalgae capacity of 300 tonnes per year of dry matter for the development of biofertilizer, bio-pesticides and feed additives via microalgae. The raw materials for the cultivation of microalgae are CO_2 and seawater, whilst the nutrients come from sewage, centrate or pig manure. The products obtained are of high value, such as aquafeed additives and bio-stimulants, as well as lower-value biofertilizers (Acien 2017).

The project is led by the University of Almeria (Spain), with some companies involved such as FCC Aqualia, GEA Westfalia and Biorizon Biotech. In total, this project has 11 partners from 5 different countries of the EU. This project aims to develop microalgae production systems to large-scale production, new products and/or application as well as the complete use of the biomass generated.

6 Conclusion

A biorefinery is a hub of technologies, which aims to convert and transform raw biomass into different products to substitute the petrol-based materials. This is being developed nowadays in some of the industries which are trying to use the whole crop or different kinds of biomass. The first definitions of the biorefining process and biorefinery came out more than 30–40 years ago. The future of the biorefineries lies in the development, implementation and combined use of the different technologies, where biological treatment plays a key role to reduce the economic cost and environmental impact.

Acknowledgements The authors are grateful for the support and advices provided by Mr. Pádraic ÓhUiginn and Dr. Tania Palmeiro (National University of Ireland, Galway). This work was supported by the Science Foundation Ireland (SFI) Research Professorship *Innovative Energy Technologies for Biofuels, Bioenergy and a Sustainable Irish Bioeconomy* (IETSBIO[3]).

References

Acien FG (2017) Sustainable integrated Algae Biorefinery for the production of bioactive compounds for Agriculture aNd Aquaculture (SABANA). 10th World Congress of Chemical Engineering, Barcelona, p 17

Additives For Polymers (2012) Metabolix offers new bio-based PHA polymeric modifiers to improve PVC performance. Addit Polym 2012(12):1–2

Additives For Polymers (2014) Metabolix dissolves German subsidiary and focuses on PHA-based additives business. Addit Polym 2014(12):7–8

Additives For Polymers (2016) Metabolix and CJ CheilJedang sign MOU to construct PHA production unit. Addit Polym 2016(6):9

Algae Biomass Organization (2015) Algae Industry Project Book 2015

Algenol (2010) Algenol integrated biorefinery for producing ethanol from hybrid algae. https://www.energy.gov/sites/prod/files/nepapub/nepa_documents/RedDont/EA-1786-FONSI-2010.pdf

Amorim H, Lopes M (2005) Ethanol production in a petroleum dependent world: the Brazilian experience. Sugar J 2005:11–14

Anastas PT, Warner JC (1998) Green chemistry: theory and practice. Oxford University Press, Oxford

Andersen M, Kiel P (2000) Integrated utilisation of green biomass in the green biorefinery. Ind Crop Prod 11(2):129–137

Avantium (2013) ALPLA joins The Coca-Cola Company and Danone in Avantium's PEF bottle development. https://www.avantium.com/press-releases/alpla-joins-coca-cola-company-danone-avantiums-pef-bottle-development/

Avantium (2018) PEF pilot phase set to be extended. Avantium, Amsterdam. https://www.avantium.com/press-releases/pef-pilot-phase-set-extended/

Bailey A (2015) Algenol ethanol plant will have capacity to produce 18 MMgy. Ethanol Producer Magazine. http://ethanolproducer.com/articles/12629/algenol-ethanol-plant-will-have-capacity-to-produce-18-mmgy

Basf (2016) Synvina: Joint venture of BASF and Avantium established. Basf, Ludwigshafen and Amsterdam. https://www.basf.com/en/company/news-and-media/news-releases/2016/10/p-16-322.html

BetaRenewables (2013) Crescentino's biorefinery grand opening. http://www.betarenewables.com/en/media-relations/press-release-detail/1?locale=en

BetaRenewables (2018) Proesa™/What is it?. http://www.betarenewables.com/en/proesa/what-is-it. Accessed April 2018

Bio-on (2015a) Agreement to realise the first France-based facility for producing PHAs biopolymers from sugar beet co-products. Bio-on, Paris and Bologna

Bio-on (2015b) World first facility for producing bioplastic from biodiesel co-product glycerol, to be realised in Italy. Industrial licence granted. Bio-on, Bologna

Bio-on (2015c) World wide project minerv SUPERTOYS. http://www.bio-on.it/immagini/minerv-supertoys/BROCHURE_SUPERTOYS_EN.pdf. Accessed April 2018

Bio-on (2017a) Bio-on's minerv bio cosmetics bioplastic obtains international NATRUE certification. http://www.bio-on.it/immagini/comunicati-finanziari/CS_37_BIO-ON_10_11_2017_NATRUE_eng.pdf

Bio-on (2017b) Construction begins of new bio-on plant for special bioplastics production. http://www.bio-on.it/immagini/comunicati-finanziari/CS_32_BIO-ON_14_09_2017_cspt_ENG.pdf

Bio-on (2017c) Project launches for sustainable low-cost production of levulinic acid, for the green chemical industry. Bio-on, Bologna

Biorizon-Biotech (2018) Biotechnology, R+D+I. http://www.biorizon.es/working-area/biotechnology-rdi/?lang=en#1489480970436-d1e9b7ee-42e9. Accessed April 2018

Bomgardner MM (2014) BASF plans pilot plant for bioacrylic acid. Chem Eng News 92(38):14

Bopp R (2012) NatureWorks® Ingeo™ polylactide: past, present and future. Biopolymers & Biocomposites Workshop, Iowa State University, Iowa
Bozell JJ, Petersen GR (2010) Technology development for the production of biobased products from biorefinery carbohydrates-the US Department of Energy's "Top 10" revisited. Green Chem 12
Brindle F (2016) Global bioenergies obtains a grant for renewable gasoline additives. Hydrocarbon Engineering. https://www.hydrocarbonengineering.com/clean-fuels/05072016/global-bio-energies-secures-grant-for-renewable-gasoline-additives-audi-engine-testing-3641/. Accessed April 2018
Brinkman MLJ, Wicke B, Faaij APC (2017) Low-ILUC-risk ethanol from Hungarian maize. Biomass Bioenergy 99:57–68
BusinessWire (2011) MINERV® PHA SC (Sugar Cane): the first 100% biodegradable plastic produced from sugar cane waste is now certified. https://www.businesswire.com/news/home/20111012005766/en. Accessed April 2018
Capaldo F (2015) GFBiochemicals starts levulinic acid output in Caserta, Italy
Cargill (2015) Cargill acquires OPX Biotechnologies' fermentation technology. Cargill, Menneapolis and Boulder. https://www.cargill.com/news/releases/2015/NA31794255.jsp
Catalysts Fo (2008a) Sekab defends technology for making ethanol from cellulose. Focus Catal 5:5
Catalysts Fo (2008b) Sekab switches to enzymes for production of cellulose ethanol. Focus Catal 9:4
Catalysts Fo (2010) Green chemicals feature high in Sekab's plans. Focus Catal 6:4
Chemicals-Technology (2012) Algenol biofuels pilot-scale integrated bio-refinery, Florida. https://www.chemicals-technology.com/projects/algenol-biofuels-integrated-bio-refinery-florida/
Chemicals-Techonology (2010) Avantium YXY pilot plant. https://www.chemicals-technology.com/projects/avantiumyxy/
Cherubini F (2010) The biorefinery concept: using biomass instead of oil for producing energy and chemicals. Energy Convers Manag 51(7):1412–1421
Cherubini F, Jungmeier G, Wellisch M, Willke T, Skiadas I, Van Ree R, de Jong E (2009) Toward a common classification approach for biorefinery systems. Biofuels Bioprod Biorefin 3:534–546
Colling Klein B, Bonomi A, Maciel Filho R (2018) Integration of microalgae production with industrial biofuel facilities: a critical review. Renew Sust Energ Rev 82:1376–1392
Corbion (2008) PURAC-SULZER develop new polymerization technology for PLA. http://www.corbion.com/media/press-releases?newsId=1785672
Corbion (2011) BASF and CSM explore a bio-based succinic acid joint venture. Corbion, Ludwigshafen and Diemen
de Paula FC, Kakazu S, de Paula CBC, Gomez JGC, Contiero J (2017) Polyhydroxyalkanoate production from crude glycerol by newly isolated Pandoraea sp. J King Saud Univ Sci 29(2):166–173
Dias M, Cavalett O, Filhob RM, Bonomi A (2014) Integrated first and second generation ethanol production from sugarcane. Chem Eng Trans 37:445–450
DiGregorio BE (2009) Biobased performance bioplastic: Mirel. Chem Biol 16(1):1–2
Dishman M (2017) U.S. Department of Energy (DOE) Bioenergy Technologies Office (BETO) 2017 Project Peer Review POET-DSM Project Liberty. https://www.energy.gov/sites/prod/files/2017/05/f34/demonstration_market_transformation_dishman_3433.pdf
Dow continues specialities shift (2009) Focus on Powder Coatings 2009(10):5–6
Duffield J, Johansson R, Meyer S (2015) U.S. ethanol: an examination of policy, production, use, distribution, and market interactions. United States Department of Agriculture, Washington, DC
DuPont (2015) DuPont celebrates the opening of the world's largest cellulosic ethanol plant. http://www.dupont.com/corporate-functions/media-center/press-releases/dupont-celebrates-opening-of-worlds-largest-cellulosic-ethanol-plant.html. Accessed April 2018
Earthrise (2015) Linablue® grand opening. Earthrise, Calipatria. http://earthrise.com/linablue-grand-opening/. Accessed 10 June 2018

Earthrise (2016) DIC announces major expansion of production capacity for Linablue®—The #1 natural blue food coloring on earth. http://earthrise.com/major-expansion-production-capacity-linablue/. Accessed 10 June 2018

Elliott DC (2016) Review of recent reports on process technology for thermochemical conversion of whole algae to liquid fuels. Algal Res 13:255–263

Erwin V, Steve D (2015) Life cycle inventory and impact assessment data for 2014 Ingeo™ polylactide production. Ind Biotechnol 11(3)

Fargione J, Hill J, Tilman D, Polasky S, Hawthorne P (2008) Land clearing and the biofuel carbon debt. Science 319(5867):1235–1238

Fava F, Totaro G, Diels L, Reis M, Duarte J, Carioca OB, Poggi-Varaldo HM, Ferreira BS (2015) Biowaste biorefinery in Europe: opportunities and research & development needs. New Biotechnol 32(1):100–108

FitoplanctonMarino (2018) Fitoplancton Marino: company. http://www.fitoplanctonmarino.com/en/company.html. Accessed 09 June 2018

Galletti GC (1991) Production and utilization of lignocellulosics: plant refinery and breeding, analysis, feeding to herbivores, and economic aspects. Springer, Dordrecht

GFBiochemicals (2015a) GFBiochemicals breakthrough levulinic acid technology ready for commercialization. http://www.gfbiochemicals.com/news/2015/03/gfbiochemicals-breakthrough-levulinic-acid-technology-ready-for-commercialization-2/. Accesed April 2018

GFBiochemicals (2015b) GFBiochemicals at BIO World Congress. http://www.gfbiochemicals.com/news/2015/07/gfbiochemicals-at-bio-world-congress/. Accesed April 2018

GFBiochemicals (2018) Greensolres process. http://www.gfbiochemicals.com/greensolres/. Accessed April 2018

Global-Bioenergies (2015) Cristal Union and Global Bioenergies have formed a joint venture to build and operate France's first bio-sourced isobutene production plant. Global-Bioenergies, Evry and Paris

Global-Bioenergies (2016a) Global Bioenergies: IBN-One awards front-end engineering design to Technip and IPSB for the first plant converting renewable resources into isobutene. Global-Bioenergies, Evry and Paris

Global-Bioenergies (2016b) The construction of the Leuna demo plant is completed. Global-Bioenergies, Evry and Leuna

Global-Bioenergies (2017) First batch delivered to L'Oreal by Global Bioenergies. Global-Bioenergies, Evry

GranBio (2014) GranBio begins producing second-generation ethanol. http://www.granbio.com.br/en/wp-content/uploads/sites/2/2014/09/partida_english.pdf

GranBio (2018) Energy cane. http://www.granbio.com.br/en/conteudos/energy-cane/. Accessed April 2018

Hazarika M (2016) Metabolix partners with CJ CheilJedang to commercially produce specialty PHA. https://www.chemicals-technology.com/news/newsmetabolix-partners-with-cj-cheiljedang-to-commercially-produce-specialty-pha-4853010/. Accessed April 2018

Hellsmark H, Frishammar J, Söderholm P, Ylinenpää H (2016a) The role of pilot and demonstration plants in technology development and innovation policy. Res Policy 45(9):1743–1761

Hellsmark H, Mossberg J, Söderholm P, Frishammar J (2016b) Innovation system strengths and weaknesses in progressing sustainable technology: the case of Swedish biorefinery development. J Clean Prod 131:702–715

ICIS (2016) BioAmber special supplement. https://www.icis.com/contact/icis-bioamber-special-supplement/

Kann Y (2016) Biobased modifiers for polyvinylchloride blends [Online]. https://patentimages.storage.googleapis.com/69/dc/59/889a183ef1f3ab/US9464187.pdf

Kawaguchi H, Hasunuma T, Ogino C, Kondo A (2016) Bioprocessing of bio-based chemicals produced from lignocellulosic feedstocks. Curr Opin Biotechnol 42:30–39

Khanra S, Mondal M, Halder G, Tiwari ON, Gayen K, Bhowmick TK (2018) Downstream processing of microalgae for pigments, protein and carbohydrate in industrial application: a review. Food Bioprod Process 110:60–84

Koukios EG (1985) Biomass refining: a non-waste approach. In: Hall DO, Myers N, Margaris NS (eds) Economics of ecosystems management. Springer, Dordrecht, pp 233–244

Kumar K, Mishra SK, Shrivastav A, Park MS, Yang J-W (2015) Recent trends in the mass cultivation of algae in raceway ponds. Renew Sust Energ Rev 51:875–885

Laird K (2011) Chemical recycling closes the LOOPLA for cradle-to-cradle PLA. https://www.plasticstoday.com/content/chemical-recycling-closes-loopla-cradle-cradle-pla/78502650816467. Accessed April 2018

Laird K (2012) Telles tale ends. https://www.plasticstoday.com/content/telles-tale-ends/26043942116953. Accessed April 2018

Laird K (2018) Metabolix signs $10 million binding LOI with CJ CheilJedang. https://www.plasticstoday.com/business/metabolix-signs-10-million-binding-loi-cj-cheiljedang/127307180225084. Accessed April 2018

Lam GP, Vermuë MH, Eppink MHM, Wijffels RH, van den Berg C (2018) Multi-product microalgae biorefineries: from concept towards reality. Trends Biotechnol 36(2):216–227

Lane J (2015) Algenol CEO exits, staff cut by 25 percent. https://www.renewableenergyworld.com/articles/2015/10/algenol-ceo-exits-staff-cut-by-25-percent.html. Accessed April 2018

Larson A, O'Brien K, Anderson A (2010) Natureworks: Green Chemistry's Contribution to Biotechnology, Innovation, Commercialization, and Strategic Positioning. Darden Case No. UVA-ENT-0089

Laurens L (2017) State of Technology review – algae bioenergy. IEA Bioenergy. ISBN: 978-1-910154-30-4

Levy PF, Sanderson JE, Kispert RG, Wise DL (1981) Biorefining of biomass to liquid fuels and organic chemicals. Enzym Microb Technol 3(3):207–215

Maeda Y, Yoshino T, Matsunaga T, Matsumoto M, Tanaka T (2018) Marine microalgae for production of biofuels and chemicals. Curr Opin Biotechnol 50:111–120

Marc D, Anissimova M Tallon R, Marliere P (2015) Production of alkenes by combined enzymatic conversion of 3-hydroxyalkanoic acid. [Online]. https://patentimages.storage.googleapis.com/f2/ea/e9/d701368578d1e2/US9193978.pdf

Marlière P, Anissimova M, Allar M (2015) Method for producing monoalkene by enzymatic conversion of an alkyl monoester. [Online]. https://patentimages.storage.googleapis.com/68/03/60/990f12c5645f83/US20150159177A1.pdf

McCarthy A (2003) Metabolix, Inc. and Tepha, Inc. Bioplastics for industry and medical devices. Chem Biol 10(10):893–894

McMahon J (2012) Why did Obama's favorite algae biofuels company break up with Dow?. https://www.forbes.com/sites/jeffmcmahon/2012/02/28/why-did-obamas-favorite-algae-biofuels-company-break-up-with-dow/#53e015327e35. Accessed April 2018

Metabolix (2012) Metabolix announces termination of telles joint venture. http://www.mirelplastics.com/wp-content/uploads/2017/03/011212-MBLXAnnoucesTerminationTellesJV-1.pdf

Metgen (2017) Shaping the future of bioenergy and renewable chemicals. http://www.metgen.com/wp-content/plugins/pdf-viewer/beta/web/viewer.html?file=http://www.metgen.com/wp-content/uploads/2017/02/MetGen-Presentation-March-2017.pdf. Accessed

Metgen (2018) Metgen invents novel chemo-enzymatic route to FDCA. http://www.metgen.com/metgen-invents-new-technology-renewable-chemicals/

MIRACLE (2017) Multi-product integrated biorefinery of algae: from carbon dioxide and light energy to high-value specialties. Netherlands, Spain, Norway, Chile, Belgium, Portugal, Greece and Germany, FP7-KBBE-Specific Programme "Cooperation": Food, Agriculture and Biotechnology

Morais AR, Bogel-Lukasik R (2013) Green chemistry and the biorefinery concept. Sustain Chem Process 1(1):18

Moreno-Garcia L, Adjallé K, Barnabé S, Raghavan GSV (2017) Microalgae biomass production for a biorefinery system: Recent advances and the way towards sustainability. Renew Sust Energ Rev 76:493–506

Mossberg J, Söderholm P, Hellsmark H, Nordqvist S (2018) Crossing the biorefinery valley of death? A role-based typology for understanding actor networks' ability to overcome barriers in sustainability transitions. Environ Innov Soc Transit 27:83–101
Mudombi S, Nyambane A, von Maltitz GP, Gasparatos A, Johnson FX, Chenene ML, Attanassov B (2018) User perceptions about the adoption and use of ethanol fuel and cookstoves in Maputo, Mozambique. Energy Sustain Dev 44:97–108
Nandy SK, Srivastava RK (2018) A review on sustainable yeast biotechnological processes and applications. Microbiol Res 207:83–90
NatureWorks (2012) NatureWorks broadens Ingeo product portfolio with Sulzer proprietary production equipment. https://www.natureworksllc.com/news-and-events/press-releases/2012/09-05-12-sulzer-equipment-for-increased-ingeo-production
Nickel Institute (2014) Really green ethanol. https://www.nickelinstitute.org/en/NickelMagazine/MagazineHome/AllArchives/2014/Volume29-2/InUseReallyGreenEthanol.aspx. Accessed 26 April 2018
Nizami AS, Rehan M, Waqas M, Naqvi M, Ouda OKM, Shahzad K, Miandad R, Khan MZ, Syamsiro M, Ismail IMI, Pant D (2017) Waste biorefineries: enabling circular economies in developing countries. Bioresour Technol 241:1101–1117
Novozymes (2012) CleanStar Mozambique launches world's first sustainable cooking fuel facility. https://www.novozymes.com/en/news/news-archive/2012/05/cleanstar-mozambique-launches-worlds-first-sustainable-cooking-fuel-facility. Accessed April 2018
Novozymes (2014) BASF, Cargill and Novozymes achieved another milestone in bio-based acrylic acid. https://www.novozymes.com/es/news/news-archive/2014/09/basf-cargill-and-novozymes-achieved-another-milestone-in-bio-based-acrylic-acid. Accessed April 2018
Novozymes (2015a) Novozymes and Cargill remain committed to bringing technology to market. https://www.novozymes.com/es/news/news-archive/2015/01/novozymes-cargill-continue-bio-acrylic-acid-partnership-basf-exits. Accessed April 2018
Novozymes (2015b) Novozymes to supply enzymes to St1 Biofuels in Finland. https://www.novozymes.com/en/news/news-archive/2015/05/novozymes-to-supply-enzymes-to-st1-biofuels-in-finland
Pannonia (2017) Pannonia ethanol corporate video. http://www.pannoniaethanol.com/en/. Accessed April 2018
RFA (2017) Building partnerships and growing markets. RFA, New Orleans. http://www.ethanolrfa.org/wp-content/uploads/2017/02/Ethanol-Industry-Outlook-2017.pdf
RFA (2018) Ethanol strong. RFA, San Antonio. http://www.ethanolrfa.org/wp-content/uploads/2018/02/NECfinalOutlook.pdf
Rizwan M, Mujtaba G, Memon SA, Lee K, Rashid N (2018) Exploring the potential of microalgae for new biotechnology applications and beyond: a review. Renew Sust Energ Rev 92:394–404
Rocha GJM, Gonçalves AR, Nakanishi SC, Nascimento VM, Silva VFN (2015) Pilot scale steam explosion and diluted sulfuric acid pretreatments: comparative study aiming the sugarcane bagasse saccharification. Ind Crop Prod 74:810–816
Roux J-M, Lamotte H, Achard J-L (2017) An overview of microalgae lipid extraction in a biorefinery framework. Energy Procedia 112:680–688
Safari A, Karimi K, Shafiei M (2017) Dilute alkali pretreatment of softwood pine: a biorefinery approach. Bioresour Technol 234(Supplement C):67–76
Santos VEN, Ely RN, Szklo AS, Magrini A (2016) Chemicals, electricity and fuels from biorefineries processing Brazil's sugarcane bagasse: production recipes and minimum selling prices. Renew Sust Energ Rev 53:1443–1458
Schill SR (2010) Fagen begins construction on Hungarian plant. Ethanol Producer Magazine. http://www.ethanolproducer.com/articles/6906/fagen-begins-construction-on-hungarian-plant. Accessed April 2018
Sekab (2013) Davy Process Technology Ltd and Sekab E-technology AB form strategic partnership. http://www.sekab.com/wp-content/uploads/2013/03/SEKAB-2013-03-13.pdf

Sekab (2015) Sekab's history. http://www.sekab.com/about-us/our-history/. Accessed April 2018
Simha P (2015) A policy analysis for pelagonia ethanol: regulatory implications, compliance gaps and uncertainties. https://doi.org/10.13140/RG.2.1.2902.1201
Sims B (2011) Avantium secures funding to accelerate YXY technology. http://biomassmagazine.com/articles/7095/avantium-secures-funding-to-accelerate-yxy-technology. Accessed April 2018
Sims B (2012) OPXBio hits key production milestone for bioacrylic acid. http://biomassmagazine.com/articles/7692/opxbio-hits-key-production-milestone-for-bioacrylic-acid. Accessed April 2018
Siripuram R (2015) BASF terminates partnership with Cargill and Novozymes in bio-based acrylic-acidproject. https://www.chemicals-technology.com/news/newsbasf-terminates-partnership-with-cargill-and-novozymes-in-bio-based-acrylic-acid-project-4498045/
Speight J (2008) Synthetic fuels handbook: properties, process, and performance. McGraw-Hill Education, New York
st1 (2008) St1 opens bioethanol dehydration plant in Finland. http://www.st1biofuels.com/company/news/st1-opens-bioethanol-dehydration-plant-in-finland
st1 (2015a) Creating new business from waste-based advanced ethanol. http://www.st1biofuels.com/files/15226/St1Biofuels_Whitepaper01062015_web.pdf. Accessed April 2018
st1 (2015b) St1 built a waste-based Etanolix® ethanol production plant in Gothenburg. http://www.st1.eu/news/st1-built-a-waste-based-etanolix-ethanol-production-plant-in-gothenburg. Accessed April 2018
st1 (2018) http://www.st1biofuels.com/solutions. Accessed April 2018
Stadler T, Chauvet J-M (2018) New innovative ecosystems in France to develop the bioeconomy. New Biotechnol 40:113–118
Stephan D (2012) BASF, Cargill and Novozymes target commercial bio-based acrylic acid process. https://www.process-worldwide.com/basf-cargill-and-novozymes-target-commercial-biobased-acrylic-acid-process-a-375761/
Thyssenkrupp (2016) Reduced CO2 emissions and increased sustainability in plastics production: commercial debut for PLAneo® technology from thyssenkrupp. https://www.thyssenkrupp-industrial-solutions.com/en/press_detail_34432.html?id=637548. Accessed April 2018
Tullo AH (2013) Hunting for biobased acrylic acid. Chem Eng News 91(46):18–19
Unamunzaga Escosura C, Mantecon Galvez E (2016) Method for obtaining a biomass of a microalga of the species Tetraselmis chuii enriched in superoxide dismutase (SOD). [Online]
Vink E, Davies S, Kolstad JJ (2010) Original research: the eco-profile for current Ingeo® polylactide production. Ind Biotechnol 6(4):212–224
Vinuesa JF, Mirabel P, Ponche JL (2003) Air quality effects of using reformulated and oxygenated gasoline fuel blends: application to the Strasbourg area (F). Atmos Environ 37(13):1757–1774
Vizcaíno AJ, López G, Sáez MI, Jiménez JA, Barros A, Hidalgo L, Camacho-Rodríguez J, Martínez TF, Cerón-García MC, Alarcón FJ (2014) Effects of the microalga Scenedesmus almeriensis as fishmeal alternative in diets for gilthead sea bream, Sparus aurata, juveniles. Aquaculture 431:34–43
Vonshak A (1997) Spirulina platensis arthrospira: physiology, cell-biology and biotechnology. Taylor & Francis, London and Bristol
Ward L (2015) DOE Bioenergy Technologies Office (BETO) IBR Project Peer Review POET-DSM Project Liberty. https://www.energy.gov/sites/prod/files/2015/04/f22/demonstration_market_transformation_ward_3433.pdf
Werner L (2015) St1 a new actor on biofuels in Sweden and Finland. Faculty of engineering LTH. http://www.lth.se/fileadmin/energiportalen/Energy_Portal/Files/Werner.pdf
Werpy T, Holladay J, White J (2004) Top value added chemicals from biomass: I. results of screening for potential candidates from sugars and synthesis gas. https://doi.org/10.2172/926125
Wessberg N, Eerola A (2013) Basic value chain analysis for Etanolix® and Bionolix® bioethanol production by St1 in Finland: TOPNEST-WP3. Nordic Energy Research https://

brage.bibsys.no/xmlui/bitstream/handle/11250/276878/Value%20Chain%20Analysis%20-%20Finland%20-%20Case%20ST1%20-%20edited%20for%20TOP-NEST%20website.pdf?sequence=1&isAllowed=y. Accessed April 2018

Xie S, Lawlor PG, Frost JP, Wu G, Zhan X (2012) Hydrolysis and acidification of grass silage in leaching bed reactors. Bioresour Technol 114:406–413

Zhang Y H P (2013) Next generation biorefineries will solve the food, biofuels, and environmental trilemma in the energy-food-water nexus. Energy Science & Engineering 1 (1):27–41

Zhu L (2015) Biorefinery as a promising approach to promote microalgae industry: an innovative framework. Renew Sust Energ Rev 41:1376–1384

Biomass for Biorefineries: Availability and Costs

Niclas Scott Bentsen

1 Biomass for Energy: Bioenergy Resource Assessment

Biomass has been used for energy for millennia (Smil 2018) and is the largest contributor to renewable energy generation. Currently (2016) biomass contributes 12.8% (46.4 EJ) of the global final energy consumption, with 7.8% in the form of traditional biomass and 5% as modern bioenergy (REN21 2018). Traditional biomass covers wood, straws and dung used in general by the poorer populations in developing countries (Chum et al. 2011). Globally more than 25% of the energy used for residential space heating and cooking is provided by biomass, and it is here the 28 EJ of traditional biomass is used. In industrial heating biomass contributes 6%, in transport 3% and in electricity generation 2.1% of the global total final energy consumption (REN21 2018).

There are large geographical differences in the contribution from biomass to energy consumption. In South Saharan Africa, biomass contributed close to 75% of the final energy consumption in 2015, and also in Southeast and South Asia, biomass contributed 30–35% (WEC 2016). In the other end of the spectrum in West Asia, biomass contributed 2%, in North America 3% and Europe 6.3%. Particularly the European Union (EU) has seen a large increase in the consumption of bioenergy due to political agreements to promote renewable energy (Bentsen and Felby 2012): first with the directive on renewable electricity production[1] in 2001 followed by the

[1]Directive 2001/77/EC of the European Parliament and of the Council of 27 September 2001 on the promotion of electricity produced from renewable energy sources in the internal electricity market.

N. S. Bentsen (✉)
Department of Geosciences and Natural Resource Management, University of Copenhagen, Copenhagen, Denmark
e-mail: nb@ign.ku.dk

© Springer Nature Switzerland AG 2019

J.-R. Bastidas-Oyanedel, J. E. Schmidt (eds.), *Biorefinery*,
https://doi.org/10.1007/978-3-030-10961-5_2

directive on the promotion of the use of biofuels[2] in 2003, followed again by the Renewable Energy Directive (RED)[3] and the Fuel Quality Directive (FQD)[4] of 2009 (Bureau et al. 2010). In 2018 the revised Renewable Energy Directive (REDII) has passed the governing bodies of the EU.

Transport biofuel production amounted to 143 billion litres (3.5 EJ) in 2017 (REN21 2018). Production of liquid and gaseous fuels for transport is geographically concentrated to mainly Brazil, the USA, and the EU, which account for 80% of the total production (REN21 2018).

Fundamentally biomass is solar energy converted and stored in a chemical form. While the primary reactions on the molecular level converting photon energy to stored energy are efficient, around 35% (Mauzerall 2013), on the plant level, it is much lower. Theoretically plants can convert 4.6–6.0% of the solar energy influx to biomass (Zhu et al. 2010), but in nature conversion efficiencies are lower due to, e.g. incomplete coverage, diseases and senescence. Still, the annual production of biomass, the net primary production (NPP), by far surpasses the amount of energy used by society. Annually terrestrial plants assimilate close to 60 billion tonnes of carbon (Haberl et al. 2007) corresponding to 125 billion tonnes biomass or 2400 EJ. Global energy consumption in 2017 was 633 EJ (BP 2018). Humans already exploit ~20% of the annual NPP (Imhoff et al. 2004), the so-called human appropriation of net primary production (HANPP). Food production takes up 35% of the HANPP, materials 27% and bioenergy 37% (Imhoff et al. 2004). Global NPP may be seen as an inflexible upper limit for biomass production (Smith et al. 2012), and Running (2012) argues that annual HANPP can only sustainably increase by 5 billion tonnes C. On a local scale, however, it has been shown that productivity can be increased (DeLucia et al. 2014; Larsen et al. 2017).

Estimates of bioenergy potentials are seen to differ a lot. Not only because potentials can be defined in several ways (Resch et al. 2008), but also because underlying assumptions can influence the reported potential (Bentsen and Felby 2012). Based on an extensive literature review, Slade et al. (2014) reported that the main potential sources of future (2050) bioenergy are energy crops (22–1272 EJ $year^{-1}$), agricultural residues (10–66 EJ $year^{-1}$), forest residues (3–35 EJ $year^{-1}$), waste (120 EJ $year^{-1}$) and forestry (60–230 EJ $year^{-1}$). A critical characteristic of bioenergy potentials is that they are dependent on economic and social development and on human behaviour. Slade et al. (2014) characterised global bioenergy potentials above 600 EJ $year^{-1}$ as extreme as they would require up to 2.5 billion ha used for energy

[2] Directive 2003/30/EC of the European Parliament and of the Council of 8 May 2003 on the promotion of the use of biofuels or other renewable fuels for transport.

[3] Directive 2009/28/EC of the European Parliament and of the Council of 23 April 2009 on the promotion of the use of energy from renewable sources and amending and subsequently repealing Directives 2001/77/EC and 2003/30/EC.

[4] Directive 2009/30/EC of the European Parliament and of the Council of 23 April 2009 amending Directive 98/70/EC as regards the specification of petrol, diesel and gas-oil and introducing a mechanism to monitor and reduce greenhouse gas emissions and amending Council Directive 1999/32/EC as regards the specification of fuel used by inland waterway vessels and repealing Directive 93/12/EEC.

crops, widespread deployment of high-input farming and land-less animal production and also a widespread adoption of vegetarian diets. Potentials below 100 EJ year^{-1} could be achieved with no or moderate allocation of degraded land to energy crops, no technological revolution in agriculture and a continued development towards high-meat diets.

2 Agricultural Residue Potentials for High-Value Products

Biomass resources can be characterised in a so-called resource hierarchy (Demirbas 2011). With reference to agricultural residues, primary resources include stalk, stem, leave, stover and seed pods. Resources are not or only slightly processed, e.g. separated by the crop harvest equipment and can be collected directly from the field (Fig. 1). Secondary resources include processing residues such as husk, chaff, shell, root, bagasse, dung and manure. These resources have been further processed at the farm or industrially, and they are not collected from the field. Tertiary resources include food waste and sewage sludge. There is some overlap between different levels in the hierarchy.

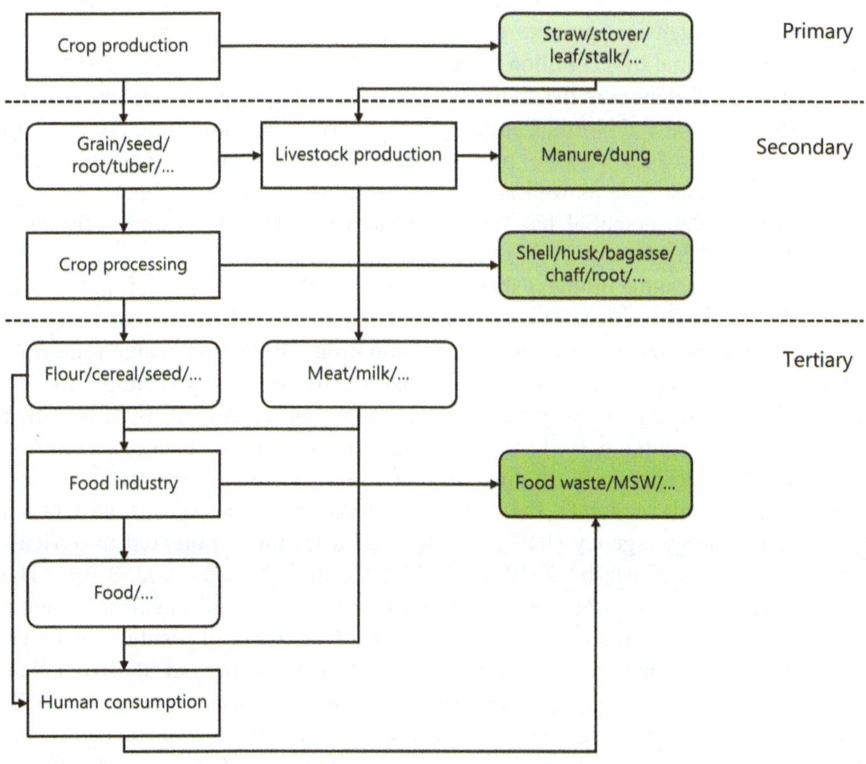

Fig. 1 Simplified flow diagram of resource streams in agriculture

2.1 Primary Resources

Primary agricultural residues constitute a relatively easily available biomass resource but also a dispersed resource with a low energy density. The resource is available as it is produced on land already under some form of management; the land is accessible, and infrastructure is established. In intensive agriculture as winter wheat production in Denmark, the energy density of the straw resource is estimated to 6.5 MJ m^{-2} (Bentsen et al. 2018). An energy crop of *Miscanthus* or willow could, in the same location, produce 18–20 MJ m^{-2} (Larsen et al. 2016) and reed canary grass 17.7 MJ m^{-2} (Peake 2018). Other crops in other locations offer higher energy density of the residue resource, e.g. sugar cane in Brazil with 11.3 MJ m^{-2} (Peake 2018). In a landscape perspective considering crop rotations and other land uses, the energy density of straw resources in Eastern Denmark has been estimated to 2.1 MJ m^{-2} and in Southern Sweden 1.2 MJ m^{-2} (Bentsen et al. 2018). The low energy density has implications for transport and cost of the resource.

The current global theoretical potential of agricultural residues has been estimated to 4.6 billion tonnes dry matter annually corresponding to 78 EJ $year^{-1}$ (Bentsen et al. 2014). Barley, maize, rice, soybean, sugar cane and wheat account for the majority of the residue potential: 3.7 billion tonnes (65 EJ $year^{-1}$). Widespread deployment of intensive agriculture can increase crop production and subsequently the residue potential by 1.3 billion tonnes annually (Bentsen et al. 2014). There are geographical differences in the additional potential through agricultural intensification. While Europe and North, Central and Southern Africa have little prospects of significantly increased residue production, a huge potential exists in South, Southeast and East Asia as well as in East Africa (Bentsen et al. 2014). A future theoretical residue potential has been estimated to 109–128 EJ $year^{-1}$ by 2100 (Daioglou et al. 2016).

Agricultural residue biomass should not be considered as waste left in the field available for biorefineries. The biomass has a number of ecological functions such as contributing to soil carbon accumulation, increasing water retention, reducing wind erosion, and supplying nutrients to new crops (Daioglou et al. 2016). Furthermore parts of the residue resource may already be used for other purposes, mainly animal feed, bedding, building material or energy generation (Smil 1999). Consequently the available potential (technical, economic or sustainable potential) is lower than the theoretical potential. The International Renewable Energy Agency (IRENA) estimated a technical potential of agricultural residues and waste by 2030 of 13–30 EJ $year^{-1}$ (Nakada et al. 2014). The special report on renewable energy by the Intergovernmental Panel on Climate Change (IPCC) reported a technical potential of agricultural residues (primary and secondary resources) by 2050 of 15–70 EJ $year^{-1}$ (Chum et al. 2011). The review by Slade et al. (2014) identified a global technical potential by 2050 between 10 and 66 EJ $year^{-1}$. By 2100 (Daioglou et al. 2016) found an ecological residue potential between 57 and 67 EJ $year^{-1}$ and an available potential between

27 and 30 EJ year^{-1}. For the EU Bentsen and Felby (2012) identified a current potential between 1 and 2 EJ year^{-1} rising to up to 5 EJ year^{-1} by 2050. Global technical potentials of primary, secondary and tertiary agricultural residues are illustrated in Fig. 2.

2.2 *Secondary Resources*

Secondary residue resources constitute a wide variety of resources from a wide variety of processes and productions, and the potential is difficult to accurately estimate. IRENA estimated the global potential of processing residues by 2030 to 18 EJ year^{-1} (Nakada et al. 2014). Chum et al. (2011) reported a potential of dung by 2050 in the range 5–50 EJ year^{-1}, emphasising that population development, diets and the setup of livestock production systems are determinants.

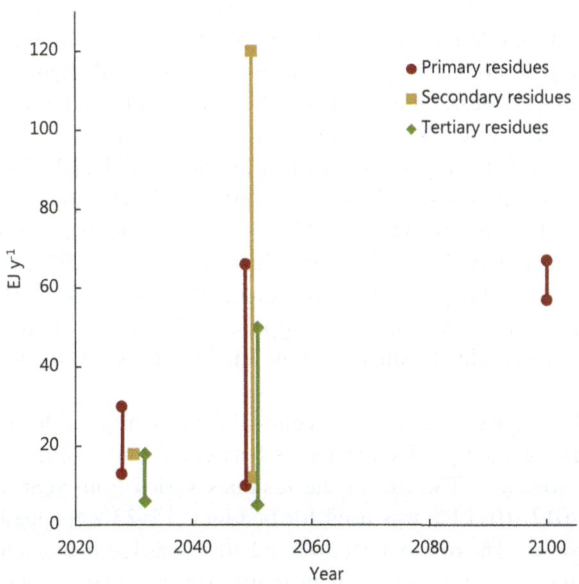

Fig. 2 Global technical/ecological resource potentials of primary, secondary and tertiary agricultural residues by 2030, 2050 and 2100. The larger variability of 2050 estimates compared to 2030 and 2100 estimates does not reflect a larger uncertainty in the estimates; it reflects that 2050 estimates summarise findings of a larger number of studies. Based on data from Chum et al. (2011), Nakada et al. (2014), Slade et al. (2014) and Daioglou et al. (2016)

2.3 Tertiary Resources

A main component of tertiary residue resources is food waste. It is estimated that 24% of food produced is lost in the food supply chain (Kummu et al. 2012). Lin et al. (2013) estimated that food waste amounted to 1.3 billion tonnes annually. Utilisation of food waste for energy purposes or in biorefineries represents a number of challenges. The resource is heterogenic in terms of its composition of carbohydrates, lipids and proteins; volumes fluctuate over the season; food waste usually has a high water content, which impairs storage and increases transport cost (Lin et al. 2013). Consequently, development of feasible, large-scale, consistent and robust industrial processes and facilities is challenging (Lin et al. 2013).

From an energy point of view, IRENA estimated the resource potential of animal and household waste to 6–18 EJ $year^{-1}$ by 2030 (Nakada et al. 2014). For food waste alone, Chum et al. (2011) estimated a 2050 potential of 5–50 EJ $year^{-1}$. As for secondary resources, projections on tertiary resources are determined by social, economic and behavioural development.

2.4 Current Use of Residue Resources

As stated above, part of the residue resources may be used already, and as such they are not readily available for biorefineries. Very few countries monitor on a regular basis how and to what extent agricultural residues are used (Bentsen et al. 2018). Attempts are made to estimate indirectly the amount of residue used for various purposes by linking a residue consumption, which is unknown, to an economic activity, which is known, e.g. the number of livestock produced or mushroom production. Such assessments are inherently uncertain. (Krausmann et al. 2008) found on a global level for the year 2000 that 24% of crop residues was harvested for various purposes. Scarlat et al. (2010) found that on average (1998–2007) in the EU ~11% of the total residue production was allocated to livestock (>10%) and mushroom (<1%) production. With a similar approach, Searle and Malins (2016) found that 8% of the total residue production in the EU in 2010 was allocated to other uses than energy.

In Denmark, national statistics report annual data on crop residue production and use (Fig. 3). The annual production varies between 5 and 6 million tonnes (fresh weight, 15% moisture). The use of the residues varies from year to year. In the period 2006–2017, 10–14% was used for bedding, 15–23% for feed and 23–32% for energy purposes. The rest, between 32 and 50%, was left in the field (Fig. 3). The comparably high exploitation rate of primary crop residues in Denmark may be attributable to cereals being a dominant crop type and a long history of using residues for energy purposes (Bentsen et al. 2018).

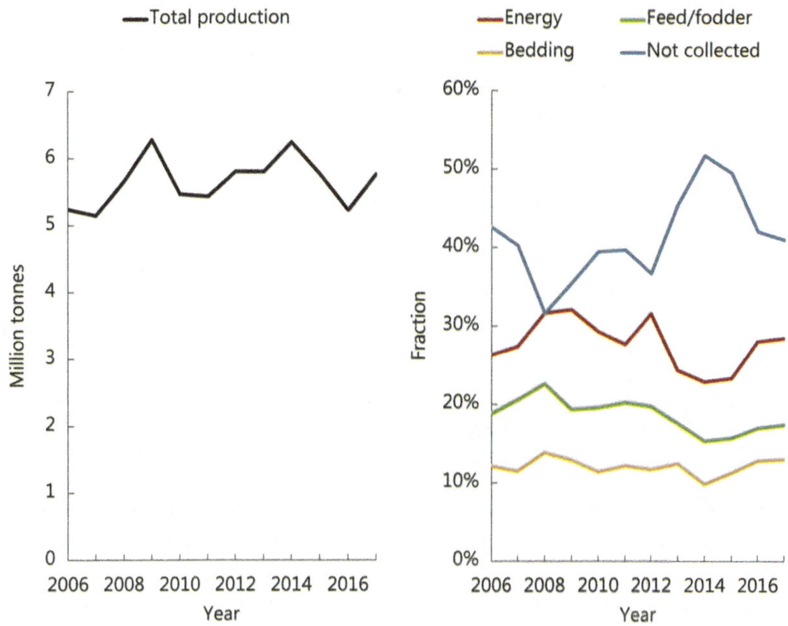

Fig. 3 Production (left panel) and relative use (right panel) of agricultural residues in Denmark from 2006 to 2017. Based on data from Statistics Denmark (www.statistikbanken.dk)

3 Cost

Residues from crop harvest and crop processing are relatively low cost compared to, e.g. wood chips and wood pellets (Bentsen et al. 2017, 2018). The absolute cost of different biomass types varies from region to region depending on, e.g. salaries, land prices, taxes, subsidies and technology cost. The cost relation between different biomass types seems to be more consistent between regions. The general picture is that industrial processing residues, manure and dung are among the cheapest resources available followed by crop residues, which again are cheaper than wood chips and even more than wood pellets (Nakada et al. 2014; Bentsen et al. 2018). Local conditions determined by specific supply options may change the competitive strength of different biomass types, particularly for those resources that are less mobile due to high transportation cost or perishable material characteristics. Also a local demand may increase the cost, which is seen with crop residues in Denmark with a comparably high supply cost to utilities of 6.5–7 USD GJ^{-1} (Bentsen et al. 2018).

IRENA estimated the supply potential and associated cost ranges for a number of biomass types (Table 1) (Nakada et al. 2014).

Some biomass types are globally traded commodities, e.g. wood pellets, wood chips and liquid biofuels (Lamers et al. 2011, 2012). Other types, such as agricultural residues, rarely cross borders, due to difficulties in handling and a low energy

Table 1 Supply cost and cost components of different biomass types

	Production	Collection	Transport to processor	Transport to end user	Total supply cost
	USD GJ^{-1} (2030)				
Energy crops	2–78			2	4–80
Crop residues	1.1	1.1	0.01–1.3		4.2–5.5
Processing residues	0	0			2–3.3
Biogas	0	0			2–3.3
Fuel wood	8–35				10–37
Logging residues	3–18				5–20
Wood waste	3–18				5–20

Adopted from Nakada et al. (2014) Table 3, page 13.

density. Nakada et al. (2014) estimated the current cost of longer-distance cross-border transport to 0.5–4 USD GJ^{-1}; the low estimate for rail transport and the higher estimates for long-distance overseas ship transport.

As can be seen from Table 1, transport is a significant cost component of supply chains based on crop residues. The generic data proposed by IRENA are to some extent supported by more detailed location-specific studies. Marchand (2015) found for Southwestern Ontario that transportation would account for more than 30% of supply costs of cornstalk to biorefineries. In Denmark transportation accounts for 25–30% of supply cost (Bentsen et al. 2016), and a similar cost structure is seen in Norway with the exception that long-distance transport does not take place in Norway (Belbo and Talbot 2014).

Traditionally agricultural residues are supplied through vertically integrated feedstock supply system relying on existing technologies and requiring pioneer biorefineries to adapt to the diversity of the feedstock (Bentsen et al. 2017). However, scaling up a biorefinery industry will require increasing feedstock volumes at decreasing costs, i.e. larger sourcing areas and lower transportation cost. It remains unclear if a conventional supply system can meet that outside highly productive regions. These supply uncertainties tend to increase the risk, which could limit the biorefinery concept from being broadly implemented (Lamers et al. 2015). Advanced supply systems emulate the current grain commodity supply system, which manages crop diversity at the point of harvest and introduces methods to reduce feedstock volume, price and quality supply uncertainties. A network of distributed biomass preprocessing centres uses one or several biomass types to generate uniform-format feedstock 'commodities'; intermediates with consistent physical and chemical characteristics. Such centres would allow subsequent supply system infrastructure to be similar for all biomass resources (Lamers et al. 2015).

4 Mobilisation of Biomass for Energy and High-Value Products

As reported above agricultural residues constitute a large but geographically dispersed biomass resource potentially available for future biorefineries. There are, however, a number of barriers that prevent mobilisation of the resource but also opportunities to overcome those barriers. Smith et al. (2017) identified four types of barriers and constraints to mobilisation of biomass resources. The strength and significance of individual barriers are context specific, depend on local conditions and vary over time.

Institutional barriers include unstable markets; uncoordinated policies across departments of energy, agriculture, forestry and environment; complicated regulation for obtaining permits; no consistent sustainability guidelines; no coherent long-term energy strategy; power of nationalised or monopolised utility companies; or ambiguous policy definition of sustainable energy. Opportunities to overcome institutional barriers include policy development and coordination; development of a cooperative organisational structure along the supply chain; and guaranteed support frameworks in the form of feed-in tariffs, subsidies or renewable energy credits (Smith et al. 2017).

Competition from cheap alternative fuels, fossil or non-fossil, is a strong *economic barrier* alongside lack of access to finances, a risky and uncertain investment profile, high capital expenditure and capacity investment in 'new technologies' and high transportation cost due to the often low energy density of biomass resources and geographically dispersed production. Some of the opportunities to overcome institutional barriers also help overcome economic barriers such as long-term political commitment to promote or support specific supply chains, which reduce the financial risk. Furthermore valuation of material or non-material by-products, e.g. CO_2 emission reductions, can help build a competitive business case (Smith et al. 2017).

Technical barriers include existing quality standards and the heterogeneity of most biomass fractions, immature technology and supply chains, seasonal variations in supply and for some biomass resources limited storability and combustion and conversion issues such as slagging, fouling, clogging and recalcitrance. Research, development and international technology transfer are key to overcome these barriers as local solutions to technical problems, e.g. corrosion from agricultural residue combustion, may be scaled up and deployed more widespread. A learning-by-doing approach to deployment can help overcome some of the chicken-and-egg problems in initiating and scaling up a new supply chain (Smith et al. 2017).

In some regions new biomass supply chains develop in a hostile environment as the climatic benefit is challenged and potential environmental impacts are highlighted. Also lack of transparency or public awareness of environmental impacts of fossil fuels and biomass-based supply chains contributes to *social barriers and constraints* to mobilisation. Successful deployment of biomass-based supply chains requires a social license to operate, and communication, knowledge transfer and a broad societal stakeholder consensus seem to help overcome some of the barriers. The role of knowledge and science is not clear as science often provide diverging evidence within the bioenergy and biofuel area (Bentsen 2017; Robledo-Abad et al. 2017).

A practical example of how different barriers and organisational structures affect the mobilisation of biomass resources can be seen in the case of the use of agricultural residues for energy in Denmark and Southern Sweden (Bentsen et al. 2018). While the climatic conditions, climate and energy policies and agricultural opportunities do not differ much between the two nations, the use of agricultural residues for energy differs a lot. In Denmark app. 1.3 million tonnes of cereal straw is used in 27 large-scale combined heat and power (CHP), 58 medium-scale district heating and close to 7000 individual heating facilities. In comparison 0.1 million tonnes of cereal straw is used in Sweden, in app. 150 small- and medium-scale local heat production facilities (Bentsen et al. 2018).

Bentsen et al. (2018) concluded that the main drivers behind the observed differences in mobilisation can be attributed to policy instruments, organisational framework and resource density. While the overall climate and energy policy goals are shared between Denmark and Sweden, the *policy instruments* differ. Sweden has chosen a technology neutral renewable energy certificate market together with Norway, whereas Denmark has applied a technology-specific support scheme and mandated the use of agricultural residues. The *organisational framework* differs, and a mature and transparent straw to energy market has been established in Denmark with the help of the large utilities and the Danish Straw Suppliers Association. No such framework exists in Sweden and the market is still developing. Although the yield of the main cereal crops is fairly similar in Denmark and Southern Sweden, the agronomic tradition differs. Cereals as wheat and barley cover a much larger fraction of the land area in Eastern Denmark than in Southern Sweden, and as such, in a landscape perspective, the residue *resource density* is higher in Denmark (Bentsen et al. 2018).

As evidenced from the experience in Denmark, establishment of viable production systems based on agricultural residues is possible when the supply chains meet the diverse criteria of its stakeholders. A consistent policy framework that supports a bioeconomy, biorefineries or production based on biomass and waste is required. Awareness about credible knowledge on process costs and sustainability aspects must be developed among stakeholders (farmers, utilities, regulating bodies, NGOs) along the supply chain. This could be in the form of expert systems to transfer knowledge, acknowledged certification systems or best management practices to ensure that residue removal does not jeopardise long-term soil health. Long-term contracts to increase business confidence in a supply chain together with economic incentives reduce the financial risk of establishing a new supply chain (Smith et al. 2017).

References

Belbo H, Talbot B (2014) Performance of small-scale straw-to-heat supply chains in Norway. Wiley Interdiscip Rev Energ Environ 3(4):400–407

Bentsen NS (2017) Carbon debt and payback time—lost in the forest? Renew Sust Energ Rev 73:1211–1217

Bentsen NS, Felby C (2012) Biomass for energy in the European Union—a review of bioenergy resource assessments. Biotechnol Biofuels 5:25

Bentsen NS, Felby C, Thorsen BJ (2014) Agricultural residue production and potentials for energy and materials services. Prog Energy Combust Sci 40:59–73

Bentsen NS, Nilsson D, Larsen S, Stupak I (2016) Agricultural residues for energy in Denmark and Sweden—differences and commonalities, IEA Bioenergy Task 43. TR2016:05

Bentsen NS, Lamers P, Lalonde C, Wellisch M, Dale VH, Bonner I, Jacobson J, Stupak I, Gan J, Girouard P (2017) Mobilisation of agricultural residues for bioenergy and higher value bio-products: resources, barriers and sustainability, IEA Bioenergy Taks 43. TR2017:01

Bentsen NS, Nilsson D, Larsen S (2018) Agricultural residues for energy—a case study on the influence of resource availability, economy and policy on the use of straw for energy in Denmark and Sweden. Biomass Bioenergy 108:278–288

BP (2018) BP statistical review of world energy. BP, London

Bureau J-C, Guyomard H, Jacquet F, Tréguer D (2010) European biofuel policy: how far will public support go? In: Khanna M, Scheffran J, Zilberman D (eds) Handbook of bioenergy economics and policy. Springer, New York, pp 401–423

Chum H, Faaij A, Moreira J, Berndes G, Dhamija P, Dong H, Gabrielle B, Eng AG, Lucht W, Mapako M, Cerutti OM, McIntyre T, Minowa T, Pingoud K (2011) Bioenergy. In: Edenhofer O, Pichs-Madruga R, Sokona Y et al (eds) IPCC special report on renewable energy sources and climate change mitigation. Cambridge University Press, Cambridge

Daioglou V, Stehfest E, Wicke B, Faaij A, van Vuuren DP (2016) Projections of the availability and cost of residues from agriculture and forestry. GCB Bioenergy 8(2):456–470

DeLucia EH, Gomez-Casanovas N, Greenberg JA, Hudiburg TW, Kantola IB, Long SP, Miller AD, Ort DR, Parton WJ (2014) The theoretical limit to plant productivity. Environ Sci Technol 48(16):9471–9477

Demirbas A (2011) Waste management, waste resource facilities and waste conversion processes. Energy Convers Manag 52(2):1280–1287

Haberl H, Erb KH, Krausmann F, Gaube V, Bondeau A, Plutzar C, Gingrich S, Lucht W, Fischer-Kowalski M (2007) Quantifying and mapping the human appropriation of net primary production in earth's terrestrial ecosystems. Proc Natl Acad Sci U S A 104(31):12942–12947

Imhoff ML, Bounoua L, Ricketts T, Loucks C, Harriss R, Lawrence WT (2004) Global patterns in human consumption of net primary production. Nature 429(6994):870–873

Krausmann F, Erb K-H, Gingrich S, Lauk C, Haberl H (2008) Global patterns of socioeconomic biomass flows in the year 2000: a comprehensive assessment of supply, consumption and constraints. Ecol Econ 65(3):471–487

Kummu M, de Moel H, Porkka M, Siebert S, Varis O, Ward PJ (2012) Lost food, wasted resources: global food supply chain losses and their impacts on freshwater, cropland, and fertiliser use. Sci Total Environ 438:477–489

Lamers P, Hamelinck C, Junginger M, Faaij A (2011) International bioenergy trade—a review of past developments in the liquid biofuel market. Renew Sust Energ Rev 15(6):2655–2676

Lamers P, Junginger M, Hamelinck C, Faaij A (2012) Developments in international solid biofuel trade—an analysis of volumes, policies, and market factors. Renew Sust Energ Rev 16(5):3176–3199

Lamers P, Tan ECD, Searcy EM, Scarlata C, Cafferty KG, Jacobson JJ (2015) Strategic supply system design—a holistic evaluation of operational and production cost for a biorefinery supply chain. Biofuel Bioprod Biorefin 9(6):648–660

Larsen S, Jaiswal D, Bentsen NS, Wang D, Long SP (2016) Comparing predicted yield and yield stability of willow and Miscanthus across Denmark. GCB Bioenergy 8(6):1061–1070
Larsen S, Bentsen NS, Dalgaard T, Jørgensen U, Olesen JE, Felby C (2017) Possibilities for near-term bioenergy production and GHG-mitigation through sustainable intensification of agriculture and forestry in Denmark. Environ Res Lett 12(11):114032
Lin CSK, Pfaltzgraff LA, Herrero-Davila L, Mubofu EB, Abderrahim S, Clark JH, Koutinas AA, Kopsahelis N, Stamatelatou K, Dickson F, Thankappan S, Mohamed Z, Brocklesby R, Luque R (2013) Food waste as a valuable resource for the production of chemicals, materials and fuels. Current situation and global perspective. Energy Environ Sci 6(2):426–464
Marchand L (2015) Cost assessment for cornstalk supply chain for bioprocessing purposes. University of Guelph, Guelph
Mauzerall D (2013) Thermodynamics of primary photosynthesis. Photosynth Res 116(2):363–366
Nakada S, Saygin D, Gielen D (2014) Global bioenergy: supply and demand projections. IRENA, Abu Dahbi
Peake S (ed) (2018) Renewable energy—power for a sustainable future. Oxford University Press, Oxford
REN21 (2018) Renewables 2018 global status report. REN21 Secretariat, Paris
Resch G, Held A, Faber T, Panzer C, Toro F, Haas R (2008) Potentials and prospects for renewable energies at global scale. Energy Policy 36(11):4048–4056
Robledo-Abad C, Althaus H-J, Berndes G, Bolwig S, Corbera E, Creutzig F, Garcia-Ulloa J, Geddes A, Gregg JS, Haberl H, Hanger S, Harper RJ, Hunsberger C, Larsen RK, Lauk C, Leitner S, Lilliestam J, Lotze-Campen H, Muys B, Nordborg M, Ölund M, Orlowsky B, Popp A, Portugal-Pereira J, Reinhard J, Scheiffle L, Smith P (2017) Bioenergy production and sustainable development: science base for policymaking remains limited. GCB Bioenergy 9(3):541–556
Running SW (2012) A measurable planetary boundary for the biosphere. Science 337(6101):1458–1459
Scarlat N, Martinov M, Dallemand J-F (2010) Assessment of the availability of agricultural crop residues in the European Union: potential and limitations for bioenergy use. Waste Manag 30(10):1889–1897
Searle SY, Malins CJ (2016) Waste and residue availability for advanced biofuel production in EU member states. Biomass Bioenergy 89:2–10
Slade R, Bauen A, Gross R (2014) Global bioenergy resources. Nat Clim Change 4(2):99–105
Smil V (1999) Crop residues: agriculture's largest harvest—crop residues incorporate more than half of the world agricultural phytomass. Bioscience 49(4):299–308
Smil V (2018) Energy and civilization: a history. MIT Press, Cambridge
Smith WK, Zhao M, Running SW (2012) Global bioenergy capacity as constrained by observed biospheric productivity rates. Bioscience 62(10):911–922
Smith CT, Lattimore B, Berndes G, Bentsen NS, Dimitriou I, Langeveld JWA, Thiffault E (2017) Opportunities to encourage mobilization of sustainable bioenergy supply chains. Wiley Interdiscip Rev Energ Environ 6:e237
WEC (2016) World energy resources—bioenergy. World Energy Council, London
Zhu X-G, Long SP, Ort DR (2010) Improving photosynthetic efficiency for greater yield. Annu Rev Plant Biol 61(1):235–261

Analysis and Optimization of Multi-actor Biorefineries

Muhammad Tahir Ashraf, Ana Inés Torres, Jens Ejbye Schmidt, and George Stephanopoulos

1 Introduction

Biorefineries convert biomass into chemicals and are commonly regarded as the biomass-based analogs of the traditional petroleum refinery. Unlike its predecessor, which evolved simultaneously with catalysis, polymer chemistry, process design methodologies and other (chemical) engineering areas, biorefineries can take advantage of the maturity of these fields and be optimally designed from the beginning to satisfy different needs. Currently, as several technologies for the production of many biomass-based chemicals are already available, how to optimally design biorefineries has itself become a very active research area (Murillo-Alvarado et al. 2013; Torres and Stephanopoulos 2016). Starting from the general idea that a

The authors Ana I. Torres and Muhammad Tahir Ashraf have equally contributed.

M. T. Ashraf
Department of Chemical Engineering, Khalifa University of Science and Technology, Abu Dhabi, United Arab Emirates
e-mail: muhammad.ashraf@ku.ac.ae

A. I. Torres
Instituto de Ingeniería Química, Facultad de Ingeniería, Universidad de la República, Montevideo, Uruguay
e-mail: aitorres@fing.edu.uy

J. E. Schmidt (✉)
SDC-Department of Chemical Engineering, Biotechnology, and Environmental Technology, University of Southern Denmark, Campusvej 55, DK-5230 Odense M, Denmark
e-mail: jesc@kbm.sdu.dk

G. Stephanopoulos
Department of Chemical Engineering, Massachusetts Institute of Technology, Cambridge, MA, USA
e-mail: geosteph@mit.edu

© Springer Nature Switzerland AG 2019

J.-R. Bastidas-Oyanedel, J. E. Schmidt (eds.), *Biorefinery*,
https://doi.org/10.1007/978-3-030-10961-5_3

biorefinery should be analogous to an oil refinery, different approaches can be used for their design. As each approach leads to the formulation of a different biorefinery design problem, it is pertinent to discuss their most salient features.

One approach could be to start from formal definitions of biorefineries, which, for example, state:

> Biorefinery is the sustainable processing of biomass into a spectrum of marketable products (food, feed, materials, chemicals) and energy (fuels, power, heat). (IEA Bioenergy Task 42 2007)

or

> Integrated biorefineries employ various combinations of feedstock and conversion technologies to produce a variety of products, with the main focus on producing biofuels. Co-products can include chemicals (or other materials), animal feed, and heat and power. (US Department of Energy 2018)

From the process design point of view, the above definitions are very open-ended, and allow the inclusion under the term biorefinery of almost every process that uses any kind of biomass as a feedstock. In addition, apart from the fact that many products should be produced, these definitions do not provide a clear connection between biorefineries and oil refineries.

From a different perspective, it could be argued that the future biomass-based chemical industry should mimic the petrochemical industry in its *structure*. This implies that the biomass to bio-products production chain should consist of a biorefinery section, analogous to the traditional oil refinery, whose task is to separate different biomass-based sources to their components, and an upgrade section whose task is to convert, through a sequence of processes, these components in a series of marketable chemicals. Although not explicitly stated, this is the path that is suggested by the highly influential DOE "Top ten chemicals report" (Werpy et al. 2004), which presents flowcharts with the current value chain for chemicals and products from petroleum and proposes an analogous biomass-based value chain. Figures 1 and 2 present a summary of the information contained in these flowcharts.

Figure 3 schematizes the current structure of the petrochemical industry by showing the participation of some refineries and petrochemical industries in the previous flowchart. As an example in the United States, BP has activities in both refining and petrochemical sector: it refines petroleum in Cherry Point, Whiting, and Toledo. It produces *p*-xylene in the Texas City facility from xylenes bought from "Gulf coast refineries" (BP 2018a). Most of this *p*-xylene is used to produce purified terephthalic acid (PTA) in the Cooper River facility. PTA and the by-product *m*-xylene are sold to other companies, not further processed by BP (2018a, b). Notice that although being under the same company, and according to BP's website, in the United States refinery and petrochemical activities are decoupled. AdvanSix (formerly Honeywell Resins and Chemicals division) key process is the production of nylon 6 via caprolactam and phenol which are also produced by the company (AdvanSix 2018).

Thus, AdvanSix is also one of the largest world producers of caprolactam and phenol. Phenol is produced from benzene and propylene that AdvanSix obtains

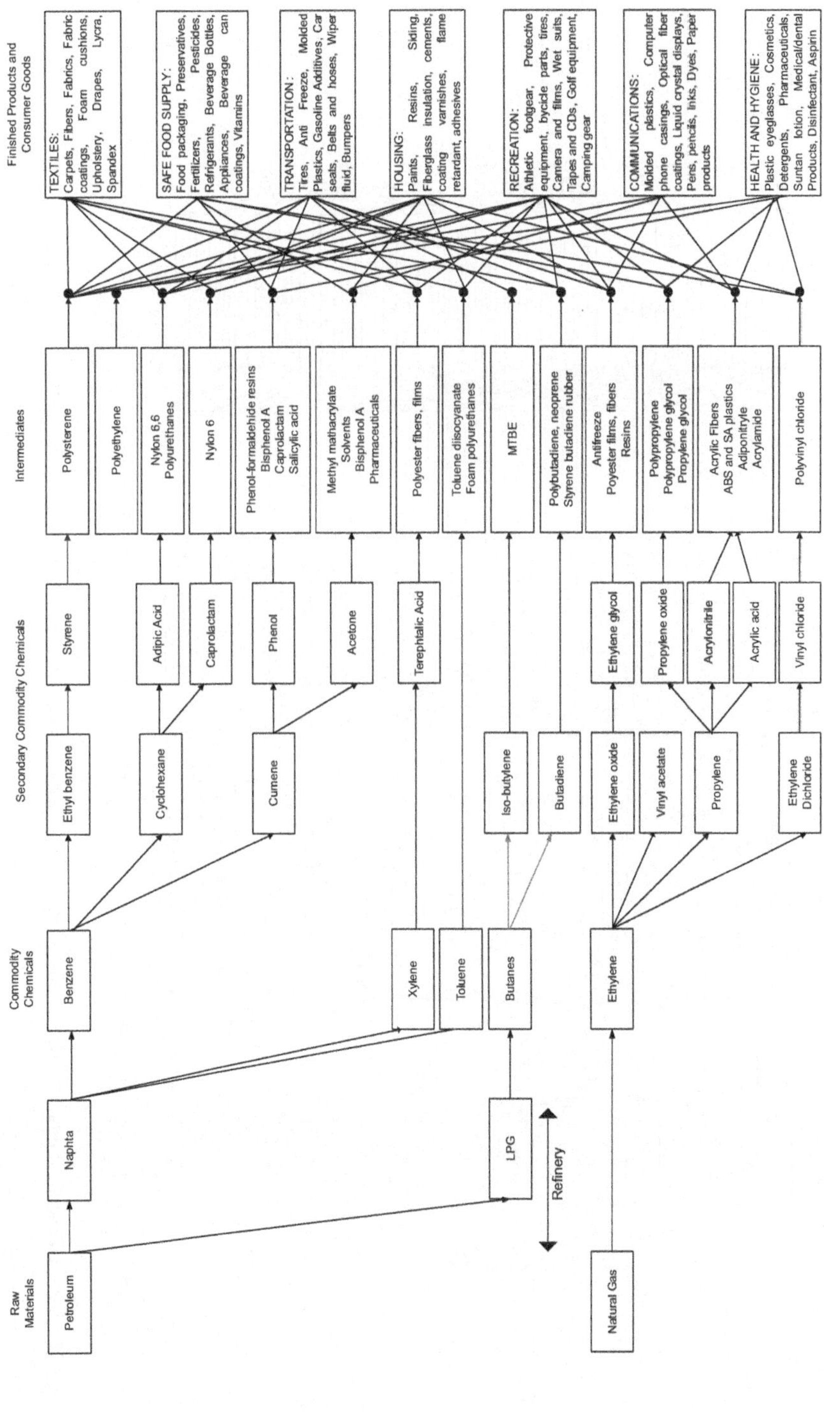

Fig. 1 Chemical product value chain from fossil-based feedstocks (selection). Adapted from Werpy et al. (2004)

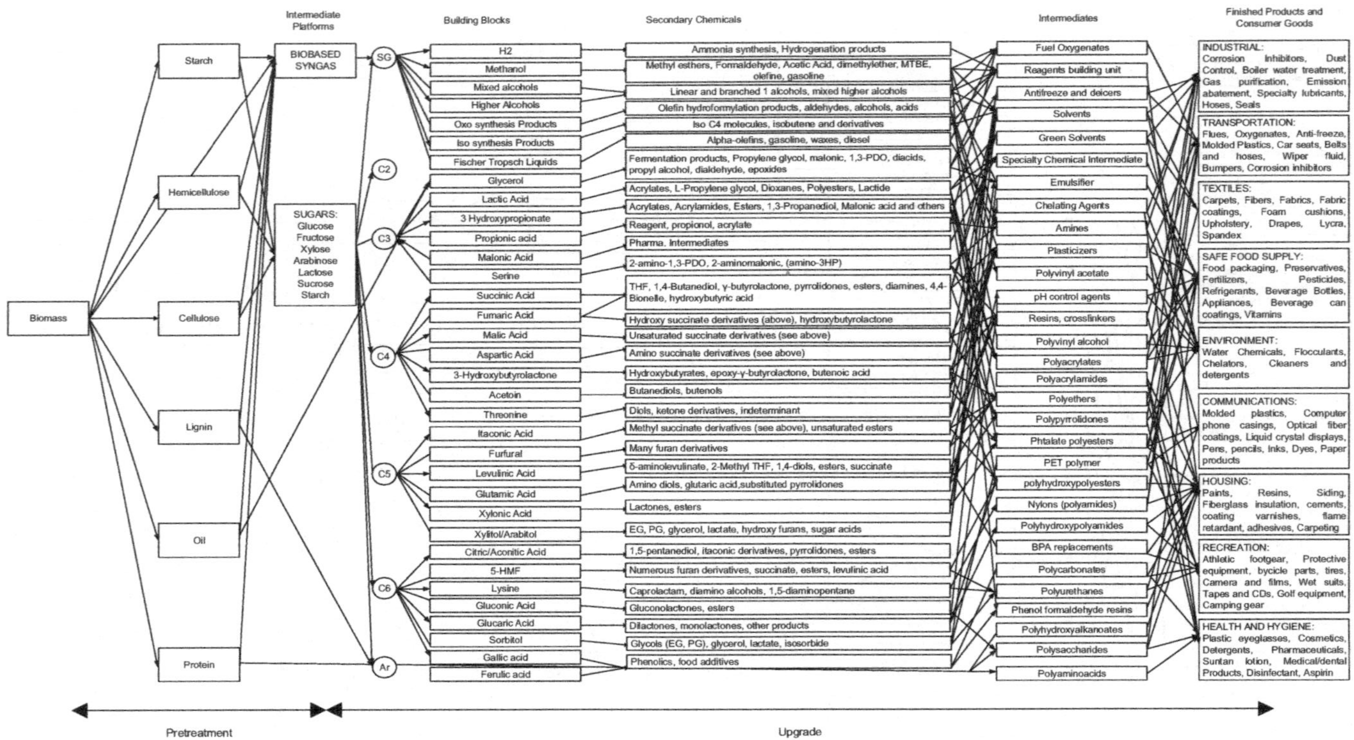

Fig. 2 Proposed chemical product value chain from biomass-based feedstocks. Adapted from Werpy et al. (2004)

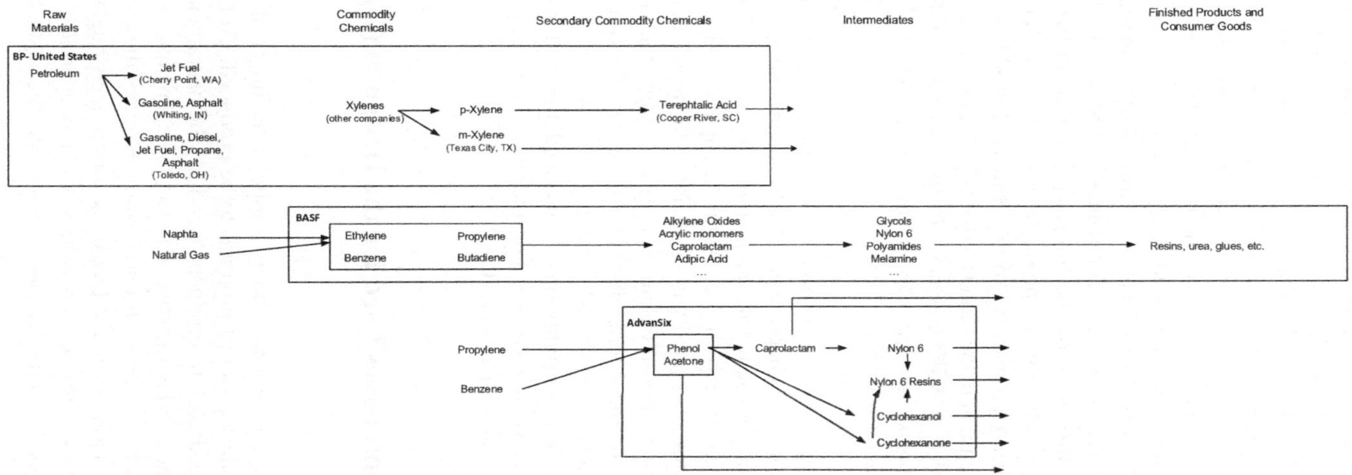

Fig. 3 Participation of three representative petrochemical industries in the value chain in Fig. 1

from other companies. Other products that are offered by the company are coproducts of the main processing line or directly related to them: acetone is a coproduct of the phenol production process; hydrogenation of phenol produces cyclohexanone and cyclohexanol, and both these products are used together with nylon 6 to produce nylon 6 resins. Overall, companies usually specialize along a portion of the value chain, not all of it. BASF could be seen as an exception to this rule as it produces more than 700 intermediates and monomers: ethylene, propylene, butadiene, benzene, alcohols, solvents, alkylene oxides, glycols, and acrylic monomers, among others. However, BASF is not involved in the refining sector and relies on the supply of naphtha or natural gas from other companies (BASF 2018). By performing a search by the product instead of by the company, it can be further noticed that the number of companies involved in the production of a particular chemical increases when moving from the "Commodity" to the "Finished products" category. As an example, nylon 6 a chemical in the "Intermediates" category is produced by approximately 150 companies, BASF, Celanese Corp., Arkema, DSM, DowDuPont, AdvanSix, and Mitsubishi Chemical Corporation, among others (MatWeb 2018).

The goal of this chapter is to discuss recent contributions to the design of optimal biorefineries which, inspired by the distributed structure of the current fossil-based industry, divides the biorefinery in a supply of intermediates section and a demand of intermediates section. The chapter starts by reviewing the approaches that have been used for designing biorefineries and motivating the need for a distributed approach. In Sect. 3.1 we discuss what constitutes a supplier and a consumer and the process design problems associated with each. Next, the mathematical formulation of a profit-based maximization problem focused on finding the optimal flow rates and exchange prices for the intermediates is presented. Section 4 introduces an alternative return-on-investment-based formulation, which has been proposed for matching production and consumption of intermediates and for estimating an exchange price in cases where the state of development of the design prevents finding a solution using the approach in Sect. 3.2.

2 Review of Approaches Used for the Design of (Optimal) Biorefineries

In a general setting, designing a biorefinery implies selecting (1) the biomass-based feedstock, (2) the product or set of products to be produced, (3) the technological pathway to connect feedstock to product(s), and (4) the operational conditions for the processing units included in the chosen technological pathways. In some design cases, all these decisions are unconstrained, while in others there exist restrictions, as, for example, in the case where the feedstocks or products might be set. The aim of this section is to provide an overview of the two main approaches that have been used for the design of biorefineries and to motivate the need for a distributed approach.

2.1 Approach 1: Traditional Process Design Applied to Biorefineries

The first approach is the one in which the biorefinery has a fairly well-specified scope. This means that the biomass feedstock is well defined, the set of desired objectives/products is also well specified, and the selection of attractive technologies has already been made. In this type of approach, the structure of the desired biorefinery does not include a number of alternative options that justify proposing a combinatorial optimization problem. Then the design of the biorefinery becomes a traditional process design problem or at most a parametric optimization one. This is the approach that most companies have followed.

As an example DuPont partnered in 2004 with Tate & Lyle creating DuPont Tate & Lyle Bio Products, LLC, a company that has designed a biorefinery that uses corn to produce 1,3 propanediol (PDO) (Dupont Tate and Lyle 2018a). PDO is one of the chemicals identified by the US DOE as a "biomass-based chemical of the future" (see Fig. 2). PDO is a substitute for oil-based PDO, ethylene, propylene and butylene glycols, and 1,4-butanediol, among others (Dupont Tate and Lyle 2018b). PDO, through its brand name Zemea©, is sold to other companies as a feedstock for their own processes, among those, it is sold to DuPont who fabricates the Sorona© fiber, a PDO-based replacement of nylon (Dupont Tate and Lyle 2018c). From the process design point of view, there is one product of interest (PDO), one feedstock of interest (corn), and one possible technology, which combines Tate & Lyle's already existing sugar-refining technology (Dupont Tate and Lyle 2018d) with DuPont's starch to PDO patented technology.

This approach is also abundant in academic research. Generally, contributions come from partnerships between experimental and process design research groups who jointly develop process flowsheets and perform a techno-economic analysis of the technology-product combinations that are being developed at laboratory scale. The goal in these cases is to "put a price" to the technology-product combinations so this price can be used as a benchmark for comparison with other technology-product options. The level of detail included in the analysis varies among the various contributions, ranging from designs that only consider the reported laboratory scale yields (Patel et al. 2010; Kazi et al. 2011), to others which consider rigorous kinetics and perform optimization of the design parameters (Torres et al. 2010, 2012).

2.2 Approach 2: Process Network Synthesis of Biorefineries

In the second type of approach, there might be multiple options for both feedstocks and products, and the connection between them might be carried out by several technological pathways. Finding the best set of feed products as well as the

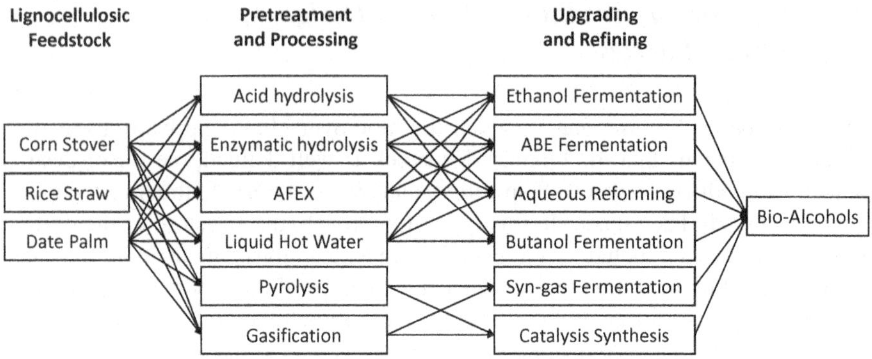

Fig. 4 A processing network for conversion of lignocellulosic feedstock to bio-alcohols. Adapted from Pham and El-Halwagi (2012)

pathways that connect them is the design objective in this case, a problem known as process network synthesis (PNS).

In this type of problem, a superstructure of all potential technologies, linking biomass feedstocks to the desired products through all possible routes is first constructed. A representative example is presented in Fig. 4 where a network of lignocellulosic feedstock conversion to bio-alcohols is depicted. This network considers three types of lignocellulosic biomass as feedstock, six alternatives for pretreatment and processing technologies, and six alternatives for upgrading and refining technologies.

PNS problems are usually solved by formulating combinatorial optimization problems in which each feedstock, technology, and product included in the superstructure is identified by a binary variable. Logical relationships among these variables become the constraints of the optimization problem, while the objective function varies depending on the goal of the problem, i.e., maximization of profit and maximization of production rate.

This is the approach that several academic studies have adopted for the optimal synthesis of biorefineries in various problem formulations (e.g., Murillo-Alvarado et al. 2013; Giuliano et al. 2016; Gong et al. 2016; Hernández-Calderón et al. 2016). A typical example is the problem formulation where one selects the best set of biomass-based products that is more suitable for market conditions of a specific region/country (Murillo-Alvarado et al. 2013). Clearly this problem formulation attempts to focus the attention to higher-level strategic decisions, when compared to the previous cases of detailed process designs, and quickly screen the broadly available options.

Models for each of the technologies under consideration are usually developed by combining literature data (obtained at laboratory scale) with off-line process flowsheet simulation, i.e., process simulation and PNS optimization are generally done in separate steps. In agreement with the macro-level of the decisions made with this approach, the level of detail in these designs is lower than in the designs used in the approach in Sect. 2.1. These simplifications allow the formulation of an optimization problem using linear input-output models for the technologies, which result in mixed integer linear programming (MILP) problems, instead of difficult-to-solve mixed integer nonlinear programming (MINLP) problems. Examples of PNS applied to the design of biorefineries that have been solved as MILP problems can be found in (Murillo-Alvarado et al. 2013; Giuliano et al. 2016; Gong et al. 2016; Hernández-Calderón et al. 2016; Cheali et al. 2014).

2.3 Distributed Approach: Motivating Examples

Although a priori dissimilar, the previous two approaches have the following aspects in common: (1) They assume an already defined set of biomasses as feedstocks. (2) They assume an already defined and, most importantly, narrow set of target products. (3) They implicitly consider that a single actor takes responsibility for the overall biomass conversion process, i.e., from biomass pretreatment until the desired product(s). Clearly, in view of the vast number of options for biomass-based chemicals, the previous approaches are not well suited for identifying economic opportunities in the initial exploratory phases, when all products in Fig. 2 are potentially attractive or when a company (interested in entering the biomass-based market) does not participate in the whole biomass processing technology chain.

A "real-life" example that falls in this category is the PlantBottle Technology© under development by The Coca-Cola Company. Given the information posted in the company's web page, the goal in this case is quite specific: to produce a "plastic beverage bottle that (…) looks and functions just like traditional PET plastic" (The Coca Cola Company 2018a). PET, polyethylene terephthalate, is traditionally made from purified terephthalic acid (PTA), which is derived from *p*-xylene and monoethylene glycol (MEG). As seen in Fig. 5 in first-generation PlantBottle packaging, oil-based MEG had already been replaced by biomass-based MEG, and then R&D efforts focused on finding oil-derived PTA replacements. The Coca-Cola Company supported R&D efforts for the production of bio-based PTA along three distinct

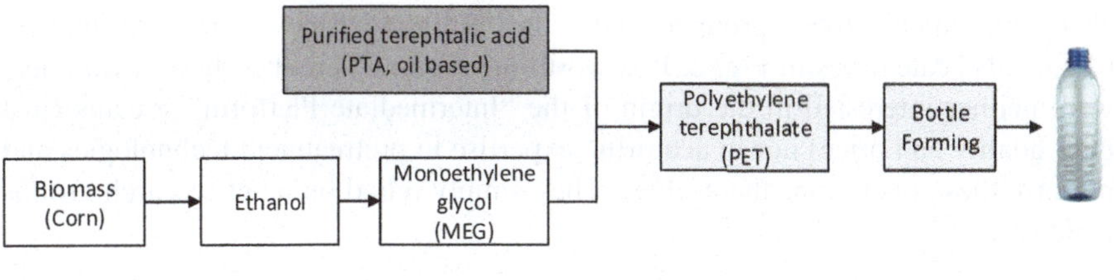

Fig. 5 First-generation PlantBottle from The Coca-Cola Company. Figure based on the diagram in The Coca Cola Company (2018b)). Bottle image from freeimages.com (plastic-bottle-1,327,399). Credit Ivan Philipov

lines (Packaging Getaway 2018): (a) direct production of *p*-xylene from biomass using thermochemical processes (collaboration with Virent), (b) indirect production through the intermediate isobutanol (collaboration with Gevo), and (c) production of PEF (polyethylene furanoate) an analog of PET (collaboration with Avantium). Figure 6 summarizes the proposed pathways. Notice that unlike the cases in Sects. 2.1 and 2.2, The Coca-Cola Company does not try to produce all the chemicals it needs in its own specially designed biorefinery. Instead, it relies on the availability of the most attractive intermediates (*p*-xylene, 5-hydroxymethylfurfural) produced by others, and it explores a number of alternative technologies leading to a polymer with the desired properties to form bottles.

Examples of distributed approaches also arise in the academic setting. Figure 7 shows the concept behind the recent Masdar Institute (now Khalifa University) UAE/MIT USA "Biorefinery: Integrated Sustainable Processes for Biomass Conversion to Biomaterials, Biofuels, and Fertilizer" joint research project. This project combined the expertise of seven researchers from both institutions with the aim of finding a biorefinery that suited UAE needs. As shown in the figure, the set of biomass feedstocks was given and corresponded to biomass streams available in the UAE. These biomass feedstocks, after pretreatment, produced a number of potential "Intermediate Platform" streams, such as cellulose, hemicellulose, lignin, lipids, and proteins, which were available in various compositions and phase characteristics, e.g., dilute or concentrated solutions, in various mixtures including only fluids or fluids and solids, with varying impurities some of which detrimental for subsequent processing, and in various chemical mixtures. Some of the research groups involved in the project, the ones shown in the left of Fig. 7, had expertise in fractionation of different types of biomass into these "Intermediate Platform" streams and subsequent conversion of the glucose stream into biofuels, mainly ethanol. These groups were interested in making their own processes more profitable by valorizing the remaining "Intermediate Platform" streams that resulted from the different pretreatments of the biomasses under their study. The objective of the collaborative project was precisely this: let researchers that have expertise in valorization of the streams take care of proposing the best processes.

These research groups, the experts in valorization of the "Intermediate Platform" streams, are the ones shown in the right of Fig. 7. Their objective was then to find technologies for upgrading the available "Intermediate Platform" streams into other chemicals, mainly those grouped under the "Building Blocks" or "Intermediate Chemicals" categories in Fig. 2. It is worth mentioning here that these researchers were neither interested in the origin of the "Intermediate Platform" streams (just their quality and price) nor in acquiring expertise in pretreatment technologies and perform these operations themselves. They simply relied on other research groups to do so.

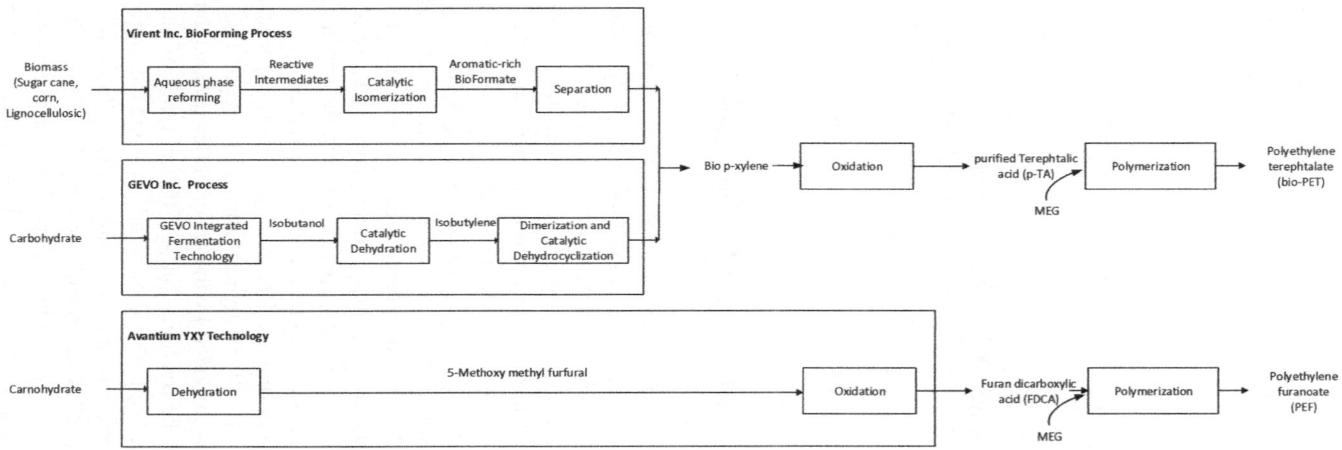

Fig. 6 Pathways for production of bio-based purified terephthalic acid (or replacement) for PlantBottle. (Peters et al. n.d.; Avantium 2018)

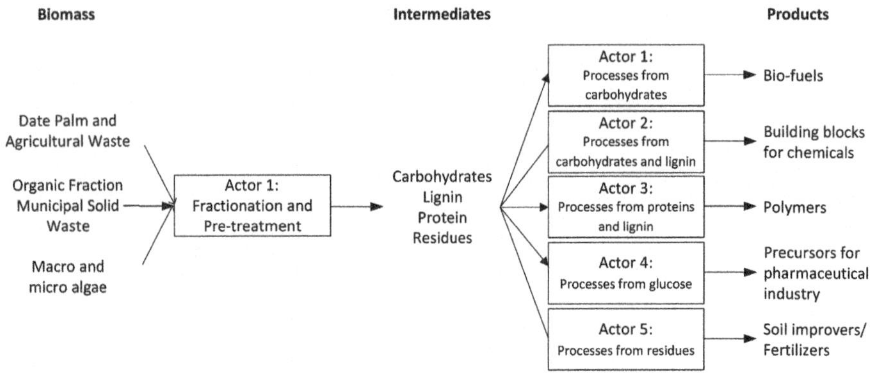

Fig. 7 The Masdar Institute (now Khalifa University)-MIT joint research project "Biorefinery: Integrated Sustainable Processes for Biomass Conversion to Biomaterials, Biofuels, and Fertilizer"

3 Foundations of the Decomposition of the Biorefinery Network

3.1 Intermediate Platform and Building Blocks

The implications arising from the cases described in Sect. 2.3, Distributed Approach: Motivating Examples, streamline the third approach for the evaluation of potential biorefineries as follows: there is a natural decomposition between the set of technologies converting the available biomass feedstocks to "Intermediate Platform" chemicals and the set of technologies that convert these "Intermediate Platform" chemicals into market products. Here, the former is referred as the *supply side* and the latter as the *demand side*. Notice that this decomposition allows an analysis of biorefineries which is in full alignment with the distributed structure of traditional refineries discussed in Sect. 1. In a general case, any of the compounds in Fig. 2 might be regarded as structure-dividing "Intermediate Platform" chemicals. Thus all actors (companies or research groups) involved in the production of these intermediates become the *suppliers*, and all actors involved in the upgrade of the intermediates become the *consumers*. However, previous experience in analysis of biorefineries (Torres et al. 2015) suggests that a first natural set of "Intermediate Platform" compounds is provided by the set of outlet streams, emanating from the operations required for the fractionation of different types of biomass. Thus actors involved in pretreatment technologies naturally become the suppliers. The next two examples illustrate the identification of "Intermediate Platform" chemicals and the processing scope of *suppliers* and *consumers*.

Figure 8a, shows an example of the compounds that are obtained when processing biomass using organosolv pretreatment. Organosolv is a type of pretreatment that fractionates high lignin biomass in a cellulose-rich solid stream and a liquid

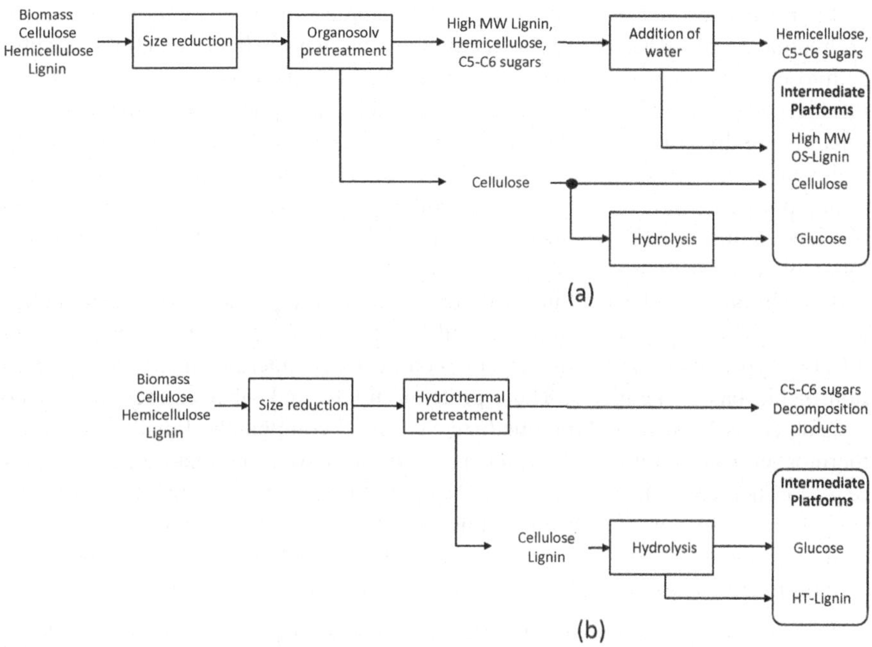

Fig. 8 Identification of "Intermediate Platform" compounds for two examples of processing trains for pretreatment of biomass. (**a**) corresponds to an organosolv-based pretreatment process, cellulose and lignin are separated in the first step, cellulose can be sold as is or further hydrolyzed to glucose, three intermediate platform compounds are obtained by this process, (**b**) corresponds to an hydrothermal pretreatment-based process; carbohydrates and lignin are separated by hydrolysis of the former, and then two intermediate platform compounds can be obtained

stream containing lignin, hemicellulose, and monomeric C5, C6 sugars (Aita and Kim 2010; Hallberg et al. 2012). *Cellulose* can be used as is for thermochemical processes; therefore it can be considered as an "Intermediate Platform" compound. With relative processing easiness, cellulose can be hydrolyzed into monomeric glucose, and this glucose can be used as a feedstock for a large number of biochemical or thermochemical processes. Therefore, *glucose* is another "Intermediate Platform" compound. The liquid stream (lignin and hemicellulose) can be sent to a separator from where a lignin-rich stream (high molecular weight lignin) is precipitated by addition of water. This lignin-rich stream could be used as a starting point for the production of polymers and aromatic compounds, among others (Ragauskas et al. 2014); thus *lignin* is another "Intermediate Platform" compound. The hemicellulose stream resulting from the lignin precipitation step could be used as is or be hydrolyzed into, more useful, monomeric C5 and C6 sugars. As this stream needs further processing (e.g., concentration, detoxification) in order to be used for current sugar upgrade technologies, the *hemicellulose* stream as is, is labeled as an "Opportunistic Intermediate," and not as a real "Intermediate Platform" compound (Torres et al. 2015).

Figure 8b shows a similar analysis, as in Example 1, for the case of hydrothermal pretreatment of biomass. With this technology, biomass is fractionated into a solid cellulose and lignin stream, and a liquid stream containing C5 sugars, organic acids, furanic, and phenolic compounds. Cellulose and lignin in the first stream are usually separated by hydrolyzing cellulose into glucose. Thus the "Intermediate Platform" compounds in this case are *lignin* and *glucose*. Here, it is worth mentioning that lignin properties are altered after hydrothermal pretreatment, hence the lignin obtained in this case (HT-lignin) is considered to be a compound different from the lignin obtained with organosolv pretreatment (OS-lignin).

In addition to the basic "Intermediate Platform" compounds, e.g., lignin, cellulose, and glucose, proteins, lipids, and other extractives can also be recovered from different types of biomasses and can be seen as critical intermediates in the production of bio-based chemicals. Thus, proteins, lipids, and other extractives can be considered as "first-line" "Intermediate Platform" compounds. Finally, in certain thermochemical biorefineries, syngas is produced. Given that mature technologies (e.g., Fischer Tropsch) already exist for upgrading syngas to valuable chemicals, syngas can be seen as a "first-line" "Intermediate Platform" compound.

The design and optimization problem for these supply-side actors can be therefore postulated as follows:

> For a given set of biomass feedstocks, select the feedstocks, sequence of pretreatment technologies and first-line intermediate platform compounds that generate the most profitable operation.

Notice that this design problem involves a superstructure of technologies. The reader is referred to Torres et al. (2015) for an example of such superstructure.

The potential products emanating from this "first-line intermediate platform compounds" correspond to the "Building Blocks" in Fig. 2. Any actor that intends to produce these "Building Blocks" using the "first-line intermediate platform compounds" is therefore regarded as an actor on the demand side. Usually these demand-side actors dominate a technology that allows the production of one or a narrow set of these "Building Blocks" products. Therefore, the design and optimization problem that they need to solve is simpler than the supply-side one.

> For a given 'Building Block' or a small set of desired 'Building Blocks,' (a) select as feedstocks the 'first-line intermediate platform compounds,' (b) design the process that converts the feedstocks to the desired 'Building Blocks,' and (c) compute the processing conditions, in order to maximize the economic benefit, e.g. annual profit, or return on investment.

After the "Building Blocks", the biomass value chain continues as follows: "Building Blocks" become the starting materials for a variety of "secondary chemicals", which in turn are the starting materials of the "intermediate chemicals", which can then be used for the production of an extremely large and diverse set of market products. There are two important notes to make at this point as follows: (1) Following the incremental number of products as one moves from feedstocks to market products, and akin to how the petrochemical industry is currently structured, we assert that the future biomass-based chemical industry will be composed of a large number of actors, each of which specializes in a technology and/or narrow set of products. (2) As one moves up the biomass value chain, the line between what is considered a supplier or a consumer becomes blurred. In fact, most of the actors will have a double role: they will consume not directly biomass but a compound derived from it and sell not a market product but a key precursor of it.

3.2 Mathematical Framework for the Analysis of a Distributed Biorefinery

Following the previous discussion on how a biorefinery should be decomposed, the optimal biorefinery is found by maximizing the profit of all the actors that take part in it. This requires formulating a multi-objective optimization problem whose objective function is a composite of the objective functions of the individual actors.

Figure 9 shows a generic actor i participating in the biorefinery: it may buy feedstocks f_i^{in} from the market at a unit price α_i and sell products f_i^{out} to the market for a unit price β_i. The exchange of intermediates of actor i with other members of the network are represented as $x_{j_p,i}$ (intermediates bought from an upstream actor j_p) and y_{i,k_q} (intermediates sold to actor k_q).

P_i, the profit of actor i, is then computed as follows:

$$P_i = \beta_i f_i^{\text{out}} + \sum_{k_q} p_{ik_q} y_{ik_q} - c_i\left(\underline{d_i}, f_i^{\text{in}}, x_{j_p i}\right) - \sum_{j_p} p_{j_p i} x_{j_p i} - \alpha_i f_i^{\text{in}} \quad (1)$$

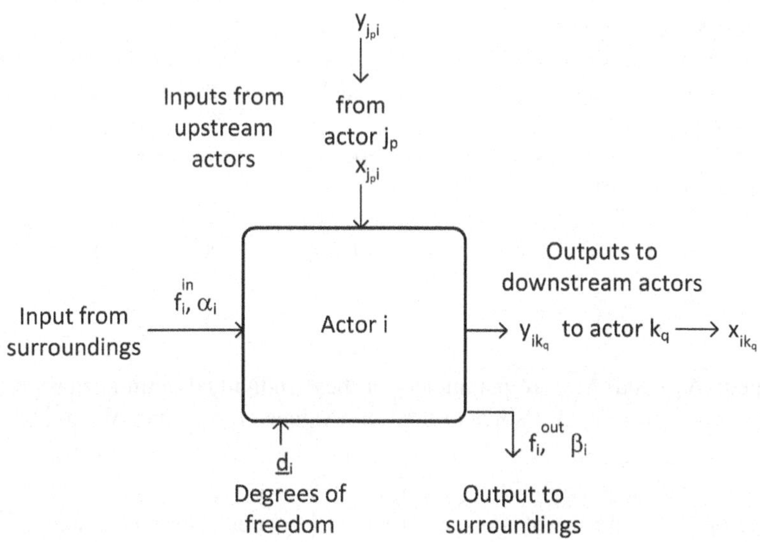

Fig. 9 Generic actor for a biorefinery network. On the basis of the original in Torres and Stephanopoulos (2016)

where

- $c_i\left(\underline{d_i}, f_i^{\text{in}}, x_{j_p i}\right)$ represents a cost function that includes annualized operating and capital costs for actor i. This cost depends on the inlet flow rates ($f_i^{\text{in}}, x_{j_p, i}$) and also the process design variables $\underline{d_i}$. These design variables also affect the input-output relations $y_{i,k_q} = y_{i,k_q}\left(\underline{d_i}, f_i^{\text{in}}, \underline{x_{ji}}\right)$.
- p_{ik_q} and $p_{j_p i}$ are the prices at which the intermediates are exchanged among the actors. Notice that these prices are not established market prices, these are prices that have to be agreed upon among the actors, and for which an optimal solution that is satisfactory for all the actors involved in the network is also required.

Then, the multi-objective optimization problem for maximizing the profit of the overall network, as a weighted sum of the individual profits, is formulated as follows:

$$\begin{aligned} &\max_{\underline{f^{\text{in}}}, \underline{d}, \underline{x}, \underline{p}} P = P_i + \sum_{t, t \neq i} s_t P_t \\ &\text{s.t.} \quad x_{ji} = y_{ji} \quad \forall i, j \\ &y_{iq} = x_{iq} \quad \forall i, q \end{aligned} \tag{2}$$

It is outside the scope of this chapter to fully discuss the mathematical derivations and solution methodologies for this optimization problem. The details can be found in Torres and Stephanopoulos (2016). In this section the most salient points and major conclusions from that work will be summarized:

1. The optimal network is achieved when all the actors have an equal weight in the network, i.e., $s_t = 1 \quad \forall\, t$.
2. In such a case, the optimization problem in Eq. 2 is separable and, under the assumption of convexity, can be solved by using the two-level Lagrangian approach.
3. In this approach, actors first solve, independently of the other actors, an optimization problem of the following form:

$$\begin{aligned} &\max_{\underline{f_i^{\text{in}}}, \underline{d_i}, x_{jp}} P_i = \beta_i f_i^{\text{out}} + \sum_{k_q} \Lambda_{ik_q} - c_i\left(d_i, f_i^{\text{in}}, x_{j_p i}\right) - \sum_{j_p} \Lambda_{j_p i} x_{j_p i} - \alpha_i f_i^{\text{in}} \\ &\text{s.t.} \quad d_i \in \mathfrak{D}_i \end{aligned} \tag{3}$$

   where Λ_{ik_q} and $\Lambda_{j_p i}$ are parameters of these individual optimization problem. The problem in Eq. 3 is solved for several values of Λ_{ik_q} and $\Lambda_{j_p i}$ to obtain the optimal response curves y_{ik_q} vs. Λ_{ik_q} and $x_{j_p i}$ vs. $\Lambda_{j_p i}$ for actor i. These curves provide the so-called first-level solutions.
4. The optimal solution of the overall network, under convexity assumptions, is found by intersecting the first-level response curves. This intersection provides the second-level solution and returns the optimal amount of intermediate to be exchanged among the actors, i.e., $y_{ik_q}^*$ and $x_{j_p i}^*$, and the equilibrium prices $\Lambda_{ik_q}^*$ and $\Lambda_{j_p i}^*$ which can be regarded as the optimal exchange price.

3.3 Illustrative Case Study

The framework of the above approach is better understood by considering the network of one supplier one intermediate one consumer. This example is reproduced from Torres and Stephanopoulos (2016) and considers the two actors shown in Fig. 10. The processes for both the supplier and consumer actors consist of a reaction and separation step, with the compound B_S being the intermediate exchanged by the actors. y_{S1} represents the flow of B_S produced by the supplier S and x_{C1} the flow of B_S required by the consumer C. p_{S1} and p_{C1}, respectively, represent the prices at which the intermediate is offered by the supplier and desired by the consumer, respectively. In this simple example, it is easy to see that $p_{S1} = p_{C1} = p_I$, but as discussed in the original text, this might not always be the case. In particular, this equality does not hold true if the supplier sells the same intermediate but at a different price to different consumers, a practice known as price differentiation.

Following the methodology in Sect. 3.2, at the first level, each actor maximizes its corresponding own profit, P_i, i.e.:

$$\begin{aligned} \max_{f_{S1} d_{S1}} \quad & P_{S1} = \Lambda y_{S1}\left(f_{S1}, d_{S1}\right) - c_{S1}\left(f_{S1}, d_{S1}\right) - \alpha f_{S1} \\ \text{s.t.} \quad & P_{S1} > 0 \\ & 0 \le f_{S1} \le f_{S1}^{\text{av}} \\ & d_{S1} \in \mathfrak{D}_{S1} \end{aligned} \tag{4}$$

$$\begin{aligned} \max_{x_{C2} d_{C2}} \quad & P_{C2} = \beta f_{C2}\left(x_{C2}, d_{C2}\right) - c_{C2}\left(x_{C2}, d_{C2}\right) - \Lambda x_{C2} \\ \text{s.t.} \quad & P_{C2} > 0 \\ & 0 \le f_{C2}\left(x_{C2}, d_{C2}\right) \le f_{C2}^{\text{de}} \\ & d_{C2} \in \mathfrak{D}_{C2} \end{aligned} \tag{5}$$

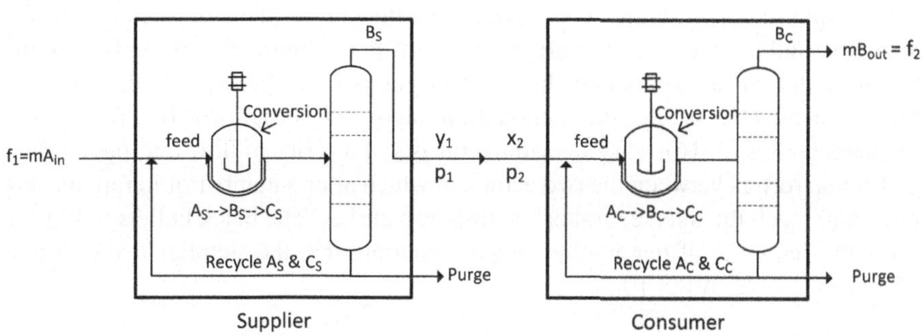

Fig. 10 Single supplier-single intermediate-single consumer case study. On the basis of the original in Torres and Stephanopoulos (2016)

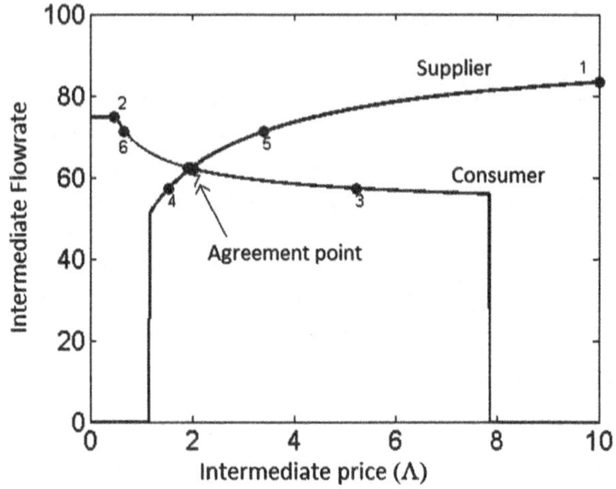

Fig. 11 Response curves for the single supplier-single intermediate-single consumer case study. On the basis of the original in Torres and Stephanopoulos (2016)

The two actors construct their corresponding response curves, i.e., x_{C1} vs. Λ for the consumer and y_S vs. Λ for the supplier. Figure 11 shows these response curves under the following assumptions:

1. The cost functions capture capital and operational costs and have the following general form:

$$c\left(m_{\text{feed}},\underline{d}\right)=\sum_{e=\text{reactor, separator}} a_e m_e^{0.6}\left(m_{\text{feed}},\underline{d}\right)+b_e m_e\left(m_{\text{feed}},\underline{d}\right) \tag{6}$$

2. The reactions follow first-order kinetics.
3. The corresponding per pass conversion, X, and recycle rate, R, are the process design variables.

The optimal solution for the overall network is obtained at the intersection of these response curves, which corresponds to a flow of 62 (units of mass intermediate/time), and a price of 1.96 (currency/units of mass intermediate). Arrival to this optimal solution can be done by direct intersection of the curves, if all the information is available, or by the negotiation algorithm described by Torres and Stephanopoulos (2016). This algorithm consists of a series of flow rate-price offers and counteroffers between the two actors, in which after a number of iterations, it is proven to reach the intersection of the response curves. The black bullets in Fig. 11 show the sequence of negotiations when assuming that the supplier makes a first offer of $y_{S1}^{(1)}=82$, $\Lambda^{(1)}=10$.

4 Distributed Processing Considering Underdevelopment Biorefinery Sections

The mathematical framework discussed in Sect. 3.2 offers a solution for analysis of the distributed biorefinery consisting of multiple independent actors, under the assumption that the response curves of such actors intersect. As discussed by Ashraf (2017) for this intersection to happen, upon optimization, the values of the design variables chosen as decision variables of the optimization problem in Eq. 3 must be sensitive to the parameter Λ.

However, in many cases there are sections of the network that are under development and do not have detailed process models to guarantee such sensitivity. Actors with underdeveloped processes usually consider experimental trials, analogy extension, or some design heuristics to develop preliminary models for assessment. The process models developed from this information satisfy mass and energy balances and can be used to estimate operating and capital costs. However, it may not suffice for the parametric optimization of design variables ($\underline{d}$). As an example consider many of the biochemical processing routes which have low solid loading, for example, dilute acid pretreatment, enzymatic hydrolysis, and ethanol fermentation. In these processes as water acts as a relatively large inert mass in the system, changes in the decision variables do not result in significant changes in the design of the equipments. Thus, the cost functions c_i discussed earlier, and the corresponding profits P_i of the individual actors, are not significantly affected.

In order to have response curves that are more sensitive to changes in process design variables, the mathematical framework presented in Sect. 3.2 was modified by Ashraf (2017) by using return on investment (ROI) as the payoff function and a minimum acceptable return on investment (MAR), instead of profit P_i, as the objective function.

4.1 ROI-Based Mathematical Framework

In this section ROI-based game theoretical framework is presented for distributed manufacturing of biorefinery processes. In the distributed manufacturing approach for a biorefinery, different processing steps are carried out by independent actors. These actors can be differentiated based on the feedstock they use, the conversion technology they employ, or the products they desire to produce. The process models for these actors may be at different level of development. These actors will interact with each other via exchange of intermediate products, which can be materials, energy, and/or services. Some of these intermediate products might be available from other sources and have an established market, for example, ethanol, while others might not have established market and price. To establish the price for these new entrants is the critical step in envisioning the distributed processing of biorefinery by multiple actors. The ROI-based game theoretical framework sets up the minimum acceptable price for these new products.

4.1.1 ROI as Objective Function

ROI is a measure of profitability or actor's payoff. Investment in a chemical processing plant by an actor is associated with a certain level of risk. For example, a new plant for an established marketable product is considered to have lower risk as compared to a plant with a new product. Similarly, a mature processing technology for a plant will have a lower risk associated with it as compared to a newer technology. Risk scenarios with different status of processing technology and product nature and their associated risk levels are discussed in detail in Peters et al. (2003). When an actor decides to make an investment, he expects a minimum return on the investment depending on the level of risk; a lower return on investment might be agreeable for a safer technology while a higher return on investment will be required for a riskier technology. In the ROI-based framework, each actor chooses its own minimum acceptable return (MAR) on investment depending on its process, technology, product, location, and economic constraints. This condition allows an actor to be a distinct entity, e.g., a distinct company in the biorefinery network.

ROI is the annual interest rate made by the profit on original investment, Eq. 7 (Seider et al. 2008). Let, t be the tax rate, c_{mat} the feedstock cost, c_{op} the annual operating cost, s the annual sales revenue, and c_{cap} the total capital investment. ROI is a gross economic profitability measure, and it does not take into account time value of money and uses straight-line depreciation. It is mostly used to compare alternative processes during the concept stage of process development (Seider et al. 2008; Douglas 1988). Applying Eq. 7 to a generic actor presented in Fig. 9 results in Eq. 8.

$$\text{ROI} = \frac{(1-t)\left(s - c_{\text{mat}} - c_{\text{op}}\right)}{c_{\text{cap}}} \tag{7}$$

$$\text{ROI}_i = \frac{(1-t_i)\left(\beta_i f_i^{\text{out}} + \sum_{k_q} p_{ik_q} y_{ik_q} - \sum_{j_p} p_{j_p i} x_{j_p i} - \alpha_i f_i^{\text{in}} - c_{op_i}\left(\underline{d_i}, f_i^{\text{in}}, x_{j_p i}\right)\right)}{c_{\text{cap}_i}\left(\underline{d_i}, f_i^{\text{in}}, x_{j_p i}\right)} \tag{8}$$

4.1.2 Foundations of the Framework

The solution approach is similar to the one presented in Sect. 3.2. The only difference is that at first level ROI is used to generate response strategy such that each actor's objective is to meet its minimum acceptable return (MAR) on investment (Eq. 9):

$$\text{ROI}_i - \text{MAR}_i = 0 \tag{9}$$

It is assumed in that the actors act independently in setting their MAR targets and act rationally in selecting their strategy with the aim of meeting their targets of MAR. Both supply and demand-side actors are assumed here to have similar payoff function; however, due to opposing place in the game, their response curves are opposite. Price and flow of the intermediate products determine the revenue for the supply-side actor, while they represent a cost for the demand-side actor. Hence, a higher value for these variables is suitable for actors on supply side, while lower values are desirable for the demand-side actors.

In the solution procedure the steps followed by the supply and demand-side actors are also symmetric. In the first level, each actor finds its feasible response value of flow at given prices. At the second level, intersection of the response curves is searched to find the feasible value of intermediates price and flow. This approach is demonstrated below using two cases of lignocellulosic biorefinery.

4.2 Illustrative Case Study for ROI-Based Framework

A lignocellulosic biorefinery network using a biochemical conversion pathway is taken here as example. It consists of supply-side actors employing the hydrothermal pretreatment and enzymatic hydrolysis to produce monomeric sugars as an intermediate product stream. On the demand side, it has actors utilizing the intermediate sugars to produce marketable products. Two case studies are presented:

- **Case 1**: Select feasible supply-side actor for the intermediate sugar production.
- **Case 2**: Find a feasible price for the intermediate sugars, when multiple actors are producing and utilizing it.

The process and cost modeling for the two cases are detailed below.

Supply-side actors Actors 1, 2, 3, and 4 are considered to be on the supply side. Process simulation and economic models for these actors are adopted from Ashraf and Schmidt (2018). They differ in the lignocellulosic feedstock, as a result their processing costs and product yields are different. Actor 1 uses Bermuda grass (BG), Actor 2 uses jasmine hedges (JH), Actor 3 uses date palm (DP) fronds as feedstock, and Actor 4 uses a mixture of the aforementioned feedstock in the following ratio BG:JH:DP::1:1:2. They all are based on hydrothermal pretreatment and enzymatic hydrolysis technology for the conversion of lignocellulosic feedstock to monomeric sugars.

Demand-side actors Actors 5 and 6 are considered here to exist on the demand side that utilizes monomeric sugars to produce value added chemicals. Actor 5 produces citric acid via fermentation of monomeric sugars using *Aspergillus niger*. Citric acid is an organic acid used in the food and beverage industries to preserve and enhance flavor. Actor 6 produces lysine from the intermediate sugar stream. Lysine is a building block for muscle proteins and is used as an additive in the animal feed. Process simulation and economic models are adopted from the SuperPro Designer v10 examples library (Intelligen Inc. 2017). Details about the process inputs and assumption can be found in Ashraf and Schmidt (2018) and Intelligen Inc. (2017).

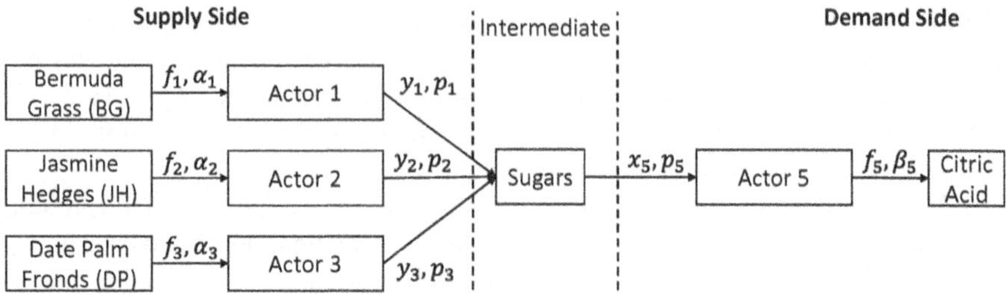

Fig. 12 Process network for the lignocellulosic biorefinery Case 1 consisting of three supply-side actors competing for the production of monomeric sugars as the intermediate product to supply to the demand-side actor producing citric acid

4.2.1 Case 1

In Case 1 the ROI-based game theoretical framework is used to find the best supply-side actor for citric acid production. The network considered in this case consists of Actors 1, 2, and 3 on the supply side and Actor 5 (citric acid production) on the demand side, as shown in Fig. 12. Actors 1, 2, and 3 use same technology; however, their feedstocks are different: Bermuda grass, jasmine hedges, and date palm fronds, respectively. Hence the solution of this problem solves the process synthesis problem for citric acid production from the three different options in the lignocellulosic feedstock.

For Actors 1, 2, and 3, the MAR is taken to be 30%, considering that the technology to produce sugars from these feedstock is new and must be associated with high risk. It is assumed that the maximum lignocellulosic feedstock available ($f_{1,2,3}^{\mathrm{av}}$) is 150 MT/h (metric tonne) at price of 20 USD/MT. For Actor 5 MAR of 10% is assumed on the basis that the citric acid fermentation technology is mature and the product has a market demand. It is assumed that maximum citric acid market demand (f_5^{de}) is 36 MT/h. The selling price of citric acid is taken as 2.0 USD/kg. Feasible response price of sugar and its flow rate are calculated by each of these actors by applying Eqs. 7–9 and using the C_{TCI}, C_{OP}, and product yield data generated from process simulations discussed in the previous section.

The feasible response curves of the actors are drawn in Fig. 13. It shows that the minimum acceptable prices of sugar for Actors 1, 2, and 3 are 0.43, 0.83, and 0.50 USD/kg, respectively—below these price values, they cannot meet their preset MAR of 30% at any flow rate of the sugar. This shows that Actor 1 is most economically competent, as it can offer a lower price for its product. Actor 2 is least economically competent as its product price is highest among the three supply-side actors. Similarly, the maximum price that Actor 5 on the demand side can afford for the sugars is 0.58 USD/kg. A price higher than 0.58 USD/kg renders the Actor 5 unable to meet its MAR of 10%.

Using the framework, Actor 5 finds the feasible price of the intermediate sugars with each of the actor on supply side separately. The intersection points are drawn over the response curves as shown in Fig. 13. Actor 5 finds equilibrium point with

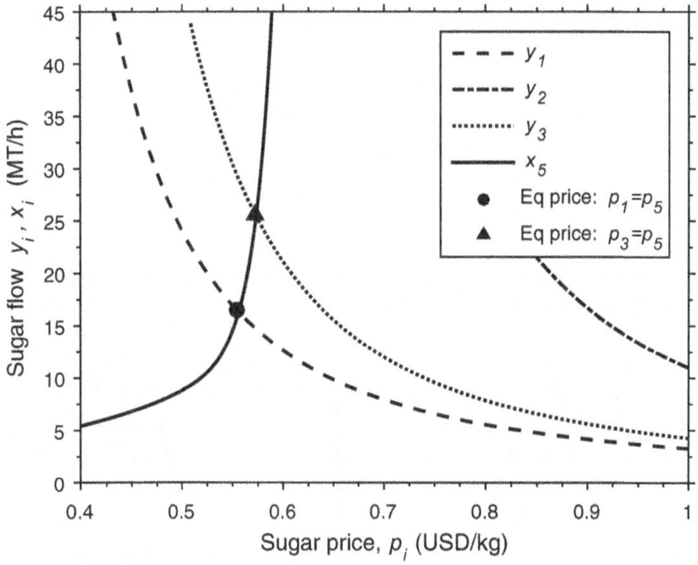

Fig. 13 Results of the negotiation algorithm applied to the lignocellulosic biorefinery network presented in Case 1 (Fig. 12]). The feasible response from the actors is plotted as lines. The equilibrium point that reached between Actors 1 and 5 is shown as bullet point (•), and the equilibrium point that reached between Actors 3 and 5 is shown as triangle (▲)

Actors 1 and 3; however, it did not find equilibrium point with Actor 2. This is due to the fact that their response curves do not intersect; the maximum price Actor 5 can pay is 0.58 USD/kg, while the minimum price that Actor 2 requires is 0.83 USD/kg. Equilibrium price with Actor 1 is $p_1 = p_5 = 0.55$ USD/kg and with Actor 3 is $p_3 = p_5 = 0.57$ USD/kg. This shows that Actor 1 offers a lower price for the sugar intermediate. A lower price of the intermediate sugar means a better payoff for demand-side actor; hence Actor 1 or Bermuda grass feedstock is selected as the economically better supply-side process to produce citric acid.

4.2.2 Case 2

In this case, we take three supply-side actors (Actors 1, 3, and 4) producing the sugar intermediate and two demand-side actors (Actors 5 and 6) competing for it, as shown in Fig. 14. It is assumed in this case that all the actors are existing at a time and ROI-based framework is used to determine the minimum acceptable price of the intermediate sugar stream.

Actor 4 here uses a mixed feedstock in the ratios BG:JH:DP::1:1:2. Actor 6 is producing lysine from sugars. Same as in Case 1 for supply-side actors, a MAR of 30% is selected for the Actors 1, 3, and 4. The lignocellulosic feedstock is assumed to be available at 20 USD/MT and a maximum of 150 MT/h is available to each actor.

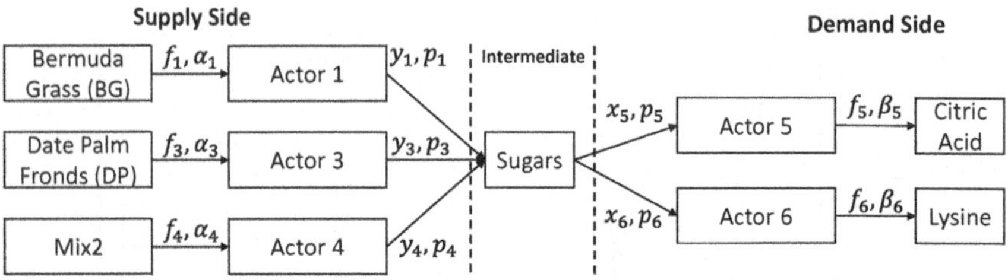

Fig. 14 Process network for the Case 2 consisting of three supply-side actors and two demand-side actors; here all actors are competing for the single sugar intermediate product which is considered as a new entrant to the market

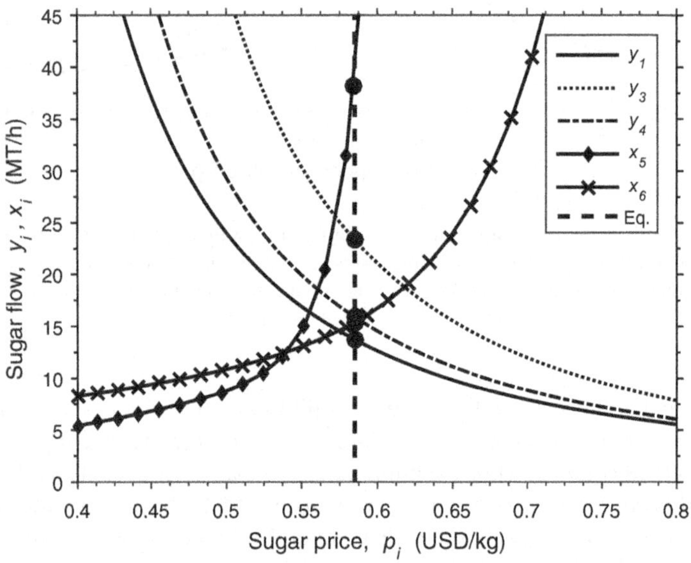

Fig. 15 Results for the negotiation algorithm applied to the network presented in Case 2 (Fig. 14) of the lignocellulosic biorefinery case studies. The feasible response from the actors is plotted as lines. The location of equilibrium price agreed among the actors is shown as a vertical dashed line (▪▪), its intersection with the actor's response curves is shown as bullet point (•), and it shows the location equilibrium flow of the actors

On the demand side, it assumed that the technologies are mature, and 10% and 15% of MAR are set for Actors 5 and 6, respectively. The selling price of marketable products is taken as 2.0 USD/kg for citric acid and 4.0 USD/kg for lysine. The maximum market demand is assumed as 35 MT/h and 24 MT/h for citric acid and lysine, respectively.

Using the cost data (C_{TCI}, C_{OP}) and product yield data from the process simulations and applying Eq. 9, the feasible response of each actor is generated. The feasible response curves are plotted in Fig. 15. The minimum price of sugar below which they cannot meet their MAR of 30% is 0.43, 0.50, and 0.45 USD/kg for Actors 1, 3, and 4, respectively. This shows that Actor 1 and Actor 4 are more economically competent than Actor 3 as they can offer a lower price for the sugar intermediate.

Similarly, on the demand side, the maximum prices that Actors 5 and 6 can take for sugar are 0.58 and 0.71 USD/kg, respectively. This shows that on the demand side, Actor 6 is more economically competent than Actor 5, as it can afford a higher price for its feed of sugar intermediate.

The framework is used to establish the minimum price of the intermediate sugars when all of these actors are assumed to exist at a time. The framework converges to equilibrium point in 30 iterations, where mass and cash balance errors are ≤ 0.1. The minimum price established for the sugar intermediate is $p_1 = p_3 = p_4 = p_5 = p_6 = 0.58$ USD/kg. The location of equilibrium price is shown with a vertical dashed line (--) in Fig. 15. The intersection of this line with the feasible response curves graphically represent the equilibrium flow for each actor, shown as bullet points (•). The equilibrium sugar flow for each actor is as follow: $y_1 = 13.7$, $y_3 = 23.3$, $y_4 = 15.8$, $x_5 = 38.1$, and $x_6 = 15.3$ MT/h. It is to be noted that the sum of supply-side flow and the sum of demand-side flow are same, 51.6 MT/h, which established minimum market size of sugar intermediate. This is due to the fact that mass balance limit is included in the framework, and it is assumed that all the actors exist at a time.

5 Summary and Concluding Remarks

The design of an "optimal biorefinery" implies the definition of the set of marketable products and the technologies to produce them from a given set of biomass feedstocks. These technologies usually consist of a series of distinct processing steps, which could be carried out by a single actor or distributed among several different actors. The latter is the case of the traditional petrochemical industry, and this chapter discusses that biorefineries should be designed analogously.

After establishing a decomposition of the proposed biomass-based chemical value chain in sections that supply intermediate platform chemicals and sections that process these intermediates to marketable products (consumers), two mathematical frameworks are discussed for establishing the flow rates and prices at which the intermediates should be exchanged between the actors. These frameworks consider that actors behave as independent entities and that they do not share proprietary information such as technologies used, processing details, or economic models. Each actor establishes the sets of flows and prices of the intermediate products that allow its participation in the network by using its own protocols. Then, upon negotiation with the other members of the network, an agreement on flows and prices that is satisfying for all members is achieved. The first framework has its basis in optimization theory and results in a distribution of flow rates and prices that is optimal for the overall network as well as for the individual actors. The second ROI-based framework has as an advantage—the ability of being applicable to process models with limited level of development and available knowledge.

These frameworks can be used for collaboration by different entities with expertise in niche areas of biorefinery process or by multidisciplinary teams developing biorefinery processes. One of the main advantages of the frameworks is that they provide an estimate of the price of intermediate products that are not currently sold in the market. Another advantage is their flexibility as both of them allows testing the feasibility of coupling new processing sections within the existing processing network, without the need of solving large optimization problems. In a general setting, the methodologies discussed in this chapter can also be applied to other distributed manufacturing processes where distinct operating sections or companies are interacting via exchange of intermediate products or services.

References

AdvanSix: Products (2018) Chemical intermediates. https://www.advansix.com/chemicalintermediates/products/. Accessed 17 May 2018

Aita GM, Kim M (2010) Pretreatment technologies for the conversion of lignocellulosic materials to bioethanol. Chapter 8. In: Sustainability of the sugar and sugar–ethanol industries, vol 1058, American Chemical Society, pp 117–145

Ashraf MT (2017) Valorization of mixed lignocellulosic residues in the UAE using a biorefinery concept. Khalifa University of Science and Technology, Abu Dhabi

Ashraf MT, Schmidt JE (2018) Process simulation and economic assessment of hydrothermal pretreatment and enzymatic hydrolysis of multi-feedstock lignocellulose—separate vs combined processing. Bioresour Technol 249:835–843

Avantium (2018) YXY technology. https://www.avantium.com/yxy/yxy-technology/. Accessed 17 May 2018

BASF (2018) Products and industries: chemicals. Chemicals. https://www.basf.com/en/products-and-industries/chemicals.html. Accessed 17 May 2018

BP (2018a) Petrochemicals. Texas City Chemicals, Texas. https://www.bp.com/en_us/bp-us/what-we-do/petrochemicals/texas-city.html. Accessed 17 May 2018

BP (2018b) Petrochemicals. Cooper River Chemicals, Cooper River. https://www.bp.com/en_us/bp-us/what-we-do/petrochemicals/cooper-river.html. Accessed 17 May 2018

Cheali P, Gernaey KV, Sin G (2014) Toward a computer-aided synthesis and design of biorefinery networks: data collection and management using a generic modeling approach. ACS Sustain Chem Eng 2:19–29

Douglas JM (1988) Conceptual design of chemical processes. McGraw-Hill, New York

Dupont Tate & Lyle: Susterra (2018a) Life cycle analysis overview—Susterra® propanediol. http://www.duponttateandlyle.com/sites/default/files/Susterra(R)Propanediol-LCA Overview.pdf. Accessed 17 May 2018

Dupont Tate & Lyle: FAQS (2018b) 1,3-propanediol FAQ. http://www.duponttateandlyle.com/faqs. Accessed 17 May 2018

Dupont Tate & Lyle: Story (2018c) Our Story—DuPont and Tate & Lyle bio products. http://www.duponttateandlyle.com/our_story. Accessed 17 May 2018

Dupont Tate & Lyle (2018d) History. https://www.tateandlyle.com/about-us/history. Accessed 17 May 2018

Giuliano A, Cerulli R, Poletto M, Raiconi G, Barletta D (2016) Process pathways optimization for a lignocellulosic biorefinery producing levulinic acid, succinic acid, and ethanol. Ind Eng Chem Res 55:10699–10717

Gong J, Garcia DJ, You F (2016) Unraveling optimal biomass processing routes from bioconversion product and process networks under uncertainty: an adaptive robust optimization approach. ACS Sustain Chem Eng 4:3160–3173

Hallberg C, O'Connor D, Rushton M, Pye EK, Gjennstad G, Berlin A, et al (2012) Continuous counter-current organosolv processing of lignocellulosic feedstocks. US008772427B2

Hernández-Calderón OM, Ponce-Ortega JM, Ortiz-Del-Castillo JR, Cervantes-Gaxiola ME, Milán-Carrillo J, Serna-González M et al (2016) Optimal design of distributed algae-based biorefineries using CO2 emissions from multiple industrial plants. Ind Eng Chem Res 55:2345–2358

IEA (2007) IEA bioenergy Task 42 on biorefineries: co-production of fuels, chemicals, power and materials from biomass. Minutes Third Task Meet

Intelligen Inc. (2017) SuperPro(R) designer

Kazi FK, Patel AD, Serrano-Ruiz JC, Dumesic JA, Anex RP (2011) Techno-economic analysis of dimethylfuran (DMF) and hydroxymethylfurfural (HMF) production from pure fructose in catalytic processes. Chem Eng J 169:329–338

MatWeb: Material Property Data (2018) Nylon manufacturers, suppliers, and distributors. http://www.matweb.com/reference/Manufacturers.aspx?MatGroupID=17. Accessed 20 July 2018

Murillo-Alvarado PE, Ponce-Ortega JM, Serna-Gonzalez M, Castro-Montoya AJ, El-Halwagi MM (2013) Optimization of pathways for biorefineries involving the selection of feedstocks, products, and processing steps. Ind Eng Chem Res 52:5177–5190

Packaging Getaway (2018) Coca-Cola's 100% plant-based bottle. http://www.packaging-gateway.com/projects/coca-cola-plant-based-bottle/. Accessed 5 July 2018

Patel AD, Serrano-Ruiz JC, Dumesic JA, Anex RP (2010) Techno-economic analysis of 5-nonanone production from levulinic acid. Chem Eng J 160:311–321

Peters MS, Timmerhaus KD, West R (2003) Plant design and economics for chemical engineers, 5th edn. McGraw-Hill, New York

Peters MW, Taylor JD, Jenni M, Manzer LE, Henton DE (n.d.) Integrated process to selectively convert renewable isobutanol to P-xylene

Pham V, El-Halwagi M (2012) Process synthesis and optimization of biorefinery configurations. AICHE J 58:1212–1221

Ragauskas AJ, Beckham GT, Biddy MJ, Chandra R, Chen F, Davis MF et al (2014) Lignin valorization: improving lignin processing in the biorefinery. Science 344:1246843

Seider WD, Seader JD, Lewin DR, Widagdo S (2008) Product and process design principles: synthesis, analysis and design, 3rd edn. Wiley, Hoboken

The Coca Cola Company (2018a) PlantBottle. http://www.coca-colacompany.com/our-company/plantbottle. Accessed 17 May 2018

The Coca Cola Company (2018b) PlantBottle basics. https://www.coca-colacompany.com/stories/plant-bottle-basics

Torres AI, Stephanopoulos G (2016) Design of multi-actor distributed processing systems: a game-theoretical approach. AICHE J 62:3369–3391

Torres AI, Daoutidis P, Tsapatsis M (2010) Continuous production of 5-hydroxymethylfurfural from fructose: a design case study. Energy Environ Sci 3:1560–1572

Torres AI, Tsapatsis M, Daoutidis P (2012) Biomass to chemicals: design of an extractive-reaction process for the production of 5-hydroxymethylfurfural. Comput Chem Eng 42:130–137

Torres AI, Cybulska I, Fang C, Thomsen MH, Schmidt JE, Stephanopoulos G (2015) A novel approach for the identification of economic opportunities within the framework of a biorefinery. Comput Aided Chem Eng 37:1175–1180

US-DOE (2018) Integrated biorefineries. https://energy.gov/eere/bioenergy/integrated-biorefineries. Accessed 17 May 2018

Werpy T, Petersen G, Aden A, Bozell J, Holladay J, White J, et al (2004) Top value added chemicals from biomass (volume 1: results of screening for potential candidates from sugars and synthesis gas). 1

Part II
Thermochemical Processes

Pyrolysis and Gasification of Lignocellulosic Biomass

Maria Puig-Arnavat, Tobias Pape Thomsen, Giulia Ravenni, Lasse Røngaard Clausen, Zsuzsa Sárossy, and Jesper Ahrenfeldt

1 Introduction

Thermochemical conversion processes (combustion, gasification, and pyrolysis) are widely used to recover energy from biomass. In a biorefinery context, pyrolysis and gasification are especially interesting as both platforms offer high feed and product flexibility, providing the possibility to convert many different biogenic feedstocks into a wide variety of products such as heat, electricity, chemicals, transport fuels, and high-value ash and char products.

Pyrolysis is a thermal decomposition process of carbonaceous materials in the absence of air/oxygen. The cracking of chemical bonds leads to the formation of molecules with a lower molecular weight. Different product fractions are obtained: a solid (char), a liquid/condensed (tars), and a non-condensable gaseous fraction. Depending on the heating rate and solid residence time, biomass pyrolysis can be divided into three main types: slow pyrolysis, fast pyrolysis, and flash pyrolysis. Slow pyrolysis (typically <20 °C/min and retention times >15 min) has been conventionally used for the production of charcoal and to maximize the solid yields. Fast pyrolysis and flash pyrolysis processes are often applied in systems with focus

M. Puig-Arnavat (✉)
Department of Chemical and Biochemical Engineering, Technical University of Denmark, DTU Risø Campus, Roskilde, Denmark

Biomass Gasification Group, Chemical and Biochemical Engineering Department, Technical University of Denmark, Roskilde, Denmark
e-mail: mpar@kt.dtu.dk

T. P. Thomsen · G. Ravenni · Z. Sárossy · J. Ahrenfeldt
Department of Chemical and Biochemical Engineering, Technical University of Denmark, DTU Risø Campus, Roskilde, Denmark

L. R. Clausen
Department of Mechanical Engineering, Section of Thermal Energy, Technical University of Denmark, Kgs. Lyngby, Denmark

© Springer Nature Switzerland AG 2019

J.-R. Bastidas-Oyanedel, J. E. Schmidt (eds.), *Biorefinery*,
https://doi.org/10.1007/978-3-030-10961-5_4

on bio-oil production and typically involve much higher heating rates and shorter residence times. In fast pyrolysis the heating rate is typically around 10–200 °C/s and the residence times around 0.5–10 s but typically <2 s (Demirbas and Arin 2002). Bio-oil yield from optimized fast pyrolysis processes can be as high as 50–70 wt% on dry biomass basis. The flash pyrolysis process is characterized by extremely high heating rates of >1000 °C/s and even shorter residence times (<0.5 s), resulting in very high bio-oil yields which can achieve up to 75–80 wt% (Jahirul et al. 2012; Bridgwater et al. 1999). Bio-oils from pyrolysis have a great potential to be used in refineries to synthesize chemicals and fuels. However, more research is needed to overcome the problems they present like strong corrosiveness (pH = 2–4), high viscosity, immiscibility with conventional fuels, and poor chemical stability with polymerization of components on storage (Milina et al. 2014).

The gasification platform adds to the pyrolysis a char conversion process where carbon in the char reacts with a gasification agent such as steam or carbon dioxide at elevated temperatures. The gasification reactions are endothermic, and the heat is either supplied externally (allothermal gasification) or by supplying air/oxygen to the gasifier. The main product from the gasification process is a non-condensable gas product (CO_2, CO, H_2, H_2O, and other gaseous hydrocarbons), and by-products can include smaller quantities of char, ash, and several condensable compounds (tars and oils). Biomass gasification involves a sequence of several stages occurring at different temperatures: drying (100–200 °C), pyrolysis (200–700 °C), partial combustion (>800 °C), and reduction or gasification (700–900 °C). These stages often overlap depending on the specific gasifier design. The quality of the gas produced varies according to the gasifying agent used, feedstock, bed material, operational conditions (temperature, pressure, air to fuel ratio), and gasification technology. The gas obtained covers a wide range of calorific values from 4 to 7 MJ/m^3 when using air to 12–28 MJ/m^3 when pure oxygen is used. Thermal gasification can play an important role in the future energy system because it offers a flexible and efficient platform that can meet a variety of needs. However, despite the many advantages that biomass gasification presents, it has not yet been able to consolidate its role and become a mature technology in other areas than small-scale CHP from wood with mediocre thermal efficiencies. Around 1000 units of this type exist in Germany and surrounding countries. However, in other areas, great expectations have led to great disappointment because insufficient resources have been allocated for immediate and future operational issues related to:

- Handling and feeding the biomass feedstock
- Optimizing the energy efficiency to high moisture content feedstocks
- Addressing variations in biomass fuel properties
- Unrealistic fuel flexibility expectations
- Upscaling
- Ash-related problems including sintering, agglomeration, deposition, erosion, and corrosion
- Tar-related issues, e.g., condensation at lower temperatures, which can lead to fouling and plugging of the plant pipelines, filters, catalyst units, or engines
- Problems with gas cleaning trains and impurities such as sulfur compounds (e.g., H_2S, COS), hydrogen chloride, alkali, and ammonia

Despite past experiences with these challenges in several projects and plants, there is a firm belief in academia, industry, and among energy political consultants and advisors that thermal gasification of biomass will play a crucial role in future energy systems. Thus, in order to increase the utilization of biomass pyrolysis and gasification and make it commercially interesting in future energy systems, new integrated biorefinery designs with optimized thermal concepts and combinations of different technologies are required to maximize product yield and value, increase the overall process efficiency, and improve the economic viability.

The present chapter provides an insight on the versatility and potential of biomass pyrolysis and gasification processes and their products. It also describes some new concepts and solutions like process integration schemes, polygeneration strategies, biochar uses, and tar abatement strategies that can help to overcome the operational challenges that the technology is facing.

2 Thermal Pyrolysis and Gasification of Secondary Resources and Fuel Mixes

Pyrolysis and gasification platforms are often designed for conversion of one or several conventional biomass types such as wood, straw, macroalgae, and *Miscanthus* grass (Trinh et al. 2013; Saleh et al. 2014; Ahrenfeldt et al. 2013). However, in a biorefinery context, it may be valuable to extend the potential range of organic material fractions converted in thermal processes to cover more of the biogenic materials in the category *organic secondary resources*, not only the classic biomass fractions. Secondary resources refer, in this chapter, to a category of organic residues and degraded organic materials, which often have a low or negative price due to undesirable circumstances and characteristics, e.g.:

- Difficulties in collection, transport, storage, or handling
- Low energy density per unit of mass or volume due to a high moisture content or ash content
- Very inhomogeneous, requiring flexible and resource consuming handling
- Highly polluted, polluting, and/or harmful
- Very volatile, difficult to contain and generates local odor or dust problems

If a resource is sufficiently problematic, management of the resource will be costly, and, in many cases, the price of the resource will be low. A low price does not necessarily entail a low value of a given resource. The category of secondary organic resources includes many different subgroups. Some of the most important groups include:

- Agricultural by-products and residues, e.g., crop residues (stalk, leaf, cob, etc.), manure fibers, muck and bedding, and fibers from biogas slurry
- Municipal by-products and residues, e.g., source-segregated organic waste, used textiles, grass and cuttings from road and park maintenance, beach cleaning waste, and sewage sludge

- Industrial by-products and residues, e.g., residues from breweries, food packaging, food retail, and food preparation or residues from production of nonfood products based on the partial conversion of animal and vegetable raw materials

Optimized treatment of secondary resources has great potentials from an economic, environmental, and resource political point of view. The positive effects of thermal pyrolysis or gasification of organic secondary resources can include (1) increased non-fossil energy production, (2) reduced emissions of greenhouse gases from storage and direct application of unstable fractions, (3) carbon sequestration and soil enhancement, (4) nutrient recovery and recycling, (5) increased security of supply, and (6) reduced risks of terrestrial toxicity in soil management (Thomsen et al. 2016). For these reasons, it may prove beneficial to develop biorefineries and waste refineries to convert or co-convert secondary resources. Finally, application of thermal processes developed for low-quality fuels will also increase the technical potential for waste and by-product valorization within the biorefinery concept and add to the energy and mass integration of the system.

Recent research in thermal valorization of secondary resources in systems with pyrolysis- or gasification-based processes have indicated a huge potential for optimizing the present management systems. However, the thermal conversion of secondary resources has often proven problematic due to undesirable fuel characteristics (e.g., high moisture contents, low energy density, low melting points, and particle bridging). To reduce the impact of varying and problematic fuel characteristics, it may become useful—or necessary—to mix two or more different fuels in optimized ratios to match the criteria of the conversion technology as well as the potential end use of the produced products. In this way, the practical impact of problematic fuel characteristics can be reduced, and potential synergies obtained. Potential benefits that could be obtained from proper fuel mixing include:

- Reduced requirements for drying by mixing wet and dry fuels
- Standardized fuel characteristics such as energy density, proximate composition, and char reactivity
- Thermal purification of valuable elements in problematic resources like waste, manure, and sludge
- Optimized ash composition for fertilizer use by mixing fuels with, e.g., high P content with fuels with, e.g., high K content
- Potentially increased P fertilizer quality and decreased heavy metal content in ashes by co-gasification of P-rich secondary resources with fuels with high K, Na, Cl, or Mg content (Herzel et al. 2016; Nanzer et al. 2014; Nowak et al. 2012; Stemann et al. 2015; Vogel et al. 2013; Adam et al. 2009)

A small study published in 2016 (Thomsen et al. 2016) illustrates the potential benefits and drawbacks of mixing different fuels in different thermal processes (Fig. 1) by investigating, in laboratory conditions, which fuel mix characteristics were linear combinations of the characteristics of the individual fuels and which deviated substantially from such linearity.

The screening included proximate composition; higher heating value of fuels and chars; bulk density of fuels, ashes, and chars; char reactivity; char deposition

Fig. 1 (Left) Mix 1—palm kernel shells and rice husks, 50/50 w/w. (Center) Mix 2—palm kernel shells and animal meat and bone meal C1, 50/50 w/w. (Right) Mix 3—rice husks and animal meat and bone meal C1, 50/50 w/w. All weights on as-received basis

and formation of char agglomerates during pyrolysis; ash deposit sticking and vaporization; and P extractability of incubated ash and char samples. Results showed (1) good agreement between measured and calculated values of proximate composition, higher heating values (fuel and char), and fuel bulk densities; (2) inconsistent, but substantial, deviations from the linear predictions within char reactivity; (3) generally lower char and ash bulk densities of the mixes than predicted by linear sums of the char and ash bulk densities of the involved fuels; (4) substantial improvement of char deposition and char agglomeration during pyrolysis, especially for the animal meat and bone meal samples; (5) huge improvements also in ash deposit sticking as well as ash sintering in incineration or gasification processes, especially in samples with palm kernel shells and animal meat and bone meal; and (6) tremendous increase in the Mix 1 ash sample and Mix 3 char sample P fertilizer quality compared to the predicted values (Thomsen et al. 2016). In this context, mixing biomass into animal meat and bone meal before pyrolysis was previously found to increase extractability of bone meal char P (Zwetsloot et al. 2015). Similarly, positive experiences were obtained in studies with thermochemical treatment of waste and sludge with Na- and Cl-rich additives, and it is therefore expected that the high Na and Cl content of the palm kernel shell sample could have a positive influence on the P extractability of the mixed chars and ashes (Herzel et al. 2016; Nanzer et al. 2014). Based on these results and other studies recently published in the open literature, e.g., (Zwetsloot et al. 2015; Jiang et al. 2016; Pettersson et al. 2008; Lin and Ma 2012; Ding and Jiang 2013; Ong et al. 2015; Rong et al. 2015; Seggiani et al. 2012a, b; Skoglund et al. 2013; Ren and Li 2015; Li et al. 2013; Kern et al. 2012; Manara and Zabaniotou 2012), it is anticipated that further investigations into fuel mixing and thermal co-conversion of problematic residual resources could contribute greatly to increase the economy and the level of sustainability in the agricultural and waste handling sectors.

3 Fuels and Chemicals

The use of renewable resources such as biomass to produce synthetic fuels, chemicals, and other high-value products has attracted a lot of interest worldwide. One of the possible routes is synthesis of fuels and chemicals from syngas, produced via biomass gasification. Hydrogen and carbon monoxide, the major components of clean and conditioned syngas, are the basic building blocks of a number of products, including fuels and chemicals. Figure 2 shows the huge potential of biomass gasification to produce bio-products. In addition, a gasification plant can be designed to produce more than one product at a time (polygeneration), such as electricity and chemicals.

The production efficiency of synthetic fuels and biofuels depends on the type of fuel and the production pathway. Theoretically, LHV efficiencies of 84, 82, and 78% can be obtained for production of methanol, DME, and Fischer-Tropsch diesel using natural gas as feedstock (Van der Drift and Boerrigter 2005). However, for biofuels produced via biomass gasification, the total biomass to fuel efficiency is lower. In principle, biofuel production from biomass-derived syngas is not different from processes where fossil syngas is used. There are, however, a few biomass-related challenges that influence the conversion efficiency. One of those is the scale of operation. A large-scale plant will suffer less heat loss per unit of product, whereas fossil-based syngas plants are typically huge; it is believed that biomass plants may not be larger than a few hundred MW_{th} due to the limited availability of biomass feedstock at a negotiated price. Another difference between biofuel synthesis and the fossil fuel-based processes relates to the additional conversion steps required in the biomass conversion pathway and the related need for additional process integration. All stages in the biomass conversion pathway lead to potential losses as heat or by-product losses. Therefore, biomass-based plants require a biorefinery approach and have to be integrated with other technologies to increase the process efficiency and the internal use or external valorization of by-products.

Several pilot-scale plants for synthesis of biofuels can be found in research institutions and universities, but full-scale plants are scarcer. Bioenergy2020+ GmbH maintains a database of facilities for the production of advanced liquid and gaseous biofuels for transport (http://demoplants.bioenergy2020.eu). In the present chapter, three remarkable facilities are presented:

- GoBiGas: BioSNG production in Sweden
- Fortum Otso® bio-oil plant in Joensuu, Finland
- Enerkem Alberta Biofuels plant producing methanol and ethanol in Canada

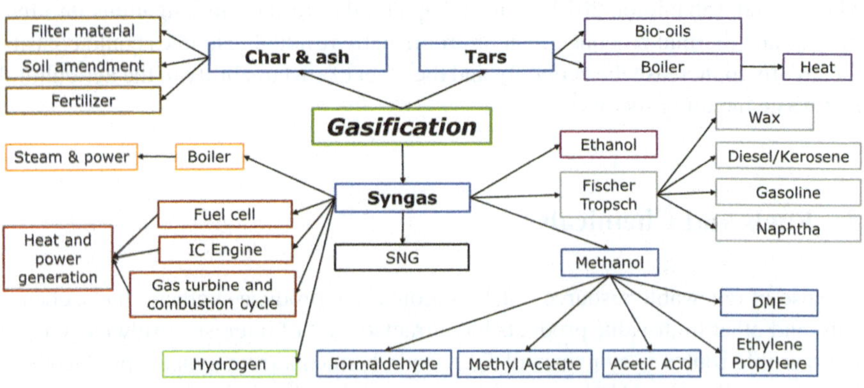

Fig. 2 Potential bio-products that can be obtained from biomass gasification

In addition to these three, it is also worth mentioning the Piteå BioDME pilot plant in Sweden producing around 4 tons of BioDME from black liquor from an adjacent pulp mill. The plant is based on oxygen-blown high-pressure gasification of the black liquor followed by gas conditioning, methanol synthesis, and dehydration to DME.

GoBiGas (Gothenburg Biomass Gasification Project) in Sweden is the world's largest plant, producing high-quality bio-methane from gasification of wood pellets and forest residues. It produces 20 MW of BioSNG, 5 MW of district heating, and 6 MW of heat to heat pumps. The plant was originally designed as a two-phased project. Phase I of the project, which involved the construction of the 20 MW gas demonstration plant, was inaugurated in March 2014 and became fully operational in December 2014 when it started injecting BioSNG to the grid. The capacity of the demonstration plant is big enough to supply BioSNG to approximately 15,000 cars or 400 buses a year. Phase II of the project consisted of a 80–100 MW gas commercial plant, scheduled for 2016, but that was cancelled due to the large amount of biomass needed and because the plant lacks the economy of scale achieved in coal-based plants. The 20 MW demonstration plant is currently looking for new owners/investors. The process (Fig. 3) involves indirect gasification of forest residues. The feedstock is fed into the circulating fluidized bed gasifier and gasified at approximately 850 °C by steam injection from a separate combustion chamber, producing nitrogen-free and low-tar syngas. The synthesis gas is purified by removing tar, sulfur, and carbon dioxide and upgraded in a methanation plant, where bio-methane (BioSNG) with a methane content of more than 95% is produced. The TREMP methanation technology including catalyst for the process reactors and engineering for the gas cleaning facilities was provided by Haldor Topsøe, whereas the gasification technology for the project was provided by Repotec in collaboration with Metso Power.

Other BioSNG projects and plants include the "Go Green Gas" 1 MW_{BioSGN} pilot plant connected to an industrial waste gasification unit in the UK, the AMBIGO 4 MW_{th}/2.8 MW_{BioSGN} demonstration project in Alkmaar in the Netherlands, and the 600 kW_{th} Gaya project in France.

Fortum Otso® bio-oil plant in Joensuu, Finland, is the first CHP-integrated pyrolysis plant in commercial size. It produces bio-oil from forest residues, wood chips, and saw dust. The bio-oil plant is integrated with Fortum's Joensuu combined heat and power (CHP) plant. Joensuu's bio-oil plant is based on fast pyrolysis technology (Fig. 4) and was commissioned in autumn 2013. The annual production capacity of the Joensuu bio-oil plant is 50,000 tons, which is equivalent to the heating needs of more than 10,000 single-family homes. Fortum Otso bio-oil can be used as a replacement for heavy and light fuel oil in heat production plants or in the production of industrial steam. In the future, bio-oil could also be used as a raw material for various biochemicals or transport fuels. Fortum uses the bio-oil at the heat plants in Espoo and Joensuu (Finland). Savon Voima, a Finnish energy company, has also started to use bio-oil at their plant in Iisalmi. According to Fortum, the bioliquid is acidic (pH 2–3) and sulfur free (<0.05%) and has a LHV of 15 MJ/kg, a density of 1.2 t/m^3, and a viscosity "between heavy and light fuel oil." The development and

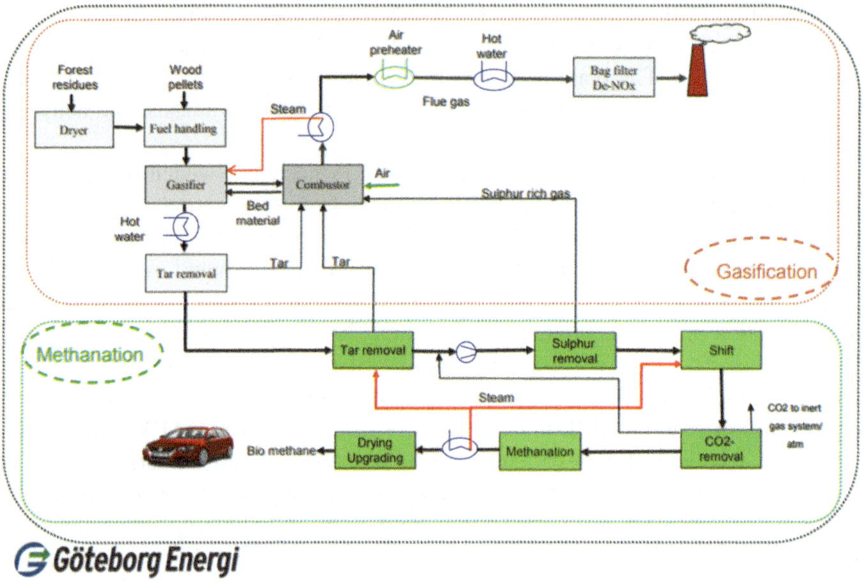

Fig. 3 Schematic of the GoBiGas BioSNG production plant. Copyright Göteborg Energi

conceptualization of the new technology were done collaboratively between Fortum, Metso, UPM, and VTT Technical Research Centre of Finland. The research was part of Tekes – the Finnish Funding Agency for Technology and Innovation's Biorefine program.

Enerkem Alberta Biofuels in Canada is the first of its kind to convert non-recyclable, non-compostable municipal solid waste into liquid biofuels. It was commissioned in 2014 and built adjacent to the Edmonton Waste Management Centre. The facility produces methanol and ethanol and contributes to the City of Edmonton's goal to divert up to 90% of household waste from the landfill. This commercial-scale facility has the capacity to process 100,000 metric tons of solid waste annually, which includes items like textiles, non-recyclable plastics, or soiled food containers, to produce over 40 million liters of biofuels. The Enerkem Alberta Biofuels plant initiated the commercial production of biomethanol in the summer of 2015.

Enerkem's gasification technology (Fig. 5) is based on a bubbling fluidized bed reactor with a front-end feeding system that is capable of handling fluffy material with no need to pelletize it. The fuel is continuously fed into a reactor where an inert heat carrier (i.e., sand) is fluidized under relatively low temperatures (700–750 °C) and moderate pressures of ~2 atm. Slurries or liquids can also be fed into the gasifier through appropriately designed injectors. Oxygen and steam are used as fluidizing gases and gasification agents. The produced gas is drawn from the top of the gasifier and goes through a cleaning and conditioning system. This process, including cyclones to remove particles and scrubbers and absorption to remove impurities, upgrades the gas to a chemical-grade syngas that can be synthesized into liquid

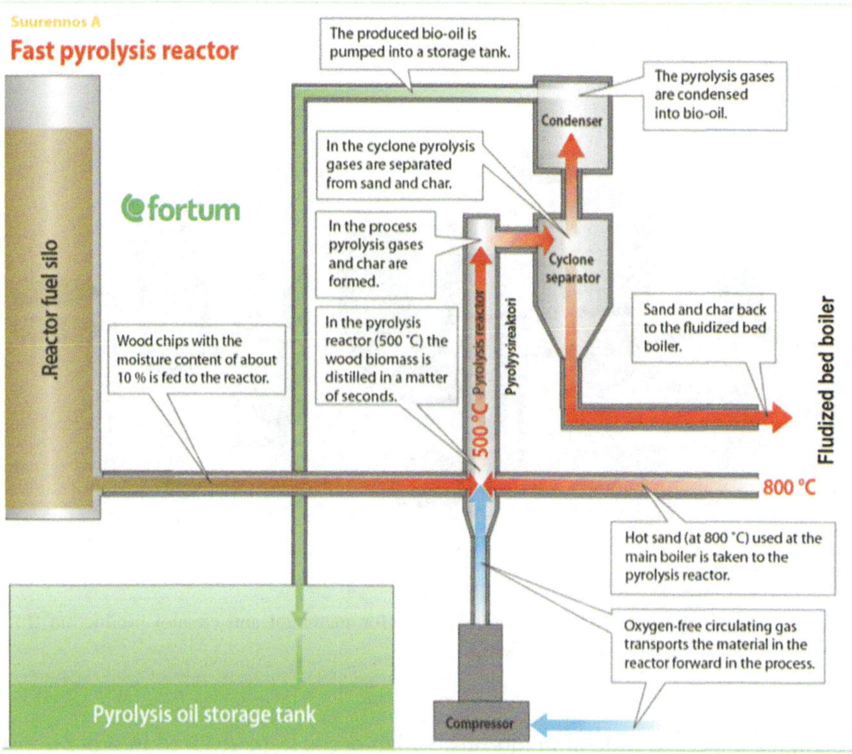

Fig. 4 Schematic of the Fortum Otso® bio-oil plant. Copyright Fortum

fuels and chemicals. A portion of the syngas reacts with a commercially available catalyst to produce methanol, which can either be sold as an end product or used as a chemical intermediate to form other products. To produce ethanol, methanol reacts with carbon monoxide from the syngas with a commercially available catalyst to produce methyl acetate. The final conversion step in the ethanol production process entails splitting the methyl acetate by inserting a hydrogen molecule that is also extracted from the produced syngas. The resulting ethanol is then distilled in a final refining step to improve product quality.

3.1 Bio-Oil Production, Separation, and Upgrading

Pyrolytic vapors upon their condensation in condensers form bio-oil. Bio-oils have a dark brown color and a complex structure and may contain a wide range of organic compounds including aromatic hydrocarbons, phenol derivatives, ketones, esters, ethers, sugars, amines, alcohols, furans, and water. The bio-oil has a H/C molar ratio higher than 1.5 (Guedes et al. 2018; Isahak et al. 2012). This composition results in

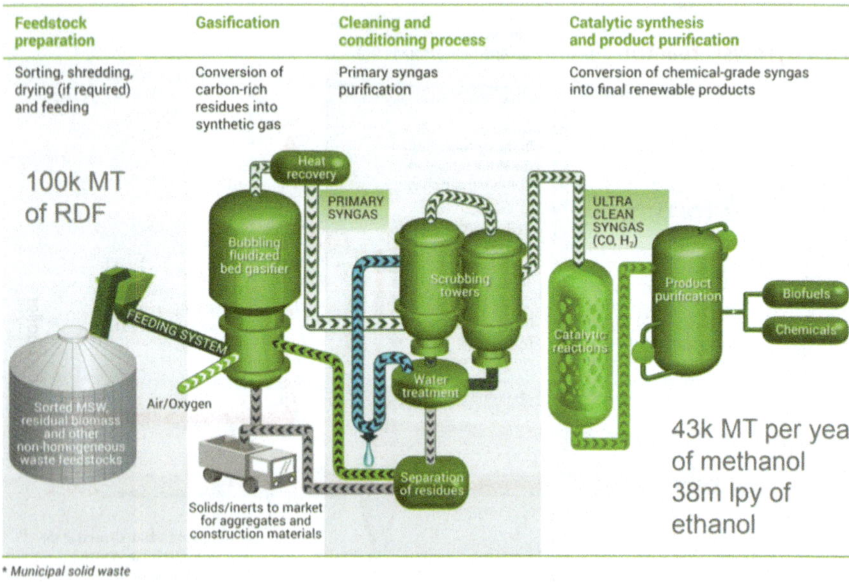

Fig. 5 Schematic Enerkem Alberta Biofuels plant for methanol and ethanol production from municipal solid waste. Copyright Enerkem Inc.

a significant oxygen content in bio-oils, which needs to be reduced in order to use the oil product for most applications. Bio-oils can have many application areas including their direct use as fuels in boilers. After upgrading, the oil can be used to produce fuels and bulk chemicals (Guedes et al. 2018). Partly deoxygenated bio-oil may be used as a feed for gasoline and diesel production in a conventional oil refinery and thereby provide a renewable transportation fuel.

Several types of condenser units have been tested for bio-oil collection, including a simple condenser chamber filled with isopropanol (Trinh et al. 2013), ethanol, ethyl acetate (Asadieraghi and Wan Daud 2015), but also two condensers or even a condenser train (Zhang et al. 2012; Jae et al. 2014). More advanced condensing systems may include spray condensers, where cooled recirculated condensed bio-oil can be used to condense the pyrolysis oil. Droplets escaping from the different condensers can be captured and collected with an electrostatic precipitator (ESP) (Palla et al. 2015; Hossain et al. 2013; Mante and Agblevor 2011). Wet electrostatic precipitation has proved itself over many years at the gasification plant in Harboøre, Denmark (Ahrenfeldt et al. 2013), and so has oil scrubbers, which have been validated in the OLGA process by ECN (Boerrigter et al. 2005). It has also been demonstrated that the bio-oil composition can be controlled by the condensation temperature. Thus sequential condensation may result in a fractionation of the produced bio-oil and carry a positive effect in terms of energy efficiency (Chang et al. 2012; Westerhof et al. 2011).

The direct application of untreated bio-oil within existing infrastructure is impeded by its high oxygen content (17–50 wt%) and acidity (pH = 2.5–3). This composition results in undesirable properties such as low heating value, immiscibility with hydrocarbon fuels, thermal and chemical instability, high viscosity, and corrosiveness (Milina et al. 2014). Upgrading of bio-oils and reduction of the rather high oxygen content are hence required and can be done in a number of ways, such as hydrotreatment and catalytic deoxygenation, while other upgrading methods involve supercritical water processes or catalytic cracking and may be used to increase heating value or for the production of chemicals (Isahak et al. 2012).

Upgrading of hot pyrolysis vapors at atmospheric pressure using in situ catalytic measures is one of the most promising processes to produce enhanced bio-oils from biomass pyrolysis. Zeolite cracking is one of the attractive processes within this category due to process simplicity, no requirement for pressurized operation, and no need for add-in of hydrogen or other compounds (Mortensen et al. 2011). By coupling a pyrolysis unit directly to a hot catalytic reactor based on a zeolite, a relatively simple system can be provided, which can produce a high-quality low-oxygen hydrocarbon liquid product. Studies on atmospheric catalytic biomass pyrolysis combined with zeolite upgrading have been performed using different setups. Microscale fixed bed batchwise laboratory equipment using a mixture of catalyst and biomass (Mochizuki et al. 2013; Yu et al. 2012; Ma et al. 2014), fluidized bed with a feed mixed with catalyst (Jae et al. 2014), and continuous pyrolysis reactors directly coupled with a separate reactor for catalytic gas-phase deoxygenation (Zhou et al. 2016; Carlson et al. 2009; Patwardhan et al. 2011). The latter system type will probably be the most efficient for commercial systems. Recently there has also been development into zeolite-based upgrading of gasification producer gas, but this development has not progressed as far as pyrolysis gas upgrading.

Several other methods for deoxygenation of pyrolysis oil are also under development. However, most of these methods are based on processes operating at high pressure such as the hydrodeoxygenation (HDO) process (more than 20 bars). They often consume large amounts of hydrogen (Mortensen et al. 2011) which in return results in higher bio-oil yields. Jointly, HDO and zeolite cracking are referred to as catalytic bio-oil upgrading, and these could become routes for production of second-generation biofuels in the future (Zwetsloot et al. 2015).

4 Char and Ash

Pyrolysis and gasification generate gaseous, liquid, and solid products, namely, chars and ashes. In general, char is considered as the main product of pyrolysis, whereas gas is the main product of gasification, but in both cases all fractions are present in variable amounts. In the optic of a circular economy, all the process products and by-products should find use in order to minimize the waste streams and increase product and process value. Particularly, chars and ashes offer several options for being recycled in an efficient and sustainable way thanks to their chemical and structural properties. For example, they maintain most of the inorganics contained in the feedstock, including valuable elements such as phosphorus. Depending on the process conditions, char can have interesting surface properties such as developed porosity and large specific surface area, suggesting various technical applications. It is important to mention that residual char from pyrolysis and gasification could as well be used as biochar for soil amendment, remediation, and carbon sequestration, if the required quality and toxicity standards are satisfied, as specified by the European biochar guidelines or the International Biochar Initiative (European Biochar Foundation (EBC) and Arbaz 2016; International Biochar Initiative 2015).

The difference between pyrolysis char and gasification char or ash is the reduction in carbon content induced by the additional gasification reactions taking place during the last stage of the gasification process (Fig. 6).

During pyrolysis the feedstock is carbonized: volatiles are released, and a large part of the mass and energy potential is left as mineralized (fixed) carbon. During gasification, fixed carbon reacts with a gasification agent (e.g., H_2O or CO_2) at elevated temperatures. In most gasification processes, the carbon conversion is incomplete, and the solid residues appear black, attesting the presence of a significant carbon fraction.

The characteristics and possible applications of chars and ashes are very process-specific, as they are heavily influenced by the operating conditions of the feedstock conversion (e.g., temperature, pressure, residence time). With this in mind, repurposing of char and ash can be conceived in two ways: on-site, in the frame of a biorefinery, a system which integrates thermal conversion of feedstock with other processes, or as a commodity to be sold and used elsewhere. In the following, some of the possible solutions for the recycling of the solid residues of pyrolysis and gasification are described and discussed.

4.1 Direct (On-Site) Valorization of Char

A particularly convenient application of residual char within the system is the treatment of gasification producer gas for the removal of tar and other contaminants. The advantages would be manifold: char is continuously produced on-site and readily available and could offer a convenient alternative to costly metal-based catalysts. If

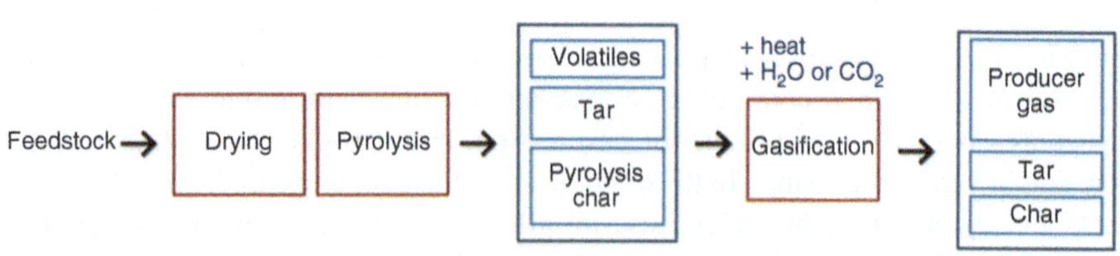

Fig. 6 Schematic of pyrolysis and gasification processes and products

the material gets deactivated as effect of poisoning or coking or adsorption saturation, it is possible to gasify it along with fresh feedstock.

Char-based tar removal can be carried out by physical adsorption from the gas phase (physisorption), or by using the char as a catalyst or tar reforming enhancer. In the case of physisorption, char is able to clean producer gas through filtering in the temperature range 150–250 °C. Higher temperatures would reduce the adsorption capacity of char, whereas lower temperatures could cause unwanted condensation of the tar species.

Mastral et al. (2001, 2003, 2004) investigated thoroughly the adsorption of PAHs from gas phase by using a variety of different activated carbons (ACs). They concluded that the porous structure of the adsorbent is determinant: the optimal AC should be rich in micropores (pores smaller than 2 nm), but the presence of larger pores, up to 50 nm (mesopores), is also important, especially for the adsorption of larger molecules (phenanthrene, pyrene). Similar conclusions were found also by Hu et al. (2007). Thus, if char is to be used as adsorbent, its porous structure should be somehow optimized for the tar mixture, which is to be removed.

The effectivity of biomass chars in the decomposition of tars has been observed in several studies (Boroson et al. 1989; Dabai et al. 2014; Matsuhara et al. 2010; Al-Rahbi et al. 2016; Zhang et al. 2015) within the temperature range 450–850 °C. If the contact between the char surface and the producer gas takes place at temperatures higher than 600 °C, cracking and reforming of tars into stable gases (H_2, CH_4) can be assisted. The most accredited mechanism for the decomposition of aromatics over the surface of char involves dehydrogenation and carbon deposition over the char surface (coking) (reaction 1).

$$C_nH_m = C_nH_x + m - x/2H_2 \quad (1)$$

The reaction occurs mainly in the micropores, which can easily be blocked by coking in absence of a gasifying agent. However, in the presence of H_2O or CO_2, gasification reactions of solid carbon (2 and 3) are able to "clean up" or regenerate the micropores, maintaining the activity of char for a longer time.

$$C_{(s)} + H_2O = CO + H_2 \quad (2)$$

$$C_{(s)} + CO_2 = 2CO \quad (3)$$

Indeed, according to Hosokai et al. (2008), the activity of char for the decomposition of aromatics can be maintained if the gasification rate is higher than the rate of carbon deposition. With this in mind, it is possible to imagine an actual integration of a char-based cleaning system in gasification, where char is used as a substrate for gas cleaning and upgrading. However, it is important to design and dimension the gas cleaning step so to guarantee its sustainability: the char bed should not be quickly deactivated nor quickly gasified away.

As an example, the two-stage gasifier (known as "Viking") developed at DTU Risø is able to produce an almost tar-free gas, thanks to the passage of the producer gas through a bed of hot char (Brandt et al. 2000).

The use of char could also assist the removal of other undesired substances present in producer gas such as sulfur compounds. Hervy et al. (2018) studied the adsorption of H_2S on the surface of pyrolysis chars obtaining promising results, which were significantly improved after steam activation of the original chars.

In the frame of a biorefinery concept, the properties of biochar could also be exploited to adsorb ammonium (NH_4^+), ammonia (NH_3), and phosphates for manure/sewage sludge/biogas fiber treatment plants. Naturally occurring characteristics of biochar make it very suitable for the adsorption of organic molecules, but it might need to undergo functionalization to be used as an effective adsorbent for cationic and anionic pollutants, such as ammonium and phosphate.

Negatively charged oxygen groups on the surface of chars are responsible for ammonium retention. According to Wang et al. (2015a), oxidation of biochar (preferably through natural aging) is useful to improve the adsorption capacity of ammonium and cations in general. On the other hand, recent studies focused on impregnation of biochars with metals to improve the adsorption of anions (Wan et al. 2017; Wang et al. 2016). Biochar modification methods for optimized adsorption of various contaminants have been comprehensively reviewed by Sizmur et al. (2017).

Biochar has been found suitable for uptaking NH_3 (Asada et al. 2006; Taghizadeh-Toosi et al. 2012a), with reported uptakes ranging from <1 to over 60 mg/g biochar (Seredych and Bandosz 2007).

The application of biochar for adsorption of nutrients is, of course, useful to control eutrophication but could also help the recycling of nitrogen and phosphorus. Indeed, Taghizadeh-Toosi et al. (2012b) demonstrated that nitrogen taken up by biochar is plant available. Similar findings were reported for phosphorus (Zhang et al. 2016). As a consequence, biochar could be used to prevent nutrient leaching and then applied as a slow-release fertilizer (Vikrant et al. 2017), concurrently acting as a carbon sequester.

However, it is important to keep in mind that the physical and chemical properties of char—and biochar—are highly variable. Thus, to take advantage of these multiple benefits, it will be necessary to carefully "design" char production and if necessary post-treat it with impregnation or oxidation to guarantee the characteristics required by specific applications.

4.2 By-Products with Added Value: Repurposing Char and Ash to Active Carbon and Fertilizers

Chars and ashes from pyrolysis and gasification may also be sold as renewable substitutes for industrial products such as fertilizers for agriculture and active carbon for industry and remediation.

4.2.1 Fertilizer Value of Ashes and Chars from Thermal Biorefinery Processes

In an ideal biorefinery encompassing thermal processing of nutrient-rich feedstocks, the majority of the organic fractions of the converted biogenic material are utilized to produce, e.g., chemicals, fuels, or materials, while nutrients are preserved in a plant-available form in the ashes or chars. This will allow for recycling of the nutrients back into the system where the biogenic material originated from, closing nutrient loops, and increasing the level of long-term sustainability. In addition, the thermal process may allow for removal or reduction of heavy metals and destruction of organic xenobiotics. The most valuable nutrient to recover in this way is phosphorous (P). P is an essential macronutrient, and the main source for P fertilizer is mined phosphate rock, which is a critical nonrenewable globally demanded resource. There is an increasing concern about the commercial availability and cost of phosphate rock in the near future (Cordell and White 2014). With the proper match between thermal process design and operation, fuel characteristics and end use, it may be feasible to dispatch biorefinery and waste refinery technology to close, e.g., phosphorus loops in modern society while also produce high-quality products and reduce toxicity and risk issues as illustrated in Fig. 7 (Zwetsloot et al. 2015; Thomsen et al. 2017a, b; Klinglmair et al. 2015; Wang et al. 2015b).

The characteristics and potential fertilizer value of chars and ashes vary with the characteristics of the converted fuel, the thermal process design, the operational parameters, the post-process treatment (if any), the end use, and the overall match between these different aspects. An illustration of such solid P-rich residuals is provided in Fig. 8, showing variation in color, morphology, and particle size distribution of six potential fertilizer substrates originating from six different thermal treatments of the same sewage sludge sample (Thomsen et al. 2016, 2017b).

When considering application of chars or ashes as fertilizer and/or soil enhancer, it is essential to optimize key characteristics including nutrient content and composition, organic/inorganic toxicity, pH, nutrient fertilizer quality, carbon content, and other biochar-related characteristics (e.g., water retention).

The nutrient content and composition varies greatly with the fuel composition and temperature profile of the thermal process. In general, low temperatures, low heating rates, and short retention times will increase nutrient recovery in ash and char fractions.

Content of organic pollutants in pyrolysis char and gasification ashes in the form of PAHs have been investigated several times. Total PAH content in cyclone ashes from low-temperature gasification of straw, dry chicken manure fibers, dry pig manure fibers, digested pig manure fibers, and sewage sludge has previously been found to range from 0.2 to 6.2 mg/kg (Thomsen et al. 2016; Nielsen 2007; Stoholm et al. 2002). In a study on the influence of the thermal process design on PAH content in chars and ashes from sewage sludge conversion, it was found that no PAHs persisted in bottom ash from low-temperature gasification, while small amounts persisted in the cyclone ashes from the same process and in incineration ashes from a full-scale fluid bed sludge incineration facility. Significantly larger amounts were found in ashes from two-stage downdraft gasification and slow pyrolysis of the same material (Thomsen et al. 2016).

Heavy metal content also varies with the composition of the converted material, the design of the thermal process, the operation parameters, and the type of ash product. Filter ash will often contain more volatile heavy metals than cyclone ash, fly ash, and bottom ash, while bottom ash may contain more thermally stable heavy metals than the other ash fractions (Thomsen et al. 2017a). Heavy metal concentrations in a set of different ashes and chars from conversion of biomass and sewage sludge are provided in Table 1.

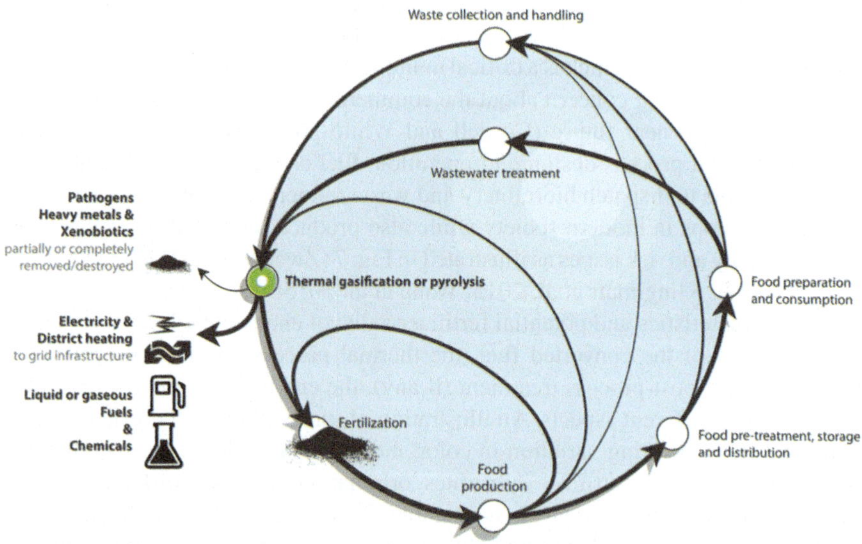

Fig. 7 Closing phosphorus loops in a modern society using biorefinery and waste refinery technology encompassing thermal processes (Thomsen et al. 2016)

Fig. 8 Char and ash fertilizer substrates from thermal conversion of the same sewage sludge sample. Adapted from Thomsen et al. (2016)

Table 1 Content of selected heavy metals in ashes from thermal conversion of various fuels in a range of different conversion technologies

Fuel and reference	Ash/char product	Thermal plant	Cu	Zn	Cd	Ni	Cr	Pb
			mg kg^{-1}					
Wheat straw (Müller-Stöver et al. 2012)	Cyclone/fly ash	Low-temperature gasification	31	160	<1	48	100	<10
Citrus peel fiber residues (Müller-Stöver et al. 2012)	Cyclone/fly ash		12	60	<1	34	<100	20
Digested pig manure fibers (Nielsen 2007)	Cyclone/fly ash		350	1900	2	57	22	13
Danish sewage sludge (Thomsen et al. 2017b)	Cyclone/fly ash		380	1906	4.5	158	182	106
Danish sewage sludge (Thomsen et al. 2017b)	Bottom ash		591	1636	2	67	182	159
Danish sewage sludge (Thomsen et al. 2017b)	Grate ash	Two-stage downdraft gasifier	804	2226	1	124	165	47
Danish sewage sludge (Thomsen et al. 2017b)	Char	Slow pyrolysis	458	1617	4	47	135	109
Danish sewage sludge (Thomsen et al. 2017b)	Fly ash	Full-scale fluid bed incineration	769	2567	6	105	127	210

Db dry basis

In addition to toxicity, the fertilizer quality of the nutrients in the char or ashes is also very important. Fertilizer quality relates to the mobility and plant availability of the nutrients. This is usually determined by plant growth experiments in pots, small field plots, or large field plots. Approximations of the mobility of nutrients can also be attempted by incubation of soil and fertilizer substrate with subsequent nutrient extraction using a variety of procedures (Wüenscher et al. 2015; Wünscher 2013). Proper quantification of fertilizer quality is a very complex issue, and several factors influence the results of the assessment. These factors include (but are not limited to):

- The chemical composition and nutrient speciation in the substrate (Thomsen et al. 2017b)
- The liming effect of the substrate (Jakobsen and Willett 1986; Li et al. 2017)
- The structure, composition, and pH of the soil (Li et al. 2017)
- The particle size distribution of the substrate (Thomsen et al. 2016)
- The type of plant applied (plant growth experiments) (Pearse et al. 2007; Kalaji et al. 2014)
- The type of extraction method used to extract nutrients from soil or plant after the incubation or growth period (Wüenscher et al. 2015; Wünscher 2013)
- The temporal scope of the investigation (Thomsen et al. 2016)
- Nutrient dosage (Thomsen et al. 2017a; Müller-Stöver et al. 2012; Li et al. 2017; Six et al. 2012; Mackay et al. 2017)

– Limiting effects of deficient levels of macro- and micronutrients not examined in the study (Schmidt et al. 2016; Guo et al. 2016; Wang et al. 2015c)

Despite the complexity of plant-substrate-soil-climate interactions, it is possible to obtain useful indications about a substrate fertilizer quality with relatively simple measures. This may be done in a screening measure by incubating substrate/soil mixtures with high moisture levels and subsequently extract the investigated nutrients (e.g., P) from the incubated mixture using different methods (Wüenscher et al. 2015; Six et al. 2012; Nuernberg et al. 1998; Sibbesen 1983). An example of such screening of P fertilizer quality in a range of pyrolysis char and incineration ash substrates is provided in Fig. 9 (Thomsen et al. 2016).

The results (Fig. 9) show P fertilizer quality compared to a commercial P fertilizer with 100% mineral P. The variation in the results is substantial among different fuels as well as between char and ashes from the same fuels. In general, the different substrates show a substantially lower P fertilizer quality than the commercial P fertilizers with only two exceptions (beet seed ash and tomato residue ash). However, the results are from a 1-week incubation study and not representative for the full fertilizer potential obtained in full plant growth cycles. Short-term incubations are good for relative comparison between comparative samples, but the method often underestimates the fertilizer potential of char and ash substrates compared to mineral references. This is due to a double effect where the mineral fertilizer is fixed over time by soil particles in real plant growth systems while immediately inaccessible P substrates are solubilized over time in the same systems. In combination with other results from the published literature, it is emphasized that fuel characteristics and design and operation of the thermal process as well as the design of the

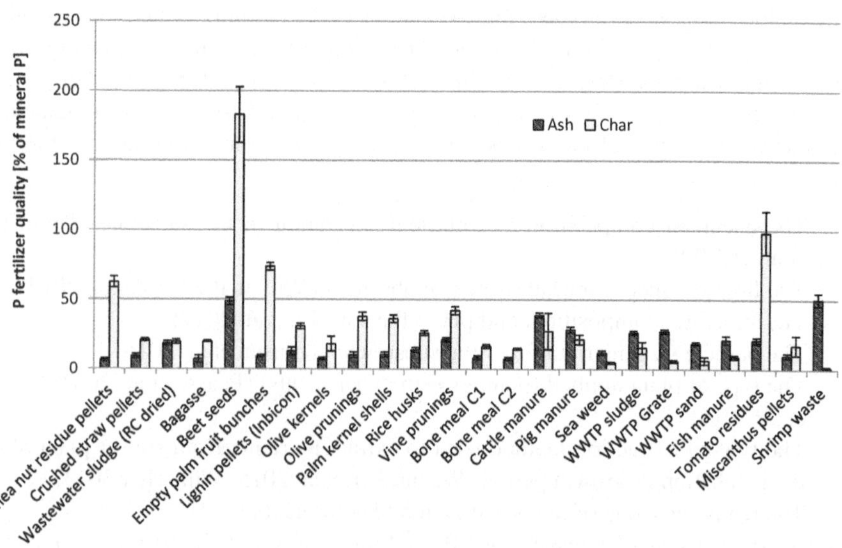

Fig. 9 Anion exchange resin extractability of phosphorus in incubated char and ash samples from the low-temperature gasification fuel screening

plant growth system severely influence the fertilizer quality measures of solid residuals from thermal biorefinery processes. For further results, see, e.g., Zwetsloot et al. (2015), Thomsen et al. (2017a, b), Müller-Stöver et al. (2012), Jakobsen and Willett (1986), Qian and Jiang (2014), Mellbye et al. (1982), Bierman and Rosen (1994), Kumpiene et al. (2016), Hossain et al. (2015), Song et al. (2014), Hansen et al. (2016), Kuligowski et al. (2012), and Kuligowski (2009).

Biochar contains a very large organic fraction, consisting primarily of highly recalcitrant carbon. Some of the most important physical characteristics of biochar are the total surface area, porosity (nano- and macro-), density, particle size, stability, mineral content, residual oils and tars, and surface chemistry and sorption properties. There have been substantial amounts of research done on biochar production, characteristics, and use in recent years, and much of it is available in the published literature, e.g., Bruun et al. (2014), Kizito et al. (2015), Van Wesenbeeck et al. (2014), Zielińska et al. (2015), Hossain et al. (2011), and Lehmann et al. (2011). In addition to potential nutrient release, chars or carbon-rich ashes also hold the potential to enhance soil structure and soil quality and sequester carbon for a very long time. In this way, these residual products may contribute to the mitigation of climate change while also enhancing the productivity and quality of the soil in which the char (biochar) is amended.

When amending biochar or gasification biochar, the recalcitrant carbon fraction remains unconverted in the soil for a very long time, and this positively affects the carbon balance. In a LCA study by Sigurjonsson et al. (2015), it was concluded that due to the fertilizing effect and content of recalcitrant carbon in cyclone ash from low-temperature gasification of cereal straw, a system with straw gasification and recycling of the ashes to soil could deliver carbon-neutral and even carbon-negative energy, depending on the carbon content in the ashes (Sigurjonsson et al. 2015). The stability of carbon in gasification biochar from straw gasification was investigated in a study by Hansen et al. (2015). It was found that after 110 days of incubation of ashes and dry straw in soil, about 3% of the ash carbon was respired as CO_2, while 80% of the carbon from a straw reference was respired. The structure of the ash was also investigated and was found to have a high porosity and specific surface area, which was proposed as key quality parameters in regard to improvements of soil structure and the soil ability to retain nutrients and water (Hansen et al. 2015).

4.2.2 Activated Carbon

Activated carbons (ACs) are industrially produced through carbonization followed by chemical and physical activation. The activation process is controlled and optimized to produce the desired pore structure and surface area for specific adsorption applications. The specific surface area of ACs ranges from 500 to 2000 m^2/g and the pore volume from 0.20 to 0.60 cm^3/g (Marsh and Rodríguez-Reinoso 2006). ACs are largely used for removing contaminants from liquids and gases: their global demand can be expected to reach 2.1 million metric tons in 2018 (Maneerung et al. 2016). As a consequence, cheaper precursors or substitutes for ACs such as residual biomass char could help decrease the costs and encourage the utilization of this material. Pyrolysis and gasification are not optimized for producing AC; however, under certain operating conditions, the properties of these chars can be comparable to AC in terms of surface area and pore volume.

Surface properties of pyrolysis char depend on the charring temperature, on the retention time, and on the feedstock. As an example, Keiluweit et al. (2010) observed that the surface area of wood char produced at either 100 or 700 °C changed from 1.6 to 347 m^2/g, with significantly higher values in comparison with straw char, which ranged from 1.8 to 139 m^2/g at the same charring temperatures. Liu et al. (2011) focused on the effect of retention time and found it to be a determining factor for the development of a large surface area and a consequently improved phenol adsorption capacity.

Pyrolysis chars do not undergo gasification reactions thus have generally a lower surface area and a less developed porosity in comparison with gasification chars. The fundamental steps of the gasification process (drying, pyrolysis, and gasification) are somehow comparable to the production phases of AC. However, the properties of gasification chars are variable and strongly dependent on the process conditions. Hérnandez et al. (2016) operated a lab-scale gasification reactor under different conditions and always obtained a solid residue with a specific surface area below 70 m^2/g. Klinghoffer et al. (2012) produced gasification char with a surface area ranging from 429 to 687 m^2/g, which increased with temperature and reaction time. The specific surface area of residual chars from four different small-scale gasifiers was found to range between 78 and 352 m^2/g (Benedetti et al. 2018). On the other hand, residual char from pilot-scale, two-stage gasification of wood was reported to be around 1027 m^2/g (Hansen et al. 2015). According to Benedetti et al. (2018), char with a larger specific surface area is produced in two-stage gasifiers, where pyrolysis and gasification are performed separately, and the burn-off of the pyrolyzed material is better controlled. If the conversion process is designed to produce a solid product of reasonable quality, it could be possible to use it directly as a substitute for active carbon, without further activation. Indeed, Runtti et al. (2014) tested gasification char with and without chemical activation for adsorption of metals (iron, copper, and nickel ions). In all cases, chars performed better than commercial ACs.

Nonetheless, oftentimes gasification char does not have a specific surface area as large as industrial AC, but it can be a good precursor for AC production. Promising results in this sense were reported by Kilpimaa et al. (2015) and Maneerung et al. (2016). They tested physically activated char from gasification for adsorption of nitrate and phosphate and dye (Rhodamine B), respectively, from aqueous solutions. Galhetas et al. (2014) focused on adsorption of caffeine and acetaminophen on K_2CO_3-activated gasification char, which performed comparably or even better than commercial AC.

ACs are often employed for wastewater treatment, thanks to their good adsorption capacity toward a number of contaminants. Again, pyrolysis and gasification chars could substitute AC produced for this purpose: Ahmad et al. (2014) and

Rosales et al. (2017) are authors of comprehensive reviews on the application of char as sorbent for contaminants in soil and water.

Water purification can also be performed by using biologically active carbon (BAC). Conventional and advanced water treatment systems use AC filtration: granular AC gradually becomes saturated with microorganisms and organic/inorganic matter, developing a rough and porous surface that can be favorable for bacterial colonization forming a biologically active film (biofilm). The biofilm is capable of biodegrading a significant fraction of waterborne nutrients, organic matter, minerals, and microorganisms (Simpson 2008). In a recent study by Dalahmeh et al. (2018), biochar filters with active biofilm were found to be more efficient than sand filters in removal of organic matter and nitrogen; in addition, biochar was able to efficiently remove pharmaceutically active compounds in sewage facilities.

Porous materials such as AC and similar materials can also be used for carbon capture and storage (CCS) to mitigate the risks and impacts of climate change. Indeed, large specific surface area and a microporous structure of char can be favorable for the adsorption of CO_2 emissions. For this application, Madzaki et al. (2016) tested residual char from sawdust gasification with positive results, measuring capture capacities of 0.47 kg CO_2/kg biochar and 0.30 kg CO_2/kg biochar at 30 and 70 °C, respectively. Gasification char was also tested by Benedetti et al. (2017) for CO_2 capture.

5 System Integration

Gasification and pyrolysis of biomass can be integrated with a number of processes in larger systems or biorefineries to improve the overall system performance. A biorefinery approach typically improves performance parameters related to energy efficiency and carbon efficiency but also system flexibility in terms of, e.g., feedstock flexibility or product flexibility (polygeneration).

The following two integration opportunities will be discussed in this section, as they could be very important in future energy systems based on renewable energy:

- Integration with water/steam electrolysis
- Integration with anaerobic digestion

The integration of pyrolysis and/or gasification with water/steam electrolysis in biorefineries or biofuel production plants enables storage of electricity from renewables as chemical energy bound in the produced fuel, feed, or chemical. Furthermore, the integration typically enables a doubling of the product output per biomass input, because a hydrogen deficit is usually limiting the production and the required hydrogen can be provided from the electrolysis cells (Clausen 2015).

The integration with anaerobic digestion has many benefits. In this section, focus is put on the system integration part, showing that even very wet digestate from anaerobic digestion can be converted in systems based on gasification or pyrolysis with high energy efficiency.

5.1 Integration with Water/Steam Electrolysis

By integrating water/steam electrolysis in a biorefinery based on gasification or pyrolysis, a highly flexible and energy-efficient system can be made. An overall diagram of such a system or biorefinery can be seen on Fig. 10. Such a biorefinery could be highly relevant in a future renewable energy system because it can (1) perform electricity grid balancing by storing electricity as chemical energy when the electricity demand is low and then produce electricity from input biomass when the electricity demand is high; (2) produce renewable fuels for the transportation system, providing also an indirect electrification of the transportation sectors where direct electrification is difficult, such as aviation, shipping, and heavy goods transport; and (3) produce bioash/biochar for agriculture in order to recycle nutrients and provide soil improvement and carbon sequestration.

Besides providing electricity storage, the integration with electrolysis can circumvent the need of an oxygen production plant, as by-product oxygen from electrolysis can be used for gasification (Fig. 11a). In biorefineries, steam electrolysis will typically be preferred over liquid water electrolysis from an energy perspective, because waste heat generated within the biorefinery can be used to raise steam for electrolysis (Fig. 11b). Furthermore, when also integrating biomass steam drying (Fig. 11c), the system will be able to handle biomasses with a very high water content such as mechanically dewatered sludge or manure without a decrease in energy efficiency (Clausen 2017).

Besides the advantages with steam electrolysis highlighted above, it is important to note that when using solid oxide electrolysis cells (SOEC) for steam electrolysis, the same cells can be used for reversed operation as solid oxide fuel cells (SOFC) (Jensen et al. 2015; Gadsbøll et al. 2017). The system displayed in Fig. 11b could in this way operate as shown on Fig. 12.

Combining biomass gasification with solid oxide cells (SOEC/SOFC) and biofuel synthesis is therefore one way of compiling a biorefinery that has the characteristics shown in Fig. 10. Another promising way of integrating electricity from fluctuating renewable energy sources in biorefineries is by high-temperature electric heating of, e.g., gasification or reforming processes (Spagnolo et al. 1992). If carefully integrated, high-temperature electric heating can be a cheaper and more

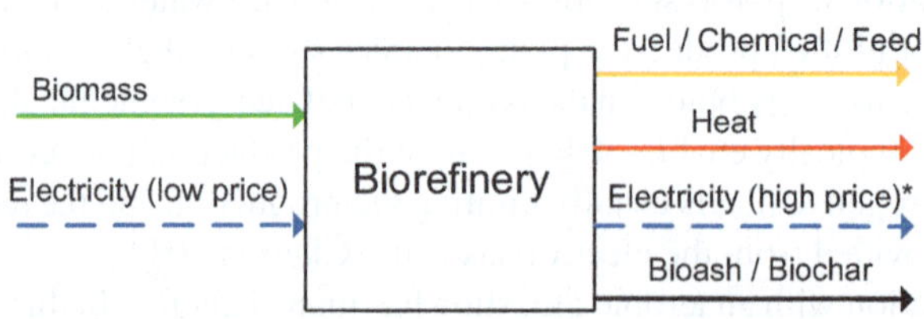

Fig. 10 Diagram of a biorefinery based on gasification or pyrolysis integrated with water/steam electrolysis. (Asterisk) The electricity output can be produced by a heat engine or by a reversible electrolysis system (e.g., solid oxide cells)

energy-efficient way of integrating electricity. The potential electric input that the biorefinery can absorb efficiently is however typically lower than by electrolysis because only energy is supplied by electric heating, whereas electrolysis supplies a mass flow of hydrogen.

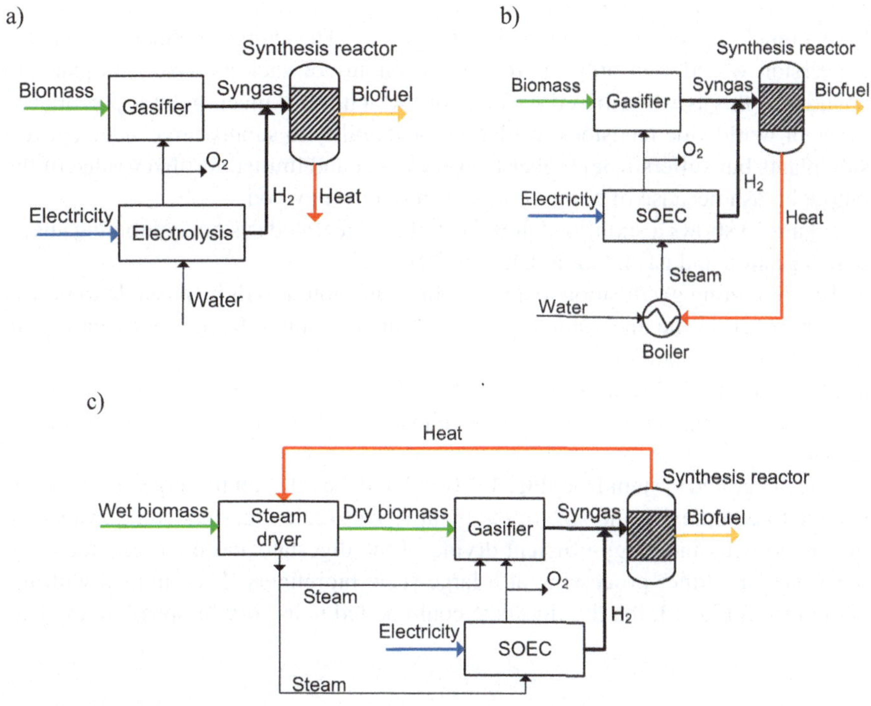

Fig. 11 Simplified biorefinery flowsheets based on biomass gasification and water/steam electrolysis. (**a**) Integration with liquid water electrolysis, (**b**) integration with steam electrolysis (SOEC = solid oxide electrolysis cells), (**c**) integration with steam electrolysis and steam drying of biomass to enable energy-efficient conversion of wet biomass. Note that the gasifier produces bioash/biochar although not shown on the figures

Fig. 12 Simplified flow sheet showing how the system from Fig. 11b could operate to produce electricity by operating the SOEC as an SOFC

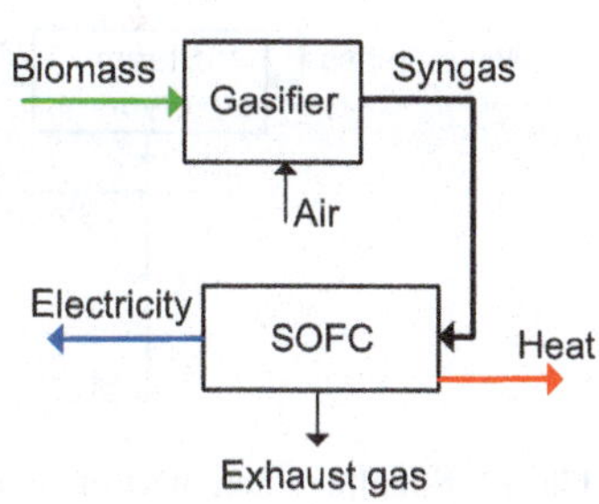

5.2 Integration with Anaerobic Digestion

An energy-efficient integration of gasification and anaerobic digestion could turn out to be an important stepping-stone for biomass gasification because a cheap or perhaps free fuel in the form of digestate is available. If the gasification process can produce a high-quality bioash or biochar product for agriculture, this product could have a much higher value than the input digestate. The energy product from such a conversion would then only improve the economy of such a conversion plant. In contrast to this is wood-based gasification, which has a much greater potential in terms of worldwide feedstock availability and ability to supply large-scale conversion plants but suffers from higher feedstock cost and limited fertilizer value of the output bioash because of a low content of nutrients in wood.

Figure 13 shows a simplified flowsheet of an integrated system combining anaerobic digestion and gasification (Clausen 2017).

By combining gasification and anaerobic digestion, it will be possible to mineralize nitrogen in the anaerobic digester and thereby minimize the nitrogen loss in the gasifier. The main nitrogen flow will leave as ammonia with the liquid fraction after the mechanical dewatering (Fig. 13). However, recovering nitrogen as fertilizer is not as important as recovering phosphorous, as nitrogen is not a limited resource.

The integrated system from Fig. 13 would only be relevant for large-scale digesters due to complexity and size of the plant. Small-scale digesters could instead be integrated with an energy-efficient drying of the digestate and then send the dried digestate for further processing at a large-scale biorefinery. If combined with the system from Fig. 13, the dry digestate could be fed in as "dry biomass" to the gas-

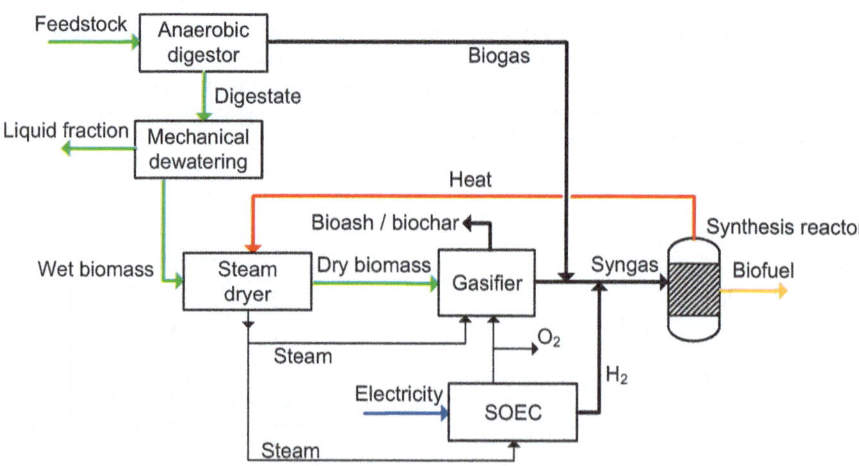

Fig. 13 Simplified flowsheet of an integrated system based on anaerobic digestion and gasification. The system is based on the system shown in Fig. 11c

ifier. An energy-efficient drying could be achieved by an integrated system combining steam drying and mechanical vapor recompression (MVR) (see Fig. 14).

6 Conclusions

Biomass pyrolysis and gasification processes are versatile processes which may provide substantial contributions to modern biorefinery concepts and energy systems. The purpose of this chapter has been to give an introduction to the current state-of-the-art, recent research and development efforts and classic challenges related to integration and operation of these technologies. The chapter focuses on key aspects and novel solutions which may become relevant in order to increase the utilization of biomass pyrolysis and gasification and make it commercially interesting in future energy systems.

The versatility of the thermochemical conversion platform is essential in a biorefinery context, especially when moving from high-quality biomass resources into low-quality organic secondary resources. Highly efficient utilization of these globally available resources is commercially and politically interesting because of its potential value from an economic, environmental, and resource political point of view. The combination of intelligent fuel design and specially designed thermochemical co-conversion systems for valorization of problematic residual resources could contribute greatly to increase the economy and the level of sustainability in the agricultural sector, waste handling sector, and transport sector.

Future development of thermochemical biorefinery processes will be fostered by the desire for a circular economy. All products and by-products from pyrolysis and thermal gasification should find use in order to minimize the waste streams and

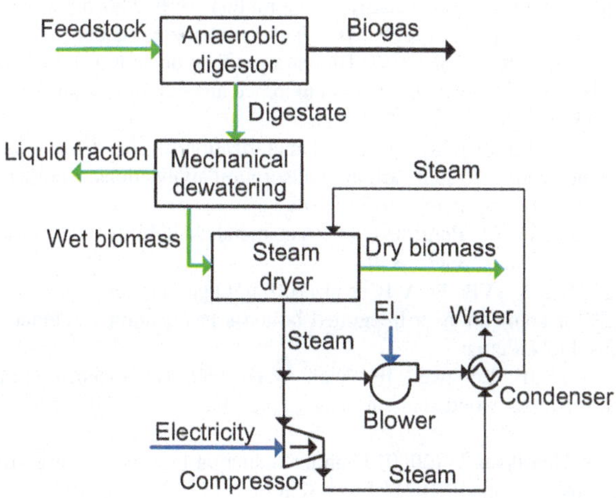

Fig. 14 Flowsheet of an integrated system for drying of digestate using steam drying and mechanical vapor recompression (MVR)

increase product and process value. Particularly, chars and ashes offer several options for being recycled in an efficient and sustainable way thanks to their chemical and structural properties. Some of the options mentioned in the chapter include the use of biochar as active carbon or for soil amendment, remediation, and carbon sequestration.

New biorefinery concepts with pyrolysis and gasification stages may benefit from further development in enhanced process integration schemes, combinations of different technologies and new polygeneration strategies. Such development is required to maximize total system product yield and value, increase the overall process efficiency, improve the economic viability, and overcome the operational challenges that the technology is facing in individual systems.

References

Adam C, Peplinski B, Michaelis M et al (2009) Thermochemical treatment of sewage sludge ashes for phosphorus recovery. Waste Manag 29:1122–1128. https://doi.org/10.1016/j.wasman.2008.09.011

Ahmad M, Upamali A, Eun J et al (2014) Biochar as a sorbent for contaminant management in soil and water: a review. Chemosphere 99:19–33. https://doi.org/10.1016/j.chemosphere.2013.10.071

Ahrenfeldt J, Thomsen TP, Henriksen U, Clausen LR (2013) Biomass gasification cogeneration—a review of state of the art technology and near future perspectives. Appl Therm Eng 50:1407–1417. https://doi.org/10.1016/j.applthermaleng.2011.12.040

Al-Rahbi AS, Onwudili JA, Williams PT (2016) Thermal decomposition and gasification of biomass pyrolysis gases using a hot bed of waste derived pyrolysis char. Bioresour Technol 204:71–79. https://doi.org/10.1016/j.biortech.2015.12.016

Asada T, Ohkubo T, Kawata K, Oikawa K (2006) Ammonia adsorption on bamboo charcoal with acid treatment. J Health Sci 52:585–589. https://doi.org/10.1248/jhs.52.585

Asadieraghi M, Wan Daud WMA (2015) In-situ catalytic upgrading of biomass pyrolysis vapor: using a cascade system of various catalysts in a multi-zone fixed bed reactor. Energy Convers Manag 101:151–163. https://doi.org/10.1016/j.enconman.2015.05.008

Benedetti V, Patuzzi F, Baratieri M (2017) Biomass gasification char as a low-cost adsorbent for CO2 capture. In: 25th European biomass conference and exhibition, pp. 760–764. https://doi.org/10.5071/25thEUBCE2017-2CV.3.14

Benedetti V, Patuzzi F, Baratieri M (2018) Characterization of char from biomass gasification and its similarities with activated carbon in adsorption applications. Appl Energy 227:92–99. https://doi.org/10.1016/j.apenergy.2017.08.076

Bierman PM, Rosen CJ (1994) Phosphate and trace metal availability from sewage-sludge incinerator ash. J Environ Qual 23:822–830

Boerrigter H, van Paasen SVB, PCA B et al (2005) "Olga" tar removal technology proof-of-concept (PoC) for application in integrated biomass gasification combined heat and power (CHP) systems. ECN, Petten

Boroson ML, Howard JB, Longwell JP, Peters WA (1989) Heterogeneous cracking of wood pyrolysis tars over fresh wood char surfaces. Energy Fuel 3:735–740. https://doi.org/10.1021/ef00018a014

Brandt P, Larsen E, Henriksen U (2000) High tar reduction in a two-stage gasifier. Energy Fuel 14:816–819. https://doi.org/10.1021/ef990182m

Bridgwater AV, Meier D, Radlein D (1999) An overview of fast pyrolysis of biomass. Org Geochem 30:1479–1493. https://doi.org/10.1016/S0146-6380(99)00120-5

Bruun EW, Petersen CT, Hansen E et al (2014) Biochar amendment to coarse sandy subsoil improves root growth and increases water retention. Soil Use Manag 30:109–118. https://doi.org/10.1111/sum.12102

Carlson TR, Tompsett GA, Conner WC, Huber GW (2009) Aromatic production from catalytic fast pyrolysis of biomass-derived feedstocks. Top Catal 52:241–252. https://doi.org/10.1007/s11244-008-9160-6

Chang CC, Wu SR, Lin CC et al (2012) Fast pyrolysis of biomass in pyrolysis gas: fractionation of pyrolysis vapors using a spray of bio-oil. Energy Fuel 26:2962–2967. https://doi.org/10.1021/ef201858h

Clausen LR (2015) Maximizing biofuel production in a thermochemical biorefinery by adding electrolytic hydrogen and by integrating torrefaction with entrained flow gasification. Energy 85:94–104. https://doi.org/10.1016/j.energy.2015.03.089

Clausen LR (2017) Energy efficient thermochemical conversion of very wet biomass to biofuels by integration of steam drying, steam electrolysis and gasification. Energy 125:327–336. https://doi.org/10.1016/j.energy.2017.02.132

Cordell D, White S (2014) Life's bottleneck: sustaining the World's phosphorus for a food secure future. Annu Rev Environ Resour 39:161–188. https://doi.org/10.1146/annurev-environ-010213-113300

Dabai F, Paterson N, Millan M et al (2014) Tar formation and destruction in a fixed bed reactor simulating downdraft gasification: effect of reaction conditions on tar cracking products. Energy Fuel 28:1970–1982. https://doi.org/10.1021/ef402293m

Dalahmeh S, Ahrens L, Gros M et al (2018) Potential of biochar filters for onsite sewage treatment: adsorption and biological degradation of pharmaceuticals in laboratory filters with active, inactive and no biofilm. Sci Total Environ 612:192–201. https://doi.org/10.1016/j.scitotenv.2017.08.178

Demirbas A, Arin G (2002) An overview of biomass pyrolysis. Energy Sources 24:471–482. https://doi.org/10.1080/00908310252889979

Ding H-S, Jiang H (2013) Self-heating co-pyrolysis of excessive activated sludge with waste biomass: energy balance and sludge reduction. Bioresour Technol 133:16–22. https://doi.org/10.1016/j.biortech.2013.01.090

European Biochar Foundation (EBC), Arbaz S (2016) Guidelines for a sustainable production of biochar. Eur Biochar Found. https://doi.org/10.13140/RG.2.1.4658.7043

Gadsbøll RØ, Thomsen J, Bang-Møller C et al (2017) Solid oxide fuel cells powered by biomass gasification for high efficiency power generation. Energy 131:198–206. https://doi.org/10.1016/j.energy.2017.05.044

Galhetas M, Mestre AS, Pinto ML et al (2014) Chars from gasification of coal and pine activated with K2CO3: acetaminophen and caffeine adsorption from aqueous solutions. J Colloid Interface Sci 433:94–103. https://doi.org/10.1016/j.jcis.2014.06.043

Guedes RE, Luna AS, Torres AR (2018) Operating parameters for bio-oil production in biomass pyrolysis: a review. J Anal Appl Pyrolysis 129:134–149. https://doi.org/10.1016/j.jaap.2017.11.019

Guo W, Nazim H, Liang Z, Yang D (2016) Magnesium deficiency in plants: an urgent problem. Crop J 4:83–91. https://doi.org/10.1016/j.cj.2015.11.003

Hansen V, Müller-Stöver D, Ahrenfeldt J et al (2015) Gasification biochar as a valuable by-product for carbon sequestration and soil amendment. Biomass Bioenergy 72:300–308

Hansen V, Hauggaard-Nielsen H, Petersen CT et al (2016) Effects of gasification biochar on plant-available water capacity and plant growth in two contrasting soil types. Soil Tillage Res 161:1–9. https://doi.org/10.1016/j.still.2016.03.002

Hernández JJ, Lapuerta M, Monedero E (2016) Characterisation of residual char from biomass gasification: effect of the gasifier operating conditions. J Clean Prod 138:83–93. https://doi.org/10.1016/j.jclepro.2016.05.120

Hervy M, Pham Minh D, Gérente C et al (2018) H2S removal from syngas using wastes pyrolysis chars. Chem Eng J 334:2179–2189. https://doi.org/10.1016/j.cej.2017.11.162

Herzel H, Krüger O, Hermann L, Adam C (2016) Sewage sludge ash—a promising secondary phosphorus source for fertilizer production. Sci Total Environ 542:1136–1143. https://doi.org/10.1016/j.scitotenv.2015.08.059

Hosokai S, Kumabe K, Ohshita M et al (2008) Mechanism of decomposition of aromatics over charcoal and necessary condition for maintaining its activity. Fuel 87:2914–2922. https://doi.org/10.1016/j.fuel.2008.04.019

Hossain MK, Strezov V, Chan KY et al (2011) Influence of pyrolysis temperature on production and nutrient properties of wastewater sludge biochar. J Environ Manag 92:223–228

Hossain MM, Scott IM, McGarvey BD et al (2013) Toxicity of lignin, cellulose and hemicellulose-pyrolyzed bio-oil combinations: estimating pesticide resources. J Anal Appl Pyrolysis 99:211–216. https://doi.org/10.1016/j.jaap.2012.07.008

Hossain MK, Strezov V, Nelson PF (2015) Comparative assessment of the effect of wastewater sludge biochar on growth, yield and metal bioaccumulation of cherry tomato. Pedosphere 25:680–685. https://doi.org/10.1016/S1002-0160(15)30048-5

Hu X, Hanaoka T, Sakanishi K et al (2007) Removal of tar model compounds produced from biomass gasification using activated carbons. J Japan Inst Energy 86:707–7011. https://doi.org/10.1093/bioinformatics/btu635

International Biochar Initiative (2015) Standardized product definition and product testing guidelines for biochar that is used in soil, pp. 1–61. https://www.biochar-international.org/wp-content/uploads/2018/04/IBI_Biochar_Standards_V2.1_Final.pdf

Isahak WNRW, Hisham MWM, Yarmo MA, Yun Hin T (2012) A review on bio-oil production from biomass by using pyrolysis method. Renew Sust Energ Rev 16:5910–5923. https://doi.org/10.1016/j.rser.2012.05.039

Jae J, Coolman R, Mountziaris TJ, Huber GW (2014) Catalytic fast pyrolysis of lignocellulosic biomass in a process development unit with continual catalyst addition and removal. Chem Eng Sci 108:33–46. https://doi.org/10.1016/j.ces.2013.12.023

Jahirul MI, Rasul MG, Chowdhury AA, Ashwath N (2012) Biofuels production through biomass pyrolysis—a technological review. Energies 5:4952–5001. https://doi.org/10.3390/en5124952

Jakobsen P, Willett IR (1986) Comparisons of the fertilizing and liming properties of lime-treated sewage sludge with its incinerated ash. Fertil Res 9:187–197. https://doi.org/10.1007/BF01050345

Jensen SH, Graves C, Mogensen M et al (2015) Large-scale electricity storage utilizing reversible solid oxide cells combined with underground storage of CO_2 and CH_4. Energy Environ Sci 8:2471–2479. https://doi.org/10.1039/C5EE01485A

Jiang LB, Yuan XZ, Li H et al (2016) Co-pelletization of sewage sludge and biomass: thermogravimetric analysis and ash deposits. Fuel Process Technol 145:109–115. https://doi.org/10.1016/j.fuproc.2016.01.027

Kalaji HM, Oukarroum A, Alexandrov V et al (2014) Identification of nutrient deficiency in maize and tomato plants by invivo chlorophyll a fluorescence measurements. Plant Physiol Biochem 81:16–25. https://doi.org/10.1016/j.plaphy.2014.03.029

Keiluweit M, Nico PS, Johnson M, Kleber M (2010) Dynamic molecular structure of plant biomass-derived black carbon (biochar). Environ Sci Technol 44:1247–1253. https://doi.org/10.1021/es9031419

Kern S, Halwachs M, Kampichler G et al (2012) Rotary kiln pyrolysis of straw and fermentation residues in a 3MW pilot plant—influence of pyrolysis temperature on pyrolysis product performance. J Anal Appl Pyrolysis 97:1–10. https://doi.org/10.1016/j.jaap.2012.05.006

Kilpimaa S, Runtti H, Kangas T et al (2015) Physical activation of carbon residue from biomass gasification: novel sorbent for the removal of phosphates and nitrates from aqueous solution. J Ind Eng Chem 21:1354–1364. https://doi.org/10.1016/j.jiec.2014.06.006

Kizito S, Wu S, Kipkemoi Kirui W et al (2015) Evaluation of slow pyrolyzed wood and rice husks biochar for adsorption of ammonium nitrogen from piggery manure anaerobic digestate slurry. Sci Total Environ 505:102–112. https://doi.org/10.1016/j.scitotenv.2014.09.096

Klinghoffer N, Castaldi MJ, Nzihou A (2012) Catalyst properties and catalytic performance of char from biomass gasification. Ind Eng Chem Res 51:13113–13122. https://doi.org/10.1021/ie3014082

Klinglmair M, Lemming C, Jensen LS et al (2015) Phosphorus in Denmark: national and regional anthropogenic flows. Resour Conserv Recycl 105:311–324. https://doi.org/10.1016/j.resconrec.2015.09.019

Kuligowski K (2009) Utilization of ash from thermal gasification of pig manure as p-fertilizer and soil amendment. Aalborg University, Aalborg

Kuligowski K, Gilkes RJ, Poulsen TG, Yusiharni BE (2012) Ash from the thermal gasification of pig manure-effects on ryegrass yield, element uptake, and soil properties. Soil Res 50:406–414. https://doi.org/10.1071/SR12075

Kumpiene J, Brännvall E, Wolters M et al (2016) Phosphorus and cadmium availability in soil fertilized with biosolids and ashes. Chemosphere 151:124–132. https://doi.org/10.1016/j.chemosphere.2016.02.069

Lehmann J, Rillig MC, Thies J et al (2011) Biochar effects on soil biota—a review. Soil Biol Biochem 43:1812–1836. https://doi.org/10.1016/j.soilbio.2011.04.022

Li L, Ren Q, Li S, Lu Q (2013) Effect of phosphorus on the behavior of potassium during the co-combustion of wheat straw with municipal sewage sludge. Energy Fuel 27:5923–5930

Li X, Rubæk GH, Müller-Stöver DS et al (2017) Plant availability of phosphorus in five gasification biochars. Front Sustain Food Syst 1:1–12. https://doi.org/10.3389/fsufs.2017.00002

Lin H, Ma X (2012) Simulation of co-incineration of sewage sludge with municipal solid waste in a grate furnace incinerator. Waste Manag 32:561–567. https://doi.org/10.1016/j.wasman.2011.10.032

Liu W-J, Zeng F-X, Jiang H, Zhang X-S (2011) Preparation of high adsorption capacity biochars from waste biomass. Bioresour Technol 102:8247–8252. https://doi.org/10.1016/j.biortech.2011.06.014

Ma Z, Custodis V, van Bokhoven JA (2014) Selective deoxygenation of lignin during catalytic fast pyrolysis. Cat Sci Technol 4:766. https://doi.org/10.1039/c3cy00704a

Mackay JE, Cavagnaro TR, Jakobsen I, Macdonald LM, Grønlund M, Thomsen TP, Müller-Stöver DS (2017) Evaluation of phosphorus in thermally converted sewage sludge: P pools and availability to wheat. Plant Soil 418(1–2):307–317

Madzaki H, Azlina W, Ab W (2016) Carbon dioxide adsorption on sawdust biochar. Proc Eng 148:718–725. https://doi.org/10.1016/j.proeng.2016.06.591

Manara P, Zabaniotou A (2012) Towards sewage sludge based biofuels via thermochemical conversion—a review. Renew Sust Energ Rev 16:2566–2582. https://doi.org/10.1016/j.rser.2012.01.074

Maneerung T, Liew J, Dai Y et al (2016) Activated carbon derived from carbon residue from biomass gasification and its application for dye adsorption: kinetics, isotherms and thermodynamic studies. Bioresour Technol 200:350–359. https://doi.org/10.1016/j.biortech.2015.10.047

Mante OD, Agblevor FA (2011) Catalytic conversion of biomass to bio-syncrude oil. Biomass Convers Biorefin 1:203–215. https://doi.org/10.1007/s13399-011-0020-4

Marsh H, Rodríguez-Reinoso F (2006) Activated carbon. Elsevier Science Ltd, Oxford

Mastral AM, García T, Callén MS et al (2001) Removal of naphthalene, phenanthrene, and pyrene by sorbents from hot gas. Environ Sci Technol 35:2395–2400. https://doi.org/10.1021/es000152u

Mastral AM, García T, Murillo R et al (2003) PAH mixture removal from hot gas by porous carbons. From model compounds to real conditions. Ind Eng Chem Res 42:5280–5286

Mastral AM, García T, Murillo R et al (2004) Development of efficient adsorbent materials for PAH cleaning from AFBC hot gas. Energy Fuel 18:202–208. https://doi.org/10.1021/ef030058+

Matsuhara T, Hosokai S, Norinaga K et al (2010) In-situ reforming of tar from the rapid pyrolysis of a brown coal over char. Energy Fuel 24:76–83. https://doi.org/10.1021/ef9005109

Mellbye ME, Hemphill DD, Volk VV (1982) Sweet corn growth on incinerated sewage sludge-amended soil. J Environ Qual 11:160–163

Milina M, Mitchell S, Pérez-Ramírez J (2014) Prospectives for bio-oil upgrading via esterification over zeolite catalysts. Catal Today 235:176–183. https://doi.org/10.1016/j.cattod.2014.02.047
Mochizuki T, Chen SY, Toba M, Yoshimura Y (2013) Pyrolyzer-GC/MS system-based analysis of the effects of zeolite catalysts on the fast pyrolysis of Jatropha husk. Appl Catal A Gen 456:174–181. https://doi.org/10.1016/j.apcata.2013.02.022
Mortensen PM, Grunwaldt JD, Jensen PA et al (2011) A review of catalytic upgrading of bio-oil to engine fuels. Appl Catal A Gen 407:1–19. https://doi.org/10.1016/j.apcata.2011.08.046
Müller-Stöver D, Ahrenfeldt J, Holm JK et al (2012) Soil application of ash produced by low-temperature fluidized bed gasification: effects on soil nutrient dynamics and crop response. Nutr Cycl Agroecosys 94:193–207. https://doi.org/10.1007/s10705-012-9533-x
Nanzer S, Oberson A, Berger L et al (2014) The plant availability of phosphorus from thermo-chemically treated sewage sludge ashes as studied by 33P labeling techniques. Plant Soil 377:439–456. https://doi.org/10.1007/s11104-013-1968-6
Nielsen RG (2007) Optimering af Lav Temperatur Cirkulerende fluid bed forgasningsprocessen til biomasse med højt askeindhold (Optimizing low temperature circulating fluidized bed gasification for high alkali biomass). Technical University of Denmark, Lyngby
Nowak B, Frías Rocha S, Aschenbrenner P et al (2012) Heavy metal removal from MSW fly ash by means of chlorination and thermal treatment: influence of the chloride type. Chem Eng J 179:178–185. https://doi.org/10.1016/j.cej.2011.10.077
Nuernberg NJ, Leal JE, Sumner ME (1998) Evaluation of an anion-exchange membrane for extracting plant available phosphorus in soils. Commun Soil Sci Plant Anal 29:467–479. https://doi.org/10.1080/00103629809369959
Ong Z, Cheng Y, Maneerung T et al (2015) Co-gasification of woody biomass and sewage sludge in a fixed-bed downdraft gasifier. AICHE J 61:2508–2521. https://doi.org/10.1002/aic.14836
Palla VSKK, Papadikis K, Gu S (2015) Computational modelling of the condensation of fast pyrolysis vapours in a quenching column. Part A: hydrodynamics, heat transfer and design optimisation. Fuel Process Technol 131:59–68. https://doi.org/10.1016/j.fuproc.2014.11.007
Patwardhan PR, Brown RC, Shanks BH (2011) Understanding the fast pyrolysis of lignin. ChemSusChem 4:1629–1636. https://doi.org/10.1002/cssc.201100133
Pearse SJ, Veneklaas EJ, Cawthray G et al (2007) Carboxylate composition of root exudates does not relate consistently to a crop species' ability to use phosphorus from aluminium, iron or calcium phosphate sources. New Phytol 173:181–190. https://doi.org/10.1111/j.1469-8137.2006.01897.x
Pettersson A, Åmand LE, Steenari BM (2008) Leaching of ashes from co-combustion of sewage sludge and wood-part II: the mobility of metals during phosphorus extraction. Biomass Bioenergy 32:236–244. https://doi.org/10.1016/j.biombioe.2007.09.006
Qian T, Jiang H (2014) Migration of phosphorus in sewage sludge during different thermal treatment processes. Sustain Chem Eng 2:1411–1419
Ren Q, Li L (2015) Co-combustion of agricultural straw with municipal sewage sludge in a fluidized bed: role of phosphorus in potassium behavior. Energy Fuel 29:4321–4327. https://doi.org/10.1021/acs.energyfuels.5b00790
Rong L, Maneerung T, Ng JC et al (2015) Co-gasification of sewage sludge and woody biomass in a fixed-bed downdraft gasifier: toxicity assessment of solid residues. Waste Manag 36:241–255. https://doi.org/10.1016/j.wasman.2014.11.026
Rosales E, Meijide J, Pazos M, Sanromán MA (2017) Challenges and recent advances in biochar as low-cost biosorbent: from batch assays to continuous-flow systems. Bioresour Technol 246:176–192. https://doi.org/10.1016/j.biortech.2017.06.084
Runtti H, Tuomikoski S, Kangas T et al (2014) Chemically activated carbon residue from biomass gasification as a sorbent for iron(II), copper(II) and nickel(II) ions. J Water Process Eng 4:12–24. https://doi.org/10.1016/j.jwpe.2014.08.009
Saleh SB, Flensborg JP, Shoulaifar TK et al (2014) Release of chlorine and sulfur during biomass torrefaction and pyrolysis. Energy Fuel 28:3738–3746
Schmidt SB, Jensen PE, Husted S (2016) Manganese deficiency in plants: the impact on photosystem II. Trends Plant Sci 21:622–632. https://doi.org/10.1016/j.tplants.2016.03.001

Seggiani M, Vitolo S, Puccini M, Bellini A (2012a) Cogasification of sewage sludge in an updraft gasifier. Fuel 93:486–491. https://doi.org/10.1016/j.fuel.2011.08.054
Seggiani M, Puccini M, Raggio G, Vitolo S (2012b) Effect of sewage sludge content on gas quality and solid residues produced by cogasification in an updraft gasifier. Waste Manag 32:1826–1834. https://doi.org/10.1016/j.wasman.2012.04.018
Seredych M, Bandosz TJ (2007) Mechanism of ammonia retention on graphite oxides: role of surface chemistry and structure. J Phys Chem C 111:15596–15604. https://doi.org/10.1021/jp0735785
Sibbesen E (1983) Phosphate soil tests and their suitability to assess the phosphate status of soil. J Sci Food Agric 34:1368–1374. https://doi.org/10.1002/jsfa.2740341209
Sigurjonsson HÆ, Elmegaard B, Clausen LR, Ahrenfeldt J (2015) Climate effect of an integrated wheat production and bioenergy system with low temperature circulating fluidized bed gasifier. Appl Energy 160:511–520. https://doi.org/10.1016/j.apenergy.2015.08.114
Simpson DR (2008) Biofilm processes in biologically active carbon water purification. Water Res 42:2839–2848. https://doi.org/10.1016/j.watres.2008.02.025
Six L, Pypers P, Degryse F et al (2012) The performance of DGT versus conventional soil phosphorus tests in tropical soils—an isotope dilution study. Plant Soil 359:267–279. https://doi.org/10.1007/s11104-012-1192-9
Sizmur T, Fresno T, Akgül G et al (2017) Biochar modification to enhance sorption of inorganics from water. Bioresour Technol 246:34–47. https://doi.org/10.1016/j.biortech.2017.07.082
Skoglund N, Grimm A, Marcus O, Bostro D (2013) Effects on ash chemistry when co-firing municipal sewage sludge and wheat straw in a fluidized bed: influence on the ash chemistry by fuel mixing. Energy Fuel 27:5725–5731
Song XD, Xue XY, Chen DZ et al (2014) Application of biochar from sewage sludge to plant cultivation: influence of pyrolysis temperature and biochar-to-soil ratio on yield and heavy metal accumulation. Chemosphere 109:213–220. https://doi.org/10.1016/j.chemosphere.2014.01.070
Spagnolo DA, Cornett LJ, Chuang KT (1992) Direct electro-steam reforming: a novel catalytic approach. Int J Hydrog Energy 17:839–846. https://doi.org/10.1016/0360-3199(92)90033-S
Stemann J, Peplinski B, Adam C (2015) Thermochemical treatment of sewage sludge ash with sodium salt additives for phosphorus fertilizer production—analysis of underlying chemical reactions. Waste Manag 45:385–390. https://doi.org/10.1016/j.wasman.2015.07.029
Stoholm P, Nielsen RG, Sarbæk L et al (2002) The low temperature CFB gasifier—further test results and possible applications. In: Proceedings of the 12th European biomass conference, pp 706–709
Taghizadeh-Toosi A, Clough TJ, Sherlock RR, Condron LM (2012a) A wood based low-temperature biochar captures NH 3-N generated from ruminant urine-N, retaining its bioavailability. Plant Soil 353:73–84. https://doi.org/10.1007/s11104-011-1010-9
Taghizadeh-Toosi A, Clough TJ, Sherlock RR, Condron LM (2012b) Biochar adsorbed ammonia is bioavailable. Plant Soil 350:57–69. https://doi.org/10.1007/s11104-011-0870-3
Thomsen TP, Ahrenfeldt J, Henriksen UB et al (2016) Closing the loop—utilization of secondary resources by low temperature thermal gasification. Technical University of Denmark, Lyngby
Thomsen TP, Hauggaard-Nielsen H, Gøbel B et al (2017a) Low temperature circulating fluidized bed gasification and co-gasification of municipal sewage sludge. Part 2: evaluation of ash materials as phosphorus fertilizer. Waste Manag 66:145–154. https://doi.org/10.1016/j.wasman.2017.04.043
Thomsen TP, Sárossy Z, Ahrenfeldt J et al (2017b) Changes imposed by pyrolysis, thermal gasification and incineration on composition and phosphorus fertilizer quality of municipal sewage sludge. J Environ Manag 198:308–318. https://doi.org/10.1016/j.jenvman.2017.04.072
Trinh TN, Jensen PA, Kim DJ et al (2013) Comparison of lignin, macroalgae, wood, and straw fast pyrolysis. Energy Fuel 27:1399–1409. https://doi.org/10.1021/ef301927y
Van der Drift A, Boerrigter H (2005) Synthesis gas from biomass for fuels and chemicals. In: SYNBIOS conference, Stockholm, Sweden, May 2005, pp 1–31

Van Wesenbeeck S, Prins W, Ronsse F, Antal MJ (2014) Sewage sludge carbonization for biochar applications. Fate of heavy metals. Energy Fuel 28:5318–5326. https://doi.org/10.1021/ef500875c
Vikrant K, Kim KH, Ok YS et al (2017) Engineered/designer biochar for the removal of phosphate in water and wastewater. Sci Total Environ 616–617:1242–1260. https://doi.org/10.1016/j.scitotenv.2017.10.193
Vogel C, Exner RM, Adam C (2013) Heavy metal removal from sewage sludge ash by thermochemical treatment with polyvinylchloride. Environ Sci Technol 47:563–567. https://doi.org/10.1021/es300610e
Wan S, Wang S, Li Y, Gao B (2017) Functionalizing biochar with Mg–Al and Mg–Fe layered double hydroxides for removal of phosphate from aqueous solutions. J Ind Eng Chem 47:246–253. https://doi.org/10.1016/j.jiec.2016.11.039
Wang B, Lehmann J, Hanley K et al (2015a) Adsorption and desorption of ammonium by maple wood biochar as a function of oxidation and pH. Chemosphere 138:120–126. https://doi.org/10.1016/j.chemosphere.2015.05.062
Wang Y, Lin Y, Chiu PC et al (2015b) Phosphorus release behaviors of poultry litter biochar as a soil amendment. Sci Total Environ 512–513:454–463. https://doi.org/10.1016/j.scitotenv.2015.01.093
Wang Y, Liu T, Xiao-kun LI et al (2015c) Nutrient deficiency limits population development, yield formation, and nutrient uptake of direct sown winter oilseed rape. J Integr Agric 14:670–680. https://doi.org/10.1016/S2095-3119(14)60798-X
Wang Z, Shen D, Shen F, Li T (2016) Phosphate adsorption on lanthanum loaded biochar. Chemosphere 150:1–7. https://doi.org/10.1016/j.chemosphere.2016.02.004
Westerhof RJM, Brilman DWF, Garcia-Perez M et al (2011) Fractional condensation of biomass pyrolysis vapors. Energy Fuel 25:1817–1829. https://doi.org/10.1021/ef2000322
Wüenscher R, Unterfrauner H, Peticzka R, Zehetner F (2015) A comparison of 14 soil phosphorus extraction methods applied to 50 agricultural soils from Central Europe. Plant Soil Environ 61:86–96. https://doi.org/10.17221/932/2014-PSE
Wünscher R (2013) A comparison of different phosphorus extraction methods with the phosphorus uptake of wheat. Universität für Bodenkultur, Wien
Yu Y, Li X, Su L et al (2012) The role of shape selectivity in catalytic fast pyrolysis of lignin with zeolite catalysts. Appl Catal A Gen 447–448:115–123. https://doi.org/10.1016/j.apcata.2012.09.012
Zhang H, Carlson TR, Xiao R, Huber GW (2012) Catalytic fast pyrolysis of wood and alcohol mixtures in a fluidized bed reactor. Green Chem 14:98–110. https://doi.org/10.1039/c1gc15619e
Zhang Y, Wu W, Zhao S et al (2015) Experimental study on pyrolysis tar removal over rice straw char and inner pore structure evolution of char. Fuel Process Technol 134:333–344. https://doi.org/10.1016/j.fuproc.2015.01.047
Zhang H, Chen C, Gray EM et al (2016) Roles of biochar in improving phosphorus availability in soils: a phosphate adsorbent and a source of available phosphorus. Geoderma 276:1–6. https://doi.org/10.1016/j.geoderma.2016.04.020
Zhou G, Jensen PA, Le DM et al (2016) Direct upgrading of fast pyrolysis lignin vapor over the HZSM-5 catalyst. Green Chem 18:1965–1975. https://doi.org/10.1039/C5GC01976A
Zielińska A, Oleszczuk P, Charmas B et al (2015) Effect of sewage sludge properties on the biochar characteristic. J Anal Appl Pyrolysis 112:201–213. https://doi.org/10.1016/j.jaap.2015.01.025
Zwetsloot MJ, Lehmann J, Solomon D (2015) Recycling slaughterhouse waste into fertilizer: how do pyrolysis temperature and biomass additions affect phosphorus availability and chemistry? J Sci Food Agric 95:281–288. https://doi.org/10.1002/jsfa.6716

Upgrading Bio-oil: Catalysis and Refinery

Robert M. Baldwin

1 Introduction

Fast pyrolysis (FP) is a biomass conversion technology that can produce feedstocks for upgrading to renewable drop-in transportation fuels and blendstocks for petroleum refineries (Richards 2013; Urbanchuk 2012; Green Goods and Services 2013; Production Statistics 2012). Fast pyrolysis is characterized by high throughput at short residence time (1–2 s) under moderate temperatures (400–650 °C) and pressures (c.a. 1 atm), which allows for compact reactors constructed with relatively inexpensive materials and utilizing mature reactor technology such as circulating and bubbling fluidized beds. The yields of organic liquids from FP can be as high as 75 wt%, allowing for high utilization of the renewable carbon (Bridgwater 2003). Some of the remaining light gases and char can be used for plant heat, and the char may be valuable as a soil amendment or for producing high-value coproducts such as bio-graphite and carbon nanotubes ($10–$20/kg).

The liquid produced by FP (pyrolysis oil or bio-oil) has a number of undesirable properties such as high viscosity, reactivity, immiscibility in hydrocarbons, and corrosivity (Oasmaa et al. 1997; Oasmaa and Peacocke 2001, 2010), which are largely due to the high oxygen content (~40 wt%). To address this barrier, a great deal of research has been conducted to develop upgrading processes to remove the oxygen either from the pyrolysis vapors or from the condensed liquids.

In one upgrading strategy known as catalytic fast pyrolysis (CFP), catalysts are employed to upgrade the bio-oil vapors (prior to condensation) with the optional addition of hydrogen to enhance yields. Two modes of operation are possible: in situ CFP where the catalyst is in contact with biomass in the pyrolysis reactor and ex situ CFP where pyrolysis vapors are catalytically upgraded in a separate reactor downstream from the primary pyrolysis reactor. A common application of CFP uses

R. M. Baldwin (✉)
National Bioenergy Center, National Renewable Energy Laboratory, Golden, CO, USA
e-mail: robert.baldwin@nrel.gov

© Springer Nature Switzerland AG 2019

J.-R. Bastidas-Oyanedel, J. E. Schmidt (eds.), *Biorefinery*,
https://doi.org/10.1007/978-3-030-10961-5_5

zeolite catalysts such as HZSM-5, which produces primarily olefins and aromatic molecules. Mono-ring aromatics (principally benzene, toluene, and xylenes or BTX) from this process can be used as a gasoline blendstock or as replacements for fossil-derived BTX, but the yields of low-oxygen hydrocarbon products from CFP are generally poor (Oasmaa et al. 1997), which results in unfavorable economics when compared to yields from fossil feedstocks. In addition, this approach does not form a significant amount of hydrocarbons that are suitable for diesel and jet fuels, which are considered better targets for advanced biofuels due to projections for future fuel usage (US Energy Information Administration 2016). As a result, alternative approaches to producing "drop-in" hydrocarbon fuels from biomass are being investigated, including the use of metal-based catalysts and adding hydrogen during CFP (Murugappan et al. 2016; Nolte et al. 2016; Venkatakrishnan et al. 2015). Although CFP may significantly reduce the amount of oxygen in the product, liquid-phase hydrotreating may still be required to meet specifications for introduction into a refinery or use as an intermediate or final fuel blendstock.

Hydrotreating of crude bio-oil is a viable strategy for producing hydrocarbon fuels (Jones et al. 2013), but the reactivity of the oil often leads to catalyst fouling and reactor plugging. This can be addressed with multistage upgrading, in which the oil is first catalytically stabilized by hydrotreating at a lower severity and then hydrotreated at higher severity to produce hydrocarbons (Wang et al. 2016). The hydrotreating stage needs to be conducted in two steps with increasing severity to prevent coke formation and to improve hydrogen utilization. The stabilization step targets the removal of reactive components, such as carbohydrates and carbonyl compounds, which can lead to gelation and solid formation. This stabilization step may not be required if the pyrolysis vapors are catalytically upgraded before they are condensed (Wang et al. 2016).

In this chapter we will first review fast pyrolysis for production of bio-oil and the properties of the resulting liquids. This will be followed by a review of options for upgrading bio-oil using catalysts and for integrating pyrolysis oil into a standard petroleum refinery.

1.1 Biomass Pyrolysis and Composition of Pyrolysis Oil

Biomass is composed of three main biopolymers, cellulose, hemicellulose, and lignin, and the decomposition of these three materials is largely responsible for the observed solid, liquid, and gaseous products during pyrolysis. As can be seen (Fig. 1), these biopolymers contain a significant amount of oxygen, which translates into 30–60 wt% oxygen (including oxygen in water) in the oil. Cellulose is a linear polymer consisting of repeat units of cellobiose with a degree of polymerization (dp) often exceeding 2000.

The organization of these polymers in cell walls is the subject of intense research, but it is clear that cellulose microfibrils form what appears to be a mat that provides the structure and strength for plant cell walls and the hemicellulose and lignin are

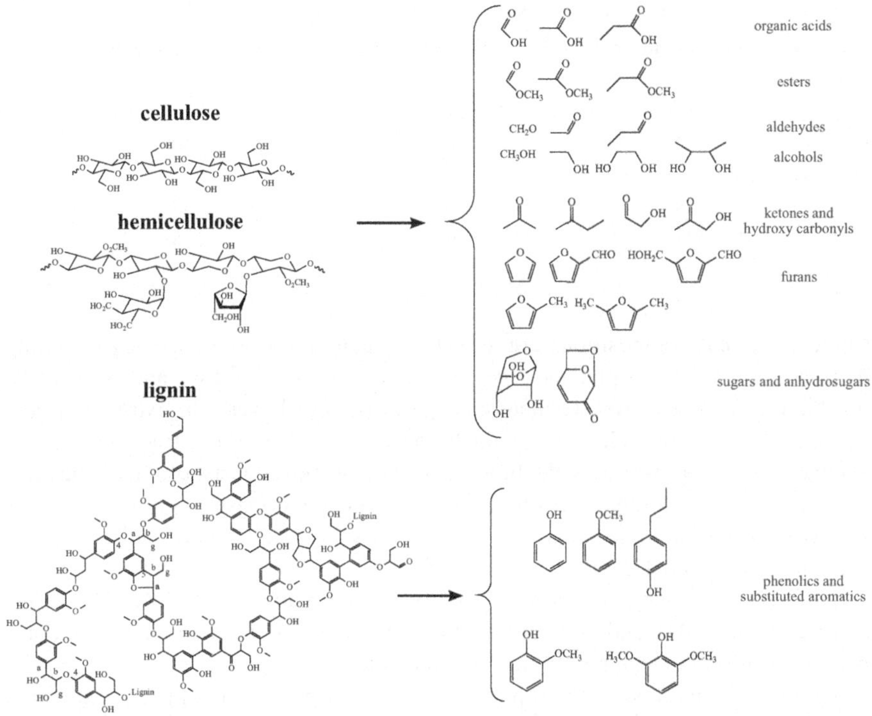

Fig. 1 Typical products formed from the pyrolysis of the biopolymers in plant cell walls

Table 1 Typical biomass composition[a]

Species	Extractives	Ash	Lignin	Hemicellulose	Cellulose
Hybrid poplar (Agblevor et al. 1992)	3.6	0.9	23.3 (24.6)	27.8 (29.3)	43.7 (46.1)
Monterey pine (Wiselogel et al. 1996)	2.7	0.3	25.9 (28.6)	23.0 (25.4)	41.7 (46.0)
Switchgrass (Johnson et al. 1993)	17.0	5.8	17.4 (23.1)	27.3 (36.1)	30.8 (40.8)
Corn stover (Qu et al. 2011)	7.6	6.8	17.2 (21.1)	26.3 (32.5)	37.8 (46.4)

[a]wt% values in parenthesis are ash- and extractive-free

intimately intertwined with the cellulose microfibrils. Although these biopolymers in biomass largely determine the yields and composition of the observed products in biomass pyrolysis oil, small levels of inorganic constituents also affect the product yields (Evans and Milne 1987; Oasmaa et al. 2010a; Patwardhan et al. 2010).

The composition of biomass varies significantly from woody to herbaceous materials; data on the composition of several potential feedstocks for biofuel manufacture is available at the US Department of Energy's Alternative Fuels Data

Table 2 Products from the fast pyrolysis of select biopolymers at 500 °C (Qu et al. 2011)

Yields	Cellulose (%)	Hemicellulose (%)	Lignin (%)
Gas	20	26	13
Char	18	23	47
Liquid	62	50	40
Products in gas			
CO_2	48	60	32
CO	47	30	25
CH_4	4	7	42

Center.[1] Typical compositions are shown in Table 1 for some species including hard- and softwoods (poplar and pine), energy crops (switch grass), and agricultural residue (corn stover). Woody materials typically have lower ash, extractives, and hemicellulose and more lignin than the herbaceous species. The extractives, materials that can be removed from the biomass using solvents, consist of fatty acids, lipids, fatty alcohols, terpenes, resin acids, and terpenoids (Oasmaa et al. 2003a) for woody materials and free sugars, sugar oligomers, alditols, organic acids, and inorganic ions (Chen et al. 2007, 2010) for herbaceous feedstocks. Extractives can have a major influence on the properties of pyrolysis oil (Oasmaa et al. 2003a, b; Oasmaa and Kuoppala 2003), and pretreatment of the biomass can often impact these components resulting in changes in bio-oil composition.

During fast pyrolysis, the biopolymers found in plant cell walls are converted into non-condensable gases, liquids, and solid char. The gases are primarily CO_2, CO, H_2, and some light hydrocarbons (C_1–C_4). The liquids contain 15–30% water in an emulsion with hydrophobic and hydrophilic organic compounds, some suspended solid material, and alkali and alkaline earth metal compounds that are present as inorganic matter in the feed biomass. Depending upon the temperature of pyrolysis, the char is primarily carbon (significant amounts of inorganics are present in the char), which can be used for processing heat or for upgrading to value-added coproducts. The organic compounds in the bio-oil typically contain a wide variety of oxygen functional groups, which impart undesirable physical and chemical properties to the oil.

Determining the yields of char, gas, and liquid is very important for assessing the viability of different feedstocks for production of bio-oil. As shown in Table 2, carbohydrates make up roughly 70% of the biopolymers and have liquid yields of 50–60%, while lignin makes up 20–30% of the biopolymers and only produces 40% liquid. Thus, the carbohydrates contribute 3–4 times as much liquid as does lignin in biomass pyrolysis.

The presence of inorganic materials—especially alkali and alkaline earth metals—in biomass also impacts yields. Oasmaa et al. measured the pyrolysis yields for several feedstocks and showed that the organic materials in the liquids were directly related to the amount of ash in the feedstock. Alkali metals present in

[1] (http://www.afdc.energy.gov/biomass/progs/search1.cgi)

the ash are known to increase the yields of char, water, and gases during pyrolysis (Patwardhan et al. 2010; Abdullah and Gerhauser 2008; Abdullah et al. 2010; Hayes and Hayes 2009; Nowakowski et al. 2007; Diebold et al. 1995; Baldwin and Feik 2013), and this could lead to lower yields of liquid organic compounds; potassium is known to be particularly active (Davidsson et al. 2002; Jensen et al. 2000; Knudsen et al. 2004; Wei et al. 2005; Oasmaa and Meier 2005).

Oxygen Content Organic oxygen in bio-oil is responsible for some of the physical and chemical properties that make it problematic for direct use or a refinery feedstock. There are a significant number of studies that report on the elemental composition of biomass fast pyrolysis oils; some of these data are shown in Table 3. In this table, the elemental composition is reported on a water-free basis (water makes up roughly 20% of the bio-oil). As can be seen, the amount of oxygen varies from 32 to 48 wt%, and significant variations are found both between and within each feedstock grouping. Organic sulfur and nitrogen levels are generally very low, which is a potential advantage since fuel standards require low levels of sulfur in finished fuels and both organic N and S compounds can be catalyst poisons. Potassium, sodium, and chloride can all be catalyst poisons; chloride can also contribute to corrosion.

Molecular Composition The chemical composition of bio-oils produced from process conditions that maximize liquid yields is very complex, and complete analysis of those oils requires the combined use of several analytical techniques. A precise description of bio-oil composition has not yet been achieved, and even with considerable analytical efforts, about 20% of the composition still remains unknown. Water is the single most abundant component of bio-oil, accounting for 15–30 wt% of the whole oil (Meier 1999). Major organic compound classes identified in bio-oil are hydroxy aldehydes, hydroxy ketones, sugars, carboxylic acids, and phenolics (Piskorz et al. 1988), with most of the phenolic compounds present as oligomers with molecular weights ranging from 900 to 2500 AMU (Meier and Scholtze 1997). GC/MS analysis has been used extensively to identify and quantify the volatile components of bio-oils (Garcìa-Pérez et al. 2007; Azeez et al. 2010). The most abundant organic components of bio-oils that have been reported studies are generally hydroxyacetaldehyde, acetic acid, formic acid, acetol, glyoxal, levoglucosan, and cellobiosan. Information on bio-oil composition from ^{13}C NMR analysis is shown in Table 4 for several bio-oils produced in an auger pyrolyzer (Ingram et al. 2008).

Two-dimensional gas chromatography coupled with time of flight mass spectrometry (2D GCxGC/TOFMS) and two-dimensional flame ionization detection (2D GCxGC/FID) has been recently applied to more comprehensively characterize bio-oil (Talmadge et al. 2014). Figure 2 presents information on major functional groups have been identified in crude bio-oil.

Table 3 Elemental composition of fast pyrolysis oil on a dry basis

Feedstock	Weight %					ppm	
	C	H	O[a]	N	S[b]	K-Na[b]	Cl[b]
Pine							
VTT (Oasmaa and Peacocke 2010)	55.8	5.8	38.2	0.1	0.02	20	30
Dynamotive (Oasmaa and Peacocke 2010)	52.6	7.53	39.52	0.09	0.0197		
BTG (Oasmaa and Peacocke 2010)	53.7	6.0	40.0	0.3			
Fortum (Oasmaa and Peacocke 2010)	57.1	6.4	36.4	0.1			
PNNL (Elliott 1994)	51.2	7.5	41.1	0.1		10	
PNNL (Elliott et al. 2012)	53.0	6.4	40.5	0.1	0.003		
Poplar							
NREL (NREL poplar bio-oil composition n.d.)	57.3	6.3	36.2	0.18	0.02	10	8
NREL (NREL poplar bio-oil composition n.d.)	60.5	6.7	32.6	0.23	0.02	12	8
Waterloo (Scott et al. 1985)	54.7	6.7	38.3				
Waterloo (Scott and Piskorz 1982)	51.8	6.7	41.3				
Waterloo (Scott and Piskorz 1984)	57.3	6.29	36.4				
Oak							
Dynamotive (Oasmaa and Peacocke 2010)	47.2	4.5	48.0	0.12	0.022		
PNNL (Oasmaa and Peacocke 2010)	56.0	6.8	37.2			57	
NREL (Baldwin and Feik 2013)	59.6	6.0	34.2	0.11	0.01	100	
Corn stover							
UMinn (Yu et al. 2007)	60.66	7.70		2.2	0.15		
USDA (Mullen et al. 2010)	53.97	6.92	37.94	1.18			
Iowa St (Shah et al. 2012)	58.4	5.2	30.9	0.5			
Straw							
VTT (Oasmaa and Peacocke 2010)	55.3	6.6	37.7	0.4	0.05	2	330
Waterloo (Scott and Piskorz 1984)	55.55	6.39					
Switchgrass							

Feedstock	Weight %					ppm	
	C	H	O[a]	N	S[b]	K-Na[b]	Cl[b]
NREL (Oasmaa and Peacocke 2010; Scott and Piskorz 1984)	55.8	6.9	36.3	0.79	0.03	128	1900
PNNL (Elliott 1994)	46.6	8.0	45.4			165	
USDA (Boateng et al. 2007)	46.0	6.7	42.6	0.3			

[a]By difference
[b]Some of the studies did not report sulfur, alkali, and chloride

Physicochemical Properties of Oils Table 5 shows specifics for pyrolysis oil properties that are relevant to catalytic upgrading and/or introduction into a refinery (Talmadge et al. 2014); these properties are discussed in more detail below.

2 Catalytic Upgrading of Bio-oil

2.1 Vapor-Phase Upgrading

CFP of biomass has been studied for nearly 30 years using biomass, biopolymers, and model compounds from the microscale up to pilot scale (Diebold et al. 1988; Evans et al. 1988; Horne and Williams 1995). Use of commercial-scale CFP to produce hydrocarbon fuels was recently attempted in the United States, and Anellotech, Inc. is currently scaling up in situ CFP technology for production of BTX from biomass (Anthrop 2013). In addition, the development and scale-up of the two-stage catalytic hydropyrolysis IH^2 technology is ongoing (Turriff 2014).

Table 4 Distribution in bio-oils produced in an auger reactor (Ingram et al. 2008)

	Carbon content (% of all bio-oil carbon)			
Type of carbon	Pine wood oil	Pine bark oil	Oak wood oil	Oak bark oil
Carbonyl	11.8	0.5	18.1	2.4
Aromatic	48.4	43.9	40.1	35.3
Carbohydrate	5.8	1.4	10.3	2.1
Methoxy/hydroxy	16.1	20.8	16.1	12.5
Alkyl	17.9	33.4	15.5	47.7

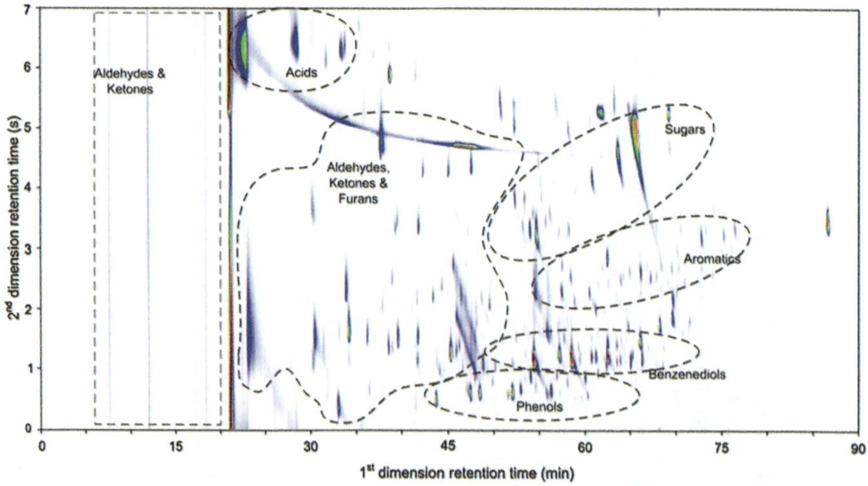

Fig. 2 Group-type analysis of the crude bio-oil by two-dimensional GCxGC

There have been several recent reviews of CFP (Liu et al. 2014; Martín-Aranda and Čejka 2010; Ruddy et al. 2014; Perego and Bosetti 2011; Taarning et al. 2011; Bulushev and Ross 2011; Asadieraghi et al. 2015; Rezaei et al. 2014); some aspects of advances in technology are highlighted in the following section.

Upgrading Using HZSM-5 HZSM-5 has been the most extensively used zeolite catalyst for CFP and is effective for removing oxygen from pyrolysis vapors, at a cost of reduced carbon yields for condensed organic oil. Typical product distributions from CFP of woody biomass with HZSM-5 compared to FP are shown in Table 6. As can be seen, the oxygen content of the oil can be significantly reduced from close to 40% for FP to 4% in one instance for CFP (Iisa et al. 2017). Oxygen reduction is accompanied by a decrease in the organic oil yield, which has a strong negative impact on the cost of fuel production; total liquid yields (including water) are reduced from ~70 wt% for FP to less than 55 wt% for CFP. FP oil typically is an emulsion with 15–30 wt% water (Czernik and Bridgwater 2004), while CFP oil

Table 5 Properties of biomass pyrolysis oil

Property	Notes
Water	15–30 wt%
Viscosity	13–80 cSt @ 50 °C
Solids content	0.01–1 wt%
Miscibility in organic solvents	Poor
Stability	Oil components polymerize, particularly at elevated temperatures
Corrosivity	pH 2.0–3.7, TAN 50–200
Distillation	30–50% residue
Density	1.2 g cm^{-3}

Table 6 Product distributions (g/g feed, %) from CFP over HZSM-5

	CFP[a]	CFP[a]	CFP[b]	CFP[c]	CFP[d]	CFP[e]	FP[f]
Liquids	**32**	**33**	**52**	**45**	**43**	**51**	**67**
Organic	9	11	28	26		32	
Aqueous	27	22	24	19		19	
Gas	**34**	**33**	**29**	**36**	**40**	**21**	**18**
Solids	**19**	**16**	**19**	**25**	**18**	**27**	**12**
Oxygen, wt%[g]	4	14	18		19	22	36
Bio:Cat, g/g	0.5	1.5	0.06	0.17	3.0		

[a]Iisa et al. (2017), ex situ, pine, 500 °C
[b]Vasalos et al. (2016) in situ, beechwood, 482 °C
[c]Jae et al. (2011), in situ, pine, 500 °C
[d]Mante and Agblevor (2014), in situ, pine, 550 °C
[e]Paasikallio et al. (2014) (VTT), in situ, pine, 520 °C
[f]Howe et al. (2015), pine, 500 °C; note that in some cases liquids are single-phase, hence no separate organic and aqueous fractions
[g]Oxygen content in the oil on a dry basis

typically has a separate aqueous layer, which is 20–25 wt% of the biomass, and the organic liquid yield is low (9–32 wt%; water-free basis). Lower organic yield in CFP is concomitant with higher gas and solid (i.e., coke) formation.

The chemical composition of condensable products from CFP with HZSM-5 shows a strong dependence on the ratio of the mass of the pyrolysis vapor processed to the mass of the catalyst. Direct measurements of the compositional change of CFP oil as a function of biomass-to-catalyst ratio (B:C) show that the oxygenated species increase with B:C (Iisa et al. 2017; Czernik and Bridgwater 2004; Mukarakate et al. 2014; Lappas et al. 2002). Figure 3 shows that increasing B:C leads to increased selectivity for phenols, indenols, naphthols, methoxyphenols, levoglucosan, acids, and carbonyls; these compounds are intermediates formed by incomplete deoxygenation and not simply due to breakthrough of raw pyrolysis products. These compounds are largely responsible for the corrosivity of bio-oils and fouling problems during hydrotreating but can be greatly reduced by application of CFP.

Microscale experiments have been conducted as a function of B:C and show that at low B:C the products from CFP with HZSM-5 are almost exclusively oxygen-free aromatics and olefins (Mukarakate et al. 2014). As shown in Fig. 4, increasing B:C leads to the formation of oxygenated intermediates (primarily phenols and other hydroxylated aromatic molecules and furans) followed by the breakthrough of the oxygenated molecules found in the raw pyrolysis vapors at higher B:C. This change in product composition is likely due to the buildup of carbon on the catalyst, which reduces deoxygenation reactions and deactivates the catalyst. The difference

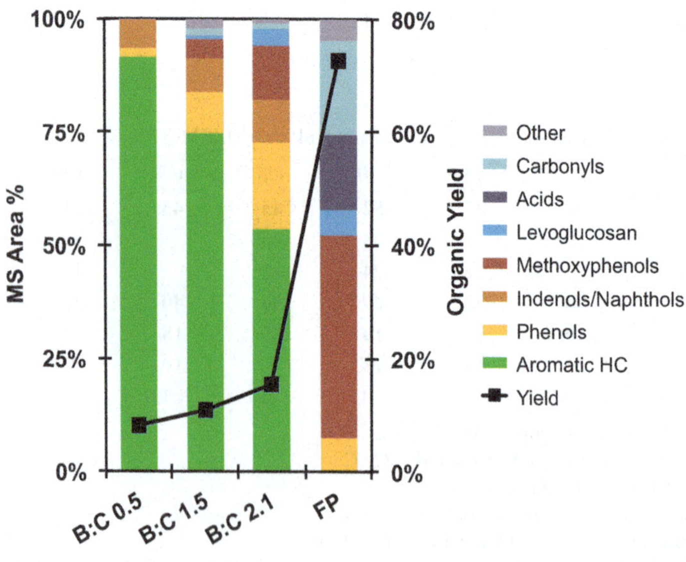

Fig. 3 Measured composition of fast pyrolysis (FP) oil and the organic oils from CFP at different biomass-to-catalyst ratios (B:C) (Knudsen et al. 2004)

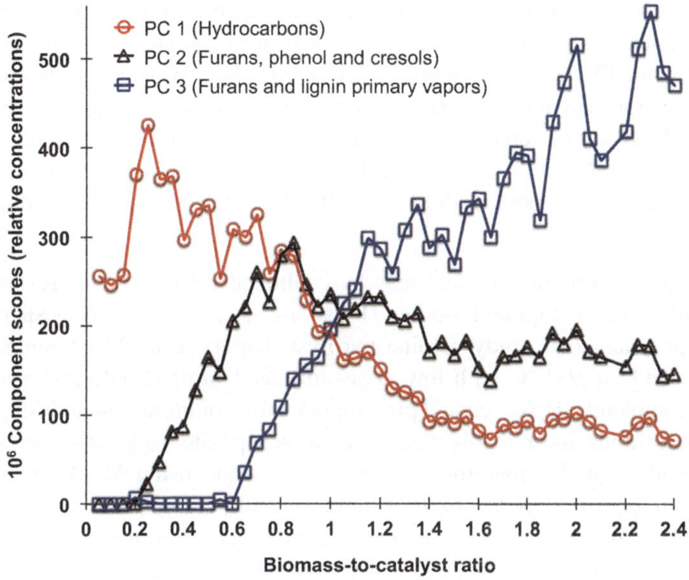

Fig. 4 Deactivation of HZSM-5 as a function of the amount of biomass vapors introduced into a fixed bed (Mukarakate et al. 2014). The mass spectra at each point were analyzed using multivariate analysis, and the red trace (PC1) is the principal component (PC) containing aromatic hydrocarbons, the blue trace (PC3) is the principal component containing the molecules of raw pyrolysis vapor, and the black trace (PC2) is the principal component containing oxygenated intermediates, primarily phenols

in chemical composition of CFP oils relative to the raw FP oils is marked by lower carbonyls and acids and explains why CFP oils are amenable to distillation (Agblevor et al. 2010) and do not require a stabilization step needed for hydrotreating raw pyrolysis oils (Agblevor et al. 2016).

Other Zeolites Other zeolites used for CFP show a similar product suite as HZSM-5 but are typically not as efficient and often lead to enhanced coke formation (Mullen and Boateng 2013). This is likely due to the micropore size and shape; smaller pores cannot pass aromatic molecules, while larger pores produce more coke (Carlson et al. 2009).

Amorphous Silicates Amorphous silicates such as MCM-41 or SBA-15 are highly structured silica materials with regular mesopores and amorphous silica walls (Ciesla and Schuth 1999). These have been studied as catalysts for vapor-phase upgrading—their weak acid sites enable greater selectivity toward oxygenated products than HZSM-5. There are a number of microscale experiments that suggest these weak acid sites contribute to the formation of oxygenated intermediates, such as furans, phenols, and ketones (Antonakou et al. 2006; Suzuki et al. 2008).

Metals Oxides of metals such as Ti, Sn, Zr, Ce, and Mo have acid and reducible metal sites that can deoxygenate biomass pyrolysis vapor when used in either the in situ or ex situ CFP mode. Many microscale studies have been conducted that show high conversion and the formation of oxygenates, such as furans, phenols, and ketones, which can be separated for chemicals or coupled to produce diesel range renewable fuels (Nolte et al. 2016; Agblevor et al. 2010; Doornkamp and Ponec 2000; Lu et al. 2010, 2009; Budhi et al. 2015; Donar and Sınağ 2016; Mante et al. 2015).

Deoxygenation by metal oxides can be facilitated with added H_2 as shown by a number of model compound studies (Prasomsri et al. 2013, 2014; Shetty et al. 2015). One microscale study of pine pyrolysis vapors using MoO_3 supported on TiO_2 and ZrO_2 at 500 °C with low pressure (c.a. 1 atm) H_2 (Ruddy et al. 2014) reported completely deoxygenated pine vapors until a biomass-to-catalyst ratio of 1, after which furans and phenols were formed. A separate study (Zhou et al. 2016) collected oil from the upgrading of lignin and wood using MoO_3. The organic phases from these studies consisted of furans and alkyl furans, ketones, cyclopentanones, phenol and alkyl phenols, naphthols, and indenols. Similar oxygenated species such as furan and alkyl furans, C_4–C_5 ketones, and cyclopentanone were also observed in the aqueous phase.

Another approach for upgrading biomass pyrolysis vapors in the presence of added hydrogen is to promote breaking of C–O bonds over C–C bonds and to reject oxygen as water. Catalysts used for this purpose have both metal sites for hydrogenation and, depending upon the support, acid sites that facilitate hydrodeoxygenation (HDO) and other reactions catalyzed by solid acids. Pd/SiO_2 was used for upgrading of *m*-cresol to form 3-methylcyclohexanone with 65% selectivity, toluene with 27% selectivity, and 3-methylcyclohexanol with 8% selectivity (de Souza et al. 2014). Similar results were also observed with Pt/SiO_2 and metals (e.g., Pt) supported on carbon, which also has low acidity (Gao et al. 2014). Upgrading of guaiacol using Ru/SiO_2 was found to favor formation of phenol by hydrogenation, while Ru/Al_2O_3 favored formation of catechol by HDO (Boonyasuwat et al. 2013). Recently there has been significant interest in carbide catalysts for ex situ CFP (Lee et al. 2014, 2015; Sullivan and Bhan 2016; Sullivan et al. 2015). Mo_2C has both metallike sites for hydrogenation and acid sites to facilitate HDO (Sullivan et al. 2016).

3 Bio-oil Hydrotreating

Hydroprocessing Hydrotreating of bio-oils takes place in the presence of a catalyst at high hydrogen partial pressures (500–2000 psi). Oxygen removal may take place by HDO, decarboxylation, decarbonylation, or dehydration reactions; other reactions such as saturation of aromatic and C–C double bonds, cracking of molecules into smaller ones, and repolymerization take place simultaneously. The literature on catalytic hydrotreating of FP bio-oil is very extensive, and several reviews of hydrotreating of pyrolysis oils have been published on both the process and catalysts (Elliott 2007; Zacher et al. 2014; Wang et al. 2013a, b; He and Wang 2012; Ma and van Bokhoven 2014; Furimsky 2000).

The main technical barrier to hydrotreating of pyrolysis oils is fouling of the hydrotreating catalyst due to carbonaceous deposits from thermally induced polymerization of reactive components such as aldehydes and sugars. At low temperatures, these compounds can be hydrogenated, but at the higher temperatures required for HDO of less reactive components, they rapidly form cross-linked polymers. A two-stage hydrotreating process with a first stage operating at a low temperature to prevent coking followed by a second, higher-temperature stage was developed early on to combat polymerization issues (Baker and Elliott 1988; de Miguel Mercader et al. 2011). More recently, a third stabilizer stage has been added to the process such that a total proposed process consists of a low-temperature stabilizer followed by two-stage hydrotreating (Jones et al. 2013). Reduction of sugars and hydrogenation of aldehydes during the stabilization step have been shown to lead to improved hydrotreating performance (Olarte et al. 2016); however, some deactivation of the stabilization catalyst was still observed, and poisoning of the catalyst by sulfur-containing species was identified as the major cause for deactivation. Recently, 1000 h of FP hydrotreating has been demonstrated (Abdullah n.d.). The FP oil was ion exchanged and filtered to remove heteroatoms responsible for deactivation of the stabilization catalyst, and coke from the stabilization stage was removed by solvent rinse.

Traditional sulfided CoMo and NiMo catalysts have been found to be efficient for hydrotreating of pyrolysis oils. Precious metal catalysts on carbon support are another group of catalysts that have been assessed for pyrolysis oil hydrotreating (French et al. 2014; Elliott et al. 2014), and sulfided Ru/C has been used for the first stage of hydrotreating in several studies (Olarte et al. 2016; Bui et al. 2011).

For deoxygenation of compounds such as phenols, the traditional sulfided Mo-based catalysts have been reported to favor direct deoxygenation by removal of the hydroxyl group without ring hydrogenation (Ma and van Bokhoven 2014; Wang et al. 2013b). In contrast, noble metal catalysts are efficient for hydrogenation and favor first hydrogenating the aromatic ring, followed by deoxygenation. Direct deoxygenation is preferable because of the lower hydrogen consumption. Therefore, sulfided CoMo and NiMo have been the catalyst of choice for the second hydrotreating stage (de Miguel Mercader et al. 2011). Sulfided NiMo and CoMo catalysts are typically supported on Al_2O_3, whereas the noble metal catalysts are usually supported on carbon. Al_2O_3 may not be a good support due to poor hydrothermal stability and resistance to organic acids, and other, more stable supports have been tested, including carbon extrudates, TiO_2, and ZrO_2 (Zacher et al. 2014).

Hydrotreating CFP Oil CFP oils are partially upgraded, but they still contain oxygen in significant quantities and may require hydrotreating if the objective is to produce finished transportation fuels or refinery blendstocks. CFP oils contain a variety of oxygenates including phenols, which are of low reactivity; hence, it can be expected that the final temperature required for hydrotreating is similar to that for hydrotreating of non-catalytic FP oils that have been stabilized. CFP may, however, remove the more reactive oxygenated compounds, such as aldehydes and ketones which are responsible for catalyst bed plugging during FP oil hydrotreating and hence require multiple hydrotreating stages at increasing severity for FP oils.

There is limited information on hydrotreating of CFP oils, but CFP oil with oxygen content as high as 25 wt% has been successfully hydrotreated in a single stage (Agblevor et al. 2016). Table 7 highlights differences in CFP oil and FP oil hydrotreating. As shown, CFP oils could be hydrotreated in a single stage for times on stream exceeding 300 h, whereas the two FP oils hydrotreated in the same equipment as the first CFP oil led to bed plugging or lost catalyst activity in less than 100 h. Limited or no catalyst deactivation was observed with the CFP oil hydrotreated at 400 °C, but the catalyst tested at 290 °C deactivated rapidly, and the oil oxygen contents increased significantly during the experiment. The yields of the hydrocarbon product are significantly higher for CFP oils than for FP oils, and oil carbon yields can exceed 90% (French et al. 2014). The higher yields are partially a direct consequence of the initially lower oxygen contents but also of the types of oxygen functional groups remaining in the CFP oil, e.g., phenols, which are likely to lose oxygen via dehydration as opposed to the release of CO and CO_2, which lead to mass and carbon losses.

The composition of the CFP oil impacts the hydrotreating. Figure 5 shows GCMS analysis of the product oils from hydrotreating of CFP oils with different oxygen contents produced over ZSM-5. Phenols were the most persistent oxygenates and were detected at low levels in all hydrotreated CFP oils (Iisa et al. 2017). In addition, low quantities of methoxy groups and carbonyls were detected in the product oils with higher oxygen contents. Aromatic hydrocarbons became partially

Table 7 Comparison of hydrotreating FP and CFP bio-oil

	FP[a]	FP[a]	CFP[b]	CFP[c]
Feed oil O, wt% dry basis	40.5	40.5	24.8	19.5
Temperature, °C	170/400	170/400	–/400	–/290
Catalyst	Sulfided Ru/C/ sulfided NiMo	Sulfided Ru/C/ sulfided CoMo/C	Sulfided CoMo/ZrO_2	Sulfided HDO catalyst
Pressure, bar	138	138	138	138
Oil yield, g/g dry feed oil	35–45%	35–43%	67–79%	68–83%
Normalized C yield, g C in product/g C in feed oil	~82%	~82%	87–96%	80–93%
Product oil O, wt% dry basis	0.2–0.3	0.3–2.7	1.0–1.2	0.5–11
H_2 consumption, g H_2/g feed oil	0.043–0.050	0.025–0.044	0.067–0.074	0.069–0.073
Time on stream (h)	90	90–100	>300	>300

[a]Elliott et al. (2014)
[b]Agblevor et al. (2016)
[c]Mante et al. (2015)

saturated; one-ring aromatics had relatively low conversions (up to 15%) to cyclohexanes; multiring compounds typically retained one aromatic ring and consisted, for example, of tetrahydronaphthalenes. The degree of saturation of the aromatic bonds decreased as hydrotreating temperature was increased, which can be attributed to thermodynamic limitations.

Even though the oxygen contents and H:C molar ratios in the products from hydrotreating of FP and CFP oils may be similar, differences in the chemical composition of the product oils can be expected. Upgrading of pyrolysis vapors during CFP in most cases results in cracking of the heavier molecules; consequently, both the CFP oils and the hydrotreated CFP oils will likely be of lower molecular weight than the corresponding FP oils. Simulated distillation (SIMDIST) analysis of hydrotreated CFP and FP oils has indeed shown higher gasoline fraction for the hydrotreated CFP oil (Agblevor et al. 2010). Many CFP catalysts, such as HZSM-5, produce aromatic rings, including monoaromatics (e.g., BTX) and bi- and multiring compounds.

The monoaromatic compounds in CFP oils have high octane numbers and can provide high-quality gasoline. For both hydrotreated FP and CFP oils, low levels of oxygen remaining in the products will be mainly in the form of phenols, but the molecular weights and side chains are likely different.

An advantage for CFP oils is separation of the liquid products into organic and aqueous phases. Only the organic fraction will be hydrotreated, which reduces the volume of the liquid to be processed and thus reduces cost. In addition, the water content of the CFP organic phase is lower than that of FP oils (generally <5 wt%), and this may enable the use of catalyst supports that are less hydrothermally stable.

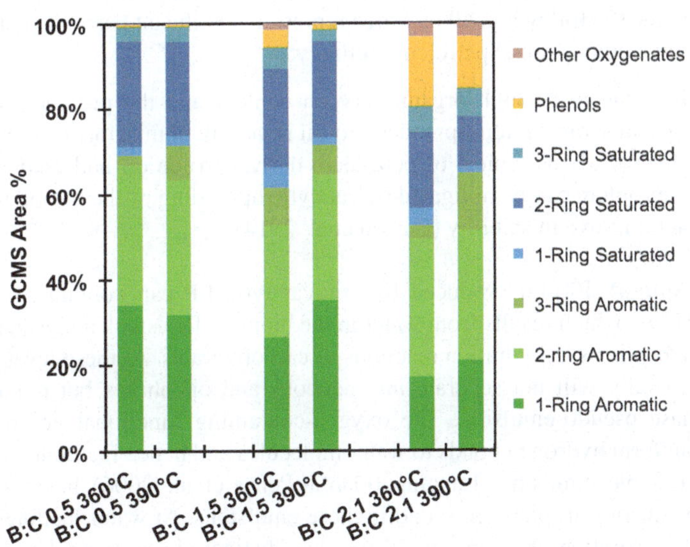

Fig. 5 Measured composition of hydrotreated CFP oils prepared at different biomass-to-catalyst ratios (B:C) and hydrotreated at either 360 or 390 °C over sulfided $CoMo/Al_2O_3$ (Iisa et al. 2017). The CFP oil prior to hydrotreating had oxygen contents of 4, 14, and 18% (dry basis)

4 Refinery Integration

4.1 Integrating Pyrolysis Oil into Standard Refineries

A potentially very attractive option for introducing biomass-derived materials into the fuel marketplace would be to use bio-oil or bio-crude (derived via hydrothermal liquefaction) as a feedstock and/or blendstock in a standard petroleum refinery, either replacing or supplementing fossil-derived materials with biomass-derived materials (Fig. 6). This would, in principle, facilitate the introduction of renewable carbon into the fuel infrastructure and would economically advantage the biofuel industry by using the multitrillion-dollar refining and distribution infrastructure already in place. Simultaneous processing of bio-oil with fractions from fossil petroleum in existing refinery unit operations is known as co-processing; a recent review article summarizes opportunities (Bezergianni et al. 2018). From the refiner's perspective when evaluating potential refinery feedstocks, important properties include the boiling range distribution obtained from the main crude oil fractionator and the hydrocarbon types (PONA) and heteroatom (sulfur, nitrogen, oxygen) contents of each of the resulting primary distillation and process-derived intermediate fractions. The boiling range distribution from the main fractionator impacts all of the major downstream unit operations, which are in turn designed to optimize the refinery product slate to produce the most profit per barrel of feedstock.

Using bio-oil as a refinery blendstock and/or feedstock introduces several potential problems due to differences in physicochemical properties of bio-oil relative to petroleum crude oil. Important properties of bio-oil as they relate to refinery feedstocks include the following, all of which represent significant barriers to utilization of bio-oil in a conventional petroleum refinery:

Miscibility Due to its high organic oxygen content and the presence of highly polar oxygenates, raw or non-upgraded bio-oil is largely immiscible in aliphatic and aromatic hydrocarbons typical of petroleum-derived crude oil and crude oil fractions. This problem can be mitigated by catalytic upgrading of the bio-oil to reduce oxygen and improve miscibility (French et al. 2014).

Water Content Bio-oil produced by non-catalytic FP can contain as much as 30 wt% H_2O—which results from water in the biomass feedstock and water formed during pyrolysis by dehydration reactions of carbohydrates. At these levels pyrolysis oil generally will not separate into aqueous and oil phases but remains as a single-phase pseudo-emulsion. The oxygen-containing functional groups on the bio-oil can form hydrogen bonds to water molecules and hence facilitate the formation of a stable emulsion. Imaging (Garcìa-Pérez et al. 2006) has shown that 5–10 μm aqueous droplets are found in these emulsions. At water loadings greater than 30%, separation of water and oil into two distinct phases can take place. The impact of water content on downstream upgrading and refinery operations is a key issue that impacts use as a refinery feedstock. Water will decrease the viscosity of

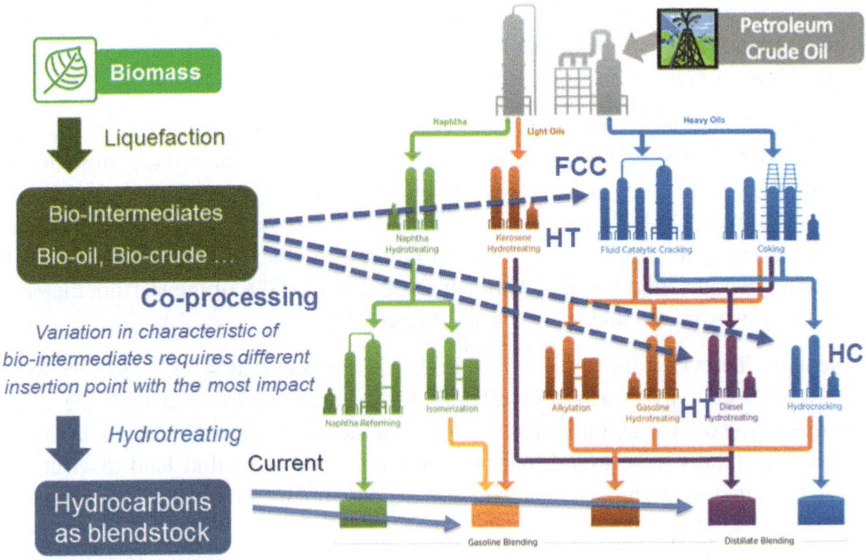

Fig. 6 Schematic biorefinery

bio-oil, but the presence of water can be damaging to some catalysts used in the downstream unit operations that lack hydrothermal stability. Production of a partially upgraded bio-oil by either in situ or ex situ catalytic fast pyrolysis normally results in a two-phase product with an organic phase that is much lower in water content than raw pyrolysis oil.

Stability and Aging Fast pyrolysis reaction conditions of rapid heating and quenching produce a bio-oil condensate that is not at thermodynamic equilibrium at ambient conditions. During storage, the chemical composition of the bio-oil shifts toward thermodynamic equilibrium resulting in changes in the viscosity, molecular weight, and co-solubility of its many compounds. Aging of biomass pyrolysis oil has been extensively studied; investigations have shown that chemical reactions in the oil lead to increased water content, the evolution of light gases, greater tendency toward phase separation, increased molecular weight, and increased viscosity. The rate of aging is strongly dependent upon temperature (Oasmaa and Sipilä 1996; Czernik 1994) which also has important implications for introduction into refinery unit operations. Studies have found viscosity increases by roughly 50% in 80 days at 37 °C, while the viscosity doubles in 15 h at 90 °C. The associated increase in molecular weight suggests that polymerization reactions are occurring during aging. The reactions that occur in pyrolysis oil during storage at ambient conditions largely involve the oxygen functionalities, including:

- Esterification: The reaction of organic acids with alcohols to form esters and water

- Condensation reactions involving aldehydes and ketones and water or alcohols
- Condensation reactions of aldehydes with phenols
- Condensation reactions involving furfurals

Many of these reactions can be catalyzed by solid material in the bio-oil, such as inorganics from the biomass. Studies where the solids are filtered out show much slower changes in viscosity (Meier and Scholtze 1997). The reactivity of pyrolysis oil can be particularly problematic for distillation operations. Polymerization and molecular weight growth during distillation lead to the formation of very high molecular weight solid residue. This can result in up to 50% of the starting material remaining as solid residue (Oasmaa and Meier 1999).

Viscosity The viscosity of bio-oil as produced (measured at 40 °C) can vary from as low as 25 cP to as high as 1000 cP or more depending on the feedstock, the water content of the oil, the amount of light ends, and the extent to which the oil has aged. Other researchers have found that polymerization reactions that lead to viscosity increases are accelerated at higher storage temperatures, and it has been shown that the rate of change in viscosity can increase from 0.009 cP/day when stored at −20 °C to more than 300 cP/day at 90 °C (Diebold 2000). Bio-oil is more viscous than crude oil at room temperature; however, its viscosity is very similar to that of crude oil in a temperature range of 35–45 °C. To transport bio-oil in pipelines, the temperature of the pipeline should be maintained in the range of 35–45 °C to keep the viscosity similar to that of crude oil (Pootakham and Kumar 2010).

Simple methods such as adding polar solvents and diesel or other fuels can address some of these undesirable bio-oil physicochemical characteristics. Polar solvents, such as methanol or ethanol, can improve the volatility and heating value and decrease the viscosity and acidity (Zheng and Kong 2010), while blending diesel or other fuels can positively impact oil viscosity. Finally, application of hot gas filtration prior to condensation offers a relatively simple method to produce a bio-oil with low particulate content (essentially zero ash) and improved stability with respect to increases in viscosity (Baldwin and Feik 2013).

Acidity The acidity of petroleum crude oil is usually very low, and acidic components present in crude oil generally represent naphthenic acids. These components are corrosive to mild steel at high temperatures. The refining industry has long since determined mechanisms[2] for mitigating the impact of corrosion imparted by naphthenic acids including blending (industry standard is that the TAN of the blend must be <0.6 (Marker 2005)), use of corrosion inhibitors, and upgraded materials of construction. The corrosivity of pyrolysis oil is primarily due to its acidity, which is derived mainly from strong acids (carboxylic acids) and weak acids (phenolic compounds). Measurements of total acid number (TAN) of bio-oil samples show that TAN values in the 90–100 range are fairly common, with a pH typically in the range 2–3; TAN values of 150–200 are not uncommon. Other groups of compounds in fast

[2] http://www.setlaboratories.com/nac/tabid/79/Default.aspx

pyrolysis liquids that influence acidity include phenolics (5–10%) and fatty and resin acids (<5%) (Oasmaa et al. 2010b).

This level of acidity has been shown to cause corrosion problems for many materials. In particular, high corrosion rates for carbon steel (AISI01) have been observed (Oasmaa and Peacocke 2010)—this is particularly problematic as many refinery unit operations are constructed from carbon steel. Further, corrosivity increases significantly at higher temperatures (Aubin and Roy 1980). Measured corrosion rates for carbon steel alloys were much more than could be tolerated in any system that was expected to last for years. Several metal and polymer materials have been tested (Das et al. 2004; Fuleki 1999; Soltes and Lin 1984; Jay et al. 1995; Darmstadt et al. 2004; Keiser 2013) for resistance to corrosivity from bio-oil. Stainless steel specimens showed minimal weight change. Hydroprocessing can address the high acidity problem; however this requires significant CAPEX investment and normally is accompanied by substantial yield losses of distillate-range product (up to 50%). Torrefying biomass has resulted in 25% less acetic acid in the produced oil; however torrefaction also results in a loss of carbon efficiency.

Organic Oxygenates Most crude oils contain very small amounts of organic oxygen, in general less than 1 wt% and often less than 0.1 wt% (Speight 1991). Accordingly, catalysts and processes used in the refinery unit operations for hydroprocessing intermediates and upgrading fossil petroleum to finished fuels are not designed to effectively process or utilize oxygen-containing molecules. The presence of organic oxygenates and oxygen functional groups can impart very different properties to feedstocks in the refinery when compared to streams that are predominately hydrocarbon. Physical properties such as density, viscosity, and storage stability can be altered, and oxygenates can also degrade elastomers used in engine parts. The presence of organic oxygen leads to changes in volatility, which impacts unit operations used for separating intermediates or producing final products such as distillation. Chemical properties are also altered by the presence of organic oxygen, which can influence the reaction chemistry of important conversion and fuel synthesis processes. Catalysts, which are effective for converting crude fractions in the absence of oxygenates, may function entirely differently when oxygenated compounds are present, leading to changes in catalyst activity and selectivity.

Alkali and Alkaline Earth Metals and Other Particulates in the Oil Crude oil has a very small inorganic content, which is generally comprised of salt and porphyrins of vanadium and nickel. Salts and suspended solids in crude oil are readily removed in a desalting unit operation prior to the initial fractionation. Refiners have developed strategies to mitigate and/or passivate the impact of vanadium and nickel contamination on refining catalysts, but introducing bio-oil brings a new suite of potential catalyst poisons from silicon, calcium, potassium, and other alkali and alkaline earth metals. These metals could act as poisons for cracking and hydrotreating catalysts, leading to changes in selectivity and/or permanent activity loss. Particulate matter is generated by pyrolysis, and in spite of attempts to filter this material with cyclones, some of the solid material is often carried into the pyrolysis oil (Garcìa-Pérez et al. 2006). In addition, there is evidence that particulate matter—ranging in size from a few nanometers to micrometers—is formed during bio-oil condensation and aging of pyrolysis oil (Oasmaa and Peacocke 2001). The solids could be from char formed during pyrolysis, sand or other heat transfer material, polymerized pyrolysis products, or inorganic material from the biomass. Studies by the National Renewable Energy Laboratory (NREL) have shown that application of hot gas filtration to pyrolysis vapors prior to condensation is effective for producing a bio-oil with very low particulate content and with very low concentrations of alkali and alkaline earth metals (Diebold et al. 1995; Baldwin and Feik 2013). As these metals are known to catalyze condensation reactions, the stability of the hot gas filtered oil was also found to be greatly improved when compared to unfiltered oil (Baldwin and Feik 2013).

4.2 Refinery Integration Studies: Co-processing

In a comprehensive investigation on incorporation of bio-renewables into the petroleum refinery, Marker (Marker 2005) examined several opportunities for including bio-oil into standard petroleum refinery unit operations. In addition to an examination of utilizing waste fats and greases as refinery feedstocks and hydrogen production from the aqueous fraction of bio-oil, this study looked at:

- Hydroprocessing pyrolytic lignin to produce aromatics and gasoline
- Co-processing bio-oil with vacuum gas oil (VGO) in the fluid catalytic cracker (FCC)

In these studies, co-processing whole raw bio-oil and pyrolytic lignin in a laboratory ACE system was carried out for blends of up to 20 wt% raw bio-oil. These experiments were accompanied by tests on the catalytic cracking of a hydrotreated whole bio-oil and for VGO alone. Results showed that all three biomass-derived oils gave greatly increased yields of coke (16 and 27 wt% for the bio-oil/VGO blends) when compared to catalytic cracking of VGO alone. The bio-oil/VGO blends were found to increase the "crackability" of the feedstock when compared to VGO alone and to increase the yields for light-end products, which is potentially an economically attractive outcome.

A comprehensive investigation of opportunities for incorporating biomass-derived materials in the petroleum refinery was carried out under the auspices of the BIOCOUP project (BIOCOUP 2011). This study concluded that the best strategy for co-processing bio-oil was the FCC using a partially deoxygenated bio-oil containing up to 20% organic oxygen as the feedstock.

A study was conducted by a team with members from NREL, PNNL, GEMI, and Valero (Christensen et al. 2011a; Arbogast et al. 2017a, b) on the impact of hydroprocessing on several of the important refinery-relevant properties of bio-oil. These properties included acidity, boiling range distribution, elemental composition (including total oxygen), and hydrocarbon and oxygenate types in streams that represent important refinery intermediates. These data were correlated with hydroprocessing severity (principally temperature, pressure, liquid hourly space velocity or LHSV) and total oxygen content of the upgraded oil. Three levels of hydroprocessing severity were analyzed, consisting of reaction conditions required to produce an oil with low oxygen content (LOC; organic oxygen = 0.4 wt% on a water-free basis), medium oxygen content (MOC; water-free organic oxygen = 4.9 wt%), and high oxygen content (HOC; water-free organic oxygen = 8.2 wt%). Information on the distillate fractions and elemental analysis of the upgraded bio-oil fractions are shown in Table 8. As can be seen, hydrotreating results in a gradual shift of the distillate product slate toward lighter fractions (naphtha + light ends) with a reduction primarily in the gas oil fraction as hydroprocessing severity increases. For the HOC and MOC oils, an additional 10 wt% of the starting oil comprised a nonvolatile residue. These data show that organic oxygen is concentrated in the lighter cuts for the HOC oil, while the opposite trend is found for the MOC oil with organic oxygen concentrated in the heavier fractions. For the LOC oil, the organic carbon content of all fractions was basically the same.

In addition to elemental composition and boiling fraction distribution, the acidity of the fractions is extremely important to the refiner. Table 9 presents data on the relationship between acidity and hydroprocessing severity for each distillate fraction obtained from this study.

These acidity data are presented in terms of total acid number (TAN) and carboxylic acid number (CAN). The intent here was to indicate the relative proportion of strong acids (carboxylic acids) in the TAN, as corrosion issues associated with carboxylic acids are anticipated to be quite problematic. As shown, for the HOC oil, TAN is still very high, and most of the total acids consist of carboxylic acids, indicating that the weak acids (phenolics) have been removed by hydrotreating. At higher hydrotreating severities, the TAN and CAN are both significantly reduced.

A paraffin, olefin, naphthene, aromatic (PONA) analysis of the LOC, MOC, and HOC fractions is shown in Table 10. Oxygenated compounds were present in all of the HOC fractions. In the light ends and naphtha fractions, these were primarily C_5, C_6, and C_7 cyclic and noncyclic ketones, esters of C_6 and shorter carboxylic acids, methyl-substituted tetrahydrofurans, and aliphatic alcohols. Some acetic acid was present in these fractions. In the jet fraction, oxygenates were primarily methyl- and ethyl-substituted phenols, with some methoxy phenols and C_6 and C_7 cyclic ketones. Ketones in the jet fraction were less than in the lighter fractions, but phenols were much greater in this fraction. The MOC fractions contained much lower levels of oxygenates compared to the HOC fractions; detected compounds consisted primarily of alkyl phenols and aryl ethers. Fractions from the LOC oil contained no oxygenates in the lights and naphtha (below detection limits), with a small amount of alkyl-substituted phenols in the jet fraction.

Results for the PONA analysis showed increasing hydroprocessing severity decreased aromatics and increased paraffins and naphthenes in the light and naphtha fractions. For the LOC fraction, the data indicate low aromatic content and moderate isoparaffin content leading to the relatively low research octane number (RON) and motor octane number (MON) for these fractions. Benzene content in the light and naphtha fractions was found to be below the limits set by the EPA for motor gasoline in all fractions analyzed. While it is clear from this study that hydropro-

Table 8 Elemental analysis of boiling range fractions (Christensen et al. 2011a)

Oil sample	Fraction	Distillate fraction, % w/w	C, % w/w	H, % w/w	N, % w/w	S, ppm	O, % w/w
HOC	Lights	5.3	72.8	11.9	0.01	25	14.2
	Naphtha	19.7	73.7	11.5	0.01	19	14.4
	Jet	18.7	77.8	11.0	0.03	23	11.9
	Diesel	17.2	82.4	10.7	0.09	101	7.5
	Gasoil	30.3	84.6	10.4	0.14	354	5.3
MOC	Lights	4.6	85.6	13.6	0.02	8	0.5
	Naphtha	17.7	84.5	11.9	0.05	8	3.9
	Jet	23.1	83.9	10.1	0.14	12	6.6
	Diesel	18.3	85.7	10.2	0.32	21	4.4
	Gasoil	32.6	87.8	9.9	0.40	116	2.5
LOC	Lights	13.9	85.9	14.6	0.01	2	0.3
	Naphtha	30.2	86.3	13.3	0.02	2	0.3
	Jet	22.0	87.0	12.3	0.02	12	0.7
	Diesel	20.6	88.4	11.4	0.02	310	0.5
	Gasoil	13.5	88.6	11.5	0.03	243	0.4

Table 9 Acidity of hydrotreated fractions (Christensen et al. 2011a)

	HOC		MOC		LOC	
mg KOH/g	CAN[a]	TAN[b]	CAN	TAN	CAN	TAN
Lights	102	102	BD	14	BD	BD
Naphtha	123	123	BD	100	BD	2
Jet	67	154	BD	199	BD	14
Diesel	20	20	BD	0.3	0.1	0.1
Gasoil	9	9	BD	BD	0.4	0.4

BD below detection limit
[a]Carboxylic acid number (strong acids)
[b]Total acid number (strong plus weak acids)

cessing is very effective in improving the quality of bio-oil, the CAPEX and OPEX associated with oxygen removal to low levels (<1%) remain a significant challenge.

Integration of Bio-oil in the FCC The FCC is the single most important unit operation in the modern petroleum refinery that has been optimized for producing motor gasoline; accordingly, a great deal of interest has been focused on processing bio-oil in the FCC—either by itself or as a blend with petroleum-derived gas oil/vacuum gas oil (co-processing). It has been speculated that decarboxylation via FCC could provide a more cost-effective route for producing transportation fuels from biomass when compared to deoxygenation by hydroprocessing (Butler et al. 2011). Further, the FCC is a flexible refinery unit operation that can, in principle, be readily tuned to accommodate different feedstocks by modifying catalysts and/or operating conditions. Several potentially viable strategies exist for integrating bio-oil into the fluid catalytic cracking unit of an existing petroleum refinery.

Co-processing Whole Bio-oil in the FCC A simple and straightforward method for integrating bio-oil into an existing refinery would be to use whole bio-oil without pre-treatment or fractionation as a blendstock with petroleum-derived GO or VGO and direct feed the blend to the FCC (Watkins et al. 2008). In one study, mixtures of model compound oxygenates (acetone, acetic acid, 2-propanol) and iso-octane as a surrogate for gas oil were cracked over an industrial equilibrium catalyst (E-cat) in a fixed-bed laboratory reactor (Domine et al. 2008). In general, selectivity to light gases and olefins was reduced, and coke was found to be dramatically increased by adding oxygenates. In a different study, blends of model oxygenates such as acetic acid, hydroxyacetone, and phenol with petroleum-derived gas oil were processed under standard FCC conditions in a lab-scale reactor using both an E-cat and a mixture of E-cat and ZSM-5 (Graca et al. 2009). Adding the oxygenates increased overall conversion, reduced the coke yield, and increased the yield of fuel gas, LPG, and gasoline. Overall conversion of the gas oil was not significantly altered.

Fluid catalytic cracking mixtures of petroleum-derived gas oil with whole bio-oil has been reported by Fogassy et al., who investigated co-processing VGO and whole bio-oil over a standard FCC catalyst, H-Y zeolite, and HZSM-5 in a laboratory reactor. These studies found that introducing bio-oil resulted in lower rates for formation of cracked products except for coke and aromatics (Fogassy et al. 2011). These researchers also investigated partitioning fossil carbon and biomass-derived carbon in products from co-processing bio-oil with petroleum gas oil. Using carbon-14, they were able to determine that both coke and light gases were richer in ^{14}C than the gasoline from the FCC, suggesting that biomass-derived components react preferentially to undesirable products under cat cracking conditions (Fogassy et al. 2012). In a laboratory cracking reactor (ACE system), Agblevor et al. (2012) were able to produce fuel-range products by co-processing bio-oil with gas oil in a ratio of 15/85 (wt/wt). The product yields were almost identical to that for cracking gas oil alone, and the products were found to contain negligible amounts of oxygen. Similar results were reported for co-processing a mixture of 10 wt% bio-oil and 90 wt% vacuum gas oil using an E-cat in a laboratory ACE system (De Almeida

Table 10 PONA analysis of the distillate fractions as a function of hydrotreating severity (Christensen et al. 2011a)

	LOC		MOC	HOC	
Vol %	Lights	Naphtha	Lights	Lights	Naphtha
Paraffins	28.3	15.4	13.6	7.9	5.9
Isoparaffins	14.9	26.8	25.9	32.8	38.8
Naphthenes	51.3	46	47.8	31.8	20.3
Aromatics	5.6	11.8	5.2	10.9	27.0
Olefins	0.07	0.01	7.54	16.7	8.3
Benzene	0.5	0.4	0.6	0.3	0.8
RON	64	71	73	79	88
MON	61	68	72	77	87

2008); however in this case, up to 1500 ppm phenols were found in the liquid products. In a recent pioneering pilot-scale study by NREL and Petrobras, co-processing 10 wt% bio-oil and 90 wt% vacuum gas oil indicated substantial differences in the yields of coke and liquid products, and the products contained significant organic oxygen content (Pinho et al. 2017). Similar findings were reported by these same researchers when cracking a feedstock containing 100% bio-oil in a laboratory ACE unit.

Co-processing whole bio-oil with hydrogen-rich materials other than petroleum has been studied (Bezergianni et al. 2018; Chang et al. 1976; Chen et al. 1988). These investigations found that mixing bio-oil (a hydrogen-deficient material with a low effective hydrogen index or EHI) with a hydrogen-rich material (such as methanol) dramatically improved the conversion of bio-oil to hydrocarbons during catalytic cracking in the vapor phase over HZSM-5. It was reported that a mixture with a combined EHI of 1.0 or greater resulted in a greater than 300% increase in C5+ hydrocarbon yield accompanied by a 32 wt% reduction in coke-on-catalyst (water-free basis) when compared to vapor-phase cracking of whole bio-oil alone. Petrobras has applied this concept to the catalytic cracking of petroleum-derived hydrocarbons with ethanol to produce ethylene in high yields (Pinho et al. 2011).

Findings from these studies show that whole bio-oil and model compounds representing the major oxygenated compounds in whole bio-oil produce large amounts of coke and light gases when processed over acid catalysts typical of those used in a conventional FCC unit. Catalyst deactivation was found to be rapid, and alkali and alkaline earth metals present in the whole oil caused severe and irreversible poisoning of the catalysts. Other factors, including the acidity and high water content of whole bio-oil, make whole bio-oil a particularly difficult feedstock for the cat cracker. FCC units are generally not made from high alloy steel, and the corrosivity of whole bio-oil would present severe operational difficulties. Similarly, the high water content of whole bio-oil is deleterious to catalyst integrity in the FCC unit. Finally, it is unlikely that production facilities for bio-oil will be able to supply sufficient quantities of product. Typical modern petroleum refineries process upward of 200,000 barrels/day of crude; a significant fraction of that amount is fed to the FCC. Single biorefineries based on pyrolysis will initially produce bio-oils at a rate of only about 8000 BBL/D,[3] which is insufficient to satisfy the demand for the FCC in even one small- to medium-sized refinery. Accordingly, integration strategies based on processing whole bio-oil without blending with refinery feedstocks and/or intermediates do not appear to be technically or commercially feasible (Diebold et al. 1995). A blend of up to 10 wt% whole (untreated) bio-oil was suggested to be a suitable feed for the FCC unit in a conventional petroleum refinery (Melero et al. 2012). Problems associated with co-processing whole bio-oil can be partially addressed by upgrading the whole oil prior to blending with gas oil. Thermal and catalytic hydrotreating of FP pyrolysis oil and use of in situ and ex situ catalytic fast pyrolysis (CFP) have all been investigated as upgrading strategies to improve properties with respect to co-processing of bio-oil in the FCC. Both low-severity thermal

[3]Assuming a single biorefinery processing 2000 metric tons/day lignocellulosic biomass.

(e.g., non-catalytic) hydrotreating and catalytic hydrotreating have been investigated (Samolada et al. 1998); it was found that the heavy fraction from thermal hydrotreating could be successfully co-processed with light cycle oil in the FCC. Co-processing hydrotreated bio-oil in the FCC has been studied by several investigators (Lappas et al. 2009; Mercader et al. 2010; Fogassy et al. 2010). Using a laboratory FCC system, Mercader et al. found that co-processing HDO bio-oil in the FCC with long residue and light cycle oil produced products that were almost free of organic oxygen without excessive coke formation. Fogassy et al. found that blending HDO bio-oil and VGO at a level as high as 20% gave comparable yields for the gasoline fraction when compared to cracking VGO alone. A common thread in many of these studies is that removing oxygen in the FCC consumes hydrogen from the hydrocarbon feedstock, resulting in the production of more olefins and aromatics in the products.

Co-processing partially upgraded bio-oil produced by catalytic fast pyrolysis (CFP) has been compared to co-processing an HDO bio-oil by Thegarid et al. (2014). This study showed that product distributions were similar but that the CFP oil could eliminate the need for upstream hydrodeoxygenation. Organic carbon efficiency of the CFP/FCC strategy was found to be significantly better than the HDO/FCC strategy. Co-processing upgraded bio-oil in the FCC provides a technical solution to some of the more problematic issues associated with using biomass-derived liquids in the refinery. However, the economics of these strategies are dominated by the high capital and operating costs associated with hydroprocessing. These costs are present, in part, because the strategies being employed involve high-severity hydrodeoxygenation and then co-processing the whole bio-oil. This results in high CAPEX due to large reactor volumes and high OPEX due to the hydrogen demand for hydroprocessing/deoxygenation of the whole oil. A different strategy, which in principle could circumvent some of these problems, is shown in Fig. 7. This scheme involves first *mildly* deoxygenating the whole bio-oil to the point where the oil can be distilled followed by fractionation. This mild deoxygenation step could be done either by hydrotreating, or by generating the bio-oil using catalytic fast pyrolysis. Conditions could be adjusted to allow for water removal as a separate phase during this initial step as mild hydrotreating has been shown to be effective in facilitating this separation (Fogassy et al. 2011). Depending on distillation characteristics and boiling range, the bio-derived fractions could then be sent to the appropriate unit operation in the refinery (e.g., bio-naphtha to the reformer hydrotreater, bio-diesel to the diesel hydrotreater) for blending with petroleum-derived material and further processing into finished fuels.

In this scheme, high severity hydroprocessing associated with the HDO step would be reserved for that fraction of the bio-oil that requires more severe processing to reduce acidity and improve miscibility. This would result in improved hydrogen utilization efficiency and savings in both capital and operating costs when compared to the whole-oil strategies discussed above. This strategy of selective hydrotreating also applies when the bio-oil is available from CFP. In the context of the scheme shown above, CFP is used to provide a partially upgraded bio-oil that can be fractionated, perhaps removing the need for hydrotreating the whole oil prior to fractionation and improving the overall economics and carbon efficiency accordingly. Conditions and catalysts for CFP required to produce a bio-oil that can be fractionated have not been widely investigated to date.

Strategies for optimizing bio-oil deoxygenation for CFP vs. hydrotreating have been recently proposed (Iisa et al. 2018). As shown in Fig. 8, the carbon efficiency of hydrotreating is high at low bio-oil oxygen levels, while CFP evidences high carbon efficiencies at high bio-oil oxygen contents. The dashed green line represents the efficiency of the combined process; the shaded zone indicates the target bio-oil oxygen content to maximize the synergy between the two deoxygenation strategies.

Co-processing in Hydrotreaters Co-processing bio-oil with petroleum-derived materials in hydrotreaters has not been extensively investigated to date but is a very important component in overall strategies for refinery utilization of bio-oil or bio-crude. Bui et al. (2009) investigated co-processing straight-run gas oil with guaiacol as the surrogate for bio-oil in a laboratory hydrodesulfurization (HDS) reactor using a standard $CoMo/Al_2O_3$ HDS catalyst and found a competition existed between HDS and HDO with a decrease in HDS activity under certain conditions. Pinheiro et al. (2009) used model oxygenates blended with straight-run gas oil (SRGO) to investigate the impact of bio-oil on HDS, HDN, and aromatic ring saturation. These studies showed no impact of 2-propanol, cyclopentanone, anisole, and guaiacol on HDS, HDN, or ring saturation; propanoic acid and ethyldecanoate were found to inhibit all three hydrotreating functions. In a separate study (Pinheiro et al. 2011), these same investigators found that CO and CO_2 formed during hydropro-

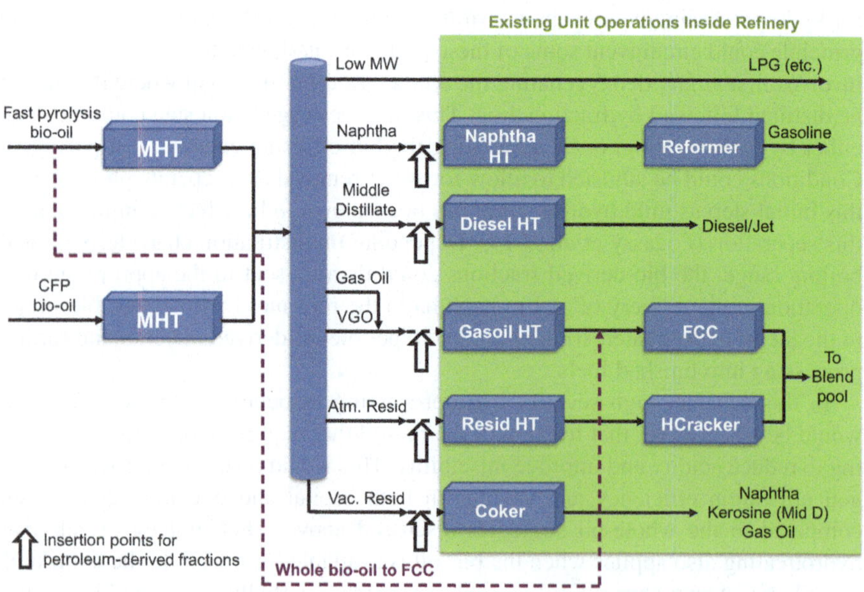

Fig. 7 Alternate refinery integration schemes

cessing also inhibited HDS and HDN for hydrotreating SRGO. One of the few studies of co-processing authentic bio-oil with petroleum-derived material in a hydrotreater was conducted by Mercader et al. (2011). These investigators processed HDO bio-oil with SRGO under typical HDS conditions and also found competition between HDS and HDO; the product from co-processing contained substantially higher levels of sulfur when compared to HDS of the SRGO alone. Catalyst activity for HDS was not reduced by co-processing with bio-oil as indicated by a return to the original low sulfur levels in the product when the bio-derived material was removed from the feed. Product yields were the same for SRGO and when SRGO was co-processed with bio-oil.

4.3 Co-processing in Other Refinery Unit Operations: Coker

Advanced pyrolysis approaches such as fast pyrolysis and catalytic fast pyrolysis can be used to produce feedstocks for transportation fuels and bio-based chemicals, but the high molecular weight fraction is difficult and costly to convert and thus represents a low value stream. An alternate approach to valorizing this material is to remove the "heavies" via fractional distillation and then upgrade the remaining bio-oil to fuel feedstocks and/or blendstocks. When bio-oil is used for co-processing in the refinery, removal of heavy fractions provides an upgraded distillate material that contains fewer coke precursors and hence represents a premium feedstock for refinery co-processing applications such as fluid catalytic cracking.

A portion of these heavy materials can be used to produce functional replacements for high-value carbon and graphite that are traditionally made from petroleum coke, coal tar, and natural mineral graphite. An especially important application

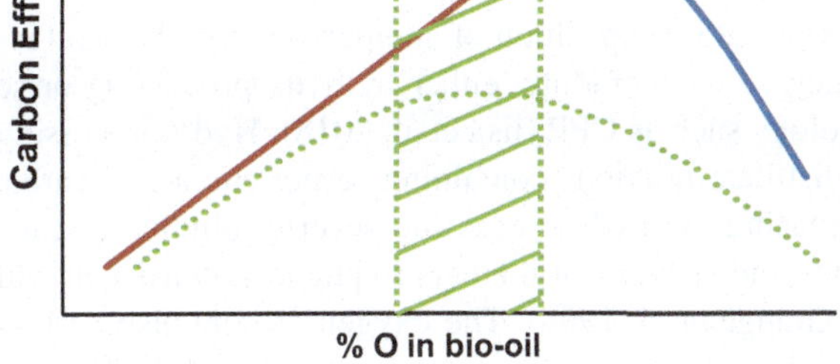

Fig. 8 Hydrotreating and CFP carbon efficiencies as a function of bio-oil oxygen content (Iisa et al. 2018)

for this material is for advanced energy storage devices such as lithium-ion batteries or carbon capacitors. Current pathways for production of anode-grade graphite depend on upgrading of mineral or petroleum-derived materials, which contain significant amounts of contaminants that must be reduced or eliminated. These processes are very energy intensive and environmentally unfriendly. Replacing fossil graphite with bio-graphite will provide a less energy-intensive pathway to a renewable and sustainable source for these and similar materials. Alternate strategies for valorizing bio-oil residuum are shown in Fig. 9.

4.4 Biomass-Derived Oxygenates in Finished Fuels

Because of the high oxygen content of bio-oils, there is a strong economic incentive to leave much of this oxygen in the finished fuel product to the extent that government regulations and product quality standards will allow. Arbogast et al. (2017a) have highlighted the high hydrotreating costs required to reduce oxygen content to the 2–3 wt% range. These costs increase exponentially as the oxygen content goes below approximately 2 wt%. Accordingly, it is important to investigate the potential for oxygenates in bio-oil to become components of drop-in fuels.

As described above, the three components of biomass (cellulose, hemicellulose, and lignin) produce different oxygenated products during pyrolysis. Cellulose and hemicellulose form low molecular weight (C_4 and smaller) ketones, aldehydes, acids, esters, ethers, and alcohols that cannot easily be directly incorporated into gasoline or diesel fuel. Hydrogenation of these compounds leads to low molecular weight hydrocarbons, suggesting that some form of oligomerization to increase molecular weight is necessary if this bio-derived carbon is to be incorporated into fuel. Cellulose and hemicellulose can also produce furanic compounds such as furfural, furfuryl alcohol, and furoic acid that, upon hydrogenation, can yield methyl furans because of the relative recalcitrance of the furan ring structure (Grange et al. 1996). Sugars and anhydrosugars have also been observed in the pyrolysis products, with hydrogenation producing 5- and 6-carbon alcohols. Pyrolysis of lignin, on the other hand, produces phenols and alkyl phenols, methyl aryl ethers, and guaiacols. Ethers are generally converted to phenolics by hydrotreating at adequately severe conditions.

The actual oxygenate composition of an upgraded pyrolysis oil is highly dependent upon the degree of upgrading, either by hydroprocessing or some advanced pyrolysis technology such as CFP (Iisa et al. 2018). Hydroprocessing to 8–10 wt% oxygen yields distillate fractions containing carboxylic acids, carbonyls, phenols, and ethers. Increasing the hydroprocessing severity eliminates carbonyl and carboxyl compounds and converts aryl ethers to phenols, consistent with model compound studies (Grange et al. 1996). The oxygen present also varies with distillate fraction. At roughly 8–10 wt% oxygen, the light and naphtha fractions will primarily contain carbonyl, carboxyl, and ether groups. The jet and diesel fractions will contain these functional groups at lower levels but will also contain phenolic compounds. Upgrading further, under more severe conditions, to roughly 5% oxygen leads to fractions containing almost exclusively phenolic compounds which may be acceptable in gasoline and diesel fuel but would never be accepted (under current standards) in jet fuel.

4.4.1 Properties of Biomass-Derived Oxygenates

With the exception of jet fuel, certain oxygen functional groups present in bio-oil are unlikely to be acceptable in fuel products. While carboxylic acids are used in fuels as corrosion inhibitors at very low levels (Peyton 2002), at higher levels, they cause corrosion and are potentially poorly soluble in hydrocarbons at cold temperatures. Aldehydes and ketones may undergo condensation reactions leading to the formation of gums. Esters, ethers, and alcohols have all been used successfully in fuels (biodiesel, methyl tert-butyl ether or MTBE, and ethanol, respectively)—with the caveat that MTBE's poor biodegradability in groundwater ultimately led to its removal from the US market (McCarthy and Tiemann 2006). Table 11 shows property data for a number of oxygenated compounds that have been observed in raw and upgraded pyrolysis oils. For gasoline, the boiling point must be between about 25 and the 225 °C; the end point limit is set in standard specifications (ASTM D4814). Additionally, the 90% volume boiling point is limited to 185 or 190 °C, depending on volatility class (time of year). Therefore, only limited amounts of compounds boiling above about 185 °C can be blended. Examination of the data in Table 11 indicates that the furans, as well as anisole and methyl anisole, boil in the acceptable range and also have high octane number and very low water solubility. Christensen et al. (2011b) have described the properties of dimethyl furan and 2-methyl furan blends with gasoline; and these oxygenates have many desirable properties, including little effect on vapor pressure. Singerman described the use of methyl aryl ethers as gasoline blend components in the early 1980s (Singerman

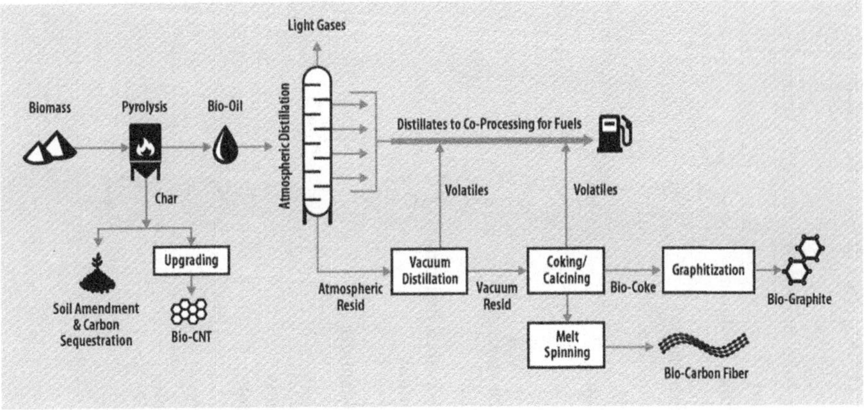

Fig. 9 Strategies for valorizing bio-oil residuum

Table 11 Properties of model biomass-derived oxygenates

Compound class/name	Molecular structure	Research and motor octane numbers	Boiling point (°C)	Solubility in water at 20 °C (wt%)	Solubility in hydrocarbon
Furans					
2,5-Dimethylfuran	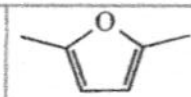	153, 109	94	0.26	Miscible
2-Methylfuran	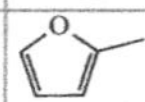	155, 92	65	0.3	Miscible
Ethers					
Anisole	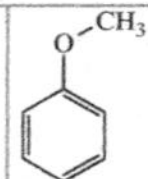	Unknown	154	Insoluble	Miscible
4-Methylanisole	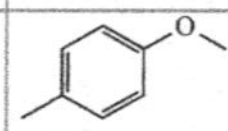	166,148	174	Insoluble	Miscible
1,2 Dimethoxybenzene (veratrole)	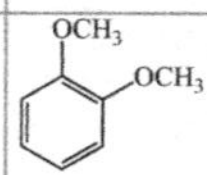	Unknown	206	Insoluble	Miscible
Propylanisole	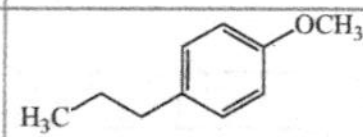	Unknown	215	Insoluble	Miscible
Phenols					
Phenol	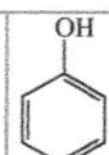	Unknown	181.7	8.3	Soluble, not miscible
p-Cresol	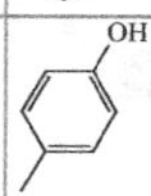	153, 149	202	1.9	Miscible

Compound class/name	Molecular structure	Research and motor octane numbers	Boiling point (°C)	Solubility in water at 20 °C (wt%)	Solubility in hydrocarbon
2,4-Xylenol	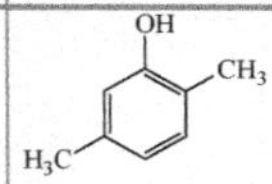	140, 113	211	0.5	Miscible
Guaiacol	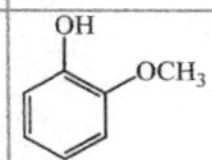	Unknown	205	1.7	Soluble, not miscible
Syringol	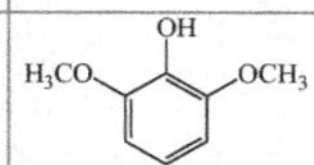	Unknown	261	1.7	Soluble, not miscible
4-Propylphenol	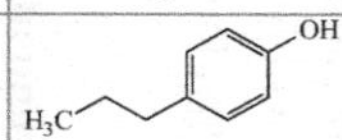	Unknown	232	Insoluble	Miscible
4-Propylguaiacol	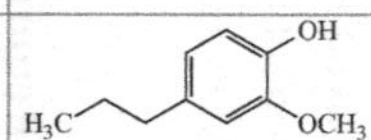	Unknown	250	Insoluble	Miscible

1981) and reported that these compounds improved octane number without degrading other gasoline properties. An important caveat is that gasoline aromatics have been linked to fine particle emissions for emerging gasoline direct injection engines (Khalek and Jetter 2012) and to the formation of secondary organic aerosol in the atmosphere (von Stackelberg et al. 2013). Both types of fine particles have been shown to have negative health effects. The US Environmental Protection Agency currently limits benzene in gasoline to an average of 0.62 vol%, not to exceed a maximum of 1.30 vol% (US Environmental Protection Agency 2007). It is unknown if furans or aryl ethers show the same effect on atmospheric fine particles. Phenol also has a boiling point just in the acceptable range; however, it also has high water solubility and poor solubility in hydrocarbon at cold temperatures and is corrosive. Other ethers and phenols have boiling points too high to be used in gasoline as blend components, although low residual levels (below roughly 1000 ppm oxygen) may be tolerable.

Diesel fuels boil between either 200 and 350 °C (No. 2 Grade) or 145–300 °C (No. 1 Grade). No. 1 grade or blends of No. 1 and No. 2 are used predominantly in cold climate, wintertime environments. Thus, the oxygenates in Table 23 that boil at too high a temperature for use in gasoline could be used in diesel fuels based on boiling point. Additionally, as C/O ratio increases, the phenolics become less soluble in water and more soluble in hydrocarbon. However, because these oxygenates are all aromatic compounds, they have a very low cetane number, significantly limiting the amount that could be economically blended. Their impact on precipitate formation at cold temperatures is also unknown. Potentially, these oxygenates could be tolerated in diesel fuels as residual components up to an oxygen content of roughly 1000 ppm. Very little research has been published on the potential for biomass-derived oxygenates to be present in fuels at these low levels.

Jet engine fuels boil between 180 and 300 °C and have a freezing point below −40 °C. However, quality standards and regulatory requirements for jet engine fuels are necessarily more strict. Jet engines require clean, low-soot formation combustion, and so the sooting tendency of jet fuels (measured as smoke number) is limited in ASTM standard D1655. The presence of aromatic compounds can lead to high sooting tendency, and so the standard also limits aromatics to 25 vol%. Oxygenated compounds other than specifically approved fuel additives are not permitted. Given these requirements, pyrolysis oil components will need to be fully hydrogenated to alkanes before their use in jet fuel could be considered.

4.4.2 Regulatory and Commercial Requirements

New transportation fuels cannot simply be produced and then introduced into the fuel marketplace. There are many federal and state regulatory, commercial, and consequent testing requirements that must be met. The exact requirements will depend on the chemical makeup of the new fuel. If it is demonstrably hydrocarbon (primarily hydrogen and carbon with less than perhaps 1000 ppm of sulfur, nitrogen, and oxygen), then requirements for market introduction are likely to be less than if the fuel is an oxygenate (such as ethanol, butanol, or biodiesel). Compliance with the

Clean Air Act is mandatory for fuels sold in the United States. If new fuels are to be used in existing engines designed for gasoline or diesel, the Environmental Protection Agency (EPA) will require demonstration that the emission performance of existing engines/vehicles is not degraded for the full useful life of the engine/vehicle (120,000 mi for a car, as much as 435,000 mi for a heavy duty truck). Potentially, this could involve testing many vehicles.

Since most, if not all, new fuels will be blended with petroleum fuels and utilized in existing engines, acceptance of the new fuel by the petroleum and auto/engine industries is critical. If the petroleum distribution industry refuses to distribute the blended fuel because they are uncomfortable handling it for safety or environmental reasons, because it cannot be obtained with consistent quality, or because they feel they accept too much liability for engine operating problems, then the new fuel will not be distributed. While these requirements are not directly legislated, they are an important aspect of consumer protection. Primarily, this involves developing data to show that the new material can be safely handled and that it is compatible with existing engines and vehicles. This compatibility is different than the emission performance mandated by the Clean Air Act. The primary way that this is accomplished is through development of an ASTM specification for the new fuel, which may take the form of a blendstock specification (such as exists for ethanol and biodiesel), adding the new fuel to existing specifications (such as those for gasoline or diesel fuel), or a new specification for a blended fuel (such as that for B6 to B20 blends). ASTM specification development requires data on a variety of issues, and what testing must be done ultimately depends on the properties of the new fuel. But it should be clear that this is a nontrivial requirement and may take 2–5 years to complete.

5 Summary

Bio-oil is currently being produced and sold as a boiler fuel and is increasingly finding niche markets for use as a transportation fuel. Use of bio-oil as a refinery feedstock for co-processing applications is expanding, with a number of demonstration projects either completed or planned. While upgrading of bio-oil to a drop-in replacement for fossil gasoline and diesel fuel remains economically challenging, introducing biomass pyrolysis oils into existing petroleum refineries offers an opportunity to accelerate the use of lignocellulosic bio-oils for production of renewable biofuels. Because raw pyrolysis oils have physical and chemical properties that make direct insertion into refinery unit operations challenging, technology development is currently underway for co-processing strategies that circumvent some of these issues by either blending or partial upgrading to reduce oxygen and water content. Finally, research and development on valorizing "the bottom of the barrel" could lead to entirely new concepts that could economically advantage pyrolysis-derived bio-oils either as refinery feedstocks or as products from stand-alone bio-refineries.

Acknowledgments and Disclaimers This work was authored by the National Renewable Energy Laboratory, operated by Alliance for Sustainable Energy, LLC, for the US Department of Energy (DOE) under Contract No. DE-AC36-08GO28308. Funding is provided by the US Department of Energy Office of Energy Efficiency and Renewable Energy Bioenergy Technology Office. The views expressed in the article do not necessarily represent the views of the DOE or the US Government. The US Government retains and the publisher, by accepting the article for publication, acknowledges that the US Government retains a nonexclusive, paid-up, irrevocable, worldwide license to publish or reproduce the published form of this work, or allow others to do so, for US Government purposes.

References

Abdullah Z (n.d.) Upgrading of biomass fast pyrolysis oil (Biooil), DOE Bioenergy Technology Office (BETO) 2015 peer review. https://www.energy.gov/sites/prod/files/2015/04/f21/thermochemical_conversion_abdullah_231401.pdf

Abdullah N, Gerhauser H (2008) Bio-oil derived from empty fruit bunches. Fuel 87:2606–2613

Abdullah N, Gerhauser H, Sulaiman F (2010) Fast pyrolysis of empty fruit bunches. Fuel 89:2166–2169

Agblevor FA, Chum HL, Johnson DK (1992) Compositional analysis of NIST biomass standards from the IEA whole feedstock round Robin, energy from biomass and wastes XVI: proceedings of the Institute of Gas Technology Conference, Orlando, Florida, 2–6 March 1992, pp 395–421

Agblevor FA, Mante O, Abdoulmoumine N, McClung R (2010) Production of stable biomass pyrolysis oils using fractional catalytic pyrolysis. Energy Fuel 24:4087–4089

Agblevor FA, Mante O, McClung R, Oyama ST (2012) Co-processing of standard gas oil and biocrude oil to hydrocarbon fuels. Biomass Bioenergy 45:130–137

Agblevor FA, Elliott DC, Santosa DM, Olarte MV, Burton SD, Swita M, Beis SH, Christian K, Sargent B (2016) Red mud catalytic pyrolysis of pinyon juniper and single-stage hydrotreatment of oils. Energy Fuel 30:7947–7958

Anthrop DF (2013) Gasoline from trees. Oil Gas J 111(1):18

Antonakou E, Lappas A, Nilsen MH, Bouzga A, Stöcker M (2006) Evaluation of various types of Al-MCM-41 materials as catalysts in biomass pyrolysis for the production of bio-fuels and chemicals. Fuel 85:2202–2212

Arbogast S, Bellman D, Paynter D, Wykowski J, Baldwin RM (2017a) Integration of pyrolysis oil in petroleum refineries, part I hydrocarbon processing

Arbogast S, Bellman D, Paynter D, Wykowski J, Baldwin RM (2017b) Integration of pyrolysis oil in petroleum refineries, part II hydrocarbon processing

Asadieraghi M, Ashri Wan Daud WM, Abbas HF (2015) Heterogeneous catalysts for advanced bio-fuel production through catalytic biomass pyrolysis vapor upgrading: a review. RSC Adv 5(28):22234–22255

Aubin H, Roy C (1980) Study on the corrosiveness of wood pyrolysis oils. Fuel Sci Technol Int 8:77–86

Azeez AM, Meier D, Odermatt J, Willner T (2010) Fast pyrolysis of African and European lignocellulosic biomasses using Py-GC/MS and fluidized bed reactor. Energy Fuel 24:2078–2085

Baker EG, Elliott DC (1988) Research in thermochemical biomass conversion. Elsevier Science, Barking, p 883

Baldwin RM, Feik CJ (2013) Bio-oil stabilization and upgrading by hot gas filtration. Energy Fuel 27:3224–3238

Bezergianni S, Dimitriadis A, Kikhtyanin O, Kubicka D (2018) Refinery co-processing of renewable feeds. Prog Energy Combust Sci 68:29–64
BIOCOUP (2011) Publishable final activity report. VTT, Espoo
Boateng AA, Daugaard DE, Goldberg NM, Hicks KB (2007) Bench-scale fluidized-bed pyrolysis of switchgrass for bio-oil production†. Ind Eng Chem Res 46:1891–1897
Boonyasuwat S, Omotoso T, Resasco DE, Crossley SP (2013) Conversion of guaiacol over supported Ru catalysts. Catal Lett 143:783–791
Bridgwater AV (2003) Renewable fuels and chemicals by thermal processing of biomass. Chem Eng J 91:87–102
Budhi S, Mukarakate C, Iisa K, Pylypenko S, Ciesielski PN, Yung MM, Donohoe BS, Katahira R, Nimlos MR, Trewyn BG (2015) Molybdenum incorporated mesoporous silica catalyst for production of biofuels and value-added chemicals via catalytic fast pyrolysis. Green Chem 17:3035–3046
Bui VN, Toussaint G, Laurenti D, Mirodatos C, Geantet C (2009) Co-processing of pyrolysis bio oils and gas oil for new generation of bio-fuels: hydrodeoxygenation of guaiacol and SRGO mixed feed. Catal Today 143:172–178
Bui VN, Laurenti D, Delichere P, Geantet C (2011) Hydrodeoxygenation of guaiacol: part II: support effect for CoMoS catalysts on HDO activity and selectivity. Appl Catal B 101:246–255
Bulushev DA, Ross JRH (2011) Catalysis for conversion of biomass to fuels via pyrolysis and gasification: a review. Catal Today 171(1):1–13
Butler E, Devlin G, Meier D, McDonnell K (2011) A review of recent laboratory research and commercial developments in fast pyrolysis and upgrading. Renew Sust Energ Rev 15:4171–4186
Carlson TR, Tompsett GA, Conner WC, Huber GW (2009) Aromatic production from catalytic fast pyrolysis of biomass-derived feedstocks. Top Catal 52:241
Chang CD, Lang WH, Silvestri AJ (1976) US patent 3,998,898. Accessed 21 Dec 1976
Chen NY, Walsh DE, Koenig LR (1988) Fluidized-bed upgrading of wood pyrolysis liquids and related-compounds. In: Pyrolysis liquids from biomass, ACS symposium series: American Chemical Society, pp 264–275
Chen S-F, Mowery RA, Scarlata CJ, Chambliss K (2007) Compositional analysis of water-soluble materials in corn stover. J Agric Food Chem 55:5912–5918
Chen S-F, Mowery RA, Sevcik RS, Scarlata CJ, Chambliss CK (2010) Compositional analysis of water-soluble materials in switchgrass. J Agric Food Chem 58:3251–3258
Christensen ED, Chupka GM, Luecke J, Smurthwaite T, Alleman TL, Iisa K, Franz JA, Elliott DC, McCormick RL (2011a) Analysis of oxygenated compounds in hydrotreated biomass fast pyrolysis oil distillate fractions. Energy Fuel 25:5462–5471
Christensen E, Yanowitz J, Ratcliff M, McCormick RL (2011b) Renewable oxygenate blending effects on gasoline properties. Energy Fuel 25:4723–4733
Ciesla U, Schuth F (1999) Ordered mesoporous materials. Microporous Mesoporous Mater 27:131–149
Czernik S (1994) Storage of biomass pyrolysis oils. In: Proceedings from biomass pyrolysis oil: properties and combustion, Estes Park, CO, 26–28 September 1994, NREL CP-430-7215, pp 67–76
Czernik S, Bridgwater AV (2004) Overview of applications of biomass fast pyrolysis oil. Energy Fuel 18:590–598
Darmstadt H, Garcia-Perez M, Adnot A, Chaala A, Kretschmer D, Roy C (2004) Corrosion of metals by bio-oil obtained by vacuum pyrolysis of softwood bark residues. An X-ray photoelectron spectroscopy and auger electron spectroscopy study. Energy Fuels 18:1291–1301
Das P, Sreelatha T, Ganesh A (2004) Bio oil from pyrolysis of cashew nut shell-characterisation and related properties. Biomass Bioenergy 27:265–275
Davidsson KO, Stojkova BJ, Pettersson J (2002) Alkali emission from birchwood particles during rapid pyrolysis. Energy Fuel 16:1033–1039
De Almeida MBB (2008) MSc thesis. Federal University of Rio de Janeiro
de Miguel Mercader F, Groeneveld MJ, Kersten SRA, Geantet C, Toussaint G, Way NW, Schaverien CJ, Hogendoorn KJA (2011) Hydrodeoxygenation of pyrolysis oil fractions: pro-

cess understanding and quality assessment through co-processing in refinery units. Energy Environ Sci 4:985–997
de Souza PM, Nie L, Borges LEP, Noronha FB, Resasco DE (2014) Role of oxophilic supports in the selective hydrodeoxygenation of m-cresol on Pd catalysts. Catal Lett 144:2005–2011
Diebold JP (2000), A review of the chemical and physical mechanisms of the storage stability of fast pyrolysis bio-oils. NREL Report, SR-570-27613, pp 1–59
Diebold J, Scahill J (1988) Biomass to gasoline—upgrading pyrolysis vapors to aromatic gasoline with zeolite catalysis at atmospheric pressure. In: Soltes EJ, Milne TA (eds) Pyrolysis oils from biomass, producing, analyzing, and upgrading. American Chemical Society, Washington, DC, pp 264–276
Diebold JP, Scahill JW, Czernik S, Phillips SD, Feik CJ (1995) Progress in the production of hot-gas filtered biocrude oil at NREL. NREL technical report, pp 1–16
Domine ME, van Veen AC, Schuurman Y, Mirodatos C (2008) Coprocessing of oxygenated biomass compounds and hydrocarbons for the production of sustainable fuel. ChemSusChem 1:179–181
Donar YO, Sinag A (2016) Catalytic effect of tin oxide nanoparticles on cellulose pyrolysis. J Anal Appl Pyrolysis 119:69–74
Doornkamp C, Ponec V (2000) The universal character of the Mars and Van Krevelen mechanism. J Mol Catal A Chem 162:19–32
Elliott DC (1994) Water, alkali and char in flash pyrolysis oils. Biomass Bioenergy 7:179–185
Elliott DC (2007) Historical developments in hydroprocessing bio-oils. Energy Fuel 21:1792–1815
Elliott DC, Hart TR, Neuenschwander GG, Rotness LJ, Olarte MV, Zacher AH, Solantausta Y (2012) Catalytic hydroprocessing of fast pyrolysis bio-oil from pine sawdust. Energy Fuel 26:3891–3896
Elliott DC, Wang H, French R, Deutch S, Iisa K (2014) Hydrocarbon liquid production from biomass via hot-vapor-filtered fast pyrolysis and catalytic hydroprocessing of the bio-oil. Energy Fuel 28:5909–5917
U.S. Environmental Protection Agency (2007) Control of hazardous air pollutants from mobile sources. Fed Regist, 72(37):8428–8570
Evans R, Milne T (1987) Molecular characterization of the pyrolysis of biomass 1. Energy Fuel 1:123–137
Evans RJ, Milne TA (1988) Molecular-beam, mass-spectrometric studies of wood vapor and model compounds over an HZSM—5 catalyst. In: Soltes EJ, Milne TA (eds) Pyrolysis oils from biomass, producing, analyzing, and upgrading. American Chemical Society, Washington, DC, pp 311–327
Fogassy G, Thegarid N, Toussaint G, van Veen AC, Schuurman Y, Mirodatos C (2010) Biomass derived feedstock co-processing with vacuum gas oil for second-generation fuel production in FCC units. Appl Catal B Environ 96:476–485
Fogassy G, Thegarid N, Schuurman Y, Mirodatos C (2011) From biomass to bio-gasoline by FCC co-processing: effect of feed composition and catalyst structure on product quality. Energy Environ Sci 4:5068
Fogassy G, Thegarid N, Schuurman Y, Mirodatos C (2012) The fate of bio-carbon in FCC co-processing products. Green Chem 14:1367
French RJ, Stunkel J, Black S, Myers M, Yung MM, Iisa K (2014) Evaluate impact of catalyst type on oil yield and hydrogen consumption from mild hydrotreating. Energy Fuel 28:3086–3095
Fuleki D (1999) Bio-fuel system material testing. PyNE Newsletter, Issue no. 7. Aston University, Bio-Energy Research Group, Birmingham
Furimsky E (2000) Catalytic hydrodeoxygenation. Appl Catal A 199:147–190
Gao D, Schweitzer C, Hwang HT, Varma A (2014) Conversion of guaiacol on noble metal catalysts: reaction performance and deactivation studies. Ind Eng Chem Res 53:18658–18667
Garcìa-Pérez M, Chaala A, Pakdel H, Kretschmer D, Rodrigue D, Roy C (2006) Multiphase structure of bio-oils. Energy Fuel 20:364–375
Garcìa-Pérez M, Chaala A, Pakdel H, Kretschmer D, Roy C (2007) Characterization of bio-oils in chemical families. Biomass Bioenergy 31:222–242

Graca I, Ribeiro FR, Cerqueira HS, Lam YL, de Almeida MBB (2009) Catalytic cracking of mixtures of model bio-oil compounds and gasoil. Appl Catal B Environ 90:556–563
Grange P, Laurent E, Maggi R, Centeno A, Delmon B (1996) Hydrotreatment of pyrolysis oils from biomass: reactivity of the various categories of oxygenated compounds and preliminary techno-economical study. Catal Today 29:297–301
Green Goods and Services (GGS) (2013) Bureau of Labor Statistics U.S. Department of Labor
Hayes DJ, Hayes MHB (2009) The role that lignocellulosic feedstocks and various biorefining technologies can play in meeting Ireland's biofuel targets. Biofuels Bioprod Biorefin 3:500–520
He Z, Wang X (2012) Hydrodeoxygenation of model compounds and catalytic systems for pyrolysis bio-oils upgrading. Catal Sustain Energy 1:28–52
Horne PA, Williams PT (1995) The effect of zeolite ZSM-5 catalyst deactivation during the upgrading of biomass-derived pyrolysis vapours. J Anal Appl Pyrol 34:65–85
Howe D, Westover T, Carpenter D, Santosa D, Emerson R, Deutch S, Starace A, Kutnyakov I, Lukins C (2015) Field-to-fuel performance testing of lignocellulosic feedstocks: an integrated study of the fast pyrolysis–hydrotreating pathway. Energy Fuel 29:3188–3197
Iisa K, French RJ, Orton KA, Dutta A, Schaidle JA (2017) Production of low-oxygen bio-oil via ex situ catalytic fast pyrolysis and hydrotreating. Fuel 207:413–422
Iisa K, Robichaud DJ, Watson MJ, ten Dam J, Dutta A, Mukarakate C, Kim S, Nimlos MR, Baldwin RM (2018) Improving biomass pyrolysis economics by integrating vapor and liquid phase upgrading. Green Chem 20:567
Ingram L, Mohan D, Bricka M, Steele P, Strobel D, Crocker D, Mitchell B, Mohammad J, Cantrell K, Pittman CUJ (2008) Pyrolysis of wood and bark in an auger reactor: physical properties and chemical analysis of the produced bio-oils. Energy Fuel 22:614–625
Jae J, Tompsett GA, Foster AJ, Hammond KD, Auerbach SM, Lobo RF, Huber GW (2011) Investigation into the shape selectivity of zeolite catalysts for biomass conversion. J Catal 279:257–268
Jay DC, Rantanen O Sipilä K, Nylund NO (1995) Wood pyrolysis oil for diesel engines. In: Proceedings of the 1995 fall technical conference, Milwaukee, Wisconsin, 24–27 Sept 1995. ASME, Internal Combustion Engine Division, New York
Jensen PA, Frandsen FJ, Dam-Johansen K, Sander B (2000) Experimental investigation of the transformation and release to gas phase of potassium and chlorine during straw pyrolysis. Energy Fuel 14:1280–1285
Johnson DK, Adam P, Ashley P, Chum H, Deutch S, Fennell J, Wiselogel A (1993) Study of compositional changes in biomass feedstocks upon storage (results), storage and drying of woody biomass: proceedings of the international energy agency/bioenergy agreement task IX activity 5 workshop, New Brunswick, Canada, May 19, 1993. Swedish University of Agricultural Sciences, Department of Forest Products, New Brunswick, pp 28–52
Jones SB, Tan E, Jacobson I, Meyer P, Dutta A, Cafferty K, Snowden-Swan L, Padmaperuma A (2013) Process design and economics for the conversion of lignocellulosic biomass to hydrocarbon fuels. Fast pyrolysis and hydrotreating bio-oil pathway. U. S. Department of Energy, Washington, DC
Keiser JR (2013) Bioenergy technology. Office of the Department of Energy Program Review, Washington, DC
Khalek I, Jetter J (2012) Effect of commercially available gasoline fuels properties on particle emissions from a 2010 vehicle equipped with gasoline direct injection engine. In: 22nd CRC real world emissions workshop, San Diego, CA, 2012
Knudsen JN, Jensen PA, Dam-Johansen K (2004) Transformation and release to the gas phase of Cl, K, and S during combustion of annual biomass. Energy Fuel 18:1385–1399
Lappas AA, Samolada MC, Iatridis DK, Voutetakis S, Vasalos IA (2002) Biomass pyrolysis in a circulating fluid bed reactor for the production of fuels and chemicals. Fuel 81:2087–2095
Lappas AA, Bezergianni S, Vasalos IA (2009) Production of biofuels via co-processing in conventional refining processes. Catal Today 145:55–62
Lee WS, Wang ZS, Wu RJ, Bhan A (2014) Selective vapor-phase hydrodeoxygenation of anisole to benzene on molybdenum carbide catalysts. J Catal 319:44–53

Lee W-S, Kumar A, Wang Z, Bhan A (2015) Chemical titration and transient kinetic studies of site requirements in Mo_2C-catalyzed vapor phase anisole hydrodeoxygenation. ACS Catal 5:4104–4114

Liu C, Wang H, Karim AM, Sun J, Wang Y (2014) Catalytic fast pyrolysis of lignocellulosic biomass. Chem Soc Rev 43(22):7594–7623

Lu Q, Xiong WM, Li WZ, Guo QX, Zhu XF (2009) Catalytic pyrolysis of cellulose with sulfated metal oxides: a promising method for obtaining high yield of light furan compounds. Bioresour Technol 100:4871–4876

Lu Q, Zhang Y, Tang Z, Li W-z, Zhu X-f (2010) Catalytic upgrading of biomass fast pyrolysis vapors with titania and zirconia/titania based catalysts. Fuel 89:2096–2103

Ma Z, van Bokhoven J (2014) Thermal conversion of biomass–pyrolysis and hydrotreating. Catalysis 26:249–272

Mante OD, Agblevor FA (2014) Catalytic pyrolysis for the production of refinery-ready biocrude oils from six different biomass sources. Green Chem 16:3364–3377

Mante OD, Rodriguez JA, Senanayake SD, Babu SP (2015) Catalytic conversion of biomass pyrolysis vapors into hydrocarbon fuel precursors. Green Chem 17:2362. 60

Marker L (2005) Opportunities for biorenewables in oil refineries: final technical report (DE-FG36-05GO15085). US Department of Energy

Martín-Aranda RM, Čejka J (2010) Recent advances in catalysis over mesoporous molecular sieves. Top Catal 53(3):141–153

McCarthy JE, Tiemann M (2006) MTBE in gasoline: clean air and drinking water issues, congressional research service reports. Paper 26. http://digitalcommons.unl.edu/crsdocs/26

Meier D (1999) New methods for chemical and physical characterization and round robin testing. In: Bridgwater AV, Czernik S, Diebold J, Meier D, Oasmaa A, Peacocke C, Piskorz J, Radlein D (eds) Fast pyrolysis of biomass: a handbook, pp 92–101

Meier D, Scholtze B (1997) Fast pyrolysis liquid characteristics. In: Kaltschmitt M, Bridgwater AV (eds) Biomass gasification and pyrolysis, state of the art and future prospects. CPL Scientific, Newbury, pp 431–441

Melero JA, Iglesias J, Garcia A (2012) Biomass as renewable feedstock in standard refinery units. Feasibility, opportunities and challenges. Energy Environ Sci 5:7393

Mercader FD, Groeneveld MJ, Kersten SRA, Way NWJ, Schaverien CJ, Hogendoorn JA (2010) Production of advanced biofuels: co-processing of upgraded pyrolysis oil in standard refinery units. Appl Catal B Environ 96:57–66

Mercader FD, Groeneveld MJ, Kersten SRA, Geantet C, Toussaint G, Way NWJ, Schaverien CJ, Hogendoorn KJA (2011) Hydrodeoxygenation of pyrolysis oil fractions: process understanding and quality assessment through co-processing in refinery units. Energy Environ Sci 4:985–997

Mukarakate C, Zhang X, Stanton AR, Robichaud DJ, Ciesielski PN, Malhotra K, Donohoe BS, Gjersing E, Evans RJ, Heroux DS, Richards R, Iisa K, Nimlos MR (2014) Real-time monitoring of the deactivation of HZSM-5 during upgrading of pine pyrolysis vapors. Green Chem 16:1444–1461

Mullen CA, Boateng AA (2013) Accumulation of inorganic impurities on HZSM-5 zeolites during catalytic fast pyrolysis of switchgrass. Ind Eng Chem Res 52:17156–17161

Mullen CA, Boateng AA, Goldberg NM, Lima IM, Laird DA, Hicks KB (2010) Bio-oil and biochar production from corn cobs and stover by fast pyrolysis. Biomass Bioenergy 34:67–74

Murugappan K, Mukarakate C, Budhi S, Shetty M, Nimlos M, Roman-Leshkov Y (2016) Supported molybdenum oxides as effective catalysts for the catalytic fast pyrolysis of lignocellulosic biomass. Green Chem 18:5548–5557

Nolte MW, Zhang J, Shanks BH (2016) Ex situ hydrodeoxygenation in biomass pyrolysis using molybdenum oxide and low pressure hydrogen. Green Chem 18(1):134–138

Nowakowski DJ, Jones JM, Brydson RMD, Ross AB (2007) Potassium catalysis in the pyrolysis behaviour of short rotation willow coppice. Fuel 86:2389–2402

NREL poplar bio-oil composition (n.d.)

Oasmaa A, Kuoppala E (2003) Fast pyrolysis of forestry residue. 3. Storage stability of liquid fuel. Energy Fuel 17:1075–1084

Oasmaa A, Meier D (1999) Analysis, characterization and test methods of fast pyrolysis liquids. In: Proceeding of biomass a growth opportunity in green energy and value-added products biomass a growth opportunity in green energy and value-added products, at proceedings of the 4th biomass conference of the Americas, Oakland, CA, II, pp 1229–1234

Oasmaa A, Meier D (2005) Norms and standards for fast pyrolysis liquids: 1. Round robin test. J Anal Appl Pyrol 73:323–334

Oasmaa A, Peacocke C (2001) A guide to physical property characterisation of biomass—derived fast pyrolysis liquids, vol 450. VTT, Espoo, pp 1–102

Oasmaa A, Peacocke C (2010) Properties and fuel use of biomass-derived fast pyrolysis liquids. A guide, vol 731. VTT, Espoo, pp 1–134

Oasmaa A, Sipilä K (1996) Pyrolysis oil properties: use of pyrolysis oil as fuel in medium-speed diesel engines. In: Bio-oil production and utilization, conference proceedings, pp 175–185

Oasmaa A, Leppamaki E, Koponen P, Levander J, Tapola E (1997) Physical characterisation of biomass-based pyrolysis liquids, vol 306. VTT, Espoo, pp 1–87

Oasmaa A, Kuoppala E, Gust S, Solantausta Y (2003a) Fast pyrolysis of forestry residue. 1. Effect of extractives on phase separation of pyrolysis liquids. Energy Fuel 17:1–12

Oasmaa A, Kuoppala E, Solantausta Y (2003b) Fast pyrolysis of forestry residue. 2. Physicochemical composition of product liquid. Energy Fuel 17:433–443

Oasmaa A, Solantausta Y, Arpiainen V, Kuoppala E, Sipilä K (2010a) Fast pyrolysis bio-oils from wood and agricultural residues. Energy Fuel 24:1380–1388

Oasmaa A, Elliott DC, Korhonen J (2010b) Acidity of biomass fast pyrolysis bio-oils. Energy Fuel 24:6548–6554

Olarte MV, Zacher AH, Padmaperuma AB, Burton D, Job HM, Lemmon TL, Swita MS, Rotness LJ, Neuenschwander GN, Frye JG, Elliott DC (2016) Stabilization of softwood-derived pyrolysis oils for continuous bio-oil hydroprocessing. Top Catal 59:55–64

Paasikallio V, Lindfors C, Kuoppala E, Solantausta Y, Oasmaa A, Lehto J, Lehtonen J (2014) Product quality and catalyst deactivation in a four day catalytic fast pyrolysis production run. Green Chem 16:3549–3559

Patwardhan PR, Satrio JA, Brown RC, Shanks BH (2010) Influence of inorganic salts on the primary pyrolysis products of cellulose. Bioresour Technol 101:4646–4655

Perego C, Bosetti A (2011) Biomass to fuels: the role of zeolite and mesoporous materials. Microporous Mesoporous Mater 144(1–3):28–39

Peyton K (2002) Nalco fuel field manual, Revised ed. McGrawHill, New York

Pinheiro A, Hudebine D, Dupassieux N, Geantet C (2009) Impact of oxygenated compounds from lignocellulosic biomass pyrolysis oils on gas oil hydrotreatment. Energy Fuel 23:1007–1014

Pinheiro A, Dupassieux N, Hudebine D, Geantet C (2011) Impact of the presence of carbon monoxide and carbon dioxide on gas oil hydrotreatment: investigation on liquids from biomass cotreatment with petroleum cuts. Energy Fuel 25:804–812

Pinho AD, Ramos JA, Leite LF (2011) US patent 7,867,378 B2, 11 Jan 2011

Pinho AR, Almeida M, Mendes FL, Casavechia LC, Talmadge MS, Kinchin CM, Chum HL (2017) Fast pyrolysis oil from pinewood chips co-processing with vacuum gas oil in a FCC unit for second generation fuel production. Fuel 188:462–473

Piskorz J, Scott DS, Radlein D (1988) Composition of oils obtained by fast pyrolysis of different woods. In: Soltes EJ, Milne TA (eds) Pyrolysis oils from biomass: producing, analyzing, and upgrading, ACS symposium series, vol 376. ACS, Washington, DC, pp 167–178

Pootakham T, Kumar A (2010) Bio-oil transport by pipeline: a techno-economic assessment. Bioresour Technol 101:7137–7143

Prasomsri T, Nimmanwudipong T, Román-Leshkov Y (2013) Effective hydrodeoxygenation of biomass-derived oxygenates into unsaturated hydrocarbons by MoO3 using low H2 pressures. Energy Environ Sci 6:1732–1738

Prasomsri T, Shetty M, Murugappan K, RománLeshkov Y (2014) Insights into the catalytic activity and surface modification of MoO3 during the hydrodeoxygenation of lignin-derived model

compounds into aromatic hydrocarbons under low hydrogen pressures. Energy Environ Sci 7:2660–2669
Production Statistics (2012) National Biodiesel Board
Qu T, Guo W, Shen L, Xiao J, Zhao K (2011) Experimental study of biomass pyrolysis based on three major components: hemicellulose, cellulose, and lignin. Ind Eng Chem Res 50:10424–10433
Rezaei PS, Shafaghat H, Daud WMAW (2014) Production of green aromatics and olefins by catalytic cracking of oxygenate compounds derived from biomass pyrolysis: a review. Appl Catal A Gen 469:490–511
Richards E (2013) Careers in biofuels. U.S. Bureau of Labor Statistics, Washington, DC
Ruddy DA, Schaidle JA, Ferrell Iii JR, Wang J, Moens L, Hensley JE (2014) Recent advances in heterogeneous catalysts for bio-oil upgrading via "ex situ catalytic fast pyrolysis": catalyst development through the study of model compounds. Green Chem 16(2):454–490
Samolada M, Baldauf W, Vasalos I (1998) Production of a bio-gasoline by upgrading biomass flash pyrolysis liquids via hydrogen processing and catalytic cracking. Fuel 77:1667–1675
Scott D, Piskorz J (1982) The flash pyrolysis of aspen poplar wood. Can J Chem Eng 60:666–674
Scott D, Piskorz J (1984) The continuous flash pyrolysis of biomass. Can J Chem Eng 62:404–412
Scott D, Piskorz J, Radlein D (1985) Liquid products from the continuous flash pyrolysis of biomass. Ind Eng Chem Proc Des Dev 24:581–588
Shah A, Darr MJ, Dalluge D, Medic D, Webster K, Brown RC (2012) Physicochemical properties of bio-oil and biochar produced by fast pyrolysis of stored single-pass corn stover and cobs. Bioresour Technol 125:348–352
Shetty M, Murugappan K, Prasomsri T, Green WH, Román-Leshkov Y (2015) Reactivity and stability investigation of supported molybdenum oxide catalysts for the hydrodeoxygenation (HDO) of m-cresol. J Catal 331:86–97
Singerman GM (1981) Methyl aryl ethers from coal liquids as gasoline extenders and octane improvers. SAE International, Warrendale
Soltes EJ, Lin J-CK (1984) Hydroprocessing of biomass tars for liquid engine fuels. In: Tillman DA, Jahn EC (eds) Progress in biomass conversion. Academic Press, New York, pp 1–69
Speight JG (1991) The chemistry and technology of petroleum, 2nd edn. Marcel-Dekker, New York
Sullivan MM, Bhan A (2016) Acetone hydrodeoxygenation over bifunctional metallic–acidic molybdenum carbide catalysts. ACS Catal 6:1145–1152
Sullivan MM, Held JT, Bhan A (2015) Structure and site evolution of molybdenum carbide catalysts upon exposure to oxygen. J Catal 326:82–91
Sullivan MM, Chen C-J, Bhan A (2016) Catalytic deoxygenation on transition metal carbide catalysts. Cat Sci Technol 6:602–616
Suzuki K, Aoyagi Y, Katada N, Choi M, Ryoo R, Niwa M (2008) Acidity and catalytic activity of mesoporous ZSM-5 in comparison with zeolite ZSM-5, Al-MCM-41 and silica–alumina. Catal Today 132:38–45
Taarning E, Osmundsen CM, Yang X, Voss B, Andersen SI, Christensen CH (2011) Zeolite-catalyzed biomass conversion to fuels and chemicals. Energy Environ Sci 4(3):793–804
Talmadge MS, Baldwin RM, Biddy MJ, McCormick RL, Beckham GT, Ferguson GA, Czernik S, Magrini-Bair KA, Foust TD, Metelski PD, Hetrick C, Nimlos MR (2014) A perspective on oxygenated species in the refinery integration of pyrolysis oil. Green Chem 16(2):407–453
Thegarid N, Fogassy G, Schuurman Y, Mirodatos C, Stefanidis S, Iliopoulou EF, Kalogiannis K, Lappas AA (2014) Second-generation biofuels by co-processing catalytic pyrolysis oil in FCC units. Appl Catal B Environ 145:161–166
Turriff SL (2014) Pulp and Paper Canada 115(3):11–13
Urbanchuk JM (2012) Contribution of the ethanol industry to the economy of the United States. Cardno Entrix, New Castle
US Energy Information Administration (2016) International Energy Outlook 2016. U.S. Energy Information Administration, Washington, DC
Vasalos IA, Lappas AA, Kopalidou EP, Kalogiannis KG (2016) Biomass catalytic pyrolysis: process design and economic analysis. Wiley Interdiscip Rev Energy Environ 5:370–383

Venkatakrishnan VK, Delgass WN, Ribeiro FH, Agrawal R (2015) Oxygen removal from intact biomass to produce liquid fuel range hydrocarbons via fast-hydropyrolysis and vapor-phase catalytic hydrodeoxygenation. Green Chem 17(1):178–183

von Stackelberg K, Buonocore J, Bhave PV, Schwartz JA (2013) Public health impacts of secondary particulate formation from aromatic hydrocarbons in gasoline. Environ Health 12:19

Wang Z, McDonald AG, Westerhof RJM, Kersten RA, Cuba-Torres CM, Ha S, Pecha B, Garcia-Perez M (2013a) Effect of cellulose crystallinity on the formation of a liquid intermediate and on product distribution during pyrolysis. J Anal Appl Pyrolysis 100:56–66

Wang H, Male J, Wang Y (2013b) Recent advances in hydrotreating of pyrolysis bio-oil and its oxygen-containing model compounds. ACS Catal 3:1047–1070

Wang H, Lee S-J, Olarte MV, Zacher AH (2016) Bio-oil stabilization by hydrogenation over reduced metal catalysts at low temperatures. ACS Sustain Chem Eng 4(10):5533–5545

Watkins B, Olsen C, Sutovich K, Petti N (2008) New opportunities for co-processing renewable feeds in refinery processes. WR Grace Catalagram 103

Wei XL, Schnell U, Hein K (2005) Behaviour of gaseous chlorine and alkali metals during biomass thermal utilisation. Fuel 84:841–848

Wiselogel AE, Agblevor FA, Johnson DK, Deutch S, Fennell JA, Sanderson MA (1996) Compositional changes during storage of large round switchgrass bales. Bioresour Technol 56:103–109

Yu F, Deng S, Chen P, Liu Y, Wan Y, Olson A, Kittelson D, Ruan R (2007) Physical and chemical properties of bio-oils from microwave pyrolysis of corn stover. Appl Biochem Biotechnol 137:957–970

Zacher AH, Olarte MV, Santosa DM, Elliott DC, Jones SB (2014) A review and perspective of recent bio-oil hydrotreating research. Green Chem 16:491–515

Zheng J-L, Kong Y-P (2010) Spray combustion properties of fast pyrolysis bio-oil produced from rice husk. Energy Convers Manag 51:182–188

Zhou G, Jensen PA, Le DM, Knudsen NO, Jensen AD (2016) Atmospheric hydrodeoxygenation of biomass fast pyrolysis vapor by MoO_3. ACS Sustain Chem Eng 4:5432–5440

Lignocellulosic Thermochemical Pretreatment Processes

Iwona Cybulska, Tanmay Chaturvedi, and Mette Hedegaard Thomsen

1 Introduction

Various pretreatment methods currently applied or under research and development to process lignocellulosic biomass into bioproducts. Ongoing research is focusing on optimizing and improving these technologies in order to reduce energy demands, the use of chemicals, the formation of by-products, and, more importantly, finding applications for the coproducts produced during the lignocellulosic ethanol production process (creating a complete and economically feasible biorefinery) (Wyman 1996). A desired pretreatment method should be as simple as possible and applicable to a wide variety of feedstock and should ensure purity of all products obtained (Chandra et al. 2007). Other important factors include catalyst use, catalyst recovery, and waste treatment (Zheng et al. 2009).

Efficient delignification should be the number one goal of an effective pretreatment method, since the presence of lignin determines cellulose accessibility and, lignin is also a valuable coproduct of lignocellulosic biomass processing. Most of the delignification methods developed to date produce lignin that is contaminated with hemicellulose and oftentimes with chemical residues—such as those used in the paper industry, which utilize sulfite or kraft processes (Wyman 1996). Most of the pretreatments studied are preceded by a size reduction of the feedstock. The most common particles size is 3 mm or less (Wyman 1996). Hydrothermal and organosolv processes have been chosen in this chapter as examples of the most

I. Cybulska (✉)
Faculty of Bioscience Engineering, Earth and Life Institute, Université Catholique de Louvain, Louvain-la-Neuve, Belgium
e-mail: iwona.cybulska@uclouvain.be

T. Chaturvedi · M. H. Thomsen
Department of Energy Technology, Aalborg University, Esbjerg, Denmark
e-mail: mht@et.aau.dk

© Springer Nature Switzerland AG 2019
J.-R. Bastidas-Oyanedel, J. E. Schmidt (eds.), *Biorefinery*,
https://doi.org/10.1007/978-3-030-10961-5_6

promising, optimized, and commercially feasible thermochemical pretreatment methods applied to lignocellulosic biomasses.

2 Hydrothermal Pretreatment

2.1 General Principles

Hydrothermal pretreatment (liquid hot-water treatment, high-temperature water treatment) or hot-compressed water (HCW) treatment is also known as vapocracking. Hydrothermal process is based on a principle that water heated under isochoric conditions has different physical and chemical properties from those at ambient temperature and atmospheric pressure. An exemplary diagram of a hydrothermal pretreatment process is illustrated in Fig. 1.

Under high-temperature/pressure conditions, the water properties as a solvent are modified such that (1) the density decreases and hydrogen bonds become weaker and the dielectric constant decreases, which leads to a decrease in polarity, mimicking organic solvents; (2) the translational and rotational movement of molecules are weaker because of the reduction in hydrogen bond strength, leading to an increase in diffusivity; and (3) the ionic strength increases through a higher number of H_3O^+ and OH^- ions. These changes also support ionic reactions, polar nonionic reactions, and radical reactions (Peterson et al. 2008; Akiya and Savage 2002). Moreover, hydrothermal technologies involve water autocatalysis, being organic-solvent free and limiting corrosion or the use of toxic compounds, which is in agreement with the principles of green chemistry (Anastas and Warner 2000). Autohydrolysis/hydrothermal processes of biomass follow similar mechanisms as dilute acid

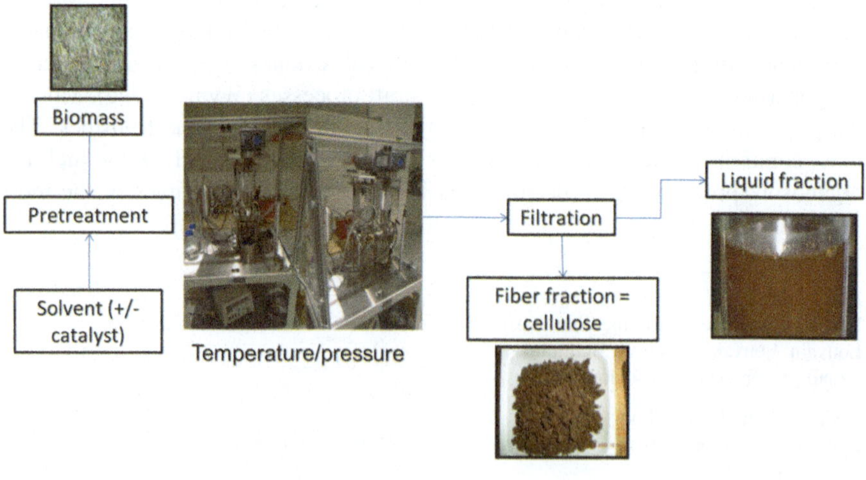

Fig. 1 Typical hydrothermal pretreatment diagram

hydrolysis: both being hydrolysis processes catalyzed by hydronium ions (H_3O^+). In an autohydrolysis process, in which water is the only reagent, the hydronium ions coming from water autoionization catalyze the depolymerization of polysaccharides by selective hydrolysis of both glycosidic linkages and acetyl groups. In a second step, hydronium ions from generated acetic acid also act as catalysts, improving reaction kinetics (Carvalheiro et al. 2008). Therefore, when submitted to hydrothermal treatment, hemicelluloses initially form shorter chains as well as oligosaccharides by hydrolysis of glycosidic bonds of polysaccharide chains (Carvalheiro et al. 2008). The hemicellulose oligomers in solution are partially further hydrolyzed to monomeric sugars and sugar degradation products that include hydroxymethylfurfural (HMF) formed from hexoses and 2-furfuraldehyde (2-F) obtained from pentoses and uronic acids (Chen et al. 2010), both by dehydration reactions (Chheda et al. 2007). These compounds can further undergo decomposition reactions, yielding formic and levulinic acids (Carvalheiro et al. 2008) as well as aldehydes or insoluble carbon compounds (Moreira 2008; Antal Jr et al. 1990), which are also considered valuable bioproducts. Furthermore, it has been observed that furan concentrations below potential inhibitory levels can be actually beneficial to yeast metabolism (Klinke et al. 2004). Lignin fraction is also being depolymerized in water and converted into phenolic compounds; however repolymerization process takes place as well. Repolymerized lignin precipitates on cellulose and is generally called pseudolignin (Negro et al. 2003).

Research approaches have shown the merits of water as a pretreating agent. Pressure cooking of plant materials using hot water was found to maximize physical changes and minimize hydrolysis of cellulose and therefore sugar degradation products during pretreatment, while making the pretreated cellulose highly reactive for subsequent enzymatic hydrolysis to achieve maximal glucose yield (Lei et al. 2008; Kohlmann et al. 1993; Weil et al. 1998; Walch et al. 1992). Physical changes by hydrothermal pretreatment that improve enzymatic hydrolysis of cellulose are well known and include an increase in pore size to enhance enzyme penetration and an increase in accessible cellulose by decreasing its crystallinity and association with lignin (Mok and Antal 1992; Grethlein 1985; Ladisch et al. 1983; Lee 2000). The effect of hydrothermal treatment on cellulose fibers can be seen using a scanning electron microscope (SEM), shown in Fig. 2.

Hydrothermal pretreatment has features of potentially optimal method, since it produces highly digestible cellulose (~90% hydrolysis glucose yield), and the highest results are achieved at temperatures above 200 °C (Mok and Antal 1992; Cybulska et al. 2009; Cara et al. 2007; Weil et al. 1997). This treatment uses no chemicals and moderate processing temperature (Chandra et al. 2007). Low (acidic) pH generated in water at high temperatures promotes lignin-carbohydrates ether bonds cleavage, thus using mineral acids (most commonly sulfuric) has been widely used for lignocellulose pretreatment prior to ethanol production (Wyman 1996; Zheng et al. 2009; Aita and Kim 2010; Schell et al. 2003). Lignin polymer alteration and partial removal of hemicellulose occur during the treatment (Aita and Kim

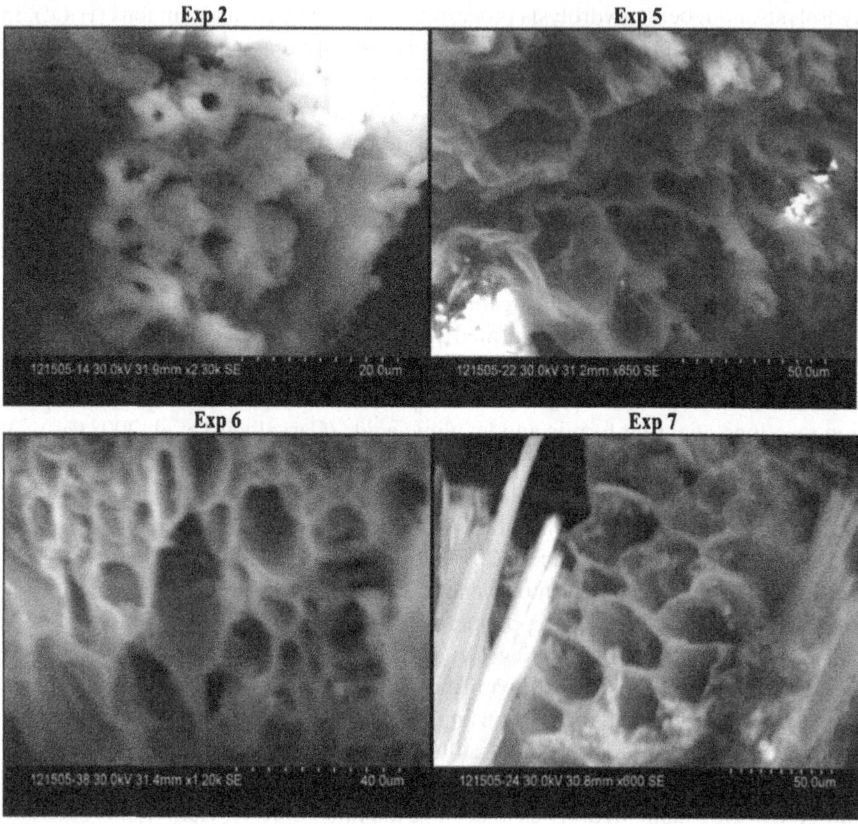

Fig. 2 Scanning electron microscopy (SEM) photographs of samples pretreated under various conditions: 210 °C/10 min (Experiment 2), 161.72 °C/15 min (Experiment 5), 218 °C/15 min (Experiment 6), and 190 °C/15 min (Experiment 7)

2010; Schell et al. 2003; Alvira et al. 2010; Carvalheiro et al. 2009). High temperatures and pressures modify lignin structure by melting, coagulation, and subsequent reprecipitation on the cellulose fibers (Aita and Kim 2010). Despite the fact that only small amounts of lignin are dissolved during the hydrothermal treatment, lignin alteration results in its decreased ability to interact with enzymes during cellulose hydrolysis and leaves cellulose fibrils more accessible to enzymes (Wyman 1996). High pressures generated during the treatment physically increase pore volume of the cellulose and therefore its surface area, which has been proven to improve enzymatic hydrolysis yields (Wyman 1996; Cybulska et al. 2009; Aita and Kim 2010).

2.2 State of the Art

Generally it is established that hydrothermal treatment does not involve explosion, but the reactor is slowly heated and then cooled after the process. The process is gaining interest as a pretreatment method for the ethanol industry, since it does not require any chemicals and is simple in operation. It has been implemented in pilot scale at DONG Energy facility in Skærbek, Denmark. The hydrothermal treatment has been applied in a screw-conveying reactor. A particle pump prevents the biomass from splashing due to pressure release while exiting the reactor. Splashing is a common problem occurring at the location of biomass removal from the reactor (in either continuous or a batch mode) (Larsen et al. 2008). Hydrothermal treatment may be followed by a liquid and solid fraction separation step, or the effluent can be fed into the next step in the form of slurry (Wyman 1996).

This technology can be applied to a variety of lignocellulosic feedstocks, including agricultural residues or energy crops (both woody and herbaceous). A study using olive tree residues revealed that the highest glucose yields obtained in enzymatic hydrolysis were found in samples pretreated at temperatures between 200 and 210 °C, while the highest xylose recovery occurred at lower temperatures (~170 °C) (Cara et al. 2007). Glucose yield of 90% occurred during enzymatic hydrolysis of hydrothermally pretreated (at 240 °C) yellow poplar sawdust. However, the inhibitory compounds formed during the treatment resulted in a low ethanol yield (50%) (Weil et al. 1997). According to Mok and Antal (1992), the efficiency of the treatment does not depend on the reaction conditions but mainly on the feedstock type, especially when woody and herbaceous materials are being compared. However, the variability among the feedstock type was not found to be significant. It was also observed that compressed hot-water percolation can remove 76–100% of hemicellulose (depending on the material type) and up to 60% of lignin at a severity parameter (log R_0) equal to 4.1 (230 °C and 2 min). Good hemicellulose solubility (64%) was also achieved in a study using hydrothermally treated wheat straw (with severity factor equal to 3.96, corresponding to 215 °C). Hemicellulose removal resulted in enriching the solids in cellulose up to 61%. The amount of inhibitors generated was acceptable (Carvalheiro et al. 2009).

Studies on prairie grasses (prairie cordgrass) have demonstrated high efficiencies in generating digestible cellulose, which was easily converted to glucose and further to ethanol (Cybulska et al. 2009). Applying typical conditions to woody municipal residues (landscaping trimmings of palm trees in Abu Dhabi) has shown high efficacy as well. There was a 60% enhancement of glucan-to-glucose conversion by pretreatment at 210 °C/10 min (100% conversion) compared with untreated leaflets (40% conversion). As for palm rachis, the pretreatment enhancements of glucan-to-glucose conversion (52% increase) were lower than for pretreated leaflets. The increase of the fiber digestibility has been related to structural changes, specifically destruction of the cell wall (Fang et al. 2015). The highest theoretical ethanol yield was observed at 210 °C/10 min for both leaflets (183.6 kg/t dry biomass) and rachis (235.0 kg/t dry biomass). Liquid-solid separation was found to sufficiently eliminate the inhibition caused by the pretreatment by-products (Torres et al. 2015). Generally, the most beneficial configuration for the hydrothermal treatment reactor would be continuous flow-through. This would ensure constant removal of hemicellulose from the material (Alvira et al. 2010). Other configurations include cocurrent and countercurrent flow reactors (Mosier et al. 2005).

2.3 *Special Case: Steam Explosion*

Steam explosion is a type of hydrothermal treatment and one of the most common and efficient methods of lignocellulosic biomass pretreatment. In this process, steam is injected into a reactor along with biomass feedstock, resulting in swelling of the lignocellulosic structure. Steam temperature, which also generates high pressures, is usually in the range of 180–200 °C. After a specified reaction time, usually between 5 and 30 min, the steam is released suddenly through a release valve, causing the biomass structure to explode. The explosion rapidly disrupts the linkages between cellulose, lignin, and hemicellulose. During this treatment, due to high temperature and pressure, some of the hexoses and pentoses from hemicellulose fraction are degraded to aldehydes and organic acids, which are inhibitory to fermenting microorganisms (Kosarie et al. 2001). The treatment does not involve any chemicals nor harsh processing conditions (Chandra et al. 2007). Steam explosion efficiency is affected by the particle size of the feedstock (Zheng et al. 2009).

For example, a steam explosion at 210 °C, 4 min residence time applied to poplar resulted in a 60% glucose enzymatic hydrolysis yield, 60% ethanol yield obtained during the simultaneous saccharification and fermentation (SSF), and 41% xylose recovery in the liquid fraction (Negro et al. 2003). This pretreatment method is one of the few that have been demonstrated on both a pilot scale and a commercial scale (Wyman 1996; Zheng et al. 2009). One of the industrial demonstration-scale steam explosion facilities is operated by Iogen Corporation in Canada. Its production capacity is 340 L of ethanol per tonne of fiber (Zheng et al. 2009). Another example of industrial-scale steam explosion pretreatment is the Masonite batch process used for the production of fiberboard and other products in the early twentieth century (Mosier et al. 2005). This pretreatment was applied to wood chips and used steam at pressures up to 90 atm and residence time between 1 and 10 min. A continuous mode of the Masonite process, called Stake II, used an extruder as the reactor. Another example of a continuous industrial-scale process is rapid steam hydrolysis (RASH). In this process, the liquid fraction is continuously drained from the reactor, which reduces generation of inhibitory by-products (Young 1998; Garrote et al. 1999).

Addition of H2SO4 (or SO2) or CO2 in steam explosion can improve enzymatic hydrolysis, decrease the production of inhibitory compounds, and lead to more complete removal of hemicellulose (Morjanoff and Gray 1987). The optimal conditions of steam explosion pretreatment of sugarcane bagasse have been found to be as follows: 220 °C; 30 s residence time; water-to-solid ratio, 2; and 1% H2SO4

(Morjanoff and Gray 1987). The advantages of steam explosion pretreatment include the low energy requirement compared to mechanical comminution and no recycling or environmental costs (Sun and Cheng 2002). Limitations of steam explosion include incomplete disruption of the lignin-carbohydrate matrix, destruction of a portion of the xylan fraction, and generation of compounds that may be inhibitory to microorganisms used in downstream processes (Mackie et al. 1985).

3 Organosolv Treatment

3.1 General Principles

By choosing an appropriate chemical solvent, certain fractions of the lignocellulosic biomass can be solubilized and removed from the biomass structure, leaving the remaining fractions as a solid. It is possible to remove cellulose as a liquid, leaving the lignin as the solid fraction, or to extract lignin as the liquid fraction, leaving cellulose as the solid. Solvents such as alcohols, ketones and aldehydes, organic acids, and alkalis have been used for lignin removal. Cellulose removal into the solution has been practiced using various acids (sulfuric, hydrochloric, and phosphoric). Another example of a cellulose solvent is cadoxen, used in biomass studies in the 1970s. Cadoxen is an aqueous alkaline solution of ethylene diamine and cadmium oxide, which extracts cellulose into the liquid phase, from which it can be precipitated by addition of excess water (Kosarie et al. 2001; Chang et al. 1981). Another cellulose-dissolving agent that has been tested is zinc chloride. Cellulose-dissolving processes have gained less interest by researchers than lignin-dissolving processes (Wyman 1996). Materials pretreated by cellulose-dissolving agents can yield glucose release levels of 80–90% during enzymatic hydrolysis in as little as 5–10 h (Chang et al. 1981).

Overall, solvent application is an expensive (compared to aqueous treatment) but more selective alternative for biomass fractionation. In addition to the fractionation, solvent treatment also has the potential to reduce cellulose crystallinity (Kosarie et al. 2001). However, when it comes to using organic solvents, this method is not very efficient at low temperatures, and introducing a physical aspect of temperature and pressure as part of the treatment is often necessary.

The most important reaction responsible for breaking down the lignin polymer and therefore delignification is cleavage of ether bonds. The α-aryl ether bonds are easier to hydrolyze, and so their cleavage is the most extensive. However, cleavage of β-aryl ether bonds, which are more difficult to break, also occurs to some degree. The cleavage of α-aryl ether bonds occurs most easily on structural units with a hydroxyl group in the para-position (such as syringil units). Also, the reaction is catalyzed by hydronium ions, coming either from an acidic catalyst or hemicellulose acetyl group degradation which occurs at high temperatures. The intermediate formed in this reaction can undergo an undesired condensation reaction, which limits the delignification process. Carbohydrate-lignin bonds are found to be mainly ether linkages involving α-carbon of lignin side chains, which are easily hydrolyzed in acidic conditions of organosolv treatment. Therefore, organosolv lignin is usually free of carbohydrate contamination.

3.2 *State of the Art*

3.2.1 Lignol Process

The Lignol Biorefinery Technology process is a pure organosolv treatment process using an ethanol-water mixture for softwood fractionation into three relatively pure streams: lignin, cellulose, and hemicellulose. This treatment results in more than 90% enzymatic hydrolysis glucose yield (Zhao et al. 2009). The process is divided into two main stages: organosolv treatment with multiple coproduct recovery and cellulose saccharification. Contrary to other pretreatment methods, this process is not focused only on ethanol production. It refines biomass with generation of other valuable coproducts, especially relatively pure lignin. Hemicellulose-originating products such as xylose, furfural, and acetic acid are also considered as a potential revenue source. This makes ethanol only 25–30% of the predicted revenue (Arato et al. 2005).

The organosolv cooking takes place at temperatures between 185 and 190 °C with residence time between 30 and 90 min, LSR from 4:1 to 10:1. The resulting pH is usually between 3.0 and 2.0. Cellulose is hydrolyzed and fermented to ethanol. Lignin recovered from the Lignol process represents about 75–80% of initial Klason lignin of the substrate. The process is more efficient when applied to hardwoods. The main solvent (ethanol) used in the cooking liquor is also the main product, so any losses of the solvent can be made up for without external sources. The process has been demonstrated on a pilot scale in batch mode, but the predictions show that countercurrent continuous mode would be more beneficial (Arato et al. 2005).

3.2.2 Acetic Acid Pulping (Acetosolv, Acetocell, and Biodyne Processes)

In order to reduce the temperature needed for the generation of hydronium ions that promote the delignification process, low amounts of organic or mineral acids can be added to the organosolv cooking. One of the alternatives is using acetic acid with or without mineral acid and with or without other organic solvents (Young 1998).

The Acetosolv system was invented in Hamburg, Germany, in the 1980s and employs a solvent mixture composed of 93% acetic acid and 0.5–3% hydrochloric acid. The optimal temperature for the digestion was found to be 110 °C and time varied from 0.5 to 5 h. Delignification rates were found to be higher than the ones obtained using the Kraft process; however, economic aspects of the process have not been specified. The biggest disadvantage of the process was the presence of chloride ions, which created a need for special corrosion-resistant equipment. Many modifications of the Acetosolv pulping have been investigated, including utilization of formic (e.g., FORMACELL process), propionic, or sulfonic acids (Young 1998).

The Acetocell process originated from the Acetosolv system, modified by eliminating the hydrochloric acid and using higher temperatures. Concentrations of acetic acid used in this process were in the range of 80–90%, and pulping was performed at 170–190 °C for 2–3 h. Acetylation of pulp fibers and lignin occurs to some extent when using acetic acid as a pulping solvent, and often deacetylation has to be employed after digestion.

Acetic acid in the mixture with ethyl acetate and water was proposed for use in pulping and biomass pretreatment at the University of Wisconsin and was licensed to Biodyne Chemicals Inc. The process was based on the fractionation principle and started with digestion at high temperatures (up to 200 °C), which was followed by phase separation of the spent liquor into organic phase (with dissolved lignin) and aqueous phase (containing hemicellulose). The proposed solvent mixture was found to produce good delignification rates for both hardwoods and softwoods. Ethyl acetate is miscible with both acetic acid and water, but it also enables the phase separation for chemical recovery (Young 1998; Baierl et al. 1987).

3.2.3 Current Research

The difficulty in valorizing lignin fraction originates from its recalcitrance which is due to a complex polymeric structure built of repeating units of three building blocks, coumaryl, coniferyl, and sinapyl alcohols, bound together by C-C and ether bonds. While C-C bonds are difficult to disrupt without causing a strong defunctionalization of the biomolecule, ether bonds are more easily broken and have therefore been targeted by researchers for lignin depolymerization (Van den Bosch et al. 2015; Verboekend et al. 2016). Disassembly of the biopolymer leads to the production of a low molecular weight lignin oil, mainly composed of monomers (alkyl phenols) and oligomers, highly functionalized intermediates which can be further upgraded by dealkylation or conversion to alkyl cyclohexanols (Schutyser et al. 2016) and cyclohexanones (Schutyser et al. 2015a) to ultimately manufacture valuable polymers. In addition, alternative routes to valorize lignin oil include the production of biofuels (Van den Bosch et al. 2015; Abdullah et al. 2017) or the exploitation of their antioxidant activity for food and pharmaceutical applications (Vinardell and Mitjans 2017).

Among the different techniques that have been studied to fractionate lignocellulose into its main components, organosolv delignification is a method that offers the possibility to simultaneously extract lignin from lignocellulose while making cellulose more digestible (Cybulska et al. 2017). The principle of organosolv treatment is to dissolve the lignin component in an organic solvent, while the cellulose and the hemicellulose fractions remain in the solid (Salapa et al. 2017). In some processes, water is added to the solution in order to extract hemicellulose components and by-products and to obtain higher purity of the desired cellulose and lignin streams.

Moreover, an acid or base is sometimes added to enhance lignin depolymerization (Van den Bosch et al. 2017).

The strong advantage of organosolv delignification, with respect to harsher pretreatment methods aimed at increasing lignocellulose enzymatic digestibility, is offering the opportunity to fully utilize biomass potential and creating a holistic biorefinery via less common "lignin-first" approach. Recent findings by Van den Bosch et al. (2015) have shown that lignin depolymerization can be achieved from the unprocessed biomass through direct precious metal catalysis in an organic solvent, purifying fiber fraction at the same time, with a process referred to as "reductive catalytic fractionation" (Schutyser et al. 2015a). According to this study, conversion of birch sawdust leads to the production of a small group of monomers (mainly 4-n-propylguaiacol and 4-n-propylsyringol) with a yield of about 50%, together with a delignification efficiency of nearly 85% and a carbohydrate retention in the pulp of over 90%. The same research team also inspected the influence of various organic solvents on the delignification yield and carbohydrate retention achieved during birchwood processing (Schutyser et al. 2015b), as well as the effects on lignin oil composition deriving from the employment of different commercial metal catalysts, contributing to building the foundations of a model for lignin depolymerization under reductive catalytic conditions (Van den Bosch et al. 2017). Process optimization is still ongoing, with the goal of bringing the "lignin-first" biorefinery concept to the industry.

4 Summary

Despite the fact that currently there are no perfect pretreatment methods for the lignocellulosic biomass, during the last decade, advancements have been made in optimizing already discovered techniques and modifying old industrial processes (pulp and paper industry) for new applications (ethanol fermentation and lignin utilization).

Hydrothermal processing possesses features of a perfect pretreatment for ethanol production purposes. It is usually short (minutes), does not require any chemicals, and produces good hydrolysis and fermentation results. It cleaves the bonds between lignin and carbohydrates and alters lignin structure to the point where it does not interfere with the enzymes. Moderate energy demand and water usage are also favorable characteristics of this process.

Organosolv fractionation is basically the only method that generates relatively pure lignin in addition to highly digestible cellulose and is low in inhibitory by-products and recoverable hemicellulose sugars. The organosolv method provides an opportunity to implement the biorefinery concept, producing a wide variety of value-added products in addition to fuel from lignocellulosic biomass feedstock. These two treatments have also low environmental impact in comparison to other methods.

References

Abdullah, B., Muhammad, S. A. F. A. S., & Mahmood, N. A. N. (2017). Production of biofuel via hydrogenation of lignin from biomass. In New Advances in Hydrogenation Processes-Fundamentals and Applications. InTech. https://doi.org/10.5772/66108

Aita GM, Kim M (2010) Pretreatment technologies for the conversion of lignocellulosic materials to bioethanol. In: Eggleston G (ed) Sustainability of the sugar and sugar ethanol industries. American Chemical Society, Washington, p 117

Akiya N, Savage PE (2002) Roles of water for chemical reactions in high-temperature water. Chem Rev 102:2725

Alvira P, Tomas-Pejo E, Ballesteros M, Negro MJ (2010) Pretreatment technologies for an efficient bioethanol production process based on enzymatic hydrolysis: a review. Bioresour Technol 101:4851

Anastas PT, Warner JC (2000) Green chemistry: theory and practice. Oxford University Press, New York

Antal MJ Jr, Mok WS, Richards GN (1990) Mechanism of formation of 5-(hydroxymethyl)-2-furaldehyde from D-fructose and sucrose. Carbohydr Res 199:91

Arato C, Pye EK, Gjennestad G (2005) The lignol approach to biorefining of woody biomass to produce ethanol and chemicals. In: Davison BH, Evans BR, Finkelstein M, McMillan JD (eds) Twenty-sixth symposium on biotechnology for fuels and chemicals. Humana Press, New York, p 871

Baierl, S., Young, K. W., Young, R. A., & Timothy, R. (1987). Process for digesting lignocellulosic material. Biodyne Chemicals, Inc. European Patent Application. Appl. number EP19860305606.

Cara C, Romero I, Oliva JM, Sáez F, Castro E (2007) Liquid hot water pretreatment of olive tree pruning residues. In: Mielenz JR, Klasson KT, Adney WS, McMillan JD (eds) Applied biochemistry and biotechnology. Humana Press, New York, p 379

Carvalheiro F, Duarte LC, Gírio FM (2008) Hemicellulose biorefineries: a review on biomass pretreatments. J Sci Ind Res 67:849

Carvalheiro F, Silva-Fernandes T, Duarte LC, Girio FM (2009) Wheat straw autohydrolysis: process optimization and products characterization. Appl Biochem Biotechnol 153:84

Chandra R, Bura R, Mabee W, Berlin A, Pan X, Saddler J (2007) Substrate pretreatment: the key to effective enzymatic hydrolysis of lignocellulosics? In: Olsson L (ed) Biofuels. Springer, Berlin/Heidelberg, p 67

Chang M, Chou T, Tsao G (1981) Structure, pretreatment and hydrolysis of cellulose. In: Bioenergy. Springer, Berlin/Heidelberg, p 15

Chen X, Lawoko M, van Heiningen A (2010) Kinetics and mechanism of autohydrolysis of hardwoods. Bioresour Technol 101:7812

Chheda JN, Huber GW, Dumesic JA (2007) Liquid-phase catalytic processing of biomass-derived oxygenated hydrocarbons to fuels and chemicals. Angew Chem Int Ed 46:7164

Cybulska I, Lei H, Julson J (2009) Hydrothermal pretreatment and enzymatic hydrolysis of prairie cord grass. Energy Fuel 24:718

Cybulska I, Brudecki GP, Zembrzuska J, Schmidt JE, Lopez CG-B, Thomsen MH (2017) Organosolv delignification of agricultural residues (date palm fronds, Phoenix dactylifera L.) of the United Arab Emirates. Appl Energy 185:1040

Fang C, Schmidt JE, Cybulska I, Brudecki GP, Frankær CG, Thomsen MH (2015) Hydrothermal pretreatment of date palm (Phoenix dactylifera L.) leaflets and rachis to enhance enzymatic digestibility and bioethanol potential. BioMed Res Int 2015:216454

Garrote G, Dominguez H, Parajo JC (1999) Hydrothermal processing of lignocellulosic materials. Holz Roh Werkst 57:191–202

Grethlein HE (1985) The effect of pore size distribution on the rate of enzymatic hydrolysis of cellulosic substrates. Nat Biotechnol 3:155

Klinke HB, Thomsen AB, Ahring BK (2004) Inhibition of ethanol-producing yeast and bacteria by degradation products produced during pre-treatment of biomass. Appl Microbiol Biotechnol 66:10

Kohlmann K, Westgate P, Weil J, Ladisch MR (1993) Biological based systems for waste processing. In: ICES meeting

Kosarie N, Sukan-Vardar F, Pieper HJ, Senn T (2001) The biotechnology of ethanol. Classical and future applications. Wiley-VCH Verlag GmbH, Wien

Ladisch MR, Lin KW, Voloch M, Tsao GT (1983) Process considerations in the enzymatic hydrolysis of biomass. Enzym Microb Technol 5:82

Larsen J, Østergaard Petersen M, Thirup L, Wen Li H, Krogh Iversen F (2008) The IBUS process—lignocellulosic bioethanol close to a commercial reality. Chem Eng Technol 31:765

Lee YY (2000) Enhancement of dilute-acid total-hydrolysis process for high-yield saccharification of cellulosic biomass. Department of Chemical Engineering. Auburn University, Auburn

Lei H, Hennessey K, Liu Y, Lin X, Wan Y, Ruan R (2008) Optimization of hydrothermal pretreatment of corn stover. In: ASABE annual international meeting, Providence

Mackie KL, Brownell HH, West KL, Saddler JN (1985) Effect of sulphur dioxide and sulphuric acid on steam explosion of aspenwood. J Wood Chem Technol 5:405–425

Mok WSL, Antal MJ (1992) Uncatalyzed solvolysis of whole biomass hemicellulose by hot compressed liquid water. Ind Eng Chem Res 31:1157

Moreira L (2008) An overview of mannan structure and mannan-degrading enzyme systems. Appl Microbiol Biotechnol 79:165

Morjanoff PJ, Gray PP (1987) Optimization of steam explosion as a method for increasing susceptibility of sugarcane bagasse to enzymatic saccharification. Biotechnol Bioeng 29:733–741

Mosier N, Wyman C, Dale B, Elander R, Lee YY, Holtzapple M et al (2005) Features of promising technologies for pretreatment of lignocellulosic biomass. Bioresour Technol 96:673

Negro MJ, Manzanares P, Ballesteros I, Oliva JM, Cabanas A, Ballesteros M (2003) Hydrothermal pretreatment conditions to enhance ethanol production from poplar biomass. Appl Biochem Biotechnol 105–108:87

Peterson AA, Vogel F, Lachance RP, Fröling M, Antal MJ Jr, Tester JW (2008) Thermochemical biofuel production in hydrothermal media: a review of sub-and supercritical water technologies. Energy Environ Sci 1:32

Salapa I, Katsimpouras C, Topakas E, Sidiras D (2017) Organosolv pretreatment of wheat straw for efficient ethanol production using various solvents. Biomass Bioenergy 100:10

Schell DJ, Farmer J, Newman M, McMillan JD (2003) Dilute-sulfuric acid pretreatment of corn stover in pilot-scale reactor: investigation of yields, kinetics, and enzymatic digestibilities of solids. Appl Biochem Biotechnol 105–108:69

Schutyser W, Van den Bosch S, Dijkmans J, Turner S, Meledina M, Van Tendeloo G et al (2015a) Selective nickel-catalyzed conversion of model and lignin-derived phenolic compounds to cyclohexanone-based polymer building blocks. ChemSusChem 8:1805

Schutyser W, Van den Bosch S, Renders T, De Boe T, Koelewijn S-F, Dewaele A et al (2015b) Influence of bio-based solvents on the catalytic reductive fractionation of birch wood. Green Chem 17:5035

Schutyser W, Van den Bossche G, Raaffels A, Van den Bosch S, Koelewijn S-F, Renders T et al (2016) Selective conversion of lignin-derivable 4-alkylguaiacols to 4-alkylcyclohexanols over noble and non-noble-metal catalysts. ACS Sustain Chem Eng 4:5336

Sun Y, Cheng J (2002) Hydrolysis of lignocellulosic materials for ethanol production: a review. Bioresour Technol 83:1–11

Torres, A. I., Cybulska, I., Fang, C. J., Thomsen, M. H., Schmidt, J. E., & Stephanopoulos, G. (2015) A novel approach for the identification of economic opportunities within the framework of a biorefinery. Comput Aided Chem Eng 37:1175–1180. Elsevier.

Van den Bosch S, Schutyser W, Vanholme R, Driessen T, Koelewijn S-F, Renders T et al (2015) Reductive lignocellulose fractionation into soluble lignin-derived phenolic monomers and dimers and processable carbohydrate pulps. Energy Environ Sci 8:1748

Van den Bosch S, Renders T, Kennis S, Koelewijn S-F, Van den Bossche G, Vangeel T et al (2017) Integrating lignin valorization and bio-ethanol production: on the role of Ni-Al 2 O 3 catalyst pellets during lignin-first fractionation. Green Chem 19:3313

Verboekend D, Liao Y, Schutyser W, Sels BF (2016) Alkylphenols to phenol and olefins by zeolite catalysis: a pathway to valorize raw and fossilized lignocellulose. Green Chem 18:297

Vinardell MP, Mitjans M (2017) Lignins and their derivatives with beneficial effects on human health. Int J Mol Sci 18:1219

Walch E, Zemann A, Schinner F, Bonn G, Bobleter O (1992) Enzymatic saccharification of hemicellulose obtained from hydrothermally pretreated sugar cane bagasse and beech bark. Bioresour Technol 39:173

Weil J, Sarikaya A, Rau SL, Goetz J, Ladisch CM, Brewer M et al (1997) Pretreatment of yellow poplar sawdust by pressure cooking in water. Appl Biochem Biotechnol 68:21

Weil J, Sarikaya A, Rau S-L, Goetz J, Ladisch C, Brewer M et al (1998) Pretreatment of corn fiber by pressure cooking in water. Appl Biochem Biotechnol 73:1

Wyman CE (1996) Handbook on bioethanol. Production and utilization. Taylor & Francis, Washington

Young RA (1998) Environmentally friendly technologies for the pulp and paper industry. Wiley, New York

Zhao X, Cheng K, Liu D (2009) Organosolv pretreatment of lignocellulosic biomass for enzymatic hydrolysis. Appl Microbiol Biotechnol 82:815

Zheng Y, Pan Z, Zhang R (2009) Overview of biomass pretreatment for cellulosic ethanol production. Int J Agric Biol Eng 2:51

Part III
Biodiesel and Water/Ethanol Separation

Biodiesel: Use of Green Feedstocks and Catalysts

Hanifa Taher

1 Introduction

The continuous growth of the global population and industrialization has resulted in an energy demand increase. Fossil fuel resources supply the majority of the energy demanded, which has several influences on the ecosystem. The combustion of fossil fuels causes several issues, such as greenhouse gas emissions. Burning of fossil fuels leads to a huge increase in greenhouse gas (GHG) emissions, which contains 70% CO_2 (Rahman et al. 2017). Recent studies showed that the current average CO_2 concentration in the atmosphere is close to 400 ppm, which is above the save level of 350 ppm (Wennersten et al. 2015). Besides worries about fossil fuel environmental impacts, the world energy consumption of fossil fuels is growing from limited resources. The continuous reserve dwindle has made it necessary to look for alternative "green" and more "sustainable" sources of energy. Bio-based fuels are the most promising alternative for the transportation. Biodiesel, produced from oil-rich source, has rapidly gained interest over petroleum-based diesel due to its valuable characteristics and ability to lower the dependency on fossil fuels. It also reduces the emissions of greenhouse gas to the atmosphere (Eshton et al. 2013), as produced from renewable biomass and therefore CO_2 neutral. Carbon dioxide that is emitted when biodiesel is burned is simply taken into plants by photosynthesis.

Today, biodiesel is produced at industrial scale in several counties, including China, Brazil, and Thailand, but several issues were raised, mainly from with feedstock supply, catalysts performance, and technology cost. The main obstacle in biodiesel production and its commercialization through oil transesterification is its production cost. Efforts have been carried out to use different feedstocks, catalysts, and technologies, and challenges are present.

H. Taher (✉)
Chemical Engineering Department, Masdar Institute, A Part from Khalifa University of Science and Technology, Abu Dhabi, United Arab Emirates
e-mail: hanifa.alblooshi@ku.ac.ae; healblooshi@masdar.ac.ae

© Springer Nature Switzerland AG 2019

J.-R. Bastidas-Oyanedel, J. E. Schmidt (eds.), *Biorefinery*,
https://doi.org/10.1007/978-3-030-10961-5_7

2 History of Biodiesel and Its Properties

Biodiesel is a mixture of fatty acid methyl esters (FAMEs) derived from oil-rich feedstocks, such as vegetable oils and animal fats. The major components of plant-based oils and animal fats are triglycerides (Marchetti 2012). It has been verified as an environmental benign fuel, since emissions are much lower than petroleum-based diesel. It has been reported that usage of biodiesel (in its pure form) reduces CO_2 emissions by 50–75% (Schröder et al. 2013). Having said so, the CO_2 emission from biodiesel is considered neutral and does not add any carbon to the atmosphere. This is because CO_2 released by biodiesel burning is balanced by plants through photosynthesis (Farahani et al. 2011). Moreover, biodiesel results to reduce unburned and aromatic and hydrocarbon emissions (Aresta et al. 2005; Demirbas 2007). The emission of sulfur dioxide, which is the primary cause of acid rain, is also minimized when biodiesel is used, as it has almost very low sulfur content reaching 0.02 compared to 0.59 (wt%) in diesel fuel (Basha et al. 2009; Khoobbakht et al. 2016).

Direct use of oil, namely, peanut oil, in diesel engine was initially tested by the inventor of the diesel engine, Rudolf Diesel (Murugesan et al. 2009). Though the oil could be used directly, it was not easy as the high viscosity and low volatility of selected oil lead to incomplete combustion carbon deposits (Fan and Burton 2009; Helwani et al. 2009). Considerable efforts have been made to derivatize oils to products with properties similar to petroleum diesel. Transesterification is the most commonly accepted method, where short-chain alcohols are used to break the oil molecules to esters and glycerol in the presence of catalyst (Al-Zuhair 2007; Lai et al. 2005). The latter is a by-product.

3 Current Status of Biodiesel Production

3.1 Feedstocks

There are several triglyceride feedstocks for biodiesel production that include canola, palm, and sunflower seeds (Antunes et al. 2008; Dubé et al. 2007; Al-Zuhair and Taher 2016). Namely, feedstocks with fatty acid chain length in range of C_{14} to C_{22} with low unsaturation level are the most suitable one for biodiesel production (Williams and Laurens 2010). The fuel properties that are affected by oil composition are cetane number, heat of combustion, cold flow, stability, viscosity, and lubricity (Knothe 2005). It has been reported that low fuel cetane numbers are usually related to the level of unsaturated fatty acids, and oils that contain high levels of polyunsaturated acids have high iodine values (Ramos et al. 2009). Table 1 summarizes the fatty acid profile of different tested oils for biodiesel production (Verma et al. 2016).

Table 1 Fatty acid composition of lipids from common feedstocks tested for biodiesel production (Becker 2007; Demirel et al. 2006; Kamal-Eldin and Andersson 1997; Al-Zuhair and Taher 2016)

	% wt composition									
Fatty acids	C14:0	C16:0	C16:1	C18:0	C18:1	C18:2	C18:3	C20:0	C20:1	C22:0
Canola oil	–	4	–	2	61	22	10	–	1	–
Soybean oil	–	11	–	4	23	54	8	–	–	–
Sunflower oil	–	6	–	5	29	58	1	–	–	1
Cotton oil	1	23	1	2	17	56	–	–	–	–
Linseed oil	–	5.6	–	3.2	17.7	15.7	57.8	–	–	–
Palm oil	1	45	–	4	39	11	–	–	–	–
Olive oil	–	13.8	1.4	2.8	71.6	9	1	–	–	–
Sesame oil	–	9.6	0.2	6.7	41.1	41.2	0.7	–	–	–
Corn oil	–	11	–	2	28	58	1	–	–	–
Chicken fat	1	25	8	6	41	18	1	–	–	–
Lamb fat	2	19		26	44	2	4	–	–	–
Madhuca longifolia	–	18	–	14	46	18	–	–	–	–
Pongamia pinnata	–	9	–	8	66	12	–	1	1	3
Spirulina platensis	0.7	45.5	9.6	1.3	3.8	14.5	21.4	–	–	–
Spirulina maxima	0.3	45.1	6.8	1.4	1.9	14.6	20.6	–	–	–
Scenedesmus obliquus	0.6	16	8	0.3	8	6	28	–	–	–
Chlorella vulgaris	0.9	20.4	5.8	15.3	6.6	1.5	–	–	–	–
Dunaliella bardawil	–	41.7	7.3	2.9	8.8	15.1	20.5	–	–	–

Although vegetables are available in abundant, cultivation of edible plants for fuels production has negative effects on food security (Liu et al. 2011) and results to an increase in food price. In addition, plantation of oil-rich feedstocks requires lands and freshwater, which are limited resources. It has been reported that the cost of vegetable oils used in biodiesel production accounts for more than 60% of the fuel production cost (Al-Zuhair 2007). This proposes the use of non-edible oils, such as those from *Jatropha*, castor, and neem that can be cultivated on wastelands. Although this can be seen as possible, they also require freshwater, fertilizers, and land for their cultivation (González et al. 2015). The use of waste oils and fats has been also suggested, where their uses can be also a waste management strategy (Phan and Phan 2008). Nevertheless, these contain large amount of free fatty acids and water that require biodiesel post-purification and therefore increasing production cost. Moreover, such oil supply is not stable and cannot satisfy the global need of diesel fuel (Predojević 2008; Taufiqurrahmi et al. 2011; Zhang et al. 2003). Therefore, it is hard to depend on them in industrialization. Finding alternatives had encouraged researchers to look at microorganisms. Thus, significant works were carried on testing the feasibility of using microalgae lipids in biodiesel production. Microalgae have attracted the attention due to their ability to grow in several environments and to remediate effluents. In addition to their high growth rate and considerable amount of lipids in their cell, microalgae have been utilized as source of other fuels as well, including bioethanol.

3.2 Catalysts

Oils can be converted into biodiesel by different approaches, which include dilution, microemulsion, catalytic cracking, or transesterification. Transesterification is the most commonly used method at industrial scale, where a reaction is generally carried out between the oil and acyl acceptor, such as methanol, in the presence of catalyst. Conventionally, the reaction is catalyzed by alkali catalysts, and high yields are determined within few hours of reaction. Sodium hydroxide (NaOH) and potassium hydroxide (KOH) are the more common used catalysts, due to their low cost and high achievable yields of more than 98% within hours at reasonable temperature of 60 °C (Atadashi et al. 2012; Fukuda et al. 2001). Produced biodiesel is similar to petrol-diesel and has almost similar characteristics. The quality is usually compared using ASTM D6751 (American) and EN 14,214 (European) standards.

Biodiesel production yields are affected by several process variables, including oil quality and reaction conditions. Reaction conditions are alcohol to oil ratio, catalyst type and concentration, reaction time, and temperature. Table 2 lists some of common chemical-based catalysts used with different oil sources for biodiesel production at different process conditions. The effect of reaction temperature and alcohol to oil molar ratios are the most significant. From the stoichiometry of the reaction, each mole of oil reacts with three moles of alcohol; however higher molar ratios are usually required to shift the reaction toward more biodiesel production (Jain and Sharma 2010). Nevertheless, high methanol concentrations result to emulsify reaction products and decrease biodiesel yield. Increasing reaction temperature speeds up reactions and reduces oils viscosity, allowing more mixing between the immiscible reactants and faster separation of products (Noureddini and Zhu 1997). Catalysts increase reaction rate; however, the presence of chemical-based catalysts in biodiesel production complicates biodiesel purification and increases the overall process cost. This calls to test the effectiveness of using alcohols above their critical points (Demirbas 2007). Supercritical production of biodiesel is a one-step process, where reactants heated up or pressurized above the critical temperature and pressure of the alcohol used. This is usually carried out at high temperatures, reaching 400 °C, and high pressures reaching 150 bars and results to produce biodiesel within 15 min of reaction without any need to purify the oil prior its use of biodiesel at end of reaction. The use of supercritical methanol has been tested, and high yield reaching above 95% was achieved from rapeseed and sunflower oils (Demirbaş 2003; Madras et al. 2004). Although this seems fast, high energy is required to reach the critical point of the alcohol. This points back to use catalysts to speed up reactions.

Alkaline-based biodiesel processes are simple, but not that feasible with oils containing high free fatty acids and water contents, such as that in waste oils. This is mainly because the presence of fatty acids and water in the oil promotes soap

Table 2 Biodiesel production from different oil sources at different operating conditions using homogeneous chemical alkali and acid catalysts

Catalyst	Feedstock	Temperature (°C)	Molar ratio (alcohol/oil)	Reaction time	Yield (%)	References
NaOH	Sunflower oil	65	6:1	–	86.7	Vicente et al. (2004)
NaOH	Canola oil	70	6:1	15 min	93.5	Leung and Guo (2006)
NaOH	Frying oil	60	7:1	20 min	88.8	Leung and Guo (2006)
NaOH	Soybean oil	45	6:1	20 min	100	Ji et al. (2006)
NaOH	Sunflower oil	60	6:1	1.5 h	97.1	Rashid et al. (2008)
NaOH	Sunflower oil	50	12:1	10 min	99	Marjanović et al. (2010)
NaOH	Soybean oil	70	12:1	1 h	97.2	Kucek et al. (2007)
KOH	Karanja oil	65	6:1	3 h	97–98	Meher et al. (2006)
KOH	Soybean oil	70	12:1	–	95.6	Kucek et al. (2007)
KOH	Sunflower oil	65	6:1	–	91.6	Vicente et al. (2004)
KOH	Fish oil	60	6:1	30 min	98	Armenta et al. (2007)
KOH	Waste cooking oil	65	6:1	1 h	98.16	Refaat et al. (2008)
H_2SO_4	Waste cooking oil	95	20:1	20 h	90	Ji et al. (2006)
H_2SO_4	Used sunflower oil	65	30:1	69 h	99	Freedman et al. (1984)
H_2SO_4	Waste frying oil	70	245:1	4 h	99	Zheng et al. (2006)
H_2SO_4	Rice bran oil	60	5:1	12	99	Zullaikah et al. (2005)

formation, which decreases biodiesel yield and affects its quality (Ma et al. 1998; Sivasamy et al. 2009). Pre-treatment of oil prior transesterification by acid esterification has been therefore suggested, where sulfuric acid (H_2SO_4) is commonly adopted. Acid esterification is a very slow step and requires large amount of alcohols to be used, besides acids are corrosive chemicals (Akoh et al. 2007; Al-Zuhair 2007; Marchetti et al. 2007).

As mentioned earlier, not all oil-rich feedstocks can be used. This is mainly because the free fatty acids (FFAs) present in the oil influence significantly the produced biodiesel quality and therefore properties. Fatty acid profile of many microalgae species shows to be in range of C12 to C22, and unsaturation level in some of them was found to be high (Williams and Laurens 2010). Therefore, alkaline catalysts cannot be used with feedstocks containing free fatty acid contents higher than 1%, where soaps usually form. In addition, the saturation level and fatty acid chain length affect biodiesel stability, as confirmed by Issariyakul and Dalai (2012) when comparing the stability of biodiesel produced from palm and mustard oils in the presence of an alkaline catalysts, namely, potassium hydroxide.

4 Future Prospects in Biodiesel Production

4.1 Bio-based Feedstocks

As previously mentioned, oils should be refined before being used for biodiesel production, and many of the current used crops are used in human food and affect food security. To avoid such limitations, the direction nowadays is directed toward using microalgae lipids. Microalgae were specifically selected due to their remarkable characteristics, such as their high growth rate, oil content, and oil productivity (Adamczak et al. 2009; Chisti 2007; Spolaore et al. 2006; Vijayaraghavan and Hemanathan 2009; Sheehan et al. 1998). In addition, cultivation of microalgae cells does not require land neither freshwater, where several algae strains can grow in saltwater and wastewaters. It has been reported that by using microalgae, demanded land for biomass cultivation would decrease by more than 111 times (Singh and Sharma 2012). The oil contents of many microalgae strains were found to exceed 20% and in some cases might reach 50 and 80%. Moreover, growing microalgae cells would decrease CO_2 level at atmosphere, which have been revealed in several studies. The leftover, residue after lipid extrication, can be also utilized for other applications, such as animal feeding. Proteins can be also used for biomethane production while pigments in pharmaceutical applications.

Although lipid contents are reported by several algae strains, microalga biodiesel production cost is still not at commercial stage. This is mainly bottlenecked by the limited growth rate, not exceeding 8 g m^{-2} day^{-1}, and the need of enhancing lipid contents through environmental stress (Lu et al. 2011). The two main biomass cultivation systems are open ponds and photobioreactors. Open ponds are commonly used due to their low installation, operation, and maintenance costs, compared to photobioreactors. Poor light utilization, evaporation, and exposure to microbial contamination are common raised issues with the use of open ponds. The recent studies also focus on microalgae biomass production as biofilms, to simplify the harvesting process, where microalgae grow on walls of a rotary reactor to form a biofilm during their cultivation. In such systems, the biomass is more concentrated (Christenson and Sims 2012), reaching to a solid content of 10–20% in dry basis compared to 0.05% in suspension growth (Gross et al. 2013). This would result in a reduction in the costs of downstream harvesting and dewatering processing. In addition, the rate of nutrient uptake from cultivation media is commonly enhanced (Schnurr and Allen 2015). In biofilm systems, algal cells stick through hydrophobic interactions and grow on a surface of the material to produce the biomass that can be easily scraped off from the surface. Biomass production and biofilm generation mainly depend on strain selection, cultivation conditions, and characteristics of the material used as surface. Several works are considered using algae biofilms for wastewater treatment (Christenson and Sims 2012; Johnson and Wen 2010), where the rotating biofilm reactor was shown to provide high biomass productivity with efficient nutrient uptakes. In this system, the growth material (substratum) is attached to a rotating cylinder, which is partially submerged in the media and used to rotate the biofilm in and out of the wastewater, providing the sufficient nutrient while submerged in the medium and subjecting the biofilm to light and CO_2 when exposed to air phase, at which O_2 can also transfer out of the biofilm.

4.2 Bio-based Catalysts

Recently, the use of bio-based materials is becoming predominant to overcome the issues associated with chemical catalysts. Enzymes, which are biocatalyst, have been used for biodiesel production. They are available in abundant and can be produced at mild operating conditions, in range of 45–50 °C. Their uses in biodiesel production processes would require less energy, compared to chemical-based catalysts. In addition, enzymes, known as lipases, are insensitive to the FFA content of the oil, and their uses generate high-purity biodiesel. Among the several available enzymes, the non-specific lipases (EC 3.1.1.3) received a great attention for biodiesel production, due to their ability to act on different ester bonds and work at mild operating conditions with less energy, pre-treatments, and purification needs (Marchetti et al. 2007). Lipases from *Candida antarctica* (Nelson et al. 1996; Shimada et al. 1999; Watanabe et al. 2000, 2002; Samukawa et al. 2000; Fedosov et al. 2013; Taher et al. 2011, 2014; Al-Zuhair et al. 2012), *Pseudomonas fluorescens* (Guldhe et al. 2015; Devanesan et al. 2007), *Pseudomonas cepacia* (Noureddini et al. 2005), *Candida rugosa* (Moreno-Pirajàn and Giraldo 2011; Tan et al. 2014; Lee et al. 2011), and *Rhizomucor miehei* (Huang et al. 2012, 2014) are commonly tested.

Compared to chemical catalyst, the produced hydrophilic glycerol limits high yield biodiesel production. Generally, glycerol accumulation in the reactor increases reaction viscosity and deposit on immobilized lipase surfaces, preventing reacting substrates to reach enzyme active site (Dossat et al. 1999; Xu et al. 2011), and requires continuous removal of glycerol from the reaction mixture (Chen et al. 2011; Azócar et al. 2014). Methanol also negatively affects enzyme activity. Above certain alcohol concentration, commonly 1.5 molar equivalents, it becomes insoluble in oils and results in stripping-off the hydration layer of water from the lipase surface, which is required to keep the enzyme active (Du et al. 2004; Fjerbaek et al. 2009; Li et al. 2006; Zheng et al. 2009). Several works have proved lipase inhibition by methanol. It was also proposed to stepwise the addition of the alcohol to the reaction mixture (Shimada et al. 2002; Watanabe et al. 2000), and replace alcohols with acetates, as acceptors, (Du et al. 2004; Modi et al. 2006, 2007).

Although enzyme use is still promising, enzymatic biodiesel production processes are not yet commercialized because of enzyme high costs, reaching $1000 per 1 kg. Strategies have been done to improve lipase activity. In addition, cost reduction through immobilization has been considered. Through immobilization, the enzyme is confined or attached to a carrier while retaining its catalytic activity, where the stability of the lipase might be enhanced (Modi et al. 2007). Immobilization has been extensively studied using different carriers, including polymer resins, celite, silica, and ceramics. Nevertheless, limited number of carriers have been used at commercial scale. The first industrial immobilized enzyme was by Chibata and co-worker (Robinson 1997). The use of immobilized lipases has received a great interest. Several immobilization techniques and carriers have been investigated. However, there are several drawbacks, which include carriers cost and requirement of certain reactors.

For effective biodiesel production, the cost of biodiesel production should be reduced. This can be achieved by improving the immobilization techniques, optimizing the transesterification variables, designing and developing new bioreactors, and intensifying processes to one unit. Among the several immobilized lipases used in biodiesel production and available at commercial scale is Novozym®435 with a price of 1000/kg.

4.3 Green Solvents

Due to lipase sensitivity to hydrophilic alcohols and glycerol deposition, numerous organic solvents have been proposed to minimize their negative effects (Adamczak and Krishna 2004). By employment of organic solvents, the effect can be reduced. It has been reported that biodiesel production rates increase with the increase of solvent hydrophobicity (Samukawa et al. 2000; Klibanov 1997; Doukyu and Ogino 2010; Yang et al. 2004; Gorman and Dordick 1992). Nelson et al. (1996) tested the use of *n*-hexane in tallow fat transesterification with methanol catalyzed using *Mucor miehei* lipase. Biodiesel yield reached to 95% in *n*-hexane, compared to less than 20% in solvent-free system. The use of *tert*-butanol becomes dominant nowadays, as it can dissolve both methanol and glycerol, which are hydrophilic and has negative impacts on enyzyme activity (Yang et al. 2010; Demirbas 2009; Al-Zuhair et al. 2007; Peng et al. 2001; Lai et al. 2012; Royon et al. 2007). However, an additional downstream unit is essential when organic solvents are used, resulting to increase in overall biodiesel production cost. In addition, organic solvents are toxic, and their use might result to introduce some environmental issues.

Efforts have been made to use greener solvents. Thus, supercritical CO_2 and nonvolatile solvents have been suggested. Supercritical CO_2 (SC-CO_2) is a fluid at conditions above the critical points of CO_2. It assembles both liquid and gas properties, allowing it to be used in both oil extraction reaction media (Del Valle et al. 2004; Reverchon and Marrone 2001; Sovova et al. 2001). The concordance of SC-CO_2 with lipases is well known, and its employment for biodiesel production could enhance the mass transfer of reaction substrates into enzyme active sites.

Biodiesel yields exceeding 80% have been reported from microalgae lipids when transesterified with methanol in the presence of Novozym®435 and SC-CO_2 (Taher et al. 2014). Main obstacles of using SC-CO_2 are in the high pressure needed for pumping, resulting in process cost increase. The possibility of using SC-CO_2 for energy production is not evident, but combining oil extraction conversions to biodiesel in SC-CO_2 in one integrated system would be feasible, and the additional pumping cost for energy production could be verified (Taher et al. 2014; Al-Zuhair et al. 2012; Al-Zuhair and Taher 2016).

The use of ionic liquids (ILs) has been also proposed to replace volatile organic solvents. ILs are nonvolatile solvents composed of cations and anions. The first effort to use ILs in lipase-catalyzed transesterification was with [bmim][PF_6] and [bmim][BF_4] (Madeira Lau et al. 2000). By careful selection of the cation and anion groups, the properties of designed IL can be tuned (Ohno and Yoshizawa 2002; Endres and Zein El Abedin 2006), and increasing chain branching results in an increased melting point (Xue et al. 2016; Zhang et al. 2009). Similar to the role of organic solvents, hydrophobic ILs are preferable in lipase-catalyzed biodiesel process, where hydrophilic ILs may deactivate the lipase. [bmim][PF_6] is the most common IL used in biodiesel production. It has been used in sunflower oil transesterification in the presence of Novozyme®435, and high yield reaching 98% was achieved. Overall, the main issue of using ILs at industrial scale is in their high costs associated from their synthesis. The price of the IL is almost 10 times higher than that of *n*-hexane.

5 Commercialization of Lipase-Biodiesel Systems

Although lipase-catalyzed biodiesel production looks promising, most of the reported findings are carried at laboratory scale (Trani et al. 1991). For process to be scaled up, it is very important to determine the effect of key process factors on large-scale system performance, wherein scaling strategies and design models need to be identified prior to commercialization. As mentioned earlier, reaction substrate characteristics and quality determine the type of catalyst to be used, while catalyst efficiency and characteristics determine the type and configuration of the reactor. Compared to soluble enzymes, immobilized lipase acts in different phases from the reaction mixture and thereby reduces the energy and cost needed to recover the enzyme at the end of reaction. Typically, lipase-catalyzed transesterification reaction for biodiesel production involves two steps, where the immobilized lipase first hydrolyzes fatty acids and then esterifies the hydrolyzed fatty acids with selected alcohol. The mechanism follows the ping-pong bi bi kinetic model, where produced glycerol and biodiesel are released between the addition of reaction substrates. Continuous types of reactors are usually used as they are easy to scale up, can enhance the reaction rate, and reduce energy input and the molar ratio of methanol to oil. This is mainly by intensification processes and in situ product separation (Qiu et al. 2010). Stirred tank reactor and packed bed reactor are the most commonly used reactors. To avoid inactivation of the enzyme in impeller-based reactors, stepwise addition of alcohol to reaction mixture is usually suggested.

Scale-up reactors involve increasing the production amount of the desired product in a very complex aspect, which depends on the interaction of mass transfer, kinetics, and hydrodynamics. The most common engineering approach used to scale such processes is the dimensionless analysis, where a set of dimensionless numbers are fixed to obtain the same behavior in the different scales. Among the several similarities, geometric similarity is used to fix the dimensional ratios between the two scale reactors. Kinematic similarity requires that Re number on the two different scales has the Reynold value to be fixed. Moreover, in such heterogeneous systems, mass transfer resistance is common to increase with scale, and similarity in mass transfer coefficient should be always maintained.

6 Conclusions

Limited sources of fossil fuels and their negative environmental impacts direct the attention to look for green alternatives. Biodiesel from oil-rich feedstocks is among the several potential fuels that can be used in diesel engines. First-generation feedstocks, however, have several drawbacks that compete with food security, whereas the second generation cannot be continuously supplied. On the other hands, algae-based fuels are on way of commercialization. Although alkali catalysts are commercially used for biodiesel production, they are very sensitive to oil quality, and acid catalysts are corrosive. Therefore, the use of immobilized enzymes becomes predominant. The technology is not yet commercialized due to the limitation raised by reaction mixture hydrophilicity that strips off the essential water needed to keep the enzyme active.

References

Adamczak M, Krishna SH (2004) Strategies for improving enzymes for efficient biocatalysis. Food Technol Biotechnol 42(4):251–264

Adamczak M, Bornscheuer UT, Bednarski W (2009) The application of biotechnological methods for the synthesis of biodiesel. Eur J Lipid Sci Technol 111:800–813

Akoh CC, Chang S-W, Lee G-C, Shaw J-F (2007) Enzymatic approach to biodiesel production. J Agric Food Chem 55(22):8995–9005

Al-Zuhair S (2007) Production of biodiesel: possibilities and challenges. Biofuels Bioprod Biorefin 1(1):57–66. https://doi.org/10.1002/bbb.2

Al-Zuhair S, Taher H (2016) Supercritical fluids technology in lipase catalyzed processes. CRC Press, Boca Raton

Al-Zuhair S, Ling FW, Jun LS (2007) Proposed kinetic mechanism of the production of biodiesel from palm oil using lipase. Process Biochem 42(6):951–960. https://doi.org/10.1016/j.procbio.2007.03.002

Al-Zuhair S, Hussein A, Al-Marzouqi AH, Hashim I (2012) Continuous production of biodiesel from fat extracted from lamb meat in supercritical CO2 media. Biochem Eng J 60:106–110. https://doi.org/10.1016/j.bej.2011.10.010
Antunes WM, Veloso CO, Henriques CA (2008) Transesterification of soybean oil with methanol catalyzed by basic solids. Catal Today 133–135:548–554. https://doi.org/10.1016/j.cattod.2007.12.055
Aresta M, Dibenedetto A, Carone M, Colonna T, Fragale C (2005) Production of biodiesel from macroalgae by supercritical CO2 extraction and thermochemical liquefaction. Environ Chem Lett 3(3):136–139. https://doi.org/10.1007/s10311-005-0020-3
Armenta RE, Vinatoru M, Burja AM, Kralovec JA, Barrow CJ (2007) Transesterification of fish oil to produce fatty acid ethyl esters using ultrasonic energy. J Am Oil Chem Soc 84(11):1045–1052. https://doi.org/10.1007/s11746-007-1129-2
Atadashi IM, Aroua MK, Abdul Aziz AR, Sulaiman NMN (2012) The effects of water on biodiesel production and refining technologies: a review. Renew Sust Energ Rev 16(5):3456–3470. https://doi.org/10.1016/j.rser.2012.03.004
Azócar L, Navia R, Beroiz L, Jeison D, Ciudad G (2014) Enzymatic biodiesel production kinetics using co-solvent and an anhydrous medium: a strategy to improve lipase performance in a semi-continuous reactor. New Biotechnol 31(5):422–429. https://doi.org/10.1016/j.nbt.2014.04.006
Basha SA, Gopal KR, Jebaraj S (2009) A review on biodiesel production, combustion, emissions and performance. Renew Sust Energ Rev 13(6):1628–1634. https://doi.org/10.1016/j.rser.2008.09.031
Becker W (2007) Microalgae in human and animal nutrition. In: Handbook of microalgal culture. Blackwell Publishing Ltd, Oxford, pp 312–351. https://doi.org/10.1002/9780470995280.ch18
Chen H-C, Ju H-Y, Wu T-T, Liu Y-C, Lee C-C, Chang C, Chung Y-L, Shieh C-J (2011) Continuous production of lipase-catalyzed biodiesel in a packed-bed reactor: optimization and enzyme reuse study. J Biomed Biotechnol 2011:6. https://doi.org/10.1155/2011/950725
Chisti Y (2007) Biodiesel from microalgae. Biotechnol Adv 25(3):294–306. https://doi.org/10.1016/j.biotechadv.2007.02.001
Christenson LB, Sims RC (2012) Rotating algal biofilm reactor and spool harvester for wastewater treatment with biofuels by-products. Biotechnol Bioeng 109(7):1674–1684. https://doi.org/10.1002/bit.24451
Del Valle JM, Rivera O, Mattea M, Ruetsch L, Daghero J, Flores A (2004) Supercritical CO2 processing of pretreated rosehip seeds: effect of process scale on oil extraction kinetics. J Supercrit Fluids 31(2):159–174. https://doi.org/10.1016/j.supflu.2003.11.005
Demirbaş A (2003) Biodiesel fuels from vegetable oils via catalytic and non-catalytic supercritical alcohol transesterifications and other methods: a survey. Energy Convers Manag 44(13):2093–2109. https://doi.org/10.1016/S0196-8904(02)00234-0
Demirbas A (2007) Importance of biodiesel as transportation fuel. Energy Policy 35(9):4661–4670. https://doi.org/10.1016/j.enpol.2007.04.003
Demirbas A (2009) Progress and recent trends in biodiesel fuels. Energy Convers Manag 50:14–34
Demirel G, Ozpinar H, Nazli B, Keser O (2006) Fatty acids of lamb meat from two breeds fed different forage: concentrate ratio. Meat Sci 72(2):229–235. https://doi.org/10.1016/j.meatsci.2005.07.006
Devanesan MG, Viruthagiri T, Sugumar N (2007) Transesterification of Jatropha oil using immobilized Pseudomonas fluorescens. Afr J Biotechnol 6(21):2497–2501
Dossat V, Combes D, Marty A (1999) Continuous enzymatic transesterification of high oleic sunflower oil in a packed bed reactor: influence of the glycerol production. Enzym Microb Technol 25(3–5):194–200. https://doi.org/10.1016/S0141-0229(99)00026-5
Doukyu N, Ogino H (2010) Organic solvent-tolerant enzymes. Biochem Eng J 48(3):270–282. https://doi.org/10.1016/j.bej.2009.09.009
Du W, Xu Y, Liu D, Zeng J (2004) Comparative study on lipase-catalyzed transformation of soybean oil for biodiesel production with different acyl acceptors. J Mol Catal B Enzym 30(3–4):125–129

Dubé MA, Tremblay AY, Liu J (2007) Biodiesel production using a membrane reactor. Bioresour Technol 98(3):639–647. https://doi.org/10.1016/j.biortech.2006.02.019

Endres F, Zein El Abedin S (2006) Air and water stable ionic liquids in physical chemistry. Phys Chem Chem Phys 8(18):2101–2116. https://doi.org/10.1039/b600519p

Eshton B, Katima JHY, Kituyi E (2013) Greenhouse gas emissions and energy balances of Jatropha biodiesel as an alternative fuel in Tanzania. Biomass Bioenergy 58:95–103. https://doi.org/10.1016/j.biombioe.2013.08.020

Fan X, Burton R (2009) Recent development of biodiesel feedstocks and the applications of glycerol: a review. Open Fuels Energy Sci J 1:100–109

Farahani M, Pagé DJYS, Turingia MP (2011) Sedimentation in biodiesel and Ultra Low Sulfur Diesel Fuel blends. Fuel 90(3):951–957. https://doi.org/10.1016/j.fuel.2010.10.046

Fedosov SN, Brask J, Pedersen AK, Nordblad M, Woodley JM, Xu X (2013) Kinetic model of biodiesel production using immobilized lipase Candida antarctica lipase B. J Mol Catal B Enzym 85–86:156–168. https://doi.org/10.1016/j.molcatb.2012.09.011

Fjerbaek L, Christensen KV, Norddahl B (2009) A review of the current state of biodiesel production using enzymatic transesterification. Biotechnol Bioeng 102(5):1298–1315

Freedman B, Pryde EH, Mounts TL (1984) Variables affecting the yields of fatty esters from transesterified vegetable oils. J Am Oil Chem Soc 61(10):1638–1643. https://doi.org/10.1007/bf02541649

Fukuda H, Kondo A, Noda H (2001) Biodiesel fuel production by transesterification of oils. J Biosci Bioeng 92(5):405–416

González LE, Díaz GC, Aranda DAG, Cruz YR, Fortes MM (2015) Biodiesel production based in microalgae: a biorefinery approach. Nat Sci 7:358–369

Gorman LAS, Dordick JS (1992) Organic solvents strip water off enzymes. Biotechnol Bioeng 39(4):392–397. https://doi.org/10.1002/bit.260390405

Gross M, Henry W, Michael C, Wen Z (2013) Development of a rotating algal biofilm growth system for attached microalgae growth with in situ biomass harvest. Bioresour Technol 150:195–201. https://doi.org/10.1016/j.biortech.2013.10.016

Guldhe A, Singh B, Rawat I, Permaul K, Bux F (2015) Biocatalytic conversion of lipids from microalgae Scenedesmus obliquus to biodiesel using Pseudomonas fluorescens lipase. Fuel 147:117–124. https://doi.org/10.1016/j.fuel.2015.01.049

Helwani Z, Othman MR, Aziz N, Fernando WJN, Kim J (2009) Technologies for production of biodiesel focusing on green catalytic techniques: a review. Fuel Process Technol 90(12):1502–1514

Huang D, Han S, Han Z, Lin Y (2012) Biodiesel production catalyzed by Rhizomucor miehei lipase-displaying Pichia pastoris whole cells in an isooctane system. Biochem Eng J 63:10–14. https://doi.org/10.1016/j.bej.2010.08.009

Huang J, Xia J, Yang Z, Guan F, Cui D, Guan G, Jiang W, Li Y (2014) Improved production of a recombinant Rhizomucor miehei lipase expressed in Pichia pastoris and its application for conversion of microalgae oil to biodiesel. Biotechnol Biofuels 7(1):1–11. https://doi.org/10.1186/1754-6834-7-111

Issariyakul T, Dalai AK (2012) Comparative kinetics of transesterification for biodiesel production from palm oil and mustard oil. Can J Chem Eng 90(2):342–350. https://doi.org/10.1002/cjce.20679

Jain S, Sharma MP (2010) Kinetics of acid base catalyzed transesterification of Jatropha curcas oil. Bioresour Technol 101(20):7701–7706. https://doi.org/10.1016/j.biortech.2010.05.034

Ji J, Wang J, Li Y, Yu Y, Xu Z (2006) Preparation of biodiesel with the help of ultrasonic and hydrodynamic cavitation. Ultrasonics 44(Suppl 1):e411–e414

Johnson MB, Wen Z (2010) Development of an attached microalgal growth system for biofuel production. Appl Microbiol Biotechnol 85(3):525–534. https://doi.org/10.1007/s00253-009-2133-2

Kamal-Eldin A, Andersson R (1997) A multivariate study of the correlation between tocopherol content and fatty acid composition in vegetable oils. J Am Oil Chem Soc 74(4):375–380. https://doi.org/10.1007/s11746-997-0093-1

Khoobbakht G, Najafi G, Karimi M, Akram A (2016) Optimization of operating factors and blended levels of diesel, biodiesel and ethanol fuels to minimize exhaust emissions of diesel engine using response surface methodology. Appl Therm Eng 99:1006–1017. https://doi.org/10.1016/j.applthermaleng.2015.12.143
Klibanov AM (1997) Why are enzymes less active in organic solvents than in water? Trends Biotechnol 15(3):97–101. https://doi.org/10.1016/S0167-7799(97)01013-5
Knothe G (2005) Dependence of biodiesel fuel properties on the structure of fatty acid alkyl esters. Fuel Process Technol 86(10):1059–1070. https://doi.org/10.1016/j.fuproc.2004.11.002
Kucek KT, César-Oliveira MAF, Wilhelm HM, Ramos LP (2007) Ethanolysis of refined soybean oil assisted by sodium and potassium hydroxides. J Am Oil Chem Soc 84(4):385–392. https://doi.org/10.1007/s11746-007-1048-2
Lai C-C, Zullaikah S, Vali SR, Ju Y-H (2005) Lipase-catalyzed production of biodiesel from rice bran oil. J Chem Technol Biotechnol 80(3):331–337. https://doi.org/10.1002/jctb.1208
Lai J-Q, Hu Z-L, Wang P-W, Yang Z (2012) Enzymatic production of microalgal biodiesel in ionic liquid [BMIm][PF6]. Fuel 95:329–333. https://doi.org/10.1016/j.fuel.2011.11.001
Lee JH, Kim SB, Kang SW, Song YS, Park C, Han SO, Kim SW (2011) Biodiesel production by a mixture of Candida rugosa and Rhizopus oryzae lipases using a supercritical carbon dioxide process. Bioresour Technol 102(2):2105–2108. https://doi.org/10.1016/j.biortech.2010.08.034
Leung DYC, Guo Y (2006) Transesterification of neat and used frying oil: optimization for biodiesel production. Fuel Process Technol 87(10):883–890
Li L, Du W, Liu D, Wang L, Li Z (2006) Lipase-catalyzed transesterification of rapeseed oils for biodiesel production with a novel organic solvent as the reaction medium. J Mol Catal B Enzym 43(1–4):58–62. https://doi.org/10.1016/j.molcatb.2006.06.012
Liu J, Huang J, Sun Z, Zhong Y, Jiang Y, Chen F (2011) Differential lipid and fatty acid profiles of photoautotrophic and heterotrophic Chlorella zofingiensis: assessment of algal oils for biodiesel production. Bioresour Technol 102(1):106–110. https://doi.org/10.1016/j.biortech.2010.06.017
Lu J, Sheahan C, Fu P (2011) Metabolic engineering of algae for fourth generation biofuels production. Energy Environ Sci 4(7):2451–2466. https://doi.org/10.1039/C0EE00593B
Ma F, Clements LD, Ha MA (1998) Biodiesel fuel from animal fat. Ind Eng Chem Res 37(9):3768–3771
Madeira Lau R, Van Rantwijk F, Seddon KR, Sheldon RA (2000) Lipase-catalyzed reactions in ionic liquids. Org Lett 2(26):4189–4191. https://doi.org/10.1021/ol006732d
Madras G, Kolluru C, Kumar R (2004) Synthesis of biodiesel in supercritical fluids. Fuel 83(14–15):2029–2033. https://doi.org/10.1016/j.fuel.2004.03.014
Marchetti JM (2012) A summary of the available technologies for biodiesel production based on a comparison of different feedstock's properties. Process Saf Environ Prot 90(3):157–163. https://doi.org/10.1016/j.psep.2011.06.010
Marchetti JM, Miguel VU, Errazu AF (2007) Possible methods for biodiesel production. Renew Sust Energ Rev 11:1300–1311
Marjanović AV, Stamenković OS, Todorović ZB, Lazić ML, Veljković VB (2010) Kinetics of the base-catalyzed sunflower oil ethanolysis. Fuel 89(3):665–671. https://doi.org/10.1016/j.fuel.2009.09.025
Meher LC, Dharmagadda VSS, Naik SN (2006) Optimization of alkali-catalyzed transesterification of Pongamia pinnata oil for production of biodiesel. Bioresour Technol 97(12):1392–1397
Modi MK, Reddy JRC, Rao BVSK, Prasad RBN (2006) Lipase-mediated transformation of vegetable oils into biodiesel using Propan-2-ol as acyl acceptor. Biotechnol Lett 28(9):637–640
Modi MK, Reddy JRC, Rao BVSK, Prasad RBN (2007) Lipase-mediated conversion of vegetable oils into biodiesel using ethyl acetate as acyl acceptor. Bioresour Technol 98(6):1260–1264
Moreno-Pirajàn JC, Giraldo L (2011) Study of immobilized candida rugosa lipase for biodiesel fuel production from palm oil by flow microcalorimetry. Arab J Chem 4(1):55–62. https://doi.org/10.1016/j.arabjc.2010.06.019

Murugesan A, Umarani C, Subramanian R, Nedunchezhian N (2009) Bio-diesel as an alternative fuel for diesel engines—a review. Renew Sust Energ Rev 13(3):653–662. https://doi.org/10.1016/j.rser.2007.10.007

Nelson LA, Foglia TA, Marmer WN (1996) Lipase-catalyzed production of biodiesel. J Am Oil Chem Soc 73(9):1191–1195

Noureddini H, Zhu D (1997) Kinetics of transesterification of soybean oil. J Am Oil Chem Soc 74(11):1457–1463. https://doi.org/10.1007/s11746-997-0254-2

Noureddini H, Gao X, Philkana RS (2005) Immobilized Pseudomonas cepacia lipase for biodiesel fuel production from soybean oil. Bioresour Technol 96(7):769–777. https://doi.org/10.1016/j.biortech.2004.05.029

Ohno H, Yoshizawa M (2002) Ion conductive characteristics of ionic liquids prepared by neutralization of alkylimidazoles. Solid State Ionics 154–155:303–309. https://doi.org/10.1016/S0167-2738(02)00526-X

Peng W, Wu Q, Tu P (2001) Pyrolytic characteristics of heterotrophic *Chlorella protothecoides* for renewable bio-fuel production. J Appl Phycol 13(1):5–12. https://doi.org/10.1023/a:1008153831875

Phan AN, Phan TM (2008) Biodiesel production from waste cooking oils. Fuel 87(17–18):3490–3496

Predojević ZJ (2008) The production of biodiesel from waste frying oils: a comparison of different purification steps. Fuel 87(17–18):3522–3528. https://doi.org/10.1016/j.fuel.2008.07.003

Qiu Z, Zhao L, Weatherley L (2010) Process intensification technologies in continuous biodiesel production. Chem Eng Process Process Intensif 49(4):323–330. https://doi.org/10.1016/j.cep.2010.03.005

Rahman FA, Aziz MMA, Saidur R, Bakar WAWA, Hainin MR, Putrajaya R, Hassan NA (2017) Pollution to solution: capture and sequestration of carbon dioxide (CO2) and its utilization as a renewable energy source for a sustainable future. Renew Sust Energ Rev 71:112–126. https://doi.org/10.1016/j.rser.2017.01.011

Ramos MJ, Fernández CM, Casas A, Rodríguez L, Pérez Á (2009) Influence of fatty acid composition of raw materials on biodiesel properties. Bioresour Technol 100(1):261–268. https://doi.org/10.1016/j.biortech.2008.06.039

Rashid U, Anwar F, Moser BR, Ashraf S (2008) Production of sunflower oil methyl esters by optimized alkali-catalyzed methanolysis. Biomass Bioenergy 32(12):1202–1205

Refaat AA, Attia NK, Sibak HA, Sheltawy ST, ElDiwani GI (2008) Production optimization and quality assessment of biodiesel from waste vegetable oil. Int J Environ Sci Technol 5(1):75–82. https://doi.org/10.1007/bf03325999

Reverchon E, Marrone C (2001) Modeling and simulation of the supercritical CO_2 extraction of vegetable oils. J Supercrit Fluids 19(2):161–175. https://doi.org/10.1016/S0896-8446(00)00093-0

Robinson GK (1997) Methods in biotechnology—immobilization of enzymes and cells; Edited by G F Bickerstaff. pp 367. Humana Press, New Jersey. 1996 ISBN 0-89603-386-4. Biochem Educ 25(4):232–232. https://doi.org/10.1016/S0307-4412(97)87538-3

Royon D, Daz M, Ellenrieder G, Locatelli S (2007) Enzymatic production of biodiesel from cotton seed oil using t-butanol as a solvent. Bioresour Technol 98(3):648–653. https://doi.org/10.1016/j.biortech.2006.02.021

Samukawa T, Kaieda M, Matsumoto T, Ban K, Kondo A, Shimada Y, Noda H, Fukuda H (2000) Pretreatment of immobilized Candida antarctica lipase for biodiesel fuel production from plant oil. J Biosci Bioeng 90(2):180–183. https://doi.org/10.1016/S1389-1723(00)80107-3

Schnurr PJ, Allen DG (2015) Factors affecting algae biofilm growth and lipid production: a review. Renew Sust Energ Rev 52:418–429. https://doi.org/10.1016/j.rser.2015.07.090

Schröder O, Bünger J, Munack A, Knothe G, Krahl J (2013) Exhaust emissions and mutagenic effects of diesel fuel, biodiesel and biodiesel blends. Fuel 103:414–420. https://doi.org/10.1016/j.fuel.2012.08.050

Sheehan J, Dunahay T, Benemann R, Roessler G, Weissman C (1998) A look back at the U.S. Department of Energy-Aquatic Species Program: biodiesel from algae

Shimada Y, Watanabe Y, Samukawa T, Sugihara A, Noda H, Fukuda H, Tominaga Y (1999) Conversion of vegetable oil to biodiesel using immobilized Candida antarctica lipase. J Am Oil Chem Soc 76(7):789–793. https://doi.org/10.1007/s11746-999-0067-6

Shimada Y, Watanabe Y, Sugihara A, Tominaga Y (2002) Enzymatic alcoholysis for biodiesel fuel production and application of the reaction to oil processing. J Mol Catal B Enzym 17(3–5):133–142

Singh RN, Sharma S (2012) Development of suitable photobioreactor for algae production—a review. Renew Sust Energ Rev 16(4):2347–2353. https://doi.org/10.1016/j.rser.2012.01.026

Sivasamy A, Cheah KY, Fornasiero P, Kemausuor F, Zinoviev S, Miertus S (2009) Catalytic applications in the production of biodiesel from. ChemSusChem 2:278–300

Sovova H, Zarevucka M, Vacek M, Stransky K (2001) Solubility of two vegetable oils in supercritical CO_2. J Supercrit Fluids 20(1):15–28. https://doi.org/10.1016/S0896-8446(01)00057-2

Spolaore P, Joannis-Cassan C, Duran E, Isambert A (2006) Commercial applications of microalgae. J Biosci Bioeng 101(2):87–96

Taher H, Al-Zuhair S, AlMarzouqui A, Hashim I (2011) Extracted fat from lamb meat by supercritical CO2 as feedstock for biodiesel production. Biochem Eng J 55(1):23–31. https://doi.org/10.1016/j.bej.2011.03.003

Taher H, Al-Zuhair S, Al-Marzouqi AH, Haik Y, Farid M (2014) Enzymatic biodiesel production of microalgae lipids under supercritical carbon dioxide: process optimization and integration. Biochem Eng J 90:103–113. https://doi.org/10.1016/j.bej.2014.05.019

Tan Z, Fang M, Du H, Song L, Ren D, Tang X, Han X, Liang X (2014) Production of biodiesel catalyzed by Candida rugosa lipase at interface of w/o microemulsion system. J Braz Chem Soc 25:1704–1711

Taufiqurrahmi N, Mohamed AR, Bhatia S (2011) Production of biofuel from waste cooking palm oil using nanocrystalline zeolite as catalyst: process optimization studies. Bioresour Technol 102(22):10686–10694. https://doi.org/10.1016/j.biortech.2011.08.068

Trani M, Ergan F, André G (1991) Lipase-catalyzed production of wax esters. J Am Oil Chem Soc 68(1):20–22. https://doi.org/10.1007/bf02660302

Verma P, Sharma MP, Dwivedi G (2016) Evaluation and enhancement of cold flow properties of palm oil and its biodiesel. Energy Rep 2:8–13. https://doi.org/10.1016/j.egyr.2015.12.001

Vicente G, Martínez M, Aracil J (2004) Integrated biodiesel production: a comparison of different homogeneous catalysts systems. Bioresour Technol 92(3):297–305

Vijayaraghavan K, Hemanathan K (2009) Biodiesel production from freshwater algae. Energy Fuel 23(11):5448–5453. https://doi.org/10.1021/ef9006033

Watanabe Y, Shimada Y, Sugihara A, Noda H, Fukuda H, Tominaga Y (2000) Continuous production of biodiesel fuel from vegetable oil using immobilized Candida antarctica lipase. J Am Oil Chem Soc 77(4):355–360. https://doi.org/10.1007/s11746-000-0058-9

Watanabe Y, Shimada Y, Sugihara A, Tominaga Y (2002) Conversion of degummed soybean oil to biodiesel fuel with immobilized Candida antarctica lipase. J Mol Catal B Enzym 17(3–5):151–155. https://doi.org/10.1016/S1381-1177(02)00022-X

Wennersten R, Sun Q, Li H (2015) The future potential for Carbon Capture and Storage in climate change mitigation—an overview from perspectives of technology, economy and risk. J Clean Prod 103:724–736. https://doi.org/10.1016/j.jclepro.2014.09.023

Williams PJB, Laurens LML (2010) Microalgae as biodiesel & biomass feedstocks: review & analysis of the biochemistry, energetics & economics. Energy Environ Sci 3(5):554–590. https://doi.org/10.1039/B924978H

Xu Y, Nordblad M, Nielsen PM, Brask J, Woodley JM (2011) In situ visualization and effect of glycerol in lipase-catalyzed ethanolysis of rapeseed oil. J Mol Catal B Enzym 72(3–4):213–219. https://doi.org/10.1016/j.molcatb.2011.06.008

Xue L, Gurung E, Tamas G, Koh YP, Shadeck M, Simon SL, Maroncelli M, Quitevis EL (2016) Effect of alkyl chain branching on physicochemical properties of imidazolium-based ionic liquids. J Chem Eng Data 61(3):1078–1091. https://doi.org/10.1021/acs.jced.5b00658

Yang L, Dordick JS, Garde S (2004) Hydration of enzyme in nonaqueous media is consistent with solvent dependence of its activity. Biophys J 87(2):812–821. https://doi.org/10.1529/biophysj.104.041269

Yang Z, Zhang K-P, Huang Y, Wang Z (2010) Both hydrolytic and transesterification activities of Penicillium expansum lipase are significantly enhanced in ionic liquid [BMIm][PF6]. J Mol Catal B Enzym 63(1–2):23–30. https://doi.org/10.1016/j.molcatb.2009.11.014

Zhang Y, Dube MA, McLean DD, Kates M (2003) Biodiesel production from waste cooking oil: 1. Process design and technological assessment. Bioresour Technol 89(1):1–16. https://doi.org/10.1016/S0960-8524(03)00040-3

Zhang S, Lu X, Zhou Q, Li X, Zhang X, Li S (2009) Ionic liquids: physicochemical properties. Elsevier Science, Amsterdam

Zheng S, Kates M, Dubé MA, McLean DD (2006) Acid-catalyzed production of biodiesel from waste frying oil. Biomass Bioenergy 30(3):267–272. https://doi.org/10.1016/j.biombioe.2005.10.004

Zheng Y, Quan J, Ning X, Zhu L-M, Jiang B, He Z-Y (2009) Lipase-catalyzed transesterification of soybean oil for biodiesel production in tert-amyl alcohol. World J Microbiol Biotechnol 25(1):41–46. https://doi.org/10.1007/s11274-008-9858-4

Zullaikah S, Lai C-C, Vali SR, Ju Y-H (2005) A two-step acid-catalyzed process for the production of biodiesel from rice bran oil. Bioresour Technol 96(17):1889–1896. https://doi.org/10.1016/j.biortech.2005.01.028

Applications of Ionic Liquids and Deep Eutectic Solvents in Biorefinery-Biodiesel Production

I. Wazeer, M. K. Hadj-Kali, and I. M. AlNashef

Abbreviations

[BMIM][BF_4]	1 butyl-3-methyl imidazolium tetrafluoroborate
[BMIM][$InCl_4$]	1-*n*-butyl-3-methylimidazolium tetrachloro-indate
[BMIM][$N(CN)_2$]	1-butyl-3-methyl imidazolium dicyanamide
[BMIM][PF_6]	1-butyl-3-methyl imidazolium hexafluorophosphate
[BMIM][Tf_2N]	1 butyl-3-methyl imidazolium bis(trifluoromethyl sulfonyl)imide
[BMIM]Im	1-butyl-3-methylimidazolium imidazolide
[BMIM]OH	1-butyl-3-methylimidazolium hydroxide
[BMMIM][Tf_2N]	1-butyl-2,3-dimethylimidazolium bis(trifluoromethyl sulfonyl)imide
[BSPy][CF_3SO_3]	1-(4-sulfonic acid) butylpyridinium trifluoromethane sulfonate
[BSPy][HSO_4]	1-(4-sulfonic acid) butylpyridinium hydrogen sulfate
[C_{16}MIM][Tf_2N]	1-hexadecyl-3-methylimidazolium bis(trifluoromethyl sulfonyl)imide
[C_{18}MIM][Tf_2N]	1-octadecyl-3-methylimidazolium bis(trifluoromethyl sulfonyl)imide
[EMIM][OAc]	1-ethyl-3-methylimidazolium acetate
[EMIM][PF_6]	1-ethyl-3-methyl imidazolium hexafluorophosphate
[EMIM][TfO]	1-ethyl-3-methylimidazolium trifluoromethanesulfonate
[EMIM]BF_4	1-ethyl-3-methyl imidazolium tetrafluoroborate
[EMIM]Cl	1-ethyl-3-methyl imidazolium chloride
[EMIM]DEP	1-ethyl-3-methyl imidazolium diethylphosphate

I. Wazeer · M. K. Hadj-Kali
Chemical Engineering Department, King Saud University, Riyadh, Saudi Arabia

I. M. AlNashef (✉)
Department of Chemical Engineering, Khalifa University of Science and Technology, Masdar City Campus, Abu Dhabi, United Arab Emirates
e-mail: enashf@masdar.ac.ae

© Springer Nature Switzerland AG 2019

J.-R. Bastidas-Oyanedel, J. E. Schmidt (eds.), *Biorefinery*,
https://doi.org/10.1007/978-3-030-10961-5_8

[EMIM]EtOSO$_3$	1-ethyl-3-methyl imidazolium ethyl sulfate
[EMIM]SCN	1-ethyl-3-methyl imidazolium thiocyanate
[HMIM][BF$_4$]	1-hexyl-3-methyl imidazolium tetrafluoroborate
[HMIM][N(CN)$_2$]	1-hexyl-3-methyl imidazolium dicyanamide
[HMIM][Tf$_2$N]	1-hexyl-3-methyl imidazolium bis(trifluoromethyl sulfonyl)imide
[HMMIM][Tf$_2$N]	1-hexyl-2,3-dimethylimidazolium bis(trifluoromethyl sulfonyl)imide
[NMP][CH$_3$SO$_3$]	*N*-methyl-2-pyrrolidonium methyl sulfonate
[OMIM][PF$_6$]	1-methyl-3-octylimidazolium hexafluorophosphate
[OMIM][Tf$_2$N]	1-methyl-3-octylimidazolium bis(trifluoromethyl sulfonyl)imide
[OMMIM][Tf$_2$N]	1-octyl-2,3-dimethylimidazolium bis(trifluoromethyl sulfonyl)imide
[OMPY][BF$_4$]	1-octyl-3-methylpyridinium tetrafluoroborate
[SBP][HSO$_4$]	(2-(4-sulfobutyl) pyrazolium hydrogensulfate)
[SMIM][HSO$_4$]	1-(4-sulfonic acid) butyl-3-methylimidazolium hydrogen sulfate
[SO$_3$H-(CH$_2$)$_3$-HIM][HSO$_4$]	1-(propyl-3-sulfonate) imidazolium hydrogen sulfate
[SPyr][HSO$_4$]	1-(4-sulfonic acid) butylpyridinium hydrogen sulfate
ACPO	Acidic crude palm oil
BAO	Bitter apple oil
ChAc	Choline acetate
ChCl	Choline chloride
DBT	Dibenzothiophene
DESs	Deep eutectic solvents
DMC	Dimethyl carbonate
FAME	Fatty acid methyl esters
Gly	Glycerol
HBA	Hydrogen bond acceptor
HBD	Hydrogen bond donor
HPyrBr	1-hexylpyridinium bromide
ILs	Ionic liquids
IMC$_2$OH	Bis-(3-methyl-1-imidazole)-ethylene dihydroxide
IMC$_3$OH	Bis-(3-methyl-1-imidazole)-propylene dihydroxide
IMC$_4$OH	Bis-(3-methyl-1-imidazole)-butylene dihydroxide
IMC$_5$OH	Bis-(3-methyl-1-imidazole)-pentylene dihydroxide
IMC$_6$OH	Bis-(3-methyl-1-imidazole)-hexylene dihydroxide
KOH	Potassium hydroxide
LGCPO	Low-grade crude palm oil
MIL	Magnetic ILs
MTBE	Methyl tertiary butyl ether
MTPPB	Methyl triphenyl phosphonium bromide
MW	Microwaves
PTSA	*p*-toluenesulfonic acid
scCO$_2$	Supercritical carbon dioxide
TFA	Total fatty acid

1 Introduction

The rapidly growing global energy demand, the exploitation of renewable energy sources, environmental pollution and global warming due to the emission of greenhouse gases, and the diminution of fossil fuels are the crucial factors that have fuelled both researchers and governments to explore alternative sources of renewable energy (Shahbaz et al. 2011a). Some prominent alternative renewable energy sources that are capable of replacing fossil fuels include water, wind energy, solar energy, and biofuels (Atadashi et al. 2011). At present, around 100% of the energy required for the transportation and 86% of the world energy consumption are delivered by fossil fuels (Dorian et al. 2006). Currently, the only alternative renewable energy source able to substitute transport fuel in vehicles without concerning significant modifications to vehicle engines is biofuel (Kaygusuz 2009). Biofuels have attained a great deal of interest and are the focus of current energy research. Due to the large and unexpected of fluctuations of prices of crude oil and its derivatives, the production of biofuel is becoming exceedingly attractive with the aim of elevating decarburization of fuel for transportation, boosting fuel supply sources and depleting harmful gaseous emission. Currently, many countries including Austria, Australia, Italy, Germany, and the Unites States are already producing biofuels in the forms of biodiesel and bioethanol (Atadashi et al. 2011).

Biochemical (e.g., sugar fermentation) and thermochemical processes (e.g., pyrolysis and gasification) have been used to process biofuel resources (Hoekman 2009). These routes yield bioethanol, syngas, and other fuel types. Cellulosic biomass and lipid, microbes and algae, and soybeans are the most promising biofuel feedstocks.

Biodiesel, an alternative to diesel fuel, is synthesized using four different routes: (1) transesterification, (2) thermal cracking, (3) microemulsions, and (4) direct use and blending of raw oils (Shahbaz et al. 2011a). Among these approaches, the most common route is transesterification for producing biodiesel. In transesterification, vegetable oils and animal fats react with primary aliphatic alcohol (ethanol or methanol) in the existence of a catalyst to synthesize fatty acid alkyl esters and glycerol (Gly) as a by-product. As one of the alternative renewable energy sources, biodiesel is becoming a promising renewable, biodegradable, and potentially CO_2-neutral fuel for heating systems and diesel engines. The fuel economy of biodiesel is comparable to petroleum-based diesel, and it can also minimize the releases of contaminating substances, such as carbon monoxide, particulate matter, and hydrocarbon (Zhao and Baker 2013).

The production of biodiesel through conventional chemical methods has various limitations such as emulsification and corrosion, energy-intensive procedures, and high waste treatment. Enzyme-catalyzed transesterification is a greener alternative to chemical transesterification because of mild operating conditions, energy-efficient operation, low waste treatment, and use of green catalysts (lipases) and permits a small quantity of water to be present in substrates. However, high cost and inactivation of lipase by lipase-compatible nonaqueous solvents (methanol) are the major bottleneck associated with enzyme-catalyzed transesterification.

Before reaching the market, biodiesel needs to meet specific characterizations depending on regional standards. Commonly, three approaches have been used for the purification of biodiesel, viz., water washing, dry washing, and membrane separation (Shahbaz et al. 2011a). Water washing is the most traditional method being used in the industry for biodiesel purification. However, this technique has several drawbacks such as increased production time and substantial product loss due to retention in the water phase. In dry washing method, the impurities are removed using an ion-exchange resin or a magnesium silicate powder. The disadvantage of this method is the increased production cost due to added material expenses. The third route to purify biodiesel is the use of a membrane. This method increases the final production cost and has smaller throughput (Leung et al. 2010). To develop "green" method for the production of biodiesel, green alternative low-cost solvents and promising catalysts are both required.

To overcome the hurdles of the current biodiesel production processes, a number of novel technologies have been actively applied. Over the last two decades, the growth of ionic solvent system known as ionic liquids (ILs) has attracted great interest because of their special physical traits such as non-flammability, very low vapor pressure, easy recycling, and excellent dissolving strength for variety of organic and inorganic compounds (Welton 1999). More importantly, it is possible to choose a particular IL based on the selection of cation and anion to enhance its physical properties (such as viscosity, hydrophobicity, polarity, and hydrogen bond basicity).

ILs are organic salts composed of organic cations and organic/inorganic anions. Owing to their variable structure, ILs can dissolve different compounds, including polar and apolar organic substances and polymeric and inorganic molecules (Gamba et al. 2008). ILs have attracted substantial attention in various fields, e.g., electrochemistry, biotechnology, organic/inorganic transformations, electrolysis, catalysis, organic synthesis, and extraction/separation processes (Potdar et al. 2015). The common choices of IL cations include nitrogen-containing cations (e.g., alkylammonium, *N*, *N*'-dialkylimidazolium, *N*-alkylpyridinium, and pyrrolidinium). Typical IL anions are halides BF_4^-, PF_6^-, $CF_3CO_2^-$, NO_3^-, and Tf_2N^-. The most common ILs are the salts comprised of 1,3-dialkylimidazolium cations (Andreani and Rocha 2012). Figure 1 displays the common anions and cations used to form ILs.

The development of a new class of ionic solvent system recognized as deep eutectic solvents (DESs) has recently attracted much attention due to their environmentally friendly nature and superior chemical and physical properties (Zhang et al. 2012). DES usually consists of an organic salt, e.g., choline chloride (ChCl), and a hydrogen bond donor (HBD) component, e.g., alcohols, amines, amides, and carboxylic acids, which combine via hydrogen bonding to form a eutectic mixture that has a melting point lower than that of each compound (Wazeer et al. 2018).

DESs exhibit several substantial advantages over organic solvents and conventional ILs including low cost (e.g., the price of ethaline is 50 $/kg, while the average

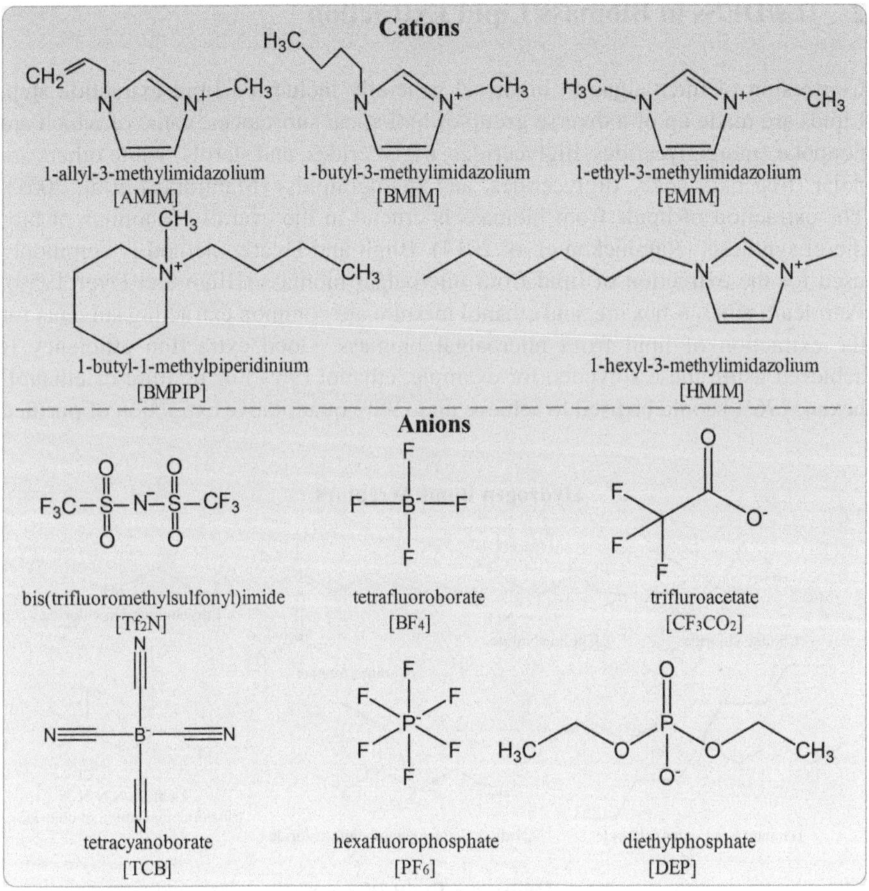

Fig. 1 The structures of common cations and anions used to form ILs

price of IL is not less than 500 $/kg), negligible vapor pressure, non-flammability, simple preparation, low viscosity (e.g., the viscosity of ChCl/ethylene glycol (1:2) DES is 37 cP at room temperature) (D'Agostino et al. 2011), high biodegradability, and compatibility with water (Abbott et al. 2004). Owing to their unique solvation properties, DESs have been utilized as solvents in electrochemistry, enzyme catalysis, material chemistry, chemical reactions, separation and purification processes, and polymer synthesis (Wazeer et al. 2018). The DESs have superior biocompatibility and biodegradability compared with conventional reaction systems and organic and aqueous solvents. Furthermore, the preparation of DES is straightforward, not requiring purification, compared to conventional ILs. The most common salt used to prepare DESs with different HBDs is ChCl. Recently, DESs have been introduced in many applications for the biorefinery such as the production of 5-hydroxymethylfurfural (Zuo et al. 2018), the conversion of carbohydrates (Zuo et al. 2017), and enzymatic reactions for biodiesel production (Hayyan et al. 2014). Figure 2 shows some common salts and HBDs used to form DESs.

2 ILs/DESs in Biomass Lipid Extraction

Conversion of microalgae to biodiesel generally includes a lipid extraction step. Lipids are made up of a diverse group of biological substances, some of which are nonpolar (monoglycerides, diglycerides, triglycerides, and sterols) while others are polar (free fatty acids, diglycerides, and sphingolipids) (Manirakiza et al. 2001). The extraction of lipids from biomass is crucial in the overall economics of biodiesel synthesis (Ramluckan et al. 2014). Bligh and Dyer's method is commonly used for the extraction of lipid from microalgal biomass (Bligh and Dyer 1959). Petroleum ether, *n*-hexane, and ethanol mixture are common extracting solvents for the extraction of lipid from microalgal biomass. Good extraction efficiency is achieved using these solvents, for example, ethanol (96%) or mixture of ethanol-hexane (96%) could be used to achieve up to 98% quantitative extraction of purified

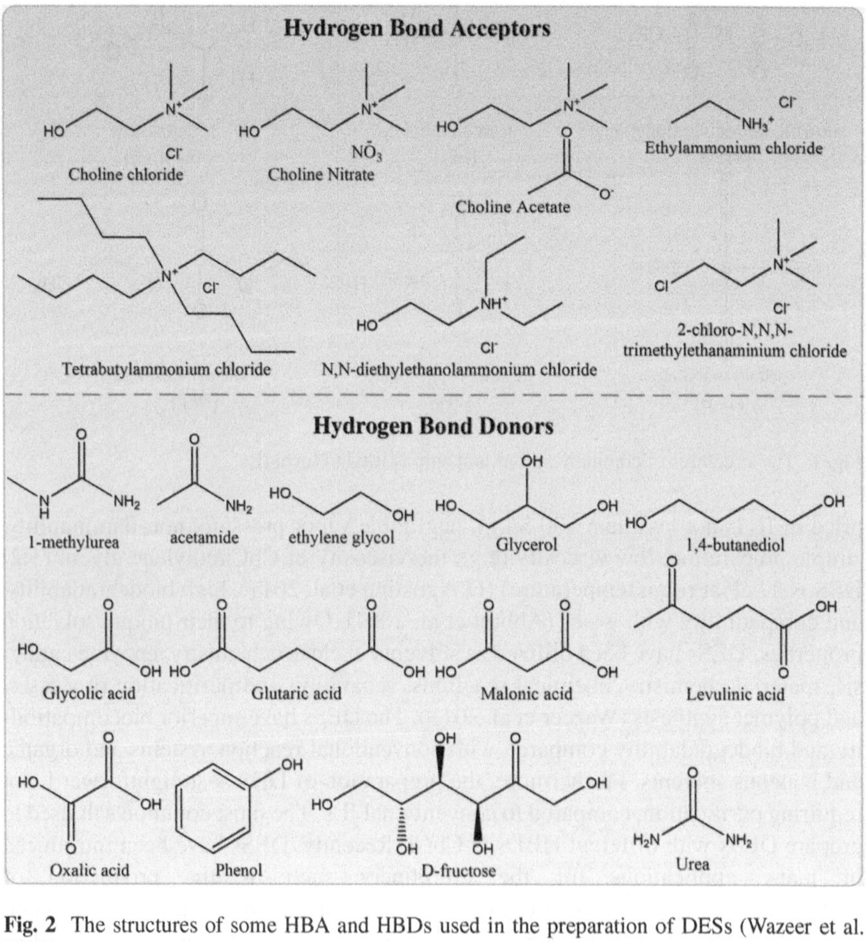

Fig. 2 The structures of some HBA and HBDs used in the preparation of DESs (Wazeer et al. 2018)

fatty acids (Richmond 2013). However, these solvents are volatile, flammable, and toxic. Another major drawback of using these solvents (e.g., ethanol) is that they also extract undesirable proteins, sugars, and pigments (Mata et al. 2010). Numerous ILs have been used for the extraction of lipids from biomass. Choi et al. (2014) investigated the effect of 12 ILs on lipid extraction from microalgae *Chlorella vulgaris*. The yield of lipid extraction using single ILs was compared with the yield attained with organic solvents and IL mixtures. The yield using hexane-methanol solvent (185.4 mg/g cell) was lower than some of the single ILs (the lipid extraction yields using [EMIM][OAc], [EMIM]DEP, [EMIM]BF_4, and [EMIM]Cl were higher than 20%). While some ILs such as [EMIM]$EtOSO_3$ and [EMIM]SCN showed lower yields (6.05 and 4.2 7%, respectively), however, the yield of their mixture (1:1 weight ratio) was increased to 15.8 2% cell.

On the other hand, some researchers have used the mixture of methanol and ILs to extract lipids from algal biomass. Young et al. (2010) proposed IL-methanol cosolvent system to extract lipids. The lipids were auto-partitioned to a separate immiscible phase instead of dissolving in the cosolvent system. Kim et al. (2012) also used IL-methanol cosolvent system to extract lipids from algal biomass. The total contents of lipids derived from cultivated and commercial *Chlorella vulgaris* were 11.1% and 10.6%, respectively, by the Bligh and Dyer's method, while a mixture of IL (1-butyl-3-methylimidazolium trifluoromethanesulfonate) and methanol extracted 19% and 12.5% of the lipids, respectively. Protic ILs have also been used for lipid extraction from biomass. Chiappe and co-authors (2016) used low-cost protic ILs for lipid extraction. Wet microalgae (85% water) was used to investigate the use of different protic ILs, and high extraction yields (up to 88%) were achieved. The direct transformation of fatty acids and triglycerides was also obtained using switchable protic ILs.

Recently, an innovative method was proposed for extracting wet algae directly (Yang et al. 2017). Gas exposure approach and liquid-liquid extraction were employed to recycle the solvents while consuming less amount of energy. An extraction yield of 85% was achieved by using the amphiphilic amine solvent to extract wet docosahexaenoic acid algae. Furthermore, an algal lipid recycling yield of 77% was obtained using 1-butyl-3-methylimidazolium bis(trifluoromethylsulfonyl) imide and *N,N*-dimethylcyclohexylamine. DESs have also been employed for the treatment of microalgae cells. Pan et al. (2017) studied one- and two-step methods for the extraction of lipid from wet and unbroken *Chlorella* sp. and *Chlorococcum* sp. The effect of DES on the destruction of microalgae cell was analyzed by using methanol and hexane as cosolvents. In comparison with two-step method, one-step method with DES treatment improved the total content of FAME by 30%. Very recently, Tommasi et al. (2017) developed a new protocol for the extraction of lipid from the diatom *Phaeodactylum tricornutum*. Microwaves (MW) and DESs were examined as pretreatments for environmentally benign solvent extractions using supercritical carbon dioxide ($scCO_2$) and dimethyl carbonate (DMC). Various DESs formed by mixing ChCl with different HBD (oxalic acid, levulinic acid, ethylene glycol, urea, and sorbitol) were investigated in combination with DMC extraction. Highest selectivity and total fatty acid (TFA) extraction yield of DMC were achieved using ChCl/carboxylic acid DES. The combination of DESs and MW followed by

DMC extraction resulted in a TFA yield and fatty acid profile comparable to those of Bligh and Dyer extraction method to be reached, along with a much higher selectivity (88% vs 35%). The extraction efficiency of $scCO_2$ was also significantly improved due to this pretreatment; the yield of TFA was increased by a factor of 20, and highly purified triglyceride extracts were obtained.

2.1 *Extraction of Free Fatty Acids from Biomass*

Manic et al. (2011) studied the use of two ILs (i.e., Ammoeng 100 and 1-butyl-3-methylimidazolium dicyanamide) and poly(ethyleneglycol)s (with various molar masses) as alternative solvents for the extraction of FFA (e.g., linoleic acid) from soybean oil. Liquid-liquid phase equilibrium data revealed that the alternative solvents (ILs and poly(ethyleneglycol)) are completely miscible with the FFA, but not miscible with soybean oil. Ammoeng 100 as extraction solvent yielded the highest values of linoleic acid distribution coefficient. This study indicated that the proposed solvents showed high potential for biodiesel deacidification than short-chain alcohols.

Recently, a vacuum headspace single-drop microextraction method based on the use of magnetic ILs (MIL) was presented for the extraction of short-chain FFA (Trujillo-Rodríguez et al. 2017). The proposed method was more effective and faster than the method at atmospheric pressure. The most suitable MIL for the proposed method was trihexyl(tetradecyl)phosphonium tris(hexafluoroacetylaceto) manganate(II). In another work (Grimes and Kewcharoenwong 2017), a mixed IL system (dual extraction and catalytic functions) was developed for the extraction and conversion of FFA from waste cooking oils to biodiesel. The dual IL system consisted of the miscible ILs 1-HPyrBr and one of the Brønsted ILs ([SPyr][HSO_4] or 1-(4-sulfonic acid) butyl-3-methylimidazolium hydrogen sulfate [SMIM] [HSO_4]). A summary of the results is shown in Table 1.

3 ILs/DESs as Reaction Media

3.1 *IL Solvents*

ILs offer an ideal media for many biocatalytic reactions, enzyme immobilization, and separations in biotechnology due to their stability over an extended period during the reaction (Jain et al. 2005). In comparison with organic solvents, ILs impart increased stability, no inactivation of enzymes, faster rates, and higher selectivity in biocatalytic reactions (Park and Kazlauskas 2003). It is known that polar organic solvents inactivate enzymes; notably, ILs do not inactivate enzymes; this feature simplifies reactions and increases the solubility of polar substrates such as sugars (Park and Kazlauskas 2001). Furthermore, it is also possible to recycle enzyme when using ILs as solvents for the synthesis of biodiesel.

Table 1 Yield of FFA methyl esters derived from 1-hexylpyridinium bromide (HPyrBr)/[SMIM][HSO_4]-catalyzed esterification with methanol (Grimes and Kewcharoenwong 2017)

IL catalyst	Acid	Lipid number (C:D)	Temp (°C)	Atmosphere	Product	Yield (%)
[SMIM][HSO_4]	Palmitic acid	C16:0	80	Air	Methyl palmitate	90
HPyrBr/[SMIM][HSO_4]	Palmitic acid	C16:0	80	Air	Methyl palmitate	90
HPyrBr/[SMIM][HSO_4]	Stearic acid	C18:0	80	Air	Methyl stearate	91
HPyrBr/[SMIM][HSO_4]	Oleic acid	C18:1	80	Air	Methyl oleate	83
HPyrBr/[SMIM][HSO_4]	Linoleic acid	C18:2	20	Nitrogen	Methyl linoleate	67
HPyrBr/[SMIM][HSO_4]	Linoleic acid	C18:2	30	Nitrogen	Methyl linoleate	67
HPyrBr/[SMIM][HSO_4]	Linoleic acid	C18:2	60	Nitrogen	Methyl linoleate	70
HPyrBr/[SMIM][HSO_4]	Linoleic acid	C18:2	80	Nitrogen	Methyl linoleate	73
HPyrBr/[SMIM][HSO_4]	Acid mixture		30	Nitrogen	Mixed FAME	80

Several groups demonstrated the enzymatic transesterification of vegetable oils to produce biodiesel in ILs. 23 ILs were screened as solvents by Ha et al. (2007) for biodiesel production from soybean oil using *Candida antarctica* lipase as catalyst. Among the tested ILs, the highest biodiesel yield of 80% was achieved using 1-ethyl-3-methylimidazolium trifluoromethanesulfonate ([EMIM][TfO]) IL as a solvent. As compared to the conventional solvent-free system, the yield was eight times higher. 4:1 was found to be the optimum substrate molar ratio of methanol to soybean oil for biodiesel production in the same IL. In another work (Neto et al. 2007), 1-butyl-3-methylimidazolium tetrachloro-indate ([BMIM][$lnCl_4$]) IL was used as reaction media for the methanolysis of soybean oil to biodiesel in the presence of tin-based catalyst [Sn(3-hydroxy-2-methyl-4-pyrone)$_2$(H_2O)$_2$], and 83% biodiesel yield was achieved under reflux condition in 4 h. Electrospray ionization tandem mass spectrometry screening proposed that a cationic specie was formed via the substitution of pyrone ligand by alcohol, accompanied by the coordination of carboxylate compound to tin. Liu et al. (2011) studied 19 ILs for the production of biodiesel catalyzed by *Burkholderia cepacia*. They studied the influence of anion/cation structures of ILs on the production of biodiesel. The authors reported that the choice of anion had a much larger influence on the conversion of biodiesel than that of the cation. 1-octyl-3-methylpyridinium tetrafluoroborate ([OMPY][BF_4]) produced the highest yield (82.2 ± 1.2%) of biodiesel in 12 h reaction time. Low biodiesel yields were achieved in ILs with strong water miscible properties.

Four ILs (two hydrophobic and two hydrophilic ILs) were studied to investigate their effectiveness as reaction media for the methanolysis of sunflower oil using

Candida antarctica lipase as a catalyst (Sunitha et al. 2007). High fatty acid methyl ester (FAME) yield (98–99%) was achieved in hydrophobic ILs (1-butyl-3-methyl imidazolium hexafluorophosphate ([BMIM][PF_6]) and 1-ethyl-3-methyl imidazolium hexafluorophosphate ([EMIM][PF_6])). On the other hand, very poor yields of FAME were achieved in hydrophilic ILs (1-hexyl-3-methyl imidazolium tetrafluoroborate ([HMIM][BF_4]) and 1 butyl-3-methyl imidazolium tetrafluoroborate ([BMIM][BF_4])). The catalyst activity of an enzyme for transesterification process is maintained when using hydrophobic ILs because they protect lipase from being deactivated by methanol, while the hydrophilic ILs were not able to prevent the deactivation of lipase by methanol which justify the low yield of FAME.

Novozym 435 lipase is commonly used as a catalyst for the production of biodiesel using enzyme-catalyzed transesterification. Imidazolium ILs with different anions such as $[Tf_2N]^-$, $[BF_4]^-$, and $[PF_6]^-$ and different alkyl chain lengths (C_{10}–C_{18}) are used as efficient cosolvents in the Novozym 435 lipase-catalyzed production of biodiesel from methanol/triolein mixtures (Troter et al. 2016). For example, [BMIM][PF_6] IL and methyl acetate were used as cosolvent with Novozym 435 catalyst for the production of biodiesel (Ruzich and Bassi 2010). Eighty-three percent of FAME yield was achieved using a small-scale reaction. Diego et al. (2011) used two hydrophobic ILs with $[Tf_2N]^-$ anion, namely, 1-hexadecyl-3-methylimidazolium bis(trifluoromethyl sulfonyl)imide ([C_{16}MIM][Tf_2N]) and 1-octadecyl-3-methylimidazolium bis(trifluoromethyl sulfonyl)imide ([C_{18}MIM] [Tf_2N]), for the enzymatic methanolysis of lipase (Novozym 435) to produce FAME. Excellent biodiesel yield (98%) was obtained after 6 h of reaction at 60 °C. ILs consisting of *N, N* dialkylimidazolium cations, and different anions were scrutinized in alcoholysis reactions to analyze lipase stability and activity (Qin et al. 2016). It was found that the main factor affecting enzyme stability and activity was the structure of the anion of IL. The initial activity of the enzyme in $[Tf_2N]^-$- and $[PF_6]^-$-based ILs was much higher than in ILs containing $[N(CN)^2]^-$ anion as shown in Fig. 3. The initial enzyme activities were in an increasing order of $[N(CN)^2]^- < [PF_6]^- < [Tf_2N]^-$.

3.2 *DES Solvents*

Choline-based DESs are the most commonly used DES solvents for the production of biodiesel (Merza et al. 2018; Zhao et al. 2013). Zhao et al. (2013) used choline-based DESs as solvents for the enzymatic synthesis of biodiesel from soybean oil. Under optimized conditions such as 7:3 (v/v) of DES (ChCl/Gly (1:2)) and methanol, 0.2% water (v/v), 40 mgmL Novozym 435/mL, 50 °C, and 24-h reaction time, a high conversion (88%) of triglyceride was achieved. Without losing much activity, the enzyme could be reused for at least four times. The lipase showed high stability in the eutectic mixture of DES and methanol for an extended period (24 h) at 50 °C. In another work (Gu et al. 2015), the same choline-based DES was used as cosolvent for the transesterification of rapeseed oil to biodiesel catalyzed by sodium

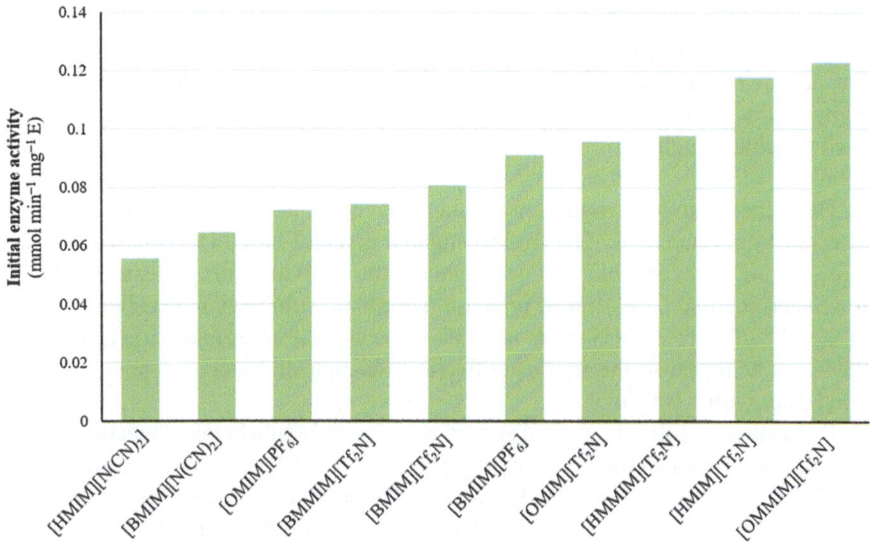

Fig. 3 The initial enzyme activity of alcoholysis reaction in ten different ILs

hydroxide. Response surface methodology and Box-Behnken Design were used to optimize the yield of FAME. At optimum conditions (6.95 methanol/oil molar ratio, 9.27 wt% DES concentration, and 1.34 wt% catalyst concentration), FAME yield of up to 98% was obtained. Addition of DES in the transesterification reaction enabled a straightforward biodiesel synthesis and purification and reduced the side reactions such as saponification. Zhang et al. (2016) studied 11 DESs as reaction media for the production of biodiesel from yellow horn seed oil through transesterification using immobilized enzyme Novozym 435. Among 11 DESs, ChCl/Gly (1:2) was most efficient in producing biodiesel where 95% conversion yield was achieved under optimum conditions. Enzymes were recovered and used four times in the reaction without any significant loss of activity. DES enabled the enzymes to retain their activity, and the conversion yield was also increased.

Some researchers have investigated DESs as promising solvents for lipase-catalyzed reactions due to their high activity and selectivity. Durand et al. (2012) studied the limitations and advantages of few DESs as green solvents for biotransformation using immobilized *Candida antarctica* lipase B as a catalyst. As compared with organic media, the DESs composed of ChCl combined with urea or glycerol exhibited the best initial specific activity of the lipase. Side reactions were observed when using dicarboxylic acid-based DES in the alcoholysis reactions. However, other DESs such as ChCl/Gly and ChCl/urea showed high selectivity and activity for lipase-catalyzed reactions. The use of DESs in the synthesis of biodiesel with free lipases starting from low-quality cooking oils and refined rapeseed oil was also evaluated. Free lipases Lipozyme CALB L and Lipozyme TL 100 L were dissolved in DES in order to develop a two-step-one-pot enzymatic transesterification reaction, in which pure ethyl ester with low acid value was synthesized. Ninety-seven percent yield of ethyl ester was obtained after a single refinement step. Zhao et al. (2011a) explored novel and biodegradable eutectic mixtures based upon choline acetate (ChAc) combined with glycerol for the enzymatic transesterification of triglycerides with methanol. Under optimal conditions, ChAc/Gly (1:1.5) DES was used to achieve high conversions (82–97%) of Miglyol oil using Novozym 435 as a catalyst.

DESs have also been used as promising solvents in the protease-catalyzed transesterification. Zhao et al. (2011b) investigated glycerol-based DESs based on choline salt (acetate or chloride) to evaluate the activities of protease-catalyzed transesterification. Cross-linked protease (subtilisin) exhibited an excellent activity in the ChCl/Gly (1:2) DES containing 3% (v/v) water. Furthermore, a selectivity of 98% was achieved in the transesterification reaction of *N*-acetyl-L-phenylalanine ethyl ester with 1-propanol.

Deep eutectic solvents were also used as solvents in biochemical reactions other than biodiesel production. For example, Maugeri et al. (2013) investigated protease-catalyzed peptide synthesis in different ChCl-based DESs combined with urea, glycerol, xylitol, and isosorbide. Under optimal conditions, high productivities of 20 g L^{-1} h^{-1} were reached with ChCl/Gly and 10–30% water. Furthermore, α-chymotrypsin could be reused over several cycles in the DES solution before being deactivated.

Epoxide hydrolases also find promising applications in DESs containing systems. Until now, there are limited publications using DES in epoxide hydrolase-catalyzed reactions. Gorke and co-authors (2008) reported excellent catalytic activity of hydrolases in DESs. The conversion of styrene oxide catalyzed by epoxide hydrolase AD1 increased from 4.6 to 92% due to the addition of 25% ChCl/Gly (v/v). Furthermore, no change in the enantioselectivity was noticed with the increase of conversion. Lindberg et al. (2010) also studied different DESs in enzyme-catalyzed hydrolysis of a chiral (1,2)-trans-2-methylstyrene oxide. Among different ChCl-based DESs, ChCl/Gly exhibited superior solvent property. Due to the presence of glycerol or ethanol diol (HBD), the regioselectivity in hydrolysis was altered to favor epoxide ring opening at the benzylic carbon, improving the regioselectivity detected in buffer-only systems. Few applications of DESs and ILs as solvents in the production of biodiesel are listed in Table 2.

4 ILs/DESs as Catalysts for Biodiesel Production

4.1 *ILs as Catalysts*

There are several types of acidic and basic ILs being prepared and used as catalysts in one- or two-step processes of biodiesel synthesis. ILs have proven to be auspicious catalysts in biodiesel synthesis. Sodium hydroxide and potassium hydroxide (KOH) are the most commonly used basic catalysts for the production of commercial biodiesel (Shahid and Jamal 2011). However, such catalytic systems have several drawbacks: (1) sensitive process to free fatty acids and residual water; (2) catalysts cannot be recycled; (3) FAME separation becomes difficult due to the formation of stable emulsions. In comparison with conventional solid/liquid catalysts, ILs offer excellent stability, higher catalytic activity, easy product isolation, and environmental benefits (Wu et al. 2007).

4.1.1 Brønsted Acidic ILs

The efficient catalytic activity of acidic ILs based on pyrazolium, pyrrolidinium, and pyridine rings was reported by several researchers. Li et al. (2010) examined ILs containing pyridine rings as catalysts for esterification and transesterification of Jatropha oil. High biodiesel yield (92%) was obtained using 1-(4-sulfonic acid) butylpyridinium trifluoromethanesulfonate ([BSPy][CF_3SO_3]). After the reaction, the product was easily removed, and the catalyst maintained its activity even after seven cycles. Pyrazolium-based IL ((2-(4-sulfobutyl) pyrazolium hydrogen sulfate), [SBP][HSO_4]) showed excellent catalyst activity in the methanolysis of bitter apple oil (BAO) to produce biodiesel (Elsheikh 2014). The highest yield (89.5%) of esters was achieved when the reaction was carried out under the conditions of 5.2 wt% of [SBP][HSO_4], molar ratio of BAO to methanol of 1:15, 170 °C, and 800 rpm for 6 h.

Zhang et al. (2009) synthesized biodiesel using Brønsted acidic IL *N*-methyl-2-pyrrolidonium methyl sulfonate ([NMP][CH_3SO_3]) as a catalyst. Good catalytic efficiency and reusability were achieved under mild conditions using this IL. The authors achieved 95.3% yield of fatty acid alkyl esters after 8 h at 70 °C, and the catalytic system could be reused eight times, keeping the conversion above 90%. Wu et al. (2007) investigated the cottonseed oil transesterification with methanol to produce biodiesel using Brønsted acidic ILs containing sulfonic group in cations. Among different ILs evaluated by the authors, 1-(4-sulfonic acid) butylpyridinium hydrogen sulfate ([SPyr][HSO_4]) IL presented the highest Brønsted acidity and was identified as the best catalyst. An optimized condition (oil/methanol/IL = 1:12:0.057 (molar ratio) for 5 h at 170 °C) was used to obtain the highest biodiesel yield (92%).

Both the acidity of the anions and length of the carbon chain in the cations affect the acidity of ILs. Longer carbon chain ILs stimulate efficient esterification because these long carbon chains expedite mass transfer in the reaction. 3,3′-(Octane-1,8-

Table 2 Applications of ILs and DESs as solvents in biodiesel production

IL/DES	Feedstock	Alcohol	T (°C)	Yield (%)	Refs.
[EMIM][TfO]	Soybean oil	Methanol	50	80	Ha et al. (2007)
[OmPy][BF_4]	Soybean oil	Methanol	40	82.2	Liu et al. (2011)
[EMIM][PF_6]	Sunflower oil	Methanol	58–60	98–99	Sunitha et al. (2007)
[C_{18}MIM][Tf_2N]	Vegetable oil	Methanol	60	98	De Diego et al. (2011)
ChCl/Gly (1:2)	Soybean oil	Methanol	50	88	Zhao et al. (2013)
ChCl/Gly (1:2)	Seed oil	Methanol	50	95	Zhang et al. (2016)
ChAc/Gly (1:2)	Miglyol® oil	Methanol	50	97	Zhao et al. (2011a)

diyl)-bis(4-sulfobenzyl-1H-imidazol-3-ium) hydrogen sulfate showed the best catalytic efficiency in the esterification of oleic acid with methanol (Fig. 4) because of the highest space and lowest hindrance between sulfonic groups for approaching of fatty acid to the active sites (Aghabarari et al. 2014). The catalytic activity of the IL is dependent on the steric hindrance and the number of Brønsted acidic functions.

Wu et al. (2014) synthesized silica-coated Fe_3O_4 magnetic nanoparticle-supported dual Brønsted acidic IL 1-(propyl-3-sulfonate) imidazolium hydrogen sulfate ([SO_3H-$(CH_2)_3$-HIM][HSO_4]) (Fig. 5). The prepared catalyst exhibited good performance in the production of biodiesel with ethanol and oleic acid. Zhang et al. (2017) prepared a Brønsted-Lewis acidic IL and immobilized on the surface of Fe_3O_4@SiO_2. The catalyst was used to produce biodiesel from nonedible feedstock *Koelreuteria integrifoliola*. Under optimized conditions, the FAME yield of 93.7% was achieved. The catalyst demonstrated a heterogeneous catalyst behavior and could be reused for at least five times with no significant loss of activity.

4.1.2 Brønsted Basic ILs

Several Brønsted acidic ILs have been used as catalysts in the production of biodiesel; however, studies concerning the applications of basic ILs as catalysts in biodiesel synthesis are rare. These basic ILs that have been used are commonly choline- and imidazolium-based ILs. Zhou et al. (2012) studied imidazolium hydroxide-based basic ILs as recycle catalysts for the transesterification of glycerol trioleate with methanol. They obtained high biodiesel yield of 87.2% for 8 h at 120 °C. Liang et al. (2010) used five basic binuclear imidazolium-based ILs as catalysts in the production of biodiesel via transesterification from cottonseed oil. The results of the transesterification of cottonseed oil to biodiesel over various Brønsted basic ILs are given in Fig. 6. Best catalytic performance was exhibited using IMC_2OH IL. The same authors reported that the activity of catalysts was affected by the length of the carbon chain in the cations. Luo et al. (2013) obtained high biodiesel yield (up to 95%) using 1-butyl-3-methylimidazolium imidazolide ([BMIM]Im) IL as a catalyst. They used different vegetable oils, such as sunflower oil, soybean oil, and rapeseed oil. The catalytic efficiency of [BMIM]Im remained excellent after reusing for several times. Few applications of ILs as catalysts in the production of biodiesel are summarized in Table 3.

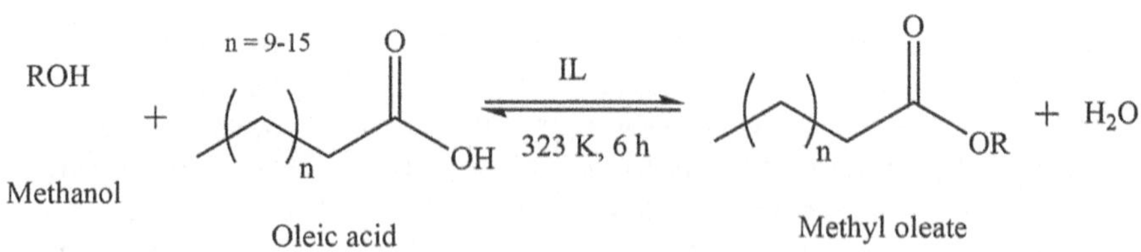

Fig. 4 Esterification reaction in the presence of IL (Aghabarari et al. 2014)

Fan et al. (2013) used choline-based basic IL (choline hydroxide) for the catalytic production of biodiesel from soybean oil at atmospheric pressure. The authors concluded that choline hydroxide exhibited better catalytic activity compared with other basic IL catalysts. It was also reported that better basicity was obtained when using IL catalyst in methanol solution. Quasi in situ infrared spectroscopy was used to elucidate the interaction between methoxyl group and the carbonyl group of the triglyceride.

4.2 *DESs as Catalysts*

Various DESs have been prepared and used as catalysts for the production of biodiesel. For example, $ChCl/ZnCl_2$ (1:2) DES was used as a Lewis acidic catalyst in the soybean oil methanolysis; 55% conversion was achieved under optimum

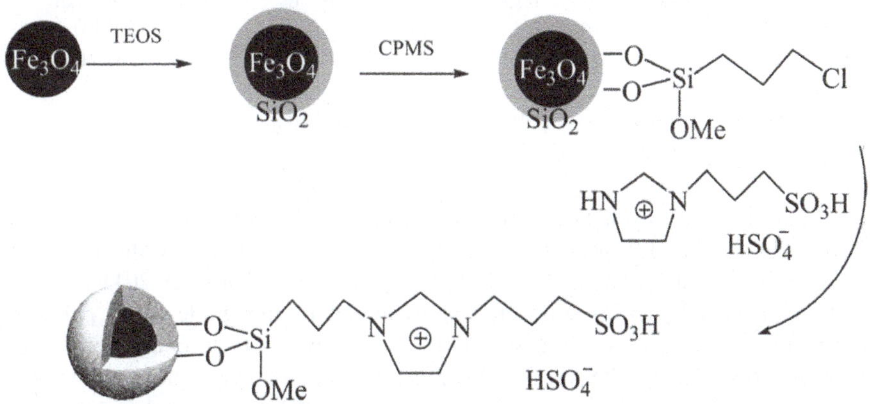

Fig. 5 Synthesis route to IL-$Fe_3O_4@SiO_2$ (Wu et al. 2014)

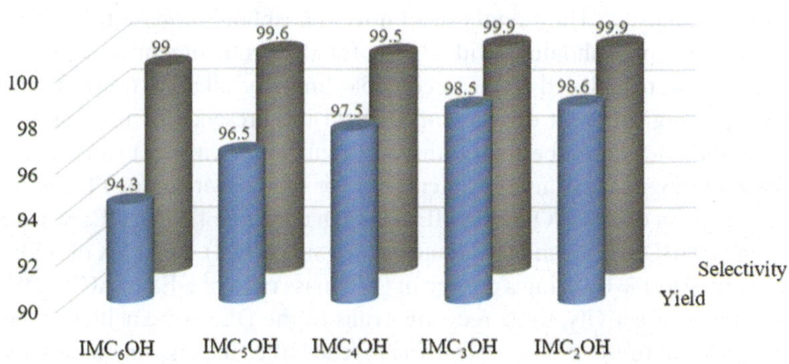

Fig. 6 Performance of various imidazolium-based ILs in the transesterification of cottonseed oil with methanol to biodiesel (Liang et al. 2010)

Table 3 Applications of ILs in the transesterification reaction of oil with methanol to produce biodiesel

IL	Oil	Oil/alcohol (molar ratio)	*T* (°C)	Time (h)	Yield (%)	Ref.
$[BSPy][CF_3SO_3]$	Jatropha oil	1:10	100	5	92.0	Li et al. (2010)
$[BSPy][CF_3SO_3]$	Jatropha oil	1:10	100	8	90.3	Li et al. (2010)
$[BSPy][HSO_4]$	Jatropha oil	1:10	100	6	88.6	Li et al. (2010)
[BSPy][p-TSA]	Jatropha oil	1:10	100	5	75.5	Li et al. (2010)
$[BSPy]_3PW_{12}O_{40}$	Jatropha oil	1:10	120	5	80.1	Li et al. (2010)
$[BSPy]_3SiW_{12}O_{40}$	Jatropha oil	1:10	120	5	79.5	Li et al. (2010)
$[SBP][HSO_4]$	BAO	1:15	170	6	89.5	Elsheikh (2014)
$[SPyr][HSO_4]$	Cottonseed oil	1:12	170	5	92.0	Wu et al. (2007)
[BMIM]OH	Glycerol trioleate	1:9	120	8	87.2	Zhou et al. (2012)

conditions (10% catalyst, methanol/oil molar ratio 16:1, 70 °C, and 72 h) (Tao et al. 2010). $ChCl/FeCl_3$ (1:2) was used as catalyst in the palm oil transesterification reaction to produce ester, and yield of 67.4% was achieved (Isahak et al. 2011).

Hayyan et al. (2013a) conducted pretreatment of low-grade crude palm oil (LGCPO) using ammonium-based DES as catalyst which comprised of *p*-toluenesulfonic acid (PTSA) monohydrate and *N*,*N*-diethyl ethanol ammonium chloride. After optimization, the FFA content of LGCPO was reduced from 9.5% to less than 1%. Four recycling runs of the DES were used without losing its activity. The same group (Hayyan et al. 2013b) explored the possibility of using a two-stage process to produce LGCPO, in which an alkali- and a phosphonium-based DES were used as catalysts. They synthesized novel phosphonium-based DESs by mixing allyltriphenylphosphonium with PTSA for the pretreatment of LGCPO. The levels of FFA were reduced to the acceptable limit for alkaline transesterification reaction by using different dosages of DES in the presence of methanol. It was found that the DES could be reused three to four times without losing its activity. ChCl-based DES was also used as a catalyst for the conversion of FFA content in acidic crude palm oil (ACPO) to FAME (Hayyan et al. 2014). The DES was formed by mixing an HBD (PTSA monohydrate) with a salt (ChCl). The FFA of ACPO was reduced from 9 to 1% by using a dosage of 0.75 mass ratio of DES to ACPO. Without significant loss of activity, three recycling runs of the DES were achieved. Ninety-two percent yield of the final product was reached after the alkaline transesterification with 0.07% FFA and 96 mol% FAME content. Table 4 displays the applications of choline-based DESs catalysts in biodiesel production.

A phosphonium-based DES was also utilized as catalyst in the esterification reaction of glycerol with oleic acid (Williamson et al. 2017). The DES was prepared by mixing allyltriphenylphosphonium bromide (hydrogen bond acceptor, HBA) and PTSA monohydrate (HBD) in a molar ratio of 1:3 (HBA to HBD). The authors investigated the effect of DES catalyst concentration (1, 3, and 5 wt%) and temperature (120, 150, and 180 °C) on the esterification efficiency of fatty acid conversion. The lowest activation energy of 54.64 KJ/mol was achieved using 5 wt% of DES catalyst. Consequently, 95% FFA conversion was attained under optimum reaction conditions (150 °C temperature and 5 wt% DES catalyst). Very recently, ChCl-based DESs combined with carboxylic acids were utilized as low-cost, effective bifunctional catalysts to catalyze epoxidation of soybean oil with peroxyformic acid (Wang et al. 2017). ChCl/oxalic acid exhibited high selectivity (93.68%) and conversion (88.8%) at 50 °C for 8 h reaction time. Moreover, ChCl/oxalic acid catalyst obtained a good catalyst activity (89.98% selectivity) after five successive runs.

5 IL/DES Biodiesel Purification

ILs and DESs have also been used as extraction solvents in the purification of biodiesel. Several applications of these solvents in the purification of biodiesel include (a) removal of lipids from biomass, (b) extraction of FFA and unsaturated fatty acid, (c) extraction of glycerol from biodiesel after reaction, and (d) extraction of residual catalyst from crude biodiesel.

Table 4 Applications of choline-based DESs as catalysts in biodiesel production

DES	DES molar ratio	Feed	Reaction	Oil/alcohol (molar ratio)	Time (h)	Yield, conversion (%)	Ref.
ChCl/ $ZnCl_2$	1:2	Soybean oil, methanol	Transesterification of soybean oil to biodiesel	1:10–1:30	72	54.52	Tao et al. (2010)
ChCl/ Gly	1:2	Soybean oil, methanol	Enzymatic preparation of biodiesel		24	88	Zhao et al. (2013)
ChCl/ Gly	1:2	Rapeseed oil, methanol	Cao-catalyzed transesterification of rapeseed oil	1:14.28	3	91.9	Huang et al. (2013)
ChCl/ $FeCl_3$	1:2	Palm oil, methanol	Transesterification of soybean oil to biodiesel	1:10–1:15	4	67.4	Isahak et al. (2011)
ChAc/ Gly	1:2	*M. pinnata* seed oil, methanol	Transesterification reaction	1 mg:3 mL	48	54.8	Huang et al. (2014)
ChCl/ PTSA	1:3	Palm oil, methanol	Pretreatment esterification of ACPO	1:1–1:20	0.5	97	Hayyan et al. (2014)

5.1 Extraction of Glycerol and Leftovers

Different DESs have been employed as solvents for the removal of glycerol from the transesterification biodiesel product. Abbott et al. (2007) used a 1:1 mixture of quaternary ammonium salts (such as ChCl, [ClEtMe$_3$N]Cl, and [EtNH$_3$]Cl) and glycerol to extract excess glycerol from biodiesel. ChCl/Gly DES resulted in 51 wt% of glycerol removal in a continuous separation process (Hayyan et al. 2010). The efficiency of extraction process was investigated by analyzing the effect of DES to biodiesel ratio and the composition of DES. The optimum ratio of biodiesel/DES was 1:1 and DES composition was 1:1 (HBD/salt). Shahbaz and co-authors (2010) improved the work mentioned above by using two different ChCl-based DESs. DESs were formed by combining ChCl with ethylene glycol or 2,2,2-trifuracetamide in different molar ratios. All synthesized DESs were capable of separating all free glycerol, successfully. The optimum molar ratio for all DESs was found to be 1:1 (biodiesel/DES).

The DESs based on methyl triphenyl phosphonium bromide (MTPPB) have also been employed as solvents for the removal of glycerol from palm oil-based biodiesel (Shahbaz et al. 2011b). Glycerol, ethylene glycol, and triethylene glycol were used as HBD. These DESs were able to reduce the content of diglycerides and monoglycerides. Triethylene glycol-based DESs at a molar ratio of 1:3 (biodiesel/DES) showed maximum removal efficiency of total glycerol. A neural network-based model has also been proposed to predict the removal of total glycerol from biodiesel (Shahbaz et al. 2012). The purpose of this model was to optimize the extraction process in terms of time and cost required for experimentations. The model revealed that glycerol-based DESs have lower removal efficiencies in comparison with those synthesized with other HBD.

The residual catalyst such as KOH must be removed at the end of alkali-catalyzed transesterification reaction. Shahbaz et al. (2011a) examined 18 DESs based on ChCl and MTPPB as solvents for the separation of residual KOH from palm oil-based biodiesel. It was found that the biodiesel/DES molar ratio has a significant influence on the removal efficiency of residual catalyst. The MTPPB/Gly and ChCl/Gly DESs showed maximum removal efficiency of 97.57% and 98.59%, respectively. Furthermore, the amount of water in biodiesel was also reduced, and DESs could be used five times, without any significant loss in their efficiencies.

6 Recovery, Regeneration, and Recycling of ILs/DESs

As mentioned in previous sections, ILs and DESs have attracted much attention in both academics and industries as promising solvents for a diverse range of applications. However, there were a few industrial processes employing ILs or DESs due to the economical and efficient use of ILs and contamination. The economic efficiency can be improved by recycling and reuse of ILs. Several attempts have been made by the researchers for recovery and recycling of ILs. Mai et al. (2014) reported a comprehensive summary on the methods used for recovery and recycling of ILs. The possibilities of recovering ILs and DESs lead to reducing the amount of waste produced for a particular application. Since not all ILs are nontoxic, discharging them into the environment could contaminate soils and groundwater, thus affecting human health and other living organisms as well. From an economical point of view, recovery of used ILs could reduce their cost dramatically. The used IL could either be reused without treatment or be regenerated by undergoing recovery processes so that it can be used for the next process. Recovery operations should be done when the used ILs have been applied in the process for several times and thus their capacities and performances would decrease. Idaham et al. (2014) found that the feasibility of using water to recover ILs from dibenzothiophene (DBT) is quite high, which was 100% for butyl-methylimidazolium methyl sulfate ([BMIM][CH_3SO_4]). However, the water content in the recovered ILs was higher than that in the pure one. This means that the physical properties of the IL may change with the increased concentration of water. In this case, water can be separated from IL by moderate heating. However, this means an increase in the cost of the process.

Kuzmina (2016) discussed in detail the possible ways to recover and reuse ILs from waste streams. Kuzmina indicated that the terms recovery, solvent regeneration, and recycling are frequently used incorrectly in the literature. All of these processes together are needed for the recycling of the ILs. The purification stage is needed to avoid the deterioration of ILs, while regeneration is needed if a compound cannot be reused directly. During the processing stages and continued reuse, some impurities can accumulate, and the concentration of IL in the final extract can be significantly lower than expected. Most impurities found in ILs are traces of unreacted starting materials. However, after the processing of the IL, the waste stream can contain traces of the new reaction products, other chemicals, and their impurities. Without recycling or removal, ILs can become a persistent hazard in the aquatic environment because their low volatility prevents them from being released into the atmosphere (Haerens et al. 2010).

Secondary waste streams can be produced at the stage of IL recovery. The appropriate waste treatment can significantly decrease production costs, which define the overall price of the final product and the benefits for the producer. During product usage, the IL ends up as a waste with very little possibility of reuse. The best way to decrease the final cost of waste disposal is to reduce the amount of waste produced. A specific example of application of ILs in closed production cycle proposed for biomass pretreatment by Dibble et al. (2011) is shown in Fig. 7. The IL reclamation step is one of the most difficult and least explored aspects of this process. After precipitation of pretreated corn stover from the IL solution, several biomass components remain dissolved in the IL and wash solvent solution. The components, or residual solutes, include low molecular weight sugars, a portion of the original lignin, and nonpolar biomass components. Recovery of these residual solutes reduces their accumulation with IL reuse and the corresponding potential for reduction in IL pretreatment efficacy. The authors described a novel process for reclaiming the used IL. Dibble et al. utilized the interaction strength of hydrophilic ILs with water to

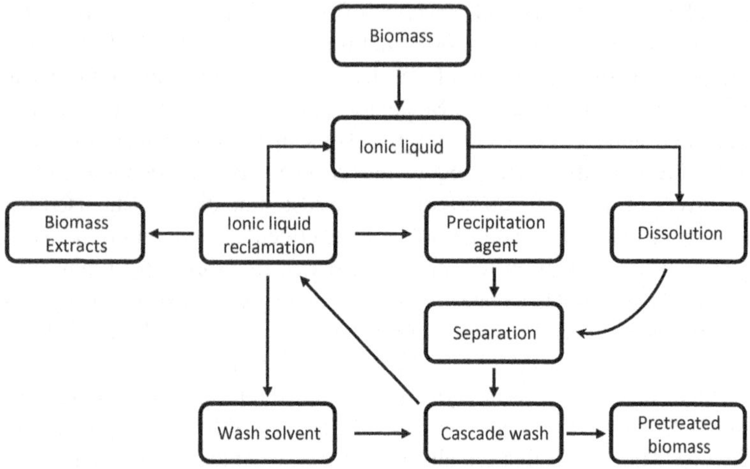

Fig. 7 Simplified diagram of an ionic liquid/biomass pretreatment process (Dibble et al. 2011)

drive phase separation and biomass fractionation, as shown in Fig. 7. Dry acetone and the hydrophilic IL are miscible, but acetone containing only 1.0 wt% water reduces the solubility of the IL to less than 0.05 wt%. Ethanol and other alcohols are good cosolvents for this system, restoring the solution to a single-phase state when added in similar quantities to the amount of water present. This phase-switchable quaternary system of IL-water-ketone-alcohol provides a very sensitive and convenient means of controlling separations. There are four basic operations in their proposed IL recovery: distillation of the combined precipitation and wash solvents, extraction of nonpolar substances from the IL, concentration of IL solutes by a second extraction that removes IL and water, and precipitation and reduction of IL content in the recovered solutes by washing. The separate products of the final three operations can then be "dried" by thermal removal of undesired water or solvents.

The possible ways to reduce the amount of industrial waste streams containing ILs were suggested by Siedlecka et al. (2011) (Fig. 8). Generally, removal of the ILs from waste streams, their regeneration, and recovery need to be implemented wherever and whenever possible. The recovery processes are limited to the concentration of contaminants. The exceeding of a reasonable limit of impurities prevents IL use and turns it to IL waste. From this point of view, the recovery processes for IL wastes are of special interest.

The main impurities that can be found in the postreaction mixture containing ILs include but are not limited to the aqueous solution (or water), traces of unreacted starting materials (mainly cellulose derivatives and organics), the residues after galvanic treatment, and catalysts. They all have different natures and properties and require unique methods for IL separation.

Despite it being clearly stated in the literature earlier that ILs are recyclable, their purification, especially for those ILs soluble in water, has proved to be more difficult than expected. The first method that comes to mind when thinking about separation

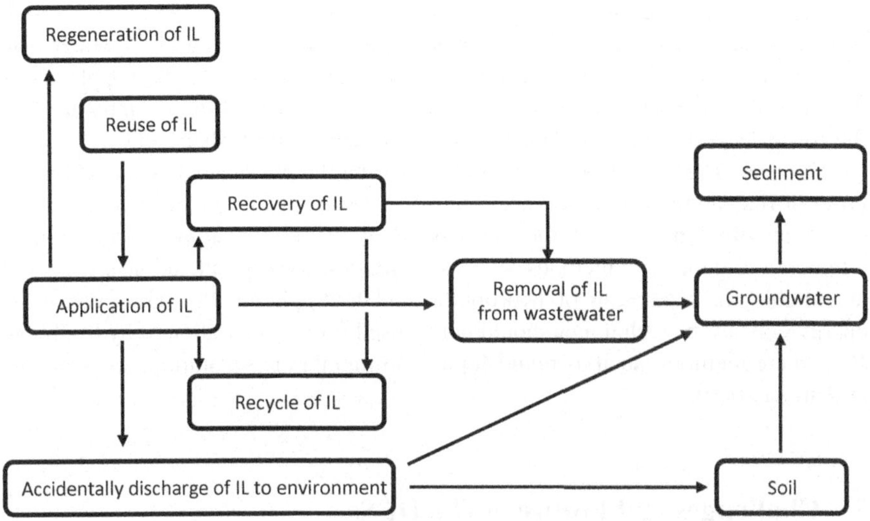

Fig. 8 Reduction of wastes containing ILs in industrial applications and potential sources of ILs released to the environment (Siedlecka et al. 2011)

of ILs from water is evaporation. However, this method requires high temperatures and therefore is very energy-consuming. The lowest-energy consumption separation methods are barrier separation methods such as membrane separation and nanofiltration. Nanofiltration can be applied to IL separation from other solvents because ILs consist only of ions. Charged and neutral compounds or mono- and divalent ions can be separated by nanofiltration membranes. The charge and size of the ions or molecules and the choice of membrane influence the retention (Schäfer et al. 2005). This method is especially promising for separation of nonvolatile compounds or, at low concentration, ILs from waste streams when distillation methods are not applicable. The principle of nanofiltration is that the membranes are permeable to the ILs but not to the other products in the waste stream. Another barrier method is pervaporation, which is more energy-consuming in comparison with nanofiltration but is often used as a less energy-demanding alternative to vacuum or extractive distillation. Some of the main difficulties of nanofiltration and pervaporation are the viscosity of ILs and the requirement of large membrane areas.

Methods such as evaporation of volatiles under vacuum, extractions with VOC solvents, supercritical CO_2 extraction, and distillation/stripping of the solute from the thermally stable ILs have been known for some time and have been applied to ILs (Kanel 2003).

The question of how to understand the recyclability of ILs was raised by Wu et al. (2009). Methods for separation and recycling of ILs were discussed in a mini review combined with their own experience on separation of ILs from their "working" environment. The authors stated that the choice of the appropriate separation method should be based on the specific properties of each system. Also, they questioned the need to obtain absolutely pure ILs, which makes sense only for academic researchers, who do not want even negligible amounts of impurities to have an effect on their results. Usually grades of purity are greater than 95% "for synthesis," greater than 99% "high purity," and greater than 99.9% "ultrapure." However, some applications do not need such purity, and additional purification steps would significantly increase the cost of ILs when produced on a commercial scale (Abu-Eishah 2011).

All these broad and deep research investigating the physical and chemical properties of ILs and their behavior in solution and waste streams are needed in order to create an efficient method for the separation of ILs to support recyclability. Temperature-dependent methods, such as distillation and crystallization, are marked as suitable for separation of hydrophilic and hydrophobic ILs. However, these energy-consuming techniques should not be used for the separation of hydrophobic ILs, where methods, such as liquid-liquid and membrane separation, are available (Kuzmina 2016).

7 Challenges and Future of ILs/DESs

The use of ILs and DESs for biorefinery treatment is gaining momentum globally. The unique properties of ILs/DESs and their convergence with current treatment technologies present great opportunities to improve biomass treatment and biofuel production and purification. Although all applications of ILs discussed in this chapter are still in the laboratory research stage and none has made their way to pilot testing, there is an ample chance for some of these applications to reach even full-scale application.

The most important challenges that face the transfer of the use of ILs from lab scale into commercial applications include cost, commercial availability, toxicity, recyclability, and compatibility with the existing infrastructure.

The challenges faced by using ILs/DESs in biorefineries are important, but many of these challenges are temporary, including technical hurdles, high cost, and potential environmental and human risk. To overcome these barriers, collaboration between research institutions, industry, government, and other stakeholders is important. It is believed that advancing applications of ILs/DESs by carefully steering its direction while avoiding unintended consequences can continuously provide robust solutions to these challenges.

References

Abbott AP, Boothby D, Capper G, Davies DL, Rasheed RK (2004) Deep eutectic solvents formed between choline chloride and carboxylic acids: versatile alternatives to ionic liquids. J Am Chem Soc 126(29):9142–9147

Abbott AP, Cullis PM, Gibson MJ, Harris RC, Raven E (2007) Extraction of glycerol from biodiesel into a eutectic based ionic liquid. Green Chem 9(8):868–872

Abu-Eishah SI (2011) Ionic liquids recycling for reuse. In: Ionic liquids-classes and properties. InTech, Rijeka

Aghabarari B, Ghiaci M, Amini SG, Rahimi E, Martinez-Huerta M (2014) Esterification of fatty acids by new ionic liquids as acid catalysts. J Taiwan Inst Chem Eng 45(2):431–435

Andreani L, Rocha J (2012) Use of ionic liquids in biodiesel production: a review. Braz J Chem Eng 29(1):1–13

Atadashi I, Aroua M, Aziz AA (2011) Biodiesel separation and purification: a review. Renew Energy 36(2):437–443

Bligh EG, Dyer WJ (1959) A rapid method of total lipid extraction and purification. Can J Biochem Physiol 37(8):911–917

Chiappe C, Mezzetta A, Pomelli CS, Iaquaniello G, Gentile A, Masciocchi B (2016) Development of cost-effective biodiesel from microalgae using protic ionic liquids. Green Chem 18(18):4982–4989

Choi S-A, Oh Y-K, Jeong M-J, Kim SW, Lee J-S, Park J-Y (2014) Effects of ionic liquid mixtures on lipid extraction from Chlorella vulgaris. Renew Energy 65:169–174

D'Agostino C, Harris RC, Abbott AP, Gladden LF, Mantle MD (2011) Molecular motion and ion diffusion in choline chloride based deep eutectic solvents studied by 1 H pulsed field gradient NMR spectroscopy. Phys Chem Chem Phys 13(48):21383–21391

De Diego T, Manjón A, Lozano P, Iborra JL (2011) A recyclable enzymatic biodiesel production process in ionic liquids. Bioresour Technol 102(10):6336–6339

Dibble DC, Li C, Sun L, George A, Cheng A, Çetinkol ÖP, Benke P, Holmes BM, Singh S, Simmons BA (2011) A facile method for the recovery of ionic liquid and lignin from biomass pretreatment. Green Chem 13(11):3255–3264

Dorian JP, Franssen HT, Simbeck DR (2006) Global challenges in energy. Energy Policy 34(15):1984–1991

Durand E, Lecomte J, Baréa B, Piombo G, Dubreucq E, Villeneuve P (2012) Evaluation of deep eutectic solvents as new media for Candida antarctica B lipase catalyzed reactions. Process Biochem 47(12):2081–2089

Elsheikh YA (2014) Optimization of novel pyrazolium ionic liquid catalysts for transesterification of bitter apple oil. Process Saf Environ Prot 92(6):828–834

Fan M, Huang J, Yang J, Zhang P (2013) Biodiesel production by transesterification catalyzed by an efficient choline ionic liquid catalyst. Appl Energy 108:333–339

Gamba M, Lapis AA, Dupont J (2008) Supported ionic liquid enzymatic catalysis for the production of biodiesel. Adv Synth Catal 350(1):160–164

Gorke JT, Srienc F, Kazlauskas RJ (2008) Hydrolase-catalyzed biotransformations in deep eutectic solvents. Chem Commun (10):1235–1237

Grimes SM, Kewcharoenwong P (2017) Dual-functionality ionic liquid mix for extraction and esterification of fatty acids as a step towards increasing the efficiency of conversion of waste cooking oils to biodiesel. J Chem Technol Biotechnol 92(8):2098–2105

Gu L, Huang W, Tang S, Tian S, Zhang X (2015) A novel deep eutectic solvent for biodiesel preparation using a homogeneous base catalyst. Chem Eng J 259:647–652

Ha SH, Lan MN, Lee SH, Hwang SM, Koo Y-M (2007) Lipase-catalyzed biodiesel production from soybean oil in ionic liquids. Enzym Microb Technol 41(4):480–483

Haerens K, Van Deuren S, Matthijs E, Van der Bruggen B (2010) Challenges for recycling ionic liquids by using pressure driven membrane processes. Green Chem 12(12):2182–2188

Hayyan M, Mjalli FS, Hashim MA, AlNashef IM (2010) A novel technique for separating glycerine from palm oil-based biodiesel using ionic liquids. Fuel Process Technol 91(1):116–120

Hayyan A, Hashim MA, Hayyan M, Mjalli FS, AlNashef IM (2013a) A novel ammonium based eutectic solvent for the treatment of free fatty acid and synthesis of biodiesel fuel. Ind Crop Prod 46:392–398

Hayyan A, Hashim MA, Mjalli FS, Hayyan M, AlNashef IM (2013b) A novel phosphonium-based deep eutectic catalyst for biodiesel production from industrial low grade crude palm oil. Chem Eng Sci 92:81–88

Hayyan A, Hashim MA, Hayyan M, Mjalli FS, AlNashef IM (2014) A new processing route for cleaner production of biodiesel fuel using a choline chloride based deep eutectic solvent. J Clean Prod 65:246–251
Hoekman SK (2009) Biofuels in the US–challenges and opportunities. Renew Energy 34(1):14–22
Huang W, Tang S, Zhao H, Tian S (2013) Activation of commercial CaO for biodiesel production from rapeseed oil using a novel deep eutectic solvent. Ind Eng Chem Res 52(34):11943–11947
Huang ZL, Wu BP, Wen Q, Yang TX, Yang Z (2014) Deep eutectic solvents can be viable enzyme activators and stabilizers. J Chem Technol Biotechnol 89(12):1975–1981
Idaham NF, Bustam Khalil MA, Kait CF (2014) Study on the Recovery of 1-butyl-3-methylimidazolium-based Ionic Liquids. Appl Mech Mater 625:144–147. Trans Tech Publications
Isahak W, Ismail M, Jahim JM, Salimon J, Yarmo M (2011) Transesterification of palm oil by using ionic liquids as a new potential catalyst. Trends Appl Sci Res 6(9):1055
Jain N, Kumar A, Chauhan S, Chauhan S (2005) Chemical and biochemical transformations in ionic liquids. Tetrahedron 61(5):1015–1060
Kanel JS (2003) Overview: industrial application of ionic liquids for liquid extraction. In: Chemical industry vision 2020 technology partnership workshop, New York
Kaygusuz K (2009) Bioenergy as a clean and sustainable fuel. Energy Sources A 31(12):1069–1080
Kim Y-H, Choi Y-K, Park J, Lee S, Yang Y-H, Kim HJ, Park T-J, Kim YH, Lee SH (2012) Ionic liquid-mediated extraction of lipids from algal biomass. Bioresour Technol 109:312–315
Kuzmina O (2016) Methods of IL recovery and destruction. In: Application, purification, and recovery of ionic liquids. Elsevier, Oxford, pp 205–248
Leung DY, Wu X, Leung M (2010) A review on biodiesel production using catalyzed transesterification. Appl Energy 87(4):1083–1095
Li K-X, Chen L, Yan Z-C, Wang H-L (2010) Application of pyridinium ionic liquid as a recyclable catalyst for acid-catalyzed transesterification of Jatropha oil. Catal Lett 139(3–4):151–156
Liang J-h, Ren X-q, Wang J-t, Li Z-j (2010) Preparation of biodiesel by transesterification from cottonseed oil using the basic dication ionic liquids as catalysts. J Fuel Chem Technol 38(3):275–280
Lindberg D, de la Fuente Revenga M, Widersten M (2010) Deep eutectic solvents (DESs) are viable cosolvents for enzyme-catalyzed epoxide hydrolysis. J Biotechnol 147(3):169–171
Liu Y, Chen D, Yan Y, Peng C, Xu L (2011) Biodiesel synthesis and conformation of lipase from Burkholderia cepacia in room temperature ionic liquids and organic solvents. Bioresour Technol 102(22):10414–10418
Luo H, Fan W, Li Y, Nan G (2013) Biodiesel production using alkaline ionic liquid and adopted as lubricity additive for low-sulfur diesel fuel. Bioresour Technol 140:337–341
Mai NL, Ahn K, Koo Y-M (2014) Methods for recovery of ionic liquids—a review. Process Biochem 49(5):872–881
Manic MS, Najdanovic-Visak V, da Ponte MN, Visak ZP (2011) Extraction of free fatty acids from soybean oil using ionic liquids or poly(ethyleneglycol)s. AICHE J 57(5):1344–1355
Manirakiza P, Covaci A, Schepens P (2001) Comparative study on total lipid determination using Soxhlet, Roese-Gottlieb, Bligh & Dyer, and modified Bligh & Dyer extraction methods. J Food Compos Anal 14(1):93–100
Mata TM, Martins AA, Caetano NS (2010) Microalgae for biodiesel production and other applications: a review. Renew Sust Energ Rev 14(1):217–232
Maugeri Z, Leitner W, Domínguez de María P (2013) Chymotrypsin-catalyzed peptide synthesis in deep eutectic solvents. Eur J Org Chem 2013(20):4223–4228
Merza F, Fawzy A, AlNashef I, Al-Zuhair S, Taher H (2018) Effectiveness of using deep eutectic solvents as an alternative to conventional solvents in enzymatic biodiesel production from waste oils. Energy Rep 4:77–83
Neto BAD, Alves MB, Lapis AA, Nachtigall FM, Eberlin MN, Dupont J, Suarez PA (2007) 1-n-Butyl-3-methylimidazolium tetrachloro-indate (BMI· InCl4) as a media for the synthesis of biodiesel from vegetable oils. J Catal 249(2):154–161

Pan Y, Alam MA, Wang Z, Huang D, Hu K, Chen H, Yuan Z (2017) One-step production of biodiesel from wet and unbroken microalgae biomass using deep eutectic solvent. Bioresour Technol 238:157–163

Park S, Kazlauskas RJ (2001) Improved preparation and use of room-temperature ionic liquids in lipase-catalyzed enantio- and regioselective acylations. J Org Chem 66(25):8395–8401

Park S, Kazlauskas RJ (2003) Biocatalysis in ionic liquids–advantages beyond green technology. Curr Opin Biotechnol 14(4):432–437

Potdar MK, Kelso GF, Schwarz L, Zhang C, Hearn MT (2015) Recent developments in chemical synthesis with biocatalysts in ionic liquids. Molecules 20(9):16788–16816

Qin J, Zou X, Lv S, Jin Q, Wang X (2016) Influence of ionic liquids on lipase activity and stability in alcoholysis reactions. RSC Adv 6(90):87703–87709

Ramluckan K, Moodley KG, Bux F (2014) An evaluation of the efficacy of using selected solvents for the extraction of lipids from algal biomass by the soxhlet extraction method. Fuel 116:103–108

Richmond A (2013) Chapter 11 Biological principles of mass cultivation of photoautotrophic microalgae. In: Richmond A, Hu Q(eds) Handbook of microalgal culture: applied phycology and biotechnology, Second Edition. John Wiley & Sons (Publisher), https://doi.org/10.1002/9781118567166

Ruzich NI, Bassi AS (2010) Investigation of lipase-catalyzed biodiesel production using ionic liquid [BMIM][PF6] as a co-solvent in 500 mL jacketed conical and shake flask reactors using triolein or waste canola oil as substrates. Energy Fuels 24(5):3214–3222

Schäfer AI, Fane AG, Waite TD (2005) Nanofiltration: principles and applications. Elsevier, Oxford

Shahbaz K, Mjalli FS, Hashim M, Al-Nashef IM (2010) Using deep eutectic solvents for the removal of glycerol from palm oil-based biodiesel. J Appl Sci 10(24):3349–3354

Shahbaz K, Mjalli F, Hashim M, AlNashef I (2011a) Eutectic solvents for the removal of residual palm oil-based biodiesel catalyst. Sep Purif Technol 81(2):216–222

Shahbaz K, Mjalli F, Hashim M, AlNashef I (2011b) Using deep eutectic solvents based on methyl triphenyl phosphunium bromide for the removal of glycerol from palm-oil-based biodiesel. Energy Fuels 25(6):2671–2678

Shahbaz K, Baroutian S, Mjalli FS, Hashim MA, AlNashef IM (2012) Prediction of glycerol removal from biodiesel using ammonium and phosphunium based deep eutectic solvents using artificial intelligence techniques. Chemom Intell Lab Syst 118:193–199

Shahid EM, Jamal Y (2011) Production of biodiesel: a technical review. Renew Sust Energ Rev 15(9):4732–4745

Siedlecka EM, Czerwicka M, Neumann J, Stepnowski P, Fernández J, Thöming J (2011) Ionic liquids: methods of degradation and recovery. In: Ionic liquids: theory, properties, new approaches. InTech, Rijeka

Sunitha S, Kanjilal S, Reddy P, Prasad R (2007) Ionic liquids as a reaction medium for lipase-catalyzed methanolysis of sunflower oil. Biotechnol Lett 29(12):1881–1885

Tao L, Yuefeng D, Shucai G, Ji C (2010) Application of choline chloride· xZnCl2 ionic liquids for preparation of biodiesel. Chin J Chem Eng 18(2):322–327

Tommasi E, Cravotto G, Galletti P, Grillo G, Mazzotti M, Sacchetti G, Samorì C, Tabasso S, Tacchini M, Tagliavini E (2017) Enhanced and selective lipid extraction from the microalga P. tricornutum by dimethyl carbonate and supercritical CO2 using deep eutectic solvents and microwaves as pretreatment. ACS Sustain Chem Eng 5(9):8316–8322

Troter DZ, Todorović ZB, Đokić-Stojanović DR, Stamenković OS, Veljković VB (2016) Application of ionic liquids and deep eutectic solvents in biodiesel production: a review. Renew Sust Energ Rev 61:473–500

Trujillo-Rodríguez MJ, Pino V, Anderson JL (2017) Magnetic ionic liquids as extraction solvents in vacuum headspace single-drop microextraction. Talanta 172:86–94

Wang J, Liu Y, Zhou Z, Fu Y, Chang J (2017) Epoxidation of soybean oil catalyzed by deep eutectic solvents based on the choline chloride–carboxylic acid bifunctional catalytic system. Ind Eng Chem Res 56(29):8224–8234

Wazeer I, Hayyan M, Hadj-Kali MK (2018) Deep eutectic solvents: designer fluids for chemical processes. J Chem Technol Biotechnol 93(4):945–958
Welton T (1999) Room-temperature ionic liquids. Solvents for synthesis and catalysis. Chem Rev 99(8):2071–2084
Williamson ST, Shahbaz K, Mjalli FS, AlNashef IM, Farid MM (2017) Application of deep eutectic solvents as catalysts for the esterification of oleic acid with glycerol. Renew Energy 114:480–488
Wu Q, Chen H, Han M, Wang D, Wang J (2007) Transesterification of cottonseed oil catalyzed by Brønsted acidic ionic liquids. Ind Eng Chem Res 46(24):7955–7960
Wu B, Liu W, Zhang Y, Wang H (2009) Do we understand the recyclability of ionic liquids? Chem Eur J 15(8):1804–1810
Wu Z, Li Z, Wu G, Wang L, Lu S, Wang L, Wan H, Guan G (2014) Brønsted acidic ionic liquid modified magnetic nanoparticle: an efficient and green catalyst for biodiesel production. Ind Eng Chem Res 53(8):3040–3046
Yang H-Y, Lu W-J, Chen Y-C, Chen K-T, Teng J-C, Wan H-P (2017) New algal lipid extraction procedure using an amphiphilic amine solvent and ionic liquid. Biomass Bioenergy 100:108–115
Young G, Nippgen F, Titterbrandt S, Cooney MJ (2010) Lipid extraction from biomass using co-solvent mixtures of ionic liquids and polar covalent molecules. Sep Purif Technol 72(1):118–121
Zhang L, Xian M, He Y, Li L, Yang J, Yu S, Xu X (2009) A Brønsted acidic ionic liquid as an efficient and environmentally benign catalyst for biodiesel synthesis from free fatty acids and alcohols. Bioresour Technol 100(19):4368–4373
Zhang Q, Vigier KDO, Royer S, Jérôme F (2012) Deep eutectic solvents: syntheses, properties and applications. Chem Soc Rev 41(21):7108–7146
Zhang Y, Xia X, Duan M, Han Y, Liu J, Luo M, Zhao C, Zu Y, Fu Y (2016) Green deep eutectic solvent assisted enzymatic preparation of biodiesel from yellow horn seed oil with microwave irradiation. J Mol Catal B Enzym 123:35–40
Zhang H, Li H, Pan H, Liu X, Yang K, Huang S, Yang S (2017) Efficient production of biodiesel with promising fuel properties from Koelreuteria integrifoliola oil using a magnetically recyclable acidic ionic liquid. Energy Convers Manag 138:45–53
Zhao H, Baker GA (2013) Ionic liquids and deep eutectic solvents for biodiesel synthesis: a review. J Chem Technol Biotechnol 88(1):3–12
Zhao H, Baker GA, Holmes S (2011a) New eutectic ionic liquids for lipase activation and enzymatic preparation of biodiesel. Org Biomol Chem 9(6):1908–1916
Zhao H, Baker GA, Holmes S (2011b) Protease activation in glycerol-based deep eutectic solvents. J Mol Catal B Enzym 72(3):163–167
Zhao H, Zhang C, Crittle TD (2013) Choline-based deep eutectic solvents for enzymatic preparation of biodiesel from soybean oil. J Mol Catal B Enzym 85:243–247
Zhou S, Liu L, Wang B, Xu F, Sun RC (2012) Biodiesel preparation from transesterification of glycerol trioleate catalyzed by basic ionic liquids. Chin Chem Lett 23(4):379–382
Zuo M, Le K, Li Z, Jiang Y, Zeng X, Tang X, Sun Y, Lin L (2017) Green process for production of 5-hydroxymethylfurfural from carbohydrates with high purity in deep eutectic solvents. Ind Crop Prod 99:1–6
Zuo M, Le K, Feng Y, Xiong C, Li Z, Zeng X, Tang X, Sun Y, Lin L (2018) An effective pathway for converting carbohydrates to biofuel 5-ethoxymethylfurfural via 5-hydroxymethylfurfural with deep eutectic solvents (DESs). Ind Crop Prod 112:18–23

Ethanol-Water Separation Using Membrane Technology

Munirasu Selvaraj and Fawzi Banat

1 Introduction

The increasing energy requirements for the modern lifestyle are inseparable, and nearly for a century, fossil fuels fulfilled the demand (Munirasu et al. 2016). The finite availability and concentrations in a few geographical regions make fossil fuel dependency less desirable for the long-term energy security. It is well-understood that fossil fuel as an energy source will be exhausted in the near future (Shafiee and Topal 2009). Therefore, alternative energy source is always on demand, and renewable energies like solar, wind, etc. can be the long-term solution with an added benefit of being "green" without causing environmental pollution. However, solar and wind energies cannot provide continuous energy supply, and additionally these energies are geographic and time dependent, i.e., they are concentrated in few parts of the globe with varying availability over the different period of a year. Therefore, plant-based renewable energies, particularly ethanol, is considered as highly reliable energy source for the continuous use with the option of fuel storage (Sanchez and Cardona 2008). For a very long time, ethanol is commercially produced by the fermentation process, and so far it still remains economically competitive production method compared to the conventional chemical synthesis, and its competitiveness against fossil fuel as energy source is still an ongoing debate (Bungay 2004). Initially, sugar-based feedstocks were used for the ethanol production and designated as first-generation biofuel. The ethanol produced is called as bioethanol, and the process is also termed as biorefinery as compared with the conventional refinery of petrochemicals (Kiss et al. 2016). Since first-generation bioethanol production directly competes with food grains, the second generation of bioethanol is developed from lignocellulose- and hemicellulose-based feedstocks which are noncompetitive with food stocks (Girio et al. 2010; Naik et al. 2010). Feedstock for the second-generation ethanol production can be sourced from a wide range of renewable

M. Selvaraj · F. Banat (✉)
Department of Chemical Engineering, Khalifa University, Abu Dhabi, UAE
e-mail: fawzi.banat@ku.ac.ae

© Springer Nature Switzerland AG 2019

J.-R. Bastidas-Oyanedel, J. E. Schmidt (eds.), *Biorefinery*,
https://doi.org/10.1007/978-3-030-10961-5_9

materials like softwood, microalgae, etc. (Galbe and Zacchi 2002; Singh and Dhar 2011; Ullah et al. 2015). The conversion of lignocellulose into bioethanol passes through an intermediate step, namely, hydrolysis of cellulose and hemicelluloses into monomeric sugar, and the derived sugars are subjected to fermentation which produces bioethanol subsequently (Amelio et al. 2016). The third generation of bioethanol is generally produced from algal biomasses, and the fourth-generation bioethanol is based on genetically modified and optimized feedstocks for the fermentation process.

2 Biorefinery Process

Compared with the conventional petroleum-based refinery industries, biorefinery of ethanol production has few similarities and varies considerably. For example, biorefinery invariably has to deal with low concentrated feed and product formation, whereas the conventional refinery operates the opposite way. The main reason for the low concentration in the biorefinery process is due to aqueous-based fermentation process where microorganisms, usually bacteria, enzyme, and yeast, are present and these microorganisms are sensitive to the presence of some particular compounds at certain concentration. Therefore, the fermentation process has to be performed at dilute aqueous system compared to the conventional chemical refinery which operate at much higher concentration level. In most cases, after a certain concentration, the product of the fermentation itself acts as an inhibitor for these fermenting agents. A typical yeast fermented bioconversion, for example, *S. cerevisiae*, can produce up to 18% ethanol in the broth (Zhao and Bai 2009). The microorganisms are also sensitive to the presence of low concentrated toxins like acetate, phenols, and furan family compounds (Kiss et al. 2016). The removal of these toxin compounds from the feedstock leads to a better fermentation process. Therefore most of the fermentation process is done at a low concentration level, or the product is continuously removed in order to drive the process toward the maximum conversion of the feedstock. This is identified as one reason for high capital expenditure (CapEx) for the biorefinery compared with the conventional chemical refinery (Amelio et al. 2016). Additionally, the microorganisms are thermally sensitive, and high dilution of the reaction medium makes it one of the challenging tasks for the product separation in the biorefinery process. Thus, the critical part of bioethanol production and its competitiveness with petrochemical fuel lies with the efficient separation from the aqueous reaction medium and further purification into fuel grade ethanol. One of the earliest large-scale biorefineries is acetone-butanol-ethanol (ABE) fermentation (Ranjan and Moholkar 2012). In this process, acetone, butanol, and ethanol are produced in the ratio of 3:6:1, and subsequent separation of these compounds leads to the biofuel of alcohols (Ranjan and Moholkar 2012).

The fermentation processes are conducted in batch, fed-batch (semi-batch), and continuous mode. While the batch process is the simplest one, it has many disadvantages like repeated sterilization of reactors, solvent inhibition due to the biofuel formation causing low conversion, and wastage of time due to the nature of the batch process, i.e., cleaning the reactors, recharging the reactant, etc. (Curcio et al. 2016). A continuous fermentation process can avoid these deficiencies, but the major drawback of the continuous process is maintaining the sterile condition for a long period of time (Amelio et al. 2016). In the continuous process, the product is continuously removed from the bioreactor, thereby eliminating the ethanol inhibition toward microorganisms. A fed-batch is mainly preferred where the high concentration of substrate (starting feedstock) acts as an inhibitor and/or toxic to the microorganisms present in the fermentation process. Since feedstock is added gradually at a specified time interval and it is continuously fermented into ethanol, the inhibition and/or toxic effect of feedstock is considerably reduced (Amelio et al. 2016). In this process also, the product has to be removed in order to maintain the bioreaction within the tolerance limit of product toxicity.

3 Ethanol Separation from Fermentation Mixture

In general, the biorefinery adapted the separation and purification methods employed in the oil refinery with major component of thermal distillation along with several other separation techniques used for the ethanol recovery from the reaction mixture. For the fuel grade ethanol production, multi-column distillation and molecular sieve adsorption are the two major methods employed in large scale (Vane 2008). The major drawback of the distillation process is the requirement of large amount of energy (Singh and Rangaiah 2017). This is mainly due to the presence of low concentration of bioethanol along with the large volume of water. Since the fermentation process produces bioethanol in the range of 5–15% and the remaining majority by volume is aqueous medium, the whole liquid needs to be heated for the distillation process in order to drive the separation. The presence of higher percentage of aqueous medium makes distillation process as energy-intensive process. For example, ethanol concentration of 5–10% in aqueous medium requires about 5–10 MJ-fuel/kg-ethanol resulting 94% ethanol in the distillation process (Vane 2008). The other major separation techniques studied are adsorption (Liu et al. 2014), gas tripping (Xue et al. 2014; Qureshi et al. 2001), liquid-liquid extraction, and membrane technologies like pervaporation and reverse osmosis (Jin 2016). A pass-through distillation for the concurrent alcohol recovery and fermentation (CARAF) process has also been studied for the effective bioethanol separation. This process avoided some of the attributes of conventional distillation technique which is harmful to the continuous fermentation process (McGregor and Furlong 2017). The combination of solvent extraction followed by distillation is also a reported method. In this process, ethanol is extracted using water-immiscible organic solvent, and subsequently ethanol and extracting solvents are separated by the distillation process. This method is most suitable where the temperature is highly sensitive to the bio-organisms. Ethanol is extracted at room temperature or below and followed by high-temperature distillation. A combination of processes like vapor stripping, followed by vapor compression and pervaporation process, termed as membrane-assisted vapor stripping (MAVS), has also been studied to improve the pervaporation process (Vane et al. 2010; Vane and Alvarez 2013). Thus, the continuous ongoing research is mainly focused to reduce the cost of bioethanol separation process. It was reported that the separation and purification process alone can cost about 40–50% of the total cost of refinery process (Kiss et al. 2016). A similar, perhaps higher cost of processing conditions exists for the biorefinery process. Compared to the conventional oil refinery, biorefinery differs considerably in terms of quality and quantity of the mixture that to be separated. The major differences in biorefinery are highly diluted nature of aqueous reaction mixture, i.e., highly diluted reactants and product that has to be separated, and high sensitivity to temperature due to the presence of microorganisms which performs the chemical conversion. The presence of sensitive functional groups in the feedstock also renders them unsuitable for high-temperature processing (Kiss et al. 2016). Thus biorefinery process produces high-volume and low-value fuel, whereas the conventional refinery deals with low-volume and high-value product that need to be separated. Therefore, separation of biofuel with desired purity plays a critical role for the economic viability of the product which is currently used along with the fossil fuel (Chovau et al. 2013). Thermal distillation, extraction, and membrane technologies are the separation techniques employed in large-scale biorefinery, and other separation technologies are in the research stage that needs to be upscaled (Kiss et al. 2016). Among all the separation methods, membrane technology is considered highly efficient and economical due to less energy requirement and small footprint and does not require any additional inputs like additives (Kanehashi and Nagai 2017; Wei et al. 2014; Saha et al. 2017). The overall cost of the membrane module can be considerably small provided the membrane fouling effect is minimal. For example, in membrane-based RO process, the cost of membrane is about 10–20% of overall Capex of the plant (Baker 2004). Depending on the operational conditions, the membrane module may need to be replaced in 3–6 months to few years, and in some cases more than 5 years of membrane life-span is possible (Baker 2004). Pervaporation is the most frequently studied membrane technology so far for the separation of ethanol from the fermentation mixture in situ or ex situ, and bioethanol separated by using membrane technology can be a competitive energy sources for the conventional fuels (Vane 2005).

4 Basics of Membrane Technology

A membrane is a barrier material for the separation of two species based on exclusion depending on species size or difference in solubility and diffusivity and other potential difference among the species present in the feed mixture. The membrane material can be organic or inorganic or organic-inorganic hybrid (Munirasu et al. 2016). Majority of the water treatment membranes are based on organic materials, and the separation processes are based on size exclusion. The microfiltration and ultrafiltration are the examples of the exclusion based on size separation. The most widely used membrane separation for the ethanol purification is pervaporation, and it is based on the solubility and diffusivity difference among the species present in the mixture. Two different types of membrane pervaporation process are used for the separation and enrichment of ethanol. The ethanol selective organophilic membrane is used for the selective ethanol permeation through the organic or inorganic membrane. In most cases, the fermentation broth or filtered liquid from the broth are subjected to the pervaporation, and up to 80–90% of the enriched ethanol can be obtained in this process (Vane 2008). The dehydration of ethanol-water mixture is another membrane process widely used for the production of fuel grade ethanol with the ethanol purity of >99%. In this process, the water molecule is passed through a hydrophilic membrane as permeate, and the retentate is enriched in ethanol (Huang et al. 2006, 2013a).

5 Methods of Membrane Preparation

Various methods can be employed for the preparation of membranes. The most widely used technique for the organic membrane preparation is the film casting method. Majority of the polymer-based membrane is prepared by film casting method in industrial scale, and this is a well-established process for the majority of the polymer solution cast process. This is the simplest membrane-making process and done by casting of polymer solution on a nonporous solid-like glass plate or Teflon sheet which is shown in Fig. 1a. The polymer solution is then allowed to dry in order to have the dense polymer film (membrane) which can be used for the pervaporation process. One of the disadvantages of the dense membrane is the low

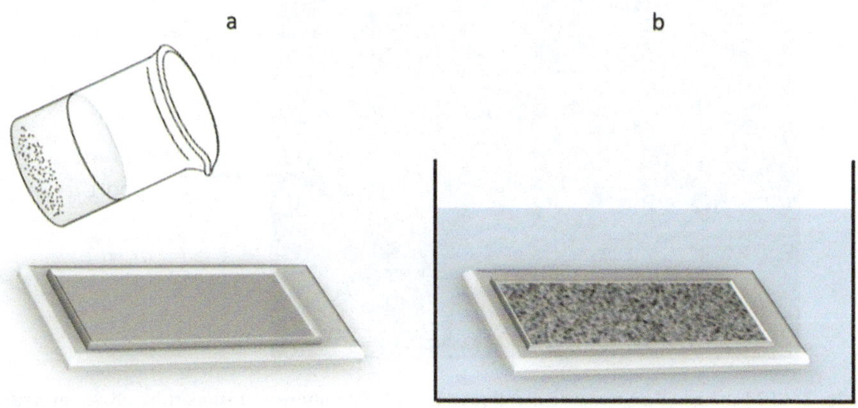

Fig. 1 (**a**) Polymer solution casting and making of dense thin film membrane and (**b**) phase inversion of the cast membrane

flux due to the slow diffusion of the permeating molecule. This drawback can be overcome by using the asymmetric membranes which are prepared using "phase inversion" method (Baker 2004; Mulder 1996). In the phase inversion method, the polymer is dissolved in suitable solvent, and the resulting polymer solution is cast on a mechanical support, generally a nonwoven support, and after a certain time interval (few seconds to minutes), it is immersed in a nonsolvent bath to make the polymer solution precipitate as a solid film (Fig. 1b). The resulting thin polymer film act as the membrane for the effective separation process. For the pervaporation process, the polymer solution can be immersed after the formation of a thin dense film on the surface of the cast polymer solution, and the immersion in the nonsolvent makes the bottom part of the film more porous compared with the dense nature of the top surface. This type of membrane is called asymmetric membrane where the cross section of the membrane is nonuniform in nature. Unlike dense film membrane where the polymer is distributed uniformly, the asymmetric membrane has two distinguished polymer layer, namely, top selective dense layer sitting on the more porous layer beneath. In the real application, asymmetric membranes are most widely used, and in most cases, a support is used for the mechanical stability of the asymmetric membrane. The support can be a nonwoven polymer or other highly porous material with adequate mechanical strength (Baker 2004; Mulder 1996).

An example of the asymmetric membrane on a support prepared by phase inversion process is shown in Fig. 2. The SEM image clearly shows the dense top selective layer followed by a highly porous layer sitting on the nonwoven support layer. The asymmetric membranes have considerable higher flux compared to the

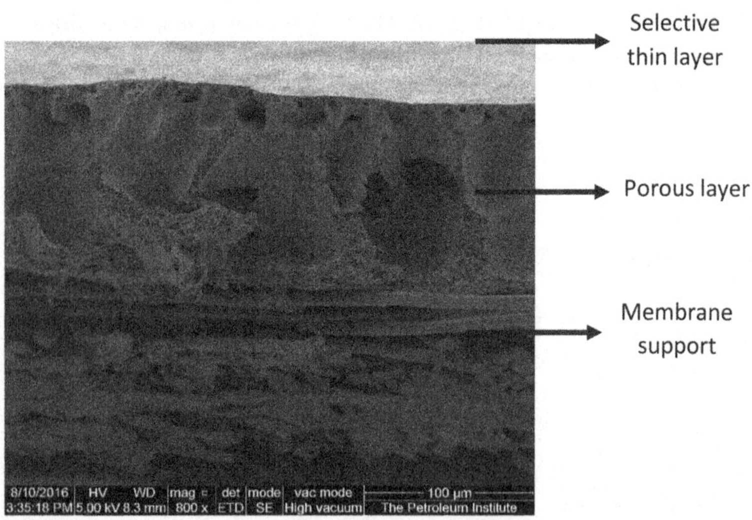

Fig. 2 SEM image of asymmetric membrane prepared on nonwoven support by phase inversion process

symmetric membranes like dense film due to less resistance. In the case of ethanol pervaporation separation process, the ethanol molecule has to diffuse through the whole thickness of dense film, whereas in the case of the asymmetric membrane, it has to pass through the thin selective layer alone resulting better flux. Therefore unless dense film membrane is required, the majority of the membrane separation is based on the asymmetric membranes (Baker 2004; Mulder 1996).

The addition of nanomaterials in membrane film makes the separation more efficient and also produces mechanically stable membrane. The addition of zeolite particle is particularly important for the pervaporation process. The addition of filler or selective nanomaterials in the polymer solution results in a composite membrane called as the mixed matrix membrane (MMM). The MMM can be prepared similar to the solution casting phase inversion process by mixing the nanomaterials with the polymer solution or spray coating the nanoparticle dispersion on the cast solution. The subsequent immersion in nonsolvent results in asymmetric mixed matric membrane and the process is schematically shown in Fig. 3.

The inorganic membranes for the pervaporation are based on the ceramic support with the seeded growth of zeolite type of materials on the surface of ceramic membranes. Alternatively, the organic polymers are also coated on the surface of the ceramic membrane which results in the high permeability of organic separation with a high mechanical stability for the ceramic membrane (Samei et al. 2013). The thin film composite (TFC) membranes can be made on the surface of the organic and ceramic membrane by interfacial crosslinking. This is widely used for the reverse osmosis of saline water to desalinated water production. By using different crosslinking, the TFC membranes can be fine-tuned for the ethanol pervaporation process (Ambrosi et al. 2018).

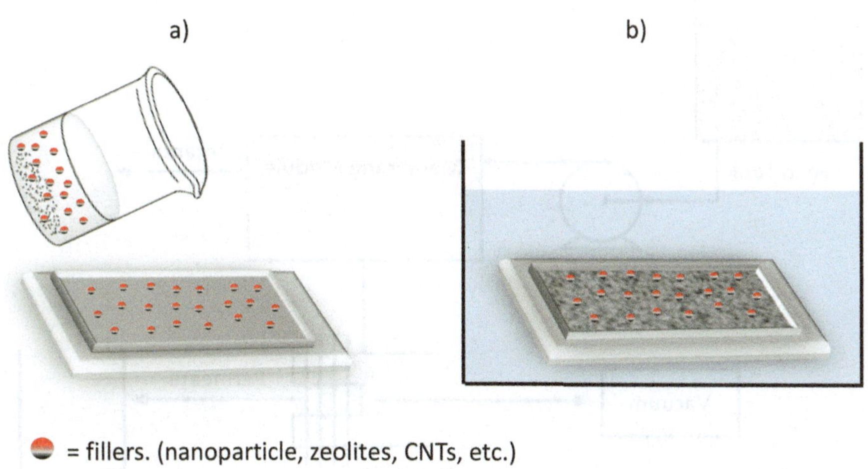

Fig. 3 Preparation of mixed matric membrane by (**a**) solution casting and (**b**) phase inversion

6 Membrane Separation by Pervaporation

The pervaporation process was first reported by Kober in the year 1917, and the term pervaporation is derived from the combination of two processes, namely, "permeation" and "evaporation" (Kanehashi and Nagai 2017). Pervaporation is a widely used membrane separation process for the liquid mixture separation (Mulder et al. 1983). Indeed, the first liquid mixture separation was demonstrated for the separation of ethanol-water in the year 1956 (Heisler et al. 1956) followed by the first commercial scale establishment in 1982 (Baker 2004). In the pervaporation separation process, the liquid mixture is fed on the membrane surface, and permeate is collected on the other side as the separated product. The schematic representation of the membrane pervaporation process is shown in Fig. 4. The solution-diffusion mechanism is used for the mass transport across the membrane, and the following three steps are involved in the separation process (Kanehashi and Nagai 2017).

1. Sorption of selective feed molecule by dissolution
2. Diffusion of the selective molecule across the membrane
3. Evaporation of the diffused molecule to the permeate side of the membrane

The diffusion is the rate-determining step in this process, and the driving force for the separation is the chemical potential gradient or partial vapor pressure difference across the membrane.

In the actual separation of ethanol from the fermentation, there are two possible configuration of membrane separation being possible. Pervaporation directly coupled with fermentation reactor where the product ethanol is continuously

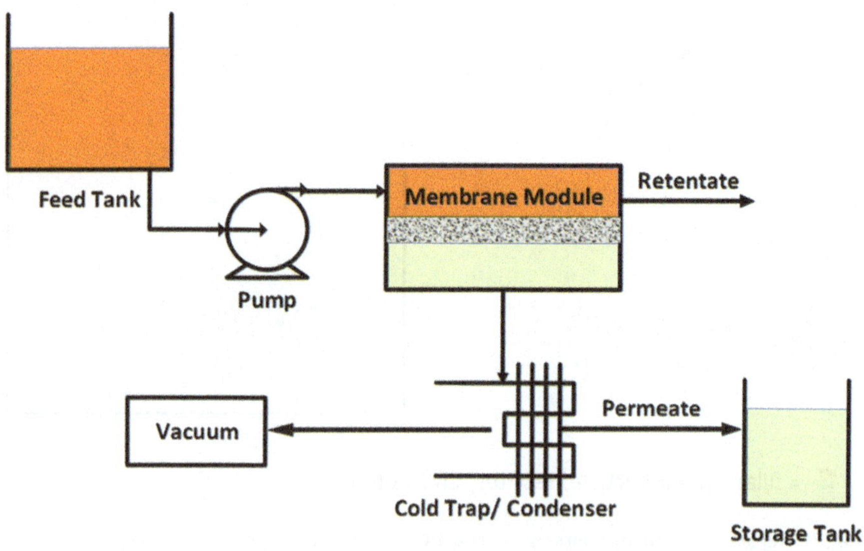

Fig. 4 General schematic representation of membrane pervaporation separation process

removed by the membrane separation. This is also sometimes called a membrane reactor (Amelio et al. 2016). The coupled process is advantageous since it continuously removes the product ethanol which is harmful to the fermenting microorganism. However, the presence of microorganism along with the starting feedstock, which is not of high quality in general, can cause severe membrane fouling. The fermentation solution at certain preheated condition is feed on the membrane top surface by the feed pump. The permeation side of the membrane is connected to the withdrawing source. The driving force for the separation can be a mild vacuum or cold trap (which creates a vacuum) or a sweep gas. Thus the product ethanol is continuously removed, and the remaining retentate solution is recirculated back to the feed tank. The continuous removal of ethanol drives the fermentation continuously, thereby avoiding the ethanol inhibition toward the microorganisms (Nakao et al. 1987).

Since membrane fouling is a severe issue for the direct coupling of the pervaporation membrane process, an intermediate membrane filtration is an attractive option to remove the particulate species from the fermentation broth. A typical membrane filtration coupled with membrane pervaporation process is shown in Fig. 5. The addition of porous membrane either microfiltration or ultrafiltration can considerably reduce the chances of microbial cell attachment on the pervaporation membrane (Pal et al. 2018).

7 Ethanol Separation by Organophilic Membrane Pervaporation

There are two different routes of membrane separation processes employed in the ethanol-water separation from the fermentation broth. The first one is the use of organophilic (hydrophobic) membranes for the selective permeation of ethanol. In this separation process, the preheated ethanol molecule selectively passes through an organic or inorganic membrane, and the collected ethanol can be enriched up to

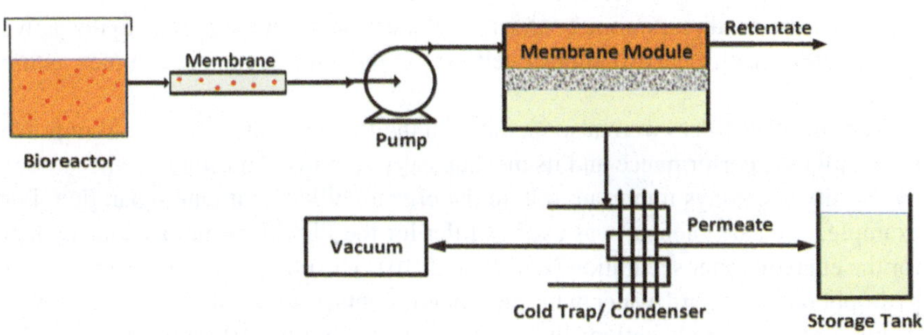

Fig. 5 General schematic representation of membrane filtration and pervaporation coupled separation process

80–90% purity. The permeate is generally collected using a vacuum or cold trap. As an alternative to the cold trap condensation, running water and air coolant were also studied for the pervaporation of ethanol-water mixture (Fan et al. 2017). The other major membrane process employed is the dehydration of enriched ethanol in to fuel grade ethanol by selectively removing water by passing through a hydrophilic (oleophobic) membrane. In this process, a highly hydrophilic membrane selectively removes the water molecule, and the retentate can be enriched up to fuel grade ethanol. In the following section, few selective membrane pervaporation processes are discussed.

As an organophilic membrane, polydimethylsiloxane (PDMS) is one of the earliest membrane used for the separation of ethanol-water by pervaporation process and still one of the dominant polymers for the membrane pervaporation process (Nakao et al. 1987). A highly flexible siloxane repeating unit capable of chemical and thermal stability along with ethanol selective permeability makes it one of the highly studied membrane material for the ethanol-water separation. Subsequently, zeolite-filled PDMS membranes were also studied for the pervaporation separation of alcohols (te Hennepe et al. 1987). Acetone-butanol-ethanol (ABE) recovery from fed-batch reactors was also reported using the silicalite-silicone membrane (Qureshi et al. 2001). In this case, an ultrafiltration membrane was used for the removal of *Clostridium acetobutylicum* (*C. acetobutylicum*) cells from the medium, and permeate was subjected to the pervaporation process of solvent recovery. These are some clear examples of the use of PDMS-based membranes for the ethanol-water separation using pervaporation process. Though PDMS is well-studied membrane material for the separation of ethanol-water, it has two distinguish drawbacks—dense nature producing low ethanol flux and the easy tendency of fouling. Various methods were employed in order to overcome these limitations, and among the most widely studied method was the use of nanofillers and PDMS-coated ceramic composite membranes. Thus, the PDMS-ceramic composite membranes were widely studied for the separation of ethanol-water (Xiangli et al. 2007). For example, PDMS/ceramic composite membrane was used for the pervaporation of fed-batch ABE fermentation and compared with and without in situ recovery. It was observed that the membrane was fouled due to the presence of active fermentation broth, and partial flux recovery was achieved by offline membrane cleaning (Wu et al. 2012). Nevertheless, the direct online connection of pervaporation membrane to the active fermentation broth clearly demonstrated the fouling issue due to the complex nature of direct feed.

Various other fillers were also included in the PDMS matrix in order to improve the membrane performance and its mechanically stability. Additionally, the surface wettability also plays important role in the organophilic membrane separation. For example, tree bark biochar was used as filler for the PDMS membrane and studied for the ethanol-water separation (Lan et al. 2016). The membrane showed the super hydrophobicity toward water with an ethanol contact angle of 27°. A separation factor of 10 with permeation flux of 0.22 kg m^{-2} h^{-1} for 10 wt% ethanol/water solution at 40 °C was obtained. A similar study demonstrating the hydrophobic effect of the superhydrophobic metal organic framework (MOF) of RHO-[Zn(eim)2]

(MAF-6) incorporated PDMS mixed matrix membrane was reported (Li et al. 2017). Compared to the conventional PDMS membrane, the new membrane showed a 1.5 times higher ethanol flux (1.2 kg m^{-2} h^{-1}) and 2.3 times better separation efficiency (factor 15). These studies clearly demonstrated the positive effect of fillers and surface wettability on the enhanced membrane performance.

In another study, the silicalite-1 was modified with various alkoxysilane groups which resulted in better dispersion of silicalite-1 in the PDMS matrix (Zhuang et al. 2016). It was observed that the silicalite-1 modified mixed matrix membrane resulted in fewer voids dense surface compared to unmodified silicalite-1 fillers. However, the modification with different silane coupling agent introduced the silane groups around silicalite-1 which reduced the separation factor. Among different silane coupling agents, vinyl-silicate-1 and PDMS provided thinnest active membrane layer which resulted in a separation factor of 34 for the dilute ethanol solution. The same mixed matrix polymer composition was used for the preparation of TFC membrane on the support of the PAN membrane (Yi and Wan 2017). The mixed matrix TFC membrane was subjected to the pervaporation of fermentation broth directly, and about 9 h operation did not show any meaningful fouling and claimed for the direct use of the membrane for the separation of ethanol directly from the fermentation broth. However, long-term stability needs to be studied in order to have better understanding of the newly developed membrane.

One of the advantages of the membrane-coupled fermentation process is the continuous removal of the solvents which facilitate the high conversion of carbohydrates into alcohols. One such coupled PDMS composite membrane showed about 90% utilization of glucose by continuous removal of ABE (Van Hecke et al. 2012). A similar study using thin-film silicalite-1 filled PDMS/PAN composite membrane showed much improved ABE productivity coupled with high glucose utilization due to the continuous removal of solvents from the fermentation broth (Li et al. 2013). The fermentation process also affects the performance of the membrane separation. It has been observed that the use of nutrients, type of yeast, and the organic by-products such as lactic acid formation severely affects the performance of silicone rubber-coated silicalite membranes (Ikegami et al. 2007). Thus, the influence of fermentation by-products on the ethanol pervaporation showed a clear trend of sorption on the selective layer and altering vapor-liquid equilibrium. The presence of sugar and salts altered the membrane performance. For example, 2,3-butanediol reduced the flux and selectivity due to strong sorption on PDMS membrane. Thus, the membrane performance will not only depend on the membrane material alone, and the fermentation process, i.e., broth media composition (starting materials and the products), also significantly influences the separation process.

One of the reasons for the membrane fouling in the direct membrane process is the presence of the microorganism in the feed, and various methods were attempted to overcome the issue. Different strategies have been widely utilized. The first one is the use of the filtration process, like microfiltration or ultrafiltration as discussed earlier, to remove the cells from the fermented broth. The second method is the immobilization of yeast cells, and after the fermentation, the fermented liquid can be easily separated by the membrane pervaporation process. For example, the yeast cells (*Saccharomyces cerevisiae*) were immobilized in alginate film and used for the fermentation process coupled with PDMS hybrid composite membrane pervaporation process (Santos et al. 2018). The immobilized yeast showed higher ethanol productivity due to the lesser ethanol presence in the vicinity of the trapped yeast in the alginate matrix. Though the concept of immobilized yeast provided better ethanol productivity, the pervaporation study was demonstrated using the physical mixture of ethanol-water, thereby not clearly demonstrating the efficacy of the real-time process. However, an innovative method of producing a high yield of ethanol using immobilized yeast with easy separation from the fermented liquid offers an attractive pathway for further study. A similar immobilization of the genus *Clostridium* cells was done in PVA cryogel (Efremenko et al. 2011). The resulting immobilized cells consistently produced high ethanol yield than free cells in the solution. The self-flocculating yeast usage in combination with the CNT-PDMS membrane was also reported (Xue et al. 2016). The self-flocculation of yeast provided better fouling resistance of the membrane, and 2 days operation did not show any noticeable fouling and decline in flux.

The other alternative way to reduce the fouling of pervaporation membrane is to combine the gas stripping followed by pervaporation process (Cai et al. 2016). In this combination of processes, the direct contact of fermentation broth with the pervaporation membrane is avoided, thereby reducing severe fouling. In a recent development, vapor permeation and gas stripping were studied and compared with the pervaporation process. It was proposed that the vapor permeation process can successfully compete with the pervaporation process with added advantages of avoiding fouling issues (Sun et al. 2017). However, it has to be seen how this process can be upscaled in noncontact membrane separation for large-scale usage of the fermentation process.

Another widely used polymer for the pervaporation process is poly[1-(trimethylsilyl)-1-propyne] (PTMSP) and it's earlier monomer family (Aoki et al. 1995; Sakaguchi et al. 2002). For example, Claes et al. reported the MMM of silica particles filled PTMSP selective layer on the PVDF membrane as support (Claes et al. 2012). A 25 wt% silica particles in PTMSP membrane with 2.4-micron thickness showed an ethanol flux of 9.5 kg m^{-2} h^{-1} with a separation factor of 104 for the 5% ethanol/water mixture at 50 °C. Apart from PDMS, the PTMSP is another promising class of polymer for the pervaporation membranes, and future developments on PTMSP can be promising for the ethanol-water separation.

The other class of polymers were also studied, and polyimide-/amide-based pervaporation membrane is an example (Jiang et al. 2009; Liu et al. 2005). The use of carbon nanotubes to poly(ether-block-amide) pervaporation membrane not only increased the mechanical strength of the membrane but also provided better operating performance for the separation of ABE. The addition of CNTs provided a 20% increase in productivity and yield with long-term membrane operational performance (Yen et al. 2012). An MMM of RHO-[Zn(eim)2] (MAF-6, Heim=2-ethylimidazole) nanoparticles and poly (ether-block-amide) (PEBA) on the ceramic hollow fiber was studied for the separation of the ethanol-water mixture (Liu et al. 2018).

The increase of water contact angle and the decrease in ethanol contact angle was observed as the loading of nanoparticles increased. Thus, the flux increased with nanoparticles concentration along with separation efficiency up to 8 wt%. A high ethanol flux of 4.446 kg m^{-2} h^{-1} with a separation factor of 5.6 was observed for 5 wt% ethanol in water at 60 °C. At the optimized condition, long-term operation of about 200 h showed stable flux and separation efficiency.

The use of organic membranes dominated the early studies of ethanol separation by pervaporation process. However, inorganic ceramic membrane as the base and various selective layers like polymer coatings, seed growth of zeolite are recently getting more attention due to its ability to perform better and withstand vigorous membrane cleaning process. Therefore recent works are carried out for the ethanol separation by pervaporation using inorganic membranes. For example, an ultrathin 0.5-micron thickness MFI membrane was prepared on the surface of the ceramic γ-Al_2O_3 membrane and used for the high flux ethanol-water separation (Korelskiy et al. 2013). A very high ethanol flux of 9 kg m^{-2} h^{-1} with a separation factor of 5 was reported for 10% ethanol/water mixture at 60 °C. The observed flux is relatively high which is presumably due to the fact of the ultralow thickness of MFI selective layer. Similarly, a high-performance silicalite-1 zeolite pervaporation membrane was prepared on porous tubular silica supports by a secondary growth method (Ueno et al. 2017). The optimized membrane showed the ethanol flux of 3 kg m^{-2} h^{-1} and separation factor of 92 for 10 wt% ethanol at 50 °C which is highest for the tubular supports.

One of the drawbacks of pure silica-based MFI zeolite membrane is the deterioration of separation during the long-time operation. It was due to the loss of silicon atom on the surface of MFI molecular sieves during the synthesis in alkaline condition, resulting in silanol (Si-OH) groups. The silanol group can react with ethanol, and other compounds present and block the narrow pore of the zeolite membrane. To overcome this limitation, the surface of the zeolite membrane was modified with dopamine and demonstrated successfully for the pervaporation performance for 180 h with a separation factor of 44 and the flux of 2.6 kg m^{-2} h^{-1} (Wu et al. 2018). The unmodified membrane showed considerable flux decline with a separation factor of 1 which is the complete loss of separation ability. Table 1 shows some selective organophilic membrane pervaporation process reported in the literature. As seen from the table, the flux and separation efficiency of the membrane operate in opposite way. If high flux is obtained, then separation factor goes down and vice versa.

8 Water Separation by Hydrophilic Membrane Dehydration

The organophilic membrane separation process can enrich ethanol from fermentation up to 80–90% purity, and the removal of remaining water content is another critical part of the separation process. Since ethanol-water forms azeotrope, the separation of the water level acceptable to fuel grade ethanol is one of the energy-consuming

Table 1 Details of hydrophobic membranes used for the ethanol pervaporation studies

Active layer/ membrane support	Feed alcohol content wt%	Temperature (°C)	Flux (kg m^{-2} h^{-1})	Separation factor	Reference
Dopamine modified MFI Zeolite	5% ethanol	60	2.6	44	Wu et al. (2018)
PDMS/polyamide	4	45	1.9	8.5	Peng et al. (2010)
Poly(1-trimethylsilyl-1-propyne) (PTMSP)—Neat film	10	50	1.1	15	González-Marcos et al. (2004)
Polymer of intrinsic microporosity-1 (PIM-1)—neat film	10	60	1.4	9	Adymkanov et al. (2008)
PVDF hallow fiber	5	50	~5	~5	Sukitpaneenit and Chung (2011)
PDMS-b-sulfone-b-4-hydroxystyrene	10	50	4.5	6.8	Okamoto et al. (1987)

processes. The removal of residual water is generally done by dehydration using membrane process or adsorption (Bolto et al. 2011). For membrane dehydration, the membrane material possesses the characteristics of high hydrophilic nature, and in most case hydrophilic polymers are crosslinked appropriately during the membrane preparation.

One of the earliest hydrophilic membranes studied for the pervaporation was polyvinyl alcohol (PVA) crosslinked with glutaraldehyde (Lee et al. 1992). In general, the hydrophilic polymers are crosslinked with suitable crosslinkers in order to avoid the swelling of the polymer molecule which results in the deterioration of the membrane separation. For example, the PVA and sodium alginate polymers were crosslinked with maleic anhydride on the surface of the polysulfone hollow fiber membrane and studied for the dehydration of various alcohols (Dong et al. 2006). A typical characteristic feature for flux and separation performance was observed for the separation of ethanol, isopropanol, *n*-butanol, and *tert*-butanol. For example, a higher degree of crosslinking provided tighter membrane with less degree of membrane swelling which translates better separation, but the flux was declined correspondingly. The water flux of the membrane can be increased by decreasing the degree of crosslinking, but the separation efficiency decreased due to the swelling of the membrane thereby contributing to the inefficient separation. A pilot scale study of ethanol dehydration using crosslinked PVA membrane was recently reported (Niemistö et al. 2013). It was observed that the activated carbon filtration was the efficient pretreatment for the feed quality used in the pervaporation process. The quality of the ethanol retentate met the requirement of the EU standard.

Polyelectrolyte membrane based on chitosan (CS) and poly(sodium vinyl sulfonate) (PVS) via the "complexation-sulfation" method was also reported for the dehydration of alcohols (Zheng et al. 2016). The CS and PVS were subjected to complexation at pH 6.15 and subsequent sulfation reaction. The resulting polymer membrane was subjected to the dehydration of ethanol. The plain CS membrane showed poor separation efficiency for 10 wt% water-ethanol mixture, whereas the newly developed membrane resulted in fuel grade ethanol (99.55%) and water flux of 2 kg m^{-2} h^{-1} at 70 °C.

One of the issues with the dehydration membrane is the presence of acidic compounds formed during the fermentation process which can have deleterious effect on the membrane. To overcome this issue, acid-resistance polyethyleneimine-sodium alginate (PEI-SA) was prepared by layer-by-layer (L*b*L) technique on PAN membrane support (Li et al. 2018). The ionic crosslinked membrane showed long-term stability at pH 3 for the separation of 90 wt% ethanol/water mixture with a permeate flux of about 1 kg m^{-2} h^{-1} and separation factor of 1542. The permeate water content was above 99% for about 5 days operational. It was observed that the membrane withstood low pH of the feed without compromising the flux or separation efficiency.

An intriguing result of 1-butanol dehydration using PVA crosslinked membrane on the Al_2O_3 ceramic membrane was reported (Peters et al. 2006). Unlike conventional membrane separation process, where flux and separation factor is operating in opposing direction, the reported membrane showed better separation as the flux was increased by increasing the feed temperature for the dehydration of 1-butanol and 2-propanol. Unfortunately, this high flux and high selectivity behavior were not observed for the dehydration of ethanol or 1-propanol, and the membrane showed a typical characteristic of high flux and low selectivity and vice versa. However, the study offered a way to develop a membrane separation process with high flux and high separation efficiency. An appropriate membrane material can be identified and optimized for the required operational conditions.

The other class of membranes used for dehydration is the zeolite-based ethanol dehydration. The zeolite is grown on the ceramic membrane or added in the polymer matrix in order to enhance the separation efficiency. For example, the surface-modified zeolitic imidazolate framework (ZIF-8) was demonstrated for the superiority of the nanoparticle distribution in the PVA membrane matrix for the dehydration of ethanol (Zhang and Wang 2016). In this case, ethylenediamine-modified and ethylenediamine-unmodified ZIF-8 as nanofillers for the PVA membrane were studied. The ethylenediamine modification of ZIF-8 allowed the particles to disperse in the polymer due to hydrogen bonding, and the flux increased with increasing amount of surface-modified ZIF-8, and separation factor also increased up to 7.5 wt% particle loading and then decreased due to particle agglomeration. At optimized condition, the MMM showed a water flux of 0.13 kg m^{-2} h^{-1} with a separation factor of 201 at 40 °C operating condition. Table 2 shows few selective membrane dehydration processes from the literature. It is clear that the flux and separation efficiency can be improved by incorporating nanofillers on the membrane matrix.

Table 2 Details of hydrophilic membranes used for the ethanol-water separation studies via dehydration mechanism

Membrane	Feed alcohol content wt%	Temperature (°C)	Flux (kg m^{-2} h^{-1})	Separation factor	Reference
Ethylenediamine modified ZIF-8	85	40	0.13	201	Zhang and Wang (2016)
Zeolite NaA	90	50	1.2	8500	Pera-Titus et al. (2008)
Zeolite TiO2	90	80	1.0	800	Sekulić et al. (2005)
TFC—polyimide	90	40	1.7	240	Kim et al. (2000)
Perfluoro-coated cellulose ester composite membrane	90	75	3	65	Huang et al. (2013b)
Polyamide-imide/ polyetherimide dual-layer hollow fiber	85	60	0.7	800	Wang et al. (2009)
Pervap® 2201	90	60	0.1	100	Van Baelen et al. (2005)
Sulfonated polysulfone	90	25	0.6	100	Chen et al. (2009)
TFC membrane on PAN support[a]	90	25	1.2	1500	Huang et al. (2010)
TFC membrane on PAN support[b]	90	25	2.2	1800	Huang et al. (2013a)
TFC membrane on PES hallow fiber support[c]	85	50	7.5	60	Sukitpaneenit and Chung (2014)

[a]Triethylenetetramine-TMC-based TFC film
[b]m-tolidine-H-TMC-based TFC film
[c]TFC film further coated with polydopamine or silicone rubber coatings

9 Summary and Outlook

The use of bioethanol as an alternating fuel is growing fast, and the immediate widespread success is depending on the competitive pricing with petroleum-based fuel. One of the major defining factors about the cost of bioethanol is the separation and purification of ethanol from the fermentation broth. There are many large-scale ethanol production units emerging to compete and compliment with the conventional fuel by blending. So far, the conventional distillation process is the first choice of the separation process for the ethanol purification. Though distillation requires high energy cost, it succeeds in its ease of operational and long-term reliability. Undoubtedly, in theory, the membrane separation is a cost-efficient separation process. However, the issue of fouling and long-term membrane stability is plaguing the widespread adaptation of the technology. Attempts were made to address these limitations and simultaneously improve the efficiency of the membrane separation. Assuming the challenges addressed in near future, the forthcoming research direction should be identifying the combination of membrane processes for the whole fermentation process instead of ethanol-water separation alone. There are many areas of pre-fermentation process in which the membrane process can be used, for example, for the starting material preparation and post-fermentation filtration. In fact, existing size exclusion-based separation of microfiltration and ultrafiltration should be widely tested for the starting material purification and intermediate use between fermentation and pervaporation process. In this direction, there are few combined membrane process that has been recently reported. The other area of interest is to combine the organophilic and hydrophilic membrane pervaporation for the ethanol-water separation. In one such case, the use of hydrophobic and hydrophilic pervaporation systems was demonstrated where hydrophobic PDMS membrane was used to concentrate ethanol to 80% from the fermentation broth (Nigiz and Durmaz Hilmioglu 2016). Subsequent use of carboxymethyl cellulose membrane for the dehydration resulted in 99% ethanol. As discussed earlier, the combination of nanofiltration and pervaporation process combination was also studied. The addition of micro and ultrafiltration as prefiltration stages can considerably increase the membrane performance. The other area of membrane development can be membrane distillation process (Banat and Shannag 2000; Munirasu et al. 2017). Compared to the solution-diffusion pervaporation process, the membrane distillation can be considerably energy efficient. However, the existing current-generation membranes have considerable limitations, and development of new membranes with superior antiwettability can lead to the new direction for the ethanol separation process in the biorefinery.

References

Adymkanov SV, Yampol'skii YP, Polyakov AM, Budd PM, Reynolds KJ, McKeown NB, Msayib KJ (2008) Pervaporation of alcohols through highly permeable PIM-1 polymer films. Polym Sci Ser A 50(4):444–450. https://doi.org/10.1134/S0965545X08040135

Ambrosi A, Al-Furaiji M, McCutcheon JR, Cardozo NSM, Tessaro IC (2018) Transport of components in the separation of ethanol from aqueous dilute solutions by forward osmosis. Ind Eng Chem Res 57(8):2967–2975. https://doi.org/10.1021/acs.iecr.7b04944

Amelio A, Van der Bruggen B, Lopresto C, Verardi A, Calabro V, Luis P (2016) Pervaporation membrane reactors. In: Membrane technologies for biorefining. pp 331–381. https://doi.org/10.1016/b978-0-08-100451-7.00014-1

Aoki T, Kobayashi H, Oikawa E (1995) Poly(p-substituted phenylacetylene) with perfluoroalkyloxydimethylsilyl side groups for oxygen and ethanol permselective membrane. Polym Bull 34(2):133–140. https://doi.org/10.1007/BF00316387

Baker RW (2004) Membrane technology and applications, 2nd edn. Wiley, West Sussex

Banat FA, Shannag MA (2000) Recovery of dilute acetone–butanol–ethanol (ABE) solvents from aqueous solutions via membrane distillation. Bioprocess Eng 23:43–649

Bolto B, Hoang M, Xie Z (2011) A review of membrane selection for the dehydration of aqueous ethanol by pervaporation. Chem Eng Process Process Intensif 50(3):227–235. https://doi.org/10.1016/j.cep.2011.01.003

Bungay HR (2004) Confessions of a bioenergy advocate. Trends Biotechnol 22(2):67–71. https://doi.org/10.1016/j.tibtech.2003.12.002

Cai D, Chen H, Chen C, Hu S, Wang Y, Chang Z, Miao Q, Qin P, Wang Z, Wang J, Tan T (2016) Gas stripping–pervaporation hybrid process for energy-saving product recovery from acetone–butanol–ethanol (ABE) fermentation broth. Chem Eng J 287:1–10. https://doi.org/10.1016/j.cej.2015.11.024

Chen S-H, Liou R-M, Lin Y-Y, Lai C-L, Lai J-Y (2009) Preparation and characterizations of asymmetric sulfonated polysulfone membranes by wet phase inversion method. Eur Polym J 45(4):1293–1301. https://doi.org/10.1016/j.eurpolymj.2008.11.030

Chovau S, Degrauwe D, Van der Bruggen B (2013) Critical analysis of techno-economic estimates for the production cost of lignocellulosic bio-ethanol. Renew Sust Energ Rev 26:307–321. https://doi.org/10.1016/j.rser.2013.05.064

Claes S, Vandezande P, Mullens S, De Sitter K, Peeters R, Van Bael MK (2012) Preparation and benchmarking of thin film supported PTMSP-silica pervaporation membranes. J Membr Sci 389:265–271. https://doi.org/10.1016/j.memsci.2011.10.035

Curcio S, De Luca G, Saha K, Chakraborty S (2016) 1—advance membrane separation processes for biorefineries. In: Figoli A, Cassano A, Basile A (eds) Membrane technologies for biorefining. Woodhead Publishing, Cambridge, pp 3–28. https://doi.org/10.1016/B978-0-08-100451-7.00001-3

Dong YQ, Zhang L, Shen JN, Song MY, Chen HL (2006) Preparation of poly(vinyl alcohol)-sodium alginate hollow-fiber composite membranes and pervaporation dehydration characterization of aqueous alcohol mixtures. Desalination 193(1–3):202–210. https://doi.org/10.1016/j.desal.2005.08.023

Efremenko EN, Stepanov NA, Nikolskaya AB, Senko OV, Spiricheva OV, Varfolomeev SD (2011) Biocatalysts based on immobilized cells of microorganisms in the production of bioethanol and biobutanol. Catal Ind 3(1):41–46. https://doi.org/10.1134/s207005041101003x

Fan S, Xiao Z, Li M, Li S, Zhou T, Hu Y, Wu S (2017) Pervaporation performance in PDMS membrane bioreactor for ethanol recovery with running water and air as coolants at room temperature. J Chem Technol Biotechnol 92(2):292–297. https://doi.org/10.1002/jctb.5004

Galbe M, Zacchi G (2002) A review of the production of ethanol from softwood. Appl Microbiol Biotechnol 59(6):618–628. https://doi.org/10.1007/s00253-002-1058-9

Girio FM, Fonseca C, Carvalheiro F, Duarte LC, Marques S, Bogel-Lukasik R (2010) Hemicelluloses for fuel ethanol: a review. Bioresour Technol 101(13):4775–4800. https://doi.org/10.1016/j.biortech.2010.01.088

González-Marcos JA, López-Dehesa C, González-Velasco JR (2004) Effect of operation conditions in the pervaporation of ethanol–water mixtures with poly(1-trimethylsilyl-1-propyne) membranes. J Appl Polym Sci 94(4):1395–1403. https://doi.org/10.1002/app.21029

Heisler EG, Hunter AS, Siciliano J, Treadway RH (1956) Solute and temperature effects in the pervaporation of aqueous alcoholic solutions. Science 124:77–79

Huang Z, Shi Y, Wen R, Guo Y-H, Su J-F, Matsuura T (2006) Multilayer poly(vinyl alcohol)–zeolite 4A composite membranes for ethanol dehydration by means of pervaporation. Sep Purif Technol 51(2):126–136. https://doi.org/10.1016/j.seppur.2006.01.005

Huang S-H, Hung W-S, Liaw D-J, Lo C-H, Chao W-C, Hu C-C, Li C-L, Lee K-R, Lai J-Y (2010) Interfacially polymerized thin-film composite polyamide membranes: effects of annealing processes on pervaporative dehydration of aqueous alcohol solutions. Sep Purif Technol 72(1):40–47. https://doi.org/10.1016/j.seppur.2009.12.026

Huang Y-H, Huang S-H, Chao W-C, Li C-L, Hsieh Y-Y, Hung W-S, Liaw D-J, Hu C-C, Lee K-R, Lai J-Y (2013a) A study on the characteristics and pervaporation performance of polyamide thin-film composite membranes with modified polyacrylonitrile as substrate for bioethanol dehydration. Polym Int 63(8):1478–1486. https://doi.org/10.1002/pi.4649

Huang Y, Baker RW, Wijmans JG (2013b) Perfluoro–coated hydrophilic membranes with improved selectivity. Ind Eng Chem Res 52(3):1141–1149. https://doi.org/10.1021/ie3020654

Ikegami T, Negishi H, Yanase H, Sakaki K, Okamoto M, Koura N, Sano T, Haraya K, Yanagishita H (2007) Stabilized production of highly concentrated bioethanol from fermentation broths by Zymomonas mobilis by pervaporation using silicone rubber-coated silicalite membranes. J Chem Technol Biotechnol 82(8):745–751. https://doi.org/10.1002/jctb.1724

Jiang LY, Wang Y, Chung T-S, Qiao XY, Lai J-Y (2009) Polyimides membranes for pervaporation and biofuels separation. Prog Polym Sci 34(11):1135–1160. https://doi.org/10.1016/j.progpolymsci.2009.06.001

Jin W (2016) Ethanol production by continous fermentation-pervaporation. In: Encyclopedia of membranes. p 721. https://doi.org/10.1007/978-3-662-44324-8

Kanehashi S, Nagai K (2017) Fundamentals and perspectives for pervaporation. In: Comprehensive membrane science and engineering II, vol 2. pp 191–225. https://doi.org/10.1016/b978-0-12-409547-2.12217-6

Kim J-H, Lee K-H, Kim SY (2000) Pervaporation separation of water from ethanol through polyimide composite membranes. J Membr Sci 169(1):81–93. https://doi.org/10.1016/S0376-7388(99)00335-X

Kiss AA, Lange J-P, Schuur B, Brilman DWF, van der Ham AGJ, Kersten SRA (2016) Separation technology—making a difference in biorefineries. Biomass Bioenergy 95:296–309. https://doi.org/10.1016/j.biombioe.2016.05.021

Korelskiy D, Leppäjärvi T, Zhou H, Grahn M, Tanskanen J, Hedlund J (2013) High flux MFI membranes for pervaporation. J Membr Sci 427:381–389. https://doi.org/10.1016/j.memsci.2012.10.016

Lan Y, Yan N, Wang W (2016) Application of PDMS pervaporation membranes filled with tree bark biochar for ethanol/water separation. RSC Adv 6(53):47637–47645. https://doi.org/10.1039/c6ra06794h

Lee YM, Nam SY, Kim JH (1992) Pervaporation of water-ethanol through poly(vinyl alcohol)/chitosan blend membrane. Polym Bull 29(3):423–429. https://doi.org/10.1007/BF00944840

Li J, Chen X, Qi B, Luo J, Zhuang X, Su Y, Wan Y (2013) Continuous acetone–butanol–ethanol (ABE) fermentation with in situ solvent recovery by silicalite-1 filled PDMS/PAN composite membrane. Energy Fuel 28(1):555–562. https://doi.org/10.1021/ef401706k

Li Q, Cheng L, Shen J, Shi J, Chen G, Zhao J, Duan J, Liu G, Jin W (2017) Improved ethanol recovery through mixed-matrix membrane with hydrophobic MAF-6 as filler. Sep Purif Technol 178:105–112. https://doi.org/10.1016/j.seppur.2017.01.024

Li J, Si X, Li X, Wang N, An Q, Ji S (2018) Preparation of acid-resistant PEI/SA composite membranes for the pervaporation dehydration of ethanol at low pH. Sep Purif Technol 192:205–212. https://doi.org/10.1016/j.seppur.2017.09.038

Liu F, Liu L, Feng X (2005) Separation of acetone–butanol–ethanol (ABE) from dilute aqueous solutions by pervaporation. Sep Purif Technol 42(3):273–282. https://doi.org/10.1016/j.seppur.2004.08.005

Liu D, Chen Y, Ding F-Y, Zhao T, Wu J-L, Guo T, Ren H-F, Li B-B, Niu H-Q, Cao Z, Lin X-Q, Xie J-J, He X-J, Ying H-J (2014) Biobutanol production in a Clostridium acetobutylicum biofilm reactor integrated with simultaneous product recovery by adsorption. Biotechnol Biofuels 7:5. https://doi.org/10.1186/1754-6834-7-5

Liu Q, Li Y, Li Q, Liu G, Liu G, Jin W (2018) Mixed-matrix hollow fiber composite membranes comprising of PEBA and MOF for pervaporation separation of ethanol/water mixtures. Sep Purif Technol. https://doi.org/10.1016/j.seppur.2018.01.050

McGregor I, Furlong S (2017) Concurrent alcohol recovery and fermentation using pass-through distillation. Ind Biotechnol 13(3):107–112. https://doi.org/10.1089/ind.2017.29081.imc

Mulder M (1996) Basic principles of membrane technology. Kluwer Academic Publishers, Dordrecht

Mulder MHV, Hendrickman JO, Hegeman H, Smolders CA (1983) Ethanol—water separation by pervaporation. J Membr Sci 16:269–284. https://doi.org/10.1016/S0376-7388(00)81315-0

Munirasu S, Haija MA, Banat F (2016) Use of membrane technology for oil field and refinery produced water treatment—a review. Process Saf Environ Prot 100:183–202. https://doi.org/10.1016/j.psep.2016.01.010

Munirasu S, Banat F, Durrani AA, Haija MA (2017) Intrinsically superhydrophobic PVDF membrane by phase inversion for membrane distillation. Desalination 417:77–86. https://doi.org/10.1016/j.desal.2017.05.019

Naik SN, Goud VV, Rout PK, Dalai AK (2010) Production of first and second generation biofuels: a comprehensive review. Renew Sust Energ Rev 14(2):578–597. https://doi.org/10.1016/j.rser.2009.10.003

Nakao S-i, Saitoh F, Asakura T, Toda K, Kimura S (1987) Continuous ethanol extraction by pervaporation from a membrane bioreactor. J Membr Sci 30(3):273–287. https://doi.org/10.1016/S0376-7388(00)80123-4

Niemistö J, Pasanen A, Hirvelä K, Myllykoski L, Muurinen E, Keiski RL (2013) Pilot study of bioethanol dehydration with polyvinyl alcohol membranes. J Membr Sci 447:119–127. https://doi.org/10.1016/j.memsci.2013.06.048

Nigiz FU, Durmaz Hilmioglu N (2016) Anhydrous fuel ethanol production by a combined hydrophobic–hydrophilic pervaporation system. Energy Sources A 38(22):3348–3353. https://doi.org/10.1080/15567036.2016.1148085

Okamoto KI, Butsuen A, Nishioka S, Tsuru S, Tanaka K, Kita H, Asakawa S (1987) Pervaporation of water-ethanol mixtures through poly-dimethylsiloxane block-copolymer membranes. Polym J 19(6):747–756. https://doi.org/10.1295/polymj.19.747

Pal P, Kumar R, Ghosh AK (2018) Analysis of process intensification and performance assessment for fermentative continuous production of bioethanol in a multi-staged membrane-integrated bioreactor system. Energy Convers Manag 171:371–383. https://doi.org/10.1016/j.enconman.2018.05.099

Peng P, Shi B, Lan Y (2010) A review of membrane materials for ethanol recovery by pervaporation. Sep Sci Technol 46(2):234–246. https://doi.org/10.1080/01496395.2010.504681

Pera-Titus M, Bausach M, Llorens J, Cunill F (2008) Preparation of inner-side tubular zeolite NaA membranes in a continuous flow system. Sep Purif Technol 59(2):141–150. https://doi.org/10.1016/j.seppur.2007.05.038

Peters T, Poeth C, Benes N, Buijs H, Vercauteren F, Keurentjes J (2006) Ceramic-supported thin PVA pervaporation membranes combining high flux and high selectivity; contradicting the flux-selectivity paradigm. J Membr Sci 276(1–2):42–50. https://doi.org/10.1016/j.memsci.2005.06.066

Qureshi N, Meagher MM, Huang J, Hutkins RW (2001) Acetone butanol ethanol (ABE) recovery by pervaporation using silicalite–silicone composite membrane from fed-batch reactor of Clostridium acetobutylicum. J Membr Sci 187(1):93–102. https://doi.org/10.1016/S0376-7388(00)00667-0

Ranjan A, Moholkar VS (2012) Biobutanol: science, engineering, and economics. Int J Energy Res 36(3):277–323. https://doi.org/10.1002/er.1948

Saha K, Uma Maheswari R, Sikder J, Chakraborty S, da Silva SS, dos Santos JC (2017) Membranes as a tool to support biorefineries: applications in enzymatic hydrolysis, fermentation and dehydration for bioethanol production. Renew Sust Energ Rev 74:873–890. https://doi.org/10.1016/j.rser.2017.03.015

Sakaguchi T, Yumoto K-i, Kwak G, Yoshikawa M, Masuda T (2002) Pervaporation of ethanol/water and benzene/cyclohexane mixtures using novel substituted polyacetylene membranes. Polym Bull 48(3):271–276. https://doi.org/10.1007/s00289-002-0041-z

Samei M, Mohammadi T, Asadi AA (2013) Tubular composite PVA ceramic supported membrane for bio-ethanol production. Chem Eng Res Des 91(12):2703–2712. https://doi.org/10.1016/j.cherd.2013.03.008

Sanchez OJ, Cardona CA (2008) Trends in biotechnological production of fuel ethanol from different feedstocks. Bioresour Technol 99(13):5270–5295. https://doi.org/10.1016/j.biortech.2007.11.013

Santos ELI, Rostro-Alanis M, Parra-Saldivar R, Alvarez AJ (2018) A novel method for bioethanol production using immobilized yeast cells in calcium-alginate films and hybrid composite pervaporation membrane. Bioresour Technol 247:165–173. https://doi.org/10.1016/j.biortech.2017.09.091

Sekulić J, Elshof AT, Blank DHA (2005) Separation mechanism in dehydration of water/organic binary liquids by pervaporation through microporous silica. J Membr Sci 254(1):267–274. https://doi.org/10.1016/j.memsci.2005.01.013

Shafiee S, Topal E (2009) When will fossil fuel reserves be diminished? Energy Policy 37(1):181–189. https://doi.org/10.1016/j.enpol.2008.08.016

Singh NK, Dhar DW (2011) Microalgae as second generation biofuel. A review. Agron Sustain Dev 31(4):605–629. https://doi.org/10.1007/s13593-011-0018-0

Singh A, Rangaiah GP (2017) Review of technological advances in bioethanol recovery and dehydration. Ind Eng Chem Res 56(18):5147–5163. https://doi.org/10.1021/acs.iecr.7b00273

Sukitpaneenit P, Chung T-S (2011) Molecular design of the morphology and pore size of PVDF hollow fiber membranes for ethanol–water separation employing the modified pore-flow concept. J Membr Sci 374(1):67–82. https://doi.org/10.1016/j.memsci.2011.03.016

Sukitpaneenit P, Chung T-S (2014) Fabrication and use of hollow fiber thin film composite membranes for ethanol dehydration. J Membr Sci 450:124–137. https://doi.org/10.1016/j.memsci.2013.08.047

Sun W, Jia W, Xia C, Zhang W, Ren Z (2017) Study of in situ ethanol recovery via vapor permeation from fermentation. J Membr Sci 530:192–200. https://doi.org/10.1016/j.memsci.2017.02.034

te Hennepe HJC, Bargeman D, Mulder MHV, Smolders CA (1987) Zeolite-filled silicone rubber membranes: part 1. Membrane preparation and pervaporation results. J Membr Sci 35(1):39–55. https://doi.org/10.1016/S0376-7388(00)80921-7

Ueno K, Negishi H, Okuno T, Saito T, Tawarayama H, Ishikawa S, Miyamoto M, Uemiya S, Sawada Y, Oumi Y (2017) High-performance silicalite-1 membranes on porous tubular silica supports for separation of ethanol/water mixtures. Sep Purif Technol 187:343–354. https://doi.org/10.1016/j.seppur.2017.06.071

Ullah K, Ahmad M, Sofia, Sharma VK, Lu P, Harvey A, Zafar M, Sultana S (2015) Assessing the potential of algal biomass opportunities for bioenergy industry—a review. Fuel 143:414–423. https://doi.org/10.1016/j.fuel.2014.10.064

Van Baelen D, Van der Bruggen B, Van den Dungen K, Degreve J, Vandecasteele C (2005) Pervaporation of water–alcohol mixtures and acetic acid–water mixtures. Chem Eng Sci 60(6):1583–1590. https://doi.org/10.1016/j.ces.2004.10.030

Van Hecke W, Vandezande P, Claes S, Vangeel S, Beckers H, Diels L, De Wever H (2012) Integrated bioprocess for long-term continuous cultivation of Clostridium acetobutylicum coupled to pervaporation with PDMS composite membranes. Bioresour Technol 111:368–377. https://doi.org/10.1016/j.biortech.2012.02.043

Vane LM (2005) A review of pervaporation for product recovery from biomass fermentation processes. J Chem Technol Biotechnol 80(6):603–629. https://doi.org/10.1002/jctb.1265

Vane LM (2008) Separation technologies for the recovery and dehydration of alcohols from fermentation broths. Biofuels Bioprod Biorefin 2(6):553–588. https://doi.org/10.1002/bbb.108

Vane LM, Alvarez FR (2013) Hybrid vapor stripping-vapor permeation process for recovery and dehydration of 1-butanol and acetone/butanol/ethanol from dilute aqueous solutions. Part 1. Process simulations. J Chem Technol Biotechnol 88(8):1436–1447. https://doi.org/10.1002/jctb.4087

Vane LM, Alvarez FR, Huangb Y, Baker RW (2010) Experimental validation of hybrid distillation-vapor permeation process for energy efficient ethanol–water separation. J Chem Technol Biotechnol 85:502–511. https://doi.org/10.1002/jctb.2318

Wang Y, Goh SH, Chung TS, Na P (2009) Polyamide-imide/polyetherimide dual-layer hollow fiber membranes for pervaporation dehydration of C1–C4 alcohols. J Membr Sci 326(1):222–233. https://doi.org/10.1016/j.memsci.2008.10.005

Wei P, Cheng L-H, Zhang L, Xu X-H, Chen H-l, Gao C-j (2014) A review of membrane technology for bioethanol production. Renew Sust Energ Rev 30:388–400. https://doi.org/10.1016/j.rser.2013.10.017

Wu H, Chen XP, Liu GP, Jiang M, Guo T, Jin WQ, Wei P, Zhu DW (2012) Acetone-butanol-ethanol (ABE) fermentation using Clostridium acetobutylicum XY16 and in situ recovery by PDMS/ceramic composite membrane. Bioprocess Biosyst Eng 35(7):1057–1065. https://doi.org/10.1007/s00449-012-0721-5

Wu Z, Zhang C, Peng L, Wang X, Kong Q, Gu X (2018) Enhanced stability of MFI zeolite membranes for separation of ethanol/water by eliminating surface Si-OH groups. ACS Appl Mater Interfaces 10(4):3175–3180. https://doi.org/10.1021/acsami.7b17191

Xiangli F, Chen Y, Jin W, Xu N (2007) Polydimethylsiloxane (PDMS)-ceramic composite membrane with high flux for pervaporation of ethanol−water mixtures. Ind Eng Chem Res 46:2224–2230

Xue C, Du G-Q, Sun J-X, Chen L-J, Gao S-S, Yu M-L, Yang S-T, Bai F-W (2014) Characterization of gas stripping and its integration with acetone–butanol–ethanol fermentation for high-efficient butanol production and recovery. Biochem Eng J 83:55–61. https://doi.org/10.1016/j.bej.2013.12.003

Xue C, Wang Z-X, Du G-Q, Fan L-H, Mu Y, Ren J-G, Bai F-W (2016) Integration of ethanol removal using carbon nanotube (CNT)-mixed membrane and ethanol fermentation by self-flocculating yeast for antifouling ethanol recovery. Process Biochem 51(9):1140–1146. https://doi.org/10.1016/j.procbio.2016.05.030

Yen HW, Chen ZH, Yang IK (2012) Use of the composite membrane of poly(ether-block-amide) and carbon nanotubes (CNTs) in a pervaporation system incorporated with fermentation for butanol production by Clostridium acetobutylicum. Bioresour Technol 109:105–109. https://doi.org/10.1016/j.biortech.2012.01.017

Yi S, Wan Y (2017) Separation performance of novel vinyltriethoxysilane (VTES)-g-silicalite-1/PDMS/PAN thin-film composite membrane in the recovery of bioethanol from fermentation broths by pervaporation. J Membr Sci 524:132–140. https://doi.org/10.1016/j.memsci.2016.11.037

Zhang H, Wang Y (2016) Poly(vinyl alcohol)/ZIF-8-NH2mixed matrix membranes for ethanol dehydration via pervaporation. AIChE Journal 62(5):1728–1739. https://doi.org/10.1002/aic.15140

Zhao XQ, Bai FW (2009) Yeast flocculation: new story in fuel ethanol production. Biotechnol Adv 27(6):849–856. https://doi.org/10.1016/j.biotechadv.2009.06.006

Zheng P-Y, Ye C-C, Wang X-S, Chen K-F, An Q-F, Lee K-R, Gao C-J (2016) Poly(sodium vinylsulfonate)/chitosan membranes with sulfonate ionic cross-linking and free sulfate groups: preparation and application in alcohol dehydration. J Membr Sci 510:220–228. https://doi.org/10.1016/j.memsci.2016.02.060

Zhuang X, Chen X, Su Y, Luo J, Feng S, Zhou H, Wan Y (2016) Surface modification of silicalite-1 with alkoxysilanes to improve the performance of PDMS/silicalite-1 pervaporation membranes: preparation, characterization and modeling. J Membr Sci 499:386–395. https://doi.org/10.1016/j.memsci.2015.10.018

Part IV
Foodwaste as Feedstock

Valorization of Agro-Industrial Waste into Bioactive Compounds: Techno-Economic Considerations

Eduardo Caballero and Carmen Soto

1 Current Context in Both Agricultural Waste (ACW) and Agro-Industrial Waste (AIW)

As definition, agricultural waste (ACW) comes from the agricultural phase of the cultivation of certain species, while agro-industrial waste (AIW) results from the industrial processing of biomass (Portugal-Pereira et al. 2015). This is the base to quantify and identify the ways to face the problem. In the European Union, taking into account both ACW and AIW, around 89 million tonnes of biomass (from the agricultural production, postharvest, and processing stages) are wasted annually (Stenmarck et al. 2016), and this value is expected to increase by 40% in 2020. However, this chapter is mainly focused in postharvest and processing stages, excluding the end of the food supply chain (retail and final consumption), which relates to retailers and consumers behavior.

If we analyzed just the AIW, this causes a serious disposal problem. For example, the juice industries produced a huge amount of waste as peels, the coffee industry produced coffee pulp as a waste, and cereal industries produced husks (Sadh et al. 2018). All over the world approximately 147.2 million metric tonnes of fiber sources are found, whereas 709.2 and 673.3 million metric tonnes of wheat straw residues and rice straws were estimated, respectively, in the 1990s (Belewu and Babalola 2009).

Specifically related to fruit processing industry, the proportion of AIW is significant, for example, depending on the location and method of harvest, the AIW is 30–50% for mango, 20% for banana, 40–50% for pomegranate, and 30–50% for citrus (Laufenberg et al. 2003; Parfitt et al. 2010). These percentages are clear and represent a relative number of wastes. However, if we think in terms of the scale of

E. Caballero (✉) · C. Soto
Centro Regional de Estudios en Alimentos Saludables (CREAS), Valparaíso, Chile
e-mail: ecaballero@creas.cl; carmensoto@creas.cl

© Springer Nature Switzerland AG 2019

J.-R. Bastidas-Oyanedel, J. E. Schmidt (eds.), *Biorefinery*,
https://doi.org/10.1007/978-3-030-10961-5_10

companies, the approaches could be different. For example, in micro and small-scale companies, processing waste is considered to be of negligible value compared to the processed material. On the contrary, for big companies the problem of waste is a priority, because they are paying to dispose their vegetable and food wastes in landfills. For example, in the case of an specific AIW (fruit processing wastes) in Europe, the disposal of 1 tonne of solid waste or 1 m^3 of effluent costs $28–60 dollars which includes a landfill tax of $10 (Gendebien et al. 2001). On the other hand, in developing countries such as India, the average transportation cost was found to be $11–15 dollars per tonne per trip (FICCI 2010) which may indicate $300 million dollars for total landfilling cost. If we add the environmental impact and the concept of corporate social responsibility, it is necessary to take into account that landfilling is also associated with risks of greenhouse gas emissions (Roggeveen 2010). For example, global food processing waste-related greenhouse gas emission was found to be the third highest contributor after total emissions for China and the USA.

On the other hand, the United Nations Food and Agriculture Organization estimates that about 815 million people of the 7.3 billion people in the world were suffering from chronic undernourishment in 2016 (FAO 2017). This is the reason why the AIW and ACW must be used in the most efficient way and be valorized through its healthy and functional potential. Regarding that, in developing countries the majority of the fruit and vegetable processing information was found to be fragmented and insufficient, focusing on the execution of projects about biogas or compost production to obtain biofertilizer, which unfortunately does not take into account the complete potential of AIW. On the contrary, in developed countries, the fruit and vegetable processing waste was found to be fifth highest contributor (8% of total food waste) to overall food waste (Fava et al. 2015), allowing carried out studies focused on AIW recovery (Pfaltzgraff et al. 2013).

Finally, a key factor to recognize the opportunities of AIW and its valorization or added value is to identify the main sources or industries of each country or region. In the case of Chile, for example, AIW are generated in companies that produce fruit pulps, avocado oil, ready-to-eat salads, ready-to-eat meals, canned fruit, canned vegetables, and fruit juice, among others. Regarding this AIW, it is important to know which types of bioactive compounds are able to be extracted from them and how much is their value. This is also a key factor to take into account when a company must decide to invest in a new process or a new product. Table 1 shows some bioactive compounds found in vegetable sources from AIW.

The focus is clear; however, the way to achieve the aforementioned goals is not simple. Some modifications are needed to stabilize the AIW before to be processed. Some researchers are investigating both aerobic and anaerobic digestions as a tool to stabilize the AIW (Fernandez-Bayo et al. 2018). On the other hand, if the goal is the functional compound from AIW, the stabilization must play the role of decreasing the biochemical reactions into the AIW and at the same time keep the benefits of the bioactive compounds. Some studies of AIW drying to recover bioactive compounds are focused in technologies as spray drying, freeze drying, and supercritical CO_2 drying (Santos-Rezende et al. 2018; Plazzotta et al. 2018). The cost of the abovementioned technologies is relatively high; however, conventional drying or

Table 1 Valorization of AIW and applications of its compounds

AIW	Products	Use/application	Patent reference	Market price
Mango peels	Pectin, polyphenols	Gelling agent, stabilizing agent in fruit juices, preservatives	Taboada and Siacor (2013)	Commercial extract 25–30 (US$/kg)[a]
Tomato processing by-products	Lycopene	Antioxidant, therapeutics	Lavecchia and Zuorro (2010)	Commercial extract 10% 55–60 (US$/kg)[b]
Citrus peels	Dietary fiber, flavorant, oil	Nutritional supplement, food additive	Nafisi-Movaghar et al. (2013)	Commercial extract 20:1[c] 60–70 (US$/kg)[b,d]
Pomegranate husk	Site-specific inhibitors of histone methyltransferase	Lead compounds to develop anti-neoplastic and anti-HIV therapeutics	Kundu et al. (2011)	Commercial extract 20:1[c] 55–70 (US$/kg)[b,d]
Pomegranate peels	Polyphenols	Treatment of prostate cancer by increasing doubling time of a prostate-specific antigen	Liker (2014)	Commercial extract 20:1[c] 55–70 (US$/kg)[b,d]
Cranberry and pomegranate extract powders	Hydrolyzable tannins as antibacterial agents	POMcran capsules (25–5000 mg)	Mackler (2014)	Commercial extract 20:1[c] 50–75 (US$/kg)[b,d]
Grape seed extract	Polyphenols	Antiaging and cancer treatment	–	200–250 (US$/kg)[e]
Grape pomace extract	Polyphenols	Antiaging and cancer treatment	–	180 (US$/kg)[e]
Vegetable feedstock	Succinic acid	Food, agricultural, and pharmaceutical uses	–	1.7–3.7 (US$/kg)[f]
Organic feedstock	Compost	Fertilizer	–	25–33 (US$/yard)[g]
Organic feedstock	Livestock feeding	Animal feeding	–	3.57–9.38 (US$/bushel)[h]

Source: Banerjee et al. (2017)
The estimated market price is considering a format up to 25 kg
[a]https://www.newdirectionsaromatics.com/products/fruit-extracts/mango-powder-fruit-extract.html
[b]Summer, China
[c]20:1, It has been used 1 kg of vegetable source to obtain 50 g of powder extract
[d]RuiHerb, China
[e]http://www.polyphenolics.com/consumer/grape-seed-extract/
[f]Shenyang East Chemical Science-Tech Co., Ltd.; Dalian Sinobio Chemistry Co., Ltd.; Guangzhou ZIO Chemical Co., Ltd
[g]https://www.improvenet.com/r/costs-and-prices/composting
[h]https://www.feednavigator.com/Article/2018/03/29/Feed-prices-forecast-to-rise

drying using equipment like roller drum driers could be a most cost-efficient alternative to stabilize AIW in order to obtain its bioactive compounds as a potential added value.

2 Valorization Approaches of AIW

Valorization of AIW is currently focused on composting, livestock feeding, and other products with low-added value, as shown in Table 1. There are many ways to decrease the amount of waste generated in this industry, some of them are reuse and recycle. However, in general both of them are options to valorize the residues into the same companies. Regarding that, this section is aimed to remark cases where the valorization is going forward, taking advantage from scientific studies to create new products with added value from AIW.

Some of the alternatives to valorize the AIW are the reuse and recycle. Reuse indicates the use of waste materials for other purposes without or with minor modification of their properties (Manzocco et al. 2016). Direct reuse of AIW for soil amendment has been investigated by Clemente et al. (2015). However, it is difficult to put into practice due to the high biological instability of waste (Ajila et al. 2012). On the other hand, recycle is a strategy based on the recovery of waste materials after a major modification of their characteristics (Williams and Anderson 2006), which offers more possibilities than its reuse. Recycle strategies can be divided into strategies in which the whole waste mass is recycled and strategies in which specific compounds are extracted. Hence, the valorization on this chapter will be focused on the recycle strategies to extract specific bioactive compounds from AIW.

2.1 Valorization from the Process

The AIW contribution related with fruit and vegetable processing is more than 0.5 billion tonnes worldwide (Banerjee et al. 2017). The potential of those AIW in terms of bioactive compounds, antioxidant activity, and some other micronutrients makes this feedstock an important source of detailed studies. Some examples are the study of resveratrol from grape seeds and peels from wine industry (Fernández-Mar et al. 2012) and the valorization of pomace fractions from tomato paste to extract lycopene (Allison and Simmons 2017).

In the case of lycopene extraction, there are many factors to be considered: drying method of the pomace, tomato varieties, and the characteristics of the cultivars, among others. In Table 2, the influence of different extraction methods in maximum lycopene recovery is presented.

Despite the abovementioned and the technologies used at laboratory level, the commercial technology to produce lycopene extract is solvent extraction using a mixture of ethanol and water. The commercial price of lycopene is 80–90 US$/kg,

Table 2 Extraction methods in lycopene recovery from pomace[a]

Raw material	Extraction method	Maximum lycopene yield (mg/g dry basis)
Raw tomato skins, separated	Acetone: hexane Soxhlet	770.8
Raw tomato skins, separated	Supercritical CO_2, 27.58 MPa, 80 °C	644.1
Tomato pomace	Chloroform Soxhlet	820
Tomato pomace	Supercritical CO_2, 40 MPa, 90 °C	459
Tomato pomace	Supercritical CO_2, 46.0 MPa, 80 °C	314
Tomato pomace	THF, methanol	734
Tomato pomace	Chloroform	24.5
Tomato pomace	Supercritical CO_2, 34.5 MPa, 86 °C	14.86
Tomato pomace	Hexane, acetone, ethanol, RT	19.8

[a]Extracted from Allison and Simmons (2017)

the commercial format is a red microencapsulated powder with 10% of lycopene, and the final powder is oil soluble.[1]

The case of grape pomace valorization has been very important in the recent years because of the growing of wine industry. Since grape pomace can be regarded as an excellent and affordable source of polyphenolic compounds, some researchers have been studying technologies and strategies to extract the antioxidants from that source (Tournour et al. 2015). Figure 1 shows the antioxidant activity (ORAC value) extracted from grape pomace from different cultivars of grapes in Portugal.

Another example of valorization is the case of producing succinic acid from natural resources (Dessie et al. 2018), which use AIW to carry out a solid-state fermentation. The succinic acid produced represents an added value product, which could be used in food, agricultural, and pharmaceutical industries. Other studies focused on the AIW valorization to produce succinic acid, explore the characteristics of the biomass feedstock and evaluate if mixed food waste is especially appealing as it represents less resource competition than more homogenous food waste fractions (Rex et al. 2017).

The AIW could be also a by-product from the process. For instance, the defatted soybean meal from the soybean oil industry could be used to obtain isolated soy protein. The high amount of protein of defatted soybean meal is interesting for the current market taking into account the food trends to plant-based diet. The protein content of defatted soybean meal is 45.7% w/w. There are several types of products derived from isolated soybean protein: dairy-based products including infant formula, beverages including liquid soy milk and fruit drinks, soups and sauces, energy bars, meat analogs including vegetarian food products, breads and pastries, breakfast

[1] http://es.hsfbiotech.com/micro-encapsulated-powders/lycopene/lycopene-10-cws.html

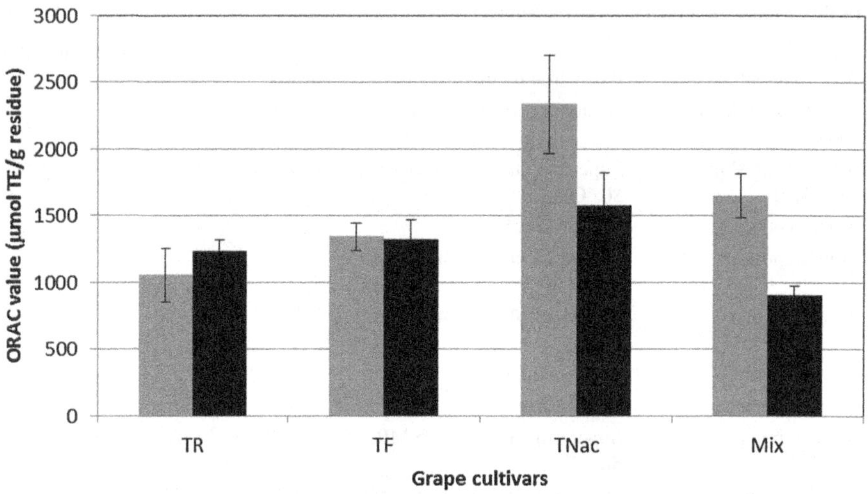

Fig. 1 Extraction of antioxidant activity from grape pomace from different grape cultivars: *TR* Tinta Roriz; *TF* Touriga Franca; *TNac* Touriga Nacional; *Mix* mix red grape pomace. Gray columns represent the ethanol/water extracts (80:20 v/v); and black columns represent the aqueous suspension (Tournour et al. 2015). ORAC (oxygen radical absorbance capacity) represents the value of antioxidant activity

cereals and other nutritional food products, and protein drink for muscle building and weight-gaining purposes (Lai et al. 2017). On the other hand, there is another type of applications of added value products from AIW. For instance, in the last few years, there is a growing interest to encompass from material as bioabsorbent (Sud et al. 2008) to rhizobial inoculant production (Ben Rebah et al. 2007).

Also, there are some bioactive compounds that are demanded for the market, and they could be included in AIW, for example, pectin, dietary fiber, polyinsaturated lipids, essential oils, flavonoids, and peptides, among others (Banerjee et al. 2017). Also structuring agents, mainly referring to colloidal polymers with interesting gelling or viscosant properties, can also be selectively extracted from AIW (McCann et al. 2011). Some recent studies relevant to bioactive extraction (e.g., carotenoids, essential oils, polyphenols, anthocyanins) from AIW using novel technologies include the use of ultrasounds, supercritical carbon dioxide, microwaves, and pulsed electric fields (Amiri-Rigi et al. 2016; Jacotet-Navarro et al. 2015; Rabelo et al. 2016; Zhou et al. 2015). For these reasons, extraction of specific compounds from AIW could be an affordable, sustainable, and profitable strategy for industries (Galanakis 2012; Laufenberg et al. 2003).

On the other hand, before to extract bioactive compounds or structuring agents from AIW, the strategy to stabilize them is drying. Flours can be obtained from drying processes, and they can be used as an ingredient for the formulation of products rich in functional compounds such as polyphenols and fiber (Ferreira et al. 2015).

The main advantage of this recycle strategy is that valuable products such as adsorbents and functional flours are obtained from low-cost raw materials. However, the main issue is the high cost required for AIW drying, due to the high water content. As a consequence, the production of AIW flour is affordable only if high value-added ingredients and products are developed (Ratti 2001). To have an estimation of the cost to dry AIW, Karam et al. (2016) have stablished the energy consumption of different drying technologies, for example, diseccant drying (6 KW/kg of AIW), freeze drying (15–20 KW/kg of AIW), and vaccum drying (5 KW/kg of AIW).

Another example of valorization is rice bran, which was previously being considered as an AIW, as the rice kernel (white rice) is an important product where the major income from rice comes from this part. Rice germ and bran are usually being considered as by-products, until researchers found that rice bran oil has good composition of monounsaturated and polyunsaturated fatty acids which turns to be health beneficial to humans (Kochhar and Gunstone 2002). Rice bran, a part of the rice kernel that contains pericarp, aleurone, and subaleurone fractions, is a by-product of rice milling. It is estimated that the world annual production of rice bran amounts to 76 million tonnes (Chiou et al. 2013). Rice bran oil has been commercialized now in India, the USA, Thailand, and many more. The benefits of rice bran oil have been discussed by many researchers, which have been summarized by Friedman (2013): anti-allergic activities, anti-cholesterol activities, anti-diabetic activities, producing liquid-solid or semisolid form of product from rice bran oil (e.g., shortening or spreads), and regulation of the immune system. These benefits are related to rice bran composition in terms of bioactive compounds, which is summarized in Table 3.

Table 3 Bioactive compounds in rice bran

Bioactive compound or extract	Amount or activity of bioactive compounds	Benefit
Phenolic acids	2101 μg/g	Antioxidant
α-Tocopherol (vitamin E)	71 mg/g	Antioxidant
Rice bran extract	Inhibition of *Vibrio cholerae*, *Vibrio vulnificus*, *Salmonella* spp., *Shigella* spp., *Escherichia coli*, and *Staphylococcus aureus*	Antibiotic activity
Pigmented rice brand extract, cycloartenyl ferulate from rice bran oil-derived	Capture immunoglobulin E (IgE)	Antiallergic an anti-inflammatory activity
MG-3 arabinoxylan rice bran	Ameliorate hepatocellular carcinoma	Anticarcinogenic effect
Tocotrienol-rich fraction of rice bran	100 mg/day during 6 weeks	Anticholesterol effect
Stabilized rice bran	20 g daily during 12 weeks	Antidiabetic activity

Source: Friedman (2013)

2.2 Valorization from the Artisanal Fishery to Food Industry

From a social point of view, fisheries and aquaculture industry provide employment to a major part of the population in several coastal provinces of many countries. This activity is physically demanding and in many cases is not well remunerated in developing countries. On the other hand, a big amount of waste is obtained from the cleaning process of fishery products, which could be used for other food applications. Hence, valorization of the underutilized and invasive species along with fishery by-products and wastes generated from these industries can enhance the industry's value chain, which can improve the profitability of the products. Also, valorization and production of various high-value pharmaceuticals/nutraceuticals (marine oils, omega-3 fatty acids, proteins, amino acids, enzymes, chitin, chitosan, and astaxanthin) from these resources can maximize the economic viability of the industry while addressing issues of waste management and environmental sustainability. Currently, polyunsaturated fatty acids mainly omega-3 fatty acids are obtained from marine fishes with high fat content. Specifically, fish oil derived from blue fish is rich of both eicosapentaenoic acid (EPA, C20:5n-3) and docosahexaenoic acid (DHA, C22:6n-3) (Ciriminna et al. 2017). High demand has motivated researchers to pursue new sources of these fatty acids, improve existing extraction methods, and develop more efficient methods for mass production and commercialization (Dave and Routray 2018).

The situation of valorizing residues in artisanal fisheries also can reach other applications. For instance, the production of seafood generates great amounts of seashells. Regarding that and in order to reduce the dependency on virgin materials for construction, efforts have been made to incorporate by-products and wastes from different industries as alternatives in concrete. Hence, an opportunity has been visualized from the fishery industry. Seashell waste, such as oyster shells, mussel shells, and scallop shells, among others, is available in huge quantities in certain regions and is usually dumped or landfilled without any reuse value. Over seven million tonnes of mollusk discards are produced every year worldwide, being mainly shells, which can be reused due to their high content of calcium carbonate (approximately 95%).[2] For example, in Europe, France has an important fishing and shellfish farming industry that produces nearly 200,000 tonnes of shells from shellfish breeding and nearly 50,000 tonnes of shellfish per year from fishing (Nguyen et al. 2013). The use of seashell waste as a partial replacement for conventional materials in concrete and other related cement-based products is a possibility. The characteristics of different types of seashell waste, as well as the effects of incorporating the seashells on the fresh and hardened properties of concrete, have been studied (Moa et al. 2018). The results of this study show that despite the reduction in the workability and strength, it is suggested that seashell waste could still be utilized as a partial aggregate at a replacement level of up to 20% for adequate workability and strength of concrete for nonstructural purposes.

[2] https://www.popsci.com/oyster-shells-alternate-uses

Chitin and chitosan are other examples of products obtained from marine waste from crustacean shells such as shrimp, crabs, lobsters, and krill. In the study of Kumari et al. (2015), chitin from marine sources is converted to chitosan and characterized through different methodologies. On the other hand, there is the case of compounds from the fishery waste industry used in applications like packaging (Leceta et al. 2015). That investigation was focused on the manufacture of sustainable films by using chitosan derived from wastes of fishery industry. Since chitosan is a hydrophilic polymer, moisture sorption behavior was analyzed in order to provide novel knowledge related to the best use conditions for chitosan as packaging films. The results revealed that chitosan films show a great potential to be used as packaging films for food products with intermediate moisture sorption. The main importance of these findings are the suitable design of chitosan-based sustainable food packaging films, valorization of waste product, maintenance of food quality, and, thus, reduction of food waste and environmental impact caused by conventional packaging systems.

3 Approaches to Optimize the Valorization of Wastes with Conventional and New Technologies

Agro-industrial wastes are often underutilized and pose a major disposal problem to the concerned parties. Food processing wastes are promising sources of valuable compounds such as dietary fiber, antioxidants, essential fatty acids, antimicrobials, and minerals because of their favorable technological, nutritional, and functional properties (Schieber et al. 2001). However, the higher-value products may be developed through various modification methods and technologies. Here, the conventional and new technologies used for valorization or extracting processes from food waste or by-products will be analyzed.

3.1 Solvent Extraction

In solvent extraction, the solvent acts as a physical carrier to transfer the target molecules between different phases of solid, liquid, and vapor (Galanakis 2012). Various compounds can be isolated using solvent extraction, which are tocopherols, flavonoids, and related compounds such as coumarins, cinnamic acid derivatives, and chalcones, phenolic diterpenes, and phenolic acids (Oreopoulou and Tzia 2007). Nonpolar solvents (hexane, petroleum ether) can be used for the recovery of tocopherols and certain phenolic terpenes. Ethyl ether and ethyl acetate are very efficient for the recovery of flavonoid aglycons, low molecular weight phenolics, and phenolic acids. Solvents of higher polarity (ethanol or ethanol water mixtures) additionally can extract flavonoid glycosides and higher molecular weight phenolics, resulting in higher yields of total extracted polyphenols. Organic solvents, such as acetone and ethyl acetate, are used for the extraction of carotenoids, and acetone results in the highest yield compared to ethanol, petroleum ether, and hexane (Calvo 2005). From the regulatory point of view, solvents permitted for use in the preparation of food ingredients in the European Union are ethanol, ethyl acetate, and acetone (Marriott 2010). Pectin extraction is accomplished by the use of mineral acids, usually hydrochloric or nitric acid (Oreopoulou and Tzia 2007). The extract is separated from the solid residue, and pectin is precipitated by the addition of ethanol or AlCl solution. Extraction of defatted soy flake with aqueous alcohol improves the flavor and color of soy protein isolate as well as markedly improved its foaming and gelling properties by removing phospholipids and other alcohol-soluble materials. Hence, it is clear that solvent extraction is a versatile way to selectively extract several types of compounds from vegetable and animal sources but is also a scalable technology and, in terms of costs, depends on the selected solvent and the downstream purification processes.

3.2 Supercritical Fluid Extraction (SFE)

Supercritical fluid extraction involves the use of gas above its critical temperature and pressure (Galanakis 2012). Supercritical fluid is defined as a state where the liquid and gas are indistinguishable from each other above its critical point or a state where the fluid is compressible and has a similar density and solvating power to liquid. Supercritical fluid extraction has several advantages over the conventional methods, including faster processing times, high selectivities (i.e., high quality extract), and high extraction yields (Farías-Campomanes et al. 2013; Herrero et al. 2006); however, energy cost to operate at high pressure is elevated. Carbon dioxide is the most commonly used fluid in SFE due to its low critical point (304.2 K/7.4 MPa), favorable environmental characteristics, and low costs. CO_2 can be easily separated by depressurization and, thus, can be recovered and reused (Farías-Campomanes et al. 2013). Numerous vegetable matrices have been used as natural sources for compressed fluid extraction. Legumes, spices, aromatic plants, and even fruit beverages, such as natural orange juice (Señoráns et al. 2001), have been processed to obtain natural antioxidant compounds.

Several applications have been developed using SFE in the recovery of value-added components from grape residues, including oil from seeds, tannins from seeds, and polyphenols from both skins and seeds (Farías-Campomanes et al. 2013). The recovery of catechin and other phenolic compounds was found to be higher during isolation of phenolic compounds from grape seeds using supercritical carbon dioxide (Murga et al. 2002; Louli et al. 2004). The use of supercritical fluid extraction after ethyl acetate extraction of wine industry by-products caused higher antioxidant activity, allowing odorless and clearer extracts.

Some authors analyzed the economic feasibility of obtaining different bioactive molecules from plant matrices. In the case of oil extraction from pre-pressed seeds, it was possible to determine that the lowest cost of manufacture (COM) is in the order of 4.08 USD/kg of oil recovered from pre-pressed seeds (by-products of the oil extraction process by cold pressing). Considering the cost of raw materials, inputs such as CO_2 and co-solvent, energy costs, manpower, among others, it allows establishing the feasibility of using this type of process for the valuation of vegetable matrices (del Valle et al. 2014). Prado et al. (2012) reported that the production cost for grape seeds depended on the operating conditions, taking into account that by increasing the scale of the process it becomes economically viable. COM values ranging from 290.17 US $/kg extract for a scale process of 5 L and operational time of 60 min, up to 11.93 US $/kg for a scale process of 500 L and 300 min of operation. It is even possible to estimate the global cost of obtaining extracts using SFE, using rapid estimation models that allow establishing that the recovery process is economically feasible (Rosa and Meireles 2005). The case of lycopene recovery from tomato processing waste have also been reported, determining that COM value is close to 1.8 M€/kg lycopene when using supercritical CO_2, while this value can be improved when it uses supercritical ethane given the higher the extraction speed, therefore the shorter the process cycle, and the higher the productivity, therefore the lower the COM (Silva et al. 2014); for its part, the recovery of oil from depleted coffee beans through supercritical fluid technology reports annual productivities up to 454 tonnes/year (extraction conditions 2 h, 300 bar, 50 ° C, 30 kg CO_2/kg raw material h) with COM of 2.4 M€ and net economic income of 56.6 M€ (Melo et al. 2014). It is important to consider that the cost of manufacture will depend basically on the raw material to be used, if there is a use of co-solvent or not, and on the scale of the extraction process and therefore of the productivity process (Veggi et al. 2014), determining, for example, variations of the COM between 1300 and 833 USD/kg crude extract rich in polyphenols from bean when CO_2 was used and considering the lower value for the greater scale of process. The data provided allows establishing the variability in the cost of manufacturing extracts from different plant matrices and therefore the potential use of AIW. This cost is associated with different variables, especially the scale and operational conditions of the process; in addition, the costs are associated with the molecule to be recovered (a crude extract or a specific compound) and are increasing with the purity required for the final product; however, it is still technically and economically feasible to take advantage of different plant matrices, with high net returns.

3.3 *Subcritical Water Extraction (SWE)*

Subcritical water extraction (SWE), an extraction which uses hot water under pressure, has recently emerged as a useful and environmentally friendly tool to replace the traditional extraction methods. Basically, the instrumentation consists of a water reservoir coupled to a high-pressure pump to introduce the solvent into the system, where the extraction cell is placed and extraction takes place and a restrictor or valve to maintain the pressure. Extracts are collected in a vial at the end of the extraction system. Subcritical water extraction has been widely used to extract different compounds from several vegetable matrices such as rosemary (Arvanitoyannis and Kassaveti 2008; Herrero et al. 2006).

Although at present there are no economic analyzes in the literature regarding the feasibility of the extraction process with subcritical water, different authors point out that it is an economically viable technology, given the fact that traditional organic solvents are replaced by water, the extraction product would not require desolventization, among other things (Zakaria and Kamal 2016; Gbashi et al. 2016; Tian et al. 2017). On the other hand, it is also necessary to consider that in general it has been reported that different green technologies, such as extraction by supercritical fluids (as seen in the previous point), microwave-assisted extraction, or electric pulse, among others, are economically feasible to be applied for the recovery of extracts with bioactive molecules (Bromberger et al. 2018).

3.4 Solid-State Fermentation (SSF)

Solid-state fermentation (SSF) is gaining wide interest these days for the production of organic acids such as citric, lactic, and oxalic acids, enzymes, and other biotechnological products, which could be extracted from AIW, which are generally considered the best substrates for this fermentation processes, especially for enzyme production (Dhillon et al. 2013). The presence of lignin and cellulose/hemicellulose acts as natural inducers, and most of these residues are rich in sugar, promoting better fungal growth and thus making the process more economical especially for the cellulo- and ligninolytic enzymes. Other potential applications of AIW through SFF technology are the production of fructooligosaccharides, bioactive compounds, and bio-insecticides, among others (Mussatto et al. 2012). For example, apple pomace undergone solid fermentation to produce organic acids, heteropolysaccharide (i.e., xanthan, chitosan), flavor and aroma compounds, bioethanol, enzymes, edible mushroom (Pleurotus ostreatus), antioxidants, and nutritional enrichment among others. Several factors making apple pomace suitable as a raw material for biotechnological products are the high content of polysaccharides (mainly cellulose, starch, and hemicelluloses); presence of mono-, di-, and oligosaccharides, citric acid, and malic acid, which can be metabolized by microorganisms; and richness in vitamins and other mineral ions which could limit the cost of nutrient supplementation for fermentation media (Dhillon et al. 2013). Also there is great potential for coffee pulp and coffee husk used as substrates to microbial aroma production by solid-state fermentation using two different strains of *C. fimbriata* (Murthy and Naidu 2012).

3.5 *Extrusion*

Extrusion combines a number of unit operations, i.e., mixing, cooking, shearing, puffing, final shaping, and drying in one energy-efficient rapid continuous process (Harper et al. 1989), and can be used to produce a wide variety of starchy foods including snacks, ready-to-eat (RTE) cereals, confectioneries, and extruded crisp breads (Suknark et al. 1997). This process of high-temperature short-time extrusion brings gelatinization of starch, denaturation of protein, modification of lipid, and inactivation of enzymes, microbes, and many antinutritional factors (Bhattacharya and Prakash 1994). Extruded foods are able to provide nutritious products, by means of combine quality ingredients and nutrients to produce processed foods that are formulated to contain specific amounts of each required nutrient (Cheftel 1986). The fruit wastes, defatted hazelnut flour, and durum clear flour can be used in combination with cereal flours for production of nutritionally balanced convenient extruded snack foods due to their valuable characteristics.

4 Conclusion: Biorefinery Concept from AIW

The use of AIW for its use as a source of bioactive compounds can be associated with two different concepts, the biorefinery and the circular economy. The processes based on agro-industrial wastes can consider different steps and technologies that allow the recovery of bioactive compounds of interest in a better way. As shown in Fig. 2, it is possible to consider a first stage of wet milling that allows reducing the particle size, facilitating both the development of later extractive processes and the release of different compounds through the syneresis process. Depending on the raw material, it is possible to remove soluble molecules (low molecular weight) such as minerals or sugars of low molecular weight, among others. In this aspect, also AIW can be an important source of macromolecules such as fibers (soluble and insoluble) and proteins, so that a fractionation of these molecules would allow a better use of waste as a raw material. In the case of proteins, these can be solubilized by the addition of an alkaline solution that is then removed by centrifugation or pressing, which allows obtaining a solid residue rich in fibers, which can contain molecules as antioxidants. In this way, the solid can be treated by different technologies for the recovery of these compounds, for example, direct extraction for the recovery of antioxidants, leaving an expended solid fraction rich in fibers; the application of SFF as a technology for the generation of enzymes or other compounds; or the application of different treatments prior to the extraction itself (such as an enzymatic treatment, or physicochemical treatments), obtaining a final residue rich in fiber that can be used by the food industry or a solid with a better metabolism capacity of the raw material and that can be used in the generation of bioenergy.

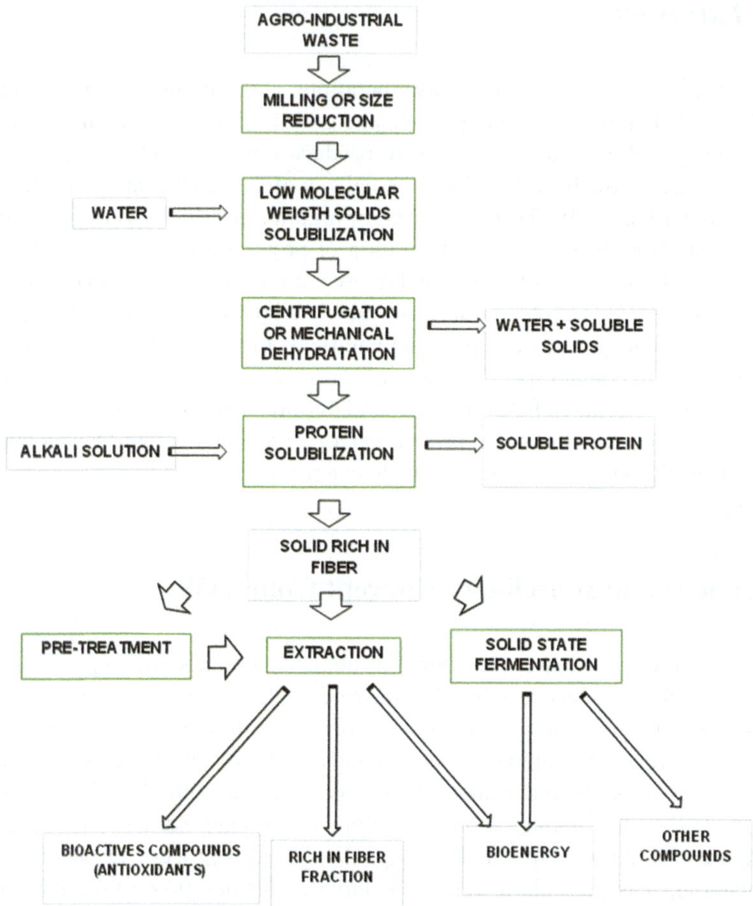

Fig. 2 AIW processing under the biorefinery concept

References

Ajila CM, Brar SK, Verma M, Prasada Rao UGS (2012) Sustainable solutions for agro processing waste management: an Overview. In: Malik A, Grohmann E (eds) Environmental protection strategies for sustainable development. Springer, Dordrecht, pp 65–109

Allison BJ, Simmons CW (2017) Valorization of tomato pomace by sequential lycopene extraction and anaerobic digestion. Biomass Bioenergy 105:331–341

Amiri-Rigi A, Abbasi S, Scanlon MG (2016) Enhanced lycopene extraction from tomato industrial waste using microemulsion technique: optimization of enzymatic and ultrasound pre-treatments. Innov Food Sci Emerg Technol 35:160–167

Arvanitoyannis IS, Kassaveti A (2008) Dairy waste management: treatment methods and potential uses of treated waste. In: Waste management for the food industries. Food science and technology, pp 801–860

Banerjee J, Singh R, Vijayaraghavan R, MacFarlane D, Patti AF, Arora A (2017) Bioactives from fruit processing wastes: Green approaches to valuable chemicals. Food Chem 225:10–22

Belewu MA, Babalola FT (2009) Nutrient enrichment of some waste agricultural residues after solid state fermentation using Rhizopus oligosporus. J Appl Biosci 13:695–699

Ben Rebah F, Prevost D, Yezza A, Tyagi RD (2007) Agro-industrial waste materials and wastewater sludge for rhizobial inoculant production: a review. Bioresour Technol 98:3535–3546

Bhattacharya S, Prakash M (1994) Extrusion of blends of rice and chick pea flours: a response surface analysis. J Food Eng 21:315–330

Bromberger M, de Marsillac L, Peixoto C (2018) Green technologies for the extraction of bioactive compounds in fruits and vegetables. CyTA J Food 16(1):400–412. https://doi.org/10.1080/19476337.2017.1411978

Calvo MM (2005) Lutein: a valuable ingredient of fruit and vegetables. Crit Rev Food Sci Nutr 45:671–696

Clemente R, Pardo T, Madejón P, Madejón E, Bernal MP (2015) Food byproducts as amendments in trace elements contaminated soils. Food Research International 73:176–189

Cheftel JC (1986) Nutritional effects of extrusion-cooking. Food Chem 20:263–283

Chiou TY, Ogino A, Kobayashi T, Adachi S (2013) Characteristics and antioxidative ability of defatted rice bran extracts obtained using several extractants under subcritical conditions. J Oleo Sci 62:1–8

Ciriminna R, Meneguzzo F, Delisi R, Pagliaro M (2017) Enhancing and improving the extraction of omega-3 from fish oil. Sustain Chem Pharmacy 5:54–59

del Valle JM, Núñez GA, Aravena RI (2014) Supercritical CO2 oilseed extraction in multi-vessel plants. 1. Minimization of operational cost. J Supercrit Fluids 92:197–207

Dave D, Routray W (2018) Current scenario of Canadian fishery and corresponding underutilized species and fishery byproducts: a potential source of omega-3 fatty acids. J Clean Prod 180:617–641

Dessie W, Zhanga W, Xina F, Donga W, Zhanga M, Maa J, Jiang M (2018) Succinic acid production from fruit and vegetable wastes hydrolyzed by onsite enzyme mixtures through solid state fermentation. Bioresour Technol 247:1177–1180

Dhillon GS, Brar SK, Kaur S, Verma M (2013) Bioproduction and extraction optimization of citric acid from *Aspergillus niger* by rotating drum type solid-state bioreactor. Ind Crop Prod 41:78–84

FAO (2017) The state of food security and nutrition in the world 2017. http://www.fao.org/state-of-food-security-nutrition/en/

Farías-Campomanes AM, Rostagno MA, Meireles MAA (2013) Production of polyphenol extracts from grape bagasse using supercritical fluids: yield, extract composition and economic evaluation. J Supercrit Fluids 77:70–78

Fava F, Totaro G, Diels L, Reis M, Duarte J, Beserra-Carioca O, Poggi-Varaldo HM, Sommer-Ferreira B (2015) Biowaste biorefinery in Europe: opportunities and research & development needs. N Biotechnol 32(1):100–108

Fernandez-Bayo JD, Yazdani R, Simmons CW, VanderGheynst JS (2018) Comparison of thermophilic anaerobic and aerobic treatment processes for stabilization of green and food wastes and production of soil amendments. Waste Manag 77:555–564

Fernández-Mar MI, Mateos R, García-Parrilla MC, Puertas B, Cantos-Villar E (2012) Bioactive compounds in wine: Resveratrol, hydroxytyrosol and melatonin: a review. Food Chem 130:797–813

Ferreira JPA, Miranda I, Gominho J, Pereira H (2015) Selective fractioning of Pseudotsuga menziesii bark and chemical characterization in view of an integrated valorization. Ind Crop Prod 74(15):998–1007

FICCI (2010) Bottlenecks in Indian Food Processing Industry. http://ficci.in/sedocument/20073/Food-Processing-bottlenecks-study.pdf

Friedman M (2013) Rice brans, rice bran oils, and rice hulls: composition, food and industrial uses, and bioactivities in humans, animals, and cells. J Agric Food Chem 61:10626–10641

Galanakis CM (2012) Recovery of high added-value components from food wastes: conventional, emerging technologies and commercialized applications. Trends Food Sci Technol 26(2):68–87

Gbashi S, Adebo OA, Piater L, Madala NE, Njobeh PB (2016) Subcritical water extraction of biological materials. Separ Purif Rev 46:21–34. https://doi.org/10.1080/15422119.2016.1170035

Gendebien R, Ferguson J, Brink H, Horth M, Davis R, Brunet HPJ (2001) Survey of wastes spread on land—final report study contract. http://ec.europa.eu/environment/waste/studies/compost/landspreading.pdf

Harper JM, Linko P, Mercier C (1989) Extrusion cooking. American Association of Cereal Chemists, St. Paul, MN

Herrero M, Cifuentes A, Ibañez E (2006) Sub- and supercritical fluid extraction of functional ingredients from different natural sources: plants, food-by-products, algae and microalgae: a review. Food Chem 98(1):136–148

Jacotet-Navarro M, Rombaut N, Fabiano-Tixier AS, Danguien M, Bily A, Chemat F (2015) Ultrasound versus microwave as green processes for extraction of rosmarinic, carnosic and ursolic acids from rosemary. Ultrason Sonochem 27:102–109

Karam MC, Petit J, Zimmer D, Djantou EB, Scher J (2016) Effects of drying and grinding in production of fruit and vegetable powders: a review. J Food Eng 188:32–49

Kochhar S, Gunstone F (2002) Sesame, rice-bran and flaxseed oils. In: Vegetable oils in food technology: composition, properties and uses. Blackwell, Oxford, pp 297–326

Kumari S, Rath P, Kumar ASH, Tiwari TN (2015) Extraction and characterization of chitin and chitosan from fishery waste by chemical method. Environ Technol Innov 3:77–85

Kundu TK, Vikru RSB, Annavarapu HK., Kempegowda M (2011) Site-specific inhibitors of histone methyltransferase (HMTASE) and process of preparation thereof. Patent No. 8,003,698, U.S.

Lai WT, Khong NMH, Lim SS, Hee YY, Sim BI, Lau KY, Lai OM (2017) A review: modified agricultural by-products for the development and fortification of food products and nutraceuticals. Trends Food Sci Technol 59:148–160

Laufenberg G, Kunz B, Nystroem M (2003) Transformation of vegetable waste into value added products: (A) the upgrading concept; (B) practical implementations. Bioresour Technol 87:167–198

Lavecchia R, Zuorro A (2010) Process for the extraction of lycopene misc. Google Patents. http://www.google.com/patents/US20100055261

Leceta I, Uranga J, Arana P, Cabezudo S, De la Caba K, Guerrero P (2015) Valorisation of fishery industry wastes to manufacture sustainable packaging films: modelling moisture-sorption behavior. J Clean Prod 91:36–42

Liker H (2014) Method of using pomegranate extracts for increasing prostate specific antigen doubling time. Patent No. US 2014/0044809, U.S.

Louli V, Ragoussis N, Magoulas K (2004) Recovery of phenolic antioxidants from wine industry by-products. Bioresour Technol 92(2):201–208

Mackler A (2014) Methods and compositions for treatment of urinary tract infections. Patent No. US 2016/9,248,154. U.S.

Manzocco L, Alongi M, Sillani S, Nicoli MC (2016) Technological and consumer strategies to tackle food wasting. Food Eng Rev 8(4):457–467

Marriott RJ (2010) Greener chemistry preparation of traditional flavour extracts and molecules. Agro Food Ind Hi Tech 21:46–48

McCann TH, Fabre F, Day L (2011) Microstructure, rheology and storage stability of low-fat yoghurt structured by carrot cell wall particles. Food Res Int 44(4):884–892

Melo MMR, Barbosa HMA, Passos CP, Silva CM (2014) Supercritical fluid extraction of spent coffee grounds: measurement of extraction curves, oil characterization and economic analysis. J Supercrit Fluids 86:150–159

Moa KH, Alengaram UJ, Jumaat MZ, Lee SC, Goh WI, Yuen CW (2018) Recycling of seashell waste in concrete: a review. Construct Build Mater 162:751–764

Murga R, Sanz MT, Beltrán S, Cabezas JL (2002) Solubility of some phenolic compounds contained in grape seeds, in supercritical carbon dioxide. J Supercrit Fluids 23(2):113–121

Murthy PS, Naidu MM (2012) Sustainable management of coffee industry by-products and value addition—a review. Resour Conserv Recycl 66:45–58

Mussatto SI, Ballesteros LF, Martins S, Teixeira JA (2012) Use of agro-industrial wastes in solid-state fermentation processes. In: Show KY (ed) Industrial waste. IntechOpen, Rijeka. https://doi.org/10.5772/36310. https://www.intechopen.com/books/industrial-waste/use-of-agro-industrial-wastes-in-solid-state-fermentation-processes

Nafisi-Movaghar K, Druz LL, Victoria CP (2013) Conversion of citrus peels into fiber, juice, naringin, and oil. Patent No. US 2013/0064947 A1, U.S.

Nguyen DH, Boutouil M, Sebaibi N, Leleyter L, Baraud F (2013) Valorization of seashell by-products in pervious concrete pavers. Construct Build Mater 49:151–160

Oreopoulou V, Tzia C (2007) Utilization of plant by-products for the recovery of proteins, dietary fibers, antioxidants, and colorants. In: Utilization of by-products and treatment of waste in the food industry. Springer, Berlin

Parfitt J, Barthel M, Macnaughton S (2010) Food waste within food supply chains: quantification and potential for change to 2050. Philos Trans R Soc Lond B Biol Sci 365(1554):3065–3081

Pfaltzgraff LA, De bruyn M, Cooper EC, Budarin V, Clark JH (2013) Food waste biomass: a resource for high-value chemicals. Green Chem 15(2):307–314

Plazzotta S, Calligaris S, Manzocco L (2018) Innovative bioaerogel-like materials from fresh-cut salad waste via supercritical-CO2-drying. Innov Food Sci Emerg Technol 47:485–492

Portugal-Pereira J, Soria R, Rathmann R, Schaeffer R, Szklo A (2015) Agricultural and agro-industrial residues-to-energy: technoeconomic and environmental assessment in Brazil. Biomass Bioenergy 81:521–533

Prado JM, Dalmolin I, Carareto NDD, Basso RC, Meirelles AJA, Vladimir Oliveira J, Batista EAC, Meireles MAA (2012) Supercritical fluid extraction of grape seed: process scale-up, extract chemical composition and economic evaluation. J Food Eng 109:249–257

Rabelo RS, Machado MTC, Martínez J, Hubinger MD (2016) Ultrasound assisted extraction and nanofiltration of phenolic compounds from artichoke solid wastes. J Food Eng 178:170–180

Ratti C (2001) Hot air and freeze-drying of high-value foods: a review. J Food Eng 49(4):311–319

Rex E, Rosander E, Røyne F, Veide A, Ulmanen J (2017) A systems perspective on chemical production from mixed food waste: the case of bio-succinate in Sweden. Resour Conserv Recycl 125:86–97

Roggeveen K (2010) Tomato journey from farm to fruit shop: greenhouse gas emissions and cultural analysis. University of Wollongong, Wollongong

Rosa PTV, Meireles MAA (2005) Rapid estimation of the manufacturing cost of extracts obtained by supercritical fluid extraction. J Food Eng 67:235–240

Sadh PK, Duhan S, Duhan JS (2018) Agro-industrial wastes and their utilization using solid state fermentation: a review. Bioresour Bioprocess 5:1. https://doi.org/10.1186/s40643-017-0187-z

Santos-Rezende RR, Pedreira-Nogueira J, Narain N (2018) Microencapsulation of extracts of bioactive compounds obtained from acerola (Malpighia emarginata DC) pulp and residue by spray and freeze drying: chemical, morphological and chemometric characterization. Food Chem 254:281–291

Schieber A, Stintzing FC, Carle R (2001) By-products of plant food processing as a source of functional compounds—recent developments. Trends Food Sci Technol 12(11):401–413

Señoráns FJ, Ruiz-Rodrıguez A, Ibañez E, Tabera J, Reglero G (2001) Optimization of countercurrent supercritical fluid extraction conditions for spirits fractionation. J Supercrit Fluids 21(1):41–49

Silva AF, de Melo MMR, Silva CM (2014) Supercritical solvent selection (CO2 versus ethane) and optimizationof operating conditions of the extraction of lycopene from tomato residues: innovative analysis of extraction curves by a response surface methodology and cost of manufacturing hybrid approach. J Supercrit Fluids 95:618–627

Stenmarck A, Jensen C, Quested T, Moates G (2016) Estimates of European food waste levels. Stockholm, Sweden. ISBN 978-91-88319-01-2

Sud D, Mahajan G, Kaur M (2008) Agricultural waste material as potential adsorbent for sequestering heavy metal ions from aqueous solutions: a review. Bioresour Technol 99:6017–6027

Suknark K, Phillips RD, Chinnan MS (1997) Physical properties of directly expanded extrudates formulated from partially defatted peanut flour and different types of starch. Food Res Int 30(8):575–583

Taboada E, Siacor FD (2013) Preparation of pectin and polyphenolic compositions from mango peels. Misc, Patent No. WO2013/141723 A1, Talamban

Tian Y, Wang Y, Ma Y, Zhu P, He J, Lei J (2017) Optimization of subcritical water extraction of resveratrol from grape seeds by response surface methodology. Appl Sci 7:321. https://doi.org/10.3390/app7040321

Tournour HH, Segundo MA, Magalhães LM, Barreiros L, Queiroz J, Cunha LM (2015) Valorization of grape pomace: extraction of bioactive phenolics with antioxidant properties. Ind Crop Prod 74:397–406

Veggi PC, Cavalcanti RN, Meireles MAA (2014) Production of phenolic-rich extracts from Brazilian plants using supercritical and subcritical fluid extraction: experimental data and economic evaluation. J Food Eng 131:96–109

Williams PJ, Anderson PA (2006) Technical bulletin: operational cost savings in dairy plant water usage. Int J Dairy Technol 59(2):147–154

Zakaria SM, Kamal SMM (2016) Subcritical water extraction of bioactive compounds from plants and algae: applications in pharmaceutical and food ingredients. Food Eng Rev 8:23–34

Zhou Y, Zhao X, Huang H (2015) Effects of pulsed electric fields on anthocyanin extraction yield of blueberry processing by-products. J Food Process Preserv 39(6):1898–1904

Industrial Food Waste Valorization: A General Overview

Juan A. Cecilia, Cristina García-Sancho, Pedro J. Maireles-Torres, and Rafael Luque

1 Introduction

The increase of the world population in the last centuries has caused a decline of the Earth reserves. Currently, fossil fuels, such as coal and crude oil, represent more than 80% of the energy production in the world. The overexploitation of nonrenewable resources leads to more unstable fuels and their depletion within a few decades. In addition, the more stringent environmental laws are demanding cleaner and sustainable processes to obtain energy and products. Several sources have been proposed to obtain energy; however, the biomass is, together with the fossil fuels, the only source that also allows to obtain chemicals.

However, the selected biomass must be a nonedible source to prevent possible economic impact on the food market. In this sense, lignocellulosic biomass is a potential source to obtain high added-value products. Another sustainable alternative is the use of biomass waste for its valorization in energy and chemicals. In this sense, the Food and Agriculture Organization (FAO) of the United Nations reports estimated that as much as 45% of the food produced is lost or wasted before and after reaching the consumer, accounting for over 1.3 billion tonnes per year of food globally produced for human consumption (Fig. 1) (Lin et al. 2013). Food waste, being high in nutritional content, putrefies on accumulation, providing breeding grounds for disease-causing organisms. This environmental issue can be limited diminishing the generation of food residues, although the most sustainable process is related with the recycling of these wastes for other uses and applications.

J. A. Cecilia (✉) · C. García-Sancho · P. J. Maireles-Torres
Dpto. Química Inorgánica, Cristalografía y Mineralogía, Facultad de Ciencias, Universidad de Málaga, Málaga, Spain
e-mail: jacecilia@uma.es

R. Luque
Departamento de Química Orgánica, Universidad de Córdoba, Córdoba, Spain

© Springer Nature Switzerland AG 2019

J.-R. Bastidas-Oyanedel, J. E. Schmidt (eds.), *Biorefinery*,
https://doi.org/10.1007/978-3-030-10961-5_11

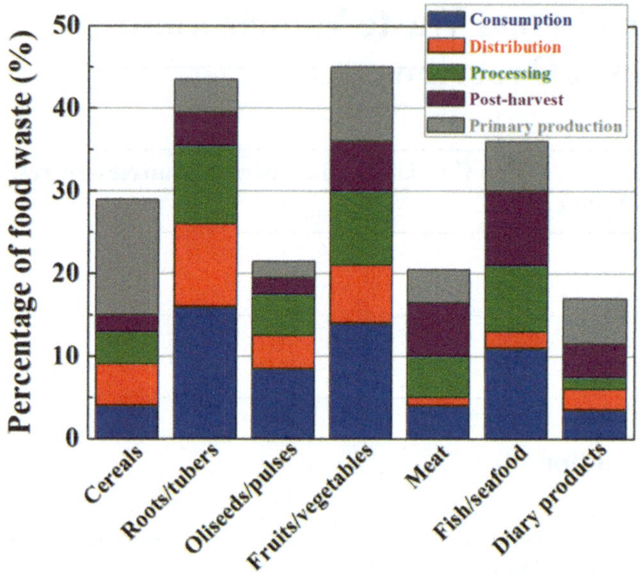

Fig. 1 Percentage of food waste along the food supply chain (Source: FAO)

Nowadays, the food waste can be valorized as animal feed, although the chemical composition of these wastes is regulated, so its use is limited (Directive 2008a). The most common use of the food waste is its composting to obtain fertilizers (Lin et al. 2013). Food waste can also be incinerated for generation of heat or energy. However, the presence of high levels of moisture in these residues can generate dioxins during their combustion, which are harmful to health (Katami et al. 2004). In addition, the incineration of the food waste can potentially cause air pollution by the CO_2 emissions as well as the formation of small particulate matter, so these residues require an appropriate treatment (Fig. 2).

Another sustainable process to valorizate the food waste is associated to the synthesis of valuable products for companies (Lin et al. 2013; Litchfield 1987). Generally, food waste is composed by a heterogeneous mixture formed by carbohydrates (starch, cellulose, hemicellulose, or lignin), proteins, lipids, organic acids, and smaller inorganic part. The main drawback of the food waste valorization is related to the wide variety of compounds and the fluctuating volumes in seasons as well as insufficient legislative and infrastructure support for its implementation at large scale (Poeschl et al. 2010).

Food waste is obtained in several points in the food chain from the harvesting until its postconsumer use (Fig. 1). It must be considered that each step of the food chain is subjected to losses. Thus, the harvesting and subsequent processing only use a proportion of the starting material. After its commercialization, a large proportion of products are wasted, as takes place in restaurant or catering services, or are perishable products that expire before consumption. A statistical study has reported that the wastes coming from the industrial processing are about 69%, while the

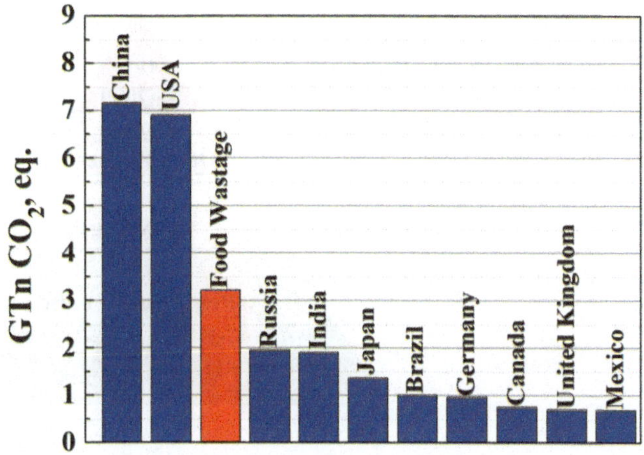

Fig. 2 Global greenhouse gas-emitting countries (CO_2 emissions in gigatonnes, GTn) versus food loss and waste (Source: FAO)

commercial wastes are around 31% in industrialize countries (ENDS 1999). In the case of the undeveloped countries, the amount of the industrial processing waste diminishes, while the commercial waste, as organic crop residue, increases. The increase of the purchasing power and the worsen management of the food have caused an increment in the generation of food waste in recent decades, so the valorization of these wastes is a key issue for the society. However, the first steps must be focused to the social awareness to manage and make a better use of food. In this sense, the legislation of more restrictive environmental policies, a formation where the world population becomes aware that the resources of the planet must be used responsibly and the bonus for those users who recycle their wastes.

2 Food Waste Reduction and Valorization Regulations

The European Union (EU) has developed several legislations aimed at regulating the production, handling, storage, transfer, treatment, and disposal of waste food with the overriding objective of avoiding as well as minimizing the negative effects of waste generation on human health and the environment (Directive 2008b). The animal by-product waste is treated by other regulations, being incinerated to avoid associated risks to animal and public health (Regulation (EC) 2009).

EU has established in the Directive 2008/98/EC the hierarchy related with the waste, where limiting the waste is prioritized (Fig. 3) (Directive 2008c). The food waste generated must be adapted for reuse or recycling to strengthen the economic value of waste. The regulations related with the biological waste, including the food waste, are based on three points: (a) the separate collection of biowaste, (b) the treatment of the biowaste, and (c) its use of environmentally safe product, such as

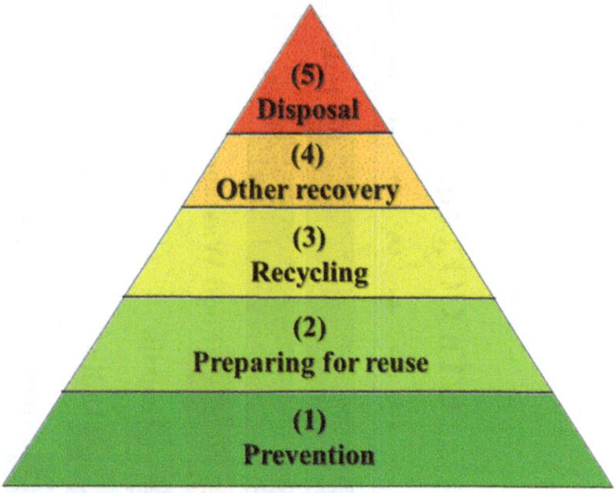

Fig. 3 Hierarchy related with the waste treatment

compost or digestate. Recently, EU is demanding waste prevention programs to limit the link between economic growth and waste generation through social awareness (Directive 2008c). The valorization of the biowaste to obtain high added-value products is a concept that was not defined by the EU until 2010 (Lin et al. 2013). Nowadays, the majority of food waste produced in the EU is either landfilled, composted, or incinerated, wherewith the valorization of these wastes in larger scale is a challenge to the governments for the next decades.

3 Classification of Food Waste

A classification of food waste types is proposed in Fig. 4, which considers the limitations and opportunities for the food waste valorization, based on the regulations (see Sect. 2) and waste properties.

3.1 Organic Crop Residues

Organic crop residues include harvested vegetables, fruits, or grains as well as processing by-products such as stones, husks, peels, straw, stover, pomace, oleochemical residues, and factory vegetable oil (Fig. 4). These residues are source of carbohydrates, sugars, lipids, and inorganic compounds (mainly silica). In addition, organic crop residues can include large proportions in phytochemicals, such as carotenoids, phenolics, and tocopherols, which display high potential in the fields of food, cosmetics, or pharmaceutical industry (Djilas et al. 2009). Organic crop residues can also be used as animal feed since the transport and treatment of the residues are economically ineffective.

Several organic crop residues, such as wheat, barley, potato, apple, or grape, are prone to microbial spoilage, so they are required a preliminary drying step. However,

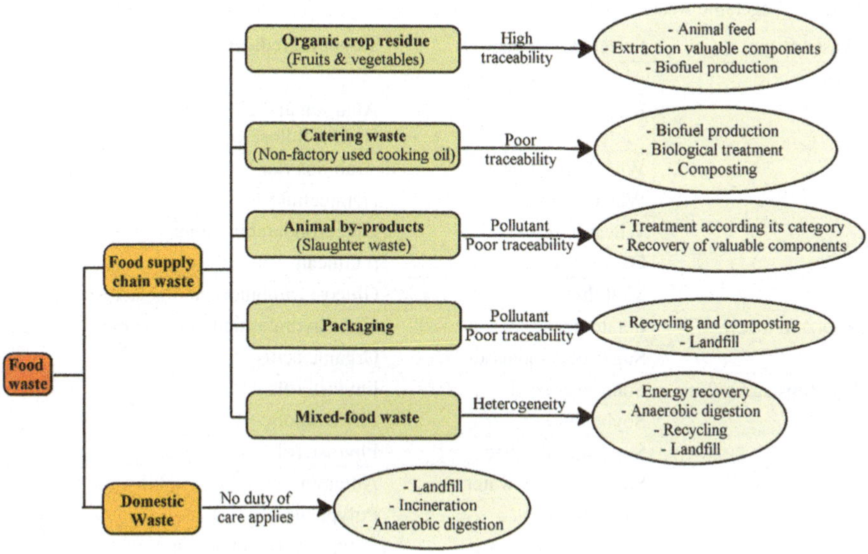

Fig. 4 Food waste classification based on their applications

other residues can be immediately treated after their generation using clean technologies which are compatible to high moisture content (Table 1).

The lignocellulosic residues of these vegetable wastes display a high potential to obtain valuable products. Thus, lignin-rich residues can be valorizated by a pyrolytic treatment to synthesize bio-oil rich in phenolic compounds, which are used in the plastic and cosmetic resins manufacturing industries as well as extracting agent. On the other hand, cellulose- and hemicellulose-rich residues can be subjected to obtain their respective monomers, which can be dehydrated leading to "building block molecules" to synthesize a wide range of high added-value products. The slag and fly ash originating from rice husk or wheat straw contains high silica content, being utilized in the synthesis of bio-derived adhesives used in bio-boards (Dodson et al. 2012).

On the other hand, sugars, peptides, amino acids, fatty acids, hydroxy acids, and lipids coming from agricultural residues and food processing wastes can lead to biocompatible and biodegradable surfactants, such as sodium lauryl sulfate, which is used in home and personal care products, (N^{α},N^{ε}-dioctanoyl lysine), and ethyl-*N*-lauroyl-L-arginate, used by its high biodegradability and antimicrobial properties, favoring its use in the cosmetics and pharmaceutical industries (Moran et al. 1999).

Table 1 Vegetable waste origin, sources, and corresponding target ingredients for recovery

Waste origin	Selected sources	Target ingredients
Plant		
Cereal	Rice bran	Albumin and globulin Hemicellulose B and fiber
	Wheat middling	Arabinoxylan
	Wheat straw	Hemicellulose
	Wheat bran	Glucuronoarabinoxylans
	Oat mill waste	β-Glucan
	Malt dust	Glucose, arabinose and galactose
Root and tubers	Potato peel	Carbohydrates and polyphenols
	Sugar beet molasses	Organic acids
Oil crops and pulses	Sunflower seed	Phytosterols
	Soybean seed	Phytosterols
	Soybean oil waste	Phytosterols
	Soybean wastewater	Albumin
	Olive pomace	Polyphenols
	Olive mill wastewater	Polyphenols and pectin
Fruits and vegetables	Cold hardy mandarin peel	Narirutin
	Orange peel	Hesperidin, apocarotenoid, and limonene
	Lemon by-product	Pectin
	Apple pomace	Pectin
	Apple skin	Polyphenols
	Peach pomace	Pectin
	Apricot kernel	Protein
	Grape pomace	Dietary fiber
	Grape skin	Phenols
	Wine lees	Calcium tartrate and enocyanin
	Banana peel	Cyanidin-3-rutinoside
	Kiwi fruits	Dietary fiber
	Carrot peel	β-Carotene and phenols
	Tomato pomace	Lycopene
	Tomato skin	Carotenoids
	Cauliflower floret and curd	Pectin

3.2 *Catering Waste and Derivatives*

Catering wastes are those generated in restaurants, coffee shops, and pubs or other wastes related with the food production no longer intended for human consumption. About 90% of the catering waste could be reused or recycled. However, they are hardly recycled due to lack of awareness or logistical problems as well as the difficulty of separating the waste from their containers. Catering waste is generally composed by mixed waste from food preparation, packaging and separated waste, organic, glass, cardboard and plastic, or used cooking oil. The valorization of these wastes are focused to the treatment of organic waste, recycling of clean packaging waste, and the utilization of used cooking oil in non-feed/technical applications, such as the synthesis of biofuels or other high added-value products (Fig. 5).

Oils and fats are products with high market volume (Fig. 6). These products are mainly used for the human consumption, although they can also be used as animal feed and the production of oleochemicals and biodiesel. Frying used cooking oil has been traditionally valorized into animal feed products, although the used cooking oil also displays interesting applications in fuel boilers, lubricants/surfactant precursors, and biodiesel production, being biodiesel the most economical from the feedstock viewpoint (ADAS UK Ltd 2007).

The synthesis of biodiesel has generated controversy in the last years since edible oils, mainly sunflower and soybean oils, were used for their production (Luque et al. 2008). The use of nonedible oils and recovered oils and fats has emerged as an alternative to attain a sustainable biodiesel production worldwide (Janaun and Ellis 2010).

Focusing on the cooking oils, these used oils are composed mono-, di-, and triglycerides with variable range of free fatty acids (5–20% w/w), which are generated during the frying process. Employed feedstocks include waste frying, olive, rapeseed, or sunflower oil, rendered animal fats as well as others sourced from food industries, restaurants, and catering facilities which are largely exposed to air, high temperatures, and moisture. These used oils can react with alcohols, mainly methanol or ethanol, to obtain fatty acid methyl esters (FAME) or fatty acid ethyl esters (FAEE), respectively, i.e., biodiesel, by transesterification reactions using a wide range of acid or basic catalysts. Although basic catalysts display higher activity than acid catalysts, additional neutralization and separation steps of the final reaction mixture are required, thus leading to a series of environmental problems related to the use of high amounts of water and energy. Nonetheless, the acid catalysts can

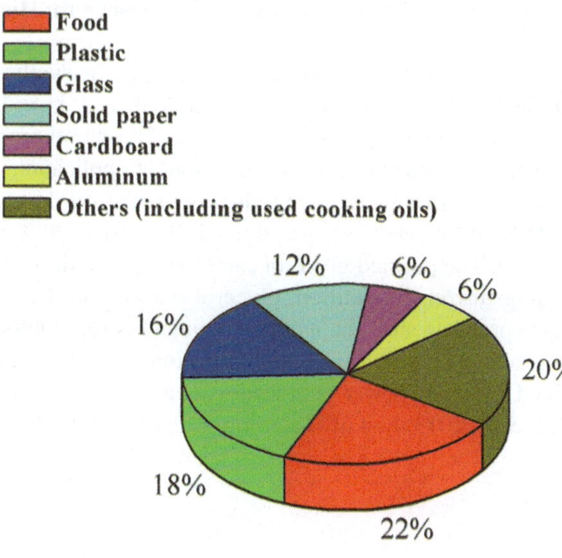

Fig. 5 Breakdown of the type of waste generated in catering facilities

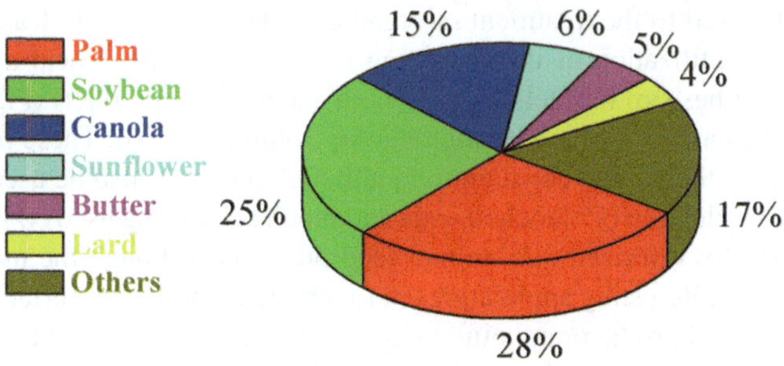

Fig. 6 World market of oils and fats. Total production 199.75 million Mt/year (Source: HIS Markit 2015)

carry out simultaneously the esterification of free fatty acids (FFAs) and transesterification of triglycerides in a single catalytic step, facilitating the use of used cooking oils as raw material (Lin et al. 2013). However, the transesterification reactions using cooking oils as source require for severe conditions in comparison to the virgin oils due to the existence of higher proportion of free fatty acids and water (Cetinkaya and Karaosmanoglu 2004). Traditionally, these transesterification reactions have been carried out by using homogeneous catalysts; however, the current trend is to change them for heterogeneous catalysts, which are more sustainable and easier to regenerate. The heterogeneous catalysts also show drawbacks related with the lower biodiesel conversion and higher deactivation of the active centers. CaO, which is the most active heterogeneous catalyst in the synthesis of biodiesel, can also be obtained from food waste as eggshell, mollusk, or crab shells after calcination step (Viriya-Empikul et al. 2012), which can decrease the cost of the process.

3.2.1 Glycerol as By-Product of Biodiesel with High Potential

Glycerol is the main by-product obtained in the synthesis of biodiesel (about 10 kg of biodiesel produce glycerol equivalent to 1 kg). The quality of this glycerol depends on the feedstock and the type of biodiesel processing employed. The glycerol can be refined to increase its range of applications, although these purification processes also increase the production costs from 11 cents kg^{-1} for a glycerol purity of 80% to 88 cents kg^{-1} for a glycerol purity of 99% (Narashimharao et al. 2007) (Fig. 7). The purified glycerol can be used as additive in food, skin care product, or as moisturizer. In addition, glycerol without purification step can also be used as land spread agent if the methanol content is low. Other applications include its use as a co-substrate for anaerobic digestion, additive in cements, combustion in combined heat and power engines, or as a co-substrate for fermentation bioreactions (Lin et al. 2013; List of Hazardous Waste, EN 2000).

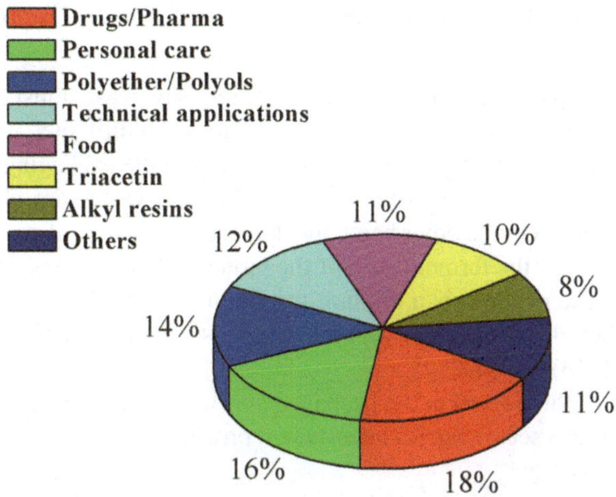

Fig. 7 Different uses of glycerol by market volumes

Table 2 Animal waste origin, sources, and corresponding target ingredients for recovery

Waste origin	Selected sources	Target ingredients
Animal		
Meat products	Chicken by-products	Proteins
	Slaughterhouse by-products	Proteins
	Bovine blood	Proteins
	Beef lung	Protein concentrates
	Sheep visceral mass	Protein hydrolyzates
Fish and seafood	Fish leftovers (skin, head, and bones)	Proteins
	Shrimp and crab shells	Chitosan/chitin, proteins
	Surimi wastewater	Proteins
Dairy products	Cheese whey	Lactose, β-lactoglobulin, α-lactalbumim

3.3 Animal By-Products

The most of food waste is attributed to meat, poultry, and fish industries (Fig. 1 and Table 2). These wastes display a variable composition such as fats, blood, or residues from intestines, which are highly polluting due to their high levels of BOD (biological chemical demands) and COD (chemical oxygen demands). As was indicated in Sect. 2, the animal by-product wastes follow other regulations (Regulation (EC) 2009) since these residues can cause risks to animal and public health. Thus, the valorization of the animal by-product wastes requires a prior treatment to remove pathogens and pollutants, which could raise the price of its valorization, so the use of these wastes from animal by-products is a challenge to the governments (Poeschl et al. 2010).

3.4 Mixed Domestic Waste and Waste Packaging

The packaging from catering and domestic waste must also undergo pretreatment to their recycling or eventually end up in landfill sites. Currently, recycling policies have caused a growing interest related to the packaging valorization (cans, plastics, glass, or papers) since these packaging wastes are about 73% of the household waste. On the other hand, the domestic waste can also be used to the biogas production by the fermentation of the biodegradable wastes. However, the composition of these residues is a mixture formed mainly by food waste and its packaging (cardboard and plastics) which hinders its treatment mainly on a large scale (Lin et al. 2013).

With regard to the packaged surplus production that cannot be sold, the food and drinks sector carries out efficient processes to reuse a large proportion of food waste generated.

3.5 Domestic Waste

Domestic waste disposal is an issue that is important to the management of any urban area. Cities without a functioning waste disposal plan face risks of disease running rampant and economic activity grinding to a halt. The developed countries use the sanitary landfill method of waste disposal, which has served fairly well for quite a while; however, in situations where space is at a premium, incineration and material-recycling-based waste disposal are more likely to come to the forefront.

The modern sanitary landfill is much more than a simple dumping ground, due to the fact that waste materials are handled in a more controlled fashion to preserve the quality of groundwater in the area. Lighter materials are placed at the bottom of the sanitary landfill, which contains the majority of toxic compounds, thereby protecting the local environment. Sanitary landfills require continuous maintenance and treatment of wastewater as well as the recovery of toxic gases, making them potentially hazardous if left derelict long enough to allow systems to fail. The main flaw of the sanitary landfill concept is that it continually consumes both land and resources to contain the waste, in addition to being potentially ecologically damaging.

Incineration is a popular garbage disposal method in locations where space is at a premium or in locations where there is no sanitation service provided by local government. Although incineration can provide relief from dealing with the bulk of material waste, it does not entirely eliminate problems. Anything in the burning incinerator that contains a toxic substance, particularly heavy-metal toxic substances, will be sent up the chimney and cast as ash all over the surrounding area. The toxins from the burning of waste then build up in the local populace, contributing to a variety of health problems ranging from asthma to heavy-metal poisoning and even cancer.

4 Current Uses of Food Supply Chain Waste

In the next section, some examples of accumulative waste and their treatment are discussed.

4.1 *Citrus Peel Residues*

Citrus fruits, such as oranges, lemon, and lime, are grown around the globe in the region of the tropics and equator, being the top producer nations such as the USA, Spain, Brazil, Italy, Morocco, or Turkey. The processing of these fruits generates about 50 wt% of waste, presenting an environmental problem (Siles Lopez et al. 2010). There is a real challenge to utilize this resource, with 15.6 million metric tonnes of waste produced from 31.2 million metric tonnes of processed citrus fruit annually (Siles Lopez et al. 2010). Waste orange peel (WOP) is composed of 20% dry matter (sugars, cellulose, hemicellulose, pectin, and D-limonene) and 80% water (Fig. 8). The transformation of these wastes into higher-value products would allow companies to increase competitiveness by generating additional profits and reducing disposal costs together with improving the resource efficiency of the citrus supply chain. Nowadays, citrus waste valorization has nearly always focused on

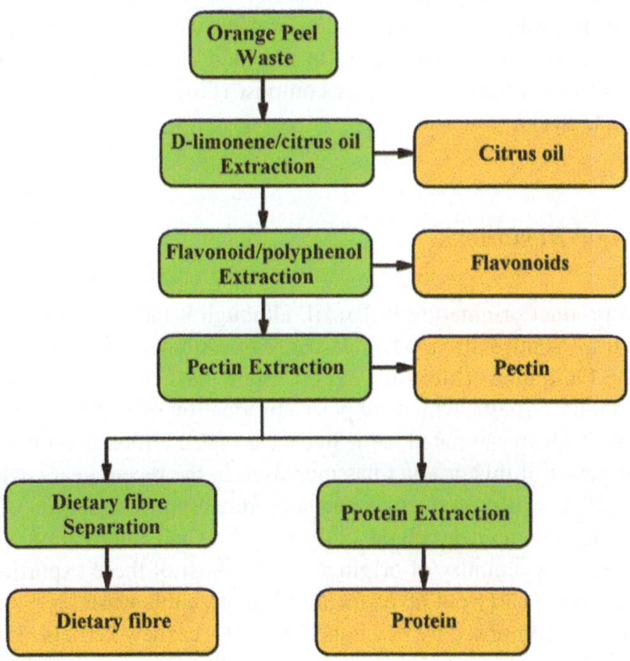

Fig. 8 Valorization of the orange peel waste

production of a single component, such as D-limonene, pectin, or bioethanol. Unfortunately, the integrated valorization of citrus wastes has hardly been reported in the literature (Siles Lopez et al. 2010).

Citrus peel is the most interesting fraction for its valorization since it is formed by a large amount of carbohydrates. The citrus peel must be dried previously to avoid its fermentation (Ferreira-Leitao and Fortes Gottschalk 2010), although this process raises the price of the treatment. Nonetheless several researches have pointed out various synthetic strategies to the citrus peel valorization such as the synthesis of activated carbon from pectin (Ma et al. 1993), its use as food additive (fiber source) (Marin et al. 2007), the synthesis of biogas (Ozmen and Aslanzadeh 2009), the synthesis of bioethanol by fermentation (Widmer et al. 2010), or the synthesis of succinic acid (Li et al. 2010). In addition, D-limonene, which can be obtained by distillation with a cost of 1–2 \$ kg^{-1} (Braddock 1999), is a compound that can be used as solvent (Kerton 2009), adhesive terpene resins, or starting material in the synthesis of valuable products such as carveol (225 USD kg^{-1}), carvone (119 USD kg^{-1}), α-terpineol (130 USD kg^{-1}), perrillyl alcohol (710 USD kg^{-1}), and perillic acid (490 USD kg^{-1}) (Braddock 1999). Other authors have pointed out that the treatment of the citrus peel with H_2SO_4 at 150 °C leads to bioethanol, methane, and D-limonene (Pourbafrani et al. 2010), while Balu et al. (2012) have established that the microwave-assisted treatment favors the formation of high added-value products such as D-limonene or α-terpineol as well as the transformation of pectin in mesoporous cellulose.

Currently, the countries with the highest citrus production have regulated the disposal of such residues. The valorization of these wastes in large scale is focused to the extraction of essential oils used in the food and cosmetic industry, animal feedstuffs, and biofuels and its use as compost (Fazlollah-Ghoreishi et al. 2007; Demirbas et al. 2011).

4.2 *Cashew Residues*

Cashew is a product originating in Brazil, although it has spread to other tropical regions, such as South Africa, Madagascar, Mozambique, Ivory Coast, India, Sri Lanka, or the Philippines (Morton 1987). Cashew nuts consist of 35–45% kernels and around 55–65% shells, which suppose a production of cashew nuts close to two million tonnes with an estimated value in excess of \$ 2 billion US dollars (Lin et al. 2013). The interest of this product has increased in the recent years, which has led to a growing of its world production and subsequently the formation of larger proportion of waste. The main drawback of the recycling and valorization of the cashew waste is related to its country of origin since the most of these exporting countries lacks of financing to carry out investments to valorize this waste.

Cashew nuts is composed by raw nuts (15–30%), cashew kernels (35–45%), and cashew nut shell liquid (55–65%) (Fig. 9). The cashew product with higher interest is its nut. Other product that can be marketed is the cashew apple, which can be used

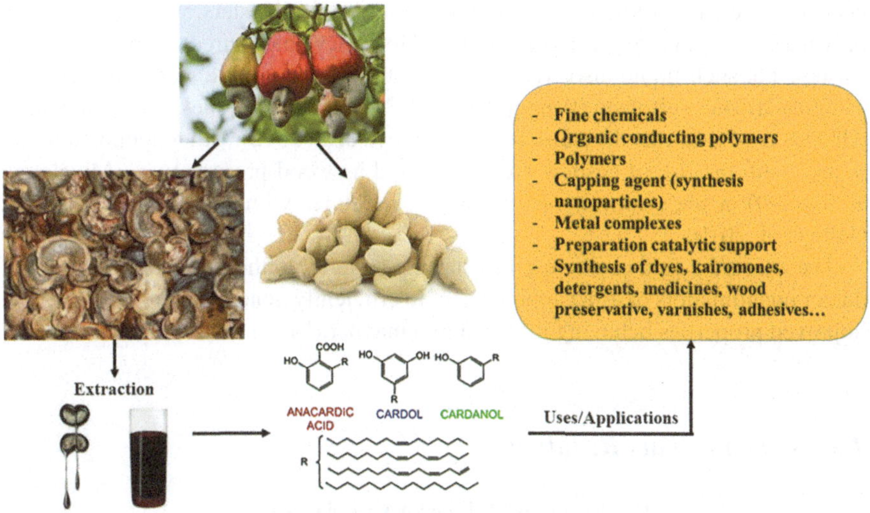

Fig. 9 Opportunities from cashew nut shell agro-waste in the production of chemicals

to produce juice, syrup, squish, pickles, jam, chutneys, candy, and canned fruit jelly. The main disadvantage of the cashew apple manufacturing is related with the fast decomposition and fermentation of this product, which causes a significant economic loss. Nonetheless, this waste could present a great potential for the synthesis of bioethanol via fermentation of the sugars present in the cashew apple.

Cashew nut shell liquid is a dark reddish-brown viscous liquid, renewable, variable composition, and with a world availability of 50,000 tonnes year^{-1}, which offers interesting opportunities for the synthesis of high added-value products and polymers. This by-product displays a wide range of applications such as paints and varnishes, rubber compounding resins, clutch disks, polyurethane-based polymers, friction linings, surfactants, epoxy resins, cashew cements, laminating resins, foundry chemicals, and intermediates for the chemical industry (Philip et al. 2007; Llomo et al. 2005; Attanasi et al. 2006).

The cashew nut shell liquid is an excellent and cheaper source to obtain unsaturated phenolic lipid (30–35 wt%), as cardol (20%), cardanol (10–15%), or anacardic acid (60–65%), which can be obtained by extraction. From cashew nut shell liquid, it has been synthesized a renewable liquid crystalline polyester as alternative to polymer fibers and films with low carbon footprint (Azam-Ali and Judge 2001). The cashew nut shell liquid can be also used as structure directing agent to the synthesis of porous silica to the adsorption of biomolecules (Hamad et al. 2011). Other authors have synthesized spherical polymeric particles by chain growth polymerization mechanisms, attaining particles with a specific surface area of 150–250 m^2 g^{-1} (Mkayula et al. 2004).

An unsaturated phenol, as cardanol, can polymerize using an Fe-salt complex as the catalyst to obtain biodegradable polyphenols (Bhunia et al. 1999). Cardanol has been also used as starting material to the synthesis of polyesters, which can replace polymer fibers and films currently produced for renewable sources (Lin et al. 2013). Cashew nut shell liquid also was used to synthesize renewable surfactants such as sodium cardanol sulfonate (Peungjitton et al. 2009) or kairomone (3-propylphenol), which is used to attract the tsetse fly (Mmongoyo et al. 2012). Other applications are related with the use of cashew nut shell liquid as wood preservatives (Mwalongo et al. 1999) or the synthesis of soft nanoarchitectures with variable morphologies (John et al. 2010).

The anacardic acid can be used as starting material to the synthesis of semiconductor quantum dots or as a capping agent to efficiently generate nanoparticles with improved properties behaving in a quantum manner (Lin et al. 2013).

4.3 Oils and Fats Residues

4.3.1 Synthesis of Biodiesel and Glycerol Valorization

As mentioned previously, the most extended way to deal food waste in developed countries is the use of composters to reduce the amount of waste and to recycle the organic matter. In the countries of the Mediterranean Sea, the production and consumption of vegetable oils (e.g., olive and sunflower) is common in all households. The olive oil is widely consumed for cooking purposes. However, the waste generated upon frying becomes a problem since until very recently this cooking oil waste was disposed without finding any alternative uses for further valorization being discharged to rivers and inland waters. Recently the residues of house and restaurant oils begin to be collected for their treatment and the generation transformation to high added-value products, mainly biodiesel. It is well known that biodiesel can be obtained by homogeneous conditions using NaOH/MeOH. Another alternative is the use of acid or basic heterogeneous catalysts or the use of enzymatic processes to obtain biodiesel, as was previously mentioned. The cooking oil wastes need a prior filtration to remove the solid food residues. The catalytic conditions to obtain biodiesel from cooking oil's waste are more severe than those required when pure edible and nonedible oils are employed by the presence of the pure edible and nonedible oils, other organic compounds, or salts incorporated in its use for cooking.

Animal fat and grease, mainly obtained from animal by-products, have been traditionally recycled to be used as fuel, cosmetic precursors, and components of pet food supplements. The most stringent regulations of biofuel production from rendered fats has recently increased in importance in the EU, obtaining high biodiesel yields from animal by-products even with 30 wt% of triglyceride (Directive 2008d). The use of wet conditions is generating growing interest because the product can be used as substrate for fermentations and feedstocks for microwave pyrolysis, which has high potential to produce biofuel and chemicals (Shuttleworth et al. 2012).

As was indicated in Sect. 3.2.1, the synthesis biodiesel produces large amounts of glycerol as by-product. The US Department of Energy has considered glycerol as one of the top ten platform chemicals for the production of a wide range of valuable chemicals (Werpy and Petersen 2004; Tuck et al. 2012). The most relevant processes developed with used cooking oil-derived glycerol is its chlorination using HCl to form epichlorohydrin (Schreck et al. 2006) and the hydrogenolysis of glycerol to Cu-based catalysts to form 1,2-propanediol (30 USD kg^{-1}) (Tuck and Tilley 2007) (Fig. 10). Another chemical obtained from glycerol with a high interest is the acrolein (11 USD kg^{-1}) since this compound is widely used as intermediate in the production of building materials, herbicides and algaecides, water treatment chemicals, and essential amino acids like methionine used as supplementary in fodder, especially for poultry. In addition, acrolein is considered as a building block molecule to obtain a wide variety of valuable chemicals with interesting applications, such as the acrylic acid in the polymer field (Katryniok et al. 2013).

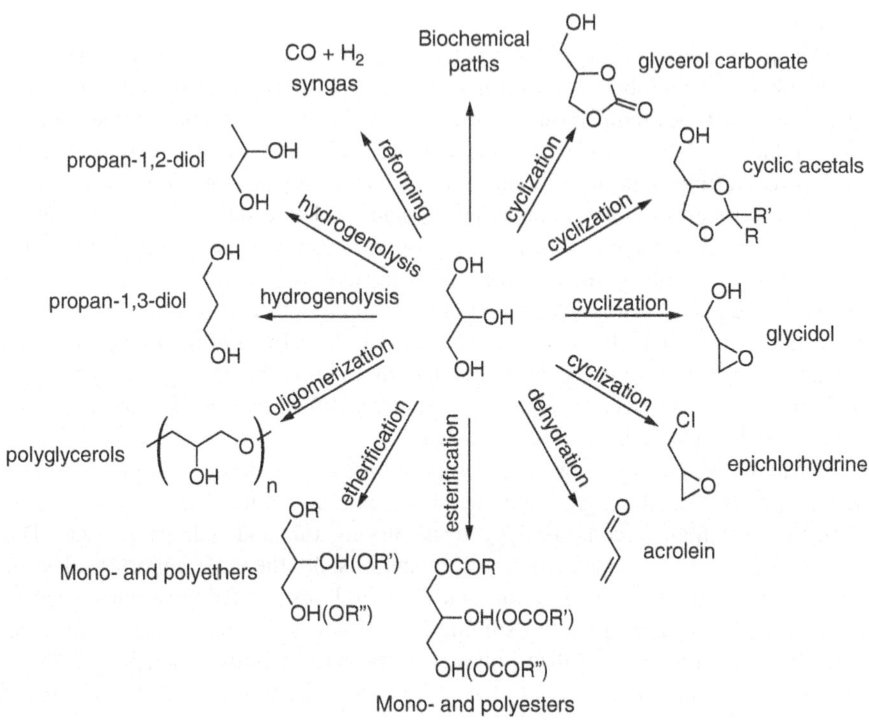

Fig. 10 Roadmap of selected glycerol valorization reactions

4.3.2 Alternative Biofuels

Apart from biodiesel, there are some other interesting catalytic routes to biofuels from waste oils. The synthesis of biofuels from oils using heterogeneous catalysts such as NiMo or CoMo sulfides is a useful methodology to remove the oxygen and to saturate the C=C double bonds (Huber et al. 2007). However, these active phases tend to suffer fast deactivation by the oxidation of the sulfide species to inactive sulfates, so alternative active phases such as noble metals or transition metal carbides and phosphides have been proposed recently (Rezaei et al. 2014). The free fatty acids and the triglycerides can lead hydrocarbons under H_2 pressure (10 bar), a temperature about 400–500 °C, using solid acid catalysts (Charusiri et al. 2006; Charusiri and Vitidsant 2005). This methodology is interesting since glycerol is not generated as a by-product, although the high H_2 pressure required raises the costs.

4.3.3 Nonfuel Applications of Used Cooking Oil

Recently, a wide range of sophisticated applications have been developed such as lubricants, coating, polymers, or surfactants (Boyde 2002; Bentley 2001). In the next section, some examples report the use of used vegetables oils and fats as starting materials to obtain valuable products with alternative applications to the biofuels.

The use of edible or nonedible oils have shown to be a promising alternative to the synthesis of biolubricants with high thermal/oxidative stability, cutting fluids, and low working temperature hydraulic fluids, being highly biodegradable (Shashidhara and Jayaram 2010; Saboya et al. 2017). The use of fats or used cooking oil as starting material to synthesize biolubricants is more limited since the obtained lubricants display poorer thermal and oxidative stabilities (Hayder et al. 2011). Nonetheless the physicochemical properties can be improved by the incorporation of some additives, such as zinc dialkyldithiophosphates as anti-wearing agent and hindered phenols or amines as antioxidants, or can be distilled to improve flash point properties (Quinchia et al. 2010; Bono et al. 2010). On the other hand, the available free fatty acid in the used cooking oils can improve their physicochemical properties, such as viscosity index or oxidative stability, by esterification reactions using branched alcohols (Saboya et al. 2017).

Used cooking oils have also been proposed for the synthesis of bio-derived varnishes, paints, and coatings, which are less toxic and safer than industrial ones. Varnishes are mainly composed by resin, drying oil, and volatile solvent. The alkyds, formed by glycerol, a dibasic acid and oils, are the most important class of resin in the coatings industry. In this sense, several unsaturated fatty acids, such as linseed, soybean, high linoleum sunflower, and castor oil, have been assayed as varnish agents after a controlled polymerization with polyols (Booth et al. 2007). The use of used cooking oils as starting material displays several disadvantages related to their relatively low shelf life and poor oxidative stability due to the increase of the free fatty acids during the frying step; however, these properties could be improved by the incorporation of other functional groups, such as carboxylic acids (Xia and Larock 2010).

5 Integration of Food Waste in Biorefineries

The valorization of food supply chain waste will be successful when these wastes can integrate within the concept of biorefinery, which produce energy and chemicals with high interest for the population from biomass. The most sustainable process to valorize the food waste into added-value products is focused toward the incorporation of these wastes to technologies already implemented, such as microwave heating or supercritical CO_2, to maximize profits of the processes.

The regional food industries can favor the decentralization of biofuel and chemical production, minimizing the transport costs. In addition, this bio-economy can significantly contribute to the future development of rural regions from the treatment of solid waste, food waste, and even effluents for the synthesis of biogas (Antizar-Ladislao and Turrion-Gomez 2010). Nowadays, food wastes are hardly implemented in biorefineries. So far, researches have mainly been carried out on a laboratory scale, so it is expected to realize studies on a larger scale or plants and its implementation in biorefineries in the coming years.

One of the methodologies used to evaluate the implementation of a process are the technology readiness levels (TRLs) since it is a method to estimate the technology maturity. These levels are determined during a technology readiness assessment (TRA) that examines program concepts and technology requirements and demonstrated technology capabilities. TRLs are based on a scale from 1 to 9 with 9 being the most mature technology. In the case of the food valorization, research techniques producing food additives (e.g., colorants, fibers, sugar, proteins), functional materials, and fine chemicals mostly have a lower TRL level (3–6) than techniques producing fuels for energy production (TRL 7–9). This can be explained because energy production techniques started (much) earlier than techniques producing chemicals and functional materials. The latter techniques are more complex and need well-developed knowledge on chemical processes and smart techniques to extract the chemical components with a high purity. This kind of science just seriously started some decades ago but now is increasing vast in number of researches and applications. EU has supported the AGRIFORVALOR project where an overview of valorization techniques for agricultural biomass is carried out (Hendriks et al. 2016), as indicated in Table 3.

Table 3 Overview of valorization techniques for agricultural biomass side streams (sorted by TRL) (adapted from Hendriks et al. (2016))

Biomass side stream	Applied in	Technique	Output	General findings	TRL
Brewer spent grain	Belgium	Extraction	Plant sterols	Food	3
Corn fiber	Hungary	Chemical fractionation, enzymatic hydrolysis, biopurification, fermentation, purifications, separations	Xylitol, arabinose ethanol, biomethane, digestion residue	Food, fine chemical, fuel, fertilizer	3
Manure and other organic waste	Hungary	Biodigestion, codigestion	Biogas	Fuel	3
Olive biomass	Spain	Steam explosion	Antioxidant, sugars	Food and fine chemical	3
Olive prunings	Spain	Hydrolysis and fermentation	Ethanol, antioxidants, oligosaccharides, lignin-derived chemicals	Fuel and fine chemical	3
Sewage sludge and manure	Sweden	Anaerobic digestion and incineration	Energy, phosphate	Fuel and fertilizer	3
Yeast	Belgium	Extraction	Squalene, phospholipids	Food, fine chemical	3
Brewer spent grain	EU	Enzyme-aided fractionation	FAX, peptides	Food	4
Leek leaves	Belgium	Fermentation	Lactic acid	Food	4
Olive crop residues	Spain	Isolation	New agrifood components	Food, feed, fine chemicals	4
Vegetable trimmings	EU	Liquefaction	Soluble colorant	Food and fine chemical	4
Agro-industrial wastewater	Spain	Microalgae and bacteria	Biofertilizer and biogas	Fertilizer and fuel	4–5
Brewery wastes	Spain	Microalgae	Nitrogen and phosphorous microalgae biomass	Fertilizer and feed	4–5
Low value biogas	Spain	Purification	Valuable biomethane	Fuel	4–5
Rape, sugar beet waste, other green plant biomass	Hungary	Acidification and gasification	Fermented biomass and energy	Fertilizer and fuel	5

(continued)

Table 3 (continued)

Biomass side stream	Applied in	Technique	Output	General findings	TRL
Vegetables, corn	Hungary	Microbial fermentation	Lactic acid, ethanol, bioenergy	Fine chemical and fuel	6
Sugar beet, pig slurry, and cow manure	Spain	Anaerobic digestion	Methane and fertilizer	Fuel and fertilizer	7
Vegetable waste	Spain	Green extraction	Food additives, biopolymers	Food and functional material	7
Vegetable waste	Spain	Green extraction	Bioactive compounds	Food and functional material	7
Olive stones	Spain	Marine organisms	Proteins and enzymes	Food and feed	7
Orange skin and other agrifood by-products	Spain	Dehydration and pelletizing	Animal feeding	Feed	7
Blood (pigs, cows)	Italy	Sterilization, coagulation, and separation	Recycled water and biogas	Fuel	8
Animal waste (bone)	Hungary	Anoxic heat treatment	Biochar	Fertilizer	8
Dairy serum	Spain	Inverse osmosis and ultrafiltration	Biogas, concentrate of proteins, concentrate of lacteal whey, feed, food, and supplements	Feed, food, fuel	9
Grape pomace	Italy	Solvent extraction	Resveratrol, anthocyanins, proanthocyanidins, quercetin, grape seed oil	Food and fine chemical	9
Olive processing waste	Italy	Solvent extraction	Polyphenol, hydroxytyrosol, oleuropein	Food	9
Sewage sludge, green waste, production residue from the food industry, straw or animal excrement	Germany	Heating and condensation	Electricity, heat gas, and oil	Fuel, fertilizer, fine chemical	9
Tomato pomace	Italy	Solvent extraction	Lycopene, fiber, seed oil, enzymes	Food	9

(continued)

Table 3 (continued)

Biomass side stream	Applied in	Technique	Output	General findings	TRL
Olive mill waste	Spain	Anaerobic digestion, catalytic reforming, and use of proton exchange membrane fuel cells	Biogas, hydrogen, energy	Fuel	9
Vineyard waste, grape seed residue	Hungary	Special pretreatment of the residues (patented application)	Fertilizer	Fertilizer	9

5.1 Integrated Biorefineries Based on Specific Food Waste

The biorefineries must adapt to the demand of the markets. The production of chemicals from food waste may not be economically feasible as low market prices of commodity products require large production capacity industrial plants, which may not be profitable in many cases since large volumes of a specific food waste can be required to synthesize a determined product. These large volumes, together with the cost of the transport and the fast deterioration of the food waste, can be important drawbacks to incorporate these wastes in the biorefineries. In this sense, the decomposition of the food waste can be minimized by the combining extraction of high-value products with subsequent fermentations to obtain a whey. This whey contains high proportion of protein and lactose, which can be hydrolyzed to obtain chemicals with interesting applications in the biochemical field (Pinto et al. 2009).

The initial steps in an integrated process must extract high-value constituents, such as proteins, oils, sugars, vitamins, waxes, colorants, and flavor and fragrance compounds, from food waste, while the majority of food waste will subsequently be processed for the production of case-specific fermentation media or treated using chemical or thermochemical methods, obtaining bioethanol or biogas as main products. Oil-rich fractions extracted from food waste could substitute for plant oils as raw materials of the chemical industry for chemical conversions and synthesis of chemically pure compounds. These oil-rich fractions could be employed for the production of biofuels, surfactants, stabilizers, fatty amines, dicarboxylic acids, resins, plasticizers, soaps, lubricants, and polyols (Carlsson 2009; Wisniewski et al. 2010).

Bioprocessing technologies could be employed for the production of bioenergy, platform chemicals, and biomaterials. Platform chemicals could be subsequently converted via clean and green chemical technologies into high added-value chemicals and polymers.

The fermentation is a biochemical process where polysaccharides and proteins contained in food wastes are hydrolyzed by an enzymatic process, leading to valuable products that can be used as substitutes for commercial nutrient supplements for the production of platform products, such as succinic, fumaric, malic,

3-hydroxypropionic, glutamic, and itaconic acids as well as sugars including xylitol and arabinitol (Table 4). These products can be converted into valuable products such as special chemicals, biofuel precursors, and bio-based polymers either through the synthesis of monomers or through direct production of biodegradable polymers (Farges-Haddani et al. 2006; Vazquez and Murado 2008). The hydrolysis of food waste by fermentation can also be used to the production of single cell oil using various oleaginous microorganisms whose composition depends on the selected organism. This cell oil could be used as a substitute for plant oils due to their similar fatty acid composition, alternatively as raw material for biodiesel and oleochemical production (Carlsson 2009; Koutinas and Papanikolaou 2011).

Table 4 Carbon source requirements for the fermentative production of various chemicals (adapted from Lin et al. (2013))

Chemical Product	Industrial applications	Worldwide production (10^6 tonnes)	Production yield (kg kg^{-1} glucose)	Quantity of glucose required (10^6 tonnes)	Quantity of food waste required
Ethanol (1.5 USD kg^{-1})	Solvent Disinfectant Building block	31	0.46	67.39	61.26 million tonnes of starch
1,3-propanediol (35 USD kg^{-1})	Polymers	0.08	0.54	0.148	0.148 million of tonnes of crude glycerol
Lactic acid (2.5 USD kg^{-1})	Additive foods Polylactic production Pharmaceuticals Personal care products	0.15	0.95	0.158	0.319 million of tonnes of waste bread
Succinic acid (2 USD kg^{-1})	Platform molecule	0.015	1.16	0.013	27 million kg of waste bread
Fumaric acid (1.5 USD kg^{-1})	Food acidulant Platform molecule	0.012	1.16	0.01	21 million kg of waste bread
PHB (2.5 USD kg^{-1})	Polymer Medical applications Biocomposites Platform molecule	0.4	0.43	0.93	1879 million kg of waste bread

6 Concluding Remarks and Future Prospects

The main efforts in the future must be aimed to raise awareness that the increase of the world population and quality of life are leading to an increase of the waste production as well as to demonstrate the potential of advanced food waste valorization to be used as inputs into other processes to recover and reuse all non-biodegradable material (Fig. 11).

The food wastes cause several environmental damages such as greenhouse gas emissions and groundwater contamination by the decomposition of the food waste in the landfill sites. The treatment of these wastes is a complex process that must be tackled by governments, policies, regulations, stakeholders, companies, products, and most importantly consumers and public opinion. Currently, the valorization of food waste is related with the recycling and composting of the residues; however, it is necessary to achieve a sufficient processing of food residues, which is limited so far.

Food waste valorization must be focused to the development of an innovative low environmental impact legislation, which should pretend the conversion of waste into value-added products by several technologies to achieve sustainable development and a circular economy. The valorization must be carried out with low environmental impact technologies using extractive or biochemical processes to obtain valuable products. These processes will be sustainable through a multidisciplinary approach to achieve a zero-waste economy and a more sustainable bio-based society. In order to obtain this goal, it is necessary cross-industry and public-private collaborations to devise a strategy with the maximum economic, social, and environmental impact. Currently, the data on food waste generation from public research are limited, so it is necessary to identify, quantify, and study the periodicity of these

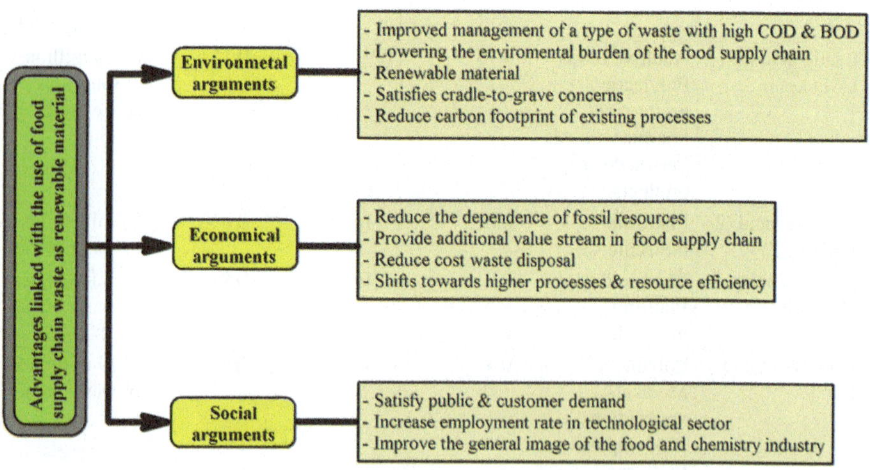

Fig. 11 Advantages linked with the use of food supply chain waste as a renewable feedstock

residues to evaluate which are the main products for their treatment toward value-added products.

Another key factor is related with the social awareness to a sustainable production and consumption of biodegradable products to minimize the formation of unnecessary waste. In addition, designed reeducation and awareness campaigns must be carried out throughout the world population in order to change the perception of food waste as a problem instead of a valuable resource to produce chemicals, materials, and fuels.

References

ADAS UK Ltd (2007) West midlands non-food crops opportunities/mapping study. Accessed 15 Oct 2007
Antizar-Ladislao B, Turrion-Gomez JL (2010) Energies 3:194–205
Attanasi OA, Berretta S, Fiani C, Filippone P, Mele G, Saladino R (2006) Tetrahedron 62:6113–6120
Azam-Ali SH, Judge EC (2001) Small-scale cashew nut processing. ITDG Schumacher Centre for Technology and Development Bourton on Dunsmore, Rugby, Warwickshire, UK, FAO
Balu AM, Budarin V, Shuttleworth PS, Pfaltzgraff LA, Waldron K, Luque R, Clark JH (2012) ChemSusChem 5:1694–1697
Bentley J (2001) In: Gunstone FD, Hamilton RJ (eds) Oleochemical manufacture and applications, 1st edn. Sheffield Academic Press, Sheffield, pp 164–193
Bhunia HP, Nando GB, Basak A, Lenka S, Nayak PK (1999) Eur Polym J 35:1381–1391
Bono A, Pin OP, Jiun CP (2010) J Appl Sci 10:2508–2515
Booth G, Delatte DE, Thames SF (2007) Ind Crop Prod 25:257–265
Boyde S (2002) Green Chem 4:293–307
Braddock RJ (1999) Handbook of citrus by-products and processing technology. Wiley-Interscience, Weinheim, p 186
Carlsson AS (2009) Biochimie 91:665–670
Cetinkaya M, Karaosmanoglu F (2004) Energy Fuel 18:1888–1895
Charusiri W, Vitidsant T (2005) Energy Fuel 19:1783–1789
Charusiri W, Yongchareon W, Vitidsant T (2006) Korean J Chem Eng 23:349–355
Demirbas FM, Balat M, Balat H (2011) Energy Convers Manage 52:1815–1828
Directive (2008a) Directive 2008/38/EC of 5th, March 2008, establishing a list of intended uses of animal feeding stuffs for particular nutritional purposes
Directive (2008b) Directive 2008/98/EC of 19 (November 2008) OJ L312, Article 3(1), pp 3–30
Directive (2008c) Directive 2008/98/EC on waste (Waste Framework Directive)
Directive (2008d) Directive 2008/98/EC of the European Parliament and of the council of the 19 November 2008 on waste and repealing certain Directives
Djilas S, Canadanovic-Brunet J, Cetkovic G (2009) Chem Ind Chem Eng Q 15:191–202
Dodson JR, Hunt AJ, Budarin VL, Matharu AS, Clark JH (2012) RSC Adv 1:523–530
ENDS (1999) ENDS Report 294, Environmental Data Services
Farges-Haddani B, Tessier B, Chenu S, Chevalot I, Harscoat C, Marc I, Goergen JL, Marc A (2006) Process Biochem 41:2297–2304
Fazlollah-Ghoreishi S, Pirmohammadi R, Teimouri-Yansari A (2007) J Anim Vet Adv 6:1074–1078
Ferreira-Leitao V, Fortes Gottschalk LM (2010) Biomass residues in Brazil: availability and potential uses. Waste Biomass Valorization 1:65–76
Hamad FB, Mubofu EB, Makame YMM (2011) Cat Sci Technol 1:444–452
Hayder AA, Rosli MY, Abdurrahman HN, Nizam MK (2011) Int J Phys Sci 6:4695–4699

Hendriks CMA, Lambrecht E, Nabuurs GJ, Gellynck X, Welck H (2016) Compendium on research results on agriculture and forest-biomass side-streams. AGRIFORVALOR Project. Grant Agreement 696394
Huber GW, O'Connor P, Corma A (2007) Appl Catal A 329:120–129
Janaun J, Ellis N (2010) Renew Sust Energ Rev 14:1312–1320
John G, Shankar BV, Jadhav SR, Vemula PK (2010) Langmuir 26:17843–17851
Katami T, Yasuhara A, Shibamoto T (2004) Formation of dioxins from incineration of foods found in domestic garbage. Environ Sci Technol 38:1062–1065
Katryniok B, Paul S, Dumeignil F (2013) ACS Catal 3:1819–1834
Kerton FM (2009) Alternative solvents for Green Chemistry, 1st edn. RSC, London, pp 97–117
Koutinas AA, Papanikolaou S (2011) In: Luque R, Campelo J, Clark JH (eds) Handbook of biofuels production—processes and technologies. Woodhead, Cambridge, pp 177–198
Li Q, Siles JA, Thompson IA (2010) Appl Microbiol Biotechnol 88:671–678
Lin CSK, Pfaltzgraff LA, Herrero-Davila L, Mubofu EB, Abderrahim S, Clark JH, Koutinas AA, Kopsahelis N, Stamatelatou K, Dickson F, Thankappan S, Mohamed Z, Brocklesby R, Luque R (2013) Food waste as a valuable resource for the production of chemicals, materials and fuels. Current situation and global perspective. Energy Environ Sci 6:426–464
List of Hazardous Waste, EN (2000) Official Journal of the European Communities L 226/3, 6.9.2000, (2000/532/EC)
Litchfield JH (1987) Food Biotechnol 1:29–57
Llomo OO, Makame YMM, Mkayula LL (2005) Bull Chem Soc Ethiop 19:9–16
Luque R, Herrero-Davila L, Campelo JM, Clark JH, Hidalgo JM, Luna D, Marinas JM, Romero AA (2008) Energy Environ Sci 1:542–564
Ma E, Cervera Q, Sanchez GMM (1993) Bioresour Technol 44:61–63
Marin FR, Soler-Rivas C, Benavente-Garcia O, Castillo J, Perez-Alvarez JA (2007) Food Chem 100:736–741
Mkayula LL, Makame YMM, Matechi TH (2004) Tanz J Sci 30:1–10
Mmongoyo JA, Mgani QA, Mdachi SJM, Pogorzelec PJ, Cole-Hamilton DJ (2012) Eur J Lipid Sci Technol 114:1183–1192
Moran C, Perez L, Pons R, Pinazo A, Infante MR (1999) In: Kjellin M, Johansson I (eds) Surfactants from renewable resource, 1st edn. Wiley-Interscience, Weinheim, pp 85–107
Morton JF (1987) In: Morton JF (ed) Fruits of warm climates. Cashew Apple, Miami, pp 239–240
Mwalongo GCJ, Mkayula LL, Dawson-Andoh B, Mubofu EB, Shields J, Mwingira BA (1999) Green Chem 35:13–16
Narashimharao K, Lee A, Wilson K (2007) J Biobaased Mater Bioenergy 1:19–30
Ozmen P, Aslanzadeh S (2009) Master of Science, University of Boras
Peungjitton P, Sangvanich P, Pornpakakul S, Petsom A, Roengsumran S (2009) J Surfactant Deterg 12:85–89
Philip J, Buchweishaija J, Mkayula LL, Ye L (2007) J Agric Food Chem 55:8870–8876
Pinto GA, Giordano RLC, Giordano RC (2009) Bioprocess Biosyst Eng 32:69–78
Poeschl M, Ward S, Owende P (2010) Renew Sust Energ Rev 14:1782–1797
Pourbafrani M, Forgacs G, Horvath IS, Niklasson C, Taherzadeh M (2010) Bioresour Technol 101:4246–4250
Quinchia LA, Delgado MA, Valencia C, Franco JM, Gallegos C (2010) Ind Crop Prod 32:607–612
Regulation (EC) (2009) Regulation (EC) No. 1069/2009 of the European Parliament and the Council of 21 October 2009 laying down health rules as regards animal by-products and derived products not intended for human consumption, OJ L300, 14 November 2009, pp 1–33
Rezaei PS, Shafaghat H, Daud WMAW (2014) Appl Catal A Gen 469:490–511
Saboya RMA, Cecilia JA, García-Sancho C, Sales AV, de Luna FMT, Rodríguez-Castellón E, Cavalcante CL Jr (2017) Catal Today 279:274–285
Schreck D, Kruper WJ, Varjian RD, Jones ME, Campbell RM, Kearns K, Hook BD, Briggs JR, Hippler JG (2006) WO2006020234
Shashidhara YM, Jayaram SR (2010) Tribol Int 43:1073–1081

Shuttleworth P, Budarin V, Gronnow M, Clark JH, Luque R (2012) J Nat Gas Chem 21:270–274
Siles Lopez JA, Li Q, Thompson IP (2010) Crit Rev Biotechnol 30:63–69
Tuck MWM, Tilley SN (2007) International Patent WO2007010299
Tuck CO, Pérez E, Horváth IT, Sheldon RA, Poliakoff M (2012) Science 337:695–699
Vazquez JA, Murado MA (2008) Enzym Microb Technol 43:66–72
Viriya-Empikul N, Krasae P, Nualpaeng W, Yoosuk B, Faungnawakij K (2012) Fuel 92:39–244
Werpy T, Petersen G (2004) Top value added chemicals from biomass. U.S. Renewable Energy Laboratory (NREL), Oak Ridge, TN
Widmer W, Zhou W, Grohmann K (2010) Bioresour Technol 101:5242–5249
Wisniewski A, Wiggers VR, Simionatto EL, Meier HF, Barros AAC, Madureira LAS (2010) Fuel 89:563–568
Xia Y, Larock RC (2010) Green Chem 12:1893–1909

Part V
Fermentation-based Products

Bio-Products from Sugar-Based Fermentation Processes

Zsófia Kádár and César Fonseca

1 Introduction

The transition from oil-based to bio-based economy is on the way, and technologies including microbial conversion processes are taking a significant share in this path, not only to replace products generated by chemical synthesis or extracted directly from natural resources, but also to obtain molecules with new functionalities (IEA Bioenergy 2012).

Among feedstock used for sugar-based fermentation processes, dedicated plant crops (e.g., sugarcane and cereal grains) are the preferred ones, followed by agriculture residues (e.g., sugarcane bagasse, wheat straw, corn stover), forestry residues, agro-industrial by-products, and bio-wastes. This biomass can also be categorized based on their carbohydrate composition as mono-, di-, or oligosaccharides (e.g., glucose and sucrose), polysaccharides (e.g., starch, cellulose, hemicellulose), or a mixture of those.

All sugar-based biomass requires a certain level of upstream processing to make its sugar content available for efficient microbial fermentation processes (Fig. 1). Biomass containing mono- or disaccharides (e.g., sugarcane and sugar beet) needs minimal crushing/extraction upstream processing to obtain a sugar solution (Nag 2008). Starch-based biomass (e.g., cereal grains) usually undergoes milling/grinding and an additional enzymatic hydrolysis step to convert starch into mono- and disaccharides (Nag 2008). In turn, lignocellulosic biomass, composed of cellulose, hemicellulose and lignin, often requires chopping and a thermochemical pretreatment step, which loosens the recalcitrant structure of lignocellulose and, in some cases, partially hydrolyzes its components, also contributing to a better access of enzymes in a subsequent hydrolysis step (Wyman 1996).

Z. Kádár (✉) · C. Fonseca
Section for Sustainable Biotechnology, Department of Chemistry and Bioscience, Aalborg University, Copenhagen, Denmark
e-mail: zsk@bio.aau.dk; csf@bio.aau.dk

© Springer Nature Switzerland AG 2019
J.-R. Bastidas-Oyanedel, J. E. Schmidt (eds.), *Biorefinery*,
https://doi.org/10.1007/978-3-030-10961-5_12

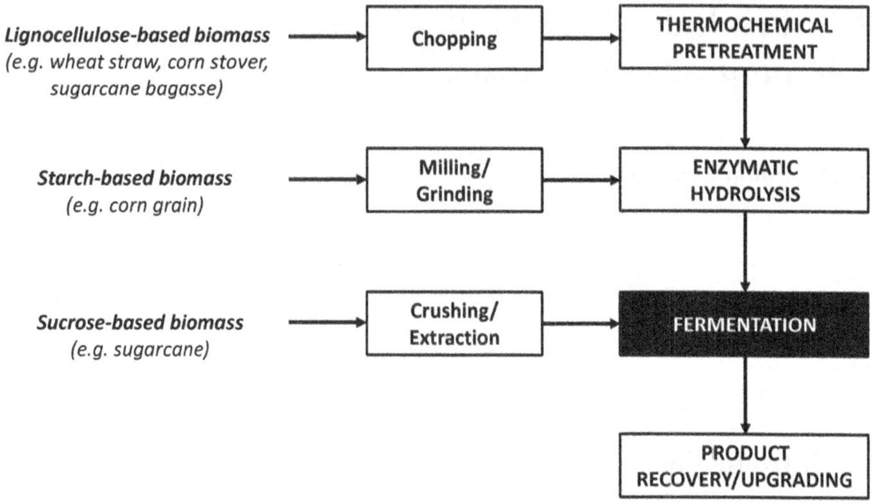

Fig. 1 Biomass processing for sugar-based fermentation processes

The aim of these upstream processes is to achieve maximal carbohydrate recovery yield from biomass in a concentrated and readily fermentable sugar solution—a sugar platform—at the minimal cost. While sucrose- and starch-based biomass has relatively easy upstream processing, the processing of lignocellulosic biomass is more complex, mainly due to the pretreatment step (Wyman and Dale 2015). Whole-crop processing is gaining interest in order to increase process efficiency, namely in terms of energy requirements for the upstream processing, even if it results in a more complex sugar mixture for the microbial fermentation process. The same challenge is faced when complex bio-wastes (e.g., organic fraction of municipal solid wastes) are used for microbial fermentation processes, especially if pure microbial cultures are applied.

Although several microorganisms are able to produce hydrolytic enzymes, most processes include the enzymatic hydrolysis of polysaccharides into a mono- and disaccharides to generate sugar solutions that are compatible with the microorganism used in the fermentation step. Starch, a homopolymer of glucose units linked via linear α-1,4 (amylose) and branched α-1,6 (amylopectin) linkages, is converted into glucose and maltose by the action of α-amylase and glucoamylase. Cellulose is also a homopolymer of glucose units, but linked by β-1,4 glycosidic bonds. Its crystalline structure makes it resistant to hydrolysis under most pretreatment processes. Endo-β-1,4-glucanases and cellobiohydrolases are applied to generate cellobiose, and β-glucosidase hydrolyzes the disaccharide into glucose. Hemicelluloses are branched heteropolysaccharides consisting of the pentose (D-xylose and L-arabinose), hexose (D-mannose, D-glucose, D-galactose), and uronic acid units. Hemicelluloses from agricultural residues are mainly arabinoxylans, i.e., a backbone of xylose units linked by β-1,4 glycosidic bonds with branches of several arabinose units (Gírio et al. 2010). The structure and composition of hemicelluloses from wood biomass is distinguished between softwood (e.g., spruce, pine) and hardwood (e.g., willow, aspen, and oak). Softwood hemicellulose has a higher mannose and glucose content, while hardwood contains a higher proportion of xylose and acetyl groups (Gírio et al. 2010). The branched and non-crystalline structure of hemicelluloses makes them more susceptible to hydrolysis during pretreatment and, in the case of agricultural residues and hardwood, hemicellulose hydrolysates rich in xylo-oligosaccharides are obtained. The complete hydrolysis of hemicellulose into monosaccharides may require the use of enzymes like endo-xylanase and β-xylosidase. Nowadays, commercial enzyme cocktails are available for the virtually complete and efficient hydrolysis of different types of biomass.

The sugar platform available from the upstream processing of different biomasses usually contains monosaccharides in the form of hexoses (often referred as C6 sugars, typically glucose) or pentoses (often referred as C5 sugars, as xylose) and/or disaccharides like sucrose or maltose. While sugarcane juice and molasses mainly contain sucrose and the hydrolysate from starch-based biomass contains maltose and glucose, the hydrolysate from lignocellulosic biomass is often composed of a mixture of C6 and C5 sugars.

The benchmarking of microbial conversion processes using sucrose-, starch-, and lignocellulose-based biomass as feedstock is usually the production of ethanol with the yeast *Saccharomyces cerevisiae*, as a long-standing biocatalyst in the food and biofuel industries (Kang and Lee 2015). This microorganism is able to efficiently convert hexoses and respective disaccharides (sucrose and maltose) into ethanol, but lacks the ability to naturally use polysaccharides and pentoses. On the contrary, other microorganisms are able to directly convert polysaccharides into bio-products, like some filamentous fungi, co-producers of enzymes (Jun et al. 2011) and organic acids. Several others are able to use hexoses and pentoses, just requiring an external source of enzymes for the hydrolysis of polysaccharides. Microbial fermentation processes requires the efficient use of all the carbohydrates from biomass and, in the past decades, major efforts have been made to engineer microorganisms towards this goal (Jansen et al. 2017).

The capacity of microorganisms to tolerate industrial settings and process specificities is also a key factor in the success of bioprocesses. The robustness of microorganisms is often challenged by upstream processing of biomass. For example, the chemicals and/or high temperature applied during pretreatment of lignocellulosic biomass often generate degradation products of cellulose, hemicellulose, and lignin. During pretreatment, the hydrolysis of hemicellulose can lead to the formation of furfural and acetic acid while lignin can release phenols, all known as microbial inhibitors (Palmqvist and Hahn-Hägerdal 2000).

Nowadays, one of the main objectives in sustainable biotechnology processes is the development of robust and tailor-made microorganisms—cell factories—for microbial fermentation processes, with efficient conversion of specific sugar mixtures into specific bio-products (IEA Bioenergy 2012). For ethanol production, the focus of strain improvement has been the substrate conversion, particularly C5 sugars by *S. cerevisiae* (Moysés et al. 2016). Conversely, *Escherichia coli* and *S. cerevisiae* have been the most common microbial chassis for metabolic engineering and synthetic biology approaches for the generation of commercially viable bio-products (Kang and Lee 2015). In fact, these microorganisms are, respectively, the prokaryote and eukaryote model organisms, and the accumulated knowledge on molecular and cell biology and the effective tools available for their genome editing, make them an excellent platform to develop a new generation of cell factories for the production of bio-products through sugar-based fermentation processes.

2 Alcohols

2.1 Ethanol

Ethanol is a two-carbon alcohol and the main liquid biofuel replacing gasoline for road transportation (Wyman 1996; Nag 2008). The research and development on biochemical conversion of biomass relied mostly on ethanol production processes. The so-called second-generation (2G) bioethanol biorefineries use lignocellulose-based substrates, in contrast to the first-generation (1G) bioethanol biorefineries, which uses sucrose- or starch-based substrates. Many microorganisms can produce ethanol as the major fermentation product from sugars, but the yeast *S. cerevisiae* is the preferred cell factory in industrial alcoholic fermentation, due to: (1) high ethanol yield and productivity; (2) robustness to harsh environments, including low pH, tolerance to inhibitors, like acids (acetic and formic) and furans (furfural and HMF); and (3) generally be regarded as safe (GRAS) (Hägerdal et al. 2007).

As the natural and most robust microorganism for ethanol production from hexoses and respective disaccharides (sucrose and maltose), *S. cerevisiae* has been engineered to face the challenges and opportunities of converting lignocellulose hydrolysates into 2G bioethanol. The metabolic and evolutionary engineering approaches included the adaptation to inhibitors generated in the pretreatment step (Almeida et al. 2007; Demeke et al. 2013), the reduction of by-product formation to increase ethanol yield (Medina et al. 2010), the hydrolysis of polysaccharides (van Zyl et al. 2007), and the fermentation of the C5 sugars obtained from hemicellulose (Gírio et al. 2010).

In the past decades, the capacity of *S. cerevisiae* to ferment C5 sugars, mainly xylose, has significantly improved, with the development of many laboratory and industrial pentose-fermenting strains. Two different metabolic pathways for xylose assimilation have been introduced in *S. cerevisiae*: (1) the "redox pathway," using xylose reductase (XR) and xylitol dehydrogenase (XDH) found in natural xylose-fermenting non-conventional yeasts (Eliasson et al. 2000), or (2) the isomerase pathway, using xylose isomerase (XI) from bacteria or from an anaerobic filamentous fungus (Kuyper et al. 2005a) (Fig. 2). Both pathways produce D-xylulose which is converted to D-xylulose 5-phosphate by xylulokinase (XK). Both the overexpression of endogenous and heterologous XK has proven to improve xylose fermentation (Eliasson et al. 2000; Jin et al. 2002). D-Xylulose 5-phosphate is an intermediate of the pentose phosphate pathway (PPP) and carbon flux can then follow glycolysis towards ethanol production through the common PPP/glycolysis intermediates, fructose 6-phosphate and glyceraldehyde 3-phosphate.

The overexpression of the endogenous PPP enzymes promoted further improvement in pentose fermentation (Karhumaa et al. 2005). Other significant improvements in xylose fermentation by *S. cerevisiae* included the reduction of by-product formation and deregulation of pentose metabolism. For example, the disruption of *GRE3*, coding an unspecific xylose reductase, reduces xylitol accumulation (Träff et al. 2001). In turn, the disruption of *PHO13*, coding a phosphatase, revealed to be

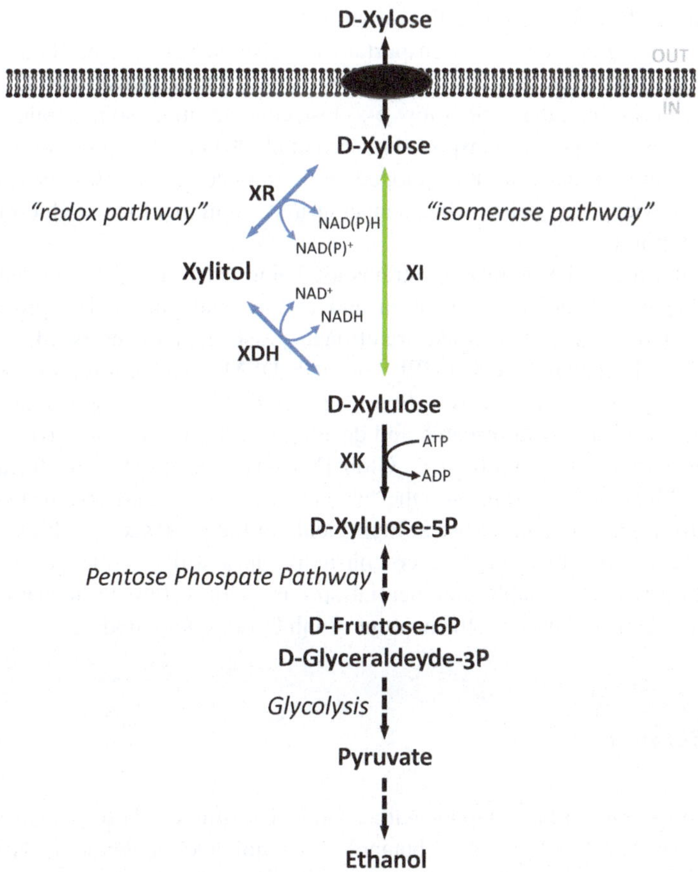

Fig. 2 Pathways for xylose fermentation in recombinant *Saccharomyces cerevisiae* (*XR* xylose reductase, *XDH* xylitol dehydrogenase, *XI* xylose isomerase, *XK* xylulokinase)

relevant to increase xylose consumption and the carbon flux through PPP, with consequent enhancement of ethanol yield and productivity (Xu et al. 2016). The metabolic engineering approaches have been often followed by evolutionary engineering protocols for further strain improvement (Kuyper et al. 2005b; Wisselink et al. 2009; Garcia-Sanchez et al. 2010). This methodology, also known as adaptive evolution, is a slow process based on natural mutations that can rationally be accelerated by appropriate selection pressure during cultivation (Mans et al. 2018). The analysis of improved C5-fermenting strains often revealed that the kinetic properties of pentose transport were altered towards increased sugar uptake fluxes when a mixed-sugar platform was used in the evolutionary engineering protocols (Kuyper et al. 2005b; Garcia-Sanchez et al. 2010). The sugars present in lignocellulose hydrolysates are often consumed by *S. cerevisiae* in a sequential mode, first glucose and then xylose, with consequences at the level of ethanol yield and, mainly, productivity. This fermentation profile is correlated to the biochemistry of sugar uptake in yeasts, which is usually dependent of nonspecific sugar transporters generally preferring glucose. Therefore, attempts have been made to develop specific transporters for xylose to overcome the inhibitory effect of glucose. The heterologous expression of a glucose/xylose transporter from *Candida intermedia* (Leandro et al. 2006) in industrial xylose-fermenting *S. cerevisiae* led to improved D-xylose uptake kinetics and revealed that, under low D-xylose concentration, some strains are limited at the level of xylose transport (Fonseca et al. 2011). Also, glucose-insensitive xylose transporters have been developed from mutated *S. cerevisiae* hexose transporters (Farwick et al. 2014), which can contribute to more efficient glucose/xylose co-consumption.

Several industrial *S. cerevisiae* strains are being used in C6/C5 fermentation in lignocellulosic ethanol demonstration and commercial plants. The providers of industrial C6/C5 yeasts include traditional yeast manufacturers like Lesaffre (CelluX™), Lallemand (C5 FUEL™), and DSM (*Saccharomyces cerevisiae* expressing xylose isomerase from *Piromyces* sp. E2) but also new players like the companies resulting from research and development performed at universities like C5 Ligno Technologies AB (C5LT), GlobalYeast (Excellulor™), and Terranol A/S (cV-110). Most of these strains use the "isomerase pathway" and are able to produce ethanol from glucose and xylose at high yields in the presence of inhibitory compounds. However, glucose/xylose co-consumption is still a challenge to be overcome in currently available commercial strains, with xylose fermentation being particularly compromised by the amount of inhibitory compounds.

2.2 *Butanol*

Butanol is a four-carbon alcohol with chemical formula $C_4H_{10}O$ which has four isomeric structures (*n*-butanol, isobutanol, 2-butanol, and *tert*-butanol). They differ in physicochemical properties and production methods but the applications are similar in some aspects. Their applications are abundant, such as: chemical intermediate for fuels and jet fuel and bio-lube oil; chemical intermediate in the production of monomers, polymeric emulsions, esters, plasticizers, glycol ethers, and amines; solvent for paints, coatings, and varnishes; extractant for antibiotics, hormones, and vitamins; perfume and cosmetics ingredient; degreasers and cleaning solutions (Schiel-Bengelsdorf et al. 2013). Compared to ethanol, *n*-butanol, and isobutanol are superior liquid fuels due to their higher energy content and lower volatility. Therefore, they are more gasoline-like and can thus be blended more easily with gasoline or even used directly in conventional internal combustion engines. Furthermore, butanol can also be blended with diesel fuels and used in jet fuels and it does not absorb moisture, so does not cause corrosion (Zhao et al. 2013).

At present, butanol and higher alcohols are mainly produced by thermochemical routes (Ndaba et al. 2015). However, the interest in the production of butanol through microbial fermentation processes has been renewed due to the general trend on the shift to renewable fuels and chemicals and the recent advances in strain and process development. Biological production of *n*-butanol has a long history (Jones and Woods 1986). Butanol fermentation process was the second largest industrial fermentation process in the world during the first part of the twentieth century. Early industrial production of *n*-butanol was based on fermentation of sugar and starch using *Clostridium* spp., typically referred to as acetone-butanol-ethanol (ABE) fermentation (Jones and Woods 1986; Sauer 2016).

ABE process historically relies on *Clostridia* spp., which are natural acetone-butanol-ethanol producers, but also known as able to generate several products which cannot be obtained through chemical synthesis (Ndaba et al. 2015). The metabolism (Fig. 3) is divided into two phases (Jones and Woods 1986; Qureshi and Ezeji 2008).

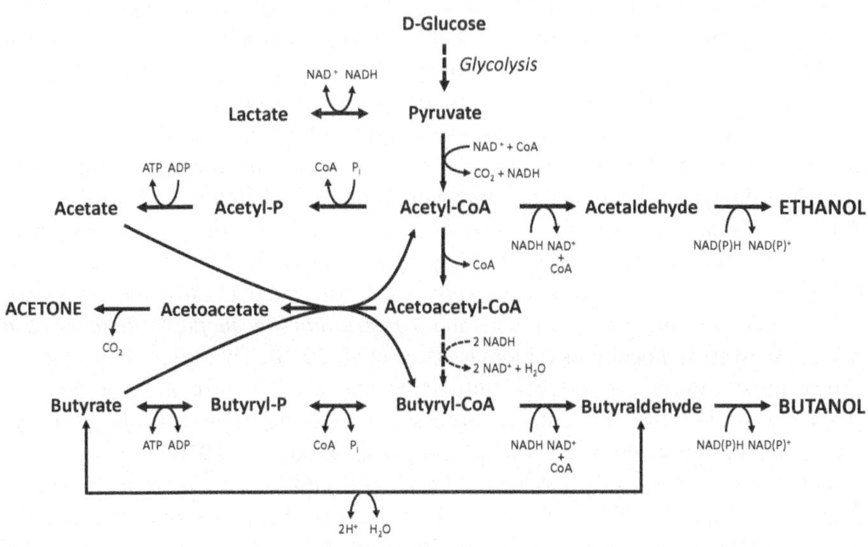

Fig. 3 Acetone Butanol Ethanol (ABE) fermentation pathway in clostridia

In the first, acidogenic phase (acidogenesis), butyrate and acetate are formed in a standard butyric acid pathway. In the second, solventogenic phase (solventogenesis), acids are converted into butanol, acetone, and ethanol. Some strains of *C. beijerinckii* are able to further reduce acetone to isopropanol (Schiel-Bengelsdorf et al. 2013). The fermentation is strictly anaerobic. The produced organic acids and alcohols above a certain titer are toxic to the cells, *n*-butanol being the most toxic (the natural tolerance is about 11–12 g/L) (Branduardi and Porro 2016). Therefore, usually *in situ* product recovery techniques are integrated in ABE fermentation (Schiel-Bengelsdorf et al. 2013).

Even though *n*-butanol is recognized as an alternative fuel, its production is still not considered economical due to several limitations, such as: (1) low *n*-butanol titers (<20 g/L) caused by inhibition during fermentation; (2) low *n*-butanol yield due to hetero-fermentative metabolism (0.28–0.33 g/g); and (3) high cost of *n*-butanol recovery from broths with low product concentration (Ndaba et al. 2015). Hence, strain improvement has been attempted to overcome the bottlenecks of *Clostridia* spp. in this process, aiming at increasing butanol yield and tolerance, but also expanding substrate utilization (e.g., xylose) and air tolerance. Some robust strains were obtained, like *C. beijerinckii BA 101* (Li and Ge 2016) and *C. acetobutylicum ATCC 55025*, reaching approximately 20 g/L *n*-butanol titers (Zhao et al. 2013). Metabolic engineering of these organisms has been challenging and the achievements on improved butanol titer, yield and productivity, enhanced butanol selectivity and increased tolerance to solvents have been mainly achieved with *C. acetobutylicum* (Lee et al. 2016; Li and Ge 2016). Still, one of the main challenges in butanol production with *Clostridia* spp. is acetone production together with butanol and ethanol. Acetone cannot be used as a fuel and reduces the yield of butanol. Therefore, metabolic engineering was also targeting eliminating acetone production pathway, but usually this resulted in reduced solvent production (Li and Ge 2016). Genome shuffling and evolutionary engineering approaches have also been applied (Li and Ge 2016). The recent development of efficient genome editing tools (Lee et al. 2016) offers great potential for further strain improvement.

Many clostridia are able to metabolize several carbohydrates, including hexoses and pentoses (Jones and Wood 1986). However, xylose utilization in mixtures with glucose is poor, due to carbon catabolite repression (Schiel-Bengelsdorf et al. 2013). Some Clostridia species are able to directly convert polysaccharides, like *Clostridium* sp. *strain NUP7*, which is able to produce butanol or isopropanol from hemicellulose (Xin et al. 2017). Some solventogenic Clostridia, such as *Clostridium thermocellum*, *Clostridium cellulolyticum*, and *Clostridium thermopapyrolyticum*, can directly convert lignocellulosic biomass (Lee et al. 2016). The different abilities of Clostridia strains in carbohydrate utilization and product formation prompt the study of mixed-culture fermentation processes in order to improve synergies in the production of lignocellulose-degrading enzymes (Baral et al. 2016).

Other cell factories, such as *S. cerevisiae*, *E. coli*, and *Pseudomonas putida* (Sauer 2016; Li and Ge 2016), have been considered as suitable chassis to introduce the pathways for *n*-butanol production. The latest results on development such strains

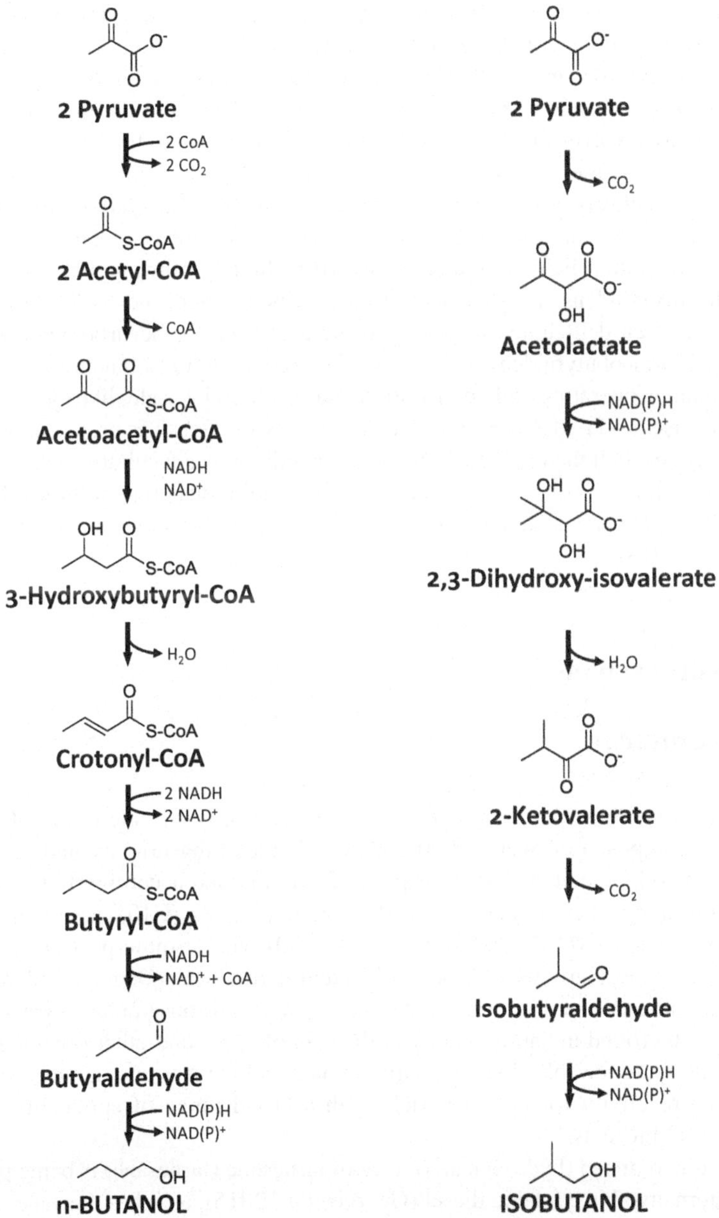

Fig. 4 Butanol production pathways in yeast

have been summarized by Li and Ge (2016). *S. cerevisiae* is not able to produce *n*-butanol, but can produce a central intermediate, acetyl-CoA, and also acetoacetyl-CoA. The "Clostridia" metabolic pathway for *n*-butanol production was introduced in *S. cerevisiae* making use of enzymes from different microorganisms (Fig. 4) (Swidah et al. 2015; Schadeweg and Boles 2016). In this pathway, two molecules of acetyl-CoA are condensed into acetoacetyl-CoA, which is reduced to 3-hydroxybutyryl-CoA, then dehydrated to crotonyl-CoA. Further reductions generate butyryl-CoA, butyraldehyde, and finally *n*-butanol (Schadeweg and Boles 2016).

An alternative route to produce butanol is the 2-keto-acid or Ehrlich pathway. This pathway involves the decarboxylation of a 2-keto-acid to form the corresponding aldehyde, and the subsequent reduction of the aldehyde to form the alcohol. The 2-keto-acid pathway was successful expressed in different chassis, like *E. coli*, *Corynebacterium glutamicum*, *Bacillus subtilis*, and *S. cerevisiae*, among others, to produce isobutanol (Felpeto-Santero et al. 2015). In this pathway, two molecules of pyruvate are condensed into 2-acetolactate, which is reduced to 2,3-dihydroxy-isovalerate, then dehydrated to 2-ketoisovalerate. Then the decarboxylation of the 2-keto-acid to isobutyraldehyde is followed by reduction to isobutanol (Fig. 4).

Companies operating at demonstration and commercial scale, like Gevo Inc. and Butamax, use modified *S. cerevisiae* in their processes. Yeast has preferentially been utilized as host cell factory since it is easy to handle, it is a facultative anaerobe and it tolerates higher alcohol concentrations. Still, continuous product removal during fermentation is part of the industrial process, allowing high yields and productivities (Ryan 2018).

3 Hydrocarbons

3.1 Farnesene

Isoprenoids (such as farnesene) are the largest and most diverse group of natural products, composed of over 50,000 compounds including primary and secondary metabolites (George et al. 2015). Isoprenoids are divided according to the number of carbon atoms: hemiterpenoids (C5), monoterpenoids (C10), sesquiterpenoids (C15), diterpenoids (C20), and triterpenoids (C30). The sesquiterpenoids (C15) are one of the largest groups of isoprenoid natural products (Demain and Martens 2017). Farnesene is a 15 carbon long-chain, branched, unsaturated hydrocarbon, which can be found in nature mainly in the skin of apple and other organic materials. It is also a renewable chemical building block which was unique physical properties and reactivity for new materials with a broad range of applications, from cosmetics to biofuels.

The fully reduced (hydrogenated) form of farnesene (farnesane) is being pursued as an alternative biosynthetic diesel (George et al. 2015), as it has a cetane number that falls within the expected range for diesel (Peralta-Yahya et al. 2012). It can be mixed directly into diesel or jet fuels without requiring any engine modifications and it is also resistance to cold. Farnesene-based renewable diesel and jet fuel is likely to hit price target and prove to be cost-efficient, which will drive its use as a drop-in replacement for synthetic fuels, lower GHG emissions and reduce particulate matter emissions, decrease pollution near airports and major metropolitan area. Amyris and Total together developed a drop-in jet fuel that contains up to 10% blends of renewable farnesane, which meets the rigorous performance requirements set for Jet A/A-1 fuel used by the global commercial aviation industry (El Takriti et al. 2017).

High-purity farnesene can be used in tire manufacturing as in polymerized form it can easily and completely react with tire rubber and unlike oil additives it can attain strong adhesion of rubber components for improving performance and shape stability. Moreover, it conveys high plasticity, maintains excellent flexibility even at low temperatures, and improves ice grip performance. Commercial tires marketed by Sumitomo Rubber Industries, Ltd. under their Dunlop brand utilizing renewable liquid-farnesene rubber was developed with Kuraray (Japan) and already launched in early 2017 (RJA 2017). Squalene is a C30 molecule formed by either biological or chemical condensation of two farnesene units. Squalane is a hydrogenation product of squalene, is used as an important moisturizing and anti-aging ingredient in the cosmetics (Beller et al. 2015). As cosmetic formulation companies prefer to procure squalane derived from biotechnology route rather than expensive and unsustainable animal sources such as ultra-refined oil or shark liver, it will continue to drive farnesene demand. The production of squalene is robust and reproducible, and along with the availability of feedstock, ensures the reliable and sustainable production of squalene both from a chemical and sensorial (i.e., odor and color) standpoint (McPhee et al. 2014).

According to Global Market Insights, Inc., the farnesene market size was estimated at over 8 kton in 2015 (https://www.gminsights.com). Cosmetics and personal care took up 37.6% of farnesene market share, followed by fuels and lubes (25.6%), while the flavors and fragrances market share was at 23.6% and performance materials (13.2%). Growing trend towards biofuel use in aviation and automobile sector to curb carbon emissions may boost farnesene market growth. Farnesene market is predicted to increase with a forecasted compound annual growth rate (CAGR) at over 40% up to 2023 (https://www.gminsights.com). The global farnesene market share is currently dominated by Amyris Inc. (California, USA).

The common biochemical precursor of all isoprenoids is the 5-carbon intermediate isopentyl pyrophosphate (IPP) and its isomer dimethylallyl pyrophosphate (DMAPP). There are two pathways for the biosynthesis of isoprenoids (Fig. 5): the mevalonate pathway (MVA) and the methylerythritol phosphate (MEP) pathway (George et al. 2015; Beller et al. 2015; Benjamin et al. 2016; Leavell et al. 2016). When IPP and DMAPP are formed, they are used for carbon chain elongation reactions to produce longer prenyl pyrophosphate precursors such as geranyl pyrophosphate (GPP), farnesyl pyrophosphate (FPP), geranylgeranyl pyrophosphate (GGPP) (Fig. 5).

Both pathways were engineered in *E. coli* and *S. cerevisiae* but the MEP pathway has turned out to be less effective than the MVA pathway (Beller et al. 2015; Benjamin et al. 2016). Amyris's trans-β-farnesene is produced through fermentation

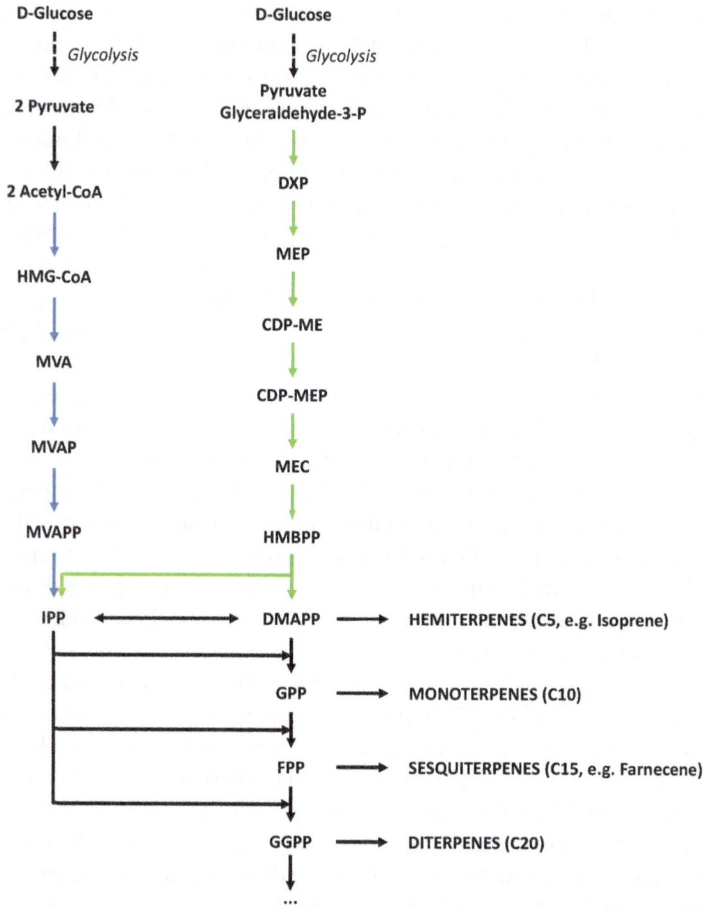

Fig. 5 Biosynthetic pathways for the production of isoprenoids, the Mevalonate pathway (MEV, in blue) and the Methylerythritol phosphate (MEP, in green). *Abbreviations*: *HMG-CoA* 3-hydroxy-3-methylglutaryl-CoA, *MVA* mevalonate, *MVAP* mevalonate-5-phosphate, *MVAPP* mevalonate-5-pyrophosphate, *IPP* isopentyl pyrophosphate, *DXP* 1-deoxy-D-xylulose-5-phosphate, *MEP* 2-C-methyl-D-erythritol-4-phosphate, *CDP-ME* 4-diphosphocytidyl-2-C-methyl-D-erythritol, *CDP-MEP* 4-diphosphocytidyl-2-C-methyl-D-erythritol 2-phosphate, MEC 2-C-methyl-D-erythritol 2,4-cyclodiphosphate, *HMBPP* hydroxymethylbutenyl-4-pyrophosphate, *DMAPP* dimethylallyl pyrophosphate, *GPP* geranyl pyrophosphate, *FPP* farnesyl pyrophosphate, *GGPP* geranylgeranyl pyrophosphate

of sugars by yeast. Target genes were selected to shift the carbon flux from ethanol to hydrocarbons. Development of the pathway for production of the antimalarial artemisinin has served as the foundation for building pathways for other terpenes (Beller et al. 2015). The first product that was developed by Amyris Inc. was artemisinin, an anti-malaria drug where the specific enzyme (amorphadiene synthase) from plant was introduced in *S. cerevisiae* to generate amorphadiene from FPP. Oxidation of amorphadiene to artemisinic acid is accomplished by the action of five plant enzymes expressed in the engineered yeast, and the final conversion of purified artemisinic acid to artemisinin is performed by organic chemistry (Benjamin et al. 2016). The flexibility of the *S. cerevisiae* chassis allowed scientists to rapidly switch from amorphadiene to β-farnesene as final product, by introducing a plant β-farnesene synthase in the MVA pathway (Benjamin et al. 2016; George et al. 2015). Since *S. cerevisiae* uses a chemically inefficient pathway for isoprenoid biosynthesis, the first attempts to produce β-farnesene resulted in low yield and productivity, with titers lower than 1 g/L (Meadows et al. 2016). Amyris Inc. generated an evolved *S. cerevisiae* strain capable of converting sugars into β-farnesene at titers as high as 100 g/L and volumetric productivities of 16.9 g/L/day (Meadows et al. 2016; Demain and Martens 2017). The microbial fermentation product is a high purity single isomer β-farnesene after distillation (Leavell et al. 2016).

Availability and cost of substrates in combination with life cycle assessment led Amyris Inc. to open a full-scale production plant in Brotas, Brazil (2013), to produce β-farnesene from sucrose (Benjamin et al. 2016; Leavell et al. 2016). The product is commercialized under the name Biofene®. Expected demand serving the polymers, nutraceuticals, and solvents markets through 2020 will require new farnesene manufacturing capacity beyond the company's current capacity. Therefore nowadays Amyris Inc. is developing an integrated scalable process with the aim to produce farnesene from cellulosic sugars at 2 USD per liter in the USA in connection to Renmatix's Plantrose® technology for cost-effective production of lignocellulosic sugars from woody feedstock (Mitrovich and Wichmann 2017). This will require further strain development for the efficient conversion of xylose from hemicellulose hydrolysates in the presence of microbial inhibitors generated during biomass processing. Alternatively, purification of sugar streams may reduce the concentration of inhibitors but those processes need to be cost-effective.

4 Organic Acids

4.1 Lactic Acid

Lactic acid (or 2-hydroxypropionic acid) is the simplest and most widely occurring natural hydroxyl acid and, similarly to ethanol, has a long history. It has an asymmetric carbon atom and is present in two optically active forms, L(+)- and D(−)-lactic acid. These isomers have the similar chemical and physical properties, making them difficult to separate with traditional techniques (Komesu et al. 2017). In humans and other mammals, only the L(+)-isomer is present (Ghaffar et al. 2014).

In fact, D(−)-lactic acid is harmful to humans since they only have L-lactate dehydrogenase. Therefore, L(+)-lactic acid is the preferred isomer in food and pharmaceutical industries (Reddy et al. 2008).

Currently, lactic acid is used in a wide variety of industrial applications, including chemicals—solvent, emulsifier, plasticizer; pharmaceuticals—implants, drug-release systems, hygiene and aesthetic products; and food—flavoring, preservative, natural product in fermented products. Furthermore, lactic acid is a precursor of several other products, like propylene oxide, acetaldehyde, acrylic acid, among others (Komesu et al. 2017). Due to the demand for bio-products from renewable resources, the biological production of lactic acid as a bulk chemical has been increasing considerably. For example, the polymerization of lactic acid into polylactic acid (PLA) generates an environment-friendly alternative to plastics derived from petrochemicals (Reddy et al. 2008; Komesu et al. 2017). Global lactic acid and PLA demand was estimated to be 714.2 and 360.8 kton in 2013, with expected annual growth of 15.5% and 18.8%, respectively, until 2020 (Abdel-Rahman and Sonomoto 2016).

While chemical synthesis from petrochemicals always generates a racemic mixture, pure lactic acid isomers can be synthetized by microbial fermentation (Komesu et al. 2017). Microbial fermentation processes utilize renewable substrates and require mild production conditions (temperature 30–45 °C, pH 5.5–6.5) and low energy consumption when compared with petroleum-based chemical synthesis (Reddy et al. 2008; Abdel Rahman and Sonomoto 2016).

Biological production of lactic acid is currently driven by lactic acid bacteria (LAB), Gram-positive, facultative anaerobes, with nutritional requirements for amino acids and vitamins (Reddy et al. 2008; Murali et al. 2017). Most common LAB genera include *Lactobacillus, Pediococcus, Aerococcus, Carnobacterium, Enterococcus, Tetragenococcus, Vagococcus, Leuconostoc, Oenococcus, Weissella, Streptococcus, and Lactococcus* (Juturu and Wu 2016). The species of LAB are usually separated into homo- and hetero-fermentative, based on their type of hexose fermentation. The homo-fermentative LAB utilizes glycolysis (Fig. 6) and produces virtually only lactic acid, with a theoretical yield of lactic acid from glucose of 1.0 g/g or 2 mol/mol. The hetero-fermentative LAB utilizes the phosphoketolase pathway and produce lactic and acetic acids, ethanol and carbon dioxide, with the theoretical yield of lactic acid from glucose reaching only 0.5 g/g or 1 mol/mol (Abdel-Rahman et al. 2011; Rooke 2003; Reddy et al. 2008).

Optically, pure lactic acid is synthesized by microbial fermentation of carbohydrates such as glucose, sucrose, lactose, and starch/maltose, which are derived from feedstocks such as sugar beet, sugarcane molasses, whey, and barley malt (Ghaffar et al. 2014). Amylolytic lactic acid bacteria (ALAB) (*Lactobacillus plantarum, Lactobacillus manihotivorans*) have been found in different tropical fermented foods (Nwankwo et al. 1989; Morlon-Guyot et al. 1998) and can contribute for the economy of the process by eliminating the two-step process of starch saccharification and lactic acid fermentation (Reddy et al. 2008). Although producing lactic

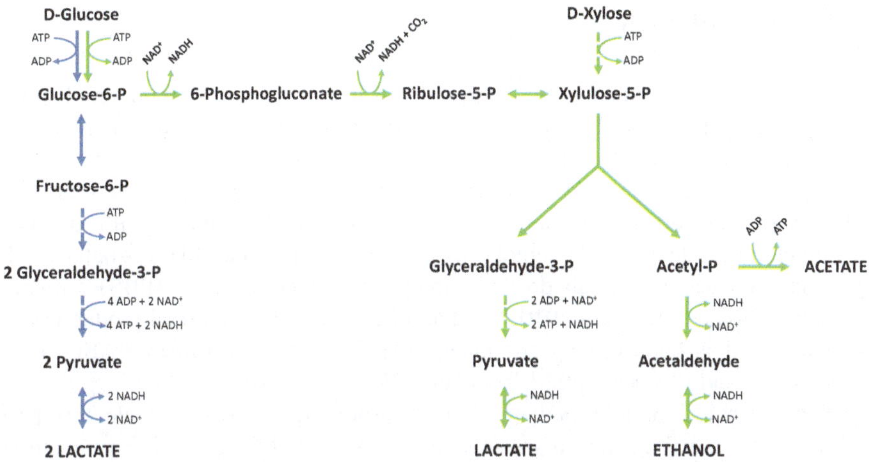

Fig. 6 Typical homolactic (in blue) and heterolactic (in green) fermentation in lactic acid bacteria

acid with high yield and productivity from different sugars, LAB are not particularly fitted to ferment lignocellulosic hydrolysates. For example, pentose fermentation mainly uses the hetero-fermentative pathway (Tan et al. 2017). Strain development for the utilization of lignocellulosic biomass is required to: (1) increase tolerance to inhibitory compounds formed during pretreatment; (2) expand the carbohydrate assimilation capacity (e.g., for direct conversion of cellulose and hemicellulose); and (3) ferment mixed-sugar streams with high lactic acid yield and productivity.

Some filamentous fungi, e.g., *Rhizopus* can also utilize glucose and produce lactic acid (Ghaffar et al. 2014), and *R. oryzae* and *R. arrhizus* can convert starch directly to L(+)-lactic acid due to their amylolytic enzyme activity (Wee et al. 2006). Fermentation by fungi has also other advantages compared to bacterial fermentation, namely in nutrient requirements (Wee et al. 2006; Tan et al. 2017).

Other microorganisms have also been considered as chassis for industrial lactic acid production such as yeasts (*S. cerevisiae, Kluyveromyces lactis, Candida boidinii*), and non-LAB bacteria (*E. coli* and *C. glutamicum*) (Abdel-Rahman et al. 2011). Microalgae and cyanobacteria (photosynthetic microorganisms) also have attracted attention because of their ability to couple CO_2 capture the potential for genetic modification (Tan et al. 2017). Furthermore, *Bacillus* spp. are mostly thermophilic, which enables simultaneous saccharification and fermentation (SSF) and reduces the risk of contamination, have low nutritional requirements and are able to ferment pentoses to lactic acid through a homo-fermentative pathway (Tan et al. 2017).

4.2 Succinic Acid

Succinic acid is a member of the family of C4-dicarboxylic acid, with a molecular formula of $C_4H_6O_4$. Succinic acid is a priority chemical of high value as food and feed ingredient and as platform chemical for replacement of the oil-based building block maleic anhydride. The applications of succinic acid include flavor additives, pharmaceuticals, detergents, and surfactants. In the chemical industry, it is a building block for producing other commodity or specialty chemicals like 1,4-butanediol, gamma-butyrolactone, tetrahydrofuran and polybutylene succinate (PBS), a biodegradable polymer (Vaswani 2010; Song and Lee 2006). Commercial production of succinic acid is deployed by companies like Myriant, BioAmber, BASF-Purac (Succinity), and Reverdia (DSM-Roquette) (Vaswani 2010).

Succinic acid is an intermediate of the tricarboxylic acid (TCA) cycle, thus part of the central metabolism of many organisms, and one of the possible fermentation end products of anaerobic metabolism. Several succinate-producing bacteria (*Actinobacillus succinogenes, Anaerobiospirillum succiniciproducens, Mannheimia succiniciproducens, Bacteroides fragilis, Enterococcus flavescens, Klebsiella pneumoniae, Basfia succiniciproducens, Succinivibrio dextrinosolvens*) and fungi (*Aspergillus niger, Paecilomyces variotii, Penicillium simplicissimum*) have been described (Li and Xing 2015).

Succinate can be generated from PEP (phosphoenolpyruvate), via oxaloacetate (OAA), malate, and fumarate, with incorporation of CO_2 and consumption of NADH. This reductive pathway is also referred as reverse TCA (rTCA) (Fig. 7). Through this pathway, succinate is often produced along with other fermentation products (lactate, acetate, ethanol, and formate), depending on the microorganism and on the cultivation conditions (McKinlay et al. 2007). Aerobically, succinate is often produced through the TCA cycle, an oxidative pathway that generates NADH and the loss of CO_2 (Fig. 7). A third option is the glyoxylate shunt, which eliminates the loss of $2CO_2$ when compared to the TCA cycle, converting isocitrate directly to succinate (Fig. 7). The co-produced glyoxylate can then be converted into malate.

The efficient production of succinate often combines the reductive pathway with the glyoxylate shunt or with the oxidative pathway, due to the requirement for reductive equivalents (in the form of NADH) by the reductive pathway (Nghiem et al. 2017). Therefore, the maximal theoretical yield of succinic acid from 6-carbon sugars, such as glucose, and CO_2 is 1.71 mol/mol sugar, or 1.12 g/g, with 57–71% of the glucose being converted through the reductive pathway and the remaining sugar used through the glyoxylate shunt or the oxidative pathway for the production of the required reducing power (Raab and Lang 2011).

During fermentation, the accumulation of succinic acid lowers the pH of the medium therefore pH control in a range suitable for the microorganism is crucial. However, the choice of the base for pH controlling will determine the produced succinate salt, which will affect its applications, or the required further process steps. Therefore, the interest has been increased significantly to isolate or engineer strains that are able to tolerate low pH.

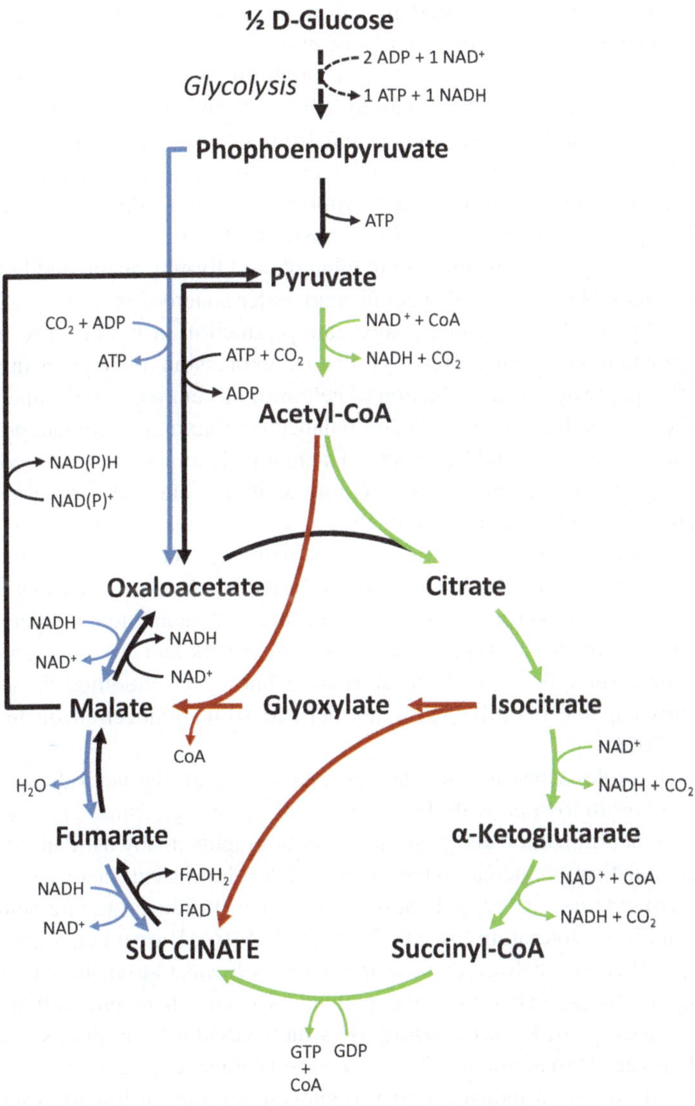

Fig. 7 Metabolic pathways for succinic acid production. Reductive pathway (in blue), a oxidative pathway (in green), and glyoxylate shunt (in red)

The first bacteria described as succinate producer was *Anaerobiospirillum succiniciproducens*, which was engineered and tested under adequate conditions to produce succinate from glucose and wood hydrolysate with yields higher than 70% of the theoretical (Guettler and Jain 1996; Lee et al. 2003). More recently, *Basfia succiniciproducens* another efficient succinate producer was described (Kuhnert et al. 2010) and the efficiency of the process was revealed to be substrate dependent

(Lange et al. 2017). PTS-dependent substrate phosphorylation of sucrose and fructose contributes to an increased pool of pyruvate and the formation of by-products (Lange et al. 2017). The presence of an alternative fructokinase allowed the disruption of the fructose-PTS, which, together with the elimination of by-product pathways, contributed to a succinate yield from sucrose higher than 70% of the theoretical (Lange et al. 2017). Succinity.technology, developed in a joint venture between BASF and Corbion, is using a proprietary strain of *Basfia succiniciproducens* in their process (Nghiem et al. 2017; Vaswani 2010).

E. coli primarily ferments glucose to ethanol, and formic, acetic and lactic acids with only detectable amounts of succinic acid under anaerobic condition (Song and Lee 2006). Approaches to promote succinate production in *E. coli* have included: (1) the removal of competing pathways; (2) overexpression of enzymes involved in the reductive pathway; (3) introduction of heterologous enzymes with superior catalytic efficiency; (4) fine tuning of redox balance for maximal succinate production (Li and Xing 2015; Song and Lee 2006). The main advantage of using recombinant *E. coli* is its fast growth rate, simple requirements for nutrients and easy genetic manipulation for high succinate yield. Recently, *E. coli* was engineered to produce succinate with a combination of the reductive pathway and the glyoxylate shunt or the oxidative pathway (Nghiem et al. 2017). In the BioAmber technology, a modified *E. coli* was used to produce diammonium succinate at ambient temperature and neutral pH, with the use of sugar and CO_2 as a feedstock and NH_3 as a neutralizing agent for the carboxylic acid. In the Myriant technology, a modified *E. coli* is also applied, this capable of utilizing sugars derived from lignocellulosic feedstocks (Ahn et al. 2016).

The bacteria described as succinate producers require the neutralization of the fermentation broth to cope with the pH needs of these organisms. The cost associated with the neutralization and consequent requirements in downstream processing for product purification increased the interest of developing cell factories capable of efficient fermentation at low pH. Several fungi can produce organic acids under aerobic conditions, tolerating low pH (Yang et al. 2016). BioAmber, at the scale up phase to commercial, realized that the applied *E. coli* was too sensitive to pH in the fermentation process. Therefore, the pathway for succinate production was re-engineered in the yeast *Pichia kudriavzevii*, which was able to produce succinic acid at a much lower pH than previously used *E. coli* (Alonso et al. 2015). *S. cerevisiae*, as robust and important industrial microorganism, tolerant of low pH values (3.0–6.0) and able to perform anaerobic fermentation, was also considered a suitable chassis for succinic acid production. The Reverdia technology, based on recombinant *S. cerevisiae* developed by DSM, is combining the reductive pathway with the glyoxylate shunt for maximal succinate yield. Moreover, it is also able to operate at low pH, thus less prone to contamination and requiring less chemical processing, equipment, and energy to convert intermediate salts into succinic acid (Nghiem et al. 2017).

5 Others

5.1 *Biosurfactants*

Surfactants are capable of reducing the surface tension and interfacial tension between individual molecule at the surface and interface. Surfactants are widely used in household detergents, industrial and agricultural chemicals (e.g., dispersants), personal care and cosmetics, pharmaceuticals, food, oilfield chemicals, among others (Nitschke and Silva 2017; Vecino et al. 2017; Souza et al. 2014; Sachdev and Cameotra 2013). Traditional surfactants are based on petrochemical resources such as ethylene, benzene, kerosene, and *n*-paraffins. Biosurfactants are produced either by microorganisms or by (bio)chemical conversion of natural products and are seen as potential alternatives to synthetic surfactants due to structural diversity, performance under different conditions and environmental performance (Scott and Jones 2000; Sajna et al. 2015). For some decades, the production of biosurfactants was limited to the use of vegetable oils, as coconut and palm oil, for the hydrophobic part of the molecule. Nevertheless, the production of these so-called first-generation biosurfactants still involves chemical synthesis. Microbial glycolipids (sophoro-, rhamno-, and mannosylerythritol lipids) (Fig. 8) are among the most promising biosurfactants for commercialization due to their technical performance, potential large-scale production through fermentation and recovery as extracellular products. These second-generation biosurfactants were only applied in niche markets until very recently (Brumano et al. 2016). Topics like sustainability and the use of bio-based home care products are getting more popular with consumers and the effect of *Green Premium* (the willingness of consumers to pay an additional price for "green materials") is expected to get more pronounced, not only in developed countries but also in the emerging regions. Also, the production costs are expected to decrease as a result of technological developments. The surfactant market is extremely big with a worldwide annual production of over 13 Mt/y expecting a CAGR (compound annual growth rate) of 5.53% during the forecast period (2018–2023) (Mordor Intelligence 2018). This demands a big input of often petrochemical resources and moreover represents a tremendous ecological load considering the large fraction used in household cleaners (over 50% of total use of surfactants) which end up in wastewater and/or directly in the environment.

Europe has taken lead in bio-based surfactant consumption and is expected to remain market leader and to enjoy 53.3% of global biosurfactants market revenue share in 2018 (Report of Transparency Market Research 2011). Owing to environmental concerns, this market is expected to grow at a promising rate in Europe in coming years. The glycolipid biosurfactants provide significant opportunities to replace chemical surfactants as sustainable alternatives, in some cases with new functionalities. Sophorolipids are mainly used in household detergents across the globe, with producers, distributers, and applicants such as Soleance (France), Ecover (Belgium), Saraya (Japan), Intobio (Korea), SyntheZyme (USA), and

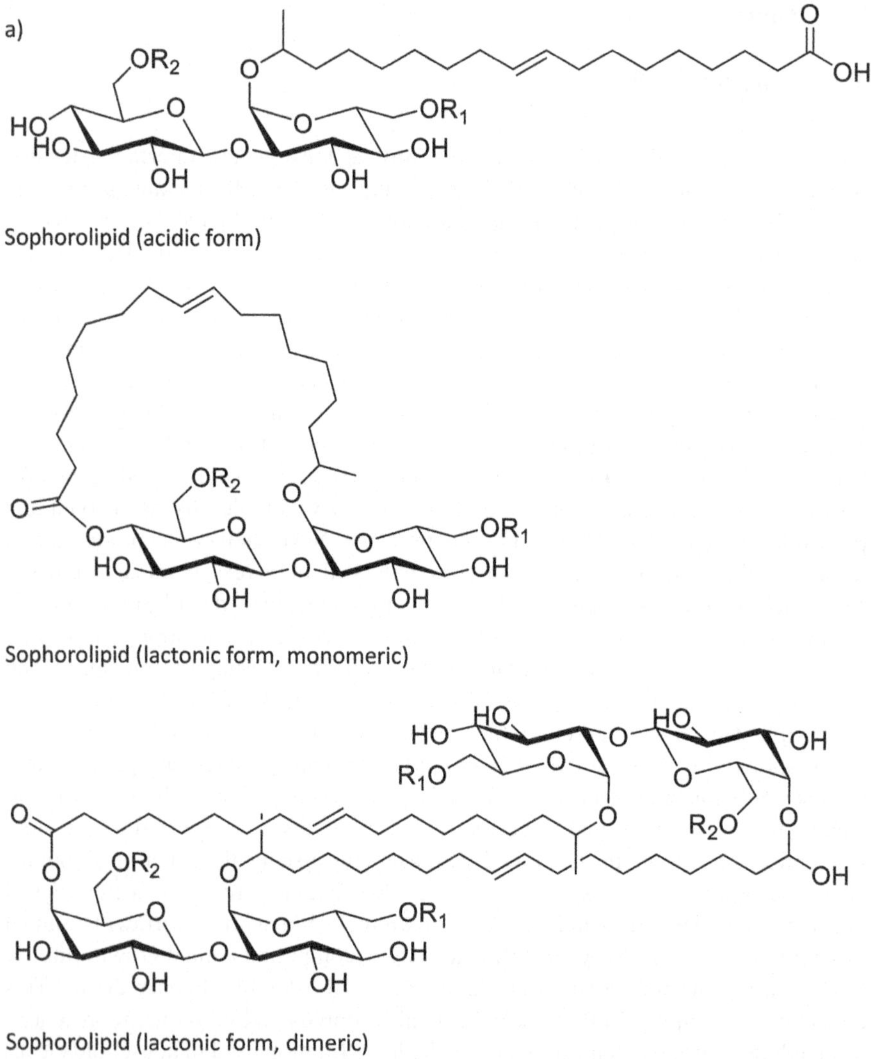

Fig. 8 Structure and variants of (**a**) sophorolipids (SL), (**b**) mannosylerythritol lipids (MEL), and (**c**) rhamnolipids (RL) (OR_1 and OR_2 represent positions for acetylation)

multinationals such as Henkel (Roelants et al. 2016). Mannosylerythritol lipids (MELs) are mainly produced and commercialized in Asia, by Toyobo (Japan) and Biotopia (South Korea), in cosmetics (Morita et al. 2015). Although the sustainability of both first as second-generation biosurfactants was expected to outperform synthetic surfactants based on fossil resources, the impact of their production is still high, mainly due to the use of vegetable oils in their production, either directly in chemical synthesis or as substrate for microbial biosynthesis.

b)

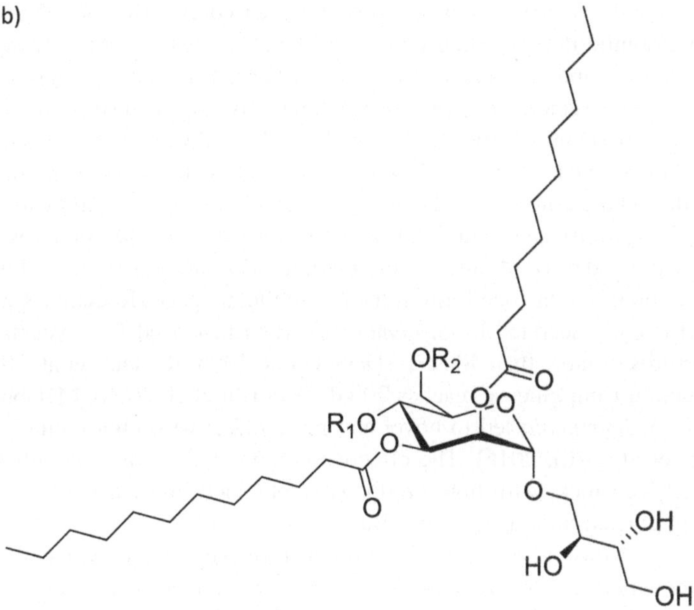

Mannosylerythritol lipid

c)

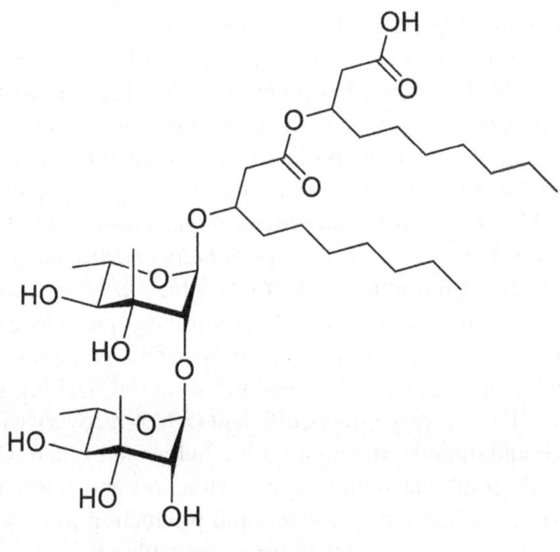

Rhamnolipid (di-rhamnolipid)

Fig. 8 (continued)

Sophorolipids (SL) are composed of sophorose (a disaccharide of glucose units) as the hydrophilic moiety, usually mono-acylated (typically with fatty acid of 18 carbons) on C-1′ and acetylated on C-6 and/or C-6′. The carboxylic group of fatty acid is either free (acidic or open form) or internally esterified (lactonic form), the later in monomeric or dimeric forms (Fig. 8a). The pathway for SL biosynthesis is described in the yeast *Starmerella bombicola* and typically involves five or six steps: hydroxylation of oleic acid at $\omega - 1$; assembly of a glucolipid by the reaction of the hydroxyl fatty acid with UDP-glucose; formation of the sophorose unit by reaction with another UDP-glucose; mono- or di-acetylation; secretion of the acidic SL; formation of the lactone form in the extracellular space (Roelants et al. 2016). SL are efficiently secreted by *S. bombicola* when produced from vegetable oils, reaching titres of more than 400 g/L (Daniel et al. 1998; Roelants et al. 2013). The SL production from glucose reaches 20 g/L (Konishi et al. 2008). Metabolic engineering of *S. bombicola* led to novel and more effective sophorolipids structures (Roelants et al. 2013, 2016). The effective commercial production still relies on oleaginous feedstock, with high productivity obtained under a fed-batch process combining glucose and rapeseed oil (Baccile et al. 2017).

Mannosylerythritol lipids (MEL) are often produced as major extracellular product by *Moesziomyces/Pseudozyma* spp. in a mixture of dozens of analogs composed of a mannosylerythritol hydrophilic moiety, usually diacylated (with fatty acids of 8–12 carbons) and di- (MEL-A), mono- (MEL-B and -C), or non-acetylated (MEL-D) on the mannosyl unit (Fig. 8b) (Morita et al. 2015). The pathway for MEL biosynthesis was first described in the fungus *Ustilago maydis* and later identified in *Moesziomyces/Pseudozyma* spp. It involves five steps: assembly of GDP-mannose and erythritol; acylation on C-2 and C-3 of the mannosyl unit to produce MEL-D; acetylation of the C-4 and/or C-6 (C-6—MEL-B, C-4—MEL-C, C-6 and C-4—MEL-A); and MMF1, for MEL export (Hewald et al. 2006) (Fig. 9). MEL can be produced by *Moesziomyces/Pseudozyma* spp. from vegetable oils at concentrations above 100 g/L (Morita et al. 2015). High production cost, related to the use of soybean oil as substrate and associated solvent-intensive recovery, is impairing their widespread application. *M. antarcticus* (former *Pseudozyma antarctica* and *Candida antarctica*) and *M. bullatus* (former *Moesziomyces aphidis and Pseudozyma aphidis*) are able to produce MEL from glucose, pentoses, glucose/xylose mixtures or directly from xylan (Faria et al. 2014a, 2015). *M. antarcticus* presents equivalent MEL yield from glucose and xylose (Faria et al. 2014a), and a process to produce MEL from cellulosic materials has been developed (Faria et al. 2014b), in which downstream process for MEL recovery is more efficient (>90% recovery with >90% purity in a single-step liquid–liquid extraction with ethyl acetate) than when produced from vegetable oils (multiple liquid–liquid extraction and lower recovery yields for the same purity), but the titers of glycolipid production from sugars are still approx. one order of magnitude lower than from vegetable oils.

Rhamnolipids (RL) are mainly produced by *Pseudomonas aeruginosa*. They are composed of one (mono-rhamnolipids) or two (di-rhamnolipids) rhamnosyl moieties linked to typically one or two beta-hydroxy fatty acids (with 8–16 carbons) (Fig. 8c). The pathway for RL biosynthesis involves the production of

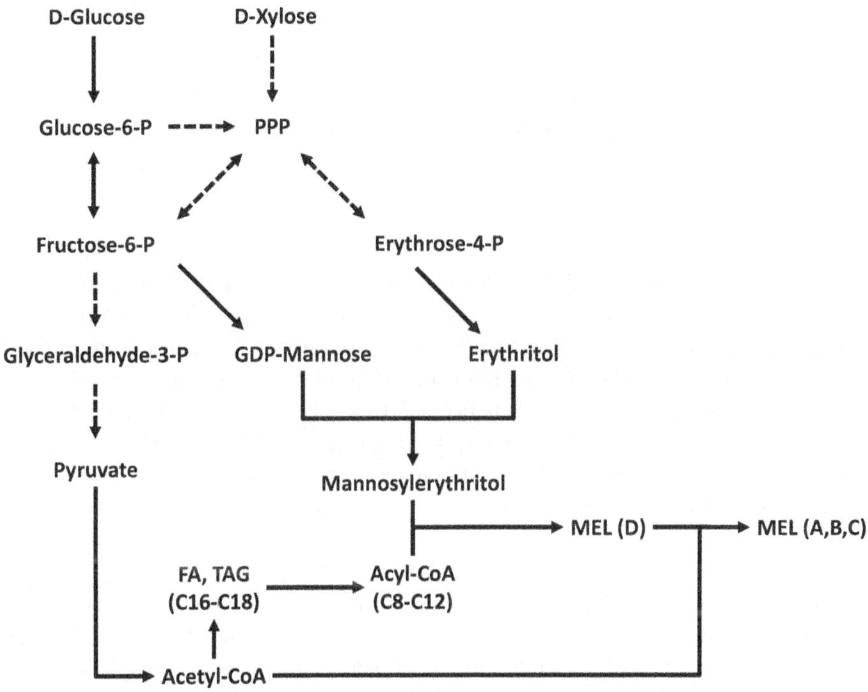

Fig. 9 Metabolic pathways for the biosynthesis of mannosylerythritol lipids (MEL) from glucose and xylose. *PPP* pentose phosphate pathway, *FA* fatty acids, *TAG* triacylglyceride

dTDP-L-rhamnose from D-glucose-1-phosphate and the assembly of beta-hydroxyalkanoyl-beta-hydroxyalkanoic acid units (Chong and Li 2017). RL titers can reach more than 100 g/L from soybean oil (Chong and Li 2017). Metabolic engineering for improved RL production was attempted in *P. aeruginosa* and in other chassis like *Pseudomonas putida, E. coli,* and *S. cerevisiae* (Beuker et al. 2016; Cabrera-Valladares et al. 2006; Bahia et al. 2018) but titers are far below those obtained with the natural producers.

5.2 Bioplastics (PHA)

Polyhydroxyalkanoates (PHA) are natural insoluble polyesters accumulated in some bacteria as energy storage. PHA is produced by metabolic transformation of carbon source under nitrogen, phosphorous, and/or sulfur-limiting conditions (Kaur and Roy 2015) although some bacteria are able to produce PHA during growth (Kourmentza et al. 2017). PHAs are composed of R(−)-3-hydroxyalkanoic acid monomers ranging from C3 to C14 carbon atoms with variety of saturated or unsaturated and linear or branched chains containing aliphatic or aromatic side groups (Fig. 10).

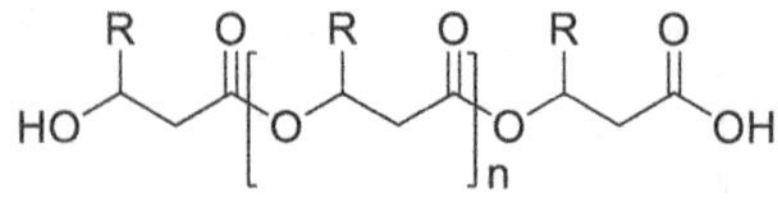

Fig. 10 General structure of PHA. If R = CH_3 – polydroxybutyrate (PHB), if R = C_2H_5 – polyhydroxyvalerate (PHV), if R = C_3H_7 –polyhydroxyhexanoate (PHH); if R = C_4H_9 – polyhydroxyoctanoate (PHO)

PHAs are a group of bioplastics that have a wide range of applications. Based on the carbon atoms comprise their monomeric units they are classified into two groups. Short-chain-length PHA (scl-PHA) consisting of 3–5 carbon atoms, and medium-chain-length PHA (mcl-PHA) consisting of 6–14 carbon atoms. The scl-PHA are mostly used for the production of disposable items and food packaging materials, while mcl-PHA are suitable for high value-added application, such as surgical sutures, implants, biodegradable matrices for drug delivery, among others (Kourmentza et al. 2017; Kootstra et al. 2017; Kaur and Roy 2015; Obruca et al. 2015).

Predictable biodegradability profile, biocompatibility, and the possibility for tailor-made structure and composition makes them attractive substitute for petrochemical plastics owed by its analogous properties (Kaur and Roy 2015). While known biopolymers such as PLA (polylactic acid) and PBS (polybutylene succinate) are produced by chemical polymerization of lactic and succinic acid, respectively, PHA polymerization is naturally performed by bacteria (Kourmentza et al. 2017).

More than 300 microorganisms are known to generate PHA (Endres and Sieber-Raths 2011). Both native and recombinant strains have been employed in PHA production. Industrial production processes for PHA have generally been developed using Gram-negative bacteria, such as *Cupriavidus necator* and *Alcaligenes latus*, mainly due to the relatively high PHA yield and the ability of some to synthesize PHA under non-limiting nutrient conditions (Jiang et al. 2016; Chen 2010). However, huge efforts have also been directed towards process development based on Gram-positive strains such as *Bacillus* sp. and *Corynebacterium glutamicum*, which can produce ideal PHA for medical applications (Kaur and Roy 2015).

The PHA production process involves a series of steps: (1) biomass growth, (2) polymer accumulation, (3) cell harvesting, (4) polymer extraction, and (5) purification. The microorganism, the respective portfolio of genes and active enzymes, and the growth conditions (medium and operation mode), influence the yield and the polymer structure (composition, molecular weight, and respective physicochemical properties) (Kaur and Roy 2015; Jiang et al. 2016).

Monosaccharides and disaccharides can be used by several microorganisms to produce PHA. The PHA biosynthesis uses acetyl-CoA as intermediate and involves three main steps: production of acetoacetyl-CoA, its reduction to (R)-3-hydroxy-

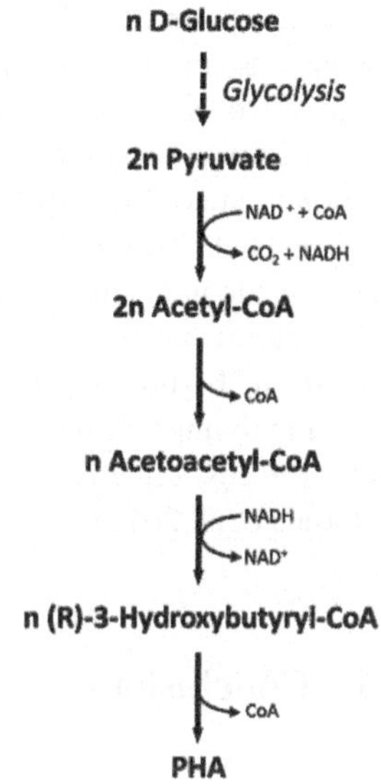

Fig. 11 Metabolic pathway for the biosynthesis of polyhydroxyalkanoates (PHA)

butyryl-CoA, and polymerization of this building block (Fig. 11). Lignocellulosic biomass and other waste materials are abundant and promising substrates for PHA production. Pentoses can be converted into PHA, but their utilization in hydrolysate contains mixtures of different carbohydrates (typically glucose and xylose) is still challenging since, depending on the applied pretreatment, inhibitors can compromise the performance of the microorganism (Jiang et al. 2016; Obruca et al. 2015).

The different composition of waste streams or by-product will significantly influence the choice of the biocatalyst. In cases where the raw material is rich in carbon and nutrients, a growth-associated PHA producer would be selected, such as *A. latus* or *Paracoccus denitrificans*. Conversely, in cases where the feedstock lacks an essential nutrient for growth (e.g., nitrogen), PHA accumulation using non-growth-associated bacteria would be preferred, i.e., *C. necator* (Kourmentza et al. 2017).

Certain bacteria (e.g., *C. necator, Protomonas extorquens, P. oleovorans*) produce PHA only when under nutrient (nitrogen or phosphorous) limitation. In this case, a two-stage process is preferred. In the first stage, growth is promoted, with limited accumulation of PHA, in a nutritionally balanced growth medium. In the second stage, an essential nutrient for growth is limited and the carbon flux is diverted from biomass production to PHA accumulation (Koller and Braunegg 2015). Other bacteria, like *A. latus*, mutant strain of *Azotobacter vinelandii* and recombinant *E. coli,* are able to accumulate PHA during exponential growth phase and are used in a one stage process (Kourmentza et al. 2017).

Industrial scale PHA production uses refined sugars as substrate (sugar beet, sugarcane, or corn) and pure cultures (*A. latus*, *C. necator*, and *P. putida*) (Jiang et al. 2016; Kourmentza et al. 2017). However, economic biotechnological polymer production is set back by (1) substrate cost, which can account for 50% of the total production cost, (2) low polymer titers, and (3) low process yield and productivity (Wang et al. 2014; Kootstra et al. 2017; Kaur and Roy 2015). Therefore, efforts have been made in (1) metabolic engineering to improve product yield and productivity, (2) using inexpensive and renewable carbon (and/or nitrogen) substrates, including waste and by-products from agriculture and industrial sources, and (3) process engineering to improve bioprocess efficiency, for maximum titer, yield and productivity, and for cost-effective product recovery (Wang et al. 2014; Kaur and Roy 2015; Obruca et al. 2015).

6 Conclusions

The European Union has the ambition of replacing at least 30% of the oil-based by bio-based chemicals in Europe by 2030. To achieve this goal and also meet the target on the reduction of greenhouse gases (GHG) emissions, the biological production of fuels and chemicals is mandatory. The deployment of a bio-based economy would not only help to reduce dependence on fossil-based products and lower GHG emissions but would also (1) create value by efficiently using and maximizing the potential of waste and residues; (2) boost the creation of rural and bio-based industrial employment; (3) revitalize industry in rural environment; (4) raise public awareness on the need for bio-based products; (5) decrease the amount of harsh chemicals and by-products.

Biological conversion or fermentation is one of the key processes for the conversion of renewable feedstock into drop-in (or ready to use) bio-products for the chemical industry, in a wide range of industries and a variety of applications. The global fermentation-based industry processes up to 200–250 million tons of carbohydrate equivalents annually from mono- and disaccharide-, starch- and lignocellulosic-based feedstock (Deloitte Report 2014). The economic feasibility of fermentation processes will be depending on the end-use and product value, cost of the feedstock and production cost, which is strongly influenced by the conversion yield and efficiency of product recovery. To increase the conversion yield some challenges remain. Most fermentation processes are still based on refined sugars (glucose or sucrose) and further development on the fermentation of mixed carbon sources (e.g., glucose/xylose) is still required. When processing lignocellulosic biomass, degradation compounds generated during pretreatment are typically highly inhibitory for the fermenting microorganism(s). Therefore, either cost-effective detoxification steps or, more relevant, the development of more robust strains is still required. Those strains should be able to cope with large-scale fermentation processes under non-sterile conditions, and compete effectively against microbial contaminants.

The recent technological advances in bioconversion processes are contributing to the reduction of production costs, making bio-based products more and more competitive against fossil-based alternatives, which, although produced by already well-established technologies, are becoming more and more costly due to the increase of oil prices. Still, the support from stakeholders and policymakers is essential for an effective deployment of fermentation-based processes within biorefineries as a relevant contribute towards a bio-based economy.

References

Abdel-Rahman MA, Sonomoto K (2016) Opportunities to overcome the current limitations and challenges for efficient microbial production of optically pure lactic acid. J Biotechnol 236:176–192

Abdel-Rahman MA, Tashiro Y, Sonomoto K (2011) Lactic acid production from lignocellulose-derived sugars using lactic acid bacteria: overview and limits. J Biotechnol 156:286–301

Ahn JH, Jang YS, Lee SY (2016) Production of succinic acid by metabolically engineered microorganisms. Curr Opin Biotechnol 42:54–66

Almeida JRM, Modig T, Petersson A, Hahn-Hägerdal B, Lidén G, Gorwa-Grauslund MF (2007) Increased tolerance and conversion of inhibitors in lignocellulosic hydrolysates by *Saccharomyces cerevisiae*. J Chem Technol Biotechnol 82:340–349

Alonso S, Rendueles M, Díaz M (2015) Microbial production of specialty organic acids from renewable and waste materials. Crit Rev Biotechnol 35(4):497–513

Baccile N, Babonneau F, Banat IM, Ciesielska K, Cuvier A-S, Devreese B, Everaert B, Lydon H, Marchant R, Mitchell CA, Roelants S, Six L, Theeuwes E, Tsatsos G, Tsotsou GE, Vanlerberghe B, Van Bogaert INA, Soetaert W (2017) Development of a cradle-to-grave approach for acetylated acidic sophorolipid biosurfactants. ACS Sust Chem Eng 5:1186–1198

Bahia FM, Almeida GC, Andrade LP, Campos CG, Queiroz LR, Silva RLV, Abdelnur PV, Corrêa JR, Bettiga M, Parachin NS (2018) Rhamnolipids production from sucrose by engineered *Saccharomyces cerevisiae*. Sci Rep 8:2905

Baral NR, Slutzky L, Shah A, Ezeji TC, Cornish K, Christy A (2016) Acetone-butanol-ethanol fermentation of corn stover: current production methods, economic viability and commercial use. FEMS Microbiol Lett 363(6)

Beller HR, Lee TS, Katz L (2015) Natural products as biofuels and bio-based chemicals: fatty acids and isoprenoids. Nat Prod Rep 32:1508

Benjamin KR, Silva IR, Cherubim JP, McPhee D, Paddon CJ (2016) Developing commercial production of semi-synthetic artemisinin, and of β-Farnesene, an isoprenoid produced by fermentation of Brazilian sugar. J Braz Chem Soc 27(8):1339–1345

Beuker J, Barth T, Steier A, Wittgens A, Rosenau F, Henkel M, Hausmann R (2016) High titer heterologous rhamnolipid production. AMB Express 6:124

Branduardi P, Porro N (2016) n-butanol: challenges and solutions for shifting natural metabolic pathways into a viable microbial production. FEMS Microbiol Lett 363:fnw070

Brumano LP, Soler MF, da Silva SS (2016) Recent advances in sustainable production and application of biosurfactants in Brazil and Latin America. Ind Biotechnol 12(1):31–39

Cabrera-Valladares N, Richardson AP, Olvera C, Treviño LG, Déziel E, Lépine F, Soberón-Chávez G (2006) Monorhamnolipids and 3-(3-hydroxyalkanoyloxy) alkanoic acids (HAAs) production using *Escherichia coli* as a heterologous host. Appl Microbiol Biotechnol 73:187–194

Chen QC (2010) Industrial production of PHA. In: Chen QC (ed) Plastics from bacteria: natural functions and applications, microbiology monographs, vol 14. Springer, Berlin, pp 121–132

Chong H, Li Q (2017) Microbial production of rhamnolipids: opportunities, challenges and strategies. Microb Cell Fact 16:137

Daniel H, Reuss M, Syldatk C (1998) Production of sophorolipids in high concentration from deproteinized whey and rapeseed oil in a two stage fed batch process using *Candida bombicola ATCC 22214* and *Cryptococcus curvatus ATCC 20509*. Biotechnol Lett 20:1153

Deloitte Report (2014) Opportunities for the fermentation-based chemical industry. An analysis of the market potential and competitiveness of North-West Europe. https://www2.deloitte.com/content/dam/Deloitte/nl/Documents/manufacturing/deloitte-nl-manufacturing-opportunities-for-the-fermentation-based-chemical-industry-2014.pdf. Accessed June 2018

Demain AL, Martens E (2017) Production of valuable compounds by molds and yeasts. J Antibiot 70:347–360

Demeke MM, Dietz H, Li Y, Foulquié-Moreno MR, Mutturi S, Deprez S, Abt TD, Bonini BM, Lidén G, Dumortier F, Verplaetse A, Boles E, Thevelein JM (2013) Development of a D-xylose fermenting and inhibitor tolerant industrial *Saccharomyces cerevisiae* strain with high performance in lignocellulose hydrolysates using metabolic and evolutionary engineering. Biotechnol Biofuels 6:89

El Takriti S, Pavlenko N, Searle S (2017) Mitigating international aviation emissions: risks and opportunities for alternative jet fuels. https://www.theicct.org/sites/default/files/publications/Aviation-Alt-Jet-Fuels_ICCT_White-Paper_22032017_vF.pdf

Eliasson A, Christensson C, Wahlbom CF, Hahn-Hägerdal B (2000) Anaerobic xylose fermentation by recombinant *Saccharomyces cerevisiae* carrying *XYL1*, *XYL2*, and *XKS1* in mineral medium chemostat cultures. Appl Environ Microbiol 66:3381–3386

Endres HJ, Siebert-Raths A (2011) Basics of PHA. Bioplast Mag 6(5):43–45

Faria NT, Santos MV, Fernandes P, Fonseca LL, Fonseca C, Ferreira FC (2014a) Production of glycolipid biosurfactants, mannosylerythritol lipids, from pentoses and D-glucose/D-xylose mixtures by Pseudozyma yeast strains. Process Biochem 49:1790–1799

Faria NT, Santos MV, Ferreira C, Marques S, Ferreira FC, Fonseca C (2014b) Conversion of cellulosic materials into glycolipid biosurfactants, mannosylerythritol lipids, by Pseudozyma spp. under SHF and SSF processes. Microb Cell Fact 13:155

Faria NT, Marques S, Fonseca C, Ferreira FC (2015) Direct xylan conversion into glycolipid biosurfactants, mannosylerythritol lipids, by Pseudozyma antarctica PYCC 5048T. Enzyme Microb Technol 71:58–65

Farwick A, Bruder S, Schadeweg V, Oreb M, Boles E (2014) Engineering of yeast hexose transporters to transport D-xylose without inhibition by D-glucose. Proc Natl Acad Sci U S A 111:5159–5164

Felpeto-Santero C, Rojas A, Tortajada M, Galán B, Ramón D, García JL (2015) Engineering alternative isobutanol production platforms. AMB Express 5:119

Fonseca C, Olofsson K, Ferreira C, Runquist D, Fonseca LL, Hahn-Hägerdal B, Lidén G (2011) The glucose/xylose facilitator Gxf1 from *Candida intermedia* expressed in a xylose-fermenting industrial strain of *Saccharomyces cerevisiae* increases xylose uptake in SSCF of wheat straw. Enzyme Microb Technol 48:518–525

Garcia-Sanchez R, Karhumaa K, Fonseca C, Sanchez N, Almeida VJR, Larsson CU, Bengtsson O, Bettiga M, Hahn-Hägerdal B, Gorwa-Grauslund MF (2010) Improved xylose and arabinose utilization by an industrial recombinant *Saccharomyces cerevisiae* strain using evolutionary engineering. Biotechnol Biofuels 3:13

George KW, Alonso-Gutierrez J, Keasling JD, Lee TS (2015) Isoprenoid drugs, biofuels, and chemicals—artemisinin, farnesene, and beyond. Adv Biochem Eng Biotechnol 148:55–89

Ghaffar T, Irshad M, Anwar Z, Aqil T, Zulifqar Z, Tariq A, Kamran M, Ehsan N, Mehmood S (2014) Recent trends in lactic acid biotechnology: a brief review on production to purification. J Radiat Res Appl Sci 7:222–229

Gírio FM, Fonseca C, Carvalheiro F, Duarte LC, Marques S, Bogel-Lukasik R (2010) Hemicelluloses for fuel ethanol: a review. Bioresour Technol 101:4775–4800

Global Market Insights, Inc.. https://www.gminsights.com/industry-analysis/farnesene-market. Accessed Jan 2018

Guettler MV, Jain MK (1996) Method for making succinic acid, *Anaerobiospirillum succiniciproducens* variants for use in process and methods for obtaining variants. U.S. patent 5521075

Hewald S, Linne U, Scherer M, Marahiel MA, Kämper J, Bölker M (2006) Identification of a gene cluster for biosynthesis of mannosylerythritol lipids in the basidiomycetous fungus *Ustilago maydis*. Appl Environ Microbiol 72:5469–5477

IEA Bioenergy Report Task 42 (2012) Bio-based chemicals: value added products from biorefineries

Jansen MLA, Bracher JM, Papapetridis I, Verhoeven MD, de Bruijn H, de Waal PP, van Maris AJA, Klaassen P, Pronk JT (2017) *Saccharomyces cerevisiae* strains for second-generation ethanol production: from academic exploration to industrial implementation. FEMS Yeast Res 17(5)

Jiang G, Hill DJ, Kowalczuk M, Johnston B, Adamus G, Irorere V, Radecka I (2016) Review carbon sources for polyhydroxyalkanoates and an integrated biorefinery. Int J Mol Sci 17:1157

Jin YS, Jones S, Shi NQ, Jeffries TW (2002) Molecular cloning of *XYL3* (D-xylulokinase) from *Pichia stipitis* and characterization of its physiological function. Appl Environ Microbiol 68:1232–1239

Jones DT, Woods DR (1986) Acetone-butanol fermentation revisited. Microbiol Rev 50(4):484–524

Jun H, Kieselbach T, Jönsson LJ (2011) Enzyme production by filamentous fungi: analysis of the secretome of *Trichoderma reesei* grown on unconventional carbon source. Microb Cell Fact 10:68

Juturu V, Wu JC (2016) Microbial production of lactic acid: the latest development. Crit Rev Biotechnol 36(6):967–997

Kang A, Lee TS (2015) Converting sugars to biofuels: ethanol and beyond. Bioengineering 2:184–203

Karhumaa K, Hahn-Hägerdal B, Gorwa-Grauslund MF (2005) Investigation of limiting metabolic steps in the utilization of xylose by recombinant *Saccharomyces cerevisiae* using metabolic engineering. Yeast 22:359–368

Kaur G, Roy I (2015) Strategies for large-scale production of polyhydroxyalkanoates. Chem Biochem Eng Q 29(2):157–172

Koller M, Braunegg G (2015) Potential and prospects of continuous polyhydroxyalkanoate (PHA) production. Bioengineering 2:94–121

Komesu A, de Oliveira JAR, da Silva Martins LH, Wolf Maciel MR, Maciel Filho RM (2017) Lactic acid production to purification: a review. BioRes 12(2):4364–4383

Konishi M, Fukuoka T, Morita T, Imura T, Kitamoto D (2008) Production of new types of sophorolipids by *Candida batistae*. J Oleo Sci 57:359–369

Kootstra M, Elissen H, Huurman S (2017) PHA's (polyhydroxyalkanoates): general information on structure and raw materials for their production. A running document for "Kleinschalige Bioraffinage WP9: PHA", Task 5. Report 727

Kourmentza C, Plácido J, Venetsaneas N, Burniol-Figols A, Varrone C, Gavala HN, Reis MAM (2017) Recent advances and challenges towards sustainable polyhydroxyalkanoate (PHA) production. Bioengineering 4:55

Kuhnert P, Scholten E, Haefner S, Mayor D, Frey J (2010) *Basfia succiniciproducens* gen. nov., sp. nov., a new member of the family Pasteurellaceae isolated from bovine rumen. Int J Syst Evol Microbiol 60:44–50

Kuyper M, Toirkens MJ, Diderich JA, Winkler AA, van Dijken JP, Pronk JT (2005a) Evolutionary engineering of mixed-sugar utilization by a xylose-fermenting *Saccharomyces cerevisiae* strain. FEMS Yeast Res 5:925–934

Kuyper M, Hartog MM, Toirkens MJ, Almering MJ, Winkler AA, van Dijken JP, Pronk JT (2005b) Metabolic engineering of a xylose-isomerase-expressing *Saccharomyces cerevisiae* strain for rapid anaerobic xylose fermentation. FEMS Yeast Res 5:399–409

Lange A, Becker J, Schulze D, Cahoreau E, Portais J-C, Haefner S, Schröder H, Krawczyk J, Zelder O, Wittmann C (2017) Bio-based succinate from sucrose: high-resolution ^{13}C metabolic flux analysis and metabolic engineering of the rumen bacterium *Basfia succiniciproducens*. Metab Eng 44:198–212

Leandro MJ, Gonçalves P, Spencer-Martins I (2006) Two glucose/xylose transporter genes from the yeast *Candida intermedia*: first molecular characterization of a yeast xylose-H+ symporter. Biochem J 395:543–549

Leavell MD, McPhee, DJ, Paddon CJ (2016) Developing fermentative terpenoid production for commercial usage. Curr Opin Biotechnol 37:114–119

Lee PC, Lee SY, Hong SH, Chang HN, Park SC (2003) Biological conversion of wood hydrolysate to succinic acid by *Anaerobiospirillum succiniciproducens*. Biotechnol Lett 25:111–114

Lee SH, Yun EJ, Kim J, Lee SJ, Um Y, Kim KH (2016) Biomass, strain engineering, and fermentation processes for butanol production by solventogenic clostridia. Appl Microbiol Biotechnol 100:8255–8271

Li Y, Ge X (2016) Advances in bioenergy, vol 1, 1st edn. Academic Press, New York

Li Q, Xing J (2015) Microbial succinic acid production using different bacteria species. In: Kamm B (ed) Microorganisms in biorefineries, Microbiology monographs, vol 26. Springer, Berlin

Mans R, Daran JG, Pronk JT (2018) Under pressure: evolutionary engineering of yeast strains for improved performance in fuels and chemicals production. Curr Opin Biotechnol 50:47–56

McKinlay JB, Shachar-Hill Y, Zeikus JG, Vieille C (2007) Determining *Actinobacillus succinogenes* metabolic pathways and fluxes by NMR and GC-MS analyses of C-13-labeled metabolic product isotopomers. Metab Eng 9:177–192

McPhee D, Pin A, Kizer L, Perelman L (2014) Squalane from sugarcane. Cosmetics & Toiletries magazine 129(6)

Meadows AL, Hawkins KM, Tsegaye Y, Antipov E, Kim Y, Raetz L, Dahl RH, Tai A, Mahatdejkul-Meadows T, Xu L, Zhao L, Dasika MS, Murarka A, Lenihan J, Eng D, Leng JS, Liu CL, Wenger JV, Jiang H, Chao L, Westfall P, Lai J, Ganesan S, Jackson P, Mans R, Platt D, Reeves CD, Saija PR, Wichmann G, Holmes VF, Benjamin K, Hill PW, Gardner TS, Tsong AE (2016) Rewriting yeast central carbon metabolism for industrial isoprenoid production. Nature 537:694

Medina VG, Almering MJ, van Maris AJ, Pronk JT (2010) Elimination of glycerol production in anaerobic cultures of a *Saccharomyces cerevisiae* strain engineered to use acetic acid as an electron acceptor. Appl Environ Microbiol 76:190–195

Mitrovich Q, Wichmann G (2017) Integrated process for production of farnesene, a versatile platform chemical, from domestic lignocellulosic feedstock. Presentation at the U.S. Department of Energy (DOE) Bioenergy Technologies Office (BETO) 2017 Project Peer Review

Mordor Intelligence (2018) Surfactants market—segmented by origin, type, application, and geography—growth, trends and forecasts (2019–2024). https://www.mordorintelligence.com/industry-reports/global-market-for-surfactants-industry

Morita T, Fukuoka T, Imura T, Kitamoto D (2015) Mannosylerythritol lipids: production and application. J Oleo Sci 64:133–141

Morlon-Guyot J, Guyot JP, Pot B, Jacobe de Haut I, Raimbault M (1998) *Lactobacillus manihotivorans* sp. nov., a new starch-hydrolyzing lactic acid bacterium isolated from cassava sour starch fermentation. Int J Syst Bacteriol 48:1101–1109

Moysés DN, Reis VCB, de Almeida JRM, de Moraes LMP, Torres FAG (2016) Xylose fermentation by *Saccharomyces cerevisiae*: challenges and prospects. Int J Mol Sci 17(3):207

Murali N, Srinivas K, Ahring BK (2017) Biochemical production and separation of carboxylic acids for biorefinery applications. Fermentation 3:22

Nag A (2008) Biofuels refining and performance. McGraw Hill, New York

Ndaba B, Chiyanzu I, Marx S (2015) n-Butanol derived from biochemical and chemical routes: a review. Biotechnol Rep 8:1–9

Nghiem NP, Kleff S, Schwegmann S (2017) Succinic acid: technology development and commercialization. Fermentation 3:26

Nitschke M, Silva SS (2017) Recent food applications of microbial surfactants. Crit Rev Food Sci Nutr 58(4):631–638

Nwankwo D, Anadu E, Usoro R (1989) Cassava fermenting organisms. MIRCEN J 5:169–179

Obruca S, Benesova P, Marsalek L, Marovaa I (2015) Use of lignocellulosic materials for PHA production. Chem Biochem Eng Q 29(2):135–144

Palmqvist E, Hahn-Hägerdal B (2000) Fermentation of lignocellulosic hydrolysates. II: inhibitors and mechanisms of inhibition. Bioresour Technol 74(1):25–33

Peralta-Yahya PP, Zhang F, del Cardayre SB, Keasling JD (2012) Microbial engineering for the production of advanced biofuels. Nature 488:320

Qureshi N, Ezeji TC (2008) Butanol, 'a superior biofuel' production from agricultural residues (renewable biomass): recent progress in technology. Biofuels Bioprod Biorefin 2:319–330

Raab AM, Lang C (2011) Oxidative versus reductive succinic acid production in the yeast *Saccharomyces cerevisiae*. Bioeng Bugs 2(2):120–123

Reddy G, Altaf M, Naveena BJ, Venkateshwar M, Vijay Kumar E (2008) Amylolytic bacterial lactic acid fermentation—a review. Biotechnol Adv 26:22–34

Report of Transparency Market Research (2011) BioSurfactants Market: Global Scenario, Raw Material and Consumption Trends, Industry Analysis, Size, Share and Forecasts, 2011 – 2018. http://www.sbdi.co.kr/cart/data/info/Transparency_Biosurfactants_Market_Sample.pdf. Accessed 31 May 2018

Roelants SL, Saerens KM, Derycke T, Li B, Lin YC, van de Peer Y, de Maeseneire SL, van Bogaert IN, Soetaert W (2013) *Candida bombicola* as a platform organism for the production of tailor-made biomolecules. Biotechnol Bioeng 110:2494–2503

Roelants SLKW, Ciesielska K, de Maeseneire SL, Moens H, Everaert B, Verweire S, Denon Q, Vanlerberghe B, van Bogaert INA, van der Meeren P, Devreese B, Soetaert W (2016) Towards the industrialization of new biosurfactants: biotechnological opportunities for the lactone esterase gene from *Starmerella bombicola*. Biotechnol Bioeng 113:550–559

Rooke JA, Hatfield RD (2003) Biochemistry of ensiling. In Silage science and technology. Edited by Buxton DR, Muck RE, Harrison JH, Madison, Wisconsin. USA: American Society of Agronomy, Crop Science Society of America, Soil Science Society of America; 95–140

Rubber Journal of Asia (RJA) (2017) Kuraray's liquid farnesene rubber used in SRI tyres for the first time. http://rubberjournalasia.com/kurarays-liquid-farnesene-rubber-used-in-sri-tyres-for-the-first-time/. Accessed Jan 2018

Ryan C (2018) An overview of Gevo's biobased isobutanol production process. https://gevo.com/wp-content/uploads/2018/02/isobutanol-process.pdf. Accessed Apr 2018

Sachdev DP, Cameotra SS (2013) Biosurfactants in agriculture. Appl Microbiol Biotechnol 97:1005–1016

Sajna KV, Höfer R, Sukumaran RK, Gottumukkala LD, Pandey A (2015) White biotechnology in biosurfactants. In: Pandey A, Hofer R, Larroche C, Taherzadeh M, Nampoothiri M (eds) Industrial biorefineries & white biotechnology. Ch. 14. Elsevier B.V., Amsterdam, pp 499–521

Sauer M (2016) Industrial production of acetone and butanol by fermentation—100 years later. FEMS Microbiol Lett 363(13)

Schadeweg V, Boles E (2016) n-Butanol production in *Saccharomyces cerevisiae* is limited by the availability of coenzyme A and cytosolic acetyl-CoA. Biotechnol Biofuels 9:44

Schiel-Bengelsdorf B, Montoya J, Linder S, Dürre P (2013) Butanol fermentation. Environ Technol 34:13–14

Scott MJ, Jones MN (2000) The biodegradation of surfactants in the environment. Biochim Biophys Acta 1508:235–251

Song H, Lee SY (2006) Review: production of succinic acid by bacterial fermentation. Enzyme Microb Technol 39:352–361

Souza EC, Vessoni-Penna TC, de Souza Oliveira RP (2014) Biosurfactant-enhanced hydrocarbon bioremediation: an overview. Int Biodeter Biodegr 89:88–94

Swidah R, Wang H, Reid PJ, Ahmed HZ, Pisanelli AM, Persaud KC, Grant CM, Ashe MP (2015) Butanol production in *S. cerevisiae* via a synthetic ABE pathway is enhanced by specific metabolic engineering and butanol resistance. Biotechnol Biofuels 8:97

Tan J, Abdel-Rahman MA, Sonomoto K (2017) Biorefinery-based lactic acid fermentation: microbial production of pure monomer product. Adv Polym Sci 279:27–66

Träff KL, Otero Cordero RR, van Zyl WH, Hahn-Hägerdal B (2001) Deletion of the *GRE3* aldose reductase gene and its influence on xylose metabolism in recombinant strains of *Saccharomyces cerevisiae* expressing the *xylA* and *XKS1* genes. Appl Environ Microbiol 67:5668–5674

van Zyl WH, Lynd LR, den Haan R, McBride JE (2007) Consolidated bioprocessing for bioethanol production using *Saccharomyces cerevisiae*. Adv Biochem Eng Biotechnol 108:205–235

Vaswani S (2010) Bio based succinic acid. Review no. 2010-14. Report by Process Economics Program, SRI Consulting

Vecino X, Cruz JM, Moldes AB, Rodrigues LR (2017) Biosurfactants in cosmetic formulations: trends and challenges. Crit Rev Biotechnol 37(7):911–923

Wang Y, Yin J, Chen GQ (2014) Polyhydroxyalkanoates, challenges and opportunities. Curr Opin Biotechnol 30:59–65

Wee YJ, Kim JN, Ryu HW (2006) Biotechnological production of lactic acid. Food Technol Biotechnol 44(2):163–172

Wisselink HW, Toirkens MJ, Wu Q, Pronk JT, van Maris AJ (2009) Novel evolutionary engineering approach for accelerated utilization of glucose, xylose, and arabinose mixtures by engineered *Saccharomyces cerevisiae* strains. Appl Environ Microbiol 75:907–914

Wyman CE (1996) Handbook on bioethanol—production and utilization. Taylor & Francis, Washington, DC

Wyman CE, Dale BE (2015) Producing biofuels via the sugar platform. American Institute of Chemical Engineers (AIChE). www.aiche.org/cep. Accessed Mar 2018

Xin F, Chen T, Jiang Y, Dong W, Zhang W, Zhang M, Wu H (2017) Strategies for improved isopropanol–butanol production by a *Clostridium* strain from glucose and hemicellulose through consolidated bioprocessing. Biotechnol Biofuels 10:118

Xu H, Kim S, Sorek H, Lee Y, Jeong D, Kim J, Oh EJ, Yun EJ, Wemmer DE, Kim KH, Kim SR, Jin YS (2016) *PHO13* deletion-induced transcriptional activation prevents sedoheptulose accumulation during xylose metabolism in engineered *Saccharomyces cerevisiae*. Metab Eng 34:88–96

Yang L, Lübeck M, Ahring BK, Lübeck P (2016) Enhanced succinic acid production in *Aspergillus saccharolyticus* by heterologous expression of fumarate reductase from *Trypanosoma brucei*. Appl Microbiol Biotechnol 100:1799–1809

Zhao J, Lu C, Chen CC, Yang ST (2013) Biological production of butanol and higher alcohols. In: Yang ST, El-Enshasy HA, Thongchul N (eds) Bioprocessing technologies in biorefinery for sustainable production of fuels, chemicals, and polymers. Wiley, New York

Biogas: Perspectives of an Old Technology

Xavier Flotats

1 Historic Perspective

One of the first descriptions of the mysterious lights and fatuous fires emerging from marshy areas is due to Pliny. For years, this phenomenon was attributed to mythical beings and other fantastic tales. Not until 1630, there is a first scientific explanation due to van Helmont, who identified a flammable gas generated during the putrefaction of organic matter, which was also contained in intestinal gases. Van Brakel (1980) considers that the beginning of the scientific history of anaerobic digestion (AD) and biogas should be dated to 1776 with Volta, who concluded that the amount of gas emitted in the marshes was a function of the amount of decomposing organic matter deposited in the bottom and that a certain proportion of this gas in the air produced an explosive mixture.

The systematic investigation of AD began in the second half of the nineteenth century. In 1884, Pasteur presented the results of his team with the fermentation of manures at 35 °C, concluding that this process could mean a new source of energy for heating and lighting. Pasteur's team received orders to build biogas plants that were not accepted, arguing that the state of the research was still preliminary (van Brakel 1980).

Two facilities are documented at the end of the nineteenth century in England and India. In 1895, the gas collected from a septic tank treating sewage fed part of the public lighting network of Exeter. In 1897, an anaerobic digester was installed in the Matunga Leper Asylum (Mumbai, India), fed with the organic waste produced. The biogas obtained was used for lighting and in 1907 also as fuel for an engine (van Brakel 1980; Deublein and Steinhauser 2008).

X. Flotats (✉)
GIRO Joint Research Unit IRTA-UPC, Department of Agrifood Engineering and Biotechnology, Universitat Politècnica de Catalunya - UPC BarcelonaTECH, Parc Mediterrani de la Tecnologia, Castelldefels, Barcelona, Spain
e-mail: xavier.flotats@upc.edu

© Springer Nature Switzerland AG 2019

J.-R. Bastidas-Oyanedel, J. E. Schmidt (eds.), *Biorefinery*,
https://doi.org/10.1007/978-3-030-10961-5_13

The development of scientific and technological knowledge of the process after 1900 was remarkable. The literature review published by van Brakel (1980) deserves special attention, as research lines, publications, technological developments, patents or governmental interests in different countries in the period 1900–1970 are referenced. It is acknowledged that France had a development in this area that led to the introduction of discontinuous digesters, identifying 50 publications between 1931 and 1962. In Germany, van Brakel (1980) documents an intense research and development of continuous anaerobic digesters (completely stirred or plug flow reactors) and references 63 publications in the period 1945–1960. Already in publications of that period, both in France and Germany, it was estimated that AD was not profitable as an energy production method, taking into account the prices of fossil fuels, but that the contribution to organic waste treatment and fertilizer production, the reduction of sewage sludge volumes in wastewater treatment plants and the fact that they could be energy-autonomous facilities gave the biogas plants a future interest of special importance. China started applying AD in 1929 to solve the fuel shortage in rural areas, developing a simple, small and low-cost design used in 38.51 million household-scale biogas digesters in rural China by the end of 2010 (Chen et al. 2012).

Checking the Scopus database of published research and looking for the terms "anaerobic digestion" or "biogas" in the title, abstract or keywords, 51 articles were found in the period 1920–1969, 433 in the period 1970–1979 and 2056 in the period 1980–989. Published research on AD has increased almost exponentially till now, with 3033 articles during year 2017 (see Fig. 1).

2 The Anaerobic Digestion Process

AD is a microbiological process following different reactions in a synergic scheme where organic matter is transformed, in the absence of oxygen, into biogas, a flammable gas constituted mainly by methane (CH_4) and carbon dioxide (CO_2), with a CH_4 content ranging from 55 to 75% by volume. Figure 2 shows a scheme of the main reactions occurring during the AD process, microorganisms catalysing them and an approach to COD (chemical oxygen demand) distribution along the process.

Since the process is performed in the absence of oxygen, COD is conservative, that is, the COD of the initial substrate is the sum of the COD of the products. This property allows the COD balance of both lab-scale and industrial facilities for characterizing the efficiency on methane production. Anaerobic biodegradability is defined as the fraction of the COD of a given substrate that is transformed to methane (0.35 Nm^3 CH_4/kg COD), volatile fatty acids and microorganisms (Holliger et al. 2016).

Relevant chemical equilibria affecting the dynamics and stability of AD are outlined in Fig. 2 too. The dynamic equilibrium between anions and cations affects the dynamics of the pH and, therefore, the overall development process since optimal

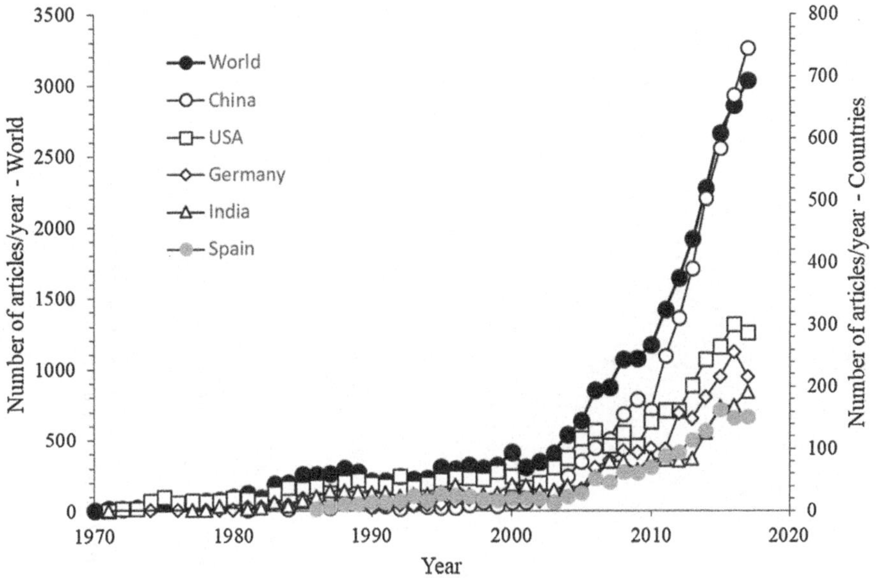

Fig. 1 Number of articles yearly published with the terms "anaerobic digestion" or "biogas" in the title, abstract or keyword, based on Scopus database (April 2018), in the world and in the 5 top countries

biomass growth is around neutral pH. All the reactions and chemical equilibria indicated in Fig. 2 are included in the Anaerobic Digestion Model number 1—ADM1—(Batstone et al. 2002), which is the current reference model and the base of further modelling developments.

2.1 *Disintegration and Hydrolysis*

Complex organic matter compounds are disintegrated into carbohydrates, proteins and lipids, and these macromolecules are hydrolysed to monosaccharides, amino acids and long-chain fatty acids (LCFA), respectively. Since the overall disintegration/hydrolysis rate depends on the ratio between concentration of hydrolytic enzymes producing biomass and the organic particles (Fernández et al. 2001), the Contois kinetics can express the behaviour of anaerobic reactors when the disintegration/hydrolysis is the rate-limiting step (Vavilin et al. 2008), thus explaining why a simplified model such as that from Chen and Hashimoto (1978), based on Contois kinetics as derived by Domenech and Flotats (1997), can be used satisfactorily in completely mixed anaerobic digesters when processing complex particulate organic matter. ADM1 (Batstone et al. 2002) adopts the first-order kinetics for both processes, which is a particular case of the Contois kinetics for high hydrolytic biomass concentration (Vavilin et al. 2008).

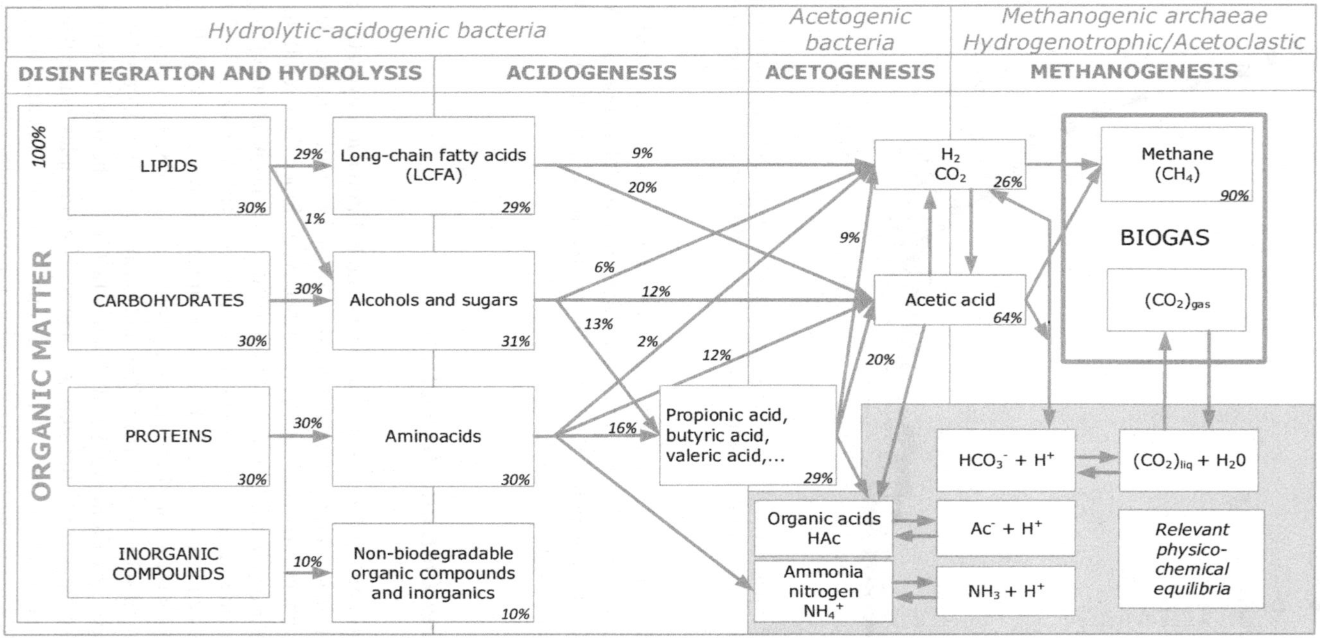

Fig. 2 Schematic representation of the anaerobic digestion process. Values (%) indicate COD flux for particulate composite constituted by 10% non-biodegradable organic matter and 30% each of carbohydrates, proteins and lipids in terms of COD, following Batstone et al. (2002)

2.2 Acidogenesis

Monosaccharides, amino acids and long-chain fatty acids (LCFA), released during hydrolysis, decompose to VFA, H_2 and NH_4^+ in the acidogenic step. The three main groups of molecules follow different pathways during acidogenesis. LCFA decompose to H_2 and acetic acid (Ac) by syntrophic acetogenic bacteria following the β-oxidation process. This process is thermodynamically favourable at low H_2 partial pressure only (Cavaleiro et al. 2016), which means that complete transformation of LCFA requires hydrogenotrophic methanogenic activity. This dependence of H_2 partial pressure can be seen in Fig. 3 for stearate, together with other acetogenic reactions. LCFA produce reversible inhibition, related to the physical adsorption of these compounds and their accumulation on the cell wall, hindering the transfer of substrates and metabolites (Pereira et al. 2005), being the acetoclastic methanogens the more affected population (Palatsi et al. 2010; Juznic-Zonta et al. 2013). This inhibition can be prevented (Palatsi et al. 2012; Affes et al. 2013), and inhibited reactors can be recovered using different strategies (Palatsi et al. 2009).

Amino acids transform into volatile fatty acids (VFA), H_2, CO_2 and NH_4^+ via oxidation-reduction reactions. Some amino acids are decomposed via an oxidative pathway and others via reductive, while others are ambivalent and follow both. In Stickland reactions, an amino acid is oxidized, while another is reduced. Stickland reactions are thermodynamically more favourable than the reductive or the oxidative

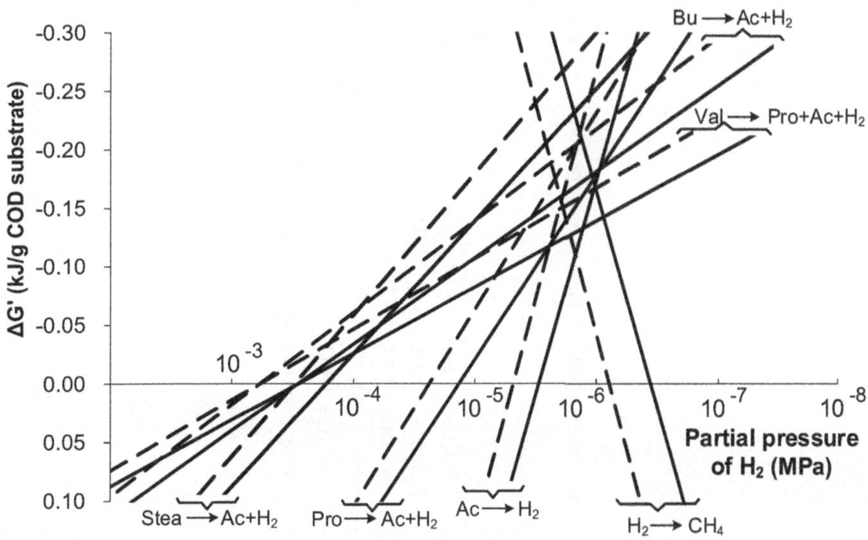

Fig. 3 Change in Gibbs free energy (kJ/g COD of substrate) depending on the partial pressure of H_2 for acetogenic reactions of stearate (Stea), valerate (Val), butyrate (Bu), propionate (Pro), oxidation of acetate (Ac) and consumption of H_2 at 35 °C (solid line) and at 55 °C (dashed line), with the following activities: acids, 1 mM; CO_2 partial pressure, 0.03 MPa; CH_4 partial pressure, 0.07 MPa; HCO_3^-, 56.2 mM (35 °C) or 85.5 mM (55 °C)

Table 1 Comparison of values of ΔG^{o}' for amino acids reactions: oxidation, reduction, Stickland and oxidation coupled to hydrogenotrophic methanogenesis

	$\Delta G^{o\prime}$ (kJ/reaction)		
	Indicated reaction	In coculture with methanogenic hydrogenotrophic archaea (with reaction 5)	Stickland reactions (with reaction 4)
1. Leucine + $3H_2O \rightarrow$ Valerate$^-$ + HCO_3^- + H^+ + NH_4^+ + $2H_2$	+ 4.2 (ox)	−63.6	−151.8
2. Valine + $3H_2O \rightarrow$ Butyrate$^-$ + HCO_3^- + H^+ + NH_4^+ + $2H_2$	+ 9.7 (ox)	−58.1	−146.3
3. Alanine + $3H_2O \rightarrow$ Acetate- + HCO_3^- + H^+ + NH_4^+ + $2H_2$	+ 7.6 (ox)	−60.2	−148.4
4. Glycine + $H_2 \rightarrow$ Acetate$^-$ + NH_4^+	−77.9 (red)		
5. $4H_2$ + HCO_3^- + $H^+ \rightarrow CH_4$ + $3H_2O$	−135.6 (red)		

$\Delta G^{o\prime}$ values calculated using free energies of formation compiled by Rittmann and McCarty (2001)

reactions, while the last ones are favoured in cultures with high hydrogenotrophic activity (Örlygsson et al. 1995; Miron et al. 2000). With the objective to propose a systematic method for establishing the overall stoichiometry of amino acids acidogenesis, Ramsay (1997) created a calculation algorithm based on Stickland reactions, the application of which is suggested by the ADM1 (Batstone et al. 2002). In a posterior study with casein, Ramsay and Pullammanappallil (2001) found that about 60% of the reactions did not follow the Stickland scheme, since amino acids must be paired for this pathway. Different available energy values for microorganisms, depending on the pathway followed (see Table 1), suggest different growth rates as function of the relative amino acids composition of the substrate (Flotats et al. 2013).

Acidogenesis of monosaccharides can lead to acetate, propionate, butyrate, H_2, lactate and ethanol, depending on different environmental conditions. The most important parameters that determine the predominant pathway are pH and H_2 partial pressure (Rodríguez et al. 2006). At low partial pressure of H_2 and at neutral pH, the predominant pathway is acetate production. As the partial pressure of H_2 rises and falls slightly the pH to 7–5.5, the pathway changes to butyrate production. Also, joint production of acetate and propionate, whose reaction is not limited by the partial pressure of H_2, has been observed and favoured at high pH. Below pH 5.6, the butyrate production decreases, and the pathway of ethanol production becomes dominant (Rodríguez et al. 2006). Since the lactic acid quickly decomposes into VFA, the lactate pathway is not usually considered, although it is included in some models (Skiadas et al. 2000).

2.3 *Acetogenesis*

VFA are converted to Ac and H_2 during this step. Under standard conditions, these reactions are not energetically feasible, and they proceed only when the reaction products are removed from the system, especially H_2 by methanogens in syntrophic association. Ac can also be synthesized by CO_2-reducing bacteria (homoacetogenic bacteria), but this pathway is thermodynamically unfavourable in methanogenic conditions, when H_2 partial pressure is low and/or Ac concentration is high (Bastidas-Oyanedel et al. 2012; Gonzalez-Cabaleiro et al. 2013). The reverse reaction, the oxidation of Ac to H_2 and CO_2 by SAOB (syntrophic acetate-oxidizing bacteria), is favoured at low partial pressure of H_2 (Hattori 2008). These bacteria are less sensitive to the NH_3 inhibition than the acetoclastic methanogenic archaea and can compete at high solid retention time, since SAOB are characterized by a high doubling time (Schnürer et al. 1994). SAOB growth conditions mentioned are the base of stable reactors operation with high ammonia content (>4 g NH_4^+-N/l) and high solid retention time (>50 days), where methane production is mainly produced via the hydrogenotrophic pathway (Schnürer and Nordberg 2008; Ruiz-Sánchez et al. 2018).

Figure 3 depicts the change in Gibbs free energy for the indicated reactions, depending on the H_2 partial pressure at mesophilic and thermophilic temperature ranges. While decreasing the partial pressure of H_2 favours the transformation of VFAs, its increase promotes the conversion of H_2 to CH_4, which is not possible for pressures below 5.5×10^{-7} MPa at 35 °C. This sets certain zones in Fig. 3 where acetogenesis or the acetate oxidation and hydrogenotrophic methanogenesis are possible simultaneously.

2.4 Methanogenesis

The final production of CH_4 in the AD process is carried out from Ac (acetoclastic methanogenesis) and H_2 (hydrogenotrophic methanogenesis). The reaction of acetoclastic methanogenesis is mainly catalysed by archaea of the genera *Methanosarcina* and *Methanosaeta*. While the former dominates at high concentrations of acetate and exhibits higher growth rates, the latter dominates at low substrate concentrations and has low growth rate. Therefore, *Methanosaeta* is mainly measured in reactors with high biomass retention time and low concentration of substrate (biofilms), and *Methanosarcina* is mainly measured in digesters with high organic loading rate (solid waste digesters). The reaction of hydrogenotrophic methanogenesis is catalysed by a number of methanogenic microorganisms and has extraordinary importance for the balance of the whole AD process, as has been shown in previous subsections. Acetoclastic archaea are very sensitive to different conventional inhibitors. The ADM1 model (Batstone et al. 2002) considers that the growth of acetoclastic archaea is inhibited by ammonia, since this population is the most sensitive (Chen et al. 2008).

3 Feedstocks for Producing Biogas

3.1 Methane Production Potentials

Almost all kinds of biodegradable organic compounds could be transformed into biogas by AD, and biodegradability is the characteristic indicating to what extent this is possible. Raw materials for producing biogas by AD are:

- Livestock manure.
- Organic fraction of municipal solid waste, including food waste from caterings, restaurants and markets.
- Waste from food industry.
- Sludge from biological urban and industrial wastewater treatment plants.
- Biodegradable wastewaters.
- Energy crops.

- Catch crops.

Catch crops are grown between two seasonal consecutive main food crops, with the objective to assimilate residual nutrients in soil, especially nitrates, in order to avoid leaching. These are suitable for the AD process without interfering with the production of food and fodder crops, but the low biomass yield has been shown as the main limiting factor for using them for biogas production (Molinuevo-Salces et al. 2015).

Difficulties in measuring COD of solid and heterogeneous organic materials (Noguerol et al. 2012) and hence anaerobic biodegradability had led to characterize the biogas productivity for a given solid or semi-solid substrate by the methane yield related to its volatile solids (VS). For a given kind of substrate, VS can present different lipids–proteins–carbohydrates ratios and even different biodegradabilities, depending on many factors. Therefore, published values can help facilities design, but updated analytical determinations must be done to optimize reactors' operation. Examples are the following:

- Low methane yields for pig manure after a long storage time before the AD process (Bonmatí et al. 2001), varying the COD/VS ratio between 1.3 and 1.8 (Palatsi et al. 2005).
- Different methane yields of the organic fraction of municipal solid waste (OFMSW) depending on the methods of sorting (Cecchi et al. 2003).
- Different methane yields of fruits and vegetable waste from central markets along the year due to seasonal variations of its composition (Edwiges et al. 2018).

Table 2 summarizes methane yields for some organic substrates usually used in AD plants. Owing to the number of factors affecting the quality and the methane yield, organic waste inventories intended for biogas plants planning should be characterized by an uncertainty degree or by intervals as was done by Lorenz et al. (2013) in a study of energy potential from biogas in the UE-27.

3.2 Pretreatments for Enhancing Methane Production

Solid substrates for which disintegration process is the rate-limiting step, or with high lignocellulosic compounds content, are difficult or very slow to biodegrade by AD. This is the case of materials constituted by cells (e.g. secondary sewage sludge) for which it is necessary to break the cell wall to increase bioavailability or fibres where lignin protects cellulose and hemicellulose against microbial attack. In these cases, some pretreatment process to increase the solid particle surface or to help particle disintegration could enhance biogas production. A number of methods, alone or combined, have been studied and successfully applied depending on the kind of substrate:

Table 2 Methane yields for different organic substrates

Organic substrate	Yield potential (Nm^3 CH_4/kg VS)	Reference
Organic fraction of municipal solid waste		
Mechanically sorted (MS)	0.160–0.370	Cecchi et al. (2003)
Separately collected (SC)	0.450–0.490	Cecchi et al. (2003)
Sorted domestic source (SS)	0.370–0.400	Cecchi et al. (2003)
Fruit and vegetable waste	0.288–0.516	Edwiges et al. (2018)
Livestock manure		
Pig manure	0.067–0.557	Bonmatí et al. (2001)
Cattle manure	0.280–0.540	Hashimoto (1986)
Poultry manure	0.228–0.390	Al Seadi (2001)
Solid fraction pig manure	0.178–0.496[a]	Jurado et al. (2013)
Industrial organic waste		
Stomach/intestinal waste	0.400–0.460	Angelidaki and Ellegaard (2003)
Slaughterhouse waste (piggery)	0.580–0.960	Rodríguez-Abalde et al. (2011)
Slaughterhouse waste (poultry)	0.460–0.480	Rodríguez-Abalde et al. (2011)
Coffee waste	0.240–0.280	Neves et al. (2006)
Citrus waste	0.314–0.548	Ruiz et al. (2016)
Fish waste	0.398–0.573	Kafle et al. (2013)
Crude glycerol (biodiesel prod.)	0.780–0.826	Baba et al. (2013)
Wastewater treatment plants		
Primary sludge (urban)	0.307–0.489	Gavala et al. (2003)
Secondary sludge (urban)	0.191–0.244	Gavala et al. (2003)
Grease waste (urban)	0.405–0.540	Silvestre et al. (2011)
Grease sludge (meat processing)	0.845–0.928	Luostarinen et al. (2009)
Energy crops		
Maize (whole crop)	0.204–0.450	Murphy et al. (2011)
Barley	0.353–0.658	Murphy et al. (2011)
Grass	0.298–0.467	Murphy et al. (2011)
Alfalfa	0.340–0.500	Murphy et al. (2011)
Miscanthus	0.179–0.218	Murphy et al. (2011)
Beet fodder	0.420–0.500	Murphy et al. (2011)
Microalgae	0.106–0.209	Passos et al. (2015)
Catch crops		
Raplanus sativus	0.274–0.474	Molinuevo-Salces et al. (2014)
Brassica napus	0.334–0.448	Molinuevo-Salces et al. (2014)
Avena sativa	0.250–0.527	Molinuevo-Salces et al. (2014)

[a]Supposing 0.7 kg VS/kg TS

- Physical processes: Maceration, grinding, low (<100 °C) or high (100–210 °C) temperature, freezing/thawing, high pressure, wet explosion, steam explosion, ultrasound, microwave and electric pulse.
- Chemical processes: Alkaline or acid pretreatment, aqueous ammonia soaking, ozonation and Fenton process.
- Biological processes: Pre-aerobic digestion, ensilage, enzymes addition, auto-hydrolytic process.

A given pretreatment process could affect differently depending on substrate composition. Thermal pretreatment at low temperature (70 °C) can significantly increase methane production from slaughterhouse waste when the carbohydrates/proteins ratio is low and the opposite when it is high, due to formation of recalcitrant compounds by Maillard reaction (Rodríguez-Abalde et al. 2011, 2013).

In a wide review of pretreatments applied to different substrates, Carlson et al. (2012) stated that pretreatment effects are complex and generally linked to substrate characteristics, pretreatment mechanisms and operation adjustments. There are different approaches to evaluate AD enhancement, and it is necessary to consider side effects: resources consumed, net energy production and environmental effects such as gaseous emissions or a modification of the digestate characteristics affecting further processes, with the corresponding overall economic balance. Carballa et al. (2011) used a LCA (live cycle assessment) method, combined with operational and economic information, to determine that, in the current state of research, the best options are mechanical or chemical pretreatments for sewage sludge.

Updated reviews about pretreatment methods can be found for sewage sludge (Neumann et al. 2016), algae and microalgae (Rodriguez et al. 2015; Passos et al. 2015), food waste (Kondusamy and Kalamdhad 2014), municipal solid waste (Jain et al. 2015) or lignocellulosic biomass (Biswas et al. 2015). In this last review, wet explosion appears to be a promising technique, adaptable to a wide range of agricultural and forestry biomass and for different biorefinery processes, allowing solubilization of some lignin compounds, making them accessible for further decomposition during AD (Ahring et al. 2015).

4 Reactor Designs for Wastes and Wastewaters

An anaerobic reactor, or anaerobic digester, is a closed volume where reactions previously described are intended to be controlled to achieve a given objective: biogas production, organic matter removal and/or some intermediate compound recovery.

Reactors can be operated at psychrophilic (<25 °C), mesophilic (25–43 °C) or thermophilic (>43 °C) temperature ranges, which correspond to three different groups of microorganism populations, with optimal temperatures around 37 and 55 °C for mesophilic and thermophilic, respectively. At thermophilic range, microorganisms growth and hydrolysis rates increase but also the concentration of some inhibitors, such as NH_3 or nonionized volatile fatty acids, and the sensitivity to small changes in temperature operation (Van Lier et al. 1993), which could lead to instabilities. Although mesophilic AD is the most widely used, thermophilic AD has been also successfully applied (De Vrieze et al. 2016).

Apart from the operation temperature, anaerobic reactors can be classified attaining the kind of biomass growth (suspended, granular or attached-biofilm), the mechanisms applied for retaining active biomass in the reactor and the flow model (completely stirred or plug flow, batch or continuous). Table 3 presents this classification for the most usual reactor types, and Fig. 4 shows some simplified schemes.

A first group of bioreactors are low-cost with undefined or variable flow model, such as the old Chinese, the plastic tubular designs (Garfí et al. 2016) or the covered anaerobic lagoons for liquid manure (Flotats et al. 2016), applied in rural areas. Stable and optimal biogas production cannot be ensured with these designs, since operation parameters are difficult to control, but they have demonstrated a wide range of environmental, social and economic benefits (Gosens et al. 2013; Garfí et al. 2016).

The CSTR (continuous stirred-tank reactor) is the most widely used digester (Fig. 4a) for substrates with less than 15% total solids (TS), corresponding approximately to the maximum TS concentration that the stirring system could maintain

Table 3 Classification of anaerobic digestion reactors

Biomass growth	Biomass retention system	Flow model	Reactor or system
Suspended/disperse	None	Undefined, not stirred	Batch, or continuous/semi-continuous
			Tubular reactors
			Covered anaerobic lagoons
		Completely mixed	CSTR
		Plug flow	PF reactor
	External settler	Completely mixed	Contact reactor
	Membrane (internal or external)	Completely mixed	AnMBR
	Internal sedimentation	Completely mixed, batch	AnSBR
Suspended/granular	Sedimentation/internal separation	Upflow (plug flow)	UASB
		Plug flow/expanded bed	EGSB
		Plug flow/expanded bed, internal recirculation	IC
Biofilm	Stationary support	Plug flow/downflow	DSFF
		Plug flow/upflow	AF
	Mobile support	Fluidized bed	FB
Suspended and biofilm	Bottom: Suspended Top: Support	Plug flow, upflow	Hybrid reactor
	Multistep and separated reactors		Several combinations

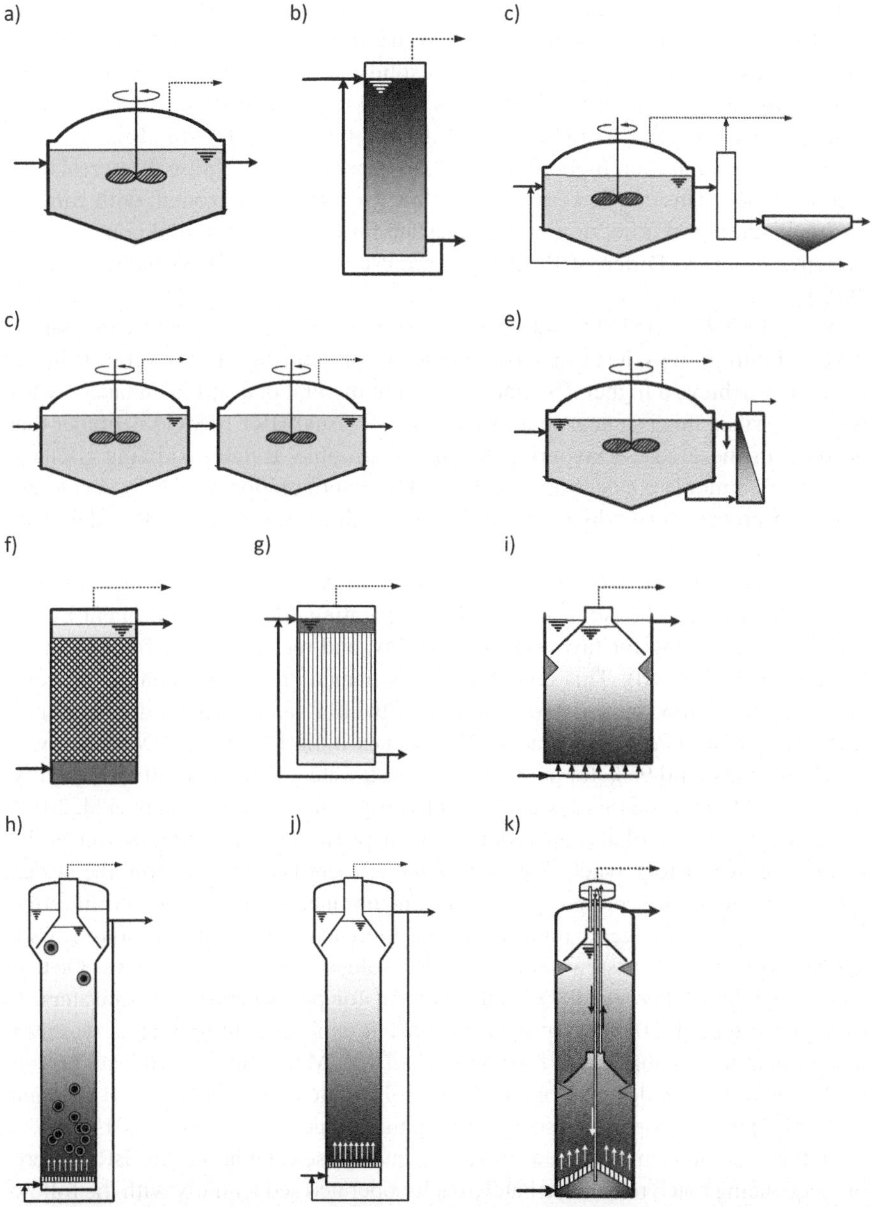

Fig. 4 Simplified schemes of the main anaerobic reactors configuration. (**a**) CSTR (continuous stirred-tank reactor), (**b**) PF (plug flow), (**c**) CSTR in series, (**d**) contact process, (**e**) AnMBR (anaerobic membrane bioreactor), (**f**) AF (anaerobic filter), (**g**) DSFF (downflow stationary fixed-film reactor), (**h**) FB (fluidized or expanded bed), (**i**) UASB (upflow anaerobic sludge blanket), (**j**) EGSB (expanded granular sludge blanket), (**k**) IC (internal circulation reactor)

without an excessive energy consumption. This could range from 30 Wh/m^3 day for low TS content (~ 1–3%TS) to 450 Wh/m^3 day for high (~ 13–15%TS), although exact values depend on the settleability of solids, volume and shape of the reactor. Usual hydraulic retention time (HRT) is 15–30 days, with same solids retention time (SRT) and organic loading rates (OLR) less than 4 kg COD/m^3 day.

For TS concentration higher than 20%, the continuous PF (plug flow) reactor is recommended. This can be vertical downflow (Fig. 4b) or horizontal, with continuous recirculation of a fraction of the digestate for ensuring an initial concentration of active biomass. Different PF designs are used for OFMSW (Vandeviere et al. 2003).

When CSTR is applied to substrates for which disintegration and hydrolysis are the rate-limiting steps, it is usual to combine two or three digesters in series (Fig. 4c) in order to achieve a higher TS concentration in the first one and to enhance hydrolysis rate, since this is a near first-order process. Usual HRT (=SRT) is higher than 50 days in these cases, favouring SAOB (syntrophic acetate-oxidizing bacteria) activity if ammonia concentration is high. This configuration confers a very stable digestion process for codigestion of manure with energy crops and/or slaughterhouse waste.

A method for increasing microbial activity in AD reactors is to increase SRT over HRT. This can be achieved in CSTR separating solids in the effluent by settling, after degassing for favouring settleability, and recirculating a fraction of the settled solids (Fig. 4d). This configuration is named "anaerobic contact process" and is applied to food industry wastewaters, with HRT values in the order of hours—few days. Nähle (1991) references OLR values until 12.9 kg COD/m^3 day with COD removals until 95% for sugar factory wastewaters. The flexibility for modifying SRT led to propose this system for biohydrogen production (Chang et al. 2011).

An improvement of the anaerobic contact process is the AnMBR (anaerobic membrane bioreactor), where liquid-digested effluent is extracted from the system through a polymeric, ceramic or metallic membrane, by micro- or ultrafiltration, and the retentate sludge is returned to the bioreactor (Fig. 4e). Although firstly applied to industrial wastewaters, with SRT values as high as 175 days, OLR of 14.6 kg COD/m^3 day and 98% COD removal for petrochemical wastewaters, as example (Lin et al. 2013), its properties make it applicable to municipal wastewaters, even at local temperature (Ozgun et al. 2013). Membrane module can be coupled to other reactor designs for enhancing solids retention (Lin et al. 2013; Ozgun et al. 2013) and can operate also at thermophilic range (Duncan et al. 2017).

Solids retention can also be achieved by internal settling in the AnSBR (anaerobic sequencing batch reactor), a batch reactor operated sequentially with the following time steps per cycle: filling, reaction (stirred), biomass settling and liquid supernatant extraction, with the appropriate settled biomass purge for maintaining the desired SRT. The successive cycles favour the enrichment of adapted biomass and a stable process (Angenent et al. 2002), only limited by the sludge settleability. Its flexibility makes this system adaptable to different substrates, such as the liquid fraction of pig manure (Massé et al. 2003) or winery wastewater (Donoso-Bravo et al. 2009), in which case Ortiz-Cabrera et al. (2018) achieved 95% total polyphenols removal.

Main reactors based on biofilm growth are the AF (anaerobic filter), the DSFF (downflow stationary fixed film) and the FB (fluidized or expanded bed) reactors. AF is an upflow reactor where biomass is attached to the support as biofilm or entrapped in the interstitial spaces (Fig. 4f). The usual problem of plugging is overcome in the downflow DSFF reactor (Fig. 4g), where support is vertically oriented, allowing until 4% TS in the influent (Kennedy and Droste 1991). Usual OLR values for the stationary biofilm reactors are 1–12 kg COD/m^3 day (Tauseef et al. 2013). In FB reactors, biofilm grows on small particles (sand, coal, resins, etc.) that are expanded or fluidized along the reactor by the upward liquid flow. Although OLR values as high as 20–30 kg COD/m^3 day were reported (Chernicharo 2007), this system is not currently applied due to operational problems (Tauseef et al. 2013).

The breakthrough in the development of high-rate AD reactors was the UASB (upflow anaerobic sludge blanket), where dense and compact granules of active biomass are retained in the reactor by settling and by a gas–liquid–solid separator device placed at the top (Fig. 4i). The system was developed by professor Lettinga and co-workers (1980) based on the observation of granules formation, for which there are different theories, related to physical, chemical and biological factors (Hulshoff Pol et al. 2004), where the interspecies H_2 transfer is favoured. Average OLR values are 4–12 kg COD/m^3 day, with HRT in the order of hours.

Improvements of the UASB system led to the development of the EGSB reactor (expanded granular sludge blanket), where granules are expanded by the upward liquid circulation (Fig. 4j), and the IC reactor (internal circulation), with two compartments, where granules and sludge that could exit the system are circulated to the bottom, allowing a complete biomass retention (Fig. 4k). Usual OLR values are between 15 and 35 kg COD/m^3 day (Tauseef et al. 2013), with maximum values that can reach 45 kg DQO/m^3 day (Van Lier 2008).

There are many combinations and variants of the previous main designs, adapted to different solid wastes and wastewaters and different operational conditions, with successful results (Tauseef et al. 2013). The technological development from the CSTR to the EGSB and IC high-rate anaerobic reactors has represented a 75-fold increase in process efficiency, in terms of treatment achieved per unit reactor volume (Van Lier 2008), which means less space and more efficiency, energy production and GHG mitigation than any aerobic biological process, stating the brilliant role to be played by the high-rate anaerobic reactors in the sanitation engineering area.

5 Anaerobic Digestion in the Context of the Circular Economy

5.1 From Waste Treatment to Resources Processing

Through the photosynthesis process photosynthetic organisms assimilate nutrients, water, inorganic carbon and solar energy in the form of organic molecules, which are the basis of biomass production and the whole trophic chain. Any solid, liquid or gaseous waste compounds from the trophic chain are derived from those initial energy and materials and, therefore, are renewable resources.

Until the beginning of the twentieth century, agricultural production depended exclusively on the natural fertility of soils and their improvement through the application of manure or mineral nitrates. With the development of the Haber-Bosch process for the production of ammonia from atmospheric N_2, and H_2 from natural gas, the synthetic production of nitrogen fertilizers made the green revolution possible since the 1960s and crops productivity increased by more than fourfold between 1900 and 2000, together with advances in breeding and crop chemical protection (Smil 2011). This author estimated a global production of about 100 MTonnes N/year for nitrogen fertilizers through the Haber-Bosch process by 2010.

Losses of nitrogen fertilizers due to volatilization, leaching, runoff or soil erosion can reduce their efficiency to 38–45% (Oenema et al. 2009). In countries with a significant weight of animal protein in the diet, produced from vegetal proteins, the global efficiency of the N in the food system can decrease below 15% (Smil 2011), livestock production being responsible for 68% of total ammonia emissions (Steinfeld et al. 2006). Only a small N fraction will be assimilated by human bodies. All these inefficiencies result in a significant release of different forms of nitrogen into the atmosphere and continental waters, as well as in livestock manure, domestic and industrial wastewaters and municipal and food industry solid waste.

Another basic nutrient for food production is phosphorus. This is a non-renewable mineral with an annual production of around 15–20 MTonnes P/year in the period 2000–2010, with the main reserves located in Morocco. At the current consumption rate, a decrease in the rate of production is estimated towards the years 2030–2040, with a production cost increase and an exhaustion into 100–150 years (Foged et al. 2012a). This nutrient is also characterized by inefficiencies in its use as fertilizer and losses along the food chain.

In a circular economy context, apart from maximizing the overall food chain efficiencies, it is necessary to recover nutrients and micronutrients from all the residual effluents, as well as their energy content. In this context, the environmental engineering area appears as strategic for future system sustainability, with changes in the objectives and nomenclature used, and moving from the concept of "treatment", which is identified as a method to reduce environmental impacts, to the concept of "processing", which is identified as a method to produce goods and services with economic value, adopting the appropriate technological strategy. I other words, to transform waste and wastewater treatment plants into new factories producing energy and valued organic compounds, recovering macro- and micronutrients, reducing dependence from fossil raw materials and contributing to the environmental quality.

5.2 Renewable Energy Production

A biogas from anaerobic digestion with 65% CH_4 and 35% CO_2 has a lower calorific value (LCV) around 6.5 kWh/Nm^3 and an upper calorific value (HCV) around 7.2 kWh/Nm^3. The simplest method to take advantage of this energy content is to produce heat or steam by a boiler, with thermal efficiencies around 90%. Combined heat and electricity production by a CHP (combined heat and power) is the most used method but with the limitation of on-site heat consumption in order to take the maximum benefit of this energy transformation. Usual electrical efficiencies of CHP units move between an average value of 31% for up to 50 kW$_e$ to an average value of 41% for 1–2.4 MWe electrical capacity (Walla and Schneeberger 2008). For these energy uses, biogas must be cleaned, decreasing the amounts of moisture, dust, H_2S, siloxanes and halogenated compounds below given values (Bailón and Hinge 2012).

Apart the classical uses of biogas for thermal or electrical energy production, the use as vehicles fuel or as natural gas substitute, after an upgrading process to produce biomethane, is gaining interest worldwide. The injection to the natural gas grid enables biomethane to be stocked and used remotely from the production site, in order to be consumed when and where the energy conversion efficiency will be higher, instead of being transformed to electricity on-site without useful and efficient recovery of the wasted heat. Upgrading consists of removing impurities and CO_2 from the biogas in order to obtain a natural gas-like quality, for which there are several applied technologies operated in large-scale plants (Beil and Beyrich 2013).

The primary energy production from around 18,000 biogas plants in Europe was 16.1 Mtoe in 2016, a 7.4-fold increase respect to 2000; the estimated electrical energy production was 62.5 TWh, which 63.7% was produced in CHP units and the rest in only electrical production units and 3.6 Mtoe consumed by thermal uses. During the same year, 497 upgrading plants injected 15.6 TWh biomethane to the natural gas grid, with 483 plants and 15.3 TWh in 9 countries (EurObserv'ER 2017). About 697 biomethane filling stations ensured the use of 160 million m^3 of biomethane as a transport fuel in 2015 (Scarlat et al. 2018). Germany is the leading country in Europe, with more than 9300 biogas plants, primary energy production close to 8 Mtoe, electrical production of 33.7 TWh (73% by CHP units) and 201 upgrading plants that injected 9.4 TWh of biomethane to the natural gas grid in 2016 (EurObserv'ER 2017). In China, 41.8 million household-scale digesters in rural areas provided 6.6 Mtoe by 2014, mainly for heating, cooking and lighting, and around 31,700 medium- and large-scale plants provided 0.75 Mtoe during the same year (Chen and Liu 2017).

Studies on biogas energy potentials in Europe indicate a value around 67 Mtoe, from which 75.5% from agriculture (34.9% from energy crops) and the rest from waste processing (Scarlat et al. 2018). In China, the energy potential from manures and crop straws was evaluated in 27.6 Mtoe, while the biogas production from these sources was 1.9% only during 2010 (Chen et al. 2012). With currently operating 2100 biogas systems in the USA, 13,000 sites have been identified with a potential of around 10 Mtoe using manure, landfill gas recovery and wastewater treatment plants, with 4.5 Mtoe from 8113 candidate farms (EPA 2014, 2018), although recent studies indicate that these potentials could be higher using industrial organic waste (Wang et al. 2018). These values indicate that biogas industry has a wide margin for growth.

5.3 *Other Roles of Anaerobic Digestion*

For years, AD was conceived as a treatment method for decreasing VS and the volume of some wastes, such as sewage sludge, and eventually to take advantage of the biogas produced, as a second objective. The biogas production was gain priority, becoming the primary objective, prompted by energy policies in several countries or because it was the only energy supply method in developing rural areas. AD can contribute to many other objectives, apart from energy production, which are presented below.

5.3.1 Greenhouse Gases Emissions (GHG) Mitigation

Poeschl et al. (2010a) reference GHG emissions values for electricity production from different renewable sources, indicating −414 g $CO_{2\,eq}$/kWh for biogas followed by +23 g $CO_{2\,eq}$/kWh for wind energy. The value of GHG emission for biogas is highly dependent on many factors and very sensible to the digestate management (Pardo et al. 2017), which should be stored in a gas-tight tank in order to avoid residual CH_4 and N_2O emissions. This is illustrated in Fig. 5 with data calculated by Giuntoli et al. (2015) for the AD of manure, corn silage and biowaste, for two options: electricity production (25% CHP electrical efficiency; CHP provides electricity and heat required by the plant) or biogas upgrading for injection of biomethane to the natural gas grid (electrical grid provides electricity for the plant, and heat is provided by a biogas boiler). Energy balance is very sensitive to the transport of digestate after storage, not included by Giuntoli et al. (2015), obtaining a negative balance for a biogas plant processing cattle manure when transport distance is higher than 22 km (Poeschl et al. 2010b), as example. Savings higher than 100% for wet manure are explained by the avoided CH_4 emission during storage before field spreading as fertilizer, suggesting that manure storages should be covered with CH_4 collection.

Although higher methane yields for maize silage, the relative low GHG savings are mainly due to equivalent emissions during cultivation and harvesting. Nevertheless, the environmental assessment done by Poeschl et al. (2012a, b) indicated a better environmental performance of corn silage than other energy crops.

Although biogas production contributes to GHG emissions mitigation, final values depend on the multitude of parameters that are often site-specific. Therefore, projects must include the whole supply chain, from the resource production to the

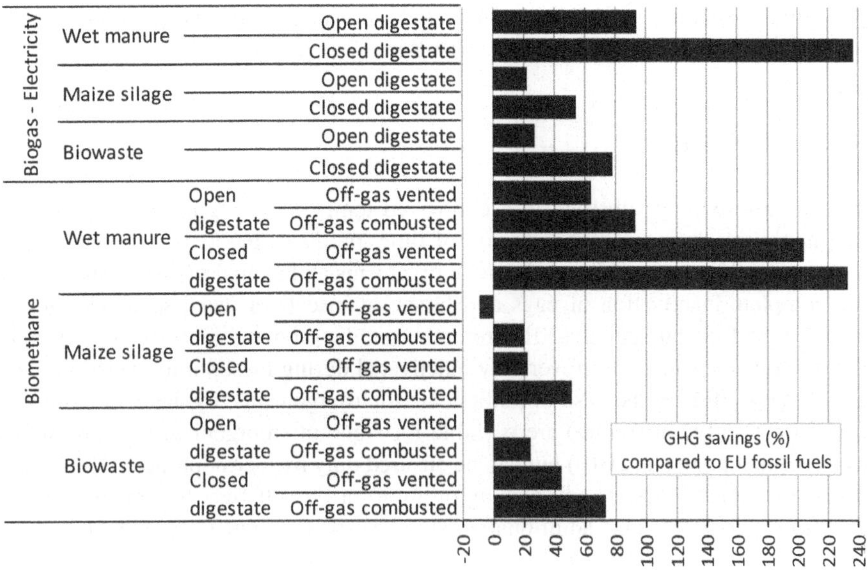

Fig. 5 GHG savings for the most representative biogas and biomethane pathways, compared to European electricity (186 g $CO_{2\,eq}/MJ_{el}$) or natural gas (72 g $CO_{2\,eq}/MJ_{NG}$). Values lower than 0% indicate systems which emit larger amounts of GHG than the compared fossil fuel. Data from Giuntoli et al. (2015)

final transformation and use of the processed products. The controversial GHG saving results for some energy crops, apart from other factors (land use, biodiversity decrease, food competing crops, etc.), explain the limits imposed in some countries, such as France where the maximum share allowed in the biogas plant feeding is 15% (ADEME 2018).

5.3.2 Better Fertilizer Value of Digestate and Odours Decrease

The anaerobic process improves the quality of manure and other organic waste for subsequent application to the soil: partial or complete stabilization of organic matter, reduced oxygen demand that avoid anoxic soil areas, mineralization, homogenization, pH increase and reduction of particles size and viscosity, which improves the infiltration in the soil when applied as fertilizer (Al Seadi et al. 2008). The concentration of ammonia nitrogen in the digested material can increase with respect to the influent, which favours the N availability to the plants if digestate is incorporated into the soil shortly after surface application to avoid volatilization (Möller and Müller 2012).

After AD there is a significant reduction of odoriferous substances (organic acids, phenolic compounds and other organic volatile substances) with respect to the initial substrates (Lukehurst et al. 2010; Powers et al. 1999), which avoids nuisances when digestate is used in agriculture.

5.3.3 Sanitation

The fact of keeping substrate in a controlled temperature regime (mesophilic or thermophilic), during a certain period of time, produces hygienization due to the decrease of thermo-sensitive pathogenic organisms and animal parasites, as well as the complete inactivation of eggs and larvae of insects or weed seeds (Johansen et al. 2013). Lukehurst et al. (2010) reference studies concluding that common fungal diseases of plants are irreversibly inactivated during mesophilic digestion with 25–30 days HRT. In the case of pathogenic microorganisms, hygienization conditions (time and temperature) are possible for some microorganisms at mesophilic range (Lukehurst et al. 2010) but can be ensured only by thermophilic AD, an additional controlled digestate composting or a pre- or post-thermal treatment. Viruses are more persistent, and additional treatments are required (Turner and Burton 1997).

5.3.4 Enhancement of Nutrient Recovery Processes

The technological strategies leading to the recovery of nitrogen and phosphorus are favoured by the previous AD due to mineralization (Flotats et al. 2012). The key to these strategies is to obtain products allowing its economic valorization (Foged et al. 2012b). Two important factors determining the optimal treatment train configuration for nutrients recovery are (a) local fertilizer markets and (b) physico-chemical characteristics of the digestate to be treated (Vaneeckhaute et al. 2018).

Phosphorous Recovery as Struvite

Struvite precipitation ($MgNH_4PO_4$ $6H_20$) is a natural process that takes place in anaerobic digesters when the concentrations of Mg^{2+}, NH_4^+ and PO_4^{3-} ions exceed the solubility product constant, causing obstructions and hydraulic problems and reducing the useful volume of digestion. Struvite is poorly soluble in water and is considered a slow-release mineral fertilizer with economic value in the fertilizers market.

The application of AD previously to struvite precipitation has the advantages of mineralization, increasing the concentration of ammonium and phosphate ions, and reducing the content of organic matter, allowing the production of large and clean struvite crystals (Cerrillo et al. 2015). Magnesium is usually found in low concentrations in manure or other organic waste, so the addition of $Mg(OH)_2$ or MgO is necessary. In order to reduce the cost of the reagents, Romero-Güiza et al. (2015)

tested the use of low-grade magnesium oxide, a by-product of magnesite calcination ($MgCO_3$), with satisfactory results.

Ammonia Recovery by Stripping and Absorption

The objective of the process is to desorb ammonia in a column by means of a countercurrent air or steam (stripping column) and its subsequent recovery in a column with an absorbing acid stream (absorption column), usually sulphuric acid. $N\text{-}NH_4^+$ is in equilibrium with NH_3, which is the gas that is entrained in the process. This equilibrium moves to NH_3 with an increase of temperature and/or pH. One objective of the process is to obtain a quality product that can be valued in the mineral fertilizers industry or other industrial uses. Therefore, the product contamination by organic matter, volatilized and entrained along with ammonia, is the main limiting factor.

Bonmatí and Flotats (2003a) compared the stripping and absorption process applied to fresh or digested pig slurry. They obtained crystals of ammonium sulphate almost free of organic matter from the digested samples, while for fresh manure the organic matter contamination did not favour crystallization. In the first case, higher desorption yields were obtained at 80 °C without modifying the pH, due to the lower alkalinity and different pH dynamics. Laureni et al. (2013) corroborated the need for prior AD to obtain a marketable product.

Ammonia Recovery with Hydrophobic Membranes

A hydrophobic and gas-permeable membrane acts as a barrier preventing direct contact and mixing of two aqueous phases (ammonia-rich feed and absorption acidic solution), separated by gas-filled pores of the membrane. The driving force for the transfer is the difference in ammonia partial pressure between digestate and the absorption solution (Lauterböck et al. 2012). Usual membranes used are fluoropolymers, such as expanded polytetrafluoroethylene (e-PTFE), but tests have been performed with polypropylene, acrylic copolymer or nylon (Simioni et al. 2011; Lauterböck et al. 2013), showing different mechanical properties and allowing design customized membranes. This method has been applied to the ammonia recovery from pig manure (García-González and Vanotti 2015) and from digestates (Wäeger-Baumann and Fuchs 2012).

The application to the continuous ammonia extraction and recovery from inside an anaerobic digester has the advantage of reducing the inhibitory effect of ammonia during the process, achieving a more stable operation and higher biogas productions for N-rich substrates. Very satisfactory results have been achieved during the AD of slaughterhouse waste (Lauterböck et al. 2014), gelatine (Ruiz-Sánchez et al. 2017) or poultry manure (Bayrakdar et al. 2018).

Thermal Concentration by Vacuum Evaporation

The objectives of the thermal concentration are the separation of water content by evaporation and to obtain a concentrated product maintaining the initial nutrients load. In order to avoid atmospheric pollution, vacuum evaporation at moderate temperatures (50–60 °C) is usually chosen, with recovery of condensates, concentrating all the nutrients for its transport and commercialization. Chiumenti et al. (2013) concluded that the thermal energy produced by a CHP unit would be sufficient to treat the entire production of digestate of a biogas plant by vacuum evaporation.

Bonmatí and Flotats (2003b) concluded that the removal of organic matter by means of a previous AD process, and an adjustment to acidic pH of the liquid fraction, allowed a reduction of the acid requirements and two orders of magnitude decrease of volatile organic matter concentration in the condensates, compared to the application of the same process to fresh pig manure. A full-scale facility with a combination of these processes, with drying of the concentrate, described by Foged et al. (2012c), concentrates around 95% of the initial total N in the pelletized final product, as well as all the phosphorous and potassium. Pellet production from digestate can be a profitable use of wasted heat, depending on dry matter content (Nagy et al. 2018). Stopp et al. (2017) propose a sequential batch evaporation of digestates, without pH modification, for recovering ammonia in the process condensates.

5.3.5 Optimization of VFA Production and Recovery

Volatile fatty acids (VFA), intermediates of the AD process, are valuable chemicals, which can be used as building blocks for the production of medium- and long-chain fatty acids, alcohols and polyhydroxyalkanoates (PHA), among others (Bastidas-Oyanedel et al. 2015). Acetic acid is used in the production of polymers, polyethylene terephthalate and solvents, and butyric acid is used in cosmetic, pharmaceutical and food industries (Jones et al. 2017). VFA recovery could be more valuable economically than methane production (Acosta and De Vrieze 2018; Bastidas-Oyanedel and Schmidt 2018).

The key is to maximize VFA production and minimize CH_4 generation, through the enhancement of hydrolysis and the control of pH, temperature and retention time (Cagnetta et al. 2016; Zhou et al. 2018). The method proposed by Acosta and De Vrieze (2018) for optimizing VFA production while recovering the remaining CH_4 potential is the temperature-phased AD (TPAD) design. This is constituted by two serial anaerobic digesters (Fig. 4c) where the first stage is operated at thermophilic conditions (50–70 °C), low HRT of 2–5 days and high OLR of 10–15 kg COD/m^3 day. In contrast, the second or mesophilic stage is operated at high HRT of 10–20 days and low OLR of 0.5–5 kg COD/m^3 day. In the first reactor, the hydrolysis is enhanced (Ge et al. 2011), and with the HRT below the generation time of methanogens, in order to avoid its growth, VFA and H_2 accumulate and as shown in Sect. 5.2, reactions depending on low partial pressure of H_2 are inhibited.

The integrated production of H_2 and VFA in the first step overcomes the limitations of schemes trying to produce biohydrogen only (Guwy et al. 2011). Also, the thermophilic stage could be operated in order to optimize the production of lactic acid from carbohydrates, with interest in the production of biodegradable plastics (Ahring et al. 2016). The second mesophilic reactor operates as a typical anaerobic digester with the objectives to optimize methane production and organic matter stabilization. Nutrient recovery techniques can be applied between the two reactors or in the second one (Acosta and de Vrieze 2018).

Production optimization of VFA (Yousuf et al. 2016; Nzeteu et al. 2018) or lactic acid (Bonk et al. 2017; Yousuf et al. 2018) has been studied at mesophilic temperature range too, using different reactor configurations.

VFA recovery can be done using different methods, such as electrodialysis (Jones et al. 2015), liquid-liquid extraction with organic solvents (Ijmker et al. 2014), adsorption on ion-exchange resins (Rebecchi et al. 2016; Yousuf et al. 2016) or gas-permeable membranes (Aydin et al. 2018), among others. In a review about VFA recovery methods, Singhania et al. (2013) concluded that VFAs in the fermented broth could be directly used for biolipids production, without previous extraction, and Spirito et al. (2014) propose some chemical chain elongation processes in the anaerobic reactor.

6 Economy of Energy Production from Biogas

Considering the broad environmental advantages of AD, to approach the profitability study of a system with only energy market prices limits the overall analysis. Other considerations must be taken into account that are site-specific. Odours abatement and improved fertilizer quality (Flotats and Gibert 2002) or a decrease on the overall manure management costs (Hjort-Gregersen 2002) could be value criteria for adopting a biogas plant, although it has extremely low internal return rates.

6.1 Electricity Production

Specific investment costs depend primarily on plant scale, or electrical production capacity, and secondarily on biogas yield of the substrate, apart from site-specific conditions such as resistance of the land for the construction or length of the electric line to evacuate the electricity produced. While specific investment cost for a plant of 200 kWe is in the range 4000–6000 €/kWe, for 1 MWe, the estimated range is 2500–3600 €/kWe (Flotats et al. 2016). A decrease on investment cost is appreciated when the biogas potential of the substrate increases, which is justified by lower reactor volume and lower input/output storage capacity per unit of energy produced (Flotats et al. 2016; IRENA 2018). Biogas yield appears as one of the key variables when analysing the economy of biogas plants.

Figure 6 shows electricity production costs from different authors, which can be compared with the EU-28 average electricity price in the period 2013–2017, without taxes, for households and non-households. Walla and Schneeberger (2008) developed a calculation model for energy crops (maize), supposing no sales of recovered heat from the CHP unit and considering no transport cost of maize and digestate or with transport cost for different substrate availabilities, showing the case of 20% availability in Fig. 6, increasing this cost for lower availabilities due to larger transport distances for a given substrate amount or specific energy output. Data from Flotats and Sarquella (2008) and from Hartmann et al. (2012) came from case studies of biogas plant models, while data from Adler et al. (2014) are real biogas plant examples, corresponding both to codigestion plants with variable amounts of manure, organic industrial waste, biowaste and energy crops. For small biogas plants, the main cost corresponds to capital cost (40–50%), while the substrate cost (production in the case of energy crops, transport and management of digestate as fertilizer for different kinds of biowaste) increases its share (>50%) for large biogas plants. Operational costs move between 14 and 23% of the total production cost (Adler et al. 2014).

Since transport and management cost of substrates and digestate increases when increasing the plant size, the optimal size can be estimated as the minimum of the curve sum of curves of cost without transport and the cost of transport and management, as done by Walla and Schneeberger (2008). The higher the substrate and transport costs, the smaller the optimal plant size. IRENA (2018) noted that transporting the amount of required feedstock in the case of waste or agriculture residues is a challenge in practice, since the required amounts are rarely available in the influence area of a plant. In an analysis of these kinds of curves, Flotats (2018)

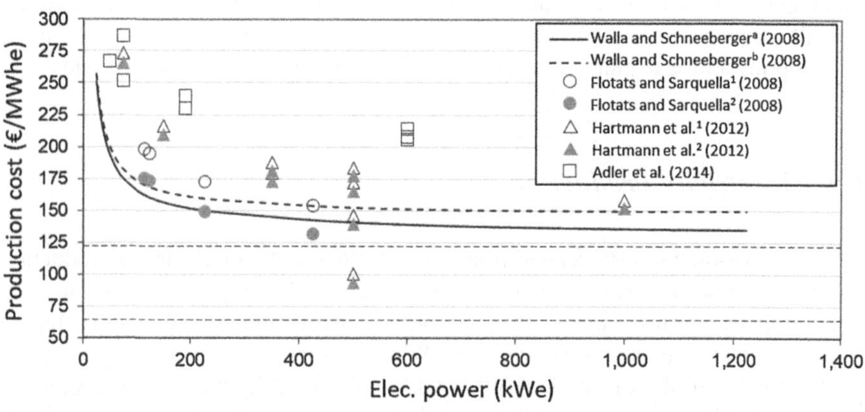

Fig. 6 Electricity production costs by CHP from biogas, estimated by the indicated authors. (**a**) Without substrate/digestate transport cost; (**b**) with substrate/digestate transport cost; (*1*) without heat sales; (*2*) with heat sales. Horizontal shaded lines are the EU-28 average electricity price 2013–2017, excluding taxes, for upper, households, and lower, non-households (EUROSTAT 2018a)

found that subsidies to investment and bonus to energy production decrease the curve without transport and favour small optimal plants.

Sales of recovered heat from CHP decrease the electricity production cost, as can be seen in Fig. 6. Flotats and Sarquella (2008) found that the studied biogas plant models offered an interesting internal return rate (IRR) in the case of heat sales only. Residual heat can be used for substituting fossil fuels in a farm-scale plant or for supplying a district heating in large-scale plants, which is applicable in cold weather but difficult in hot climate areas.

6.2 *Biomethane Production Cost*

Figure 7 shows total production cost intervals (minimum and maximum) for different plant size and substrate type, with data from different authors. Horizontal shaded lines are the EU-28 average natural gas prices in the period 2013–2017 for household and non-household users. The different hatches for every interval are distributed proportionally to the share of biogas production cost, upgrading cost and

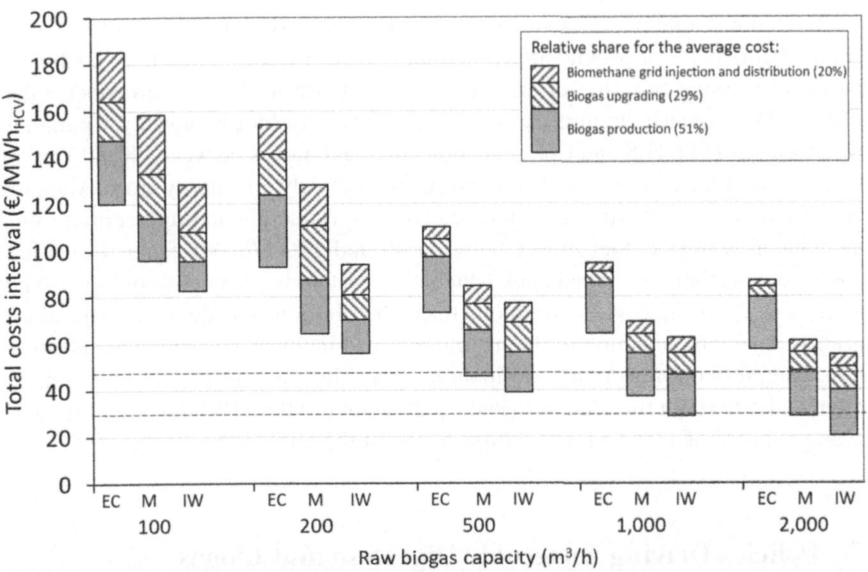

Fig. 7 Estimated intervals (minimum and maximum) of total production costs for biomethane supply through the natural gas grid from energy crops (EC), manure (M) and industrial organic waste (IW), for the indicated capacity. Hatched areas indicate the fraction of the biogas production, upgrading and injection/distribution cost for the average value of the corresponding interval (example values for M and 500 m^3/h capacity). Intervals estimation based on data from Thrän et al. (2014), Adler et al. (2014) and IRENA (2018). Horizontal shaded lines are the EU-28 average natural gas price 2013–2017, excluding taxes, for upper, households, and lower, non-households (EUROSTAT 2018b)

injection to the natural gas grid into the total costs, for the average interval cost. Values in Fig. 7 correspond to the shares for 500 m^3/h biogas processing capacity from manure. In order to uniform units from the different authors, methane HCV of 11.056 kWh/Nm^3 and 0.87 €/US$ values have been adopted. With the purpose to compare with data from Fig. 6, 100 m^3/h of biogas with 65% CH_4 corresponds to an electrical capacity of 231 kWe, and 500 m^3/h corresponds to a capacity of 1.3 MWe, taking into account the electrical CHP efficiencies proposed by Walla and Schneeberger (2008).

Biogas production cost includes capital, operational, substrate and auxiliary energy costs. Substrate costs are usually higher for energy crops than for manure and industrial organic waste (IRENA 2018). In the range of 100–2000 m^3/h biogas capacity, production costs are in the ranges of 42–75% for energy crops, 29–60% for manure and 28–56% for industrial waste of the total production costs for the average cost of every interval.

In the case of organic waste, its use can even be an income for the plant, paid by the waste producer, but the subsequent use of digestate as fertilizer can lead to an appreciable cost if not enough agricultural land is available in the surrounding area. Similar considerations can be done for manure digestate management in nutrient surplus areas, making it necessary to consider a high uncertainty degree for data of Fig. 7 for manure and industrial waste plants with capacities higher than 500 m^3/h.

Biomethane upgrading costs are not significantly different for different substrates, being between around 38 €/MWh for 100 m^3/h and 10 €/MWh for 2000 m^3/h biogas processing capacity for the average costs. IRENA (2018) references a value of 83 €/MWh for an upgrading capacity of 20 m^3/h of raw biogas capacity. Upgrading cost includes H_2O, H_2S and CO_2 reduction costs and moves between 28 and 15% of the total costs for 100 and 2000 m^3/h biogas, respectively, for energy crops. Maniatis et al. (2017) conclude that the reference cost of production and upgrading to biomethane as transport fuel moves between 40 and 120 €/MWh using wastes and energy crops from abandoned land, which is consistent with intervals of Fig. 7. This reference cost is increased when biomethane distribution is made by the natural gas grid. Biomethane injection to the natural gas grid includes measurement, odorization, compression, injection and pipeline costs. This cost moves between a minimum of 4.5 €/MWh for 2000 m^3/h and a maximum of 60 €/MWh for 100 m^3/h, with average shares of 13% and 39%, respectively, of the total costs.

7 Policies Driving Anaerobic Digestion and Biogas Production

The comparison of production cost with the electricity prices in Fig. 6 or with natural gas prices in Fig. 7, without taxes, results in the need of some kind of subsidies, feed-in tariffs, tax exemptions or green certificates and specific governments policies to overcome barriers and to exploit the benefits of AD. Substitution of natural gas by biomethane (Fig. 7) could only be competitive in some cases for manure and industrial organic waste for capacities higher than 500 m^3/h of biogas, situations where costs are characterized by a high uncertainty degree owing to the cost of digestate management.

Stakeholders' perception about manure processing techniques, and more specifically on AD, differs among countries, with visions depending on previous policies applied and results. This is the case of Spain, where policies promoting agro-industrial biogas have been scarce and farmers do consider it as a less feasible process than in other countries (Hou et al. 2018). These authors also found that the three main barriers for manure processing are (a) lack of capital for investments; (b) high cost of processing and (c) benefits return too long.

Governmental policies favouring biogas plants are key factors for the success of this industry, taking into account the investment and operational costs. In an analysis of the government policies that are capable of promoting the implementation of AD, Edwards et al. (2015) found that these are related to (1) the mitigation of climate change, (2) security in energy supply, (3) waste management and (4) regional development. These policies must be translated into economic incentives and/or into medium- and long-term stable action framework, in order to provide security for investments. To act in a coordinated and harmonized form in promoting renewable energy supply, to process organic waste for resources recovery and to promote the economic development of rural areas, thus avoiding depopulation and migration to large cities, are actions directly affecting GHG mitigation (Edenhofer et al. 2014).

Analysing the evolution of biogas plants in different European countries, the European Biogas Association (EBA 2016) concluded that financial support schemes and national legal framework play a pivotal role on the sector's development. Edwards et al. (2015) concluded that financial incentives targeted at small-medium AD plants, less than 500 kWe, have a significant positive correlation with the growth of AD industry and the number of plants, using data of the evolution of feed-in tariffs (FITs) in Germany in the period 2005–2013. In the same way, Walla and Schneeberger (2008) found that the optimal size AD plant was 250 kWe considering the economic incentives in Austria, and Zema (2017) found 300 kWe in Italy. For large plants in the agricultural sector, main limiting factors are logistics of substrates supply, or digestate management, and social aspects related to cooperation values (Flotats et al. 2009), which are site-specific and influenced by local history and development. Currently, China's policy and subsidies encourage large biogas plants, motivated by the agriculture industrialization (Chen and Liu 2017).

Foged et al. (2012b) identified different by-products from manure processing, mainly digestate and derived liquid nutrients concentrates or dried and pelletized solids, which were quite "new" for the market, suffering from a lack of infrastructure such as a standardized classification, legal requirements and price statistics. New EU legislative package related to circular economy and new fertilizers production from waste, intended to create an efficient system of control, labelling and traceability to ensure quality and safety (EP 2017), is thought to be a necessary tool for valorizing by-products obtained from digestate processing.

8 Final Remarks

AD has demonstrated to be a flexible technology with a variety of reactor designs adapted to many situations, transforming liquid and solid residual organic matter to valuable intermediates such as carboxylates, which can be recovered, and finally to biogas, which can be used for the production of heat and electricity or upgraded to biomethane. Biomethane can be used as a vehicle fuel or injected to the natural gas grid, allowing storage, transport and consumption where users can obtain the maximum benefit, as energy or as a source for chemicals production (Buelens et al. 2016). Additionally, AD enhances all the nutrients recovering processes owing to mineralization.

The focus on biogas for energy as single main product should be diversified towards creating multiple products and using biogas optimally through innovative solutions (Pfau et al. 2017). Current policies favour biogas production as an energy source, for biomethane or electricity. The potential of AD, however, exceeds the stabilization of organic waste and recovery of energy. AD should evolve from a provider of renewable energy to a beacon of bio-based chemicals and recovered nutrients, thus maximizing its potential (Acosta and De Vrieze 2018).

Probably the evolution of fossil fuel prices and the derived matter resource costs (such as ammonia) will boost the AD industry in the future, but in the meantime this industry depends on governments' vision and policies, which should prepare the transition to a new energy and a bio-based circular economy paradigm. As has been shown in the present chapter, anaerobic digestion is a key technology in this new paradigm.

References

Acosta N, De Vrieze J (2018) Anaerobic digestion as key technology in the bio-based economy. In: Stams A, Sousa D (eds) Biogenesis of hydrocarbons. Handbook of hydrocarbon and lipid microbiology. Springer, Cham

ADEME (2018) A 100% renewable gas mix in 2050? Study summary Joint study of ADEME, GrDF and GRTgaz, Angers (France), February 2018. http://www.ademe.fr/en/a-100-renewable-gas-mix-in-2050. Accessed 4 Apr 2018

Adler P, Billig E, Brosowski A, Daniel-Gromke J, Falke I, Fischer E, Grope J, Holzhammer U, Postel J, Schnutenhaus J, Stecher K, Szomszed G, Trommler M, Urban W (2014) Leitfaden Biogasaufbereitung und–einspeisung (Guideline biogas treatment and feeding). Fachagentur Nachwachsende Rohstoffe e.V. (FNR), Gulzow-Pruzen

Affes R, Palatsi J, Flotats X, Carrère H, Steyer JP, Battimelli A (2013) Saponification pretreatment and solids recirculation as a new anaerobic process for the treatment of slaughterhouse waste. Bioresour Technol 131:460–467

Ahring BK, Biswas R, Ahamed A, Teller PJ, Uellendahl H (2015) Making lignin accessible for anaerobic digestion by wet-explosion pretreatment. Bioresour Technol 175:182–188

Ahring BK, Traverso JJ, Murali N, Srinivas K (2016) Continuous fermentation of clarified corn Stover hydrolysate for the production of lactic acid at high yield and productivity. Biochem Eng J 109:162–169

Al Seadi T (2001) Good practice in quality management of AD residues from biogas production. Report made for the International Energy Agency, Task 24-Energy from Biological Conversion of Organic Waste. IEA Bioenergy and AEA Technology Environment, Oxfordshire, United Kingdom

Al Seadi T, Rutz D, Prassl H, Köttner M, Finsterwalder T, Volk S, Janssen R (2008) Biogas handbook. University of Southern Denmark, Esbjerg. http://www.lemvigbiogas.com/BiogasHandbook.pdf. Accessed 20 Mar 2018

Angelidaki I, Ellegaard L (2003) Codigestion of manure and organic wastes in centralized biogas plants; status and future trends. Appl Biochem Biotechnol 109:95–105

Angenent LT, Sung S, Raskin L (2002) Methanogenic population dynamics during startup of a full-scale anaerobic sequencing batch reactor treating swine waste. Water Res 36:4648–4654

Aydin S, Yesil H, Tugtas AE (2018) Recovery of mixed volatile fatty acids from anaerobically fermented organic wastes by vapor permeation membrane contactors. Bioresour Technol 250:548–555

Baba Y, Tada C, Watanabe R, Fukuda Y, Chida N, Nakai Y (2013) Anaerobic digestion of crude glycerol from biodiesel manufacturing using a large-scale pilot plant: methane production and application of digested sludge as fertilizer. Bioresour Technol 140:342–348

Bailón L, Hinge J (2012) Biogas and bio-syngas upgrading. Danish Technological Institute, Taastrup

Bastidas-Oyanedel JR, Schmidt JE (2018) Increasing profits in food waste biorefinery—a techno-economic analysis. Energies 11:1551

Bastidas-Oyanedel JR, Mohd-Zaki Z, Zeng RJ, Bernet N, Pratt S, Steyer JP, Batstone DJ (2012) Gas controlled hydrogen fermentation. Bioresour Technol 110:503–509

Bastidas-Oyanedel JR, Bonk F, Thomsen MH, Schmidt JE (2015) Dark fermentation biorefinery in the present and future (bio)chemical industry. Rev Environ Sci Biotechnol 14:473–498

Batstone DJ, Keller J, Angelidaki I, Kalyuzhnyi SV, Pavlostathis SG, Rozzi A, Sanders WTM, Siegrist H, Vavilin VA (2002) Anaerobic digestion model no.1 (ADM1). Scientific and Technical Report No. 13. IWA, London

Bayrakdar A, Sürmeli RÖ, Çalli B (2018) Anaerobic digestion of chicken manure by a leach-bed process coupled with side-stream membrane ammonia separation. Bioresour Technol 258:41–47

Beil M, Beyrich W (2013) Biogas upgrading to biomethane. In: Wellinger A, Murphy J, Baxter D (eds) The biogas handbook: science, production and applications, Woodhead publishing series in energy number, vol 52. Woodhead, Cambridge, pp 342–377

Biswas R, Uellendahl H, Ahring BK (2015) Wet explosion: a universal and eficient pre-treatment process for lignocellulosic biorefineries. Bioenergy Res 8:1101–1116

Bonk F, Bastidas-Oyanedel JR, Yousef AF, Schmidt JE (2017) Exploring the selective lactic acid production from food waste in uncontrolled pH mixed culture fermentations using different reactor configurations. Bioresour Technol 238:416–424

Bonmatí A, Flotats X (2003a) Air stripping of ammonia from pig slurry: characterization and feasibility as a pre- or post-treatment to mesophilic anaerobic digestion. Waste Manag 23:261–272

Bonmatí A, Flotats X (2003b) Pig slurry concentration by vacuum evaporation: influence of previous mesophilic anaerobic digestion process. J Air Waste Manage Assoc 53:21–31

Bonmatí A, Flotats X, Mateu L, Campos E (2001) Study of thermal hydrolysis as a pre-treatment to mesophilic anaerobic digestion of pig slurry. Water Sci Technol 44:109–116

Buelens LC, Galvita VV, Poelman H, Detavernier C, Marin GB (2016) Super-dry reforming of methane intensifies CO2 utilization via Le Chatelier's principle. Science 354:449–452

Cagnetta C, Coma M, Vlaeminck SE, Rabaey K (2016) Production of carboxylates from high rate activated sludge through fermentation. Bioresour Technol 217:165–172

Carballa M, Duran C, Hospido A (2011) Should we pretreat solid waste prior to anaerobic digestion? An assessment of its environmental cost. Environ Sci Technol 45:10306–10314

Carlson M, Lagerkvist A, Morgan-Sagastume F (2012) The effects of substrate pre-treatment on anaerobic digestion systems: a review. Waste Manag 32:1634–1650

Cavaleiro AJ, Pereira MA, Guedes AP, Stams AJM, Alves MA, Sousa DZ (2016) Conversion of Cn-unsaturated into Cn-2-saturated LCFA can occur uncoupled from methanogenesis in anaerobic bioreactors. Environ Sci Technol 50:3082–3090

Cecchi F, Traverso P, Pavan P, Bolzonella D, Innocenti L (2003) Characteristics of the OFMSW and behaviour of the anaerobic digestion process. In: Mata-Alvarez J (ed) Biomethanization of the organic fraction of municipal solid wastes. IWA, London, pp 141–179

Cerrillo M, Palatsi J, Comas J, Vicens J, Bonmatí A (2015) Struvite precipitation as a technology to be integrated in a manure anaerobic digestion treatment plant—removal efficiency, crystal characterization and agricultural assessment. J Chem Technol Biotechnol 90:1135–1143

Chang S, Li J, Liu F (2011) Continuous biohydrogen production from diluted molasses in an anaerobic contact reactor. Front Environ Sci Eng China 5:140–148

Chen YR, Hashimoto AG (1978) Kinetics of methane fermentation. Biotechnol Bioeng Symp 8:269–282

Chen Q, Liu T (2017) Biogas system in rural China: upgrading from decentralized to centralized? Renew Sustain Energy Rev 78:933–944

Chen Y, Cheng JJ, Creamer KS (2008) Inhibition of anaerobic digestion process: a review. Bioresour Technol 99:4044–4064

Chen L, Zhao L, Ren C, Wang F (2012) The progress and prospects of rural biogas production in China. Energy Policy 51:58–63

Chernicharo CAL (2007) Anaerobic reactors, Biological wastewater treatment series, vol 4. IWA, London

Chiumenti A, da Borso F, Chiumenti R, Teri F, Segantin P (2013) Treatment of digestate from a co-digestion biogas plant by means of vacuum evaporation: tests for process optimization and environmental sustainability. Waste Manag 33:1339–1344

De Vrieze J, Smet D, Klok J, Colsen J, Angenent LT, Vlaeminck SE (2016) Thermophilic sludge digestion improves energy balance and nutrient recovery potential in full-scale municipal wastewater treatment plants. Bioresour Technol 218:1237–1245

Deublein D, Steinhauser A (2008) Biogas from waste and renewable resources: an introduction. Wiley-VCH editors Verlag GmbH, Weinheim

Domenech PL, Flotats X (1997) A simplified mathematical model for an upflow anaerobic fixed film reactor under transient loading. Hung J Ind Chem 25:315–320. http://hdl.handle.net/2117/6744. Accessed 6 Sept 2018

Donoso-Bravo A, Rosenkranz F, Valdivia V, Torrijos M, Ruiz-Filippi G, Chamy R (2009) Anaerobic sequencing batch reactor as an alternative for the biological treatment of wine distillery effluents. Water Sci Technol 60:1155–1160

Duncan J, Bokhary A, Fatehi P, Kong F, Lin H, Liao B (2017) Thermophilic membrane bioreactors: a review. Bioresour Technol 243:1180–1193

EBA (2016) Biomethane in transport. European Biogas Association, Brussels

Edenhofer O, Pichs-Madruga R, Sokona Y, Kadner S, Minx JC, Brunner S, Agrawala S, Baiocchi G, Bashmakov IA, Blanco G, Broome J, Bruckner T, Bustamante M, Clarke L, Conte Grand M, Creutzig F, Cruz-Núñez X, Dhakal S, Dubash NK, Eickemeier P, Farahani E, Fischedick M, Fleurbaey M, Gerlagh R, Gómez-Echeverri L, Gupta S, Harnisch J, Jiang K, Jotzo F, Kartha S, Klasen S, Kolstad C, Krey V, Kunreuther H, Lucon O, Masera O, Mulugetta Y, Norgaard RB, Patt A, Ravindranath NH, Riahi K, Roy J, Sagar A, Schaeffer R, Schlömer S, Seto KC, Seyboth K, Sims R, Smith P, Somanathan E, Stavins R, von Stechow C, Sterner T, Sugiyama T, Suh S, Ürge-Vorsatz D, Urama K, Venables A, Victor DG, Weber E, Zhou D, Zou J, Zwickel T (2014) Technical summary. In: Edenhofer O, Pichs-Madruga R, Sokona Y, Farahani E, Kadner S, Seyboth K, Adler A, Baum I, Brunner S, Eickemeier P, Kriemann B, Savolainen J, Schlömer S, von Stechow C, Zwickel T, Minx JC (eds) Climate change 2014: mitigation of climate change. Contribution of Working Group III to the Fifth Assessment Report of the Intergovernmental Panel on Climate Change. Cambridge University Press, Cambridge and New York

Edwards J, Othman M, Burn S (2015) A review of policy drivers and barriers for the use of anaerobic digestion in Europe, the United States and Australia. Renew Sustain Energy Rev 52:815–828

Edwiges T, Mayer B, Frare L, Lins L, Triolo JM, Flotats X, Sarolli M (2018) Influence of chemical composition on biochemical methane potential of fruit and vegetable waste. Waste Manag 71:618–625

EPA (2014) Biogas opportunities roadmap. Voluntary Actions to Reduce Methane Emissions and Increase Energy Independence Joint USDA, EPA and DOE Report, August 2014. https://www.epa.gov/sites/production/files/2015-12/documents/biogas-roadmap.pdf. Accessed 16 Jul 2018

EPA (2018) Market opportunities for biogas recovery systems at U.S. livestock facilities. AgSTAR program, Report EPA-430-R-18-006, June 2018

EurObserv'ER (2017) Biogas Barometer. https://www.eurobserv-er.org/biogas-barometer-2017/. Accessed 2 Apr 2018

European Parliament (2017) Report on the proposal for a regulation of the European Parliament and of the Council laying down rules on the making available on the market of CE marked fertilising products and amending regulations (EC) no 1069/2009 and (EC) no 1107/2009. http://www.europarl.europa.eu/sides/getDoc.do?type=REPORT&reference=A8-2017-0270&language=EN. Accessed 12 Apr 2018

EUROSTAT (2018a) Electricity price statistics. http://ec.europa.eu/eurostat/statistics-explained/index.php?title=Electricity_price_statistics. Accessed 24 July 2018

EUROSTAT (2018b) Natural gas price statistics. http://ec.europa.eu/eurostat/statistics-explained/index.php/Natural_gas_price_statistics. Accessed 23 July 2018

Fernández B, Poirrier P, Chamy R (2001) Effect of inoculum-substrate ratio on the start-up of solid waste anaerobic digesters. Water Sci Technol 44:103–108

Flotats X (2018) Implantación de la digestión anaerobia en el sector agropecuario (Implementation of anaerobic digestion in the agricultural sector). In: Chiva S, Berlanga JG, Martínez R, Climent J (eds) Depuración de aguas residuales: digestión anaerobia. Publicacions de la Universitat Jaume I, Castellón, pp 44–66. https://doi.org/10.6035/UJI.FACSA.2018.2

Flotats X, Gibert V (2002) Mas el Cros biogas plant. Evaluation of 18 years in operation. In: Kalyuzhnyi S (ed) Proceedings of the 7th FAO/SREN workshop on "Anaerobic digestion for sustainability in waste (water) treatment and re-use", vol 1. Moscow State University, Moscow, pp 172–180

Flotats X, Sarquella L (2008) Producció de biogàs per codigestió anaeròbia (Biogas production by codigestion). Col·lecció Quadern Pràctic, número 1. Institut Català d'Energia, Barcelona. http://hdl.handle.net/2117/2265. Accessed 24 Apr 2018

Flotats X, Bonmatí A, Fernández B, Magrí A (2009) Manure treatment technologies: on-farm versus centralized strategies. NE Spain as case study. Bioresour Technol 100:5519–5526

Flotats X, Foged HL, Bonmatí A, Palatsi J, Magrí A, Schelde KM (2012) Manure processing technologies. Technical Report no. II concerning "Manure Processing Activities in Europe" to the European Commission, Directorate-General Environment. http://hdl.handle.net/2117/18944. Accessed 4 June 2018

Flotats X, Fernández B, Palatsi J (2013) Characterization of the anaerobic digestion of proteins containing animal by-products, using simultaneous batch experiments. In: Proceedings of the 13th World Congress on Anaerobic Digestion (AD13), Santiago de Compostela, 25–28 June 2013. http://hdl.handle.net/2117/97998. Accessed 15 May 2018

Flotats X, Bonmatí A, Fernández B, Sales D, Aymerich E, Irizar J, Palatsi J, Romero LI, Pérez M, Vicent T, Font X (2016) Ingeniería y aspectos técnicos de la digestión anaeróbica (Engineering and technical aspects of anaerobic digestion). Volume II.4, "De Residuo a Recurso, el camino hacia la sostenibilidad" books series. Spanish network of Composting, Ed Mundi-Prensa, Madrid

Foged HL, Flotats X, Bonmatí A (2012a) Future trends on manure processing activities in Europe. Technical Report no. V concerning "Manure processing activities in Europe" to the European Commission, Directorate-General Environment. http://hdl.handle.net/2117/18948. Accessed 4 June 2018

Foged HL, Flotats X, Bonmatí A, Palatsi J, Magrí A (2012b) End and by-products from livestock manure processing - general types, chemical composition, fertilising quality and feasibility for

marketing. Technical Report no. III concerning "Manure processing activities in Europe" to the European Commission, Directorate-General Environment. http://hdl.handle.net/2117/18945. Accessed 22 May 2018
Foged HL, Flotats X, Bonmatí A, Schelde KM, Palatsi J, Magrí A, Zonta ZJ (2012c) Assessment of economic feasibility and environmental performance of manure processing technologies. Technical Report no. IV concerning "Manure Processing Activities in Europe" to the European Commission, Directorate-General Environment. http://hdl.handle.net/2117/18947. Accessed 7 June 2018
García-González MCC, Vanotti MBB (2015) Recovery of ammonia from swine manure using gas-permeable membranes: effect of waste strength and pH. Waste Manag 38:455–461
Garfí M, Martí-Herrero J, Garwood A, Ferrer I (2016) Household anaerobic digesters for biogas production in Latin America: a review. Renew Sustain Energy Rev 60:599–614
Gavala HN, Yenal U, Skiadas IV, Westermann P, Ahring BK (2003) Mesophilic and thermophilic anaerobic digestion of primary and secondary sludge. Effect of pre-treatment at elevated temperature. Water Res 37:4561–4457
Ge HQ, Jensen PD, Batstone DJ (2011) Temperature phased anaerobic digestion increases apparent hydrolysis rate for waste activated sludge. Water Res 45:1597–1606
Giuntoli J, Agostini A, Edwards R, Marelli L (2015) Solid and gaseous bioenergy pathways: input values and GHG emissions. JRC Science and Policy Reports, European Comission, Report EUR 27215. https://doi.org/10.2790/299090
Gonzalez-Cabaleiro R, Lema JM, Rodriguez J, Kleerebezem R (2013) Linking thermodynamics and kinetics to assess pathway reversibility in anaerobic bioprocesses. Energ Environ Sci 6:3780–3789
Gosens J, Lu Y, He G, Bluemling B, Beckers TAM (2013) Sustainability effects of household-scale biogas in rural China. Energy Policy 54:273–287
Guwy AJ, Dinsdale RM, Kim JR, Massanet-Nicolau J, Premier G (2011) Fermentative biohydrogen production systems integration. Bioresour Technol 102:8534–8542
Hartmann S, Wirth B, Niebaum N, Döhler H, Keymer U, Reinhold G (2012) Economics. In: Fachagentur Nachwachsende Rohstoffe e. V. (FNR) (ed) Guide to biogas. From production to use. Abt. Öffentlichkeitsarbeit, Gülzow, pp 159–178
Hashimoto AG (1986) Ammonia inhibition of methanogenesis from cattle wastes. Agric Wastes 17:241–261
Hattori S (2008) Syntrophic acetate-oxidizing microbes in methanogenic environments. Microbes Environ 23:118–127
Hjort-Gregersen K (2002) Development and implementation of the Danish centralised biogas concept—financial aspects. In: van Ierland EC, Lansink AO (eds) Economics of sustainable energy in agriculture. Kluwer Academic, Dordrecht, pp 177–188
Holliger C, Alves M, Andrade D, Angelidaki I et al (2016) Towards a standardization of biomethane potential tests. Water Sci Technol 74:2515–2522
Hou Y, Velthof GL, Case SDC, Oelofse M, Grignani M, Balsari P, Zavattaro L, Gioelli F, Bernal MP, Fangueiro D, Trindade H, Jensen LS, Oenemab O (2018) Stakeholder perceptions of manure treatment technologies in Denmark, Italy, the Netherlands and Spain. J Clean Prod 172:1620–1630
Hulshoff Pol LW, De Castro Lopes SI, Lettinga G, Lens PNL (2004) Anaerobic sludge granulation. Water Res 38:1376–1389
Ijmker HM, Gramblicka M, Kersten SRA, van der Ham AGJ, Schuur B (2014) Acetic acid extraction from aqueous solutions using fatty acids. Sep Purif Technol 125:256–263
IRENA (2018) Biogas for road vehicles: technology brief. International Renewable Energy Agency, Abu Dhabi. https://www.irena.org. Accessed March 2018
Jain S, Jain S, Wolf IT, Lee J, Tong YW (2015) A comprehensive review on operating parameters and different pretreatment methodologies for anaerobic digestion of municipal solid waste. Renew Sustain Energy Rev 52:142–154
Johansen A, Nielsen HB, Hansen CM, Andreasen C, Carlsgart J, Hauggard-Nielsen H, Roepstorff A (2013) Survival of weed seeds and animal parasites as affected by anaerobic digestion at meso- and thermophilic conditions. Waste Manag 33:807–812

Jones RJ, Massanet-Nicolau J, Guwy A, Premier GC, Dinsdale RM, Reilly M (2015) Removal and recovery of inhibitory volatile fatty acids from mixed acid fermentations by conventional electrodialysis. Bioresour Technol 189:279–284

Jones RJ, Massanet-Nicolau J, Mulder MJJ, Premier GC, Dinsdale R, Guwy A (2017) Enhanced volatile fatty acid production via the batch electrodialysis of a dark fermentation fermenter continually fed with sucrose. Proceedings of the 15th World Congress on Anaerobic Digestion, Beijing (China), 17–20 October 2017, pp 661–664

Jurado E, Skiadas IV, Gavala HN (2013) Enhanced methane productivity from manure fibers by aqueous ammonia soaking pretreatment. Appl Energy 109:104–111

Juznic-Zonta Z, Alves MM, Flotats X, Palatsi J (2013) Modelling inhibitory effects of long chain fatty acids in the anaerobic digestion process. Water Res 47:1369–1380

Kafle GK, Kim SH, Sung KI (2013) Ensiling of fish industry waste for biogas poduction: a lab scale evaluation of biochemical methane potential (BMP) and kinetics. Bioresour Technol 127:326–336

Kennedy KJ, Droste RL (1991) Anaerobic wastewater treatment in downflow stationary fixed film reactors. Water Sci Technol 24:157–177

Kondusamy D, Kalamdhad AS (2014) Pre-treatment and anaerobic digestion of food waste for high rate methane production—a review. J Environ Chem Eng 2:1821–1830

Laureni M, Palatsi J, Llovera M, Bonmatí A (2013) Influence of pig slurry characteristics on ammonia stripping efficiencies and quality of the recovered ammonium-sulfate solution. J Chem Technol Biotechnol 88:1654–1662

Lauterböck B, Ortner M, Haider R, Fuchs W (2012) Counteracting ammonia inhibition in anaerobic digestion by removal with a hollow fiber membrane contactor. Water Res 46:4861–4869

Lauterböck B, Moder K, Germ T, Fuchs W (2013) Impact of characteristic membrane parameters on the transfer rate of ammonia in membrane contactor application. Sep Purif Technol 116:327–334

Lauterböck B, Nikolausz M, Lv Z, Baumgartner M, Liebhard G, Fuchs W (2014) Improvement of anaerobic digestion performance by continuous nitrogen removal with a membrane contactor treating a substrate rich in ammonia and sulfide. Bioresour Technol 158:209–216

Lettinga G, van Velsen AFM, Hobma SW, de Zeeuw W, Klapwijk A (1980) Use of the upflow sludge blanket (USB) reactor concept for biological wastewater treatment, especially for anaerobic treatment. Biotechnol Bioeng 22:699–734

Lin H, Peng W, Zhang M, Chen J, Hong H, Zhang Y (2013) A review on anaerobic membrane bioreactors: applications, membrane fouling and future perspectives. Desalination 314:169–188

Lorenz H, Fisher P, Schumacher B, Adler P (2013) Current EU-27 technical potential of organic waste streams for biogas and energy production. Waste Manag 33:2434–2448

Lukehurst C, Frost P, Al Seadi T (2010) Utilisation of digestate from biogas plants as biofertizer. Bioenergy Task 37—Energy from Biogas. IEA Bioenergy. http://task37.ieabioenergy.com/technical-brochures.html. Accessed 4 June 2018

Luostarinen S, Luste S, Sillanpää M (2009) Increased biogas production at wastewater treatment plants through co-digestion of sewage sludge with grease trap sludge from a meat processing plant. Bioresour Technol 100:79–85

Maniatis K, Landälv I, Waldheim L, van den Heuvel E, Kalligeros S (2017) Final report: building up the future. Sub group on advanced biofuels, sustainable transport forum, European Commission. Publications Office of the European Union, March 2017

Massé DI, Massé L, Croteau F (2003) The effect of temperature fluctuations on psychrophilic anaerobic sequencing batch reactors treating swine manure. Bioresour Technol 89:57–62

Miron Y, Zeeman G, van Lier JB, Lettinga G (2000) The role of sludge retention time in the hydrolysis and acidification of lipids, carbohydrates and proteins during digestion of primary sludge in CSTR systems. Water Res 34:1705–1713

Molinuevo-Salces B, Fernández-Varela R, Uellendahl H (2014) Key factors influencing the potential of catch crops for methane production. Environ Technol 35:1685–1694

Molinuevo-Salces B, Larsen SU, Ahring BK, Uellendahl H (2015) Biogas production from catch crops: increased yield by combined harvest of catch crops and straw and preservation by ensiling. Biomass Bioenergy 79:3–11

Möller K, Müller T (2012) Effects of anaerobic digestion on digestate nutrient availability and crop growth: a review. Eng Life Sci 12:242–257

Murphy J, Braun R, Weiland P, Wellinger A (2011) Biogas from energy crop digestion. Bioenergy Task 37—Energy from biogas. IEA Bioenergy. http://task37.ieabioenergy.com/technical-brochures.html. Accessed 18 May 2018

Nagy D, Balogh P, Gabnai Z, Popp J, Oláh J, Bai A (2018) Economic analysis of pellet production in co-digestion biogas plants. Energies 11(5):1135

Nähle C (1991) The contact process for the anaerobic treatment of wastewater: technologies, design and experiences. Water Sci Technol 24:179–191

Neumann P, Pesante S, Venegas M, Vidal G (2016) Developments in pre-treatment methods to improve anaerobic digestion of sewage sludge. Rev Environ Sci Biotechnol 15:173–211

Neves L, Oliveira R, Alves MM (2006) Anaerobic co-digestion of coffee waste and sewage sludge. Waste Manag 26:176–181

Noguerol J, Rodríguez-Abalde A, Romero E, Flotats X (2012) Determination of chemical oxygen demand in heterogeneous solid or semisolid samples using a novel method combining solid dilutions as a preparation step followed by optimized closed reflux and colorimetric measurement. Anal Chem 84:5548–5555

Nzeteu CO, Trego AC, Abram F, O'Flaherty V (2018) Reproducible, high-yielding, biological caproate production from food waste using a single-phase anaerobic reactor system. Biotechnol Biofuels 11:108

Oenema O, Witzke HP, Klimont Z, Lesschen JP, Velthof GL (2009) Integrated assessment of promising measures to decrease nitrogen losses in agriculture in EU-27. Agric Ecosyst Environ 133:280–288

Örlygsson J, Houwen FP, Svensson BH (1995) Thermophilic anaerobic amino acid degradation: deamination rates and end product formation. Appl Microbiol Biotechnol 43:235–241

Ortiz-Cabrera MA, Nayak A, Flotats X (2018) Polyphenols removal in winery wastewater using an AnSBR. XIII Latin American Workshop and Symposium on Anaerobic Digestion (DAAL XIII), Medellin (Colombia), 21–24 October 2018

Ozgun H, Dereli RK, Ersahin ME, Kinaci C, Spanjers H, van Lier JB (2013) A review of anaerobic membrane bioreactors for municipal wastewater treatment: integration options, limitations and expectations. Sep Purif Technol 118:89–104

Palatsi J, Campos-Pozuelo E, Torres M, Porras S, Flotats X (2005) Full-scale combination of anaerobic digestion and concentration by evaporation in Garrigues (Lleida, Spain): evaluation after 2 years of operation. In: Bernal MP, Moral R, Clemente R, Paredes C (eds) Proceedings of the International Conference of the FAO ESCORENA Network on Recycling of Agricultural, Municipal and Industrial Residues in Agriculture, Murcia (Spain), October 2004, vol 2, pp 155–158

Palatsi J, Laureni M, Andrés MV, Flotats X, Nielsen HB, Angelidaki I (2009) Strategies for recovering inhibition caused by long-chain fatty acids on anaerobic thermophilic biogas reactors. Bioresour Technol 100:4588–4596

Palatsi J, Illa J, Prenafeta-Boldu FX, Laureni M, Fernandez B, Angelidaki I, Flotats X (2010) Long-chain fatty acids inhibition and adaptation process in anaerobic thermophilic digestion: batch tests, microbial community structure and mathematical modeling. Bioresour Technol 101:2243–2251

Palatsi J, Affes R, Fernández B, Pereira A, Alves M, Flotats X (2012) Influence of adsorption and anaerobic granular sludge characteristics on long chain fatty acids inhibition process. Water Res 46:5268–5278

Pardo G, Moral R, del Prado A (2017) SIMSWASTE-AD–a modelling framework for the environmental assessment of agricultural waste management strategies: anaerobic digestion. Sci Total Environ 574:806–817

Passos F, Carretero J, Ferrer I (2015) Comparing pretreatment methods for improving microalgae anaerobic digestion: thermal, hydrothermal, microwave and ultrasound. Chem Eng J 279:667–672

Pereira MA, Pires OC, Mota M, Alves MM (2005) Anaerobic biodegradation of oleic and palmitic acids: evidence of mass transfer limitations caused by long chain fatty acids. Accumulation onto the anaerobic sludge. Biotechnol Bioeng 92:15–23

Pfau SF, Hagens JE, Dankbaar B (2017) Biogas between renewable energy and bio-economy policies—opportunities and constraints resulting from a dual role. Energy Sustain Soc 7:17

Poeschl M, Ward S, Owende P (2010a) Prospects for expanded utilization of biogas in Germany. Renew Sustain Energy Rev 14:1782–1797

Poeschl M, Ward S, Owende P (2010b) Evaluation of energy efficiency of various biogas production and utilization pathways. Appl Energy 87:3305–3321

Poeschl M, Ward S, Owende P (2012a) Environmental impacts of biogas deployment–part I: life cycle inventory for evaluation of production process emissions to air. J Clean Prod 24:168–183

Poeschl M, Ward S, Owende P (2012b) Environmental impacts of biogas deployment - part II: life cycle assessment of multiple production and utilization pathways. J Clean Prod 24:184–201

Powers WJ, van Horn HH, Wilkie AC, Wilcox CJ, Nordstedt RA (1999) Effects of anaerobic digestion and additives to effluent or cattle feed on odour and odorant concentrations. J Anim Sci 77:1412–1421

Ramsay IR (1997) Modelling and control of high-rate anaerobic wastewater treatment systems. PhD Thesis. Department of Chemical Engineering, University of Queensland, Brisbane

Ramsay IR, Pullammanappallil PC (2001) Protein degradation during anaerobic wastewater treatment: derivation of stoichiometry. Biodegradation 12:247–257

Rebecchi S, Pinelli D, Bertin L, Zama F, Fava F, Frascari D (2016) Volatile fatty acids recovery from the effluent of an acidogenic digestion process fed with grape pomace by adsorption on ion exchange resins. Chem Eng J 306:629–639

Rittmann BE, McCarty PL (2001) Environmental biotechnology: principles and applications. McGraw-Hill Book, New York

Rodríguez J, Kleerebezem R, Lema JM, van Loosdrecht MCM (2006) Modeling product formation in anaerobic mixed culture fermentations. Biotechnol Bioeng 93:592–606

Rodriguez C, Alaswad A, Mooney J, Prescott T, Olabi AG (2015) Pre-treatment techniques used for anaerobic digestion of algae. Fuel Process Technol 138:765–779

Rodríguez-Abalde A, Fernández B, Silvestre G, Flotats X (2011) Effects of thermal pre-treatments on solid slaughterhouse waste methane potential. Waste Manag 31:1488–1493

Rodríguez-Abalde A, Gómez X, Blanco D, Cuetos MJ, Fernández B, Flotats X (2013) Study of thermal pre-treatment on anaerobic digestion of slaughterhouse waste by TGA-MS and FTIR spectroscopy. Waste Manag Res 31:1195–1202

Romero-Güiza MS, Tait S, Astals S, del Valle-Zermeño R, Martínez M, Mata-Alvarez J, Chimenos JM (2015) Reagent use efficiency with removal of nitrogen from pig slurry via struvite: a study on magnesium oxide and related by-products. Water Res 84:286–294

Ruiz B, de Benito A, Rivera JD, Flotats X (2016) Assessment of different pre-treatment methods for the removal of limonene in citrus waste and their effect on methane potential and methane production rate. Waste Manag Res 34:1249–1257

Ruiz-Sánchez J, López A, Riau V, Prenafeta-Boldú FX, Fernández B, Flotats X (2017) Biogas production from N-rich wastes: SAOB-HM enriched digester versus hydrophobic membrane assisted digester. Proceedings of the 15th World Congress on Anaerobic Digestion, Beijing (China), 17–20 October 2017, pp 695–698

Ruiz-Sánchez J, Campanaro S, Guivernau M, Fernández B, Prenafeta-Boldú FX (2018) Effect of ammonia on the active microbiome and metagenome from stable full-scale digesters. Bioresour Technol 250:513–522

Scarlat N, Dallemand JF, Fahl F (2018) Biogas: developments and perspectives in Europe. Renew Energy 129:457–472

Schnürer A, Nordberg A (2008) Ammonia, a selective agent for methane production by syntrophic acetate oxidation at mesophilic temperature. Water Sci Technol 57:735–740

Schnürer A, Houwen FP, Svensson BH (1994) Mesophilic syntrophic acetate oxidation during methane formation by a triculture at high ammonium concentration. Arch Microbiol 162:70–74

Silvestre G, Rodriguez-Abalde A, Fernández B, Flotats X, Bonmatí A (2011) Biomass adaptation over anaerobic co-digestion of sewage sludge and trapped grease waste. Bioresour Technol 102:6830–6836

Simioni M, Kentish SE, Stevens GW (2011) Polymeric alternatives to teflon for membrane stripping. Energy Procedia 4:659–665

Singhania RR, Patel AK, Christophe G, Fontanille P, Larroche C (2013) Biological upgrading of volatile fatty acids, key intermediates for the valorization of biowaste through dark anaerobic fermentation. Bioresour Technol 145:166–174

Skiadas IV, Gavala HN, Lyberatos G (2000) Modelling of the periodic anaerobic baffled reactor (PABR) based on the retailing factor concept. Water Res 34:3691–3905

Smil V (2011) Nitrogen cycle and world food production. World Agr 2:9–13

Spirito CM, Richter H, Rabaey K, Stams AJM, Angenent LT (2014) Chain elongation in anaerobic reactor microbiomes to recover resources from waste. Curr Opin Biotechnol 27:115–122

Steinfeld H, Gerber P, Wassenaar T, Castel V, Rosales M, de Haan C (2006) Livestock long shadow; environmental issues and options. Food and Agriculture organization of the United Nations, FAO Publishing Management Service, Rome

Stopp P, Weichgrebe D, Rosenwinkel KH (2017) Innovative digestate evaporation for nutrients recovery and post-processing. Proceedings of the 15th World Congress on Anaerobic Digestion, Beijing (China), 17–20 October 2017, pp 657–660

Tauseef SM, Abbasi T, Abbasi SA (2013) Energy recovery from wastewaters with high-rate anaerobic digesters. Renew Sustain Energy Rev 19:704–741

Thrän D, Persson T, Svensson M, Daniel-Gromke J, Ponitka J, Seiffert M, Baldwin J, Kranzl L, Schipfer F, Matzenberger J, Devriendt N, Dumont M, Dahl J, Bochmann G (2014) Biomethane—status and factors affecting market development and trade. In: Junginger M, Baxter D (eds) IEA Task 40 and Task 37 Joint Study. IEA Bioenergy

Turner C, Burton CH (1997) The inactivation of viruses in pig slurry: a review. Bioresour Technol 61:9–20

Van Brakel J (1980) The Ignis Fatuus of biogas. Small-scale anaerobic digesters ("biogas plants"): a critical review of the pre-1970 literature. Delft University Press, Delft, The Netherlands. http://repository.tudelft.nl/view/ir/uuid:06a18017-8de3-4849-9337-dc41944555d6. Accessed 8 June 2018

Van Lier JB (2008) High-rate anaerobic wastewater treatment: diversifying from end-of-pipe treatment to resource-oriented conversion techniques. Water Sci Technol 57:1137–1148

Van Lier JB, Hulsbeek J, Stams AJ, Lettinga G (1993) Temperature susceptibility of thermophilic methanogenic sludge: implication for reactor start-up and operation. Bioresour Technol 43:227–235

Vandeviere P, De Baere L, Verstraete W (2003) Types of anaerobic digester for solid wastes. In: Mata-Álvarez J (ed) Biomethanization of the organic fraction of municipal solid wastes. IWA, London, pp 11–140

Vaneeckhaute C, Belia E, Meers E, Tack FMG, Vanrolleghem PA (2018) Nutrient recovery from digested waste: towards a generic roadmap for setting up an optimal treatment train. Waste Manag 78:385–392

Vavilin VA, Fernandez B, Palatsi J, Flotats X (2008) Hydrolysis kinetics in anaerobic degradation of particulate organic material: an overview. Waste Manag 28:939–951

Wäeger-Baumann F, Fuchs W (2012) The application of membrane contactors for the removal of ammonium from anaerobic digester effluent. Sep Sci Technol 47:1436–1442

Walla C, Schneeberger W (2008) The optimal size for biogas plants. Biomass Bioenergy 32:551–557

Wang S, Jenab U, Dasa KC (2018) Biomethane production potential of slaughterhouse waste in the United States. Energ Convers Manage 173:143–157

Yousuf A, Bonk F, Bastidas-Oyanedel JR, Schmidt JE (2016) Recovery of carboxylic acids produced during dark fermentation of food waste by adsorption on Amberlite IRA-67 and activated carbon. Bioresour Technol 217:137–140

Yousuf A, Bastidas-Oyanedel JR, Schmidt JE (2018) Effect of total solid content and pretreatment on the production of lactic acid from mixed culture dark fermentation of food waste. Waste Manag 77:516–521

Zema DA (2017) Planning the optimal site, size, and feed of biogas plants in agricultural districts. Biofuels Bioprod Biorefin 11:454–471

Zhou M, Yan B, Wong JWC, Zhang Y (2018) Enhanced volatile fatty acids production from anaerobic fermentation of food waste: a mini-review focusing on acidogenic metabolic pathways. Bioresour Technol 248:68–78

Nutrient and Carbon Recovery from Organic Wastes

Eric Walling, Alexandre Babin, and Céline Vaneeckhaute

1 Introduction

In 2012 the global production of municipal solid waste was of 1.3 billion tonnes per year, and it is estimated that, with the increase in population and urbanization, this amount will increase to 2.2 billion tonnes per year by 2025. Of this, approximately 40–50% is organic biodegradable matter, such as food, wood, garden and lawn clippings, and human and animal waste (Hoornweg and Bhada-Tata 2012; David 2013). These materials present a valuable source of nutrients such as phosphorus (P), potassium (K), and nitrogen (N), as well as a variety of minerals, on top of organic carbon (C). For example, food waste, similarly to most organic waste sources, is rich in carbon, usually between 40 and 55% (Adhikari et al. 2008). It can contain around 8% hydrogen (H) and 2–6% nitrogen (Vakalis et al. 2016), and other nutrients are often found below 1% (Zhang et al. 2007). However, this composition varies widely based on location and waste type (Mor et al. 2006; Siddiqui et al. 2011). With the impending nutrient depletion, it is of utmost importance to be able to recover these valuable resources from our waste (Elser and Bennett 2011). A variety of technologies currently exist to transform organic waste into value-added products, such as composting and biomethanation. This chapter will go over a variety of nutrient and carbon recovery and treatment alternatives, looking at both thermochemical and biochemical conversion processes.

E. Walling · C. Vaneeckhaute (✉)
BioEngine—Research Team on Green Process Engineering and Biorefineries, Chemical Engineering Department, Université Laval, Québec, QC, Canada

CentrEau, Centre de recherche sur l'eau, Université Laval, Québec, QC, Canada
e-mail: celine.vaneeckhaute@gch.ulaval.ca

A. Babin
BioEngine—Research Team on Green Process Engineering and Biorefineries, Chemical Engineering Department, Université Laval, Québec, QC, Canada

© Springer Nature Switzerland AG 2019
J.-R. Bastidas-Oyanedel, J. E. Schmidt (eds.), *Biorefinery*,
https://doi.org/10.1007/978-3-030-10961-5_14

In this chapter, nutrients will refer mainly to the substances used by plants to ensure their growth and maintenance. The nutrients which are necessary for proper plant life are boron (B), calcium (Ca), carbon, chlorine (Cl), copper (Cu), hydrogen, iron (Fe), magnesium (Mg), manganese (Mn), molybdenum (Mo), nickel (Ni), nitrogen, oxygen (O), potassium, sodium (Na), silicon (Si), sulfur (S), and zinc (Zn). There are a variety of ways to classify these nutrients. One form of classification is by the amount required, sorting them into either macronutrients (Ca, C, H, K, O, Mg, N, P, S) or micronutrients (B, Cl, Cu, Fe, Mn, Mo, Ni, Na, Si, Zn). They can also be classified by their roles in metabolic processes or by the location and method of their uptake (Engels et al. 2012).

2 Thermochemical Conversion

Thermochemical conversion processes are commonly used to recover energy from waste using high temperatures. The following is a short overview of thermal conversion technologies which include incineration, gasification, pyrolysis, and torrefaction.

A brief reminder of combustion can be useful to better understand the following processes. Combustion is the chemical reaction of a fuel in the presence of an oxidizer, most commonly oxygen, and heat. The fuel and the oxidizer react together to form new chemical substances, most often combinations of both, such as the combustion of hydrogen (H_2) and oxygen (O_2) to form water (H_2O). It can also include combinations of the oxidizer alone, for example, when air is used as an oxidizer, the nitrogen (N_2) and oxygen within it can react to form nitrous oxides (NO_X). Combustion can either be complete or incomplete. When there is sufficient oxidizer to allow for all the fuel to be fully oxidized, the combustion is said to be complete. Therefore, the complete combustion of hydrocarbons (C_nH_{n+2}) with oxygen will produce water and carbon dioxide (CO_2). Incomplete combustion occurs when there is not enough oxidizer to burn all the fuel. In this case, part of the fuel is only partially oxidized, which will lead to the production of hydrogen, carbon, and carbon monoxide (CO). These products can still be oxidized so, depending on the context, an incomplete combustion can either be desirable (to produce fuel) or undesirable (loss of energy).

2.1 *Incineration*

Incineration consists of oxidizing the waste source, leading to a highly exothermic process which produces flue gas, ash, and heat. The heat is used to produce steam which in turn is sent to a turbine to produce electricity or used directly as process steam. Incineration has seen widespread use throughout Europe and became popular due to restrictions on landfill use, allowing the reduction of more than 90% of the waste's volume (Chandler et al. 1997). The ash produced from the combustion process can be recovered to recycle nutrients such as phosphorus, metals, and other noncombustible materials (Zacco et al. 2014). Incineration also provides the ability to treat medical and potentially pathogenic waste. However, a wide variety of gases are released during the process, including sulfur dioxide (SO_2), NO_X, nitrous oxide (N_2O), hydrogen chloride (HCl), hydrogen fluoride (HF), CO, CO_2, dioxins, furans, polycyclic aromatic hydrocarbons (PAHs), volatile organic compounds (VOCs), and heavy metals (Jay and Stieglitz 1995; Di Maria and Micale 2015). Many of these gases have important greenhouse effects and potential health impacts; however proper flue gas and fly ash treatment should alleviate these issues. For these reasons, there still exists widespread debate over the environmental and social acceptability of these processes.

Following the incineration process, it is possible to recover some nutrients from the ash, most notably phosphorus. Phosphorus contained in organic waste may either be in organic form, which is poorly absorbed by plants, or in inorganic form, which is more easily assimilated by plants (Cieślik and Konieczka 2017). It is necessary to remove heavy metals from the ashes before, during, or after P recovery, which can be achieved by separation methods such as washing and leaching, solidification treatments, and thermal treatments (Zacco et al. 2014).

The phosphorus from the ashes can be extracted through a variety of techniques. The two main methods for phosphorus recovery include thermal conversion, which aims at enriching the phosphorus in the ash, and leaching, which aims at extracting the phosphorus from the ash (Tan and Lagerkvist 2011). Thermal conversion in the context of P recovery is notably performed in a circulating fluidized bed boiler. It can be achieved through three different methods, including biomass combustion with phosphorus in the ash as direct source, biomass combustion followed by a dry process ash for P recovery, and biomass combustion followed by a wet process ash for P recovery. Thermal conversion leads to a high recovery rate from biomass, with up to 98% of the phosphorus retained in the ashes, while other interesting macronutrients such as Ca, K, and Mg may be retained as well. Phosphorus is mainly distributed in submicron ash particles, which may also contain sulfur, silicon, and sodium, with their concentration increasing as the particle size decreases. Temperature and oxygen content also influence the phosphorus content in ashes, with higher combustion temperatures and oxygen content leading to higher P content. Increasing oxygen content also leads to higher P content in particles within the micronic scale. The main obstacle of thermal conversion is the bioavailability of the resulting phosphorus, most of which is recovered in a form that is poorly available to plants. However, bioavailability can be improved with a second thermochemical step (dry process ash), where the phosphorus is transferred into mineral phases available for plants (Adam et al. 2009). Otherwise, the phosphorus obtained from the thermal conversion can be extracted by leaching (wet process ash) (Tan and Lagerkvist 2011).

Phosphorus recovery by leaching can be done by different means, such as bioleaching, which involves microorganisms; supercritical extraction, which involves supercritical water; and chemical extraction, which involves acids such as hydrochloric acid and sulfuric acid leaching (Tan and Lagerkvist 2011).

Chemical leaching is the most common leaching method used for phosphorus recovery, where P extraction can be achieved with a fairly low consumption of chemicals, using only certain acids (Tan and Lagerkvist 2011). In general, higher temperature and acid concentration lead to higher rates of P recovery. In fact, a pH of 1 may be required to extract over 75% of the phosphorus. Solid-phase extraction of phosphorus by chemical leaching can be combined with precipitation, which is normally done by dissolution of the acidic extract containing the phosphorus, followed by a filtration (Kalmykova and Fedje 2013; Petzet et al. 2012). A series of multiple dilution-filtration cycles may be necessary to allow the extraction of every precipitation preceding the precipitation of phosphorus. An alternative to acid leaching is alkaline leaching, which uses bases such as sodium hydroxide. A two-step leaching process where acid leaching is followed by alkaline leaching may be done, which may lead to superior extraction performance under optimal conditions.

Supercritical extraction aims at enhancing the extraction of phosphorus through supercritical water oxidation (Tan and Lagerkvist 2011). In fact, supercritical water oxidation leads to higher release of phosphorus than conventional chemical leaching with normal acid concentrations. However, supercritical extraction may require pretreatments of the ashes as they may contain heteroatoms such as chlorine, sulfur, and phosphorus which may create a highly corrosive medium during supercritical water oxidation. Cost is also a concern due to the use of supercritical water and the potential need of pretreatments.

Bioleaching is unconventional as it is done prior to combustion and is only indirectly focused on the recovery of phosphorus. It involves bioacidifying microorganisms, notably sulfur-oxidizing microorganisms, for the extraction of heavy metals from soil and sludge mixtures (Shanableh and Omar 2003). Bioleaching is known to be subject to considerable losses of phosphorus and nitrogen during heavy metal removal, as it extracts a significant portion of those nutrients along with the heavy metals during the leaching. For example, 76% of phosphorus and 38% of nitrogen may be lost during bioleaching of heavy metals when using sulfur as a substrate (Shanableh and Ginige 1999). However, this problem can be attenuated by selecting the right substrate for bioleaching. For example, using $FeSO_4.7H_2O$ as a substrate reduces the loss of phosphorus to 45%, while using FeS_2 as a substrate reduces the loss of phosphorus and nitrogen to less than 6 and 15% (Wong et al. 2004). It is interesting to note that increasing solid content of biomass leads to better solubilization of heavy metals but also phosphorus (Tan and Lagerkvist 2011). Furthermore, lower sludge pH leads to faster solubilization, while higher pH tends to reduce P losses. In the context of bioleaching, using phosphate-solubilizing bacteria can also be of interest to supply plants with phosphorus from sources that are poorly available.

Another method proposed for phosphorus recovery is the fractionation method, where the ashes are agitated in an acidic solution followed by a series of centrifugations which leads to the extraction of phosphorus from the solids into the supernatant (Kleemann et al. 2017). Phosphorus can also be extracted from ashes by an electrodialytic separation process, which aims at extracting phosphates through ion exchange induced by an electric potential difference, as well as removing heavy metals (Guedes et al. 2014).

2.2 *Gasification*

Gasification is the process of transforming waste into syngas, a gas which can be used to produce fuels, chemicals, or synthetic natural gas. Whereas a complete combustion is desired during incineration, gasification comprises of an incomplete combustion using an amount of oxidant lower than the amount required for stoichiometric combustion. As a result, the waste becomes gasified into syngas, which contains unoxidized products which can be used as energy sources for later use. The resulting syngas is rich in H_2 and CO, with some methane (CH_4), but is generally contaminated and necessitates cleaning before being usable (Yan et al. 2010). The largest challenge faced by gasification processes, other than cleaning the syngas, is the need for a heterogeneous waste source (Asadullah 2014).

Furthermore, the solid by-product of gasification is biochar (Hansen et al. 2015). Biochar is rich in nutrients and has been found to increase their retention, thus decreasing leaching and increasing the nutrient availability for plants when applied to soil (Oram et al. 2014; Yuan et al. 2016). Other than its agricultural applications, biochar can also be burned to release energy or as an adsorbent for environmental applications, similarly to activated carbon (Xie et al. 2015).

As is the case for incineration, P recovery from ashes is also possible for gasification (Atienza–Martínez et al. 2014). Chemical leaching for a few hours can recover over 90% of the phosphorus with sulfuric acid, but extending the leaching duration does not significantly affect the recovery rate. A similar load of oxalic acid leads to comparable results as sulfuric acid, but may recover close to 100% of the phosphorus when leaching is done for an extended period of time.

An alternative to combustion for gasification of sewage sludge is supercritical water gasification (Acelas et al. 2014). A key advantage of supercritical water gasification of sewage sludge is that it does not require drying of the sludge, which can avoid the costs of such pretreatments. In addition to the production of biogas, a solid and a liquid residue is also produced, both containing nutrients. Almost all of the N, K, and Cl are contained in the liquid phase, while the solid phase contains almost all of the Ca, P, and Si, leading to an effective separation of nutrients. However, the process is ineffective for separation of S as it is evenly distributed between the two phases. The liquid and solid phases may be separated by filtration, and the solid phase can then undergo a chemical leaching treatment for P extraction. Increasing leaching duration of the solid residue has little impact, generally leading to a minor increase of the percentage of phosphorus recovery. Leaching temperature has a significant impact on leaching performance. A temperature of 400 °C is sufficient to ensure a recovery of at least 80% of the phosphorus, while increasing the temperature to 600 °C leads to a recovery rate close to 90%, if not higher. Once again, oxalic acid leads to a higher recovery rate than sulfuric acid. When compared to leaching of combustion ash, leaching of supercritical water gasification residue leads to higher P yields from recovery. This improvement in recovery performance may be attributed to the formation of different chemical compounds during the thermal treatment. While combustion leads to the formation of organic compounds with P-C bonds, which are inert to acid treatments, supercritical water gasification produces a higher amount of inorganic phosphorus in the form of calcium, aluminum, and iron phosphates, which are more easily dissolved during the leaching treatment.

2.3 *Pyrolysis*

Pyrolysis is the decomposition of material at high temperatures in the absence of oxygen. This process allows the conversion of waste into syngas (gas), biofuels (liquid), and biochar (solid) (Laird et al. 2009). Unlike the previous two processes, this technology is not a combustion-based process and is endothermic. By increasing the temperature with the absence of an oxidant, the chemical bonds in the organic matter are broken down. Therefore, the longer organic chains are decomposed into simpler compounds, resulting in the production of CO, H_2, and hydrocarbons (Jahirul et al. 2012). The proportion of gas, liquid, or solid product is determined by the operating conditions and the type of feedstock. Similarly to gasification, cleaning of the syngas and feedstock quality are some of the most limiting factors to the widespread use of this technology (Carpenter et al. 2014). Pyrolyzed char also benefits from a superior porosity, increasing the interest for adsorption applications (Kleemann et al. 2017).

Recovery of P from pyrolyzed biochar is a possibility, which differs significantly from P recovery from incinerated ashes in terms of performance (Kleemann et al. 2017). First of all, pyrolysis has lower P enrichment capacity relative to sludge than incineration, with P concentrations being enriched by less than three times in pyrolyzed char compared to an enrichment of about seven times in the case of ashes from incineration. Furthermore, P extraction by leaching seems to be less effective on pyrolyzed char, as a lower proportion of P extracted from the former is obtained at optimal acid concentration for incinerated ashes. However, an increase of acid concentration for the leaching process leads to an improvement in P extraction from pyrolyzed char. As a result, optimal extraction from pyrolyzed ashes by leaching requires a higher concentration of acid than for incinerated ashes. Pyrolyzed char contains a significant residual concentration of N, but contains much less heavy metals than incinerated ashes, as a large amount of them are volatilized in the pyrolysis process. The type of acid used for leaching may have an impact on P recovery performance. For example, using oxalic acid leads to higher phosphorus yields than when sulfuric acid is used (Atienza–Martínez et al. 2014).

2.4 *Torrefaction*

Torrefaction is a form of pyrolysis at lower temperatures. The aim of this technology is to improve the properties of biomass for energy generation, converting the biomass into a coal-like material. Torrefied biomass has a variety of interesting qualities, notably exhibiting a hydrophobic behavior, an inhibited biological decomposition (i.e., it doesn't rot), an improved grindability, and a higher heating value (Acharya et al. 2012). As is the case for pyrolysis, syngas and biofuels are also produced during torrefaction, but are viewed mostly as by-products, the gas usually being directed back into the process to provide heat. The biochar produced from torrefaction can then be used for combustion or gasification or reused as a soil amendment.

The following Table 1 presents a brief overview of some important technical, economic, and environmental data pertaining to these technologies.

3 Biochemical Conversion

Whereas the previously discussed technologies utilized heat to convert biomass into energy and value-added products, the following technologies use microbial processes to achieve the same goal. It is important to note, however, that these processes are limited to biodegradable waste, reducing their range of application compared to technologies such as incineration.

Table 1 Comparison of advantages and disadvantages of the thermochemical processes, adapted from Samolada and Zabaniotou (2014)

Process	Operating temperature (°C)	Advantages	Disadvantages
Incineration	850–1200[a]	– Up to 90% reduction of volume – Established technology and infrastructure – Feedstock quality is not essential	– Air pollution and GHG emissions – Low energy efficiency – Flue gas cleaning
Gasification	550–1600[a]	– High efficiency of energy recovery – Reduced emissions – Syngas and biochar production	– Complex technology – High investment and operation costs – Extensive gas cleaning of the syngas
Pyrolysis	500–800[a]	– Zero waste process – Reduced emissions – Syngas, biofuels and biochar production –Up to 90% reduction of volume	– High investment costs – Extensive gas cleaning of the syngas – Feedstock sensitivity
Torrefaction	200–300[b]		

[a]Arena (2012)
[b]Acharya et al. (2012)

3.1 Anaerobic Digestion

Anaerobic digestion is the natural process by which microorganisms break down organic material in the absence of oxygen. Examples of organic waste that can be processed by anaerobic digestion include animal manures, food scraps, fats, oils and greases, industrial organic residuals, and wastewater and sewage sludge, among others. Anaerobic digestion produces biogas, which is mainly composed of methane and carbon dioxide and can be used as an energy source or as a precursor for other chemical compounds, and a by-product known as the digestate, which can be valorized as bedding for livestock, fertilizers, and soil amendments, among others. Biogas is generally composed of around 60% methane and 40% carbon dioxide, with possible traces of hydrogen sulfide (H_2S), nitrogen, and hydrogen (H_2) and traces of oxygen, carbon monoxide, ammonia (NH_3), and certain volatile organic compounds (Noyola et al. 2006). Biogas should not be confused with syngas, the latter being composed mainly of H_2, CO, and CO_2. Anaerobic digestion is done in four key steps (Bajpai 2017). The first step is hydrolysis by microorganisms which break down carbohydrates into forms digestible by other microorganisms. The second step is known as acidogenesis, where acidogenic bacteria convert the hydrolyzed sugars into higher organic acids such as propionic acid and butyric acid. The next step is acetogenesis, where organic acids are converted into acetic acid by acetogenic bacteria, with hydrogen as a main by-product. The final step of anaerobic digestion is methanogenesis, consisting of the conversion of acetic acid into methane and carbon dioxide by methanogenic bacteria. As stated previously, the solid or liquid mixture remaining after the anaerobic digestion process is known as the digestate.

The anaerobic digestion process can be operated in three main temperature ranges, including thermophilic (50–60 °C), mesophilic (25–40 °C), and psychrophilic (15–25 °C) digestion (DeBruyn and Hilborn 2007). The higher operating temperature of the thermophilic range causes the microorganisms to break down organic matter more rapidly, with an average retention time of 3–5 days, and leads to higher production of biogas. However, more energy consumption and greater insulation are required. The mesophilic range operates at a lower temperature, meaning that more time is required for the microorganisms to break down organic material, with a retention time of 15–20 days or more. However, they are known to be more robust to temperature upsets. The psychrophilic system operates at room temperature, meaning that it has the lowest energy requirement of the three temperature ranges, but breaking down of organic matter takes longer, with retention times usually spanning over a month or two (Lettinga et al. 2001).

Prior to anaerobic digestion, substrates and feedstocks are subject to pretreatments in order to enhance performance (Ariunbaatar et al. 2014). Pretreatments include mechanical treatments to increase contact surface between substrate and bacteria; thermal treatments to remove pathogens, enhancing dewatering performance and reducing viscosity of the digestate; as well as chemical treatments for early breakdown of organic compounds and biological treatments for general performance enhancement in certain steps. Anaerobic digestion can be operated in "dry" or "wet" conditions.

3.1.1 Dry Anaerobic Digestion

Dry anaerobic digestion operates at a high total solid content, generally over 20% (Angelonidi and Smith 2015). It is mainly used for feedstocks such as leaves, grass, straw, garden waste, wood waste, as well as chicken slurry and food waste with high solid content (Steffen et al. 1998). Dry anaerobic digestion has a good number of advantages over wet digestion, including lower power and heat requirements, high tolerance to solid contaminants (sand, fibers, large particles, etc.), less critical equipment required (pumps, agitation systems, feeding equipment, etc.), less maintenance, low water consumption, arguably greater flexibility over the type of feedstock accepted, shorter retention times, more flexible management of the end product, and being less complex overall than wet anaerobic digestion (Angelonidi and Smith 2015; ADEME Bourgogne 2013). The main disadvantages of dry anaerobic digestion include the need for special technologies for loading and unloading of the digester, the need to manage the variations of biogas and heat production, uneven mixing of the substrate, lower methane production in comparison to wet anaerobic digestion, and high requirements in structure material.

Industrial dry anaerobic digesters are similar to plug flow digestors as they are generally horizontal and cylindrical, but they may be agitated and are often operated in batch (Cho et al. 2013; Guendouz et al. 2010). Dry anaerobic digesters can also be constructed in the form of a silo or a garage (ADEME Bourgogne 2013). The silo form is known to be simple, robust, and cheaper, but the garage form is easier to operate as opening and closing its doors for loading and unloading are simpler than removing and reinstalling the silo's cover to do so from above. Dry anaerobic digesters in the form of smaller and easily transportable containers have also been proposed, which would allow transportation as the digestion takes place, but installation costs are currently higher than conventional dry anaerobic digesters.

3.1.2 Wet Anaerobic Digestion

Wet anaerobic digestion operates at a low total solid content, generally less than 20% (Angelonidi and Smith 2015). The main advantages over dry anaerobic digestion include higher rate of biogas produced per tonne of waste treated, lower investment and operating costs, better energy balance, lower parasitic energy consumption, arguably greater flexibility over the type of feedstock accepted, integration of gas buffer, and biological desulfuration (Angelonidi and Smith 2015; ADEME Bourgogne 2013). Improvements in energy efficiency and biogas production rates may be attributed to optimal mixing, which is usually lacking in dry anaerobic digestion processes (Lindmark et al. 2014). The main disadvantages of wet anaerobic digestion include the need to add liquid when dry mixtures are used, the important need for mixing equipment, the significant energy requirement for pumps and agitators, and the risk of solid deposition and scum layer formation. Digestate produced from wet anaerobic digestion has lower dry matter content compared to digestate formed from dry anaerobic digestion. Digestate with low solid content can complicate storage and transportation for land application of digestate and hence require additional investments for further processing of the product, such as solid-liquid separation, drying, and/or nutrient extraction equipment. It is interesting to note, however, that wet anaerobic digestion plants may benefit from a more advantageous energy balance and economic performance overall compared to dry anaerobic digestion, despite these disadvantages (Angelonidi and Smith 2015). Since undigested sewage sludge usually has low solid content (generally ≤1–7%) (Ontario Ministry of the Environment 2008a), wet anaerobic digestion is ideal in this context. However, since higher solid content leads to lower digester sizing and energy requirements, and higher biogas production rates, dewatering the sludge prior to digestion may be desirable (Yi et al. 2014; Ontario Ministry of the Environment 2008b).

The two main types of digester used for wet anaerobic digestion include completely mixed and plug flow systems (DeBruyn and Hilborn 2007). Complete-mixed digesters consist of a large tank where fresh organic material is continuously mixed with partially digested material. This type of configuration is recommended for mixes with lower dry solid content (3–10%) such as pig and cow slurry, animal blood, whey, and fermentation slops (Chen and Neibling 2014; Steffen et al. 1998). Complete-mixed digesters can either operate in the mesophilic or thermophilic range, with a retention time generally ranging from 10 to 25 days. Plug flow digesters are continuously fed and consist of long channels in which feedstock moves through as it gets digested. This system is recommended for the treatment of material with higher dry solid content (11–14%) such as chicken slurry and food wastes, as well as pig and cow slurry with higher solid content (Steffen et al. 1998). They usually operate in the mesophilic range, with the retention time ranging from 15 to 30 days. Other key examples of wet digesters include covered lagoons and fixed-film digesters (Chen and Neibling 2014). As its name suggests, the covered lagoon is a large in-ground lagoon with a gastight cover. Covered lagoons are used for mixes with very low solid content (0.5–2%), making them ideal for raw sludge, as well as whey and fermentation slops with the lowest solid content (Steffen et al. 1998). They generally operate in the psychrophilic temperature range as they are not heated, meaning that they mostly operate in warmer regions where the temperature range can be reliably maintained. Fixed-film digesters are basically columns packed with such media as wood chips and plastic rings. The packing supports a biofilm of bacteria, with methane-forming microorganisms growing on the media. This type of digester is also used for very low solid contents (1–2%) and benefits from a very short retention time (2–6 days). However, the media is vulnerable to plugging by solids. The wet anaerobic digestion process is usually operated in continuous flow as it produces more biogas, and the operating costs are lower in comparison to a batch operation due the avoidance of the complications from restarting the system from cold (World Energy Council 2016).

3.1.3 Nutrient Recovery

Following the digestion process, it is possible to extract end products with higher nutrient concentration through implementation of nutrient recovery technologies (NRTs). For some liquid waste streams, it is possible to apply these technologies without passing through the anaerobic digestion process first, but process efficiency will be improved following digestion (Vaneeckhaute et al. 2017). The following is a quick overview of the most established NRTs to date. For detailed technical and economic information, reference is made to the literature review on NRTs for digestate treatment presented in Vaneeckhaute et al. (2017).

- Chemical crystallization: Crystallization/precipitation can be induced through a number of methods such as modification of temperature, solubility, and charges. Chemical crystallization can be used to produce struvite ($MgNH_4PO_4{\bullet}6H_2O$) and calcium phosphates ($CaHPO_4{\bullet}2H_2O$ or $Ca_5(PO_4)_3OH$) from liquid digestate. Both products are valuable synthetic fertilizer substitutes (Vaneeckhaute et al. 2017). By increasing the pH of the liquid fraction through the addition of magnesium ($MgO/MgCl_2$) or calcium hydroxide ($Ca(OH)_2$) and sodium hydroxide (NaOH), crystallization is achieved (Rahman et al. 2014). Struvite and calcium phosphates (brushite and hydroxyapatite) are of growing interest due to the depletion of rock phosphate and will likely become the chief phosphate fertilizers (Massey et al. 2007). They also contain other valuable minerals such as Mg, N, and Ca. Struvite precipitation allows for low nitrogen loss, around 5% (Kataki et al. 2016). Furthermore, struvite tends to offer some advantages over chemical phosphorus fertilizers, such as allowing for a slow release of nutrients and having a high hardness, allowing for easy transportation and storage (Latifian et al. 2012). It can also be applied to soils at greater quantities and lower frequency than conventional fertilizers (Li and Zhao 2003).
- Gas stripping and absorption: Ammonia (NH_3) stripping and absorption is one of the more established technologies for N recovery. The process involves passing the liquid fraction through a packed bed tower. Some of the ammonia is transferred from the liquid phase to the gas phase, and this gas is sent to an air scrubber. The air scrubber puts the NH_3 in contact with a liquid phase, commonly H_2SO_4, to recover a concentrated solution of ammonium sulfate ($(NH_4)_2SO_4$). Ammonium sulfate can be used as a fertilizer, being a good source of N and S (Arthington et al. 2002).
- Membrane separation: Membrane filtration involves passing the digestate through microfiltration, nanofiltration, ultrafiltration, and/or using reverse osmosis. The filtration process allows for the production of process water and a nutrient (N and K)-rich concentrate when using reverse osmosis (Vaneeckhaute et al. 2017; Drosg et al. 2015).

3.2 Fermentation

Fermentation is the use of microorganisms to convert organic matter in absence of oxygen into acids or alcohols, notably ethanol and lactic acid, as well as hydrogen. Microorganisms can also be selected to favor methane production, in which case the process is equivalent to biomethanation. Ethanol and hydrogen are clean fuels of high interest, notably for the transportation sector, while lactic acid is a versatile chemical with a long history of applications in the food, cosmetic, and pharmaceutical industries. The most important application of fermentation is for the production of bioethanol, which is usually done by yeast fermentation either in continuous or batch configuration. Although, batch fermentation is usually the preferred method due to the lower risk of contamination. The two main types of reactor used for fermentation are the continuous stirred tank reactor (CSTR) and the plug flow reactor (PFR) (Brethauer and Wyman 2010). In the CSTR, the composition in the reactor tank is homogenous if the reactor is ideally stirred, meaning that the composition is identical not only within the reactor but also to the outgoing flow. For the PFR, the reactants are pumped through a tube with the fermentation proceeding as the biomass travels through the reactor. In the case of an ideal PFR, a uniform velocity profile across the radius diffusion and negligible diffusion in the axial direction are expected. The fermentation reactor can either be single-stage or multistage, with the latter having the option of either being set up in series or in parallel. It is interesting to note that a large number of CSTRs in series may have similar performance to a single PFR.

Bioethanol production is mainly done from starchy substrates such as corn, wheat, rice, barley, and potatoes, as well as from sucrose-containing feedstocks such as sugar beet, sweet sorghum, and sugarcane (Patni et al. 2013). In fact, the world's leading sources of ethanol production are corn from the United States and sugarcane from Brazil (Crago et al. 2010). However, there is an increasing interest for the use of organic waste such as lignocellulosic biomass (dry plant matter) for bioethanol production, which would avoid the use of edible crops for bioethanol production. Lignocellulosic biomass is mainly composed of cellulose, hemicellulose, and lignin. Major sources of lignocellulosic biomass include wood, agricultural residues, herbaceous plants, municipal solid wastes, as well as algae. Bioethanol production from cellulosic feedstocks consists of four key steps (Limayem and Ricke 2012). The first step is a pretreatment consisting of enzymatic, thermal, and acid treatments. The treated biomass is then hydrolyzed into free monomer molecules readily available for fermentation conversion to bioethanol. The hydrolysis is either done through acidic or enzymatic reactions. The hydrolyzed sugars then undergo the fermentation in presence of microorganism, where they can be converted into alcohol, as well as into lactic acid or other end products, depending on the microorganisms and fermentation condition used. The final step is focused of the separation and purification of the ethanol produced from water or other by-products through a distillation process.

Nutrients such as nitrogen can be recovered from fermentation waste by electrodialysis (Lee et al. 2003). Otherwise, there is room for valorization of fermentation waste, which can be used as a metal biosorbent (Vijayaraghavan et al. 2008; Gulati et al. 2002) or as a concrete biomodifier, for example (Bolobova and Kondrashchenko 2000).

3.3 Composting

Composting is the aerobic biological decomposition of organic waste such as food or plant material by bacteria, fungi, worms, and other organisms into compost. Compost is rich in carbon and nutrients and therefore presents benefits when applied to soils by increasing the nutrient supply, sequestering carbon, acting as a pesticide, increasing crop yield, decreasing soil erosion, increasing soil workability, and increasing the nutritional quality of crops (Lazcano et al. 2014; Lairon 2011). Composting can also be undertaken in anaerobic conditions (Minale and Worku 2014), in which case it is equivalent to the anaerobic digestion treated earlier. This section will only focus on aerobic composting. Following the anaerobic digestion process described previously, the digestate can be composted to increase its suitability for use on agricultural land (Bustamante et al. 2012; Bustamante et al. 2013).

Generally, the composting process is divided into four stages: mesophilic, thermophilic, cooling, and maturation (Chen et al. 2011; Tortosa et al. 2017). Composting is started by mesophilic organisms operating between 25 and 45 °C. The decomposition of the biomass leads to a rapid increase in the temperature of the compost, reaching 50–60 °C within the first 72 h. At these higher temperatures, thermophilic organisms take the lead in decomposing the materials. This second stage lasts significantly longer than the first one (several days or weeks). The high temperatures during this stage lead to the added benefit of destroying pathogens, weed seeds, and phytotoxic compounds. However, proper temperature control is crucial. If the compost pile goes above 70 °C, there is a high risk of killing the thermophilic organisms, as well as increasing the risk of fires. Partway through this second stage, the decomposition of the compost will render less nutrients and energy available to sustain the microorganisms. The temperature will start to decrease progressively, leading into the third stage of composting: the cooling stage. During this stage, the temperature continually decreases, leading mesophilic microbial organisms to take back the reins. Once the temperature stabilizes, the final stage, known as maturation, begins. By the end of this process, which can last several months and even more than a year, the original biomass will have been decomposed into compost which can be spread over agricultural land or for home use. The first two phases are often referred to as the active phase, whereas the final two are known as the curing phase. Overall, the composting process can last between 3 months and 2 years (United States Environmental Protection Agency 2018).

Throughout the whole of this process, it is paramount that the compost receives oxygen. If proper aeration of the compost pile is not provided, the microorganisms will quickly use up all the oxygen, causing the pile to transition into an anaerobic environment. Therefore, the composting process would be drawn out and take much longer due to non-optimal operating conditions. To ensure proper aeration, there are a variety of techniques that can be used: aerated windrow composting, aerated static pile composting, and in-vessel composting (Neher et al. 2013).

Aerated windrow composting consists in making long rows, known as windrows, of organic waste and aerating the piles by turning them frequently (Kuchenrither et al. 1985). On the other hand, aerated static pile composting uses one large pile of organic waste, supplemented with bulking material such as wood chips, paper, and cardboard, to create spacing within the pile and allow air to circulate through it. Pipes can be used to draw air in or out of the pile (Ekinci et al. 2017).

For in-vessel composting, the organic waste is placed into an enclosed vessel that allows control of the air flow and temperature. The organic waste is mixed by mechanical means. This method allows for a reduction of space compared to the previous two methods, a strong control of process parameters and a control of odors, and offers the added benefit of potential integration with anaerobic digestion. Indeed, some processes have been developed wherein, following anaerobic digestion, the same reactor is used to allow aeration of the organic waste (Walker et al. 2009).

In all these cases, composting remains the process of decomposing organic waste, meaning that a significant amount of greenhouse gases (GHGs) are produced. It is therefore necessary to treat the air adequately to control these gases, most notably nitrous oxides and methane (Nigussie et al. 2016). Furthermore, a large amount of the initial nitrogen is lost during the process, being converted into NH_3, NO_x, N_2O, and N_2, which become more volatile at higher temperatures (thermophilic stage). Composting tends to have high levels of N loss, ranging from 16 to 88% depending on the waste source, therefore reducing the quality of the compost and producing undesired gases (Vu et al. 2015). It is estimated that 35–70% of this loss occurs during the thermophilic active phase (Chowdhury et al. 2014).

There exists another form of composting which allows circumventing some of these issues. Vermicomposting focuses on the use of worms for decomposition at temperatures between 12 and 25 °C with the help of microorganisms (Ndegwa and Thompson 2001). This type of decomposition takes place in a psychrophilic temperature range, avoiding many of the problems previously identified. The worms are inserted into the organic matter, which has been supplemented with bulking agents to make it more attractive to the earthworms (Lim et al. 2016), and are left to eat and digest it. The product of vermicomposting is also somewhat different from normal compost. It is known as vermicompost (or casting) and tends to be richer in nutrients, which are also more readily accessible by the plants, than compost (Lazcano et al. 2009). Vermicompost also has added benefit for soil and plant life due to some by-products excreted by the worms, such as mucus and urine which are rich in carbon and ammonium and urea, respectively (Quaik and Ibrahim 2013). By operating at lower temperatures than the thermophilic stages of composting, vermicomposting also allows for a reduction of nitrogen loss and methane emissions (Nigussie et al. 2016). However, these low temperatures also hinder the ability to destroy pathogens observed during composting (Ndegwa and Thompson 2001).

Composting can lead to the production of leachate, which arises when water passes through and leaches out some of the contents from the waste (Chatterjee et al. 2013). This liquid tends to be high in nitrogen and organic matter. Depending on regulatory constraints, this leachate can potentially be used as a fertilizer (Romero et al. 2013). It could therefore be possible to treat the leachate to produce struvite through precipitation and/or to use ammonia stripping and absorption to produce ammonium sulfate. The production of these solid fertilizers could cut down on the issues related with transportation and storage of a liquid.

Overall, composting is a relatively easy to implement method that allows for both carbon and nutrient recovery but also produces an important amount of GHGs which must be considered. The long curing time can be viewed as a drawback to this process; however, its general acceptability and ease of use when compared to anaerobic digestion make it an alluring treatment and valorization method.

3.4 Landfills with Gas Capture

A landfill is a site for the disposal of waste materials by burial. Landfills are a significant source of greenhouse gas emissions, with the organic materials decomposing in them producing a gas comprised of around 50% methane and 50% carbon dioxide with trace amounts of non-methane organic compounds (Themelis and Ulloa 2007). This gas mixture is known as landfill gas (LFG). Gas capture technologies can be integrated within the landfills to capture the greenhouse gases from LFG. Landfills with gas capture are largely focused on capturing the methane, but the carbon dioxide could technically be captured as well.

To capture the LFGs, wells are installed throughout the waste in either vertical or horizontal configurations. These wells extract the gases by producing a pressure gradient by using blowers or air compressors. The difference in pressure decreases as the distance from the well increases, meaning that proper well spacing is necessary to achieve efficient gas capture. The landfill can be covered or left open to the air, but open landfills are less efficient at capturing the gas (approximately 17% less efficient). However, methane content is reduced by around 6% in closed landfills (Powell et al. 2016).

Nutrient recovery from landfills is also a possibility (Li and Zhao 2003; Kulikowska and Klimiuk 2008). Landfill leachate, similarly to composting leachate, can be a good source of nutrients, such as nitrogen (ammonia) and phosphorus. Therefore, struvite precipitation and ammonia stripping can be excellent techniques to valorize this by-product.

4 Chemical Conversion

Biomass can be converted into biofuels using purely chemical reactions. Key reactions involved in the conversion of biomass includes hydrolysis, dehydration, isomerization, aldol condensation, reforming, hydrogenation or hydrogenolysis, and oxidation (Jiang et al. 2016). The most common reaction used in bioenergy contexts is transesterification, which converts triglycerides (oils, fats, and greases) into biodiesel. With the help of an acid or base catalyst, fatty acids from oils, fats, and greases are bonded to alcohol, which reduces the viscosity of the fatty acids and makes them combustible, thus producing biodiesel. The transesterification reaction produces glycerol as a by-product, which leaves room for valorization such as

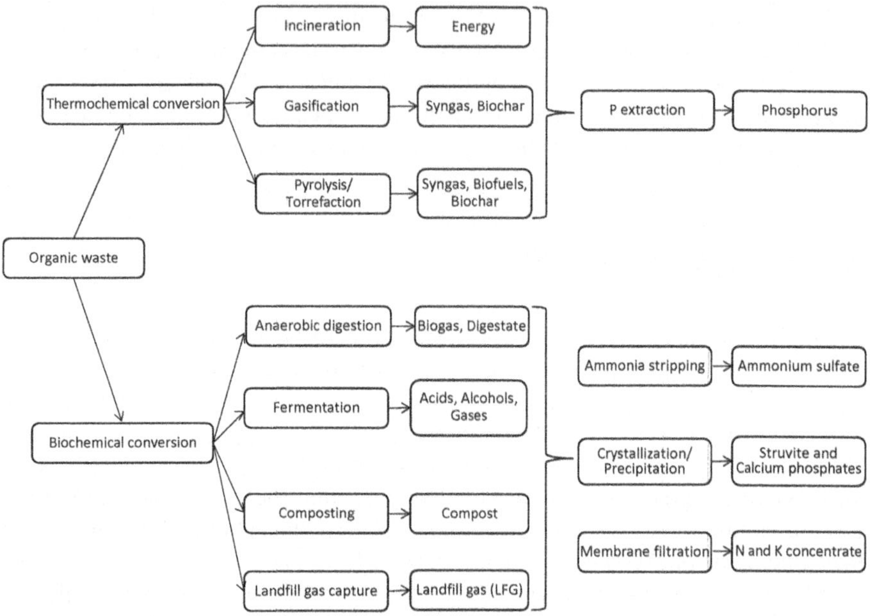

Fig. 1 Biomass conversion processes for nutrient and carbon recovery from organic waste and the main recovered end products

biofuel production by glycerol fermentation as well as hydrogen production by glycerol steam reforming.

Based on the above discussion, Fig. 1 provides a summary overview of the principal thermochemical and biochemical conversion technologies for nutrient and carbon recovery from organic waste, as well as the main recovered end products.

5 Understanding the Carbon Balance and Adding Value through Carbon Capture Technologies

Many of the technologies that have been presented in this section aim at converting biomass into energy by either burning it directly (incineration) or by using it to produce various gases which are then destined for combustion (gasification, pyrolysis, biomethanation, LFG capture). Many of these alternatives are considered to be eco-friendly or "green" when compared to simply burying waste in a landfill. However, some may wonder why this is the case; at the end of the day, the gases are still being combusted and released into the atmosphere. The answer to this can be found when examining the carbon balance of the alternatives.

The carbon cycle refers to the biogeochemical cycle of carbon on Earth. It is separated into three subcategories, which includes two types of biological carbon cycle (terrestrial and oceanic) and geological carbon cycle. The carbon balance refers to the exchange between these categories. While the carbon cycle as a whole is a self-sustainable cycle that converts or captures as much CO_2 as it produces or emits, greenhouse gas emissions from human activities create a surplus of carbon dioxide to the atmosphere (Falkowski et al. 2000). This disturbs the natural carbon balance, thereby causing a continuous increase of atmospheric CO_2 concentrations, contributing to global warming and climate change. Major sources of anthropogenic greenhouse gas emissions include energy supply, transport, industries, forestry and agriculture, as well as commercial and residential sectors (Victor et al. 2014). Although biomass actively absorbs CO_2 from the atmosphere through photosynthesis as it grows, the CO_2 is released back to the atmosphere as biodegradation of biomass takes place, combined with methane and other gases. While decomposition is largely carbon-neutral when considering biomass growth, a reduction of these emissions into the atmosphere could be valuable in the current fight against global warming. By burning syngas, biogas, or LFG, the methane which would be released to the atmosphere is converted into CO_2, which has a global warming potential much lower than methane.

It is possible to move the carbon balance to a point of "negative emissions" by integrating bioenergy with carbon capture and storage (BECCS) technologies (Biorecro 2010). Carbon capture and storage, or CCS, is mainly used for natural gas or coal power plants, consisting of capturing CO_2 emissions and storing them underground (such as in oil wells, coal seams and saline aquifers) to prevent them from reaching the atmosphere, thus causing a net reduction of greenhouse gas emissions from those energy sources (Ketzer et al. 2012). Other key applications for carbon capture and storage include cement, metal, and chemical industries. Carbon capture is mainly done by gas-liquid absorption, but other key methods proposed include adsorption, membrane separation, chemical looping combustion, cryogenic distillation, hydrate-based separation, and enzymatic capture (Leung et al. 2014). BECCS consist in implementing such technologies for the capture and storage of CO_2 emissions from combustion, fermentation, putrefaction, biodegradation, and other biological processes (Biorecro 2010). BECCS could also be implemented for the combustion of biogas produced from biomethanation. The captured CO_2 can then be valorized, which consists of using or transforming the CO_2 for other purposes, or stored under bedrock, leading to permanent storage of carbon dioxide that natural carbon sinks and valorization may not provide. Geological storage of CO_2 emissions can be combined with mineralization, which converts CO_2 into solid minerals, thus preventing any potential risk of CO_2 leakage from the storage sites (Matter et al. 2016).

CO_2 can be valorized in a number of ways. It can be used directly to supplement plant growth in greenhouses (Jaffrin et al. 2003), to aid in enhanced oil recovery by using it to displace oil (Whittaker and Perkins 2013), to act as a coolant or refrigerant (Lorentzen 1994), to act as a flame retardant for fire extinguishers (United States Department of Labor 2002), and for carbonation in food and beverage (Energy Research Centre of the Netherlands 2002), just to name a few. Otherwise, CO_2 can be converted into cements (Biello 2008), plastics, fuels, and a variety of useful chemicals, such as carboxylates, lactones, carbamates, urea, isocyanates, carbamate, alcohols, ethers, and hydrocarbons, or serve as a catalyst or a co-reactant for the conversion of organic compounds (Olajire 2013). Carbon dioxide can even be used to generate energy or electricity, thus turning unwanted greenhouse gas emissions into a renewable energy source. Energy can either be generated through electrochemical conversion of CO_2 that generates electricity (Al Sadat and Archer 2016) or in the form of supercritical CO_2 as a substitute to water for steam turbines, such as those used for natural gas, coal, nuclear, geothermal, and solar power plants (Ahn et al. 2015).

References

Acelas NY, López DP, Brilman DW, Kersten SR, Kootstra AMJ (2014) Supercritical water gasification of sewage sludge: gas production and phosphorus recovery. Bioresour Technol 174:167–175

Acharya B, Sule I, Dutta A (2012) A review on advances of torrefaction technologies for biomass processing. Biomass Convers Biorefin 2:349–369

Adam C, Peplinski B, Michaelis M, Kley G, Simon F-G (2009) Thermochemical treatment of sewage sludge ashes for phosphorus recovery. Waste Manag 29:1122–1128

ADEME Bourgogne (2013) La méthanisation agricole en voie sèche discontinue. Agence de l'Environnement et de la Maîtrise de l'Energie

Adhikari BK, Barrington S, Martinez J, King S (2008) Characterization of food waste and bulking agents for composting. Waste Manag 28:795–804

Ahn Y, Bae SJ, Kim M, Cho SK, Baik S, Lee JI, Cha JE (2015) Review of supercritical CO2 power cycle technology and current status of research and development. Nucl Eng Technol 47:647–661

Al Sadat WI, Archer LA (2016) The O2-assisted Al/CO2 electrochemical cell: A system for CO2 capture/conversion and electric power generation. Sci Adv 2:e1600968

Angelonidi E, Smith SR (2015) A comparison of wet and dry anaerobic digestion processes for the treatment of municipal solid waste and food waste. Water Environ J 29:549–557

Arena U (2012) Process and technological aspects of municipal solid waste gasification. A review. Waste Manag 32:625–639

Ariunbaatar J, Panico A, Esposito G, Pirozzi F, Lens PN (2014) Pretreatment methods to enhance anaerobic digestion of organic solid waste. Appl Energy 123:143–156

Arthington J, Rechcigl J, Yost G, McDowell L, Fanning M (2002) Effect of ammonium sulfate fertilization on bahiagrass quality and copper metabolism in grazing beef cattle 1, 2. J Anim Sci 80:2507–2512

Asadullah M (2014) Barriers of commercial power generation using biomass gasification gas: a review. Renew Sustain Energy Rev 29:201–215

Atienza–Martínez M, Gea G, Arauzo J, Kersten SR, Kootstra AMJ (2014) Phosphorus recovery from sewage sludge char ash. Biomass Bioenergy 65:42–50

Bajpai P (2017) Basics of anaerobic digestion process. In: Anaerobic technology in pulp and paper industry. Springer, Berlin

Biello D (2008) Cement from CO2: a concrete cure for global warming? Sci Am 7:61

Biorecro AB (2010) Global status of BECCS projects 2010. Global CCS Institute, Melbourne

Bolobova A, Kondrashchenko V (2000) Use of yeast fermentation waste as a biomodifier of concrete. Appl Biochem Microbiol 36:205–214

Brethauer S, Wyman CE (2010) Continuous hydrolysis and fermentation for cellulosic ethanol production. Bioresour Technol 101:4862–4874

Bustamante M, Alburquerque J, Restrepo A, De la Fuente C, Paredes C, Moral R, Bernal M (2012) Co-composting of the solid fraction of anaerobic digestates, to obtain added-value materials for use in agriculture. Biomass Bioenergy 43:26–35
Bustamante M, Restrepo A, Alburquerque J, Pérez-Murcia M, Paredes C, Moral R, Bernal M (2013) Recycling of anaerobic digestates by composting: effect of the bulking agent used. J Clean Prod 47:61–69
Carpenter D, Westover TL, Czernik S, Jablonski W (2014) Biomass feedstocks for renewable fuel production: a review of the impacts of feedstock and pretreatment on the yield and product distribution of fast pyrolysis bio-oils and vapors. Green Chem 16:384–406
Chandler AJ, Eighmy TT, Hjelmar O, Kosson D, Sawell S, Vehlow J, Van der Sloot H, Hartlén J (1997) Municipal solid waste incinerator residues. Elsevier, Amsterdam
Chatterjee N, Flury M, Hinman C, Cogger CG (2013) Chemical and physical characteristics of compost leachates. A review report prepared for the Washington State Department of Transportation. Washington State University
Chen L, Neibling H (2014) Anaerobic digestion basics. University of Idaho Extension, Moscow, p 6
Chen L, De Haro M, Moore A, Falen C (2011) The composting process: dairy compost production and use in Idaho CIS 1179. University of Idaho, Moscow
Cho S-K, Im W-T, Kim D-H, Kim M-H, Shin H-S, Oh S-E (2013) Dry anaerobic digestion of food waste under mesophilic conditions: performance and methanogenic community analysis. Bioresour Technol 131:210–217
Chowdhury MA, de Neergaard A, Jensen LS (2014) Potential of aeration flow rate and biochar addition to reduce greenhouse gas and ammonia emissions during manure composting. Chemosphere 97:16–25
Cieślik B, Konieczka P (2017) A review of phosphorus recovery methods at various steps of wastewater treatment and sewage sludge management. The concept of "no solid waste generation" and analytical methods. J Clean Prod 142:1728–1740
Crago CL, Khanna M, Barton J, Giuliani E, Amaral W (2010) Competitiveness of Brazilian sugarcane ethanol compared to US corn ethanol. Energy Policy 38:7404–7415
David A (2013) Technical document on municipal solid waste organics processing. Environment Canada
DeBruyn J, Hilborn D (2007) Anaerobic digestion basics. Ontario Ministry of Agriculture, Food and Rural Affairs, Guelph, ON, p 10
Di Maria F, Micale C (2015) Life cycle analysis of incineration compared to anaerobic digestion followed by composting for managing organic waste: the influence of system components for an Italian district. Int J Life Cycle Assess 20:377–388
Drosg B, Fuchs W, Al Seadi T, Madsen M, Linke B (2015) Nutrient recovery by biogas digestate processing. IEA Bioenergy, Dublin, pp 7–11
Ekinci K, Tosun I, Seyit Ahmet I, Memici M, Kumbul BS (2017) Design and construction of a pilot scale aerated static pile composting systems. Sci Papers Ser E Land Reclam Earth Observ Survey Environ Eng 6:7–12
Elser J, Bennett E (2011) Phosphorus cycle: a broken biogeochemical cycle. Nature 478:29
Energy Research Centre of the Netherlands (2002) Workshop on carbon capture and storage. Intergovernmental Panel on Climate Change, Geneva
Engels C, Kirkby E, White P (2012) Mineral nutrition, yield and source–sink relationships. Marschner's mineral nutrition of higher plants, 3rd edn. Elsevier, Amsterdam
Falkowski P, Scholes R, Boyle E, Canadell J, Canfield D, Elser J, Gruber N, Hibbard K, Högberg P, Linder S (2000) The global carbon cycle: a test of our knowledge of earth as a system. Science 290:291–296
Guedes P, Couto N, Ottosen LM, Ribeiro AB (2014) Phosphorus recovery from sewage sludge ash through an electrodialytic process. Waste Manag 34:886–892

Guendouz J, Buffière P, Cacho J, Carrère M, Delgenes J-P (2010) Dry anaerobic digestion in batch mode: design and operation of a laboratory-scale, completely mixed reactor. Waste Manag 30:1768–1771

Gulati R, Saxena R, Gupta R (2002) Fermentation waste of Aspergillus terreus: a potential copper biosorbent. World J Microbiol Biotechnol 18:397–401

Hansen V, Müller-Stöver D, Ahrenfeldt J, Holm JK, Henriksen UB, Hauggaard-Nielsen H (2015) Gasification biochar as a valuable by-product for carbon sequestration and soil amendment. Biomass Bioenergy 72:300–308

Hoornweg D, Bhada-Tata P (2012) What a waste: a global review of solid waste management. World Bank, Washington, DC

Jaffrin A, Bentounes N, Joan AM, Makhlouf S (2003) Landfill biogas for heating greenhouses and providing carbon dioxide supplement for plant growth. Biosyst Eng 86:113–123

Jahirul MI, Rasul MG, Chowdhury AA, Ashwath N (2012) Biofuels production through biomass pyrolysis—a technological review. Energies 5:4952–5001

Jay K, Stieglitz L (1995) Identification and quantification of volatile organic components in emissions of waste incineration plants. Chemosphere 30:1249–1260

Jiang Y, Wang X, Cao Q, Dong L, Guan J, Mu X (2016) Chemical conversion of biomass to green chemicals. Sustainable production of bulk chemicals. Springer, Berlin

Kalmykova Y, Fedje KK (2013) Phosphorus recovery from municipal solid waste incineration fly ash. Waste Manag 33:1403–1410

Kataki S, West H, Clarke M, Baruah DC (2016) Phosphorus recovery as struvite: Recent concerns for use of seed, alternative Mg source, nitrogen conservation and fertilizer potential. Resour Conserv Recycl 107:142–156

Ketzer JM, Iglesias RS, Einloft S (2012) Reducing greenhouse gas emissions with CO2 capture and geological storage. In: Handbook of climate change mitigation. Springer, Berlin

Kleemann R, Chenoweth J, Clift R, Morse S, Pearce P, Saroj D (2017) Comparison of phosphorus recovery from incinerated sewage sludge ash (ISSA) and pyrolysed sewage sludge char (PSSC). Waste Manag 60:201–210

Kuchenrither R, Martin W, Smith D, Williams D (1985) Design and operation of an aerated windrow composting facility. J Water Pollut Control Feder 57:213–219

Kulikowska D, Klimiuk E (2008) The effect of landfill age on municipal leachate composition. Bioresour Technol 99:5981–5985

Laird DA, Brown RC, Amonette JE, Lehmann J (2009) Review of the pyrolysis platform for coproducing bio-oil and biochar. Biofuels Bioprod Biorefin 3:547–562

Lairon D (2011) Nutritional quality and safety of organic food. Sustainable agriculture, vol 2. Springer, Berlin

Latifian M, Liu J, Mattiasson B (2012) Struvite-based fertilizer and its physical and chemical properties. Environ Technol 33:2691–2697

Lazcano C, Arnold J, Zaller J, Martín JD, Salgado AT (2009) Compost and vermicompost as nursery pot components: effects on tomato plant growth and morphology. Span J Agric Res 944:951

Lazcano C, Martínez-Blanco J, Christensen TH, Muñoz P, Rieradevall J, Møller J, Antón A, Boldrin A, Nuñez M (2014) Environmental benefits of compost use on land through LCA–a review of the current gaps. Proceedings of the 9th International Conference on Life Cycle Assessment in the Agri-Food Sector (LCA Food 2014), San Francisco, California, USA, 8–10 October, 2014. American Center for Life Cycle Assessment

Lee H-J, Oh S-J, Moon S-H (2003) Recovery of ammonium sulfate from fermentation waste by electrodialysis. Water Res 37:1091–1099

Lettinga G, Rebac S, Zeeman G (2001) Challenge of psychrophilic anaerobic wastewater treatment. Trends Biotechnol 19:363–370

Leung DY, Caramanna G, Maroto-Valer MM (2014) An overview of current status of carbon dioxide capture and storage technologies. Renew Sustain Energy Rev 39:426–443

Li X, Zhao Q (2003) Recovery of ammonium-nitrogen from landfill leachate as a multi-nutrient fertilizer. Ecol Eng 20:171–181

Lim SL, Lee LH, Wu TY (2016) Sustainability of using composting and vermicomposting technologies for organic solid waste biotransformation: recent overview, greenhouse gases emissions and economic analysis. J Clean Prod 111:262–278
Limayem A, Ricke SC (2012) Lignocellulosic biomass for bioethanol production: current perspectives, potential issues and future prospects. Progr Energ Combust Sci 38:449–467
Lindmark J, Eriksson P, Thorin E (2014) The effects of different mixing intensities during anaerobic digestion of the organic fraction of municipal solid waste. Waste Manag 34:1391–1397
Lorentzen G (1994) Revival of carbon dioxide as a refrigerant. Int J Refriger 17:292–301
Massey MS, Davis JG, Sheffield RE, Ippolito JA (2007) Struvite production from dairy wastewater and its potential as a fertilizer for organic production in calcareous soils. International Symposium on Air Quality and Waste Management for Agriculture, 16–19 September 2007, Broomfield, Colorado. American Society of Agricultural and Biological Engineers
Matter JM, Stute M, Snæbjörnsdottir SÓ, Oelkers EH, Gislason SR, Aradottir ES, Sigfusson B, Gunnarsson I, Sigurdardottir H, Gunnlaugsson E (2016) Rapid carbon mineralization for permanent disposal of anthropogenic carbon dioxide emissions. Science 352:1312–1314
Minale M, Worku T (2014) Anaerobic co-digestion of sanitary wastewater and kitchen solid waste for biogas and fertilizer production under ambient temperature: waste generated from condominium house. Int J Environ Sci Technol 11:509–516
Mor S, Ravindra K, De Visscher A, Dahiya R, Chandra A (2006) Municipal solid waste characterization and its assessment for potential methane generation: a case study. Sci Total Environ 371:1–10
Ndegwa P, Thompson S (2001) Integrating composting and vermicomposting in the treatment and bioconversion of biosolids. Bioresour Technol 76:107–112
Neher DA, Weicht TR, Bates ST, Leff JW, Fierer N (2013) Changes in bacterial and fungal communities across compost recipes, preparation methods, and composting times. PLoS One 8:e79512
Nigussie A, Kuyper TW, Bruun S, de Neergaard A (2016) Vermicomposting as a technology for reducing nitrogen losses and greenhouse gas emissions from small-scale composting. J Clean Prod 139:429–439
Noyola A, Morgan-Sagastume JM, Lopez-Hernandez JE (2006) Treatment of biogas produced in anaerobic reactors for domestic wastewater: odor control and energy/resource recovery. Rev Environ Sci Biotechnol 5:93–114
Olajire AA (2013) Valorization of greenhouse carbon dioxide emissions into value-added products by catalytic processes. J CO2 Utilization 3:74–92
Ontario Ministry of the Environment (2008a) Design guidelines for sewage works: sludge stabilization
Ontario Ministry of the Environment (2008b) Design guidelines for sewage works: sludge thickening and dewatering
Oram NJ, van de Voorde TF, Ouwehand G-J, Bezemer TM, Mommer L, Jeffery S, Van Groenigen JW (2014) Soil amendment with biochar increases the competitive ability of legumes via increased potassium availability. Agr Ecosyst Environ 191:92–98
Patni N, Pillai SG, Dwivedi AH (2013) Wheat as a promising substitute of corn for bioethanol production. Proc Eng 51:355–362
Petzet S, Peplinski B, Cornel P (2012) On wet chemical phosphorus recovery from sewage sludge ash by acidic or alkaline leaching and an optimized combination of both. Water Res 46:3769–3780
Powell JT, Townsend TG, Zimmerman JB (2016) Estimates of solid waste disposal rates and reduction targets for landfill gas emissions. Nat Clim Change 6:162–165
Quaik S, Ibrahim MH (2013) A review on potential of vermicomposting derived liquids in agricultural use. Int J Sci Res Publ 3:1–6
Rahman MM, Salleh MAM, Rashid U, Ahsan A, Hossain MM, Ra CS (2014) Production of slow release crystal fertilizer from wastewaters through struvite crystallization—a review. Arab J Chem 7:139–155

Romero C, Ramos P, Costa C, Márquez MC (2013) Raw and digested municipal waste compost leachate as potential fertilizer: comparison with a commercial fertilizer. J Clean Prod 59:73–78
Samolada M, Zabaniotou A (2014) Comparative assessment of municipal sewage sludge incineration, gasification and pyrolysis for a sustainable sludge-to-energy management in Greece. Waste Manag 34:411–420
Shanableh A, Ginige P (1999) Impact of metals bioleaching on the nutrient value of biological nutrient removal biosolids. Water Sci Technol 39:175–181
Shanableh A, Omar M (2003) Bio-acidification and leaching of metals, nitrogen, and phosphorus from soil and sludge mixtures. Soil Sediment Contam 12:565–589
Siddiqui Z, Horan N, Anaman K (2011) Optimisation of C: N ratio for co-digested processed industrial food waste and sewage sludge using the BMP test. Int J Chem React Eng 9:1–9
Steffen R, Szolar O, Braun R (1998) Feedstocks for anaerobic digestion. Institute of Agrobiotechnology Tulin, University of Agricultural Sciences, Vienna
Tan Z, Lagerkvist A (2011) Phosphorus recovery from the biomass ash: a review. Renew Sustain Energy Rev 15:3588–3602
Themelis NJ, Ulloa PA (2007) Methane generation in landfills. Renew Energy 32:1243–1257
Tortosa G, Castellano-Hinojosa A, Correa-Galeote D, Bedmar EJ (2017) Evolution of bacterial diversity during two-phase olive mill waste ("alperujo") composting by 16S rRNA gene pyrosequencing. Bioresour Technol 224:101–111
United States Department of Labor (2002) Portable fire extinguishers—1910.157. Occupational Safety and Health Standards
United States Environmental Protection Agency (2018) Types of Composting and Understanding the Process [Online]. Accessed May 2018
Vakalis S, Sotiropoulos A, Moustakas K, Malamis D, Vekkos K, Baratieri M (2016) Characterization of hotel bio-waste by means of simultaneous thermal analysis. Waste Biomass Valorization 7:649–657
Vaneeckhaute C, Lebuf V, Michels E, Belia E, Vanrolleghem PA, Tack FM, Meers E (2017) Nutrient recovery from digestate: systematic technology review and product classification. Waste Biomass Valorization 8:21–40
Victor D, Zhou D, Ahmed E, Dadhich P, Olivier J, Rogner H-H, Sheikho K, Yamaguchi M (2014) Climate change 2014: mitigation of climate change. Contribution of Working Group III to the Fifth Assessment Report of the Intergovernmental Panel on Climate Change: Introductory chapter. Cambridge University Press, Cambridge
Vijayaraghavan K, Lee MW, Yun Y-S (2008) Evaluation of fermentation waste (Corynebacterium glutamicum) as a biosorbent for the treatment of nickel (II)-bearing solutions. Biochem Eng J 41:228–233
Vu QD, de Neergaard A, Tran TD, Hoang HTT, Vu VTK, Jensen LS (2015) Greenhouse gas emissions from passive composting of manure and digestate with crop residues and biochar on small-scale livestock farms in Vietnam. Environ Technol 36:2924–2935
Walker L, Charles W, Cord-Ruwisch R (2009) Comparison of static, in-vessel composting of MSW with thermophilic anaerobic digestion and combinations of the two processes. Bioresour Technol 100:3799–3807
Whittaker S, Perkins E (2013) Technical aspects of CO2 enhanced oil recovery and associated carbon storage. Global CCS institute, Melbourne
Wong J, Xiang L, Gu X, Zhou L (2004) Bioleaching of heavy metals from anaerobically digested sewage sludge using FeS2 as an energy source. Chemosphere 55:101–107
World Energy Council (2016) World energy resources, waste to energy, 2016
Xie T, Reddy KR, Wang C, Yargicoglu E, Spokas K (2015) Characteristics and applications of biochar for environmental remediation: a review. Crit Rev Environ Sci Technol 45:939–969
Yan F, Zhang L, Hu Z, Cheng G, Jiang C, Zhang Y, Xu T, He P, Luo S, Xiao B (2010) Hydrogen-rich gas production by steam gasification of char derived from cyanobacterial blooms (CDCB) in a fixed-bed reactor: influence of particle size and residence time on gas yield and syngas composition. Int J Hydrogen Energy 35:10,212–10,217

Yi J, Dong B, Jin J, Dai X (2014) Effect of increasing total solids contents on anaerobic digestion of food waste under mesophilic conditions: performance and microbial characteristics analysis. PLoS One 9:e102548

Yuan H, Lu T, Wang Y, Chen Y, Lei T (2016) Sewage sludge biochar: nutrient composition and its effect on the leaching of soil nutrients. Geoderma 267:17–23

Zacco A, Borgese L, Gianoncelli A, Struis RP, Depero LE, Bontempi E (2014) Review of fly ash inertisation treatments and recycling. Environ Chem Lett 12:153–175

Zhang R, El-Mashad HM, Hartman K, Wang F, Liu G, Choate C, Gamble P (2007) Characterization of food waste as feedstock for anaerobic digestion. Bioresour Technol 98:929–935

The Future Perspectives of Dark Fermentation: Moving from Only Biohydrogen to Biochemicals

Juan-Rodrigo Bastidas-Oyanedel, Fabian Bonk, Mette Hedegaard Thomsen, and Jens Ejbye Schmidt

1 Introduction

The preservation and management of the diverse natural resources are essential for a sustainable development in the present and future. An economy based on sustainable processes requires safe and sustainable resources for industrial production, investment and finance system, ecological and health safety, and sustainability (Reddy et al. 1997; Tanaka et al. 2010). Fossil resources are not sustainable, and their availability is more than questionable in the long term. Therefore, it is essential to establish solutions that will reduce the rapid consumption of fossil resources, i.e., petroleum, natural gas, coal, and minerals. An approach is the stepwise conversion of large parts of the global economy into a sustainable biobased economy with bioenergy, biofuels, and biobased products as its main foundations (Kamm et al. 2006).

For energy generation, there is a variety of alternative renewable processes that can be established, e.g., solar, wind, tide, hydroelectric, biomass incineration,

J.-R. Bastidas-Oyanedel (✉)
Chemical Engineering Department, Khalifa University of Science and Technology, Abu Dhabi, United Arab Emirates
e-mail: juan.bastidas@ku.ac.ce

F. Bonk
Department of Environmental Microbiology, UFZ-Helmholtz Centre for Environmental Research, Leipzig, Germany

M. H. Thomsen
Department of Energy Technology, Aalborg University, Esbjerg, Denmark

J. E. Schmidt
Chemical Engineering Department, Khalifa University of Science and Technology, Abu Dhabi, United Arab Emirates

SDU-Department of Chemical Engineering, Biotechnology, and Environmental Technology, University of Southern Denmark, Campusvej 55, DK-5230 Odense M, Denmark
e-mail: jesc@kbm.sdu.dk

© Springer Nature Switzerland AG 2019

J.-R. Bastidas-Oyanedel, J. E. Schmidt (eds.), *Biorefinery*,
https://doi.org/10.1007/978-3-030-10961-5_15

nuclear fission, etc. However, products derived from petrochemical processes are ubiquitous in our present-day society, e.g., transport fuels, adhesives, and plastics, among others. Research has to be addressed toward new technologies capable of replacing/integrating both petrochemical technology and products in order to transform our society into a sustainable one (Kamm et al. 2006). Therefore, the chemical fuel industry and the industrial biotechnology will depend on the conversion of sustainable materials, i.e., will be based on biomass technologies.

Biomass technologies are a vast and interdisciplinary renewable concept in which the main role is played by solar energy recovery and CO2 transformation into biomass, i.e., photosynthesis. Under any sustainable scenario of biomass-based bioprocesses, organic residues will be produced. In this context, dark fermentation, also known as acidogenic mixed culture fermentation, appears as an attractive solution that will reduce those residues and increase the global efficiency of biomass-based production of energy and valuable chemicals, interlinked in a biorefinery concept.

As stated above, the dark fermentation of those organic residues yields energy carriers and fuels, such as hydrogen and ethanol, and chemicals like acetic acid, butyric acid, propionic acid, and lactic acid, among others. In present days, these compounds are mostly produced by petrochemical reforming, except for ethanol and lactic acid that are mainly produced by bioprocesses. The production of those compounds by dark fermentation makes it attractive as a petrochemical refinery alternative.

In recent years, dark fermentation has been extensively studied and reviewed for its capacity of producing hydrogen (Hawkes et al. 2002; Liu et al. 2002; Lin and Lay 2004; Karlsson et al. 2008; Hallenbeck 2009; Guo et al. 2010; Rittmann and Herwig 2012). Nevertheless, as is shown in Sect. 2.2, in the best-case scenario, i.e., assuming a maximum hydrogen yield of 4 molH2/molglucose, the produced hydrogen represents only a 4% of the total substrate mass consumed, while 67% corresponds to products in the liquid phase, e.g., acetic acid (Bastidas-Oyanedel et al. 2015). Hence this chapter also explores other options besides hydrogen production from dark fermentation. It aims to present and discuss the current and future status of dark fermentation in the biorefinery concept.

The first half of the review presents dark fermentation metabolic pathways, product yields and the technological importance in the present industry, microorganisms present and responsible for mixed dark fermentation, and operational parameters affecting dark fermentation, e.g., substrates, pH, temperature, and headspace composition. This section also presents the process costs of conventional dark fermentation. The second half presents and discusses the perspectives and future of dark fermentation as a core bioprocess. Links with other (bio)processes are also explored, e.g., acetone-butanol-ethanol (ABE) fermentation, syngas fermentation, biohythane, bioelectrochemical systems, and catalytic hydrocarbon synthesis.

2 Dark Fermentation: Current Status

This section presents the dark fermentation biochemistry and metabolism; product consideration as product yields, prices, and technological importance; and the operational factors that affect dark fermentation, e.g., microbial population structure, inocula sources, substrates and nutrients, pH, temperature, gas composition, and bioreactor configuration.

2.1 The Biochemistry of Dark Fermentation

Dark fermentation involves the transformation of organic compounds to various inorganic and organic products. During this process, a portion of an organic compound may be oxidized, while another portion is reduced. It is from this oxidation-reduction of organic compounds that dark fermentative microorganisms obtain their energy, producing numerous simplistic and soluble organic compounds (Gerardi 2003).

The dark fermentation process begins with bacterial hydrolysis of the fed organic materials in order to break down insoluble polymers, making them available for microorganism consumption. Acidogenic bacteria then convert the products of hydrolysis into carbon dioxide, hydrogen, ammonia, alcohols, and organic acids. Hydrolysis and acidogenesis are detailed here below.

2.1.1 Hydrolysis

Hydrolysis is the first step required for anaerobic microbial utilization of complex polymers (Aceves-Lara et al. 2008c). Hydrolytic fermentative bacteria facilitate the extracellular enzymatic hydrolysis of the initial complex organic matter formed by polymers such as polysaccharides, proteins, and fats. The hydrolases (enzymes) that catalyze these reactions are cellulase, amylase, protease, and lipase, among others. Hydrolases may be secreted to the extracellular environment or be bound to the cell surface (Pavlostathis and Gossett 1988; Mitchell and Gu 2010).

Polysaccharides are generally converted into simple monomeric or dimeric sugars. Hydrolysis of starch and cellulose yields glucose as monomeric sugar, while hemicellulose is degraded to galactose, arabinose, xylose, mannose, and glucose. Proteins are broken down into amino acids, small peptide, ammonia, and carbon dioxide by proteases. Lipids are hydrolyzed into long- and short-chain fatty acids and glycerol by lipases (Gerardi 2003; Insam et al. 2010; Mitchell and Gu 2010). The hydrolysis products then become substrates for the fermentation processes that follow.

2.1.2 Acidogenesis

Monosaccharides and amino acids, released after the hydrolysis of their respective polymers, serve as substrates to the acidogenic fermentation (Insam et al. 2010). Groups of facultative and anaerobic fermentative or anaerobic oxidizing organisms utilize these substrates yielding compounds such as ethanol, acetate, propionate, H2, and CO2 as intermediary products.

Acidogenesis is a process in which intracellular reduced cofactors such as NADH are being oxidized. The regeneration of cofactors is vital to the process as a whole, as these cofactors serve as intermediary electron acceptors in the catabolic reactions that proceed continuously within the system. The regeneration of these cofactors is done by the production of the alcohols and organic acids yielded in this process (Gujer and Zehnder 1983; Zehnder and Svensson 1986; Gerardi 2003).

2.1.3 Acidogenesis Fermentation Metabolism

In a mixed culture acidogenic fermentation, where many microbial species are present, most of the fermentative conversions explained below are possible. Further detailed dark fermentation metabolic models can be found in the literature (Bastidas-Oyanedel et al. 2008; Gonzalez-Cabaleiro et al. 2015; Zhang et al. 2013c). The fermentation patterns observed are the result of the combined effect and interaction among the microbial consortium. Acidogenic bacteria are capable of performing a variety of oxidation-reduction reactions involving organic compounds, carbon dioxide, and molecular hydrogen. Acidogenic bacteria include facultative anaerobes, aerotolerant anaerobes, and strict anaerobes. In such an ecosystem, the microorganism consortia activity depends on environmental condition changes, e.g., gas-phase composition or pH for instance. Environmental changes will conduct changes on the consortia metabolism and/or physiology, affecting the types and quantities of compounds that are produced through fermentation (Gerardi 2003).

Metabolism is the sum of all biochemical reactions performed by a living organism. The reactions have two main purposes: (1) to generate energy for the organism and (2) to build new cell material. Reactions related to generating energy are called catabolism, and all reactions leading to formation of cell material are called anabolism. Metabolism is initiated by the consumption of biodegradable substrates. The most common substrate found in organic residues is glucose. The acidogenic fermentation of glucose extracts energy in the form of ATP by substrate-level phosphorylation during oxidative substrate breakdown (Thauer et al. 1977).

The resulting reducing equivalents, in the form of NADH and reduced ferredoxin, are transferred into metabolic intermediates (Angenent et al. 2004; Zhang et al. 2013c), leading to the formation of a variety of reduced products such as H2, ethanol, and organic acids (lactic, propionic, formic, acetic, butyric acids), depending on the fermentation pathways utilized (Louis and Flint 2009). Acidogenic bacteria generally possess alternative pathways leading to the formation of these products. Hence the relative proportions of the different products formed depend on the environmental conditions.

Figure 1 illustrates the main catabolic pathways in acidogenic glucose fermentation. Glycolysis (or Embden-Meyerhof-Parnas pathway) is thought to be the archetype of a universal metabolic pathway (Kurzynski 2006). It occurs, with variations, in nearly all organisms, both aerobic and anaerobic. The wide occurrence of glycolysis indicates that it is one of the most ancient metabolic pathways known (Romano and Conway 1996). Glycolysis is a sequence of ten reactions. In Fig. 1, reaction (1), it is lumped into one reaction starting from glucose giving pyruvate. In this process, the consumption of 0.5 mol of glucose produces 1 mol of ATP, which is then used as energy source for biomass growth and maintenance processes.

The NADH resulting from glycolysis should be reconverted to NAD+ to allow glycolysis to continue. In anaerobic conditions, organisms are able to oxidase NADH back to NAD+ in several ways. One method is lactic acid fermentation, where pyruvate is converted into lactate, reaction (2). Propionic acid is formed also as a reduced product of glucose fermentation. Propionic acid can be produced through the so-called acrylate pathway (Tholozan et al. 1992) or via the succinate pathway (Schink et al. 1987). The acrylate pathway, reaction (3), reconverts extra NADH to NAD+. The succinate pathway, reactions (4) and (5), allows the reconversion of 2 NADH; this pathway also consumes CO2 for succinate synthesis. The production of ATP in the acrylate pathway and/or succinate pathway may not be possible (Seeliger et al. 2006; Zhang et al. 2013c).

Production of acetyl-coenzyme-A through pyruvate decarboxylation is a different way to regulate glucose catabolism. Two different pathways exist in anaerobic bacteria, the pyruvate formate lyase pathway and the pyruvate dehydrogenase pathway

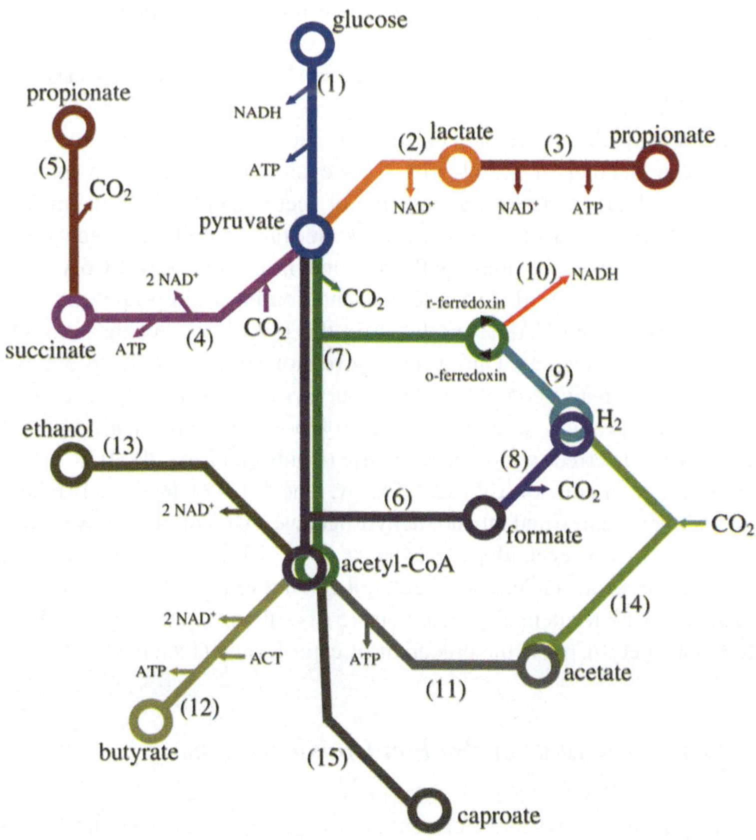

Fig. 1 Metabolic pathways of dark fermentation

(Sparling et al. 2006; Gheshlaghi et al. 2009), reactions (6) and (7), respectively. Facultative anaerobes, e.g., *Enterobacter*, *E. coli*, and *Klebsiella*, present the pyruvate formate lyase pathway (Waligorska 2012). Simultaneous presence and expression of genes encoding pyruvate formate lyase and pyruvate ferredoxin oxidoreductase can be found in members of the genus *Clostridium* (Sparling et al. 2006).

Formic acid, produced by reaction (6), is converted by the formate-hydrogen lyase, reaction (8), into H2 and CO2 (Waligorska 2012). The reduced ferredoxin, produced by reaction (7), transfers electrons to hydrogenase catalyzing the generation of H2, reaction (9) (Waligorska 2012). Reaction (10) consists on the NADH-ferredoxin oxidoreductase, which transfers electrons from reduced ferredoxin to NAD+ producing NADH (Hallenbeck 2009). This reaction does not seem to exist in many facultative anaerobes (Cai et al. 2011).

The synthesis of acetic acid proceeds from acetyl-CoA, reaction (11). This reaction produces ATP, via the phosphotransacetylase and the acetate kinase reactions. Both enzymes are present in all anaerobic bacteria that form acetyl-CoA in their energy metabolism and that use the acetyl-CoA to synthesize ATP (Thauer et al. 1977). The detection of acetate kinase activity in *C. thermolacticum* confirmed the presence of the acetate branch from acetyl-CoA and the supplementary ATP formation associated to this pathway (Collet et al. 2006). Another pathway for the generation of acetic acid is homoacetogenesis, reaction (14), which consumes H2 and CO2 (Zhao et al. 2010).

Butyric acid synthesis, reaction (12), takes place in a cycle starting with the condensation of molecules of acetyl-CoA. This cycle goes on until the formation of butyryl-CoA, which reacts with a molecule of acetic acid to yield one molecule of butyric acid and one of acetyl-CoA, which is the starting point for a new cycle. The simplification of this cycle starts with the condensation of acetyl-CoA and acetic acid to produce butyric acid. Butyrate fermentation has been proposed to not only regenerate NAD+ from NADH produced in the glycolysis, but that it might also lead to further energy gains. During the conversion of acetyl-CoA to butyric acid, one of the enzymatic complexes, the butyryl-CoA dehydrogenase electron-transferring flavoprotein, may generate a proton-motive force with a membrane-associated NADH/ferredoxin oxidoreductase (Louis and Flint 2009).

The formation of ethanol from acetyl-CoA, reaction (13), involves two enzymes, aldehyde dehydrogenase and ethanol dehydrogenase. The former catalyzes the reaction of acetyl-CoA to acetaldehyde (Thauer et al. 1977). The latter catalyzes the dehydrogenation of acetaldehyde to ethanol (Collet et al. 2006). The synthesis of caproic acid by dark fermentation, reaction (15), is still under investigation. However, it is clear that acetyl-CoA is the precursor of caproic acid (Jeon et al. 2010).

2.2 Dark Fermentation Product Considerations

This section analyzes the dark fermentation product yields. The analysis clearly shows that dark fermentation attention has not to focus only on hydrogen production but also on organic acid production. Also, here the price and market size of organic acids and ethanol are presented. Furthermore, it is described the technological importance of dark fermentation products in the present industry.

2.2.1 Product Yields and Prices

Figure 2 shows the average dark fermentation product yields of acetic acid, butyric acid, propionic acid, caproic acid, lactic acid, ethanol, formic acid, H2, CO2, and biomass, expressed as mass percentage of substrate consumed. Glucose and organic solid waste were reviewed as substrates. For glucose, the data was obtained from mixed culture dark fermentation under continuous mode with suspended microorganisms, i.e., not immobilized (Aceves-Lara et al. 2008a, b; Temudo et al. 2008a, b, 2009; Bastidas-Oyanedel et al. 2012; Zhang et al. 2012). In the case of organic solid waste, the data corresponds to the dark fermentation of organic fraction of municipal solid waste (Zahedi et al. 2013) and food waste (Jiang et al. 2013). Mass and electron balances (Bastidas-Oyanedel et al. 2008) have been applied to glucose and organic solid waste data. This was useful for the organic solid waste since it allowed estimating the CO2 and biomass average yield (data not available in the references). Both sorts of substrates have different scientific aims. Glucose has been used as a model substrate to study the metabolic changes in dark fermentation, while organic solid waste is a realistic substrate in order to apply dark fermentation in the biorefinery concept.

In both cases, organic acids and ethanol represent around 65% w/w of substrate consumed, H2 does not represent more than 3%, and CO2 and biomass represent around 35%, i.e., a large amount of the substrates, 65%, is converted to valuable products different than H2. Also, the products conforming this 65% are already

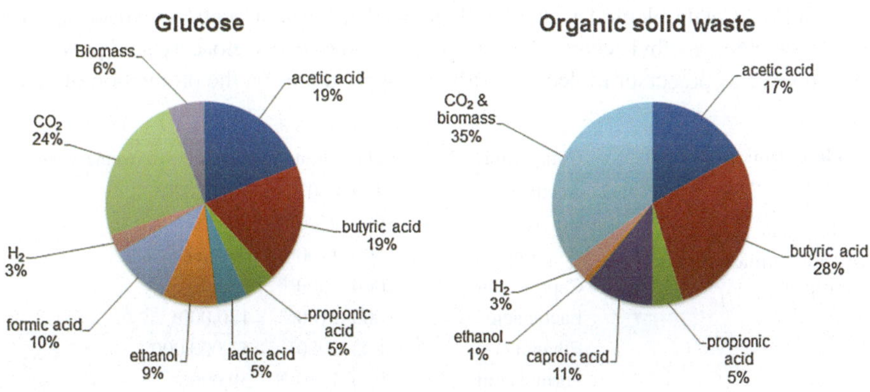

Fig. 2 Dark fermentation average product yields expressed as mass percentage of substrate consumed, i.e., glucose and organic solid waste. Glucose plot is based on experimental data of continuous mixed culture fermentation (Aceves-Lara et al. 2008a, b; Temudo et al. 2008a, b, 2009; Bastidas-Oyanedel et al. 2012; Zhang et al. 2012). Organic solid waste plot is based on Jiang et al. (2013) and Zahedi et al. (2013)

traded in the international market, while H2 is difficult to be priced since it is mainly produced in the same industries that use it.

Table 1 shows the bulk prices for acetic acid, butyric acid, propionic acid, caproic acid, lactic acid, ethanol, and formic acid. As an end product, butyric acid is by far the compound that presents the highest price, 2500 USD/ton, followed by propionic acid, 1700 USD/ton. The lowest price is for acetic acid, 800 USD/ton. In this context mixed culture dark fermentation can be controlled to maximize the production of butyric acid over the other products. Average concentrations and productivities are presented in Table 2.

2.2.2 Technological Importance of Dark fermentation Products

This section reviews the technological importance of acetic acid, butyric acid, propionic acid, caproic acid, lactic acid, ethanol, formic acid, H2, and CO2, in the current industry. Nowadays these compounds are produced by petroleum-based chemical synthesis, except for ethanol and lactic acid which are mainly produced by pure-culture bioprocess. As it was mentioned before, in a biobased economy, dark fermentation of organic residues is supposed to increase the global efficiency of biomass-based production of valuable chemicals. Hence dark fermentation appears as an alternative to petroleum-based synthesis.

Industrial acetic acid is mainly produced by petrochemical synthesis (Agreda and Zoeller 1993). Acetic acid is used in the synthesis of several derived molecules; among them, vinyl acetate represents the single largest use of acetic acid. Vinyl acetate-derived polymers are ubiquitous in modern society, which are found as part of wood panels, paper bags, cardboard boxes, labels adhesives, white glue, latex paints, high-quality paper coating, textiles, and cement additive (Agreda and Zoeller 1993; Weissermel and Arpe 1997; Lindblad et al. 2002). Acetic acid also serves as precursor of cellulose acetate, alcohol acetates, halogenated acetic acid, acetic anhydride, citrate esters, diketene, methyl acetoacetate, acetoacetamides, and acetoacetylated polymers, which serve as precursor molecules with vast applications on the production of phar-

Table 1 Bulk prices and market size of acetic acid, butyric acid, propionic acid, caproic acid, lactic acid, ethanol, formic acid, and hydrogen

Compound	Price (USD/tonne)	Market size (tonne/year)
Acetic acid	400–800[a,b]	3,500,000[a]
Butyric acid	2000–2500[a,b]	30,000[a]
Propionic acid	1500–1700[a,b]	180,000[a]
Caproic acid	2000–2500[a,b]	25,000[a]
Lactic acid	1000–2100[a,b]	120,000[a]
Ethanol	800–2000[b]	51,000,000[c]
Formic acid	950–1200[a,b]	30,000[a]
Hydrogen	600–1800[d]	

[a]Zacharof and Lovitt (2013)
[b]Alibaba (2014)
[c]Mussatto et al. (2010)
[d]Bastidas-Oyandel and Schmidt (2018)

Table 2 Concentration and productivities of acetic, butyric, and propionic acid by dark fermentation

Product	Concentration (g/L)	Calculated productivity (g/L/h)	Fermentation conditions	Ref
Acetic acid	16	0.08	Batch, 50 gCOD[a]/L (Soluble COD) initial substrate concentration, 192 h HRT[b], pH 6, 45 °C, 250 rpm stirring velocity	Jiang et al. (2013)
	3.8	0.08	Anaerobic baffle reactor, 73 gCOD/L (total COD) or 27 gCOD/L (soluble COD) initial substrate concentration, 38 h HRT	Tawfik and El-Qelish, (2012)
	0.12	0.0025	Batch, 1.9–2.5 gCOD/L (total COD) or 1–1.3 gCOD/L (soluble COD) initial substrate concentration, 48 h HRT, initial pH 6, 35 °C, 200 rpm stirring velocity	Khardenavis et al. (2013)
Butyric acid	21	0.11	Batch, 50 gCOD/L (soluble COD) initial substrate concentration, 192 h HRT, pH 6, 35 °C, 250 rpm stirring velocity	Jiang et al. (2013)
	7	0.18	Anaerobic baffled reactor, 29 gCOD/L (total COD) or 10.6 gCOD/L (soluble COD), 38 h HRT	Tawfik and El-Qelish, (2012)
	1.16	0.006	Batch, 50 gCOD/L (soluble COD) initial substrate concentration, 192 h HRT, 35 °C, 200 rpm stirring velocity	Jiang et al. (2013)
Propionic acid	18.2	0.095	Batch, 50 gCOD/L (soluble COD) initial substrate concentration, 192 h HRT, pH 6, 45 °C, 250 rpm stirring velocity	Jiang et al. (2013)
	0.6	0.0008	CSTR[c], 75 gCOD/L (total COD) or 23 gCOD/L (soluble COD), 240 h HRT, pH 5.7, 55 °C, 23 rpm stirring velocity	Zahedi et al. (2013)
	0	0	CSTR, 45 gCOD/L (total COD) or 15 gCOD/L (soluble COD), 12 h HRT, pH 4.8, 55 °C, 23 rpm stirring velocity	Zahedi et al. (2013)

[a]*COD* chemical oxygen demand
[b]*HRT* hydraulic retention time
[c]*CSTR* continuous stirred-tank reactor

maceuticals (such as aspirin, vitamin E, beta-lactam and oxacillin antibiotics, antiepileptic drugs), agrochemicals (insecticides, fungicides), dye, colorant, and polymers. Furthermore, acetic acid chemistry still offers ample opportunity for providing new discoveries in the future of material science (Agreda and Zoeller 1993).

Butyric acid is produced at industrial scale via mainly a petroleum-based chemical synthesis (Zhang et al. 2009). The main industrial application of butyric acid is in the manufacture of cellulose acetate-butyrate plastics (Rogers et al. 2006; Dwidar et al. 2012). Another well-known polymer, based on butyric acid, is the poly-3-hydroxybutyrate (PHB). Butyric acid esters, e.g., methyl, ethyl, and amyl butyrate, are used in beverages, foods, and cosmetics industries as fragrant and flavoring agents (Rogers et al. 2006; Dwidar et al. 2012). Nowadays, butyric acid attention is focused on the synthesis of fuels, since it can be converted to butanol through biological or chemical transformation (Dwidar et al. 2012).

Butyric acid also has many applications in medicine, in the field of gastroenterology (Hamer et al. 2008; Vanhoutvin et al. 2009), treatment of hematological, metabolic, and neurological pathologies (Sossai 2012). Various prodrugs (Rephaeli et al. 2000) that are derivatives of butyric acid were tried for their potential use in treatment of hypercholesterolemia (Canani et al. 2011), cancers (Canani et al. 2012), and hemoglobinopathies (Kim et al. 2009), including leukemia and sickle cell anemia, and also to protect hair follicles of radio- and chemotherapy-induced alopecia (Dwidar et al. 2012).

Propionic acid is used as antifungicide in the food industry to suppress mold growth in breads, meats, fruits, and on the surfaces of cheese and also as preservatives for grains, silage, and tobacco during storage and transport (Rogers et al. 2006). As in the case of butyric acid, propionic acid is also used in the production of cellulose-based plastics, such as cellulose acetate propionate, which are used in textiles, filters, reverse osmosis membranes, sheeting, film products, lacquers, and molding plastics (Rogers et al. 2006). Propionic acid esters and other derivatives are used as plasticizers, e.g., phenyl propionate and glycerol tripropionate, and as specialized solvents. Propionic acid esters are also used as perfumes and flavors, e.g., citronellyl propionate and geranyl propionate (Rogers et al. 2006).

The commercial use of caproic acid, also known as hexanoic acid, is primarily as precursor for the synthesis of pharmaceuticals, flavors, and other hexyl ester (Kenealy et al. 1995). Caproic acid fungicide properties have been reported as an alternative to synthetic fungicides, reducing environmental-human harm (Leyva et al. 2008). Caproic acid biosynthesis, by chain elongation of ethanol or acetic acid, is gaining interest since it can be separated with less energy input than ethanol and/or acetic acid (Vasudevan et al. 2014). In situ extraction of caproic acid has been reported from fermentation broth by water-immiscible organic solvents (Choi et al. 2013; Jeon et al. 2013).

Industrial production of lactic acid is mainly based on bioprocesses. The conventional production of lactic acid is based on pure-culture fermentation of *Lactobacillus*, achieving lactic acid concentrations of 90–180 g/L (Kwan et al. 2016; Nguyen et al. 2013). Several efforts using mixed culture fermentation have obtained lactic acid concentrations of 30–50 g/L, with 93% selectivity over other organic acids produced

(Bonk et al. 2017; Yousuf et al. 2018). Lactic acid is used in the polymer industry, as two of its polymers, the polyester syndiotactic polylactide and polylactic acid, are currently used in wide applications (www.purac.com). Other applications include its use in the food, textile, leather, dyeing, and the cosmetic industries (Kamm et al. 2006). Lactic acid is an example of a biobased molecule that is widely used in our society.

Ethanol is produced in large volumes by industrial fermentation. In 2008 the total worldwide ethanol production accounted to 51 megatonnes, of which the USA and Brazil contributed 52 and 37%, respectively (Mussatto et al. 2010). Most of the ethanol in the USA is produced from corn, while Brazilian ethanol is derived from sugar cane (Freudenberger 2009). Ethanol is also produced by petrochemical synthesis. In the chemical industry, ethanol is used as solvent, antifreeze, and fuel supplement (Mussatto et al. 2010). The major use of ethanol is as an intermediate feedstock in the synthesis of innumerable organic chemicals. Some of them include (1) diethyl ether, a solvent, extractant, and anesthetic, and (2) acetaldehyde, which is the raw material for production of a large number of organic chemicals such as butanol, acetic acid, acetic anhydride, chloral, crotonaldehyde, and ethylhexanol. These and other ethanol-derived chemicals are used in dyes, drugs, synthetic rubber, solvents, detergents, plasticizers, surface coatings, adhesives, moldings, cosmetics, explosives, pesticides, and synthetic fiber resins (Roehr 2001).

Formic acid is used in the textile industry, in tanning, in rubber processing, and in manufacturing of pharmaceuticals and is also used as food preservative agent in livestock feed (Zacharof and Lovitt 2013). Recently, it has received more attention to be used as environmental storage and transportation medium for hydrogen (Joo 2008). Hydrogen can be generated by the catabolic decomposition of formic acid. Also some researchers have demonstrated that formic acid has the potential to direct power fuel cells for electricity generation and transportation (Rice et al. 2002; Uhm et al. 2008).

Hydrogen is industrially produced by steam reforming, in which natural gas reacts with steam, releasing hydrogen. And its production costs range from 600 to 1800 USD/H_2), which depends on the natural gas market price (Bastidas-Oyandel and Schmidt 2018). These costs are the cheapest when compared to other sources, i.e., coal, electrolysis, and biomass. To illustrate this, a biomass-based biohydrogen price has been reported to reach 4700 USD/H_2 (Clarke and Alibardi 2010). The production of hydrogen by water electrolysis, by running an electrical current through it, is used where electricity is cheap and where high purity is required (Hoffmann 2001). Hydrogen uses are in energy applications and the chemical industry. Hydrogen, in combination with oxygen, is used to fuel space shuttles (Boucher 2006). Hydrogen is a nonpolluting fuel since the combustion of hydrogen with oxygen produces only water, unlike hydrocarbon internal-combustion engines that produce carbon monoxide, carbon dioxide, unburned hydrocarbons, stench, and smoke (Berry and Aceves 2005). Also, hydrogen can be used in fuel cells, producing electricity, in a flameless process which is 2.5 times more efficient than internal-combustion engines (Peighambardoust et al. 2010). Hydrogen has been thought to store renewable electricity when is less needed, e.g., during low consumption hours, through electrolysis of water (Boucher 2006). Hydrogen then can be converted back to electricity at peak hours. Hydrogen can be stored at high pressure, as an integral component in certain alloys known as hydrides, on microscopic carbon fibers. Hydrogen can be converted to formic acid in order to be stored or been directly used in fuel cells (Rice et al. 2002; Joo 2008; Uhm et al. 2008). In addition its energy applications, hydrogen is widely used in chemical industries as a raw material for hydrogenation in the production of hydrocarbons, fertilizers, dyes, drugs, and plastics. Also, it is used in the food industry for the treatment of oils and fats (Hoffmann 2001).

The most common use of carbon dioxide in industry is as supercritical CO2 in materials manufacturing industry and in food industry. Carbon dioxide appears as a nonhazardous and environmental friendly supercritical fluid. It is used as solvent in microelectronic applications (Weibel and Ober 2003), in polymer foam applications (Tomasko et al. 2009), and in the manufacture of fibers, microparticles, and films (Davies et al. 2008). Supercritical CO2 applications in food industry include extraction, fractionation, refinement, and deodorization of lipids or essential oils (Sahena et al. 2009). Another application in food industry is on dealcoholization process, i.e., reduction of ethanol content in alcoholic beverages (Ruiz-Rodriguez et al. 2010). As exposed above, the use of supercritical CO2 reduces the use of organic solvents, reducing health, environmental, and safety hazards.

2.3 *Operation Factors Affecting Dark Fermentation*

As outlined above, dark fermentation produces a mixture of organic acids, solvents, and gases. The proportion of these products depends upon the microorganisms, the substrates, and operational conditions (Cai et al. 2011). These parameters are very important for the optimization and modeling of the dark fermentation process. Dark fermentation literature has been focused on optimization of hydrogen production, including extensive reviews on the topic. Changes on dark fermentation occur in two levels: (1) microbial population structure shifts and (2) enzymatically/metabolic shifts. These two levels can be modified independently or simultaneously (Ye et al. 2007; Guo et al. 2010).

2.3.1 Dark Fermentation Microbial Population

Mixed culture dark fermentation is an autocatalytic process mediated by specified microorganisms that are able to thrive under anaerobically conditions. Mixed culture dark fermentation has two advantages over pure-culture fermentation technology, there is no need for aseptic conditions, and it can consume a vast range of substrates, which decreases the overall cost of the process (Ntaikou et al. 2010). On the other hand, pure cultures are characterized by high selectivity, yielding higher product efficiency, and well controlled by environmental parameters (Waligorska 2012).

Identified bacteria in mixed culture dark fermentation are strict anaerobes, e.g., *Clostridia* sp. (Quemeneur et al. 2011), and facultative anaerobes, e.g., *E. coli*, *Enterobacteriaceae* sp. (Aceves-Lara et al. 2008c), and *Klebsiella* sp. (Ntaikou et al. 2010). Other identified bacteria in thermophilic dark fermentation are

Caldicellulosiruptor saccharolyticus and *Thermoanaerobacter* spp. (Zhang et al. 2014). One of the phenomena related to the microbial population shift in dark fermentation is sporulation. This phenomena appears with the microbes belonging to *Clostridium* sp. Sporulation is a protection system that is activated when the environmental conditions are not favorable, e.g., increase in temperature (over 90 °C), excess or limitation of nutrients, dissolved oxygen, or a significant decrease on pH (below 3.0) (Sauer et al. 1995). This produces both advantages and disadvantages for the selection of dark fermentative microorganisms and process stability, respectively, as presented in Sect. 2.3.2. Regarding the process stability, sporulation must be controlled and/or avoided. During process, if sporulation occurs, this will decrease the substrate consumption rates and the productivity of dark fermentation and produce a washout of the spores, then losing the autocatalytic capacity of the microbes (Hawkes et al. 2002), hence increasing the dominance of bacteria populations that do not sporulate.

2.3.2 Inocula Sources

Several studies of mixed culture dark fermentation have used inocula from sewage sludge (Chen and Lin 2001), organic waste treatment sludges (Rajhi et al. 2013), animal dung (Guo et al. 2010), and fluvial (Rajhi et al. 2013) and marine (Perrotta et al. 2017) sediments. In literature, three methods have been utilized in order to select and/or enrich the dark fermentative bacteria, avoiding the presence of organisms that consume organic acids and H2, particularly methanogens (Hawkes et al. 2002). These three methods are thermal shock, pH shock, and short HRT.

Thermal shock consists of heating the inocula to 90 °C for a short time (Li and Fang 2007). Usually for reaction volumes ranging from 1 to 5 L, authors use a thermal shock time range from 10 to 20 min (Mu et al. 2007; Im et al. 2012). This treatment allows the selection of spore-forming bacteria, e.g., *Clostridium* (Hawkes et al. 2002).

The pH shock method uses pH below 5 or above 10 (Chen et al. 2002). This method uses the principle of the wide pH range that acidogenic bacteria can stand (Temudo et al. 2007), while methane-producing archaea cannot tolerate these pH conditions, being inhibited and washout of the reactors.

The third method, short HRT, uses HRT ranging from 6 to 20 h (Temudo et al. 2007; Aceves-Lara et al. 2008b; Bastidas-Oyanedel et al. 2012). This is based on the fact that acidogenic bacteria exhibit higher growth kinetics than methane-producing archaea. Methanogenic archaea is highly present in wastewater anaerobic sludge, sewage sludge, and cattle dung. These three methods can be applied independently or in a combination (Liu et al. 2002).

2.3.3 Substrates and Nutrients

Mixed culture dark fermentation has a wide range of substrate utilization (Nielsen et al. 2001), from different biomasses and organic waste streams (Li and Fang 2007; Chong et al. 2009). They are able to consume the residues from biodiesel production (Nishio and Nakashimada 2007), food waste (Han and Shin 2004; Shin et al. 2004), biomass residues from agro-industries (Hussy et al. 2003; Kaparaju et al. 2009; Guo et al. 2010; Chu et al. 2012), lignocellulosic hydrolysates (Ueno et al. 1995; Ren et al. 2009), paper/wastepaper/cardboard (Lay 2001), and pre-treated biomass (Prakasham et al. 2010). It can convert anaerobic sludge (Mu et al. 2006) and organic fraction of municipal solid waste (Lay et al. 1999; Tawfik and El-Qelish 2012; Zahedi et al. 2013). This wide range of non-sterile fermentable organic feedstock for mixed culture dark fermentation highlights its viability and flexibility as technology (Hawkes et al. 2002), compared to other bioprocesses, e.g., bioethanol production, where yeasts use a narrow range of substrates and sterile conditions are needed to avoid microbial contamination.

As mentioned previously in Sect. 2.3, substrates have a significant effect on productivities. Lin and Lay (2004) have observed metabolic changes when decreasing C/N ratio, leading to a shift from acetic acid to ethanol production. Quemeneur et al. (2011) have studied the influence of monosaccharide (glucose and fructose), disaccharide (sucrose, maltose, and cellobiose), and trisaccharide (maltotriose) on genetic diversity and product yields and in mixed culture dark fermentation.

Quemeneur et al. (2011) studied the effect of different types of carbohydrates on the microbial population structures. Glucose, fructose, sucrose, and cellobiose cultures displayed a dominant population related to *Clostridium sporogenes*. Sucrose culture showed a population related to *C. acetobutylicum*, while cellobiose showed the most diverse culture including populations of *C. cellulolyticum*, *C. acetobutylicum*, *C. saccharobutylicum*, and *C. kluyveri*. In the case of maltose and maltotriose cultures, the dominant population was related to *C. acetobutylicum*. They also found that the H2 yield decreased when increasing the chain length of carbohydrates, from 1.82 ± 0.1 to 1.38 ± 0.12 molH2/molhexose_equivalent. They also observed changes in liquid product yields. While acetic and butyric acids were the main products in all tested carbohydrates, yields were 290 ± 60 (molacetic/molhexose_equivalent) and 380 ± 90 (molbutyric/molhexose_equivalent); caproic acid only appeared when using di- and trisaccharides, with yields ranging from 10 to 40 (molcaproic/molhexose_equivalent); and ethanol yield increased dramatically when using maltotriose (trisaccharide), from 40 ± 10 (for mono- and disaccharides) to 117.5 (molethanol/molhexose_equivalent).

Minerals present on the fermentation media also affect dark fermentation. As said before, most of the literature on dark fermentation is focused on hydrogen production. In this context, the Fe2+ ion media composition has shown to be important. These ions are essential constituents of the hydrogenase active site (Aceves-Lara et al. 2008c), and the concentration of Fe2+ that maximizes H2 production was found to be around 200 mg/L, at 35 °C (Zhang and Shen 2006). Other minerals that have been identified as important for the H2 production are magnesium, sodium, and zinc (Lin and Lay 2005). In the case of phosphate, its limitation may favor solvent production over H2 and organic acids (Hawkes et al. 2002).

2.3.4 pH

Dark fermentation has a wide pH range, 3.0–11 (Chen et al. 2002; Temudo et al. 2007). The pH modifies the dark fermentation product yields. H_2 production maximization occurs in the pH range of 5–7 (Aceves-Lara et al. 2008b; Bastidas-Oyanedel et al. 2012). Acetic acid production is favored at pH over 6.5, while butyric acid production is favored at pH below 6.0 (Fang and Liu 2002). The pH range between 4.5 and 6.0 maximizes the ethanol production, while pH between 5.0 and 6.0 maximizes the propionate production (Hwang et al. 2004). Production of butanol is favored at pH below 4.3 (Kim et al. 2004). The accumulation of organic acids at pH below 5 favors the production of solvents, e.g., ethanol, propanol, or butanol (Sauer et al. 1995; Hawkes et al. 2002). Increase in the formic acid at pH below 5.5 induces the enzyme formate-hydrogen lyase, which catalyzes its decomposition into H_2 and CO_2 (Temudo et al. 2007; Bastidas-Oyanedel et al. 2012; Waligorska 2012).

2.3.5 Temperature

Dark fermentation is a biological–/biochemical-mediated process. Increase on temperature reduces enzymes' activation energy. Dark fermentation can take place in the presence of mesophilic bacteria (*Clostridium*, *Enterobacter*) with their range of temperature being around 35 °C, thermophiles (*Caldicellulosiruptor*, *Thermoanaerobacterium*) temperature around 50 °C, or hyperthermophiles (*Thermotoga*) temperature approximately 70 °C (Waligorska 2012).

In mixed culture dark fermentation, changes in temperature may produce shifts in the dominant bacteria culture. Some authors have observed shifts from *Clostridium* at mesophilic conditions to *Thermoanaerobacterium* in thermophilic conditions (Karadag and Puhakka 2010). In the case of H_2 production, their productivities are similar for mesophilic and thermophilic regions. Anyhow, H_2 yields are lower for the mesophilic temperature range (Lin and Chang 2004; Li and Fang 2007), as well as the H_2 gas fraction. This could be due to the fact that hyperthermophilic bacteria are less inhibited by the H_2 partial pressure (Waligorska 2012). Temperatures below 35 °C decrease the kinetics of dark fermentation (Lin and Chang 2004). Increasing the temperature, in the range of 33–41 °C, has demonstrated an increase in the substrate conversion and a shift from ethanol to butyrate (Mu et al. 2006).

2.3.6 Headspace Partial Pressure and Dissolved Gas Concentration (Gases Composition)

H_2 and CO_2 produced during dark fermentation generate supersaturation in the liquid phase, allowing them to be transferred to the gas phase (Pauss et al. 1990; Kraemer 2004). H_2 liquid saturation and gas partial pressure affect negatively the reactions producing H_2 and the conversion of NADH to H_2 by hydrogenases

(Tanisho et al. 1998; Mizuno et al. 2000; Hallenbeck 2005); see Sect. 2.1.3. These metabolic reactions are thermodynamically controlled (Bastidas-Oyanedel et al. 2012). The production of hydrogen using electrons from NADH is possible at very low partial pressure of hydrogen. Some authors reported H2 partial pressures below 0.09 bar (Bastidas-Oyanedel et al. 2012), while others propose the partial pressure should be less than 10^{-3} bar (Hallenbeck 2009). Other authors suggested that reduced ferredoxin is the suitable electron donor to produce H2 instead of NADH (Zhang et al. 2013c).

2.3.7 Bioreactor Configuration

Bioreactor configuration affects the hydrodynamics influencing liquid-gas mass transfer phenomena, hence producing changes on the dominant microorganism population and enzymatically/metabolic shifts (Waligorska 2012). On the other hand, since the cost of the dark fermentation bioreactors must be low, their construction has to be based on low-cost technology, e.g., anaerobic digesters used in wastewater treatment plant. The most common bioreactor used for dark fermentation is the continuously stirred-tank reactor (CSTR). In this type of reactor, the stirring must be effective in order to homogenate the biomass suspended and the substrates. This provides a good contact between the substrate and the microorganisms and enhances the mass transfer (Ntaikou et al. 2010; Show et al. 2011).

In CSTR the solid retention time (SRT) is the same as the HRT. HRT of 6–12 h is favorable for the production of acetic and butyric acids, as well as hydrogen. Also it makes the SRT short enough to promote a dominance of the microbial population by dark fermentative bacteria, preventing the mixed culture from methanogenic microorganisms, which requires SRT in the order of days. The use of HRT below 6 h is risky since it could produce biomass concentration decay, followed by a biomass washout from the reactor (Ntaikou et al. 2010; Show et al. 2011). In this context, decoupling SRT from HRT prevents biomass washout (Hafez et al. 2010). In the case of CSTR mode, several types of bioreactors have been developed in this context, based on the capacity of the microorganisms to produce biofilms, or aided by membrane technologies. In both cases, the SRT is increased independently from the HRT.

In the presence of divalent cations and an increase in carbohydrate concentration in extracellular polymeric substances (EPS), bacteria can suddenly attach to surfaces, and/or each other, creating floccules and/or granules (Jung et al. 2011b). Zhang et al. (2007) have increased the biomass concentration over 30-fold compared to a CSTR without granulation. Hafez et al. (2010) demonstrated that decoupling SRT from HRT, by biofilm formation, increases H2 yield and also allows to increase the capacity of organic waste treatment.

Membrane bioreactors (MBR) possess many advantages that include higher biomass concentration in the bioreactor; reactor volume can be reduced according to a higher substrate consumption rate, reduced production of excess sludge due to biomass decay in the reactor, and a lack of microorganisms in the effluent due to their total retention by the used membrane (Oh et al. 2004; Jung et al. 2011b). The main disadvantage limiting the use of MBR is the membrane fouling produced by EPS accumulation (Lee et al. 2008, 2010; Zheng et al. 2010). Regular backpulsing is essential in order to remove membrane fouling, hence maintaining the permeate flux through the membrane (Oh et al. 2004). Anyhow, SRT can be controlled in order to prevent or retard fouling.

Control of SRT also produces changes in the dark fermentation productivities. This has been evidenced by Lee et al. (2010), on a submerge membrane reactor, using glucose as substrate, with a HRT of 9 h. When the SRT increased from 2 to 12.5 days, the H2 productivity increased from 3.9 to 5.8 L/L/day. But using a SRT of 90 days caused a drop in H2 productivity. A similar trend was observed by (Oh et al. 2004), working on an external cross-flow membrane bioreactor, using glucose as substrate. When increasing the SRT from 3.3 to 12 h, the biomass concentration increased from 2.2 to 5.8 g/L, the glucose consumption increased from 90 to 98%, and the H2 productivity increased from 7.2 to 9.2 L/L/day. When increasing the SRT from 5 to 48 h, both biomass and glucose conversion increased, from 2.4 to 8.8 g/L and 99 to 99.5%, respectively. But H2 productivity decreased from 9.2 to 4.5 L/L/day. The authors suggest that the decrease in hydrogen productivity must have been due to changes in the physiology or composition of the microbial community in the reactor, as the SRT was increased (Oh et al. 2004).

An alternative to CSTR is upflow bioreactors. The liquid influent containing the substrates enters the reactor at the bottom, and the liquid outlet, with minimum substrates, exits at the top. In the upflow bioreactors, biomass is immobilized either in granules or in biofilms or entrapped in packed support media. The support media include sponge, granular activated carbon, expanded clay, polyethylene-octene elastomer, ceramic ball, and alginate gel. The immobilized biomass can be packed or fluidized. Upflow packed bioreactors have a higher mass transfer resistance compared to CSTR, resulting in a lower substrate conversion (Waligorska 2012). In a fluidized bed reactor, gas or liquid passes through accumulated solid matter, causing its fluidization, enhancing the mixing and mass transfer.

Upflow anaerobic sludge blanket (UASB) reactors, used commonly for methane production, consist of a gas/liquid/solid separator on top, where microbial granules are formed. The granules sediment easily creating a thick biomass blanket zone at the bottom (Hawkes et al. 2007). The main disadvantage is a long start-up period of around 5 months (Wang et al. 2007). According to (Jung et al. 2011a), the granulation rate can be increased by means of a two-stage process, consisting on a CSTR which serves as seeding to the UASB. This strategy shortened the start-up time to 20 days.

In the case of solid substrates, the current developments liquefy the solids by different means. At laboratory scale, the dark fermentation of municipal solid waste, slaughterhouse waste (Gomez et al. 2006), and household solid waste (Liu et al. 2006) has been studied in a two-stage mode, with the objective of producing biohythane (hydrogen and methane). In both experiments the solid waste has been grinded and suspended in aqueous solution.

Full-scale technologies liquefy the solid substrates by hydrolysis, trying to avoid the size reduction of the solid waste. The Aikan® technology (Aikantechnology 2015; Bonk et al. 2015; Magid 2006; Zeeshan and Hinrich 2014) has been designed for biogas and compost production from the organic fraction of municipal solid waste and green waste. It uses the hydrolytic capabilities of a preexisting biogas reactor. It consists of percolation units and biogas reactors. In the percolation units, an aqueous solution, from the biogas reactor, percolates through the waste, is collected, and is sent back to the biogas reactors. In this technology, the waste is not suspended in the fermentation broth of the biogas reactors, reducing the necessary downstream purification cost to separate or use the organic acids from the dark fermentation effluent (Bonk et al. 2015). The use of Aikan® in dark fermentation has been explored in Bonk et al. (2015).

Another full-scale technology that could be eventually adapted to dark fermentation is the REnescience process (REnescience 2015). This process uses unsorted and non-shredded municipal solid waste, which is wetted and mixed with hydrolytic enzymes. The mix is heated up for optimal enzymatic hydrolysis. The biodegradable mass is liquefied (bioliquid) by the enzymatic action. The nondegradable solids are then easily separated from the bioliquid. The bioliquid can be further processed by dark fermentation (Munster 2009).

2.4 Minimal Selling Price of Dark Fermentation Products

There is scarce literature on the economic assessment of dark fermentation systems. Granda et al. (2009) analyze the economics of the MixAlco™ process for the production of liquid fuels from biomass. In this process, biomass is converted via dark fermentation to volatile fatty acids (VFAs), i.e., acetic acid, propionic acids, and butyric acid, forming carboxylate salts. The effluent is dewatered by vapor-compression evaporation and converted to short-chain alcohols via esterification and subsequent hydrogenolysis. The MixAlco™ process minimum selling price of alcohols, with a 15% return on investment, was calculated to be around $350/ethanol. At the time of the analysis, 2009, this was economically viable where the market price of ethanol was $750/ethanol.

The organic compounds produced by dark fermentation have the potential to be the intermediates to a vast range of molecules different to alcohols, e.g., ketones, aldehydes, and olefins. They can be polymerized to produce plastics, liquid fuels, and molecules with biochemical activity. Anyhow, not all the chemical conversions of these organic compounds are economically feasible (Eggeman and Verser 2005).

A different approach could be followed. The VFAs do not need to be converted to alcohols but can be sold directly on the market if purified from the dark fermentation effluent. In a recent publication, Fasahati and Liu (2014) assess the techno-economic production and recovery of VFAs using membrane distillation followed by a methyl tert-butyl ether extraction step and a final rectification column achieving a VFA stream free from water and extracting solvent. In their assessment, it was considered the brown algae *Laminaria japonica* as the biomass converted to VFAs. The main result was the estimated minimum VFA selling price, $384/VFA, with a 10% internal rate of return after 10 years of plant operation. Still the authors have not detailed the operation cost of the VFA separation/purification steps. Bonk et al. (2015) have estimated that the production of pure VFAs from dark fermentation would be cost-efficient if the VFAs separation/purification costs do not exceed $15/m3_effluent. Yet, this estimation has to be validated. This is a possibility, considering the advancements in purification technologies.

In bioconversion of agricultural and food wastes to valuable chemicals, separation and purification of the products from the bulk liquid represent the highest percentage of the manufacturing cost (Angenent et al. 2004). Therefore, the economic feasibility of reusing wastes will strongly depend on the downstream processing efficiency, developing superior separation technologies (Agler et al. 2011).

In situ (online) product removal through diverse techniques including dialysis, distillation, adsorption, and extraction was tried for isolation of organic acids and other volatile products, and many of them can be applied for dark fermentation (Dwidar et al. 2012). In situ removal of liquid dark fermentation products enhances the productivity by decreasing the concentration of these products in the culturing medium and therefore reducing its toxic effect on the cells. In the following section, we detail the volatilization of dark fermentation products by gas stripping/flushing.

The volatilization of fermentation products by gas stripping is a known technology. In 1986, Ennis et al. (1986) used this principle on *Clostridium acetobutylicum* fermentations for separation and recovery of butanol. Further research, based on this principle, has been performed for the separation (from the fermentation broth) and recovery of acetone, butanol, and ethanol (Groot et al. 1989; Maddox et al. 1995; Ezeji et al. 2003; Ezeji and Karcher 2005). This gas stripping fermentation product recovery has been reviewed by Vane (2005), and it was also applied to separate ethanol from winery wastewater (Colin et al. 2005). Product volatilization by gas stripping, applied to acidogenic fermentation, is an attractive technology with an obvious application on biorefinery. This technology could be applied in both bioreactor and liquid outlet stream. As stated before, gas stripping directly applied in the bioreactor could increase the product yield of ethanol and organic acids, while stripping the outlet stream would allow recovery of the remaining volatile products. Organic acids are volatile in the undissociated form; thus their volatilization depends on pH. The undissociated form of an organic acid increases with pH decrease, e.g., at pH 5.0 and 4.0, the undissociated forms of both formic and lactic acids are 10% and 40%, respectively, while for both acetic and butyric acids are 40 and 80%. Thus low pH will improve the volatilization and, consequently, the product yield on acidogenic fermentation. The pH of a liquid solution decreases as a consequence of CO_2 sparging. Then CO_2 sparging is also attractive since it will allow organic acid volatilization. However, literature shows that suspended biomass acidogenic fermentation is hardly accomplished at pH lower than 4.5 (Temudo et al. 2007). Suspended biomass fermentation is a crucial system that allows the metabolic study of either pure or mixed cultures. Nevertheless, the acidogenic pH issue could be bypassed by the development of acidogenic biofilms in the reactor. Biofilms are aimed to be a technological system rather than an analysis system as suspended biomass fermentation.

3 Perspectives of Dark Fermentation in the Biorefinery Concept

As reviewed in Sect. 2, the role of dark fermentation in the bio-society is that of reduction of organic residues and increase in the global efficiency of biomass-based production of energy and valuable commodities. Hence the future of dark fermentation relies as a core bioprocess in the biorefinery concept. Here it is discussed the perspectives and future of dark fermentation. Its role as a link with other (bio)process for the production of liquid fuels, biohythane (biosyngas), and in photofermentative systems, fine chemical production, syngas fermentation, and bioelectrochemical systems is explored.

Integration of dark fermentation with other (bio)processes has the aim of reducing residues and associated operational costs and increasing revenues coming from high-value chemicals and/or bioenergy. In this context, dark fermentation metabolic capabilities and/or products can be associated to other (bio)processes. Figure 3 depicts how dark fermentation can be used as the heart of the biorefinery concept. Dark fermentation can use residues produced by different biomass-based activities, e.g., organic residues from agricultural and forestry activities (Ueno et al. 1995; Hussy et al. 2003; Kaparaju et al. 2009; Ren et al. 2009; Guo et al. 2010; Chu et al. 2012; Chaturvedi 2013; Uratani 2013), algae and microalgae activities (Sialve et al. 2009; Rashid 2013), food industry (Han and Shin 2004; Shin et al. 2004), municipal solid and liquid waste (Lay et al. 1999; Lay 2001; Li and Yu 2011; Mu et al. 2006; Nwobi 2013), and residues from other bio-industries (Nishio and Nakashimada 2007). These residues are transformed by dark fermentation into valuable chemicals that can be redirected for uses as bioenergy and biofuels, chemical and biochemical synthesis, bioplastics, and new materials.

3.1 *Liquid Fuel Production*

The dark fermentation products contained in the liquid phase can be used for the production of liquid fuels by (bio)catalytic process. Ladygina et al. (2006) have reviewed the biosynthesis of hydrocarbons by both prokaryotes and eukaryote microorganisms. In this way, organic acids, e.g., acetic or butyric acids, can be converted into C10–C35 aliphatic hydrocarbons.

Another alternative is the catalytic conversion of the organic acids into hydrocarbons. Leung et al. (1995) have demonstrated that pure saturated carboxylic acids, butyric acid among them, produce hydrocarbons when pyrolyzed at 450 °C and

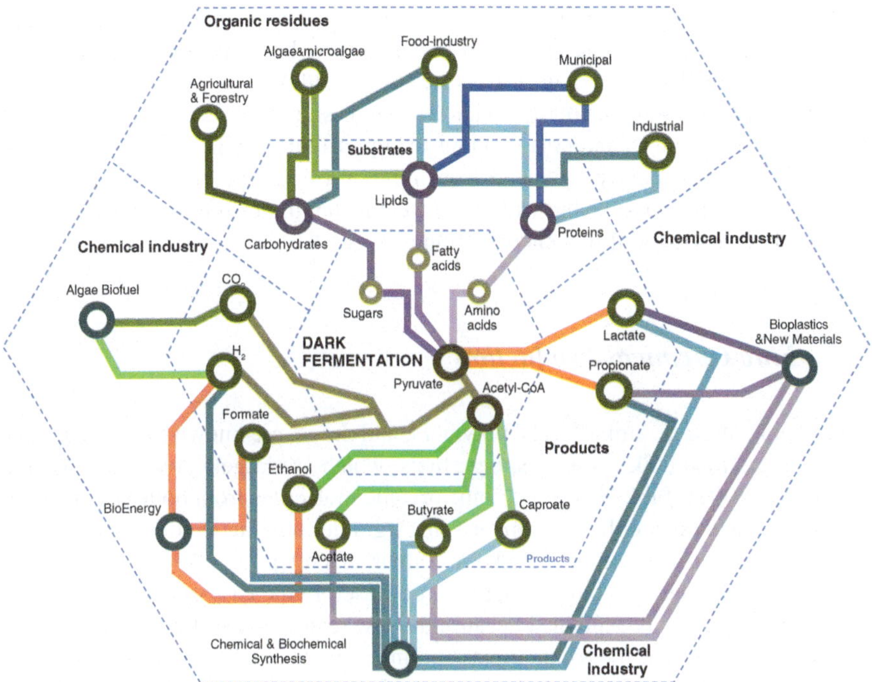

Fig. 3 Dark fermentation as a core bioprocess in the bio-society

atmospheric pressure under the presence of activated alumina as catalyzer. In this same context, Wang et al. (2012) have obtained hydrocarbons and phenol derivatives from mixed and pure acetic acid and propionic acid by catalytic cracking using silicon/aluminum as catalyzers. Ketonization of caproic acid was demonstrated by Gaertner et al. (2009) using cerium/zirconium as catalyzer at temperatures from 178 to 350 K. As demonstrated by Leung et al. (1995), ketones can be further converted into hydrocarbons, by pyrolysis.

Butyric acid, acetic acid, and ethanol produced by dark fermentation appear as carbon sources for the production of biobutanol. Biobutanol offers several advantages over ethanol as a transportation fuel, e.g., butanol provides more energy when burned, is less volatile, can replace gasoline in internal-combustion engines without any mechanical modifications, does not attract water so it can be transported in existing pipelines, is not miscible with water, and is less sensitive to colder temperatures (Dwidar et al. 2012). Despite these benefits, the main issue of fermentative bioproduction of butanol is its toxicity to the microorganisms that produce it, e.g., *Clostridia*. This toxicity results in poor concentrations in the fermentation broth and higher recovery costs. One feasible strategy to reduce the toxicity and improving the yield of butanol is to first convert biomass into organic acids by dark fermentation and then convert this downstream into butanol.

Biohydrogen, which is present in the gas-phase product streams of dark fermentation, could be used in a biohydrogen-based society; however this approach presents several drawbacks, e.g., infrastructure (Berry and Aceves 2005; Bossel 2006). In order to immediately decrease human GHG emission, biohydrogen could instead be used in the refining and/or upgrading of liquid fuels. Hydrogen used for these purposes is today mainly produced from mineral oil. Biohydrogen produced via electrolysis is also possible but with the current technologies is expensive and GHG unfriendly if using fossil fuel energy (Tran et al. 2010).

3.2 *Fine Chemical Production*

Fine chemicals, e.g., amino acids, polymers, and drugs, are known to have a higher market price than bulk chemicals, e.g., fuels, which are largely consumed (Willke and Vorlop 2004). Table 3 presents some organic molecules from both fine and bulk chemicals, as an example for comparison. Arginine, polylactic acid, and salicylic acid are well-known fine chemicals, and their market prices are in the range of 2.5–40 times higher than butanol, jet fuel, or diesel. In the case of petroleum-based bulk chemicals, this is explained because the raw material is still inexpensive compared to biomass, and the technologies and infrastructure are optimized and scaled for their respective market size. As presented in Sect. 3.1, the production of liquid fuels from dark fermentation products seems economically feasible because of two factors; mixed culture dark fermentation seems to be a low-cost process and the possibility of using organic residues. Furthermore, fermentation revenues could be increased if its products are directed to the production of fine chemicals.

As expressed above, dark fermentation could be used as an upstream process for the production of fine chemicals derived from biomass residues through (bio)catalytic processes. In the bioprocess area, dark fermentation could be linked to microbial systems, including pure cultures and engineered microbes, and/or enzymatic systems. Mixed culture systems have succeeded in the production of biopolyesters, e.g., polyhydroxyalkanoates (PHA), by using acetate and propionate as carbon sources (Sudesh and Doi 2000; Hu et al. 2005). The effect in the PHA production by using the liquid effluent of dark fermentation must be studied.

Pure culturing systems could also use the organic compounds, in the liquid effluent of dark fermentation, as substrates for the production of valuable products such as drugs, industrial chemicals, and biopolymers (Pirie et al. 2013). The case of engineered microbial pure cultures offers an expansion in number of products synthesized by microbial systems (Dhamankar and Prather 2011). The current trend is the use of biomass resources to be directly converted by engineered microbes into valuable chemicals. Pure cultures require sterile conditions and specific substrates. A direct usage of organic residues by engineered microbial systems is not advised. As exposed above, this could be done through dark fermentation in order to convert biomass residues into a mixture of organic acids and ethanol that can be therefore used as substrates for the engineered microbial systems. Alternative fractionation of the organic residues is necessary before use as substrate for engineered microbial systems.

Table 3 Comparison of some fine and bulk chemicals price and market size

	Compound	Price USD/kg	Global market size and forecast
Fine chemicals	Arginine	15–20[a]	1.2 kt in 2007[c] (in 2011 China capacity was 6.5 kt)[d]
	Polylactic acid	10–15[a]	360.8 kt in 2013 (1205 kt by 2020)[e]
	Salicylic acid	2–4[a]	95 kt in 2012 (145 kt by 2019)[f]
Bulk chemicals	Butanol	1–1.8[a]	3 Mt. in 2011 (4 Mt. by 2020)[g]
	Diesel	0.7–0.9[b]	1.2 Gt in 2009 (1.7 Gt by 2040)[h]
	Jet fuel	0.8–0.9[b]	250 Mt. in 2008 (350 Mt. by 2025)[i]

[a]Alibaba (2014)
[b]Indexmundi (2014)
[c]Demain (2007)
[d]ccminternational (2012)
[e]Grand View Research (2014)
[f]PRnewswire (2013)
[g]Market Publishers (2012)
[h]OPEC (2014)
[i]Cheze et al. (2011)

Enzymatic catalysis for synthesis of fine chemicals is already in use at industrial scale (Murzin and Leino 2008; Zhang 2014). Conventional enzymatic catalysis is carried out in organic solvents. Hence, the use of dark fermentation liquid effluent, as an upstream for enzymatic catalysis, must be treated in order to purify and extract its organic compounds that are dissolved in an aqueous solution. Also it is highly recommended to develop new approaches to directly use organic compounds from dark fermentation by enzymatic catalysis.

Another option, the chemical catalysis is being used for the conversion of biomass as a part of the so-called green chemistry (Anastas et al. 2000, 2001; Gallezot 2007; Murzin and Simakova 2011; Bui et al. 2013). The biomass compounds that have been studied as platform molecules include carbohydrates, lignocellulose, triglycerides, glycerol, and organic acids as levulinic acid. Technically there are no limitations to use dark fermentation products, e.g., acetic acid, in chemical catalysis. Still it is highly recommended to develop chemical catalysis based on this type of effluent. In this regard, as mentioned above, the direct use of dark fermentation effluent will entail the use of aqueous catalytic technologies (Cornils 1999; Ford 2001). These technologies have been reviewed for the conversion of carbohydrates (Queneau et al. 2011). Furthermore, conversion of glycerol into propylene glycol (Marinoiu et al. 2013) or polymerization of olefins (Mecking et al. 2002) has been achieved in aqueous solutions.

3.3 Algae, Cyanobacteria, and Phototrophic Bacteria Cultivation

As it was detailed in Sect. 3.1, residues from algae and microalgae cultivation and/or processing could be fed into dark fermentation. Also, products from dark fermentation can be fed to algae, cyanobacteria (blue algae), and phototrophic bacteria, in order to optimize the production of valuable chemicals, recycle nutrients, and reduce CO_2 emissions (Markou and Georgakakis 2011). Here it is explained how dark fermentation produced CO_2, organic acids, and H_2 which can be used by these three types of photosynthetic systems.

The macroalgae world market size estimative in 2004 was USD $6 9109/year, with 7.5 9106 t/year biomass harvested, while microalgae world market was USD $1.25 9109/year, 5000 t/year of biomass harvested (Pulz and Gross 2004). Valuable chemicals derived from algal products are economically attractive. Carotenoid pigments 2005 estimated marked was USD $935 9106/year (Cardozo et al. 2007). Astaxanthin, a carotenoid used as pigmentation source and antioxidant, estimated 2006 market price was 2500 USD $/kg (Jeon et al. 2006), with a market estimation of USD $150 9106/year (Pulz and Gross 2004). Another group of algal valuable chemicals are phycocolloids, where 2007 market for agar was valued at USD $200 9106/year, with 10,000 t/year of traded agar. For the same year, carrageenan market was valued at USD $200 9106 and a market size of 25,000 t/year (Cardozo et al. 2007).

Other economically attractive molecules extractable from algae are alginate, used as phycocolloid, and biochemical active molecules with use in the cosmetic industry, e.g., mycosporine-like amino acids for UV blocker sunscreen or for pharmacological industry, e.g., lectins, halogenated products, polyketide, fatty acids, sterols, and harmful algal bloom (HAB) toxins (Cardozo et al. 2007). HAB toxins have a promising potential as antibacterial and antifungal molecules (Najdenski et al. 2013). If we compare the prices of astaxanthin with acetic acid, 2500 USD/kg and 800 USD/t, respectively (see Table 1 for acetic acid details), it is clear that finding ways to convert the dark fermentation organic acids to algal valuable chemicals is economically attractive. Also the market sizes of astaxanthin and acetic acid, 60 and 3.5 9106 t/year, respectively, hint information regarding the saturation and/or expansion of their respective markets.

Literature reports that organic acids can be used for microalgae growth in mixotrophic mode, i.e., the simultaneous use of light and organic compounds for energy and carbon requirements (Markou and Georgakakis 2011) and in aerobic heterotrophic conditions, i.e., the use of organic compounds in the absence of light. The algal acetate mixotrophic growth resulted in 56–91% increase in chlorophyll content, relative to photoautotrophy, for the green algae *Chlamydomonas globosa*, *Chlorella minutissima*, and *Scenedesmus bijuga* (Bhatnagar et al. 2001). Chlorophyll content can be related to biomass production. The mixotrophic culture of *Chlorella vulgaris* in acetate increased the biomass productivity from 13 to 85 mg/L/day and the lipid productivity from 4 mg/L/day, at photoautotrophic conditions, to 28 mg/L/day (Liang et al. 2009). Another example of successful algal acetate mixotrophic growth was achieved by Kobayashi et al. (1993) and Jeon et al. (2006), where acetate enhanced both the productivity of *Haematococcus pluvialis* cell biomass and its astaxanthin formation. Mixotrophy can also support harmful algal species (HAS) (Burkholder et al. 2008), from where valuable antimicrobial toxins can be extracted (Najdenski et al. 2013). Under aerobic heterotrophic conditions, *Chlorella sorokiniana*, a single-cell green algae, specific growth rate is improved with concentrations of acetate up to 6 g/L, while it does not consume propionate (Ogbonna et al. 2000).

Organic acids produced from dark fermentation can also be used as substrates for the photofermentative H2 production. From the four groups of hydrogen producer microorganisms, i.e., green algae, cyanobacteria, phototrophic bacteria, and dark acidogenic bacteria, only the latter two can produce hydrogen from organic compounds (Cai et al. 2012).

In literature it is reported that H_2-producing phototrophic bacteria are capable of consuming the liquid products from dark fermentation, e.g., acetic or butyric, in the presence of light (Shi and Yu 2005; Argun et al. 2009; Redwood et al. 2009). The resultant dual dark/light integrated system could reach a maximum theoretical H_2 yield of 12 mol_H_2/mol_hexose. Dark fermentation maximum theoretical yield is 4 mol_H_2/mol_hexose. Experimental results using pure cultures or well-defined co-cultures of dark fermentative bacteria are promising. The integrated systems show overall H_2 yields ranging from 7 to 8 mol_H_2/mol_hexose (Yokoi et al. 2001; Kim et al. 2006). Experimental results for mixed culture dark fermentation are still missing. Shi and Yu (2006) have showed a hydrogen overall yield of 4.56 mol_H_2/mol_hexose using the effluent of an acidogenic reactor, but it is not defined if the acidogenic reactor consisted in mixed or pure culture.

Regarding the dark fermentation gas products, hydrogen can be used in a hydrogenation reaction to upgrade the algal oil/fat produced for biofuel purposes (Jang et al. 2005; Tran et al. 2010; Borowitzka and Moheimani 2013). Oil yields of microalgae, 58 to 137 m^3/ha/year, are much higher than terrestrial crops dedicated for oil, e.g., oil palm 6 m^3/ha/year (Chisti 2007; Amin 2009). Oil from microalgae can be transformed into biodiesel by a conventional catalytic methanol transesterification reaction (Demirbas 2002). The challenge is still to extract the oil from the microalgae biomass due to the content of chlorophyll. The purification of lipids from algae requires additional treatments compared with vegetable oil (Petkov et al. 2012). Also, microalgal oils are rich in polyunsaturated fatty acids, which are susceptible to oxidation during storage, and this reduces their acceptability for use in biodiesel (Chisti 2007). This issue is solved with hydrogenation converting the oils to the saturated form. Hydrogenation of algal biofuels using dark fermentative hydrogen appears as a sustainable process since hydrogenation process currently used petrochemical-based hydrogen (Tran et al. 2010).

The use of the CO_2 produced by dark fermentation can be used for the phototrophic growth of macro- and microalgae (Chung et al. 2011; Kumar et al. 2011; Singh and Ahluwalia 2013). The use of CO_2 from alcoholic fermentation has been successfully attained by Bezerra et al. (2013) for the growth of *Arthrospira platensis*. In this context, the link of dark fermentation CO_2 to algae cultivation would mitigate the CO_2 emissions of this integrated bioprocess.

3.4 Biomethane, Biohythane, and Biosyngas Production

Dark fermentation liquid effluent can be converted to biomethane by anaerobic digestion. Anaerobic digestion is a conventional process used for the treatment of organic waste. This bioprocess requires a pH of around 7 in order to maintain the activity of the methanogenic archaea, which are responsible for methane production. Fluctuations of pH could reduce, inhibit, and/or inactivate the methanogenic activity, which is why dark fermentation is used as an upstream process for methane production, i.e., conferring stability to the global system (Cavinato et al. 2012; Willquist et al. 2012).

Furthermore, the hydrogen produced in the dark fermentation step can be blended to the methane, forming biohythane. Biohythane is a combustible gas, consisting of 10% H_2, 30% CO_2, and 60% CH_4 (Cavinato et al. 2011). The use of 10% H_2 in the biohythane blend enhances combustion, thermal efficiency, and power output compared to biomethane, together with reducing hydrocarbon emissions (Sierens and Rosseel 2000; Porpatham et al. 2007).

Biohythane can be also considered as biosyngas. Syngas consists of a non-fixed mixture of CO, H_2, CO_2, and CH_4 (see Sect. 3.5). The conventional approach of production of biosyngas is the gasification of biomass (McKendry 2002b). This approach is useful for dried biomass stocks as lignocellulosic material. This approach is not advised for organic residues where the water content is above 30%, e.g., food waste, organic fraction of municipal solid waste, and algae residues, since extra energy is needed to evaporate the contained water (McKendry 2002b). The tandem system dark fermentation anaerobic digestion appears as a biogasification process of biomass (McKendry 2002a). The biosyngas produced can be used as chemical feedstock to produce liquid fuels (Wilhelm et al. 2001; Chuang 2012) and/or fine chemicals (Spivey et al. 2000; Hadipour and Sohrabi 2008).

3.5 Syngas Dark Fermentation

Syngas fermentation by mixed culture dark fermentation consists in the production of organic acids and solvents using synthesis gas (syngas). Homoacetogenesis (Sect. 2.1.3) is one example of a metabolic pathway involved in syngas dark fermentation. Syngas dark fermentation through homoacetogenesis (Zhao et al. 2010) consumes CO_2 and H_2, contained in the syngas, to produce acetic acid, whose technological importance was described in Sect. 2.2.2. Sustainable syngas, i.e., avoiding crude oil resources, could be obtained, for example, from the gasification of non-biodegradable organic solids, e.g., lignocellulosic biomass and plastics present in municipal solid waste (Arena 2012; Roddy 2013). Examples of co-gasification of lignocellulosic biomass and plastic waste are found in literature (Pinto et al. 2002; Ahmed et al. 2011). Syngas composition is not fixed (Williams et al. 2008). Pinto et al. (2002), working in co-gasification of pinewood and polyethylene, have obtained syngas compositions in the following range, in v/v %: CO 25–45, H_2

20–50, CO_2 2–15, CH_4 5–15, and hydrocarbons (CnHm) 0–15. The operational conditions used by them were temperatures ranging from 720 to 900 °C, atmospheric pressure, steam/waste ratio (0.45–0.9 w/w), and the polyethylene/pinewood ratio (0–0.4 w/w).

Syngas fermentative microorganisms have been identified and studied. Most of the work has been done in pure-culture fermentation (Heiskanen et al. 2007; Henstra et al. 2007; Mohammadi et al. 2011). Mixed culture syngas fermentation technology supposes low-cost process compared to pure-culture syngas fermentation. Gas-liquid mass transfer presumes another issue to be solved. The low solubility of H_2 in water (1.6 mg/L) in equilibrium with 1 atm of H_2 at 298 K is the main challenge for hydrogen utilization (Zhang et al. 2013a). Various methods such as vigorous stirring and gas circulation were proposed (Henstra et al. 2007). Hollow-fiber membrane bioreactors have been identified as a cheaper option to tackle this mass transfer issue (Henstra et al. 2007; Zhang et al. 2013a). In order to overcome these two pure cultures and mass transfer issues, Zhang et al. (2013a) have studied the syngas mixed culture dark fermentation in a hollow-fiber membrane bioreactor. In this kind of bioreactor, syngas permeates through the lumen of hollow-fiber membranes through the liquid phase of the bioreactor, where is directly consumed, without loss of gas through bubble gas formation, by a naturally attached mixed culture dark fermentative biofilm (enriched from an anaerobic digestion inoculum). In this work the authors have use a synthetic syngas composed by 60% H_2 and 40% CO_2 (% v/v). The bioreactor was operated at 35 °C and pH 4.5. Acetic acid represented 99% of the liquid products, reaching concentrations of 12.7 g/L (HRT: 25 days) and 3.7 g/L (HRT 9 days) in batch and continuous modes, respectively. For both bioreactor modes, their acetic acid productivities were very similar 0.02 g/L/h. Anyhow, the production of hexanoic acid and caprylic acid (medium chain fatty acids) by syngas fermentation could represent a lower operating cost and product separation when compared to acetic acid production (Agler et al. 2012; Zhang et al. 2013b).

Alternatives to syngas fermentation are the conventional catalytic syngas conversions, e.g., Fischer-Tropsch Technology (Mohammadi et al. 2011). This technology uses high temperatures ranging from 200 to 350 °C (Dry 2002) and high pressures 18–40 bar (Geerlings et al. 1999; Dry 2002), which is translated into very energy-consuming process. Also it requires metallic catalysts, e.g., iron or cobalt based. Fast heat exchange is required to avoid catalyst deactivation. Catalysts are deactivated also by syngas impurities, e.g., H2S present in municipal waste syngas (Dry 2002). Fischer-Tropsch, depending in operational conditions, produces gasoline, olefins, and waxes (Geerlings et al. 1999; Dry 2002). Deviations in the syngas composition, mainly CO/H_2 ratio, produce changes in the product selectivity. The process is fast, with retention times around 10 min (Geerlings et al. 1999).

In comparison, hollow-fiber membrane syngas dark fermentation uses temperatures around 35 °C and atmospheric pressure which results in substantial energy savings. The moderate temperature causes irreversibility of biological reactions, which results in high conversion efficiencies. This process does not require adding mineral catalysts since it is an autocatalytic process, using microorganism enzymatic systems. Hence high reaction specificity is achieved due to the enzymatic specificity. It is less sensitive to CO/H_2 ratio (Mohammadi et al. 2011). A major drawback of syngas dark fermentation is its high retention time, 9–25 days (Zhang et al. 2013a). Presently, syngas dark fermentation is still at the research level. Further study is recommended with respect to the optimization of productivities and scale-up of the process. Also long-run experiments are required in hollow-fiber membrane bioreactor syngas mixed culture dark fermentation.

4 Conclusions and Future Trends

This chapter has reviewed both the present state and future potential of dark fermentation in the (bio)-chemical industry. Dark fermentation is a viable and flexible technology since it consumes a vast range of substrates, most of which can be complex in composition, as residues streams from other bioprocess, or even municipal solid and liquid waste.

Dark fermentation contributes to the biobased society, reducing organic residues and increasing the global efficiency of biomass-based production of energy and valuable commodities. The products of dark fermentation can be purified and/or used as platform chemicals in subsequent (bio)process for the production of fuels, fine chemicals, and biosyngas. These features make dark fermentation a core bioprocess in the biorefinery concept.

References

Aceves-Lara CA, Latrille E, Bernet N, Buffiere P, Steyer JP (2008a) A pseudo-stoichiometric dynamic model of anaerobic hydrogen production from molasses. Water Res 42:2539–2550

Aceves-Lara CA, Latrille E, Buffiere P, Bernet N, Steyer JP (2008b) Experimental determination by principal component analysis of a reaction pathway of biohydrogen production by anaerobic fermentation. Chem Eng Process Process Intensif 47:1968–1975

Aceves-Lara CA, Trably E, Bastidas-Oyanedel JR, Ramirez I, Latrille E, Steyer JP (2008c) Bioenergy production from waste: examples of biomethane and biohydrogen. J Soc Biol 202:177–189

Agler MT, Wrenn BA, Zinder SH, Angenent LT (2011) Waste to bioproducts conversion with undefined mixed cultures: the carboxylate platform. Trends Biotechnol 29:70–78

Agler MT, Spirito CM, Usack JG, Werner JJ, Angenent LT (2012) Chain elongation with reactor microbiomes: upgrading dilute ethanol to medium-chain carboxylates. Energy Environ Sci 5:8189–8192

Agreda VH, Zoeller JR (1993) Acetic acid and its derivatives. Marcel Dekker, Inc, New York

Ahmed I, Nipattummakul N, Gupta A (2011) Characteristics of syngas from co-gasification of polyethylene and woodchips. Appl Energy 88:165–174

Aikantechnology (2015) Aikan technology, how it works, batch processing. http://www.aikantechnology.com/how-it-works/batch-processing.html. Accessed 12 June 2015

Alibaba (2014) Chinese e-commerce. Organic chemical bulk prices. www.alibaba.com. Accessed 24 Nov 2014

Amin S (2009) Review on biofuel oil and gas production processes from microalgae. Energy Convers Manag 50:1834–1840

Anastas P, Bartlett L, Kirchhoff M, Williamson T (2000) The role of catalysis in the design, development, and implementation of green chemistry. Catal Today 55:11–22
Anastas P, Kirchhoff M, Williamson T (2001) Catalysis as a foundational pillar of green chemistry. Appl Catal A-Gen 221:3–13
Angenent LT, Karim K, Al-Dahhan MH, Wrenn BA, Dominguez-Espinosa R (2004) Production of bioenergy and biochemicals from industrial and agricultural wastewater. Trends Biotechnol 22:477–485
Arena U (2012) Process and technological aspects of municipal solid waste gasification: a review. Waste Manag 32:625–639
Argun H, Kargi F, Kapdan I (2009) Effects of the substrate and cell concentration on bio-hydrogen production from ground wheat by combined dark and photofermentation. Int J Hydrog Energy 34:6181–6188
Bastidas-Oyandel JR, Schmidt JE (2018) Increasing profits in food waste biorefinery – a techno-economic analysis. Energies 11(6):1151. https://doi.org/10.3390/en11061551
Bastidas-Oyanedel JR, Aceves-Lara CA, Ruiz-Filippi G, Steyer JP (2008) Thermodynamic analysis of energy transfer in acidogenic cultures. Eng Life Sci 8:487–498
Bastidas-Oyanedel JR, Mohd-Zaki Z, Zeng RJ, Bernet N, Pratt S, Steyer JP et al (2012) Gas controlled hydrogen fermentation. Bioresour Technol 110:503–509
Bastidas-Oyanedel JR, Bonk F, Thomsen MH, Schmidt JE (2015) Dark Fermentation biorefinery in the present and future (bio)chemical industry. Rev Environ Sci Biotechnol 14:473–498
Berry GD, Aceves SM (2005) The case of hydrogen in a carbon constrained world. J Energ Resour ASME 127:189–194
Bezerra R, Matsudo M, Sato S, Converti A, de Carvalho J (2013) Fed-Batch Cultivation of Arthrospira platensis using carbon dioxide from alcoholic fermentation and urea as carbon and nitrogen sources. Bioenergy Res 6:1118–1125
Bhatnagar A, Chinnasamy S, Singh M, Das K (2001) Renewable biomass production by mixotrophic algae in the presence of various carbon sources and wastewaters. Appl Energy 88:3425–3431
Bonk F, Bastidas-Oyanedel JR, Schmidt JE (2015) Converting the organic fraction of solid waste from the city of Abu Dhabi to valuable products via dark fermentation—economic and energy assessment. Waste Manag 40:82–91
Bonk F, Bastidas-Oyanedel JR, Yousef AF, Schmidt JE (2017) Exploring the selective lactic acid production from food waste in uncontrolled pH mixed culture fermentations using different reactor configurations. Bioresour Technol 238:416–424
Borowitzka M, Moheimani N (2013) Sustainable biofuels from algae. Mitig Adapt Strateg Glob Change 18:13–25
Bossel U (2006) Does a hydrogen economy make sense? Proc IEEE 94:1826–1837
Boucher S (2006) La revolution de l'hydrogene, vers une energie propre et performante? Editions du Felin, France, Paris
Bui L, Luo H, Gunther W, Roman-Leshkov Y (2013) Domino reaction catalyzed by zeolites with Bronsted and Lewis acid sites for the production of gamma-valerolactone from furfural. Angew Chem Int Ed 52:8022–8025
Burkholder J, Glibert P, Skelton H (2008) Mixotrophy, a major mode of nutrition for harmful algal species in eutrophic waters. Harmful Algae 8:77–93
Cai G, Jin B, Monis P, Saint C (2011) Metabolic flux network and analysis of fermentative hydrogen production. Biotech Adv 29:375–387
Cai J, Wang G, Pan G (2012) Hydrogen production form butyrate by a marine mixed phototrophic bacterial consort. Int J Hydrog Energy 37:4057–4067
Canani R, Di Constanzo M, Leone L, Pedata M, Meli R, Calignano A (2011) Potential beneficial effects of butyrate in intestinal and extraintestinal diseases. World J Gastroenterol 17:1519–1528
Canani R, Di Constanzo M, Leone L (2012) The epigenetic effects of butyrate: potential therapeutic implications for clinical practice. Clin Epigenetics 4, 4
Cardozo K, Guaratini T, Barros M, Falcao V, Tonon A, Lopes N et al (2007) Metabolites from algae with economical impact. Comp Biochem Phys C 146:60–78

Cavinato C, Bolzonella D, Fatone F, Cecchi F, Pavan P (2011) Optimization of two-phase thermophilic anaerobic digestion of biowaste for hydrogen and methane production through reject water recirculation. Bioresour Technol 102:8605–8611

Cavinato C, Giuliano A, Bolzonella D, Pavan P, Cecchi F (2012) Biohythane production from food waste by dark fermentation coupled with anaerobic digestion process: a long-term pilot scale experience. Int J Hydrog Energy 37:11548–11555

ccminternational (2012) China's L-arginine industry enjoys rapid development. https://ccminternational.wordpress.com/2012/04/13/chinas-l-arginine-industry-enjoys-rapid-development/. Accessed 25 Nov 2014

Chaturvedi T (2013) Evaluation of bioenergy production from lignocellulosic biomass of *Salicornia Bigelovii*. MSc, Chemical Engineering, Masdar Institute of Science and Technology, Abu Dhabi, UAE

Chen CC, Lin CY (2001) Start-up of anaerobic hydrogen producing reactors seeded with sewage sludge. Acta Biotechnol 21:371–379

Chen CC, Lin CY, Lin M (2002) Acid-base enrichment enhances anaerobic hydrogen production process. Appl Microbiol Biotechnol 58:224–228

Cheze B, Gastineau P, Chevallier J (2011) Forecasting world and regional aviation jet fuel demands to the mid-term (2025). Energ Policy 39:5147–5158

Chisti Y (2007) Biodiesel from microalgae. Biotechnol Adv 25:294–306

Choi K, Jeon B, Kim B, Oh M, Um Y, Sang B (2013) Insitu biphasic extractive fermentation for hexanoic acid production from sucrose by megasphaera elsdenii NCIMB 702410. Appl Biochem Biotechnol 171:1094–1107

Chong ML, Sabaratnam V, Shirai Y, Hassan MA (2009) Bio-hydrogen production from biomass and industrial wastes by dark fermentation. Int J Hydrog Energy 34:3277–3287

Chu CY, Sen B, Lay CH, Lin YC, Lin CY (2012) Direct Fermentation of sweet potato to produce maximal hydrogen and ethanol. Appl Energy 100:10–18

Chuang S (2012) Conversion of syngas to fuels. Handbook of climate change mitigation. Springer, Berlin

Chung I, Beardall J, Mehta S, Sahoo D, Stojkovic S (2011) Using marine macroalgae for carbon sequestration: a critical appraisal. J Appl Phycol 23:877–886

Clarke WP, Alibardi L (2010) Anaerobic digestion for the treatment of solid organic waste: what's hot and what's not. Waste Manag 30:1761–1762

Colin T, Bories A, Sire Y, Perrin R (2005) Treatment and valorisation of winery wastewater by a new biophysical process (ECCFÒ). Water Sci Technol 51:99–106

Collet C, Girbal L, Peringer P, Schwitzguebel JP, Soucaille P (2006) Metabolism of lactose by Clostridium thermolacticum growing in continuous culture. Arch Microbiol 185:331–339

Cornils B (1999) Bulk and fine chemicals via aqueous biphasic catalysis. J Mol Catal A Chem 143:1–10

Davies O, Lewis A, Whitaker M, Tai H, Shakesheff K, Howdle S (2008) Applications of supercritical CO2 in the fabrication of polymer systems for drug delivery and tissue engineering. Adv Drug Deliver Rev 60:373–387

Demain A (2007) The business of biotechnology. Ind Biotechnol 3:269–283

Demirbas A (2002) Biodiesel from vegetable oils via transesterification in supercritical methanol. Energy Convers Manag 43:2349–2356

Dhamankar H, Prather K (2011) Microbial chemical factories: recent advances in pathway engineering for synthesis of value added chemicals. Curr Opin Struct Biol 21:488–494

Dry M (2002) The Fischer-Tropsch process: 1950–2000. Catal Today 71:227–241

Dwidar M, Park JY, Mitchell RJ, Sang BI (2012) The future of butyric acid in industry. Scientific World J. https://doi.org/10.1100/2012/471417

Eggeman T, Verser D (2005) Recovery of organic acids from fermentation broths. Appl Biochem Biotechnol 121–124:605–618

Ennis BM, Marshall CT, Maddox IS, Paterson AHJ (1986) Continuous product recovery by in situ gas stripping/condensation during solvent production from whey permeate using Clostridium acetobutylicum. Biotechnol Lett 8:725–730

Ezeji TC, Karcher PM (2005) Improving performance of a gas stripping-based recovery system to remove butanol from Clostridium beijerinckii fermentation. Bioprocess Biosyst Eng 27:207–214

Ezeji TC, Qureshi N, Blaschek HP (2003) Production of acetone, butanol and ethanol by Clostridium beijerinckii BA101 and in situ recovery by gas stripping. World J Microbiol Biotechnol 19:595–603

Fang HHP, Liu H (2002) Effect of pH on hydrogen production from glucose by a mixed culture. Bioresour Technol 82:87–93

Fasahati P, Liu J (2014) Techno-economic analysis of production and recovery of volatile fatty acids from brown algae using membrane distillation. Comput Aided Chem Eng 34:303–308

Ford W (2001) Catalysis by colloidal polymers in aqueous media. React Funct Polym 48:3–13

Freudenberger R (2009) A guide to small-scale ethanol, alcohol fuel, making and using ethanol as a renewable fuel. New Society Publishers, Gabriola Island

Gaertner C, Serrano-Ruiz J, Braden D, Dumesic J (2009) Catalytic coupling of carboxylic acids by ketonization as a processing step in biomass conversion. J Catal 266:71–78

Gallezot P (2007) Catalytic routes from renewables to fine chemicals. Catal Today 121:76–91

Geerlings J, Wilson J, Kramer G, Kuipers H, Hoek A, Huisman H (1999) Fischer-Tropsch technology—from active site to commercial process. Appl Catal A Gen 186:27–40

Gerardi MH (2003) The microbiology of anaerobic digesters. Wiley, Hoboken

Gheshlaghi R, Scharer JM, Moo-Young M, Chou CP (2009) Metabolic pathways of clostridia for producing butanol. Biotechnol Adv 27:764–781

Gomez X, Moran A, Cuetos MJ, Sanchez ME (2006) The production of hydrogen by dark fermentation of municipal solid wastes and slaughterhouse waste: a two-phase process. J Power Sources 157:727–732

Gonzalez-Cabaleiro R, Lema JM, Rodriguez J (2015) Metabolic energy-based modeling explains product yielding in anaerobic mixed culture fermentations. PLoS One. https://doi.org/10.1371/journal.pone.0126739

Grand View Research (2014) Global lactic acids and polylactic acid market. www.grandviewresearch.com/press-release/global-lactic-acid-poly-lactic-acid-market. Accessed 24 Nov 2014

Granda C, Holtzapple M, Luce G, Searcy K, Mamrosh D (2009) Carboxylate platform: the MixAlco process part 2: process economics. Appl Biochem Biotechnol 156:107–124

Groot WJ, van der Lans RGJM, Luyben KCAM (1989) Batch and continuous butanol fermentations with free cells: integration with product recovery by gas-stripping. Appl Microbiol Biotechnol 32:305–308

Gujer W, Zehnder AJB (1983) Conversion processes in anaerobic digestion. Water Sci Technol 15:127–167

Guo XM, Trably E, Latrille E, Carrere H, Steyer JP (2010) Hydrogen production from agricultural waste by dark fermentation: a review. Int J Hydrog Energy 35:10660–10673

Hadipour A, Sohrabi M (2008) Synthesis of some bifunctional catalysts and determination of kinetic parameters for direct conversion of syngas to dimethyl ether. Chem Eng J 137:294–301

Hafez H, Naggar MHE, Nakhla G (2010) Steady-state and dynamic modeling of biohydrogen production in an integrated biohydrogen reactor clarifier system. Int J Hydrog Energy 35:6634–6645

Hallenbeck PC (2005) Fundamentals of the fermentative production of hydrogen. Water Sci Technol 52:21–29

Hallenbeck P (2009) Fermentative hydrogen production: principles, progress and prognosis. Int J Hydrog Energy 34:7379–7389

Hamer H, Jonkers D, Venema K, Vanhoutvin S, Troost F, Brummer R (2008) Review article: the role of butyrate on colonic function. Aliment Pharmacol Ther 27:104–119

Han SK, Shin HS (2004) Biohydrogen production by anaerobic fermentation of food waste. Int J Hydrog Energy 29:567–577

Hawkes FR, Dinsdale R, Hawkes DL, Hussy I (2002) Sustainable fermentative hydrogen production: challenges for process optimisation. Int J Hydrog Energy 27:1339–1347

Hawkes FR, Hussy I, Kyazze G, Dinsdale R, Hawkes DL (2007) Continuous dark fermentative hydrogen production by mesophilic microflora: principles and progress. Int J Hydrog Energy 32:172–184

Heiskanen H, Virkajarvi I, Viikari L (2007) The effect of syngas composition on the growth and product formation of Butyribacterium methylotrophicum. Enzym Microb Technol 41:362–367

Henstra A, Sipma J, Rinzema A, Stams A (2007) Microbiology of synthesis gas fermentation for biofuel production. Curr Opin Biotechnol 18:200–206

Hoffmann P (2001) Tomorrow's energy, hydrogen, fuel cells, and the prospects for a cleaner planet. The MIT Press, Cambridge

Hu W, Sin S, Chua H, Yu P (2005) Synthesis of polyhydroxyalkanoate (PHA) from excess activated sludge under various oxidation-reduction potentials (ORP) by using acetate and propionate as carbon sources. Appl Biochem Biotechnol 121–124:289–301

Hussy I, Hawkes FR, Dinsdale R, Hawkes DL (2003) Continuous fermentative hydrogen production from a wheat starch co-product by mixed microflora. Biotechnol Bioeng 84:619–626

Hwang MH, Jang NJ, Hyun SH, Kim IS (2004) Anaerobic bio-hydrogen production from ethanol fermentation: the role of pH. J Biotechnol 111:297–309

Im WT, Kim DH, Kim KH, Kim MS (2012) Bacterial community analyses by pyrosequencing in dark fermentative H2-producing reactor using organic wastes as a feedstock. Int J Hydrog Energy 37:8330–8337

Indexmundi (2014) International prices of commodities. www.indexmundi.com. Accessed 24 Nov 2014

Insam H, Franke-Whittle I, Goberna M (2010) Microbes at work, from wastes to resources. Springer, Berlin

Jang E, Jung M, Min D (2005) Hydrogenation of low trans and high conjugated fatty acids. Compr Rev Food Sci Food Saf 1:22–30

Jeon Y, Cho C, Yun Y (2006) Combined effects of light intensity and acetate concentration on the growth of unicellular microalga *Haematococcus pluvialis*. Enzym Microb Technol 39:490–495

Jeon B, Kim B, Um Y, Sang B (2010) Production of hexanoic acid from D-galactitol by a newly isolated Clostridium sp. BS-1. Appl Microbiol Biotechnol 88:1161–1167

Jeon B, Moon C, Kim B, Kim H, Um Y, Sang B (2013) In situ extractive fermentation for the production of hexanoic acid from galactitol by Clostridium sp. BS-1. Enzym Microb Technol 53:143–151

Jiang J, Zhang Y, Li K, Wang Q, Gong C, Li M (2013) Volatile fatty acids production from food waste: effects of pH, temperature, and organic loading rate. Bioresour Technol 143:525–530

Joo F (2008) Breakthroughs in hydrogen storage—formic acid as a sustainable storage material for hydrogen. Chem Sus Chem 1:805–808

Jung K, Kim D, Shin H (2011a) A simple method to reduce the start-up in a H2-producing UASB reactor. Int J Hydrog Energy 36:1466–1473

Jung K, Kim D, Shin H (2011b) Bioreactor design for continuous dark fermentative hydrogen production. Bioresour Technol 102:8612–8620

Kamm B, Gruber PR, Kamm M (2006) Biorefineries—industrial processes and products. Wiley-VCH, Weinheim

Kaparaju P, Serrano M, Thomsen AB, Kongjan P, Angelidaki I (2009) Bioethanol, biohydrogen and biogas production from wheat straw in a biorefinery concept. Bioresour Technol 100:2562–2568

Karadag D, Puhakka J (2010) Effect of changing temperature on anaerobic hydrogen production and microbial community composition in an open-mixed culture bioreactor. Int J Hydrog Energy 35:10954–10959

Karlsson A, Vallin L, Ejlertsson J (2008) Effects of temperature, hydraulic retention time and hydrogen extraction rate on hydrogen production from the fermentation of food industry residues and manure. Int J Hydrog Energy 33:953–962

Kenealy W, Cao Y, Weimer P (1995) Production of caproic acid by cocultures of ruminal cellulolytic bacteria and *Clostridium kluyveri* grown on cellulose and ethanol. Appl Microbiol Biotechnol 44:507–513

Khardenavis A, Wang J, Ng W, Purohit H (2013) Management of various organic fractions of municipal solid waste via recourse to VFA and biogas generation. Environ Technol 34:2085–2097

Kim I, Hwang MH, Jang NJ, Hyun SH, Lee S (2004) Effect of low pH on the activity of hydrogen utilizing methanogen in biohydrogen process. Int J Hydrog Energy 29:1133–1140

Kim M, Baek J, Yun Y, Sim S, Park S, Kim S (2006) Hydrogen production from *Chlamydomonas reinhardtii* biomass using a two-step conversion process: anaerobic conversion and photosynthetic fermentation. Int J Hydrog Energy 31:812–816

Kim H, Leeds P, Chuang D (2009) The HDAC inhibitor, sodium butyrate, stimulates neurogenesis in the ischemic brain. J Neurochem 110:1226–1240

Kobayashi M, Kakizono T, Nagai S (1993) Enhanced carotenoid biosynthesis by oxidative stress in acetate-induced cyst cells of a green unicellular alga, *Haematococcus pluvialis*. Appl Environ Microbiol 59:867–873

Kraemer JT (2004) Effects of methanogenic effluent recycle on fermentative hydrogen production. PhD Thesis, University of Toronto, Toronto, Canada

Kumar K, Dasgupta C, Nayak B, Lindblad P, Das D (2011) Development of suitable photobioreactors for CO2 sequestration addressing global warming using green algae and cyanobacteria. Bioresour Technol 102:4945–4953

Kurzynski M (2006) The thermodynamic machinery of life. Springer, Leipzig

Kwan TH, Hu Y, Lin CSK (2016) Valorisation of food waste via fungal hydrolysis and lactic acid fermentation with *Lactobacillus casei* Shirota. Bioresour Technol 217:129–136

Ladygina N, Dedyukhina E, Vainshtein M (2006) A review on microbial synthesis of hydrocarbons. Process Biochem 41:1001–1004

Lay JJ (2001) Biohydrogen generation by mesophilic anaerobic fermentation of microcrystalline cellulose. Biotechnol Bioeng 74:280–287

Lay JJ, Lee YJ, Noike T (1999) Feasibility of biological hydrogen production from organic fraction of municipal solid waste. Water Res 33:2579–2586

Lee D, Li Y, Noike T, Cha G (2008) Behaviour of extracellular polymers and bio-fouling during hydrogen fermentation with a membrane bioreactor. J Memb Sci 322:13–18

Lee D, Li Y, Noike T (2010) Influence of solids retention time on continuous H2 production using membrane bioreactor. Int J Hydrog Energy 35:52–60

Leung A, Boocock D, Konar S (1995) Pathway for the catalytic conversion of carboxylic acids to hydrocarbons over activated alumina. Energy Fuel 9:913–920

Leyva M, Vicedo B, Finiti I, Flors V, Del Amo G, Real M et al (2008) Preventive and post-infection control of Botrytis cinerea in tomato plants by hexanoic acid. Plant Pathol 57:1038–1046

Li C, Fang HHP (2007) Fermentative hydrogen production from wastewater and solid wastes by mixed cultures. Crit Rev Environ Sci Technol 37:1–39

Li W, Yu H (2011) From wastewater to bioenergy and bio-chemicals via two-stage bioconversion processes: a future paradigm. Biotechnol Adv 29:972–982

Liang Y, Sarkany N, Cui Y (2009) Biomass and lipid productivities of *Chlorella vulgaris* under autotrophic, heterotrophic and mixotrophic growth conditions. Biotechnol Lett 31:1043–1049

Lin CY, Chang RC (2004) Fermentative hydrogen production at ambient temperature. Int J Hydrog Energy 29:715–720

Lin CY, Lay CH (2004) Carbon/nitrogen-ratio effect on fermentative hydrogen production by mixed microflora. Int J Hydrog Energy 29:41–45

Lin CY, Lay CH (2005) A nutrient formulation for fermentative hydrogen production using anaerobic sewage sludge microflora. Int J Hydrog Energy 30:285–292

Lindblad MS, Liu Y, Albertsson AC, Ranucci E, Karlsson S (2002) Polymers from renewable resources. In: Albertsson AC (ed) Degradable aliphatic polyesters. Springer, Berlin, pp 139–161

Liu WT, Chan OC, Fang HHP (2002) Microbial community dynamics during start-up of acidogenic anaerobic reactors. Water Res 36:3203–3210

Liu D, Liu D, Zeng RJ, Angelidaki I (2006) Hydrogen and methane production from household solid waste in the two-stage fermentation process. Water Res 40:2230–2236

Louis P, Flint HJ (2009) Diversity, metabolism and microbial ecology of butyrate-producing bacteria from the human large intestine. FEMS Microbiol Lett 294:1–8

Maddox IS, Qureshi N, Roberts-Thomson K (1995) Production of acetone-butanol-ethanol from concentrated substrates using Clostridium acetobutylicum in an integrated fermentation-product removal process. Process Biochem 30:209–215

Magid J (2006) Short-Circuit short circuiting the carbon and nutrient cycles between urban and rural districts by establishing three new systems for source separation, collection and composting of organic waste in the greater Copenhagen area. Final report presented to the EU-life program. http://orgprints.org/9230/. Accessed 21 June 2015

Marinoiu A, Cobzaru C, Carcadea E, Capris C, Tanislav V, Raceanu M (2013) Hydrogenolysis of glycerol to propylene glycol using heterogeneous catalysis in basic aqueous solutions. React Kinet Mech Catal 110:63–73

Market Publishers (2012) Global n-butanol market to reach 4 Mt by 2020. http://marketpublishers.com/lists/13495/news.html. Accessed 24 Nov 2014

Markou G, Georgakakis D (2011) Cultivation of filamentous cyanobacteria (blue-green algae) in agro-industrial wastes and wastewaters: a review. Appl Energy 88:3389–3401

McKendry P (2002a) Energy production from biomass (part 2): conversion technologies. Bioresour Technol 83:47–54

McKendry P (2002b) Energy production from biomass (part 3): gasification technologies. Bioresour Technol 83:55–63

Mecking S, Held A, Bauers F (2002) Aqueous catalytic polymerization of olefins. Angew Chem Int Ed 41:544–561

Mitchell R, Gu JD (2010) Environmental microbiology, 2nd edn. Wiley-Blackwell, Hoboken

Mizuno O, Dinsdale R, Hawkes FR, Hawkes DL, Noike T (2000) Enhancement of hydrogen production from glucose by nitrogen gas sparging. Bioresour Technol 73:59–65

Mohammadi M, Najafpour G, Younesi H, Lahijani P, Uzir M, Mohamed A (2011) Bioconversion of synthesis gas to second generation biofuels: a review. Renew Sust Energ Rev 15:4255–4273

Mu Y, Zheng XJ, Yu HQ, Zhu R (2006) Biological hydrogen production by anaerobic sludge at various temperatures. Int J Hydrog Energy 31:780–785

Mu Y, Yu HQ, Wang G (2007) A kinetic approach to anaerobic hydrogen-producing process. Water Res 41:1152–1160

Munster M (2009) Energy systems analysis of waste to energy technologies by use of energy-PLAN. Riso report, Danish Technical University. http://orbit.dtu.dk/fedora/objects/orbit:81741/datastreams/file_4069900/content. Accessed 12 June 2015

Murzin D, Leino R (2008) Sustainable chemical technology through catalytic multistep reactions. Chem Eng Res Des 86:1002–1010

Murzin D, Simakova I (2011) Catalysis in biomass processing. Catal Ind 3:218–249

Mussatto SL, Dragone G, Guimaraes PMR, Silva JPA, Carneiro LM, Roberto IC et al (2010) Technological trends, global market, and challenges of bio-ethanol production. Biotechnol Adv 28:817–830

Najdenski H, Gigova L, Illiev I, Pilarski P, Lukavsky J, Tsvetkova I et al (2013) Antibacterial and antifungal activities of selected microalgae and cyanobacteria. Int J Food Sci Technol 48:1533–1540

Nguyen CM, Choi GJ, Choi YH, Jang KS, Kim JC (2013) D- and L-lactic acid production from fresh sweet potato through simultaneous saccharification and fermentation. Biochem Eng J 81:40–46

Nielsen AT, Amandusson H, Bjorklund R, Dannentun H, Ejlertsson J, Ekedahl LG et al (2001) Hydrogen production from organic waste. Int J Hydrog Energy 26:547–550

Nishio N, Nakashimada Y (2007) Recent development of anaerobic digestion processes for energy recovery from wastes. J Biosci Bioeng 103:105–112

Ntaikou I, Antonopoulou G, Lyberatos G (2010) Biohydrogen production from biomass and wastes via drak fermentation: a review. Waste Biomass Valoriz 1:21–39

Nwobi A (2013) Techno-economic evaluation of biofuel production from organic fraction of municipal solid waste (OFMSW) generated in Abu Dhabi. MSc thesis, Masdar Institute of Science and Technology, Abu Dhabi, UAE

Ogbonna J, Yoshizawa H, Tanaka H (2000) Treatment of high strength organic wastewater by a mixed culture of photosynthetic microorganisms. J Appl Phycol 12:277–284

Oh S, Iyer P, Bruns M, Logan B (2004) Biological hydrogen production usig a membrane bioreactor. Biotechnol Bioeng 87:119–127

OPEC (2014) 2014 world oil outlook. Organization of the Petroleum Exporting Countries, Vienna

Pauss A, Andre G, Perrier M, Guiot S (1990) Liquid-to-mass transfer in anaerobic processes: inevitable transfer limitations of methane and hydrogen in the biomethanation process. Appl Environ Microb 56:1636–1644

Pavlostathis SG, Gossett JM (1988) Preliminary conversion mechanisms in anaerobic digestion of biological sludges. J Environ Eng 114:575–592

Peighambardoust SJ, Rowshanzamir S, Amjadi M (2010) Review of the proton exchange membranes for fuel cell applications. Int J Hydrog Energy 35:9349–9384

Perrotta AR, Kumaraswany R, Bastidas-Oyanedel JR, Alm EJ, Rodriguez J (2017) Inoculum composition determines microbial community and function in an anaerobic sequential batch reactor. PLoS One 12(2):e0171369. https://doi.org/10.1371/journal.pone.0171369

Petkov G, Ivanova A, Iliev I, Vaseva I (2012) A critical look at the microalgae biodiesel. Eur J Lipid Sci Technol 114:103–111

Pinto F, Franco C, Andre R, Miranda M, Gulyurtlu I, Cabrita I (2002) Co-gasification study of biomass mixed with plastic wastes. Fuel 81:291–297

Pirie C, De Mey M, Prather K, Ajikumar P (2013) Integrating the protein and metabolic engineering tollkits for next-generation chemical biosynthesis. ACS Chem Biol 8:662–672

Porpatham E, Ramesh A, Nagalingam B (2007) Effect of hydrogen addition on the performance of a biogas fuelled spark ignition engine. Int J Hydrog Energy 32:2057–2065

Prakasham RS, Sathish T, Brahmaiah P (2010) Biohydrogen production process optimization using anaerobic mixed consortia: a prelude study for use of agro-industrial material hydrolysate as substrate. Bioresour Technol 101:5708–5711

PRnewswire (2013) Global salicylic acid market is expected to reach USD 5212 million in 2019. http://www.prnewswire.com/news-releases/global-salicylic-acid-market-is-expected-to-reach-usd-5212-million-in-2019-transparency-market-research-225332251.html. Accessed 24 Nov 2014

Pulz O, Gross W (2004) Valuable products from biotechnology of microalgae. Appl Microbiol Biotechnol 65:635–648

Quemeneur M, Hamelin J, Guidici-Orticoni T, Latrille E, Steyer JP, Trably E (2011) Changes in hydrogenase genetic diversity and proteomic patterns in mixed-cultured dark fermentation of mono, di- and tri-saccharides. Int J Hydrog Energy 36:11654–11665

Queneau Y, Pinel C, Scherrmann M (2011) Some chemical transformations of carbohydrates in aqueous medium. C R Chim 14:688–699

Rajhi H, Conthe M, Puyol D, Diaz E, Sanz J (2013) Dark fermentation: isolation and characterization of hydrogen-producing strains from sludges. Int Microbiol 16:53–62

Rashid K (2013) Prospects for algal biofuels in UAE: technology and economics. MSc Thesis Masdar Institute of Science and Technology, Abu Dhabi, UAE

Reddy AKN, Williams RH, Johansson T (1997) Energy after Rio: prospects and challenges. United Nations Development Programme (UNDP), New York

Redwood M, Paterson-Beedle M, Macaskie L (2009) Integrating dark and light bio-hydrogen production strategies: towards the hydrogen economy. Rev Environ Sci Biotechnol 8:149–185
Ren N, Cao G, Xu J, Gao L (2009) Bioconversion of lignocellulosic biomass to hydrogen: potential and challenges. Biotechnol Adv 27:1051–1060
REnescience (2015) Technology and services. http://www.renescience.com/en/technology-and-services/technology. Accessed 12 June 2015
Rephaeli A, Zhuk R, Nudelman A (2000) Prodrugs of butyric acid from bench to bedside: synthetic design, mechanisms of action, and clinical applications. Drug Devlop Res 50:379–391
Rice C, Ha S, Masel RI, Waszczuk P, Wieckowski A, Barnard T (2002) Direct formic acid fuel cells. J Power Sources 111:83–89
Rittmann S, Herwig C (2012) A comprehensive and quantitative review of dark fermentative biohydrogen production. Microb Cell Factories 11:115
Roddy D (2013) A syngas network for reducing industrial carbon footprint and energy use. Appl Therm Eng 53:299–304
Roehr M (2001) The biotechnology of ethanol, classical and future applications. Wiley-VCH verlag GmbH, Weinheim
Rogers P, Chen J, Zidwick M (2006) Organic acid and solvent production. In: The Prokaryotes, 3rd edn. Springer, Berlin Heidelberg, pp, pp 511–755
Romano AH, Conway T (1996) Evolution of carbohydrate metabolic pathways. Res Microbiol 147:448–455
Ruiz-Rodriguez A, Fornan T, Hernandez E, Senorans F, Reglero G (2010) Thermodynamic modeling of dealcoholization of beverages using supercritical CO2: application to wine samples. J Supercrit Fluid 52:183–188
Sahena F, Zaidul I, Jinap S, Karim A, Abbas K, Norulaini N et al (2009) Application of supercritical CO2 in lipid extraction—a review. J Food Eng 95:240–253
Sauer U, Santangelo J, Treuner A, Buchholz M, Durre P (1995) Sigma factor and sporulation genes in Clostridium. FEMS Microbiol Rev 17:331–340
Schink B, Kremer DR, Hansen TA (1987) Pathway of propionate formation from ethanol in Pelobacter propionicus. Arch Microbiol 147:321–327
Seeliger S, Janssen PH, Schink B (2006) Energetics and kinetics of lactate fermentation to acetate and propionate via methylmalonyl-CoA or acrylyl-CoA. FEMS Microbiol Lett 211:65–70
Shi X, Yu H (2005) Optimization of volatile fatty acid compositions for hydrogen production by Rhodopseudomonas capsulata. J Chem Technol Biot 80:1198–1203
Shi X, Yu H (2006) Continuous production of hydrogen from mixed volatile fatty acids with Rhodopseudomonas capsulata. Int J Hydrog Energy 31:1641–1647
Shin HS, Younb JH, Kim SH (2004) Hydrogen production from food waste in anaerobic mesophilic and thermophilic acidogenesis. Int J Hydrog Energy 29:1355–1363
Show K, Lee D, Chang J (2011) Bioreactor and process design for biohydrogen production. Bioresour Technol 102:8524–8533
Sialve B, Bernet N, Bernard O (2009) Anaerobic digestion of microalgae as a necessary step to make microalgal biodiesel sustainable. Biotechol Adv 27:409–416
Sierens R, Rosseel E (2000) Variable composition hydrogen/natural gas mixtures for increased engine efficiency and decreased emissions. J Eng Gas Turb Power 122:135–140
Singh U, Ahluwalia A (2013) Microalgae: a promising tool for carbon sequestration. Mitig Adapt Strateg Glob Change 18:73–95
Sossai P (2012) Butyric acid: what is the future for this old substance? Swiss Med Wkly 142:w13596
Sparling R, Islam R, Cicek N, Carere C, Chow H, Levin DB (2006) Formate synthesis by Clostridium thermocellum during anaerobic fermentation. Can J Microbiol 52:681–688
Spivey J, Gangwal S, Zoeller JR, Winslow J, Srivastava RD (2000) Syngas conversion to fuels and chemicals. Catal Today 58:231
Sudesh K, Doi A (2000) Synthesis, structure and properties of polyhydroxyalkanoates: biological polyesters. Prog Polym Sci 25:1503–1555

Tanaka N, Kjorven O, Yumkella K (2010) Energy poverty, how to make modern energy access universal? International Energy Agency (IEA), United Nations Development Programme (UNDP), and United Nations Industrial Development Organization (UNIDO), Paris
Tanisho S, Kuromoto M, Kadokura N (1998) Effect of CO2 removal on hydrogen production by fermentation. Int J Hydrog Energy 23:559–563
Tawfik A, El-Qelish M (2012) Continuous hydrogen production from co-digestion of municipal food waste and kitchen wastewater in mesophilic anaerobic baffled reactor. Bioresour Technol 114:270–274
Temudo MF, Kleerebezem R, van Loosdrecht MCM (2007) Influence of the pH on (open) mixed culture fermentation of glucose: a chemostat study. Biotechnol Bioeng 98:69–79
Temudo MF, Muyzer G, Kleerebezem R, van Loosdrecht MCM (2008a) Diversity of microbial communities in open mixed culture fermentations: impact of the pH and carbon source. Appl Microbiol Biotechnol 80:1121–1130
Temudo MF, Poldermans R, Kleerebezem R, van Loosdrecht MCM (2008b) Glycerol fermentation by (open) mixed cultures: a chemostat study. Biotechnol Bioeng 100:1088–1098
Temudo MF, Mato T, Kleerebezem R, van Loosdrecht MCM (2009) Xylose anaerobic conversion by open-mixed cultures. Appl Microbiol Biotechnol 82:231–239
Thauer RK, Jungermann K, Decker K (1977) Energy conservation in chemotrophic anaerobic bacteria. Bacteriol Rev 41:100–180
Tholozan JL, Touzel JP, Samain E, Grivet JP, Prensier G, Albagnac G (1992) Clostridium neopropionicum sp-nov a strict anaerobic bacterium fermenting ethanol to propionate through acrylate pathway. Arch Microbiol 157:249–257
Tomasko D, Burley A, Feng L, Yeh S, Miyazono K, Nirmal-Kumar S et al (2009) Development of CO2 for polymer foam applications. J Supercrit Fluid 47:493–499
Tran N, Barlett J, Kannangara G, Milev A, Volk H, Wilson M (2010) Catalytic upgrading of biorefinery oil from microalgae. Fuel 89:265–274
Ueno Y, Kawai T, Sato S, Otsuka S, Morimoto M (1995) Biological production of hydrogen from cellulose by natural anaerobic microflora. J Ferment Bioeng 79:395–397
Uhm S, Chung ST, Lee J (2008) Characterization of direct formic acid fuel cells by impedance studies: in comparison of direct methanol fuel cells. J Power Sources 178:34–43
Uratani J (2013) Anerobic digestion of halophytic plant residues for biogas production and nutrient recovery. MSc Thesis, Masdar Institute of Science and Technology, Abu Dhabi, UAE
Vane LM (2005) A review of pervaporation for product recovery from biomass fermentation processes. J Chem Technol Biotechnol 80:603–629
Vanhoutvin S, Troost F, Kilkens T, Lindsey P, Hamer H, Jonkers D et al (2009) The effects of butyrate enemas on visceral perception in healthy volunteers. Neurogastroenterol Motil 21:952–e76
Vasudevan D, Richter H, Angenent L (2014) Upgrading dilute ethanol from syngas fermentation to n-caproate with reactor microbiomes. Bioresour Technol 151:378–382
Waligorska M (2012) Fermentative hydrogen production— process design and bioreactors. Chem Process Eng 33:585–594
Wang Y, Mu Y, Yu HQ (2007) Comparative performance of two upflow anaerobic biohydrogen-producing reactors seeded with different sludges. Int J Hydrog Energy 32:1086–1094
Wang S, Guo Z, Cai Q, Guo L (2012) Catalytic conversion of carboxylic acids in bio-oil for liquid hydrocarbons production. Biomass Bioenergy 45:138–143
Weibel G, Ober C (2003) An overview of supercritical CO2 applications in microelectronics processing. Microelectron Eng 65:145–152
Weissermel K, Arpe HJ (1997) Industrial organic chemistry. VCH Publishers, New York
Wilhelm D, Simbeck D, Karp A, Dickenson R (2001) Syngas production for gas-to-liquids applications: technologies, issues and outlook. Fuel Process Technol 71:139–148
Williams T, Shaddix C, Schefer R (2008) Effect of syngas composition and CO2-diluted oxygen on performance of a premixed swirl-stabilized combustor. Combust Sci Technol 180:64–88
Willke T, Vorlop K (2004) Industrial bioconversion of renewable resources as an alternative to conventional chemistry. Appl Microbiol Biotechnol 66:131–142

Willquist K, Nkemba VN, Svensson H, Pawar S, Ljunggren M, Karlsson H et al (2012) Design of a novel biohythane process with high H2 and CH4 production rates. Int J Hydrog Energy 37:17749–17762

Ye N, Lu F, Shao L, Godon J, He P (2007) Bacterial community dynamics and product distribution during pH-adjusted fermentation of vegetable wastes. J Appl Microbiol 103:1055–1065

Yokoi H, Saitsu A, Uchida H, Hirose J, Hayashi S, Takasaki Y (2001) Microbial hydrogen production from sweet potato starch residue. J Biosci Bioeng 91:58–63

Yousuf A, Bastidas-Oyanedel JR, Schmidt JE (2018) Effect of total solid content and pretreatment on the production of lactic acid from mixed culture dark fermentation of food waste. Waste Management, In Press, DOI. https://doi.org/10.1016/j.wasman.2018.04.035

Zacharof M, Lovitt R (2013) Complex effluent streams as a potential source of volatile acids. Waste Biomass Valor 4:557–581

Zahedi S, Sales D, Romero L, Solera R (2013) Hydrogen production from the organic fraction of municipal solid waste: in anaerobic thermophilic acidogenesis: influence of organic loading rate and microbial content of the solid waste. Bioresour Technol 129:85–91

Zeeshan N, Hinrich U (2014) Enhancing the hydrolysis process of a two-stage biogas technology for the organic fraction of municipal solid waste. In: Presented at the international conference on anaerobic digestion "BiogasScience 2014", Vienna, Austria

Zehnder A, Svensson B (1986) Life without oxygen: what can and what cannot? Exp Dermatol 42:1197–1205

Zhang Y (2014) Production of biofuels and biochemicals by in vitro synthetic biosystems: opportunities and challenges. Biotechnol Adv. https://doi.org/10.1016/j.biotechadv.2014.10.009

Zhang Y, Shen J (2006) Effect of temperature and iron concentration on the growth and hydrogen production of mixed bacteria. Int J Hydrog Energy 31:441–446

Zhang Z, Show K, Tay J, Liang D, Lee D, Jiang W (2007) Rapid formation of hydrogen producing granules in an anaerobic continuous stirred tank reactor induced by acid incubation. Biotechnol Bioeng 96:1040–1050

Zhang C, Yang H, Yang F, Ma Y (2009) Current progress on butyric acid production by fermentation. Curr Microbiol 59:656–663

Zhang F, Zhang Y, Chen M, Zeng RJ (2012) Hydrogen super-saturation in thermophilic mixed culture fermentation. Int J Hydrog Energy 37:17809–17816

Zhang F, Ding J, Shen N, Zhang Y, Ding Z, Dai K et al (2013a) In situ hydrogen utilization for high fraction acetate production in mixed culture hollow-fiber membrane biofilm reactor. Appl Microbiol Biotechnol 97:10233–10240

Zhang F, Ding J, Zhang Y, Chen M, Ding ZW, van Loosdrecht MCM, Zeng RJ (2013b) Fatty acids production from hydrogen and carbon dioxide by mixed culture in the membrane biofilm reactor. Water Res 47:6122–6129

Zhang F, Zhang Y, Chen M, van Loosdrecht MCM, Zeng RJ (2013c) A modified metabolic model for mixed culture fermentation with energy conserving electron bifurcation reaction and metabolite transport energy. Biotechnol Bioeng 110:1884–1894

Zhang F, Zhang Y, Ding Y, Dai K, van Loosdrecht MCM, Zeng RJ (2014) Stable acetate production in extreme-thermophilic (70 °C) mixed culture fermentation by selective enrichment of hydrogenotrophic methanogens. Sci Rep. https://doi.org/10.1038/srep05268

Zhao Y, Chen Y, Zhang D, Zhu X (2010) Waste activated sludge fermentation for hydrogen production enhanced by anaerobic process improvement and acetobacteria inhibition: the role of fermentation pH. Environ Sci Technol 44:3317–3323

Zheng H, O'Sullivan C, Mereddy R, Zeng R, Duke M, Clarke W (2010) Experimental and theoretical investigation of diffusion process in membrane anaerobic reactor for bio-hydrogen production. Int J Hydrog Energy 35:5301–5311

Biohydrogen Production Perspectives from Organic Waste with Focus on Asia

Biswarup Sen, J. Aravind, Chiu-Yue Lin, Chyi-How Lay, and Ping-Heng Hsieh

1 Introduction

With the highest energy density (143 kJ g^{-1}), hydrogen is a promising alternative energy carrier for the future. It is clean, renewable, generates harmless by-products from combustion, and does not require a separate transportation infrastructure. Gasification of hydrocarbons, water electrolysis, and fermentation of organic wastes can produce hydrogen. However, hydrogen production using organic waste, commonly termed as bio-hydrogen (bio-H_2) production, is one of the sustainable ways of energy production. Of the various bio-H_2 production routes, dark fermentation is mostly economical compared to the others (Li et al. 2012a, b; Han et al. 2016a). Mesophilic and thermophilic microorganisms primarily drive dark fermentation of organic waste, and for many industrial organic wastewaters discharged at elevated temperatures, e.g., palm mill effluent or distillery effluents, dark fermentation at thermophilic temperatures is advantageous (Lin et al. 2012; Lucas et al. 2015; Tapia-Venegas et al. 2015). Organic wastes, e.g., lignocellulosic biomass and wastewaters, contain organic matter mainly composed of glucose, xylose, starch, cellulose, and sucrose, which make organic waste the appropriate substrate for fermentation to produce bio-H_2. The main operating factors affecting dark fermentation include seed

B. Sen (✉)
School of Environmental Science and Engineering, Tianjin University, Tianjin, China
e-mail: bsen@tju.edu.cn

J. Aravind
School of Bioscience and Technology, Vellore Institute of Technology, Vellore, India

C.-Y. Lin · C.-H. Lay
Green Energy Development Center, Feng Chia University, Taichung, Taiwan

P.-H. Hsieh
Department of Environmental Engineering, National Chung Hsing University, Taichung, Taiwan

© Springer Nature Switzerland AG 2019

J.-R. Bastidas-Oyanedel, J. E. Schmidt (eds.), *Biorefinery*,
https://doi.org/10.1007/978-3-030-10961-5_16

inoculum, substrate, reactor type, nutrients (nitrogen, phosphate, trace metals, etc.), temperature, and pH. Despite low yields, production rates, and substrate conversion efficiencies, dark fermentation is relatively less energy-intensive, requires low operating costs, is technically much simpler, and is more stable and robust compared to photobiological route (Mohan and Pandey 2013).

Bio-H_2 production mechanism is complex and requires the enzymes hydrogenases and nitrogenases in a multistep fermentation process that involves three stages of biochemical reactions, namely, hydrolysis, acidogenesis, and methanogenesis (Show et al. 2012). During hydrolysis step, complex organic matters are broken down into monomers such as glucose and amino acids by the action of extracellular enzymes produced by hydrolytic bacteria. In the acidogenesis stage, the hydrolyzed monomers are converted into volatile fatty acids (VFAs) and alcohols. The higher VFAs such as propionic and butyric acids are further converted to acetic acid and H_2/CO_2 by acetogenic bacteria. In anaerobic digestion, the acetic acid and H_2/CO_2 are then converted into methane (CH_4) by methanogens; however, to recover hydrogen, the final step of fermentation—methanogenesis—must be inhibited. This is achieved by suppressing the bioactivity of methanogens in the fermenter through a heat treatment of the seed sludge. Methanogens are sensitive to heat, but hydrogen-producing bacteria are not. Other methods to subdue methanogenesis include (1) reactor operation at short HRT and (2) process operation at an acidic pH. Most dark fermentation is conducted in well-mixed systems, such as the continuous stirred-tank reactor (CSTR), carrier-induced granular sludge bed reactor (CIGSBR), anaerobic sequencing batch reactor (ASBR), membrane bioreactor (MBR), fluidized bed reactor (FBR), and upflow anaerobic sludge blanket (UASB) reactor (Lin et al. 2012).

In recent years, bio-H_2 production technologies are increasingly gaining importance worldwide (Kumar et al. 2017a). Particularly in Asia, the research on these technologies has been outstanding with the implementation of several novel- and economically feasible processes. Within Asian countries, the bio-H_2 production technologies mainly focus on the feedstock pretreatment methods, operation conditions, substrate feeding strategies, and reactor designs. Table 1 summarizes the main bio-H_2 production technologies that are developed in Asian countries. The feedstock usage in bio-H_2 production shifted from synthetic wastewater to industrial wastewater, agricultural waste, food waste, and livestock waste because Asian countries mostly rely on agriculture, animal husbandry, and manufacturing. The use of these organic wastes, to produce bio-H_2, not only removes pollutants but can also recover bioenergy. However, one common question that still remains unresolved is: "*Is bio-H_2 production process viable and poses scalability*?" which can be answered only if we consider the techno-economic evaluation, net energy ratio, energy efficiency, greenhouse gas emissions (GHG), scale-up potential, and commercial feasibility. Extensive research carried out in Asia have shown promising prospects of bio-H_2 production, and there have been substantial improvement and development in both the yield and volumetric production rates. However, for commercial scale applications that make economic sense, hydrogen yields and production rates must surpass significantly the current achievements. In addition, the technological breakthrough is desirable to pull out most of the bio-H_2 from the renewable feedstock, if not all.

Table 1 Some examples of bio-H_2 production from organic waste in Asia

Country	Feed	Reactor design	Bio-H_2 production rate	Bio-H_2 yield	Process temp. (°C)	References
Korea	Food waste	Pilot-scale two-phase hydrogen/methane fermentation system	3.88 H_2 L m^{-3} day^{-1}	1.82 moles H_2/mole glucose	30	Lee and Chung (2010)
Taiwan	Vegetable-based kitchen waste and napier grass	CSTR	0.11 ± 0.04 H_2 m^3 m^{-3} day^{-1}	14 ± 5 moles H_2/kg COD_{in}	55	Kuo et al. (2012)
Taiwan	Textile desizing wastewater	CSTR	3.8 L H_2 L-day^{-1}	–	35	Lin et al. (2017a, b)
Taiwan	Vegetable kitchen waste	I-CSTR	1.0[a]	1.7 mmoles H_2 g-COD^{-1}	55	Lee et al. (2010)
Taiwan	Starch-rich kitchen waste	I-CSTR	2.2[a]	2.1 mmoles H_2 g-COD^{-1}	55	Wang et al. (2010a, b)
India	Canteen-based composite food waste	Anaerobic SBR	–	69.95 mmoles	29	Reddy et al. (2011)
India	*Calophyllum inophyllum* oil cake	Anaerobic reactor	7.95 L L^{-1}	–	30	Arumugam et al. (2014)
India	Water hyacinth	Customized reactor	113.38 mL h^{-1}	41.38 L kg^{-1} COD	37	Varanasi et al. (2018)
China	Algal biomass	400 mL Glass bottle	4.69 mL $h^{-1}g^{-1}$ TVS	256.7 mL g^{-1} TVS	35	Cheng et al. (2014)
China	Water hyacinth	350 mL Glass bottle	84.9 mL L^{-1} h^{-1}	134.9 mL g^{-1} TVS	35	Cheng et al. (2015)
South Korea	*Gelidium amansii*	CSTR	1.3 moles $mole^{-1}$ substrate	2.7 L L^{-1} day^{-1}	53	Kumar et al. (2017a, b)
South Korea	Red algae	7 L fermenter	2.46 L g^{-1} VSS day^{-1}	2.03 moles $mole^{-1}$ galactose	35	Park et al. (2011)

(continued)

Table 1 (continued)

Country	Feed	Reactor design	Bio-H_2 production rate	Bio-H_2 yield	Process temp. (°C)	References
China	Corncob powder	Baffled photo-fermenter reactor	7.37 mmoles L^{-1} h^{-1}	512.29 mmoles L^{-1}	30	Zhang et al. (2015)
China	Waste bread	CSTR	7.4 L L^{-1} day^{-1}	109.5 mL g^{-1}	–	Han et al. (2016b)
China	Waste bread	CSTR	163.9 mL h^{-1}	52.4 mL g^{-1}	37	Han et al. (2016c)
China	Apple waste	Batch fermenter	–	111.85 mL g^{-1}TS	30.4	Lu et al. (2016)
China	*Plantanus orientalis* leaves	Batch fermenter	–	64.10 mL g^{-1}TS	35.5	Li et al. (2017)
South Korea	Korean rice straw	Serum bottle	31.77 mL L^{-1} day^{-1}	2.3 mmoles g^{-1}	–	Nguyen et al. (2010)
Taiwan	Rice straw	Serum vial	18 mL h^{-1}	4.8 mL g^{-1} TS	55	Chen et al. (2012)
South Korea	Rice husk	Serum bottle	2608 mL L^{-1} day^{-1}	473.1 mL g^{-1}	35	Gonzales and Kim (2017)
Malaysia	POME	Serum bottle	–	0.66 moles $mole^{-1}$	37	Taifor et al. (2017)
Malaysia	POME	UASB	2.5 L day^{-1}	33.48 mL g^{-1} COD	55	Krishnan et al. (2017)
Thailand	POME	CSTR-UASB	4.1 L L^{-1}POME	135 mL g^{-1}VS	55, 35	O-Thong et al. (2016)
India	Cheese whey waste	Batch reactor	139 mL $L^{-1}h^{-1}$	6.35 moles $mole^{-1}$ lactose	37	Patel et al. (2016)
China	Textile desizing wastewater	CSTR	3.8 L L^{-1} day^{-1}	–	35	Lin et al. (2017a, b)
India	Rice mill wastewater	Glass reactor	36.5 mmoles g^{-1} cell h^{-1}	1.74 moles $mole^{-1}$ sugar	33	Ramprakash and Muthukumar (2014)
India	Herbal wastewater (HWW)	Batch reactor	930 mL L^{-1} HWW	580 mL g^{-1} COD	36	Sivaramakrishna et al. (2014)

Country	Feed	Reactor design	Bio-H_2 production rate	Bio-H_2 yield	Process temp. (°C)	References
Malaysia	Sago-starch processing effluent (SSPE)	Batch reactor	57.54 mL h^{-1} L^{-1} SSPE	1.09 moles $mole^{-1}$ glucose	–	Yunus et al. (2014)
Taiwan	Mushroom cultivation waste	Batch reactor	9.50 moles H_2 kg^{-1} day^{-1}	0.29 mmoles H_2 g^{-1} TVS	55	Lin et al. (2016)
India	Sago-starch processing wastewater	Batch reactor	1340.9 mL L^{-1}	126.5 mL g^{-1} COD	30	Sen and Suttar (2012)
Iran	Rice straw	Batch reactor	–	19.73 mL g^{-1} straw	37	Asadi and Zilouei (2017)
China	Cotton stalk hyrolysate	Batch reactor	937 mL L^{-1} day^{-1}	1.44 ± 0.08 moles $mole^{-1}$ sugar consumed	37	Li et al. (2018)

[a]L-H_2 L^{-1} day^{-1}
[b]mmoles L^{-1} day^{-1}
[c]mmoles H_2 g^{-1} VSS h^{-1}
[d]mL-H_2 h^{-1}

Research addressing this challenge should be a key program of future studies on bio-based hydrogen generation. Further, scale-up of the process to the full-scale system will aid refine the technology for commercialization (Sen et al. 2013). Currently, the bio-H_2 production process is directed toward a decentralized energy production on the pilot scale and is less energy-intensive than its counterpart processes (e.g., natural gas reforming).

2 Overview of Biomass-Based Bio-H_2 Production in Asia

In the Asian region, various biomass substrates have been harnessed for the production of bio-H_2, ranging from different agricultural residues to various types of wastewaters. The wide range of substrates is broadly classified under two categories: (1) solid waste and (2) wastewaters (Table 1). Some recent case studies on biomass-based bio-H_2 are discussed below.

2.1 *Solid Waste*

2.1.1 Oilseed Cake

Among oilseed cakes, extensive research with de-oiled Jatropha waste has been conducted by Kumar et al. (2013, 2015, 2016). De-oiled Jatropha waste is generated as biodiesel industry residue which is lignocellulosic in nature. Thus, the direct utilization of this waste to hydrogen production is rate-limiting, which necessitates the use of a hydrolysis method to recover the sugars for efficient hydrogen production (Kumar et al. 2013). In other studies, the possibility of combined production of bio-H_2 and a biopolymer polyhydroxyalkanoate using *Calophyllum inophyllum* (Indian-laurel) oilseed cake as substrate and *Enterobacter aerogenes* and *Rhodobacter sphaeroides* as the inoculum was explored (Arumugam et al. 2014). The experimental setup generated 7.95 L H_2 L^{-1} and 3.23 L H_2 L^{-1} of the medium under alternative dark and photo-fermentation condition and complete dark condition alone, respectively. The fermentation was carried out at a pH of 5.1 under 30 °C temperature. The proposed method was good for the disposal of the solid waste with value addition by the production of polyhydroxyalkanoate.

2.1.2 Water Hyacinth

Su et al. (2010) initially pretreated water hyacinth with a combination of alkali, steam, and microwave heating before an enzymatic hydrolysis and subsequent hydrogen production in a two-step dark and photo-fermentation. A maximum hydrogen yield of 76.7 mL H_2 g^{-1} TVS was obtained at 20 g L^{-1} of water hyacinth under dark fermentation condition, and hydrogen yield of 522.6 mL H_2 g^{-1} TVS was obtained at 10 g L^{-1} of water hyacinth via photo-fermentation using immobilized *Rhodopseudomonas palustris*. Later, Chuang et al. (2011) determined the optimal substrate concentration (47.8 g L^{-1}) and temperature (62.5 °C) for maximum bio-H_2 production from untreated water hyacinth in batch experiments. That study demonstrated that untreated water hyacinth has good potential for bio-H_2 production and reported 31.3 $GJha^{-1}year^{-1}$. Lin et al. (2015) investigated microwave-heated alkali pretreatment to improve on the enzymatic digestibility of water hyacinth for hydrogen production and obtained a reducing sugar yield of 0.296 g g^{-1} TVS. The optimal hydrolysis conditions included a microwave temperature of 190 °C for 10 min as the duration of treatment and an enzyme dosage of 5 wt.%. Fermentative hydrogen yields from water hyacinth subjected to the microwave-assisted alkali treatment and enzymatic hydrolysis were observed to be 63.9 mL g^{-1} TVS under two-stage process and 76.7 mL g^{-1} TVS under single-stage fermentation. Further, to explore the feasibility of fermentative hydrogen production from water hyacinth, fermentative inhibitors (vanillin, furfural, and 5-hydroxymethyl furfural) were removed from the hydrolysates (Cheng et al. 2015). The water hyacinth hydrolysates were subjected to the domestication of hydrogen-producing bacteria, which resulted in an improved adaptability to fetch a hydrogen yield of 134.9 mL g^{-1} total volatile solids (TVS). Activated sludge from a marsh gas plant was utilized as the microbial source.

An integrated approach in the bioprocessing of water hyacinth for bio-H_2 production was assessed by Varanasi et al. (2018), where water hyacinth was used as the substrate in a single- to multistage process to yield maximum bioenergy in the form of hydrogen and/or methane and electricity generation. The team was able to remove 94% of the COD effectively with maximal energy recovery of 60% and claimed their current process of integrated biorefinery approach as a low-cost renewable technology to generate bioenergy, wherein a maximum cumulative bio-H_2 production of 904.24 mL L^{-1} with a yield of 41.38 L kg^{-1} COD was achieved. The authors argue that their three-stage system may prove beneficial when compared to the single- and two-stage systems.

2.1.3 Algal Biomass

Algal biomass, an abundant, clean, and renewable feedstock, is a promising source of bio-H_2. Although there are a variety of technologies for algal bio-H_2 and some laboratory- and pilot-scale systems have demonstrated a good potential for full-scale implementation (Park et al. 2011), bio-H_2 production from algal biomass is still in the early stage of development. The feasibility of bio-H_2 production from red alga *Gelidium amansii* was evaluated by Park et al. (2011) in a 7-L fermenter at a pH of 5.5 and temperature of 35 °C. The main sugar in this red alga, galactose, was readily converted to bio-H_2 via dark fermentation, and the maximum hydrogen production rate and yield were 2.46 L g^{-1} VSS day^{-1} and 2.03 moles $mole^{-1}$ $galactose_{added}$, respectively, which were higher than those of glucose (1.48 moles $mole^{-1}$). Three-stage (comprising of dark hydrogen fermentation, photo hydrogen fermentation, and methanogenesis) and combined (dark and photo-fermentation) processes with acid-domesticated hydrogenogens, sourced from activated sludge of a methane plant, were successfully tested for the efficient harnessing of the algal bloom waste from Taihu Lake, China. The domesticated hydrogenogens (*Clostridium butyricum*, *R. palustris*) in the presence of acids greatly improved the dark hydrogen production, utilizing the microwave-pretreated algae biomass, and resulted in the total hydrogen yield of 283.4 mL g^{-1} total volatile solid (TVS) in the combined system and 256.7 mL g^{-1} TVS in the three-stage process (Cheng et al. 2014). A study by Kumar et al. (2017b) probed the effectiveness of the bioconversion of *G. amansii* substrate for hydrogen production with dilute-acid pretreatment under anaerobic fermentation condition. Under mesophilic condition (temperature 35 °C, pH 5.5), the hybrid immobilized powdered sludge acted as a catalyst for hydrogen production in the continuous stirred-tank reactor. The authors reported a maximum hydrogen production rate of 1.3 moles $mole^{-1}$ of substrate and hydrogen yield of 2.7 L L^{-1} day^{-1} at a 24 h HRT.

2.1.4 Food Waste

The growing body of research suggests food waste as one of the most sustainable and easily degradable feedstock for future energy needs. With a rich content of organic matter, food waste is highly amenable to anaerobic fermentation for bio-H_2 production. In Asia, bio-H_2 production from food waste has been investigated extensively, especially in countries like South Korea, Taiwan, and China (Sen et al. 2016a).

A research was conducted for continuous hydrogen production from waste bread using anaerobic sludge, where the waste bread was initially hydrolyzed by crude enzymes of *Aspergillus awamori* and *Aspergillus oryzae* in solid-state fermentation (SSF). It was concluded that 49.78 g L^{-1} glucose and 284.12 mg L^{-1} free amino nitrogen could be generated with waste bread mass ratio of 15% (w/v). When this waste bread hydrolysate was exploited for hydrogen production using anaerobic sludge in a continuous stirred-tank reactor (CSTR), the peak hydrogen production rate of 7.4 L L^{-1} day^{-1} was observed at 6000 mg L^{-1} COD. Thus, for each g of waste bread, 0.332 g glucose could be induced that can lead to the production of 109.5 mL of hydrogen (Han et al. 2016b). In another study, food waste was initially subjected to glucoamylase and protease enzymes via SSF using *Aspergillus awamori* and *Aspergillus oryzae*; the resulting hydrolysate was then used for fermentative hydrogen production under dark condition using *Biohydrogenbacterium* R3 at 37 °C with pH maintained at 4–4.6 during the entire fermentative period. The best hydrogen yield of 52.4 mL H_2 g^{-1} food waste was achieved (Han et al. 2016c). Recently, in South Korea, food waste from a waste collection facility in Changwon was investigated for hydrogen production in an acidogenic system—a CSTR working at 37 °C, pH 6.5. The acidogenic system showed the highest hydrogen production (3.5 L day^{-1}) and yield (99.8 mL g^{-1} VS) at HRT of 8 h with OLR of 106 g VSL^{-1} day^{-1} (Paudel et al. 2017). In China, the highest apple-producing country in the world, bio-H_2 production from apple waste by photosynthetic bacteria (*Rhodospirillum rubrum*, *Rhodobacter capsulatus*, and *R. palustris*) was demonstrated (Lu et al. 2016). The peak value of 111.85 mL H_2 g^{-1} TS was obtained at the optimal conditions of pH 7.14, light intensity 3029.67 lx, temperature 30.46 °C, and material to liquid ratio of 0.21. The above case studies provide the basis for bioprospecting new and local food wastes for bio-H_2 production in Asian countries.

2.1.5 Agricultural Residues

The feasibility of various agricultural residues (soybean stalk, sorghum stalk, cotton stalk, corn stover, and corncob) as a suitable substrate for bio-H_2 production, after optimal pretreatment, was explored by Zhang et al. (2016). Based on the thermophysical properties of these agricultural residues characterized by thermogravimetric analysis (TGA), corncob (lowest residue content and E_a) was the most suitable for high hydrogen production. In a recent study, waste of *Platanus orientalis* leaves (POL) was investigated as a potential substrate for bio-H_2 production via simultaneous saccharification fermentation using photosynthetic bacteria (*R. rubrum*, *R. capsulatus*, and *R. palustris*) in a photo-fermentative batch reactor. RSM indicated that the initial pH, temperature, and inoculum load had a remarkable effect on hydrogen production. With an initial pH of 6.18, the temperature of 35.59 °C, and inoculum load of 26.29% (v/v), the peak hydrogen yield of 64.10 mL H_2 g^{-1} TS was obtained (Li et al. 2017).

Hydrogen production from sugarcane bagasse by the integration of dark and photo-fermentation using *Enterobacter aerogenes* MTCC 2822 and *Rhodopseudomonas* BHU 01, respectively, was explored by Rai et al. (2014). A novel strain *Cellulomonas fimi* was used to hydrolyze the acid-hydrolyzed cellulosic residue, and the resulting hydrolysate was subsequently subjected to simultaneous saccharification, filtration, and fermentation (SSFF) to produce hydrogen using *E. aerogenes*. Hydrogen production in dark fermentation and SSFF was found to be 1000 mL L^{-1} and 613 mL L^{-1}, respectively. Further use of the spent medium of dark fermentation and SSFF in photo-fermentation by *Rhodopseudomonas* BHU 01 resulted in the cumulative hydrogen production of 775 mL L^{-1} and 351 mL L^{-1}, respectively, in the same study. Recently, an efficient bio-H_2 production from non-detoxified sugarcane bagasse was proposed by Hu et al. (2018) who claim to have built a novel two-stage anaerobic fermentation process that can achieve a maximum hydrogen production of 6.2 L H_2 L^{-1} (277.4 mM). That study also indicated significant production of hydrogen in the second-stage fermentation (167.8 mM) when compared to the first-stage fermentation (108.6 mM).

Korean rice straw was examined as a substrate for thermophilic hydrogen fermentation by *Thermotoga neapolitana* (Nguyen et al. 2010). The study reported 29% of the substrate digestion with hydrogen yield of 2.3 mmoles g^{-1} straw in a typical fermentation of raw-rice straw. However, a combined pretreatment method (10% ammonia and 1% dilute-H_2SO_4) was more effective in increasing the digestibility of rice straw, leading to 85.4% of substrate utilization and hydrogen yield of 2.7 mmoles g^{-1} straw. Chen et al. (2012) explored bio-H_2 production using *C. pasteurianum*, *C. stercorarium*, and *Thermoanaerobacterium saccharolyticum* in a repeated-batch reactor with untreated rice straw as the substrate. A maximal hydrogen yield of 24.8 mL g^{-1} TS was obtained with rice straw at a concentration of 90 g TS L^{-1} and particle size <0.297 mm using heat-treated sludge (S1) at pH 6.5 and 55 °C under batch mode. In a recent study on the effect of acid hydrolysis and enzyme saccharification of rice husk on hydrogen fermentation, hydrogen yield was improved up to 150% by the modified hydrolysate as substrate. The highest hydrogen yield of 473.1 mL g^{-1} rice husk was achieved with an enzyme concentration of 0.75 mg protein mL^{-1} (Gonzales and Kim 2017). Sen et al. (2016b) evaluated the effects of rice straw concentration, particle size, hydrolysis time, acid concentration, $FeCl_3$, and enzyme additions in order to yield the maximum sugar monomers for hydrogen production. The study showed that 0.8–1.0 M hydrochloric acid could provide the maximum total sugar yield of 52.9% from 100 g L^{-1} rice straw with a particle size of 0.15 mm and hydrolysis time of 20 min. High volumetric hydrogen production of 771 mL L^{-1} was obtained with the pretreated rice straw.

By integrating dark fermentation with the photo-fermentation process, an enhanced hydrogen production from pretreated corncob was reported (Yang et al. 2010). The maximum hydrogen yield and rate from corncob by dark fermentation was 120.3 mL g^{-1} and 150 mL L^{-1} h^{-1}, respectively, in the first step, while in the second step, the yield was 713.6 mL g^{-1} COD from digesting the effluent of dark fermentation by photosynthetic bacteria. The highest hydrogen production rate of 7.37 mmoles L^{-1} h^{-1} with the highest cumulative hydrogen yield of 512.29 mmoles L^{-1} was achieved in a corncob fed baffled photo-fermentative bioreactor with a controlled temperature of 30 °C (Zhang et al. 2015). That system was chosen as the preferred bioreactor for continuous hydrogen production after evaluating different bioreactors with diverse mixing methods. Recently, a pilot-scale test of sequential dark and photo-fermentation was conducted in an 11 m^3 reactor (3 m^3 for dark and 8 m^3 for photo compartments) for the first time, which showed an average hydrogen production rate (HPR) of 59.7 m^3 day^{-1}. Large variation was noted for hydrogen production rate in different compartments of the tested units, underscoring the adverse effects of poor mixing, washout, and other inhomogeneity associated with large reactor operations.

2.2 *Wastewaters*

2.2.1 Palm Oil Mill Effluent (POME)

In Malaysia, POME-based hydrogen production has been extensively reported because of the abundant availability of POME. With a suspended-cell containing reactor, an optimal HPR of 0.348 L H_2 L^{-1} POME h^{-1} at 6 h HRT was obtained;

however, an immobilized-cell containing reactor exhibited a better HPR of 0.589 L H_2 L^{-1} POME h^{-1} at 2 h HRT. When the immobilized-cell containing reactor was scaled up to 5 L, the HPR increased to 0.500–0.588 L H_2 L^{-1} POME h^{-1}, which was further increased to 0.632 L H_2 L^{-1} POME h^{-1} by a thermal pretreatment (60 °C, 1 h) (Singh et al. 2013). Continuous hydrogen production from POME in a two-stage reactor at a thermophilic temperature of 55 °C exhibited 33.48 mL H_2 g^{-1} COD at the organic loading rate of 50 kg COD m^{3-1} day^{-1} (Krishnan et al. 2017). In another study, POME was used as a substrate in a two-stage thermophilic fermentation and mesophilic methanogenic process with effluent from the methanogenic stage recirculated back to the hydrogen reactor for biohythane production. The study achieved a peak hydrogen production of 4.1 L H_2 L^{-1} POME under batch conditions (O-Thong et al. 2016).

2.2.2 Cheese Whey Wastewater

Although cheese production in Asia is low compared to the USA and Europe, the application of the cheese whey wastewater in bio-H_2 production has been assessed in some Asian countries (e.g. India). As a substrate for hydrogen production in batch fermentation using *Clostridium* sp. IODB-O3, cheese whey wastewater produced 6.35 moles H_2 $mole^{-1}$ lactose. The cumulative hydrogen production observed was 3330 mL L^{-1} with an HPR of 139 mL L^{-1} h^{-1} (Patel et al. 2016). Cheese whey-based hydrogen production in batch dark fermentation using free and immobilized *Enterobacter aerogenes* MTCC 2822 followed by photo-fermentation of VFA in the spent medium by *Rhodopseudomonas* BHU 01 was investigated (Rai et al. 2012). The free and immobilized cells exhibited a cumulative hydrogen yield of 3.40 and 5.88 moles $mole^{-1}$ lactose, respectively.

2.2.3 Textile Wastewater

Some studies conducted in Taiwan on textile wastewater demonstrated that it is a feasible substrate for bio-H_2 production. With textile wastewater concentration of 13 g COD L^{-1}, a maximum HPR of 1.14 L L^{-1} day^{-1} was obtained in batch dark fermentation experiments (Lay et al. 2012). It was found that when textile wastewater was pretreated with activated carbon, it removed the bio-toxic inhibitors from the wastewater and showed increased hydrogen production (Li et al. 2012). An intermittent-flow-stirred-tank reactor fed with textile wastewater exhibited HPR of 10 L H_2 L^{-1} day^{-1} at a feeding frequency of 12 times per day with wastewater concentration of 33.1 g hexoses L^{-1} (Lay et al. 2014). Recently, continuous hydrogen production from coagulation-pretreated textile de-sizing wastewater was assessed in a stirred-tank reactor (operating conditions of pH 6.8, 35 °C, and HRTs of 8 and 4 h), and the results showed a maximum HPR of 3.8 L L^{-1} day^{-1} at an OLR of 30 g total sugar L^{-1} (Lin et al. 2017a, b).

2.2.4 Rice Mill Wastewater

In Asian countries, there are a large number of rice mills engaged in rice milling, a process of removing the husk from paddy to produce edible rice. These rice mills use water to soak the rice, which releases a large volume of wastewater when processing paddy rice, and the amount of the wastewater produced is about 1.0–1.2 L kg^{-1} paddy (Ramprakash and Muthukumar 2014). The high-COD rice mill wastewater is rich in starch, cellulose, and hemicelluloses, which make it suitable for anaerobic fermentation. A few studies conducted in India indicated the potential of bio-H_2 production from rice mill wastewater. A maximum HPR of 186 mL L^{-1} day^{-1} was noted at 57 °C, and the hydrogen yield increased with fermentation temperature in a study carried out with rice bran de-oiled wastewater (Sivaramakrishna et al. 2010). In another study, *Aspergillus niger*, *E. aerogenes*, and *C. freundii* were used for hydrogen production from acid- and enzyme-hydrolyzed rice mill wastewater (Ramprakash and Muthukumar 2014). The findings of this study indicated a maximum reducing sugar recovery of 15.8 g L^{-1} via enzymatic hydrolysis and hydrogen yield of 1.74 moles H_2 $mole^{-1}$ of reducing sugar. About 71.8% reduction in COD was also achieved at the end of 60 h of fermentation.

2.2.5 Herbal Wastewater

Herbal wastewater generated from the herbal pharmaceutical industries is an increasing concern in countries like India and China where herbal medicines are consumed in large amounts. It has high COD, BOD, and suspended solids; thus, it cannot be directly discharged into the surface water bodies. A study evaluated the hydrogen production potential of herbal wastewater under a controlled pH of 6.5 and 50 °C, which demonstrated a peak hydrogen production of 930 mL L^{-1} (Sivaramakrishna et al. 2014). That study indicated an effective hydrogen production from the herbal wastewater by thermophilic acidogenesis at the proper operational conditions.

2.2.6 Sago Starch-Processing Wastewater

Sago is an important food crop in tropical Asian countries such as Malaysia and India. The sago starch-processing wastewater generated by the starch industry is rich in carbohydrates (starch and hemicelluloses). Studies have shown that sago starch-processing effluent is an excellent substrate for fermentative hydrogen production (Sen et al. 2012). Batch experiment results using sago-processing wastewater under optimal conditions showed hydrogen yield of 126.5 mL g^{-1} COD and 456 mL g^{-1} starch. The net energy was estimated to be +2.97 kJ g^{-1} COD with a net energy value of $+2.85 \times 10^{13}$ kJ from worldwide sago-processing wastewater production. The study indicated that sago starch-processing wastewater can serve as a promising feedstock for hydrogen production with low-energy input. In another study of sago starch-processing wastewater, the peak HPR was found to be 57.54 mL H_2 h^{-1} L^{-1} using microbes sourced from sago-sludge consortia (Yunus et al. 2014).

3 High-Rate Hydrogen Fermentation Systems Developed in Asia

Among the Asian countries, Taiwan represents the leading country to develop the high-rate hydrogen production systems. These systems exhibited remarkable HPR compared to the conventional systems (Fig. 1), which laid the foundation for the future development of high-rate bio-H_2 production technology.

3.1 *Carrier-Induced Granular Sludge Bed Reactor (CIGSBR)*

Low biomass retention at short HRTs and low mass-transfer efficiency at high HRTs are the major problems associated with the long-term continuous operation of the hydrogen-producing reactor. In view of this, a CIGSBR was developed to improve the biomass retention and mass-transfer efficiency (Lee et al. 2004). The CIGSBR supplemented with calcium ions (5.4–27.2 mg Ca^{2+} L^{-1}) increased the mechanical strength of granular sludge and showed a threefold increase in the biomass concentration and a fivefold increase in HPR (up to 122.4 L L^{-1} day^{-1}). In order to enhance the mass-transfer efficiency, two reflux strategies were adopted, the liquid and gas reflux. Both the reflux strategies yielded similar HPR but differentially affected the biomass concentration; the liquid reflux produced 22 g VSS L^{-1}, whereas the gas

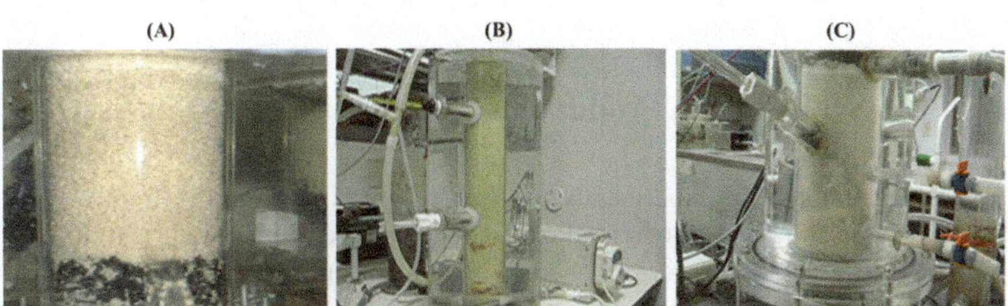

Fig. 1 Photos of the high-rate, bench-scale bio-H_2 production systems developed in Asia. (**a**) CIGSBR, (**b**) CSABR, and (**c**) AGSBR

reflux produced only 2–7 g VSS L^{-1}. Furthermore, these strategies were effective toward a stable and efficient hydrogen production for ca. 100 days (Lee et al. 2004).

3.2 *Continuously Stirred Anaerobic Bioreactor (CSABR)*

Studies on CSABR seeded with silicone-immobilized sludge demonstrated the highest HPR at 0.5–6 h of HRT using sucrose (Wu et al. 2006). At the feeding concentration range of 30–40 g COD L^{-1} and 0.5 h of HRT, the CSABR exhibited a maximum HPR of 362.2 L L^{-1} day^{-1} and an optimal hydrogen yield of 3.5 moles H_2 $mole^{-1}$ sucrose. Self-granulated sludge formation in a short HRT operation was key to the high-rate hydrogen production, which was evident from the high biomass concentration of 35.4 g VSS L^{-1} when it was operated at a very low HRT of 0.5 h. It was suggested that sludge granulation might have triggered a change of bacterial community, subsequently resulting in a two-fold increase in the specific HPR. Molecular analysis confirmed lower bacterial diversity, mostly dominated by *C. pasteurianum*, at shorter HRTs.

3.2.1 Agitated Granular Sludge Bed Reactor (AGSBR)

The AGSBR system with a working volume of 1 L included a paddle impeller agitator for a fixed agitation at 15 rpm. Activated carbon powder as the carrier was added (1 g L^{-1}) at the start of the system that was inoculated with an intertidal sludge from Tai-Xi, Taiwan. Starch wastewater was fed into the AGSBR and was operated at pH 6.0 and 5.5 under various HRTs. Interestingly, when the reactor was operated at 0.5 h of HRT, it exhibited the peak HPR (48 L L^{-1} day^{-1}) and total sludge density of 45–48 g VSS L^{-1}. *Bifidobacterium* sp. was associated with the starch degradation followed by *Clostridium* species for hydrogen production (Cheng et al. 2008). Some other examples of AGSBR include the granular activated carbon anaerobic fluidized bed reactor developed by Zhang et al. (2007, 2008) in Singapore.

4 Pilot-Scale Bio-H_2 Production Systems Developed in Asia

Conducting pilot-scale experiments to evaluate bio-H_2 production from organic waste is often too cumbersome and expensive compared to laboratory scale. It is important to evaluate the performance of the system on a pilot scale and troubleshoot the problems associated with feedstock storage, maintenance of anaerobic condition, and automation. There are several reports of pilot-scale bio-H_2 production from Asian countries, which are discussed below.

4.1 Pilot-Scale Bio-H_2 Production in Taiwan

Based on the pioneering works at laboratory scale that were conducted in Feng Chia University (FCU), Taiwan, a pilot-scale high-rate dark fermentative hydrogen production plant has been established in the FCU campus (Lin et al. 2011). The working volume of this pilot-plant system is 400 L and is composed of two feedstock storage tanks (0.75 m^3 each), a nutrient storage tank (0.75 m^3), a mixing tank (0.6 m^3), an AGSBR (working volume 0.4 m^3), a gas-liquid-solid separator (0.4 m^3), and a control panel (Sen et al. 2013). Pilot-scale studies demonstrated increased HPR with increasing OLR, but when the OLR was too high (240 kg COD m^{-3} day^{-1}), the biomass concentration decreased, which indicated overloading of the fermenter and microbial growth inhibition. The energy input (E_f) values were higher than those of ethanol production from corn, biodiesel, and sugarcane and similar to ethanol production from cellulose. Engineering strategies were further applied to increase the agitation rate and promote mass-transfer efficiency, which resulted in a HPR of 13.4 m^3 m^{-3} day^{-1} with 25–30 rpm agitation rate. Overall, the pilot-scale data indicated that engineering strategies can improve the hydrogen production efficiency and performance almost similar to that obtained from the lab-scale system.

4.2 Pilot-Scale Bio-H_2 Production in Korea

A pilot-scale study in an anaerobic-sequencing-batch reactor fed with food waste was conducted by Kim et al. (2010). The working volume of the 230-L fermenter was 150 L with a liquid depth of 770 mm and the inner diameter of 500 mm. At carbon to nitrogen (C/N) ratio of 20, the hydrogen yield was around 0.5 moles $mole^{-1}$ $hexose_{added}$, which dropped at higher C/N. At the optimal operating conditions, the pilot-scale system could account for 2.3% of energy conversion efficiency. The low performance was attributed to the co-production of lactate, propionate, and valerate, which reduced the electron flux toward hydrogen production. An alkaline shock (pH 12.5 for 1 day) further increased the hydrogen yield to 0.69 moles $mole^{-1}$ hexose added.

Another pilot-scale plant was established and operated at the solid municipal waste treatment facility in Cheonan, Korea (Lee et al. 2010). This pilot-scale system consisted of a stainless-steel hydrogen fermentation tank (working volume of 500 L) and a methane fermentation tank. The hydrogen fermentation tank was linked to a fuel cell system to recover energy from the hydrogen liberated. A maximum HPR of 3.88 L m^{-3} day^{-1} at the HRT of 21 h with 90% carbohydrate consumption was achieved by the pilot-scale system. It also generated methane over 80%. Further evaluation of the process economics underlined the greater potential of a two-stage hydrogen/methane fermentation in recovering energy than a single-stage methane fermentation.

4.3 Pilot Scale in China

A pilot-scale study of bio-H_2 production in a continuous-flow-anaerobic-fermentation reactor (available volume of 1480 L) was conducted by Ren et al. (2006). With molasses as the substrate, the pilot-scale system was operated for more than 200 days at the OLR of 3.11–85.57 kg COD m^{-3} day^{-1}. The hydrogen yield increased with OLR in the range of 3.11–68.21 kg COD m^{-3} day^{-1} but decreased at high OLR (>68.21 COD m^{-3} day^{-1}). A maximum HPR of 5.57 m^3 m^{-3} day^{-1} and hydrogen yield of 26.13 moles kg^{-1} $COD_{removed}$ was achieved at the OLR range of 35–55 kg COD m^{-3} day^{-1}.

4.4 Pilot-Scale Bio-H_2 Production in India

A pilot-scale study was conducted with sugarcane distillery effluent using co-cultures of *Citrobacter freundii* 01, *Enterobacter aerogenes* E10, and *Rhodopseudomonas palustris* P2 in a 100,000 L plant (Vatsala et al. 2008). A sequence of bioreactors was constructed with the working volumes set from 100 to 100,000 L. This is the largest dimension system reported in the literature so far for hydrogen production. At the 100,000-L scale, an average hydrogen yield of 2.76 moles $mole^{-1}$ glucose and HPR of 0.53 kg 100 m^{-3} h^{-1} was achieved. Although mixed culture is often recommended for pilot-scale systems, this study indicated that pure cultures of different genera are also suitable. Further, the application of distillery effluent in the bio-H_2 production system and simultaneous treatment of high-strength industrial effluent are highlighted.

Recently, the scaling-up of the bio-H_2 production process fed on cane molasses supplemented with groundnut de-oiled cake was studied in a 10,000-L pilot-scale reactor (Das 2017). A cumulative hydrogen production of 76,200 L was measured in 25 h at 34–37 °C operation temperature.

4.5 Pilot-Scale Bio-H_2 Production in Japan

A pilot-scale system for two-stage anaerobic digestion process was designed and constructed in Tsukuba, Japan, which comprised a slurry tank for feedstock storage, a hydrogenogenic reactor (200 L working volume), a buffer tank, a methanogenic reactor, and two gas holders. At the steady-state operation (HRT of 1.2 days for hydrogenogenesis), the system produced 5.4 m^3 H_2 m^{-3} day^{-1}. Microbial analysis of the hydrogenogenic reactor indicated that hydrogen-producing *Thermoanaerobacterium* spp. were active during the hydrogen fermentation of garbage and paper wastes. Furthermore, it was speculated that high-temperature operation supports the selective growth of the thermopiles in the inoculum and yields hydrogen fermentation efficiencies comparable to that of laboratory data. Of note, that was the first report on fermentative hydrogen production from organic waste at the pilot scale.

5 Brief Outline of the Techno-Economic Analyses of Bio-H_2 Production from Organic Waste in Asia

With intensive investment in technology and infrastructure, a transition from an oil-based economy to a hydrogen-based economy is anticipated by 2050 (Lee and Chiu 2012). Although "hydrogen roadmaps" are proposed for the USA, the EU, and other countries (Lee and Chiu 2012), they are mostly based on hydrogen from fossil fuels and renewable resources and not solely on hydrogen production from organic waste. It is expected that China will have the largest bio-H_2 market and India will see the most efficient investment in this sector. Currently, bio-H_2 production from organic waste is more expensive than other fuels (Table 2). The relatively low hydrogen yield and production rate are the two major challenges that hinder the commercialization of organic waste-based bio-H_2 production systems. Despite the in-depth understanding of the bio-H_2 fermentation technology and its operation, we are still behind the fossil fuels in terms of cost, efficiency, and reliability (Table 2). However, the energy saving company (ESCO) business model proposed by Hsu and Lin (2016) for the production and sale of hydrogen can increase the incentive to use dark fermentative hydrogen production. In such business model, the ESCO may assume the role of H_2 producer and work together with soft drink manufacturers. Additionally, ESCO can provide sewage treatment services and sign long-term energy provision and service contracts with client users, such as the hydrogen stations to guarantee successful sales. In this model, ESCO can obtain higher returns on investment by reducing the costs of H_2 fermentation, purification, and storage through additional equipment purchases. At present, more such business model are needed in order to promote new hydrogen production technologies and accelerate commercialization.

As discussed in the above sections, only a handful of pilot-scale studies are demonstrated so far. Nevertheless, these pilot-scale studies indicate that much work is further needed to resolve many of the technical issues in the laboratory prior to the scale-up and commercialization. Similar to other fermentation-based technologies, bio-H_2 production process also suffers from high maintenance costs and seasonal availability of potential organic wastes. To this end, techno-economic assessment is of prime importance to evaluate the viability and possibility of process scale-up, besides it also guides further improvement of the process.

Table 2 Cost comparison among different hydrogen production processes

Material of source	Technologies used	Production cost (USD/ kgH_2)	References
Natural gas	Steam methane reforming	0.75 (without CO_2 sequestration)	Parthasarathy and Narayanan (2014)
Nuclear	Electrolysis	2.4	Dincer (2012)
Nuclear	High temperature electrolysis	3.5	Acar and Dincer (2014)
Nuclear	Copper-chlorine	1.7	Wang et al. (2010a, b)
Solar	Electrolysis	7.7	Khan et al. (2018)
Solar	Photovoltaic electrolysis	9.1	Bhandari et al. (2014)
Solar	Photo-electrochemical	3.5	Acar et al. (2015)
Wind	Electrolysis	7.3	Wang et al. (2010a, b)
Biomass	Gasification	4.60–7.86	Koumi Ngoh and Njomo (2012)
Geothermal	Steam electrolysis	1–2.6	Yilmaz et al. (2012)
H_2O organic acids	Photobiological hydrogen	10	Saratale, et al. (2013)
Molasses	Fermentative hydrogen	10	Saratale et al. (2013)

6 Summary and Future Perspectives

Research on bio-H_2 production from organic waste has been extensively conducted in Asian countries, e.g., Taiwan, Korea, India, Japan, China, Malaysia, and Thailand. A large number of organic wastes have been evaluated, and many of them exhibited high potential for hydrogen production via dark and photo-fermentation processes. Two-stage hydrogen/methane systems are more efficient and yield higher net energy than single-stage systems. Based on the research findings, the most feasible commercial process for bio-H_2 production seems to be the on-site wastewater to bio-H_2 production (Fig. 2). Such systems can be easily operated in factories and communities, enabling stable and sufficient supply of high-organic content wastes as the feedstock. Integrating the bio-H_2 process to the conventional wastewater treatment process has many advantages such as improving the environmental compatibility of the wastewater treatment process and lowering the wastewater treatment cost by generating clean and valuable bioenergy products. The bio-H_2 produced during wastewater treatment can be fed into the boiler to reduce the fossil fuel utilization or can be converted to electricity by the fuel cell to supply the factory demand on site. Moreover, the CO_2 produced could be captured and reutilized to gain additional benefits for the factory, resulting in the reduction of capital investment. The combined process has the advantages of wastewater treatment, sludge reduction, and hydrogen production. The future of the organic waste to bio-H_2 process development lies in the selection of the appropriate biological system and technical improvements that can maximize the energy recovery and minimize the operational cost. To this end, other processes such as microbial fuel cells, microbial electrolysis cell, or microalgae

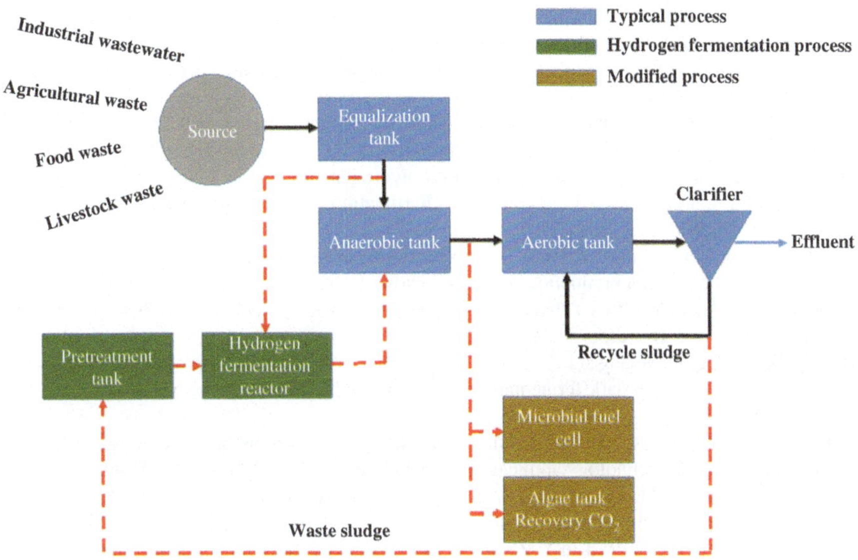

Fig. 2 A typical integrated bio-H_2 production system based on research conducted in Asia

processes should be integrated to recover the organic residue in the effluent from anaerobic reactors for increasing the energy production and reducing the pollutant.

References

Acar C, Dincer I (2014) Comparative assessment of hydrogen production methods from renewable and non-renewable sources. Int J Hydrogen Energy 39:1–12

Acar C, Ghosh S, Dincer I, Zamfirescu C (2015) Evaluation of a new continuous type hybrid photo-electrochemical system. Int J Hydrogen Energy 40:11112–11124

Arumugam A, Sandhya M, Ponnusami V (2014) Biohydrogen and polyhydroxyalkanoate co-production by *Enterobacter aerogenes* and *Rhodobacter sphaeroides* from *Calophyllum inophyllum* oil cake. Bioresour Technol 164:170–176

Asadi N, Zilouei H (2017) Optimization of organosolv pretreatment of rice straw for enhanced biohydrogen production using Enterobacter aerogenes. Bioresour Technol 227:335–344

Bhandari R, Trudewind CA, Zapp P (2014) Life cycle assessment of hydrogen production via electrolysis – a review. J Clean Prod 85:151–163

Chen C, Chuang Y, Lin C, Lay C, Sen B (2012) Thermophilic dark fermentation of untreated rice straw using mixed cultures for hydrogen production. Int J Hydrogen Energy 37:15540–15546

Cheng C-H, Hung C-H, Lee K-S, Liau P-Y, Liang C-M, Yang L-H, Lin P-J, Lin C-Y (2008) Microbial community structure of a starch-feeding fermentative hydrogen production reactor operated under different incubation conditions. Int J Hydrogen Energy 33(19):5242–5249

Cheng J, Liu Y, Lin R, Xia A, Zhou J, Cen K (2014) Cogeneration of hydrogen and methane from the pretreated biomass of algae bloom in Taihu Lake. Int J Hydrogen Energy 39:18793–18802

Cheng J, Lin R, Song W, Ao X a, Zhou J, Cen K (2015) Enhancement of fermentative hydrogen production from hydrolyzed water hyacinth with activated carbon detoxification and bacteria domestication. Int J Hydrogen Energy 40:2554–2551

Chuang YS, Lay CH, Sen B, Chen CC, Gopalakrishnan K, Wu JH, Lin C-S, Lin CY (2011) Biohydrogen and biomethane from water hyacinth (Eichhornia crassipes) fermentation: effects of substrate concentration and incubation temperature. Int J Hydrogen Energy 36(21):14195–14203

Das D (2017) A road map on biohydrogen production from organic wastes. INAE Lett 2(4):153–160

Dincer I (2012) Green methods for hydrogen production. Int J Hydrogen Energy 37:1954–1971

Gonzales RR, Kim S (2017) Dark fermentative hydrogen production following the sequential dilute acid pretreatment and enzymatic saccharification of rice husk. Int J Hydrogen Energy 42:27577–27583

Han W, Fang J, Liu Z, Tang J (2016a) Techno-economic evaluation of a combined bioprocess for fermentative hydrogen production from food waste. Bioresour Technol 202:107–112

Han W, Huang J, Zhao H, Li Y (2016b) Continuous biohydrogen production from waste bread by anaerobic sludge. Bioresour Technol 212:1–5

Han W, Ye M, Zhu AJ, Huang JG, Zhao HT, Li YF (2016c) A combined bioprocess based on solid-state fermentation for dark fermentative hydrogen production from food waste. J Clean Prod 112:3744–3749

Hsu CW, Lin CY (2016) Commercialization model of hydrogen production technology in Taiwan: dark fermentation technology applications. Int J Hydrogen Energy 41(7):4489–4497

Hu BB, Li MY, Wang YT, Zhu MJ (2018) High-yield biohydrogen production from non-detoxified sugarcane bagasse: fermentation strategy and mechanism. Chem Eng J 335:979–987

Khan MA, Ngo HH, Guo W, Liu Y, Zhang X, Guo J et al (2018) Biohydrogen production from anaerobic digestion and its potential as renewable energy. Renew Energy 129:754–768

Kim DH, Kim SH, Kim KY, Shin HS (2010) Experience of a pilot-scale hydrogen-producing anaerobic sequencing batch reactor (ASBR) treating food waste. Int J Hydrogen Energy 35(4):1590–1594

Koumi Ngoh S, Njomo D (2012) An overview of hydrogen gas production from solar energy. Renew Sustain Energy Rev 16:6782–6792

Krishnan S, Singh L, Sakinah M, Thakur S, Wahid ZA, Ghrayeb OA (2017) Role of organic loading rate in bioenergy generation from palm oil mill effluent in a two-stage up-flow anaerobic sludge blanket continuous-stirred tank reactor. J Clean Prod 142:3044–3049

Kumar G, Sen B, Lin C-Y (2013) Pretreatment and hydrolysis methods for recovery of fermentable sugars from de-oiled Jatropha waste. Bioresour Technol 145:275–279

Kumar G, Sen B, Sivagurunathan P, Lin C-Y (2015) Comparative evaluation of hydrogen fermentation of de-oiled Jatropha waste hydrolyzates. Int J Hydrogen Energy 40(34):10766–10774

Kumar G, Sen B, Sivagurunathan P, Lin C-Y (2016) High rate hydrogen fermentation of cello-lignin fraction in de-oiled jatropha waste using hybrid immobilized cell system. Fuel 182:131–140

Kumar G, Sivagurunathan P, Sen B, Mudhoo A, Davila-Vazquez G, Wang G, Kim S-H (2017a) Research and development perspectives of lignocellulose-based biohydrogen production. Int Biodeter Biodegr 119(suppl C):225–238

Kumar G, Sivagurunathan P, Anburajan P, Pugazhendhi A, Saratale GD, Choi C, Kim S (2017b) Continuous biogenic hydrogen production from dilute acid pretreated algal hydrolysate using hybrid immobilized mixed consortia. Int J Hydrogen Energy 43:11452–11459

Kuo W-C, Chao Y-C, Wang Y-C, Cheng S-S (2012) Bioaugmentation strategies to improve cellulolytic and hydrogen producing characteristics in CSTR intermittent fed with vegetable kitchen waste and napiergrass. Energy Procedia 29:82–91

Lay CH, Kuo SY, Sen B, Chen CC, Chang JS, Lin CY (2012) Fermentative biohydrogen production from starch-containing textile wastewater. Int J Hydrogen Energy 37(2):2050–2057

Lay C-H, Sen B, Kuo S-Y, Chen C-C, Lin C-Y (2014) Biohydrogen production from textile wastewater by mixed microflora in an intermittent-flow, stirred tank reactor: effect of feeding frequency. J Chin Chem Soc 61(7):791–796

Lee D-H, Chiu L-H (2012) Development of a biohydrogen economy in the United States, China, Japan, and India: with discussion of a chicken-and-egg debate. Int J Hydrogen Energy 37(20):15736–15745

Lee Y-W, Chung J (2010) Bioproduction of hydrogen from food waste by pilot-scale combined hydrogen/methane fermentation. Int J Hydrogen Energy 35:11746–11755

Lee KS, Wu JF, Lo YS, Lo YC, Lin PJ, Chang JS (2004) Anaerobic hydrogen production with an efficient carrier-induced granular sludge bed bioreactor. Biotechnol Bioeng 87:648–657

Lee ZK, Li SL, Kuo PC, Chen IC, Tien YM, Huang YJ et al (2010) Thermophilic bio-energy process study on hydrogen fermentation with vegetable kitchen waste. Int J Hydrogen Energy 35:13458–13466

Li YC, Liu YF, Chu CY, Chang PL, Hsu CW, Lin PJ, Wu SY (2012a) Techno-economic evaluation of biohydrogen production from wastewater and agricultural waste. Int J Hydrogen Energy 37(20):15704–15710

Li YC, Chu CY, Wu SY, Tsai CY, Wang CC, Hung CH, Lin CY (2012b) Feasible pretreatment of textile wastewater for dark fermentative hydrogen production. Int J Hydrogen Energy 37(20):15511–15517

Li Y, Zhang Z, Jing Y, Ge X, Wang Y, Lu C, Zhou X, Zhang Q (2017) Statistical optimization of simultaneous saccharification fermentative hydrogen production from *Platanus orientalis* leaves by photosynthetic bacteria HAU-M1. Int J Hydrogen Energy 42:5804–5811

Li Y, Zhang Q, Deng L, Liu Z, Jiang H, Wang F (2018) Biohydrogen production from fermentation of cotton stalk hydrolysate by Klebsiella sp. WL1316 newly isolated from wild carp (Cyprinus carpio L.) of the Tarim River basin. Appl Microbiol Biotechnol 102:4231–4242

Lin C-Y, Wu S-Y, Lin P-J, Chang J-S, Hung C-H, Lee K-S et al (2011) A pilot-scale high-rate biohydrogen production system with mixed microflora. Int J Hydrogen Energy 36(14):8758–8764

Lin CY, Lay CH, Sen B, Chu CY, Kumar G, Chen CC, Chang JS (2012) Fermentative hydrogen production from wastewaters: a review and prognosis. Int J Hydrogen Energy 37(20):15632–15642

Lin R, Cheng J, Song W, Ding L, Xie B, Zhou J, Cen K (2015) Characterisation of water hyacinth with microwave-heated alkali pretreatment for enhanced enzymatic digestibility and hydrogen/methane fermentation. Bioresour Technol 182:1–7

Lin CY, Lay CH, Chen CC, Sen B, Sung IY (2016) Biohydrogen production from mushroom cultivation waste by anaerobic solid-state fermentation. J Chin Chem Soc 63:199–204

Lin C, Chiang C, Nguyen TML, Lay C (2017a) Continuous biohydrogen production from coagulation-pretreated textile desizing wastewater. Int J Hydrogen Energy 42:29159–29165

Lin C-Y, Chiang C-C, Thi Nguyen M-L, Lay C-H (2017b) Enhancement of fermentative biohydrogen production from textile desizing wastewater via coagulation-pretreatment. Int J Hydrogen Energy 42(17):12153–12158

Lu C, Zhang Z, Ge X, Wang Y, Zhou X, You X, Liu H, Zhang Q (2016) Bio-hydrogen production from apple waste by photosynthetic bacteria HAU-M1. Int J Hydrogen Energy 41:13399–13407

Lucas SDM, Peixoto G, Mockaitis G, Zaiat M, Gomes SD (2015) Energy recovery from agro-industrial wastewaters through biohydrogen production: kinetic evaluation and technological feasibility. Renew Energy 75:496–504

Mohan SV, Pandey A (2013) Chapter 1 - Biohydrogen production: an introduction. In: Larroche C, Pandey A, Chang J-S, Hallenbeck PC (eds) Biohydrogen. Elsevier, Amsterdam, pp 1–24

Nguyen TD, Kim K, Kim MS, Sim SJ (2010) Thermophilic hydrogen fermentation from Korean rice straw by Thermotoga neapolitana. Int J Hydrogen Energy 35:13392–13398

O-Thong S, Suksong W, Promnuan K, Thipmunee M, Mamimin C, Prasertsan P (2016) Two-stage thermophilic fermentation and mesophilic methanogenic process for biohythane production from palm oil mill effluent with methanogenic effluent recirculation for pH control. Int J Hydrogen Energy 41:21702–21712

Park J, Yoon J, Park H, Kim YJ, Lim DJ, Kim S (2011) Feasibility of biohydrogen production from *Gelidium amansii*. Int J Hydrogen Energy 36:13997–14003

Parthasarathy P, Narayanan KS (2014) Hydrogen production from steam gasification of biomass: influence of process parameters on hydrogen yield – a review. Renew Energy 66:570–579

Patel AK, Vaisnav N, Mathur A, Gupta R, Tuli DK (2016) Whey waste as potential feedstock for biohydrogen production. Renew Energy 98:221–225

Paudel S, Kang Y, Yoo K, Seo GT (2017) Effect of volumetric organic loading rate (OLR) on H_2 and CH_4 production by two-stage anaerobic co-digestion of food waste and brown water. Waste Manag 61:484–493

Rai PK, Singh SP, Asthana RK (2012) Biohydrogen production from cheese whey wastewater in a two-step anaerobic process. Appl Biochem Biotechnol 167(6):1540–1549

Rai PK, Singh SP, Asthana RK, Singh S (2014) Biohydrogen production from sugarcane bagasse by integrating dark- and photo-fermentation. Bioresour Technol 152:140–146

Ramprakash B, Muthukumar K (2014) Comparative study on the production of biohydrogen from rice mill wastewater. Int J Hydrogen Energy 39:14613–14621

Reddy MV, Chandrasekhar K, Mohan SV (2011) Influence of carbohydrates and proteins concentration on fermentative hydrogen production using canteen based waste under acidophilic microenvironment. J Biotechnol 155:387–395

Ren NQ, Li JZ, Li BK, Wang Y, Liu SR (2006) Biohydrogen production from molasses by anaerobic fermentation with a pilot-scale bioreactor system. Int J Hydrogen Energy 31(15):2147–2157

Saratale GD, Saratale RG, Chang J-S (2013) Chapter 9 - Biohydrogen from renewable resources. In: Biohydrogen. Elsevier, Amsterdam, pp 185–221

Sen B, Suttar RR (2012) Mesophilic fermentative hydrogen production from sago starch-processing wastewater using enriched mixed cultures. Int J Hydrogen Energy 37:15588–15597

Sen B, Chu C-Y, Lin C-Y (2013) Scale-up and commercial applications of biohydrogen production processes. In: Pandey A, Chang J-S, Hallenbeck PC, Larroche C (eds) Biohydrogen, 1st edn. Elsevier, Amsterdam, p 598

Sen B, Aravind J, Kanmani P, Lay C-H (2016a) State of the art and future concept of food waste fermentation to bioenergy. Renew Sustain Energy Rev 53:547–557

Sen B, Chou YP, Wu SY, Liu CM (2016b) Pretreatment conditions of rice straw for simultaneous hydrogen and ethanol fermentation by mixed culture. Int J Hydrogen Energy 41(7):4421–4428

Show KY, Lee DJ, Tay JH, Lin CY, Chang JS (2012) Biohydrogen production: current perspectives and the way forward. Int J Hydrogen Energy 37(20):15616–15631

Singh L, Siddiqui MF, Ahmad A, Rahim MHA, Sakinah M, Wahid ZA (2013) Biohydrogen production from palm oil mill effluent using immobilized mixed culture. J Ind Eng Chem 19(2):659–664

Sivaramakrishna D, Sreekanth D, Himabindu V, Lakshmi Narasu M (2010) Thermo-acidophillic biohydrogen production from rice bran de-oiled wastewater by selectively enriched mixed culture. Int J Energy Environ 1(4):657–666

Sivaramakrishna D, Sreekanth D, Sivaramakrishnan M, Kumar BS, Himabindu V, Narasu VL (2014) Effect of system optimizing conditions on biohydrogen production from herbal wastewater by slaughterhouse sludge. Int J Hydrogen Energy 39:7526–7533

Su H, Cheng J, Zhou J, Song W, Cen K (2010) Hydrogen production from water hyacinth through dark- and photo- fermentation. Int J Hydrogen Energy 35:8929–8937

Taifor AF, Zakaria MR, Yusoff MZM, Toshinari M, Hassan MA, Shirai Y (2017) Elucidating substrate utilization in biohydrogen production from palm oil mill effluent by Escherichia coli. Int J Hydrogen Energy 42:5812–5819

Tapia-Venegas E, Ramirez-Morales JE, Silva-Illanes F, Toledo-Alarcón J, Paillet F, Escudie R et al (2015) Biohydrogen production by dark fermentation: scaling-up and technologies integration for a sustainable system. Rev Environ Sci Biotechnol 14(4):761–785

Varanasi JL, Kumari S, Das D (2018) Improvement of energy recovery from water hyacinth by using integrated system. Int J Hydrogen Energy 43:1303–1318

Vatsala TM, Raj SM, Manimaran A (2008) A pilot-scale study of biohydrogen production from distillery effluent using defined bacterial co-culture. Int J Hydrogen Energy 33(20):5404–5415

Wang Y-H, Li S-L, Chen IC, Tseng IC, Cheng S-S (2010a) A study of the process control and hydrolytic characteristics in a thermophilic hydrogen fermentor fed with starch-rich kitchen waste by using molecular-biological methods and amylase assay. Int J Hydrogen Energy 35:13004–13012

Wang ZL, Naterer GF, Gabriel KS, Gravelsins R, Daggupati VN (2010b) Comparison of sulfur–iodine and copper–chlorine thermochemical hydrogen production cycles. Int J Hydrogen Energy 35:4820–4830

Wu SY, Hung CH, Lin CN, Chen HW, Lee AS, Chang JS (2006) Fermentative hydrogen production and bacterial community structure in high-rate anaerobic bioreactors containing silicone-immobilized and self-flocculated sludge. Biotechnol Bioeng 93:934–946

Yang H, Guo L, Liu F (2010) Enhanced bio-hydrogen production from corncob by a two-step process. Bioresour Technol 101(6):2049–2052

Yilmaz C, Kanoglu M, Bolatturk A, Gadalla M (2012) Economics of hydrogen production and liquefaction by geothermal energy. Int J Hydrogen Energy 37:2058–2069

Yunus N, Jahim J, Anuar N, Rozaimah S, Abdullah S, Kofli NT (2014) Batch fermentative hydrogen production utilising sago (*Metroxylon sp.*) starch processing effluent by enriched sago sludge consortia. Int J Hydrogen Energy 39:19937–19946

Zhang Z, Tay J, Show K, Yan R, Tee Liang D, Lee D-J, Jiang W-J (2007) Biohydrogen production in a granular activated carbon anaerobic fluidized bed reactor. Int J Hydrogen Energy 32:185–191

Zhang Z-P, Show K-Y, Tay J-H, Tee Liang D, Lee D-J (2008) Enhanced continuous biohydrogen production by immobilized anaerobic microflora enhanced continuous biohydrogen production by immobilized anaerobic microflora. Energy Fuel 22(1):87–92

Zhang Z, Wang Y, Hu J, Wu Q, Zhang Q (2015) Influence of mixing method and hydraulic retention time on hydrogen production through photo-fermentation with mixed strains. Int J Hydrogen Energy 40:6521–6529

Zhang Z, He C, Sun T, Zhang Z, Song K, Wu Q, Zhang Q (2016) Thermo-physical properties of pretreated agricultural residues for bio-hydrogen production using thermo-gravimetric analysis. Int J Hydrogen Energy 41:5234–5242

Anaerobic Thermophilic Mixed Culture Fermentation Processes

Fang Zhang and Raymond Jianxiong Zeng

1 Introduction

With the ambitious undertaking to meet the needs of worldwide sustainability, governments and researchers pay much more attention to producing green biofuels, such as hydrogen and ethanol, to replace traditional fossil fuels (Dai et al. 2017; Kleerebezem et al. 2015; Pawar and Niel 2013). Meanwhile, the sustainability goal also pushes human society to use and reuse waste resources, such as lignocellulosic biomass, microalgae, activated sludge, food waste, and cattle manure, by either chemical or biological methods (Daelman et al. 2016; Khiewwijit et al. 2015; Lü et al. 2017; Tuck et al. 2012; Zamanzadeh et al. 2016; Zhang et al. 2015b). Among these methods, mixed culture fermentation (MCF) is well established as a traditional environmental biotechnology with various advantages, including the absence of sterilization requirements, an adaptive capacity to variations in feedstock or conditions, stable and continuous operation, and ability to convert the biodegradable organic wastes to biogas, chemicals, and biofuels (Bastidas-Oyanedel et al. 2015; Batstone and Virdis 2014; Kleerebezem et al. 2015; Nielsen et al. 2007).

As shown in Table 1, compared to mesophilic MCF, thermophilic MCF demonstrated several advantages, such as higher substrate degradation rate, higher hydrogen production due to favorable thermodynamics conditions, pathogen-free effluents and efficient heat utilization for treatment of hot wastewater, and so on (Gannoun et al. 2007; Labatut et al. 2014; Pawar and Niel 2013; Zhang et al. 2012). The organic waste biodegradation processes in thermophilic MCF are the cascade bioreactions, which involve three main steps, namely, (1) hydrolysis/acidogenesis, (2) acetogenesis/homoacetogenesis, and (3) methanogenesis. Additionally, different

F. Zhang (✉) · R. J. Zeng
Fujian Provincial Key Laboratory of Soil Environmental Health and Regulation, College of Resources and Environment, Fujian Agriculture and Forestry University, Fuzhou, Fujian, China
e-mail: zhfang@ysu.edu.cn; rzeng@ustc.edu.cn

© Springer Nature Switzerland AG 2019
J.-R. Bastidas-Oyanedel, J. E. Schmidt (eds.), *Biorefinery*,
https://doi.org/10.1007/978-3-030-10961-5_17

Table 1 The main advantages of thermophilic MCF over mesophilic MCF

	Fermentation conditions	References
Hydrolysis coefficient	The hydrolysis coefficient of waste-activated sludge at 60 °C (0.5 ± 0.1 day^{-1}), 65 °C (0.7 ± 0.2 day^{-1}), and 70 °C (0.8 ± 0.2 day^{-1}) was higher than mesophilic conditions (0.2 ± 0.1 day^{-1})	Ge et al. (2011)
	The maximal yields of short-chain fatty acids from sewage sludge were in the following order: thermophilic pH 8 > mesophilic pH 9 > ambient pH 10 > ambient uncontrolled pH	Zhang et al. (2010)
Hydrogen production	The hydrogen yield reached 4 mol/mol-glucose in several thermophilic bacteria (such as *Caldicellulosiruptor owensensis* and *Thermoanaerobacter tengcongensis*) and 3.2 mol/mol-glucose in thermophilic MCF, while it was around 2 mol/mol-glucose under mesophilic conditions	Straub et al. (2017), Temudo et al. (2007), Willquist et al. (2010), Zhang et al. (2014a)
Methane production	The thermophilic continuously stirred anaerobic digesters marginally outperformed the biomethane production rates (0.807 vs. 0.758 L/(L day)) and substrate stabilization (45.1% vs. 42.1%) of the mesophilic conditions	Labatut et al. (2014)
High fraction production of acetate	High fraction (>90%) of acetate production in thermophilic MCF and the concentration reached 34.4 g/L	Zhang et al. (2014b)
Pathogen-free effluents	Produce Class A biosolids in thermophilic MCF with no restrictions on crop type, harvesting, or site access for land application	Kongjan et al. (2009), Labatut et al. (2014)
Low gas solubility	The CH_4 solubility is much lower under thermophilic conditions	Smith et al. (2015)
High ethanol volatility	The boiling point of ethanol is 78.5 °C, which is easily recovered through direct evaporation and distillation in thermophilic conditions	Frock and Kelly (2012)

groups of functional microorganisms, including fermentative bacteria, acetogens, homoacetogens, and hydrogenotrophic and aceticlastic methanogens, are also involved in the processes (Kleerebezem and van Loosdrecht 2007), as shown in Fig. 1. Once the methanogens are inhibited, the intermediates, such as hydrogen, acetate, ethanol, and butyrate, notably accumulate in the reactor. Such intermediates have received more and more attention since the usage of chemicals commodity, clean fuels, and building blocks (Bastidas-Oyanedel et al. 2015; Kleerebezem et al. 2015; Temudo et al. 2007; Zhang et al. 2012).

In the present work, three main topics are covered and reviewed. First, the typical biochemical reactions shown in Fig. 1 are fundamental to understand the thermophilic MCF; accordingly, the main metabolites are covered in Sect. 2. Second, the operational conditions, such as pH, H_2 partial pressure (P_{H_2}), and reactor configuration, which may change the microbial community or the metabolic pathway in thermophilic MCF, are reviewed in Sect. 3. Lastly, the metabolites both in the headspace and in liquid solutions are always a mixture in MCF, which must be concentrated and purified before utilization. For example, the produced biogas usually consists of

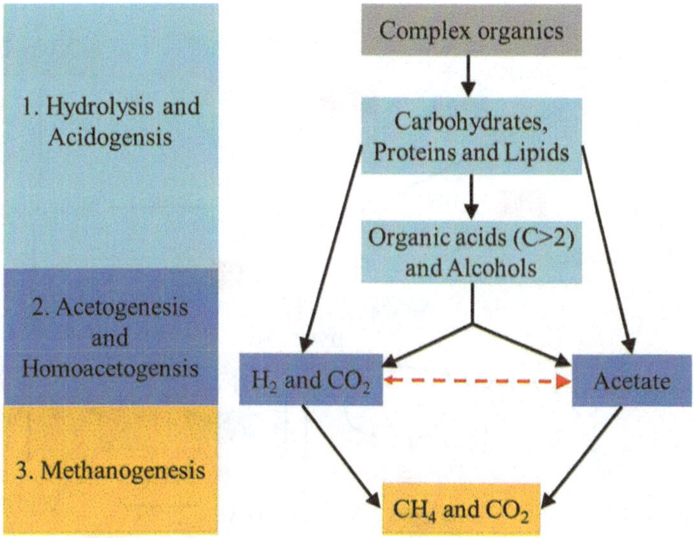

Fig. 1 Schematic of a typical mixed culture fermentation

CH_4 (40–75%), CO_2 (25–60%), and other trace gases, such as H_2S, NH_3, siloxanes, and moisture (Chen et al. 2013; Ryckebosch et al. 2011). For hydrogen production in MCF, the theoretical maximum yield is 4 mol/mol-glucose, and the practical yield is generally 1–3 mol/mol-glucose in MCF (Bastidas-Oyanedel et al. 2012; Zhang et al. 2012). The metabolites in liquid solutions are also a mixture of acetate, butyrate, ethanol, etc. Hence, the coupling of various processes with MCF, such as biogas upgrading, gas stripping, ED, and thermophilic microbial fuel cells (TMFC), has been recently proposed to deal with the separation and purification of the metabolites and is reviewed in Sect. 4. In addition, several novel technologies, such as the controlled metabolites production and medium-chain carboxylic acids production, are also evaluated in Sect. 4. Accordingly, it is anticipated that the review will promote the development and worldwide application of thermophilic MCF in the future.

2 The Metabolism and Thermodynamics in MCF

2.1 Hydrolysis and Acidogenesis

Normally, only simple substrates, such as glucose, xylose, and glycerol, can be directly utilized in the production of volatile fatty acids (VFAs), such as acetate, propionate, and butyrate, as shown in Figs. 1 and 2. Recently, the more interesting and promising substrates for MCF worldwide include industrial and agricultural wastewater, biomass, household solids, algae, and even municipal sludge (Daelman et al. 2016; Khiewwijit et al. 2015; Lü et al. 2017; Tuck et al. 2012; Zamanzadeh

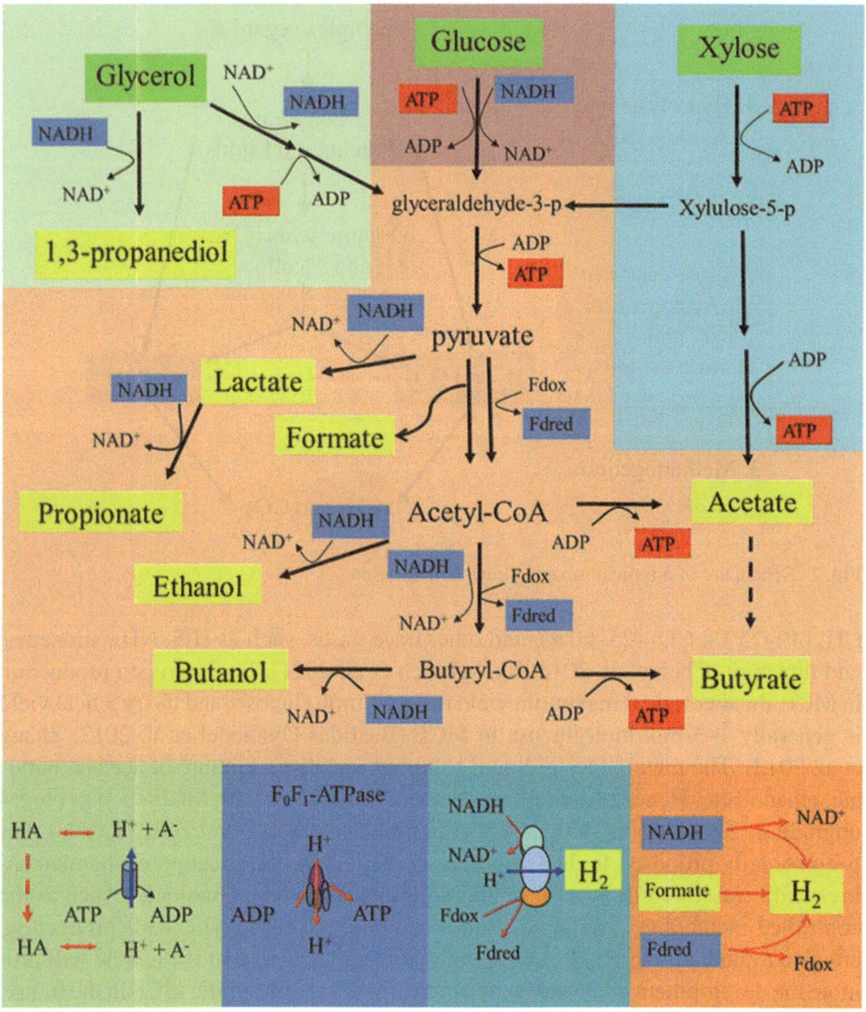

Fig. 2 The typical anaerobic acidogenic reactions in MCF

et al. 2016; Zhang et al. 2015b). However, these substrates cannot be easily directly utilized, and consequently their pretreatment and hydrolysis are necessary in MCF.

Several *Clostridium* spp., such as *Clostridium thermocellum* and *Clostridium clariflavum*, secrete cellulosomes that can degrade the cellulose and hemicellulose into soluble sugars for subsequent utilization (Artzi et al. 2015; Blumer-Schuette et al. 2014). Various chemical methods, including acid-based methods, hydrothermal processing, mild alkaline methods, oxidative methods, steam explosion, and ionic liquid solvents, have also been proposed to remove lignin and hemicelluloses in biomass (Jönsson and Martín 2016). Alternatively, the thermophilic bacteria of the *Caldicellulosiruptor* spp., such as *Caldicellulosiruptor bescii*, can directly degrade the lignocellulosic biomass to hydrogen and acetate without any pretreatments (Svetlitchnyi et al. 2013; Young et al. 2014). Moreover, the bacteria were also recently enriched in extreme-thermophilic MCF, which implied the possibility of biomass bioconversion without pretreatments (Qiu et al. 2011; Zhang et al. 2016).

On the other hand, Ge et al. (2011) reported an increase in the hydrolysis coefficient of waste-activated sludge under thermophilic pretreatment at 60 °C (0.5 ± 0.1 day^{-1}), 65 °C (0.7 ± 0.2 day^{-1}), and 70 °C (0.8 ± 0.2 day^{-1}) compared with mesophilic pretreatment (0.2 ± 0.1 day^{-1}). Alkaline pretreatment has also been proposed to promote the hydrolysis of waste-activated sludge. For example, Zhang et al. (2009) reported that when the sludge was fermented under alkaline pH, the production of short-chain fatty acids was significantly improved, and the optimal pH was 8.0 at thermophilic conditions.

In the ensuing acidogenesis step, the fermentative microorganisms convert the simple substrates into VFAs (such as acetate, propionate, and butyrate), alcohols (such as ethanol and butanol), H_2, and CO_2. The main catabolic reactions with glucose, xylose, and glycerol in MCF are shown in Fig. 2 (da Silva et al. 2009; Hoelzle et al. 2014; Lengeler et al. 1999; Temudo et al. 2009; Zhang et al. 2013c). The energy conservation mechanism includes substrate-level phosphorylation (SLP), ion-motive force mainly coupling with H^+ and Na^+, and energy conservation by electron bifurcation reaction (Lengeler et al. 1999; Peters et al. 2016; Schut and Adams 2009; Thauer et al. 2008; Turina et al. 2016).

Glucose (C6):

As shown in Fig. 2, after glucose is transported into the cytoplasm by the phosphotransferase system (PTS), glucose is mainly converted to pyruvate in the Embden-Meyerhof (EM) pathway (Eq. 1), and the percentages of other pathways, such as the Entner-Doudoroff (ED) pathway and the pentose phosphate (PP) pathway, are relatively low, for example, de Vrije et al. (2007) found that the percentage of glycolysis via the EM pathway reached to 99% in *Caldicellulosiruptor saccharolyticus*.

$$C_6H_{12}O_6 \rightarrow 2C_3H_3O_3^- + 2NADH + 2ATP \tag{1}$$

Glycerol (C3):

In glycerol fermentation, many enzymes, such as the NAD^+-dependent glycerol dehydrogenase, convert glycerol to glyceraldehyde-3-phosphate in the oxidative pathway and then it shares the same metabolic pathway of glucose to produce pyruvate, as shown in Eq. 2 (da Silva et al. 2009). Meanwhile, in the reduction pathway, glycerol and NADH are catalyzed by glycerol dehydratase and other enzymes to produce 1,3-propanediol (da Silva et al. 2009), as shown in Eq. 3.

$$C_3H_8O_3 \rightarrow C_3H_3O_3^- + 2NADH + ATP \tag{2}$$

$$C_3H_8O_3 + NADH \rightarrow C_3H_8O_2 + H_2O \tag{3}$$

Xylose (C5):
Xylose is the second most abundant carbohydrate monomer, after glucose, in agricultural wastes (Temudo et al. 2009). Xylose is initially converted to D-xylulose-5-phosphate, which is then metabolized mainly through two pathways, namely, the pentose phosphate pathway (PPP) and the phosphoketolase pathway (PKP). In the PPP pathway, D-xylulose-5-phosphate is converted to glyceraldehyde-3-phosphate (Eq. 4), whereas, in the PKP pathway, it is metabolized to glyceraldehyde-3-phosphate and then to acetyl phosphate, which is eventually converted to acetate and ATP (Eq. 5).

$$3C_5H_{10}O_5 \rightarrow 5C_3H_3O_3^- + 5NADH + 5ATP \quad (4)$$

$$C_5H_{10}O_5 \rightarrow C_3H_3O_3^- + CH_3COO^- + NADH + 2ATP \quad (5)$$

Acetate (C2):
Acetate is formed directly from acetyl-CoA, generating one molecule of ATP (Lengeler et al. 1999), *as shown in* Eq. 6. The formation of acetyl-CoA from pyruvate can occur through two pathways, namely, the pyruvate formate-lyase (PFL) pathway (Sawers 2005) and the pyruvate:ferredoxin oxidoreductase (PFOR) pathway (Lengeler et al. 1999; Madigan et al. 2002). The latter is the typical pathway in strict anaerobes, such as *Clostridium* sp., which is usually predominant in MCF (Lee et al. 2009).

$$C_3H_3O_3^- \rightarrow C_2H_3O_2^- + CO_2 + Fdred + ATP \quad (6)$$

Ethanol (C2):
The formation of ethanol is achieved through reduction by NADH via acetyl-CoA in MCF, with acetaldehyde as an intermediate, as shown in Eq. 7.

$$C_3H_3O_3^- \rightarrow C_2H_6O + CO_2 + Fdred + 2NAD^+ \quad (7)$$

Lactate (C3):
The formation of lactate from glucose occurs mainly through two pathways (Price et al. 2004), but homofermentation is the main pathway in thermophilic MCF, in which pyruvate is reduced directly to lactate by NADH, as shown in Eq. 8.

$$C_3H_3O_3^- + NADH \rightarrow C_3H_5O_3^- \quad (8)$$

Propionate (C3):
The two main pathways of propionate formation from pyruvate are the methylmalonyl-CoA pathway, which generates 2/3 ATP molecules (Eq. 9), and the acryloyl-CoA pathway, which forms lactate without ATP production (Eq. 10) (Lengeler et al. 1999; Seeliger et al. 2002).

$$C_3H_3O_3^- + 2NADH \rightarrow C_3H_5O_2^- + H_2O + \frac{2}{3}ATP \tag{9}$$

$$C_3H_3O_3^- + NADH \rightarrow C_3H_5O_2^- + H_2O \tag{10}$$

Butyrate and butanol (C4):

Two main enzyme groups control the butyrate production: phosphotransbutyrylase and butyrate kinase (Eq. 11) and butyryl-CoA:acetate CoA-transferase (Eq. 12). Butanol can be produced from the reduction of butyryl-CoA in the presence of NADH by bacteria of the genus *Clostridium*, such as *C. acetobutylicum* (Eq. 13) (Moon et al. 2016).

$$\begin{aligned}&2C_3H_3O_3^- + 3NADH \rightarrow C_4H_7O_2^- \\ &+2CO_2 + 3Fdred + H_2O + ATP\end{aligned} \tag{11}$$

$$\begin{aligned}&C_3H_3O_3^- + C_2H_3O_2^- + 3NADH \\ &\rightarrow C_4H_7O_2^- + CO_2 + 2Fdred + H_2O\end{aligned} \tag{12}$$

$$\begin{aligned}&2C_3H_3O_3^- + 4NADH \rightarrow C_4H_{10}O + \\ &2CO_2 + 3Fdred + 2H_2O + ATP\end{aligned} \tag{13}$$

Hydrogen (H_2):

Generally, the production of hydrogen is controlled via four pathways by the redox couple of NADH/NAD$^+$, Fdred/Fdox, and formate (Pawar and Niel 2013; Schut and Adams 2009; Stephen et al. 2017), as shown in Fig. 2. First, the formation of hydrogen from NADH (Eq. 14) is not possible except under rather low P_{H_2} (60 Pa) (Angenent et al. 2004). Second, since the potential of Fdred/Fdox is below −400 mV, it is thus the more suitable electron donor for hydrogen production (Eq. 15) (Temudo et al. 2007; Thauer et al. 2008). Additionally, the hydrogen production from Fdred/Fdox is also verified in *Thermoanaerobacter tengcongensis* (Soboh et al. 2004).

$$NADH + H^+ \rightarrow NAD^+ + H_2 \tag{14}$$

$$Fdred + 2H^+ \rightarrow Fdox + H_2 \tag{15}$$

$$CHO_2^- + H^+ \rightarrow H_2 + CO_2 \tag{16}$$

$$Fdred + NADH + 3H^+ \rightarrow Fdox + NAD^+ + 2H_2 \tag{17}$$

Third, hydrogen can also be produced from formate, such as in *E. coli* and *Enterobacter* sp. (Eq. 16) (Lee et al. 2009; Sawers 2005), in which the ratio of hydrogen verse (hydrogen + formate) was thermodynamically controlled (Temudo et al. 2007; Zhang et al. 2015a). The fourth pathway was found by Schut and Adams (2009), who named it the electron bifurcation reaction, in which a new type of

[FeFe] hydrogenase in the hyperthermophilic bacterium *Thermotoga maritima* simultaneously utilizes NADH and ferredoxin as electron donors to produce hydrogen (Eq. 17).

2.2 Acetogenesis and Homoacetogenesis

In the second stage of MCF, the produced VFAs and alcohols, such as propionate (Eq. 18), butyrate (Eq. 19), and ethanol (Eq. 20), are converted to acetate, H_2, and CO_2 by acetogenic bacteria (Fig. 3a) under rather low P_{H_2}. For instance, the P_{H_2} in the conversion of propionate to acetate (Eq. 12) is lower than 10 Pa (Angenent et al. 2004). Accordingly, propionate is rather hard to be consumed and is thus easily accumulated in anaerobic digester when the reactor experiences fluctuation or shock (Hori et al. 2006). For longer carbon chain organic acids, the β-oxidation pathway is dominant (Narihiro et al. 2016; Sousa et al. 2009).

$$C_3H_5O_2^- + 2H_2O \rightarrow C_2H_3O_2^- + CO_2 + 3H_2 \quad (18)$$

$$C_4H_7O_2^- + 2H_2O \rightarrow 2C_2H_3O_2^- + 2H_2 + H^+ \quad (19)$$

$$C_2H_6O + H_2O \rightarrow C_2H_3O_2^- + 2H_2 + H^+ \quad (20)$$

In contrast, high P_{H_2} favors homoacetogens (Fig. 3b), such as in *Clostridium ljungdahlii* and *Acetobacterium woodii*, to produce acetate (Eq. 21) (Bertsch and Müller 2015; Poehlein et al. 2012; Zhang et al. 2013b). For example, Poehlein et al. (2012) determined that the minimal hydrogen concentration in the reactor should be above 250 Pa to allow homoacetogens growth on hydrogen and CO_2. In addition, CO is also a substrate of *C. ljungdahlii* (Schuchmann and Muller 2014).

$$4H_2 + 2CO_2 \rightarrow C_2H_3O_2^- + H^+ + 2H_2O \quad (21)$$

2.3 Methanogenesis

Methanogens convert acetate and H_2/CO_2 to CH_4 through the processes of acetoclastic methanogenesis (Eq. 22) and hydrogenotrophic methanogenesis (Eq. 23), respectively (Thauer et al. 2008). The former reaction is carried out by *Methanosarcinaceae* and *Methanosaetaceae*, while the latter is performed by *Methanomicrobiales* and *Methanobacteriales* (Karakashev et al. 2006). Normally, methanogenesis is the rate-limiting step in anaerobic digesters because methanogens grow slowly and are susceptible to toxins and operational conditions, such as pH, NH_4^+, and temperature (Karakashev et al. 2005). Meanwhile, methanol (Eq. 24) and CO are also the common substrates (Berg et al. 2010; Thauer et al. 2008).

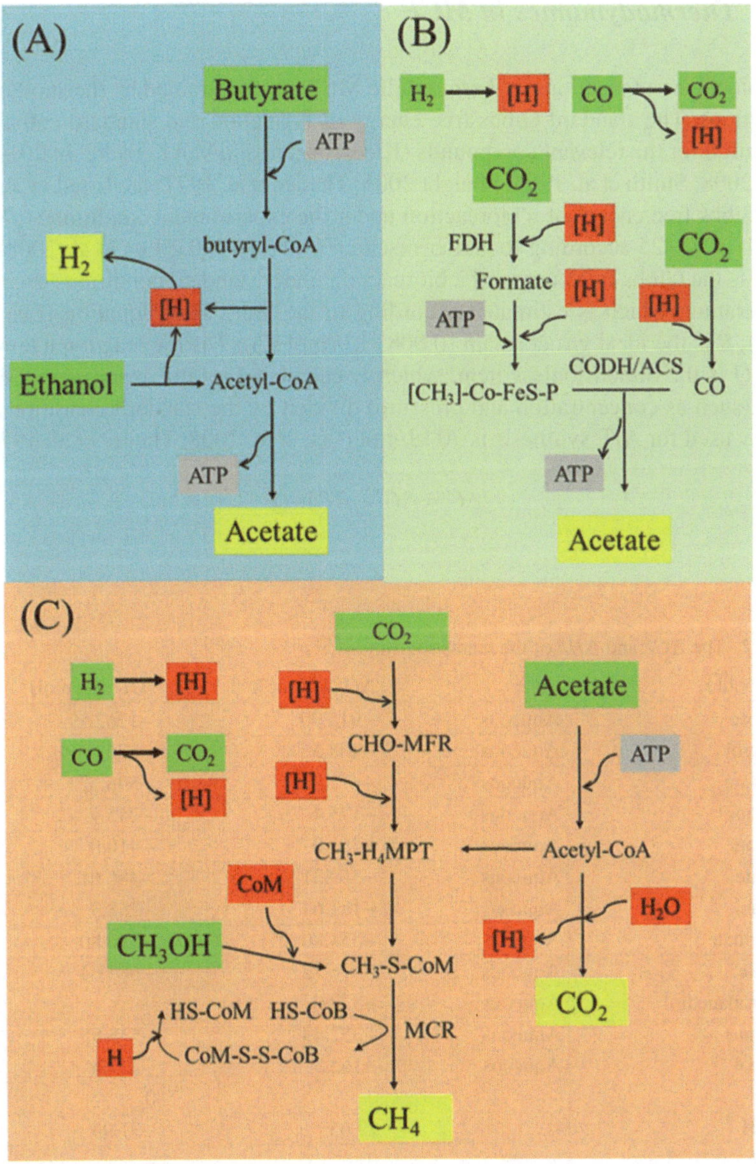

Fig. 3 The metabolic pathways of acetogenesis (**a**), homoacetogenesis (**b**), and methanogenesis (**c**) in MCF

$$C_2H_3O_2^- + H^+ \rightarrow CH_4 + CO_2 \tag{22}$$

$$4H_2 + CO_2 \rightarrow CH_4 + 2H_2O \tag{23}$$

$$CH_3OH + H_2 \rightarrow CH_4 + H_2O \tag{24}$$

2.4 Thermodynamics in MCF

The biochemical reactions in thermophilic MCF are constrained by thermodynamic conditions. The standard Gibbs free energy of formation and standard enthalpy of formation for the relevant compounds (Kleerebezem and Van Loosdrecht 2010; Lee et al. 2008; Smith et al. 1995; Speight 2005; Thauer et al. 1977) are listed in Table 2. The Gibbs free energy of a bioreaction under the experimental conditions ($\triangle G$) is shown in Eq. 25 according to earlier research (Alberty 2003; Lee et al. 2008). The $\triangle G'$ is the Gibbs free energy of a bioreaction under standard conditions except the temperature, which is estimated according to the Van't Hoff equation (Lee et al. 2008); R is the ideal gas constant (0.0083 kJ/(mol K)); T is the reactional temperature; Q is the dimensionless item, which is calculated by the experimental conditions (such as concentration and pressure) divided by the standard conditions. The energy used for ATP synthesis is 70 kJ/mol (Lee et al. 2008; Thauer et al. 1977).

$$\Delta G = \Delta G' + RT \ln Q \tag{25}$$

Table 2 The ΔG_f^0 and ΔH_f^0 of the metabolites in MCF

Metabolite	State	ΔG_f^0 (kJ/mol)	ΔH_f^0 (kJ/mol)
Glucose	Aqueous	−915.90	−1262.19
Glycerol	Aqueous	−488.5	−676.0
Pyruvate	Aqueous	−472.27	−596.22
Methanol	Aqueous	−175.4	−245.9
Formate	Aqueous	−351.0	−410.0
Acetate	Aqueous	−369.31	−486.01
Ethanol	Aqueous	−181.64	−288.3
Propionate	Aqueous	−354.80	−510.80
Lactate	Aqueous	−516.72	−686.64
1,3-propanediol	Aqueous	−327.0	–
Butyrate	Aqueous	−352.63	−535.55
Butanol	Aqueous	−162.5	−327.3
NAD^+	–	0	0
NADH	–	22.65	−31.94
Fdred	–	0	–
Fdox	–	−77.20	–
CO_2	Gas	−394.36	−393.50
CO	Gas	−137.16	−110.53
H_2	Gas	0	0
H_2O	Liquid	−237.19	−285.83
Proton	Aqueous	0	–

3 Influencing Factors in Thermophilic MCF

The operational conditions, such as pH, P_{H_2}, substrate type and concentration, and reactor configuration, may change the microbial community in thermophilic MCF or the metabolic pathway, and consequently these factors affect the reactor performances. On the other hand, the metabolites in MCF are always a mixture, and coupling processes are necessary to utilize the products. Thus, several novel technologies such as biogas upgrading, two-stage fermentation, gas stripping, ED, and microbial fuel cells are also reviewed.

3.1 H_2 Partial Pressure (P_{H_2})

As a promising clean fuel, biohydrogen has received increasing attention in the past decade, especially in thermophilic MCF (Hosseini and Wahid 2016; Kothari et al. 2012; Tapia-Venegas et al. 2015). However, due to the thermodynamic constraints, the hydrogen yield was lower than 4 mol/mol-glucose and was commonly 1–3 mol/mol-glucose (Dai et al. 2017; Zhang et al. 2015a). In addition, the P_{H_2} is an important factor to control the metabolites distribution. For example, Soboh et al. (2004) found that when H_2 accumulated during fermentation by *Thermoanaerobacter tengcongensis*, the NADH-dependent hydrogenase activity was fourfold lower, whereas the activities of aldehyde dehydrogenase and alcohol dehydrogenase were higher, and as a result the metabolites were shifted from hydrogen and acetate to ethanol.

However, in thermophilic MCF, as the P_{H_2} was decreased from 0.62 to 0.05 atm by N_2 sparging, the hydrogen yield was increased slightly from 1.9 to 2.2 mol/mol-glucose, the acetate yield was reduced from 0.58 to 0.39 mol/mol-glucose, and the butyrate yield did not change much and was between 0.48 and 0.55 mol/mol-glucose (Zhang et al. 2015a). Kim et al. (2006) reported that in mesophilic MCF, with gas sparging, the hydrogen production increased only from 0.9 to 1.2 mol/mol-glucose. Additionally, Zhang et al. (2013d) found that in extreme-thermophilic MCF (pH 8.0), with N_2 flushing, the hydrogen production increased from 0.64 to 1.2 mol/mol-glucose. Change in low hydrogen yield may be due to two reasons: (1) researchers found that hydrogen was only produced by Fdred in practical reactor, and thus the hydrogen yield was generally below 2 mol/mol-glucose (Kleerebezem and van Loosdrecht 2007); (2) the hydrogen supersaturation occurred in thermophilic MCF even with N_2 flushing and/or N_2 sparging, which resulted in high hydrogen concentrations in the liquid (Kraemer and Bagley 2006; Willquist et al. 2010; Zhang et al. 2012). However, Willquist et al. (2010) also reported that the production of hydrogen by extreme-thermophilic bacterium *Caldicellulosiruptor saccharolyticus* was close to the theoretical maximum value of 4 mol/mol-glucose in a practical reactor. Thus, the enriched bacteria that have the special hydrogenases could produce higher hydrogen yield, whereas the hydrogen production rate, not the yield, will be the reasonable factor for hydrogen production in thermophilic MCF.

3.2 Acidic pH

The toxicity of organic acids, especially at acidic pH, is a main challenge for bacteria, as the inward diffusion of organic acids through the cytoplasmic membrane leads to dissipation of the proton-motive force and bacteria have to transport out these metabolites by processes involving energy consumption in the form of ATP (Louis et al. 2004; Zhang et al. 2013c). For instance, Ueno et al. (2006) found a shift of butyrate and acetate to acetate and ethanol as the pH value increased from an acidic pH of 5.0 to a neutral pH of 7.0. In a thermophilic continuous stirred-tank reactor (CSTR) with mixed culture, the dominant metabolites of hydrogen, acetate, and butyrate did not notably change at acidic pH 4.0–6.5, whereas, at neutral pH, the main metabolites shifted to acetate, ethanol, propionate, and formate (Zhang et al. 2015a). In an extreme-thermophilic CSTR, a shift of acetate and hydrogen to ethanol was observed as the pH decreased from 7.0 to 4.0 (Zhang et al. 2014a). Thus, the removal of organic acids in thermophilic MCF is necessary.

3.3 Alkaline pH

Alkaline thermophilic MCF has been proposed to enhance waste-activated sludge hydrolysis, inhibit methanogen activity, improve VFA production, and produce pathogen-free effluent (Dai et al. 2016; Jin et al. 2016; Labatut et al. 2014; Wang et al. 2017). Researchers have therefore paid more attention to alkaline MCF, especially for use with sewage sludge. For example, Chen et al. (2017a) reported that maximum acidification of a mixed sludge was obtained at pH 8.9 and 55 °C, with a maximum production of VFAs of 423.2 mg of chemical oxygen demand (COD)/g volatile suspended solids (VSS) with no hydrogen produced. Also, Jin et al. (2016) reported that NaOH and KOH are the best alkali reagents for acidifying waste-activated sludge at 15 °C, and the percentage of acetate was about 50% of the total VFAs. At alkaline pH, the toxicity of acidic acids could be diminished; however, high free ammonia concentrations under alkaline pH inhibit bacterial activity, especially at high temperatures (Ho and Ho 2012; Wang et al. 2016; Yun et al. 2016). Moreover, Wang et al. (2016) found that when the initial P_{H_2} was 0.5 atm, in mesophilic and thermophilic conditions, the methane yield at high ammonia load (7 g NH_4^+-N/L) was 41.0% and 22.3% lower than that at low ammonia load (1 g NH_4^+-N/L), respectively. Additionally, the biomass yields under thermophilic and extreme-thermophilic alkaline MCF were also lower than that under the mesophilic condition (Dai et al. 2018). Thus, ammonia will be removed in thermophilic MCF.

3.4 Reactor Types

The reactor type also affects the metabolites yield. For example, a comparison of the hydrogen production from hemicellulose-rich wheat straw hydrolysate in a CSTR, an upflow anaerobic sludge bed (UASB) reactor, and an anaerobic filter (AF) reactor by Kongjan and Angelidaki (2010) revealed that the highest hydrogen production yield of 212.0 ± 6.6 mL-H_2/g-sugars was achieved in the UASB reactor. Moreover, lowering the hydraulic retention time (HRT) even caused biomass washout in the CSTR. Recently the anaerobic biofilms reactor has been paid more and more attention because it can completely uncouple the solids retention time from the HRT, which is especially useful for toxicity resistance and low-growing bacteria growth (Wijekoon et al. 2011). In addition, Jiang et al. (2011) proposed that *Clostridium tyrobutyricum* was immobilized in a fibrous bed bioreactor (FBB) to produce butyric acid at a high final concentration of 86.9 g/L. Also, Chaikasem et al. (2014) studied the performance of polyvinyl alcohol hydrogel (PVA-gel) beads as an effective biocarrier for VFA production in a two-stage thermophilic anaerobic membrane bioreactor (TAnMBR). However, the membrane flaws should be considered, and when the aim is to produce VFAs, the methanogens accumulation in the biofilm also needs to be prevented.

4 Coupling Technologies and Their Prospective Uses in Thermophilic MCF

4.1 Two-Stage Fermentation for Hydrogen and Methane Production

Conversion of the metabolites in broth to methane also provides an alternative for the utilization of organic acids (Li and Yu 2011). The two-stage fermentation for hydrogen and methane production is proposed through the separation of the hydrolysis/acidogenesis and methanogenesis steps (Li and Yu 2011; Ueno et al. 2007). Moreover, operating a thermophilic reactor in the first stage has attracted much more attention (Nielsen et al. 2007; Zhang et al. 2014b). For instance, Ueno et al. (2007) constructed a pilot-scale thermophilic two-stage plant for the production of hydrogen (at 60 °C) and methane (at 55 °C). Luo et al. (2011) constructed a two-stage process with HRT of 3 days for hydrogen production and 12 days for methane production and obtained 11% higher energy compared to a single-stage methanogenic process. Additionally, Pawar et al. (2013) proposed a two-step process to produce hydrogen from wheat straw using *Caldicellulosiruptor saccharolyticus* in a CSTR reactor followed by anaerobic digestion of its effluent to produce CH_4.

However, the overall operating costs after integrating the thermophilic reactors still requires further evaluation, and the extra alkali dosing also increases the operational cost. Recently, Wu et al. (2016) proposed an acidic reactor at pH 4.0 to decrease the alkali addition in the two-phase anaerobic digestion for fruit and vegetable waste treatment, and the system exhibited a low HRT (3.56 days) with a high methane yield (348.5 mL/g VS removed). However, after a long time in operation, alkali will also need to be periodically added. Therefore, novel methods to reduce operational cost are still necessary.

4.2 H_2 and CH_4 Upgrading

The H_2 and CH_4 produced in thermophilic MCF have high calorific values. The Henry's law coefficients of H_2, CH_4, CO_2, and H_2S were calculated according to Perry and Green (2008), as shown in Table 3. These data demonstrated that the solubility of H_2, CH_4, CO_2, and H_2S decreases in thermophilic conditions, which is beneficial for the gas separation/upgrading.

Recently, Sowunmi et al. (2016) showed that the energy from methane produced from biomasses in the Abu Dhabi Emirate was theoretically able to meet 6% of household electricity consumption. However, the impurities, such as CO_2, H_2S, and siloxanes, must be removed first. Water washing, membrane separation, chemical absorption, and pressure swing adsorption (PSA) are the four common CO_2 removal techniques (Deng and Hägg 2010). For example, the commercially available membranes for CO_2/CH_4 separation are generally conventional polymeric dense membranes (Deng and Hägg 2010). Bakonyi et al. (2013) proposed to use a commercial polyimide membrane module to purify H_2 and achieved the highest H_2/CO_2 gas selectivity of 1.62. In studies using chemical absorption technology, the alkali adsorption (e.g., NaOH) reduced CO_2 and H_2S simultaneously and avoided the pre-cleaning of H_2S (Petersson and Wellinger 2009) according to the ratio of dissolved to total CO_2 and H_2S (k_{CO_2} and K_{H_2S}) at 25 °C(Bastidas-Oyanedel et al. (2010)), as shown in Table 4. Thus, alkaline pH (>8) is beneficial for gas adsorption. Recently, activated carbon adsorption was suggested to be more suitable for H_2S pretreatment, and the surface acidity was found to be the key factor (Yentekakis and Goula 2017).

Siloxanes were also a major problem for biogas utilization, as the silica microparticulates formed at high temperatures can cause fouling and abrasion effects detrimental to natural gas-fueled vehicles (Cabrera-Codony et al. 2014; Yentekakis and Goula 2017). Recently, Cabrera-Codony et al. (2014) used the activated carbon for siloxanes removal, but the siloxanes polymerization cannot be avoided under

Table 3 The Henry's law coefficient of H_2, CH_4, CO_2, and H_2S

	Henry's law coefficient (1/atm)			
	25 °C	35 °C	55 °C	70 °C
H_2	1.411×10^{-5}	1.345×10^{-5}	1.308×10^{-5}	1.332×10^{-5}
CH_4	2.552×10^{-5}	2.180×10^{-5}	1.772×10^{-5}	1.633×10^{-5}
CO_2	6.116×10^{-4}	4.774×10^{-4}	3.238×10^{-4}	2.618×10^{-4}
H_2S	1.847×10^{-3}	1.500×10^{-3}	1.059×10^{-3}	8.581×10^{-4}

Table 4 The values of k_{CO_2} and K_{H_2S} at 25 °C at different pH values

pH	4	5	6	7	8	9	10
k_{CO_2}	0.996	0.959	0.699	0.189	0.023	0.002	0.0001
K_{H_2S}	0.999	0.991	0.917	0.524	0.099	0.011	0.001

long-term operation. Therefore, developing suitable methods, such as selective separating membrane technologies to remove CO_2, H_2S, and siloxanes, are prerequisites for biogas utilization.

Meanwhile, several biotechnologies have been proposed to improve the biogas purity or convert the biogas (H_2 and CO_2) to other valuable biochemicals (Luo et al. 2012; van der Ha et al. 2012; Zhang et al. 2013a, b). For example, Luo et al. (2012) proposed to convert hydrogen to methane by hydrogenotrophic methanogenesis. Also, recently, a methane-oxidizing bacterium *Methylocystis parvus* was used to convert biogas to polyhydroxybutyrate (PHB) by van der Ha et al. (2012), and the final concentration reached 295 mg-PHB/g-cell.

4.3 Gas Stripping for Alcohol Recovery and Ammonia Removal

Thanks to the high volatility under high temperature, ethanol and butanol can be easily recovered by the gas stripping technology (Frock and Kelly 2012; Xue et al. 2016). Hashi et al. (2010) used CO_2 to remove ethanol from the fermentation broth and reduced the ethanol toxicity. Frock and Kelly (2012) proposed that biofuels, such as ethanol, can be directly recovered through direct evaporation and distillation in high-temperature MCF. Xue et al. (2016) showed that much higher amounts of acetone-butanol-ethanol (ABE) (27.5 g/L of acetone, 75.5 g/L of butanol, 7.0 g/L of ethanol) were produced in the fed-batch ABE fermentation integrated gas stripping and pervaporation process. Pervaporation and membrane distillation also enabled selective separation of volatile alcohols from the fermentation broth or the purification of H_2 (Lewandowicz et al. 2011).

Gas stripping can also remove ammonia in thermophilic conditions. For instance, Serna-Maza et al. (2014) found that a high temperature (>70 °C) and a pH of 10 were needed to reduce the ammonia concentration in digesters and achieved a 48% removal of the ammonia. Although high-gas stripping rates can also cause some adverse effects, such as microbial activity inhibition and energy wasting, coupling biomass immobilization with gas stripping may be a sound method, but it still needs to be assessed in the future.

4.4 Electrodialysis for Organic Acids Removal

Given that the separation technologies, such as membrane technologies, can be achieved on a large scale, Bonk et al. (2015) showed that the purification of VFAs is economically feasible. In addition, the organic acids are disassociated from thermophilic MCF broth; thus, ED is a suitable technology to separate and concentrate these metabolites (Moresi and Sappino 2000; Zhang et al. 2011). Indeed, Meynial-Salles et al. (2008) proposed a novel three-stage continuous fermentation process, which combined an integrated membrane bioreactor-ED system to produce and concentrate succinic acid, which achieved a maximum concentration of 83 g/L. Moreover, Redwood et al. (2012) proposed an integrated hydrogen refinery of food wastes, consisting of a synergic combination of photo-fermentation, extractive fermentation, and hydrothermal hydrolysis, in which ED provided the key link in the concept of waste to energy for the selective separation of organic acids. Recently, Wang et al. constructed a device that coupled lactic acid fermentation with bipolar membrane ED (BMED) stack and achieved a lactic acid recovery rate of 86% in the batch mode (Wang et al. 2012) and 69.5% in the continuous mode (Wang et al. 2013).

Although pilot-scale experiments for VFAs separation from a real effluent of MCF also need to be conducted in the future, the overall cost is still high and membrane fouling cannot be avoided. For example, Zheng et al. (2010) reported that the membrane failed after only 24 h of operation. The addition of powdered activated carbon (PAC) was proposed to prevent the membrane from fouling. Lastly, as it is well known, except for the bacteria metabolites of organic acids and alcohols, the components of the MCF broth normally also include various kinds of inorganic salts that affect the actual separating factors. For instance, Zhang et al. (2011) examined the ion competition between organic acids (such as formate, acetate, propionate, and butyrate) and inorganic salts (such as HPO_4^{2-} and Cl^-) and found that membrane selectivity was dependent on the size, the charge, and the functional groups of the organic ions; thus the decrease of the concentrations of acetate, propionate, and butyrate was slower because there were still inorganic ions present. Accordingly, the developments of selective separating membranes for specific metabolites are urgently needed.

4.5 Thermophilic Microbial Fuel Cells for Energy Recovery

The metabolites produced in MCF can also be converted to electricity. Microbial fuel cell (MFC) is a fast-growing environmental biotechnology whereby bioconvertible substrates are consumed in the anodic chamber with simultaneous electron generation (Logan and Regan 2006; Lovley 2008). Thus, the thermophilic MCF effluent can also be the influent of TMFC to produce electricity, as well as other biochemicals. However, TMFC has seldom been reported to date (Dopson et al. 2016; Ha et al. 2012). In 2006, Jong et al. (2006) operated a TMFC, without a mediator, using acetate as the substrate and obtained open-circuit potentials of around 900 mV, while the maximum closed-circuit potential was 45 mV with an external resistance of 10 Ω. Ha et al. (2012) proposed to treat alcohol distillery wastewater with TMFC that achieved a current density of 2.3 A/m^2, and the power density reached 360 mW/m^2 with a Pt-coated cathode electrode. As shown above, the MCF under thermophilic conditions offers many advantages over mesophilic MCF; therefore, the development of TMFC could broaden the applications of thermophilic MCF.

4.6 *Other Novel Technologies*

Besides the above biotechnologies, several novel concepts have been recently proposed in thermophilic MCF. First, the production of a single metabolite in MCF can be achieved at lower cost. For example, after selective enrichment of hydrogenotrophic methanogens over acetoclastic methanogens, our group could solely produce acetate (the fraction in bulk solutions >90%) from glucose, glycerol, or tofu wastewater in extreme-thermophilic MCF (Chen et al. 2016b; Zhang et al. 2015b, 2014b). In addition, the production of high-purity propionate from glycerol or glucose in MCF induced by ammonium was recently investigated by Chen et al., who reported that the final purity of propionate was 91–100% (Chen et al. 2017b, 2016a). Recently, Bonk et al. (2017) proposed the selective production of lactic acid from food waste by in-situ product extraction using activated carbon, and their results showed that the lactic acid concentration reached 32 gCOD/L with a selectivity of 93% in a semicontinuous mode.

Second, an alternative process for anaerobic wastewater treatment with methane or acetate recovery is to elongate the carbon chain of VFAs to the medium-chain carboxylic acids, such as *n*-caproic acid from ethanol, acetate, glycerol, and syngas (Leng et al. 2017; Zhang et al. 2013b). Also, Xu et al. (2018) recently proposed a new concept that involved thermophilic lactic acid production and mesophilic chain elongation to produce organic acids of C6–C9 from acid whey via lactic acid as an intermediate. For a longer carbon chain and lower O/C ratio, the mixture of the medium-chain fatty acids produced could also be upgraded to biofuels by hydrogen reduction (Steinbusch et al. 2011; Zhang et al. 2013b). In addition, Khor et al. (2017) recently converted caproic acid to decane, an important chemical and biofuel, by the Kolbe electrolysis.

Lastly, the energy cost is the main obstacle to apply thermophilic MCF. However, Kongjan et al. (2009) reported that the energy cost of Danish thermophilic biogas plants was about 10% of the energy produced, which indicates that the extra energy cost (1–2%) for operating at a thermophilic temperature is marginal. Moreover, according to the regulations in Denmark, a sanitation step was necessary in the biogas plants, whereby specific categories of biomasses are required to be heated up to 70 °C for 1 h in Denmark, in order to ensure the sanitation of the effluents. Consequently, the combination of sanitation and fermentation step will also reduce the total operational cost. Accordingly, thermophilic alkaline MCF should be a promising technology to produce biochemicals. The utilization of industrial waste hot water to heat thermophilic MCF could further reduce the cost.

5 Conclusions

The typical metabolic pathways in hydrolysis/acidogenesis, acetogenesis/homoacetogenesis, and methanogenesis of thermophilic MCF are reviewed in this article. The operational conditions, such as pH, P_{H_2}, and reactor configuration, may change the metabolites production in thermophilic MCF and were discussed in the second section. Lastly, the metabolites, both in the headspace and in liquid solutions, are always a mixture in MCF, which must be concentrated and purified before use. In view of that, several conventional technologies to separate the metabolites and recover energy, including biogas upgrading, two-stage fermentation, gas stripping, electrodialysis, and microbial fuel cells, are described. In addition, novel technologies, such as controlled metabolites production and medium-chain carboxylic acids production, are also reviewed. Therefore, the coupling process and the development of novel technologies are necessary in MCF to promote its worldwide application.

Acknowledgments The authors would like to acknowledge the financial support from the National Natural Science Foundation of China (51408530, 50978244, and 51478447), Natural Science Foundation of Hebei Province (E2015203306), Foundation of Hebei Education Department (BJ2017014), and the Program for Changjiang Scholars and Innovative Research Team in University.

References

Alberty RA (2003) Thermodynamics of biochemical reactions. Wiley, New York

Angenent LT, Karim K, Al-Dahhan MH, Wrenn BA, Domíguez-Espinosa R (2004) Production of bioenergy and biochemicals from industrial and agricultural wastewater. Trends Biotechnol 22(9):477–485

Artzi L, Morag E, Barak Y, Lamed R, Bayer EA (2015) Clostridium clariflavum: key cellulosome players are revealed by proteomic analysis. MBio 6(3):e00411–15

Bakonyi P, Kumar G, Nemestóthy N, Lin CY, Bélafi-Bakó K (2013) Biohydrogen purification using a commercial polyimide membrane module: Studying the effects of some process variables. Int J Hydrogen Energy 38(35):15092–15099

Bastidas-Oyanedel J-R, Mohd-Zaki Z, Pratt S, Steyer J-P, Batstone DJ (2010) Development of membrane inlet mass spectrometry for examination of fermentation processes. Talanta 83(2):482–492

Bastidas-Oyanedel J-R, Mohd-Zaki Z, Zeng RJ, Bernet N, Pratt S, Steyer J-P, Batstone DJ (2012) Gas controlled hydrogen fermentation. Bioresour Technol 110:503–509

Bastidas-Oyanedel J-R, Bonk F, Thomsen M, Schmidt J (2015) Dark fermentation biorefinery in the present and future (bio)chemical industry. Rev Environ Sci Biotechnolgy 14(3):473–498

Batstone DJ, Virdis B (2014) The role of anaerobic digestion in the emerging energy economy. Curr Opin Biotechnol 27:142–149

Berg IA, Kockelkorn D, Ramos-Vera WH, Say RF, Zarzycki J, Hügler M, Alber BE, Fuchs G (2010) Autotrophic carbon fixation in archaea. Nat Rev Microbiol 8(6):447–460

Bertsch J, Müller V (2015) CO metabolism in the acetogen Acetobacterium woodii. Appl Environ Microbiol 81(17):5949–5956

Blumer-Schuette SE, Brown SD, Sander KB, Bayer EA, Kataeva I, Zurawski JV, Conway JM, Adams MWW, Kelly RM (2014) Thermophilic lignocellulose deconstruction. FEMS Microbiol Rev 38(3):393–448

Bonk F, Bastidas-Oyanedel J-R, Schmidt JE (2015) Converting the organic fraction of solid waste from the city of Abu Dhabi to valuable products via dark fermentation – economic and energy assessment. Waste Manag 40:82–91

Bonk F, Bastidas-Oyanedel J-R, Yousef AF, Schmidt JE (2017) Exploring the selective lactic acid production from food waste in uncontrolled pH mixed culture fermentations using different reactor configurations. Bioresour Technol 238:416–424

Cabrera-Codony A, Montes-Morán MA, Sánchez-Polo M, Martín MJ, Gonzalez-Olmos R (2014) Biogas upgrading: optimal activated carbon properties for siloxane removal. Environ Sci Technol 48(12):7187–7195

Chaikasem S, Abeynayaka A, Visvanathan C (2014) Effect of polyvinyl alcohol hydrogel as a biocarrier on volatile fatty acids production of a two-stage thermophilic anaerobic membrane bioreactor. Bioresour Technol 168(suppl C):100–105

Chen M, Zhang F, Zhang Y, Zeng RJ (2013) Alkali production from bipolar membrane electrodialysis powered by microbial fuel cell and application for biogas upgrading. Appl Energy 103:428–434

Chen Y, Wang T, Shen N, Zhang F, Zeng RJ (2016a) High-purity propionate production from glycerol in mixed culture fermentation. Bioresour Technol 219:659–667

Chen Y, Zhang F, Wang T, Shen N, Yu Z-W, Zeng RJ (2016b) Hydraulic retention time affects stable acetate production from tofu processing wastewater in extreme-thermophilic (70 °C) mixed culture fermentation. Bioresour Technol 216:722–728

Chen Y, Jiang X, Xiao K, Shen N, Zeng RJ, Zhou Y (2017a) Enhanced volatile fatty acids (VFAs) production in a thermophilic fermenter with stepwise pH increase – Investigation on dissolved organic matter transformation and microbial community shift. Water Res 112:261–268

Chen Y, Shen N, Wang T, Zhang F, Zeng RJ (2017b) Ammonium level induces high purity propionate production in mixed culture glucose fermentation. RSC Adv 7(1):518–525

da Silva GP, Mack M, Contiero J (2009) Glycerol: a promising and abundant carbon source for industrial microbiology. Biotechnol Adv 27(1):30–39

de Vrije T, Mars A, Budde M, Lai M, Dijkema C, de Waard P, Claassen P (2007) Glycolytic pathway and hydrogen yield studies of the extreme thermophile Caldicellulosiruptor saccharolyticus. Appl Microbiol Biotechnol 74(6):1358–1367

Daelman MRJ, Sorokin D, Kruse O, van Loosdrecht MCM, Strous M (2016) Haloalkaline bioconversions for methane production from microalgae grown on sunlight. Trends Biotechnol 34(6):450–457

Dai X, Li X, Zhang D, Chen Y, Dai L (2016) Simultaneous enhancement of methane production and methane content in biogas from waste activated sludge and perennial ryegrass anaerobic co-digestion: the effects of pH and C/N ratio. Bioresour Technol 216:323–330

Dai K, Wen J-L, Zhang F, Zeng RJ (2017) Valuable biochemical production in mixed culture fermentation: fundamentals and process coupling. Appl Microbiol Biotechnol 101(17):6575–6586

Dai K, Zhang F, Zhang Y, Zeng RJ (2018) The chemostat metabolite spectra of alkaline mixed culture fermentation under mesophilic, thermophilic, and extreme-thermophilic conditions. Bioresour Technol 249:322–327

Deng L, Hägg M-B (2010) Techno-economic evaluation of biogas upgrading process using CO2 facilitated transport membrane. Int J Greenhouse Gas Control 4(4):638–646

Dopson M, Ni G, Sleutels THJA (2016) Possibilities for extremophilic microorganisms in microbial electrochemical systems. FEMS Microbiol Rev 40(2):164–181

Frock AD, Kelly RM (2012) Extreme thermophiles: moving beyond single-enzyme biocatalysis. Curr Opin Chem Eng 1(4):363–372

Gannoun H, Othman NB, Bouallagui H, Moktar H (2007) Mesophilic and thermophilic anaerobic co-digestion of olive mill wastewaters and abattoir wastewaters in an upflow anaerobic filter. Ind Eng Chem Res 46(21):6737–6743

Ge H, Jensen PD, Batstone DJ (2011) Temperature phased anaerobic digestion increases apparent hydrolysis rate for waste activated sludge. Water Res 45(4):1597–1606

Ha PT, Lee TK, Rittmann BE, Park J, Chang IS (2012) Treatment of alcohol distillery wastewater using a bacteroidetes-dominant thermophilic microbial fuel cell. Environ Sci Technol 46(5):3022–3030

Hashi M, Tezel FH, Thibault J (2010) Ethanol recovery from fermentation broth via carbon dioxide stripping and adsorption. Energy Fuel 24(9):4628–4637

Ho L, Ho G (2012) Mitigating ammonia inhibition of thermophilic anaerobic treatment of digested piggery wastewater: use of pH reduction, zeolite, biomass and humic acid. Water Res 46(14):4339–4350

Hoelzle RD, Virdis B, Batstone DJ (2014) Regulation mechanisms in mixed and pure culture microbial fermentation. Biotechnol Bioeng 111(11):2139–2154
Hori T, Haruta S, Ueno Y, Ishii M, Igarashi Y (2006) Dynamic transition of a methanogenic population in response to the concentration of volatile fatty acids in a thermophilic anaerobic digester. Appl Environ Microbiol 72(2):1623–1630
Hosseini SE, Wahid MA (2016) Hydrogen production from renewable and sustainable energy resources: promising green energy carrier for clean development. Renew Sustain Energy Rev 57:850–866
Jiang L, Wang J, Liang S, Cai J, Xu Z, Cen P, Yang S, Li S (2011) Enhanced butyric acid tolerance and bioproduction by Clostridium tyrobutyricum immobilized in a fibrous bed bioreactor. Biotechnol Bioeng 108(1):31–40
Jin B, Wang S, Xing L, Li B, Peng Y (2016) Long term effect of alkali types on waste activated sludge hydrolytic acidification and microbial community at low temperature. Bioresour Technol 200:587–597
Jong BC, Kim BH, Chang IS, Liew PWY, Choo YF, Kang GS (2006) Enrichment, performance, and microbial diversity of a thermophilic mediatorless microbial fuel cell. Environ Sci Technol 40(20):6449–6454
Jönsson LJ, Martín C (2016) Pretreatment of lignocellulose: formation of inhibitory by-products and strategies for minimizing their effects. Bioresour Technol 199:103–112
Karakashev D, Batstone DJ, Angelidaki I (2005) Influence of environmental conditions on methanogenic compositions in anaerobic biogas reactors. Appl Environ Microbiol 71(1):331–338
Karakashev D, Batstone DJ, Trably E, Angelidaki I (2006) Acetate oxidation is the dominant methanogenic pathway from acetate in the absence of methanosaetaceae. Appl Environ Microbiol 72(7):5138–5141
Khiewwijit R, Temmink H, Labanda A, Rijnaarts H, Keesman KJ (2015) Production of volatile fatty acids from sewage organic matter by combined bioflocculation and alkaline fermentation. Bioresour Technol 197:295–301
Khor WC, Andersen S, Vervaeren H, Rabaey K (2017) Electricity-assisted production of caproic acid from grass. Biotechnol Biofuels 10(1):180
Kim D-H, Han S-K, Kim S-H, Shin H-S (2006) Effect of gas sparging on continuous fermentative hydrogen production. Int J Hydrogen Energy 31(15):2158–2169
Kleerebezem R, van Loosdrecht MCM (2007) Mixed culture biotechnology for bioenergy production. Curr Opin Biotechnol 18(3):207–212
Kleerebezem R, Van Loosdrecht MCM (2010) A generalized method for thermodynamic state analysis of environmental systems. Crit Rev Environ Sci Technol 40(1):1–54
Kleerebezem R, Joosse B, Rozendal R, Loosdrecht MCM (2015) Anaerobic digestion without biogas? Rev Environ Sci Biotechnol 14(4):787–801
Kongjan P, Angelidaki I (2010) Extreme thermophilic biohydrogen production from wheat straw hydrolysate using mixed culture fermentation: effect of reactor configuration. Bioresour Technol 101(20):7789–7796
Kongjan P, Min B, Angelidaki I (2009) Biohydrogen production from xylose at extreme thermophilic temperatures (70 °C) by mixed culture fermentation. Water Res 43(5):1414–1424
Kothari R, Singh DP, Tyagi VV, Tyagi SK (2012) Fermentative hydrogen production - an alternative clean energy source. Renew Sustain Energy Rev 16(4):2337–2346
Kraemer J, Bagley D (2006) Supersaturation of dissolved H_2 and CO_2 during fermentative hydrogen production with N_2 sparging. Biotechnol Lett 28(18):1485–1491
Labatut RA, Angenent LT, Scott NR (2014) Conventional mesophilic vs. thermophilic anaerobic digestion: a trade-off between performance and stability? Water Res 53:249–258
Lee HS, Salerno MB, Rittmann BE (2008) Thermodynamic evaluation on H_2 production in glucose fermentation. Environ Sci Technol 42(7):2401–2407
Lee HS, Krajmalinik-Brown R, Zhang HS, Rittmann BE (2009) An electron-flow model can predict complex redox reactions in mixed-culture fermentative BioH(2): microbial ecology evidence. Biotechnol Bioeng 104(4):687–697

Leng L, Yang P, Mao Y, Wu Z, Zhang T, Lee P-H (2017) Thermodynamic and physiological study of caproate and 1,3-propanediol co-production through glycerol fermentation and fatty acids chain elongation. Water Res 114:200–209

Lengeler JW, Drews G, Schlegel HG (1999) Biology of the prokaryotes. Georg Thieme, Stuttgart

Lewandowicz G, Białas W, Marczewski B, Szymanowska D (2011) Application of membrane distillation for ethanol recovery during fuel ethanol production. J Membr Sci 375(1–2):212–219

Li W-W, Yu H-Q (2011) From wastewater to bioenergy and biochemicals via two-stage bioconversion processes: a future paradigm. Biotechnol Adv 29(6):972–982

Logan B, Regan J (2006) Microbial fuel cells-challenges and applications. Environ Sci Technol 40(17):5172–5180

Louis P, Duncan SH, McCrae SI, Millar J, Jackson MS, Flint HJ (2004) Restricted distribution of the butyrate kinase pathway among butyrate-producing bacteria from the human colon. J Bacteriol 186(7):2099–2106

Lovley DR (2008) The microbe electric: conversion of organic matter to electricity. Curr Opin Biotechnol 19(6):564–571

Lü F, Chai L, Shao L, He P (2017) Precise pretreatment of lignocellulose: relating substrate modification with subsequent hydrolysis and fermentation to products and by-products. Biotechnol Biofuels 10:88

Luo G, Xie L, Zhou Q, Angelidaki I (2011) Enhancement of bioenergy production from organic wastes by two-stage anaerobic hydrogen and methane production process. Bioresour Technol 102(18):8700–8706

Luo G, Johansson S, Boe K, Xie L, Zhou Q, Angelidaki I (2012) Simultaneous hydrogen utilization and in situ biogas upgrading in an anaerobic reactor. Biotechnol Bioeng 109(4):1088–1094

Madigan M, Martinko J, Bender K, Buckley D, Stahl D (2002) The Brock biology of microorganisms (Global ed). Pearson Education Limited, Harlow

Meynial-Salles I, Dorotyn S, Soucaille P (2008) A new process for the continuous production of succinic acid from glucose at high yield, titer, and productivity. Biotechnol Bioeng 99(1):129–135

Moon HG, Jang YS, Cho C, Lee J, Binkley R, Lee SY (2016) One hundred years of clostridial butanol fermentation. FEMS Microbiol Lett 363(3):fnw001

Moresi M, Sappino F (2000) Electrodialytic recovery of some fermentation products from model solutions: techno-economic feasibility study. J Membr Sci 164(1–2):129–140

Narihiro T, Nobu MK, Tamaki H, Kamagata Y, Sekiguchi Y, Liu W-T (2016) Comparative genomics of syntrophic branched-chain fatty acid degrading bacteria. Microb Environ 31(3):288–292

Nielsen HB, Mladenovska Z, Ahring BK (2007) Bioaugmentation of a two-stage thermophilic (68°C/55°C) anaerobic digestion concept for improvement of the methane yield from cattle manure. Biotechnol Bioeng 97(6):1638–1643

Pawar S, Niel EJ (2013) Thermophilic biohydrogen production: how far are we? Appl Microbiol Biotechnol 97(18):7999–8009

Pawar SS, Nkemka VN, Zeidan AA, Murto M, van Niel EWJ (2013) Biohydrogen production from wheat straw hydrolysate using Caldicellulosiruptor saccharolyticus followed by biogas production in a two-step uncoupled process. Int J Hydrogen Energy 38(22):9121–9130

Perry RH, Green DW (2008) Perry's chemical engineers' handbook. McGraw-Hill, New York

Peters JW, Miller AF, Jones AK, King PW, Adams MW (2016) Electron bifurcation. Curr Opin Chem Biol 31:146–152

Petersson A, Wellinger A (2009) Biogas upgrading technologies–developments and innovations, vol 37. IEA-Task, p 20

Poehlein A, Schmidt S, Kaster A-K, Goenrich M, Vollmers J, Thürmer A, Bertsch J, Schuchmann K, Voigt B, Hecker M, Daniel R, Thauer RK, Gottschalk G, Müller V (2012) An ancient pathway combining carbon dioxide fixation with the generation and utilization of a sodium ion gradient for ATP synthesis. PLoS One 7(3):e33439

Price ND, Reed JL, Palsson BO (2004) Genome-scale models of microbial cells: evaluating the consequences of constraints. Nat Rev Microbiol 2(11):886–897

Qiu C, Wen J, Jia X (2011) Extreme-thermophilic biohydrogen production from lignocellulosic bioethanol distillery wastewater with community analysis of hydrogen-producing microflora. Int J Hydrogen Energy 36(14):8243–8251
Redwood MD, Orozco RL, Majewski AJ, Macaskie LE (2012) An integrated biohydrogen refinery: synergy of photofermentation, extractive fermentation and hydrothermal hydrolysis of food wastes. Bioresour Technol 119:384–392
Ryckebosch E, Drouillon M, Vervaeren H (2011) Techniques for transformation of biogas to biomethane. Biomass Bioenergy 35(5):1633–1645
Sawers RG (2005) Formate and its role in hydrogen production in Escherichia coli. Biochem Soc Trans 33(Pt 1):42–46
Schuchmann K, Muller V (2014) Autotrophy at the thermodynamic limit of life: a model for energy conservation in acetogenic bacteria. Nat Rev Microbiol 12(12):809–821
Schut GJ, Adams MWW (2009) The iron-hydrogenase of thermotoga maritima utilizes ferredoxin and NADH synergistically: a new perspective on anaerobic hydrogen production. J Bacteriol 191(13):4451–4457
Seeliger S, Janssen P, Schink B (2002) Energetics and kinetics of lactate fermentation to acetate and propionate via methylmalonyl-CoA or acrylyl-CoA. FEMS Microbiol Lett 211(1):65–70
Serna-Maza A, Heaven S, Banks CJ (2014) Ammonia removal in food waste anaerobic digestion using a side-stream stripping process. Bioresour Technol 152:307–315
Smith ET, Blamey JM, Zhou ZH, Adams MWW (1995) A variable-temperature direct electrochemical study of metalloproteins from hyperthermophilic microorganisms involved in hydrogen production from pyruvate. Biochemistry 34(21):7161–7169
Smith AL, Skerlos SJ, Raskin L (2015) Anaerobic membrane bioreactor treatment of domestic wastewater at psychrophilic temperatures ranging from 15 °C to 3 °C. Environ Sci Water Res Technol 1(1):56–64
Soboh B, Linder D, Hedderich R (2004) A multisubunit membrane-bound [NiFe] hydrogenase and an NADH-dependent Fe-only hydrogenase in the fermenting bacterium Thermoanaerobacter tengcongensis. Microbiology 150(7):2451–2463
Sousa DZ, Smidt H, Alves MM, Stams AJM (2009) Ecophysiology of syntrophic communities that degrade saturated and unsaturated long-chain fatty acids. FEMS Microbiol Ecol 68(3):257–272
Sowunmi A, Mamone RM, Bastidas-Oyanedel J-R, Schmidt JE (2016) Biogas potential for electricity generation in the Emirate of Abu Dhabi. Biomass Convers Biorefin 6(1):39–47
Speight JG (2005) Lange's handbook of chemistry. McGraw-Hill, New York
Steinbusch KJJ, Hamelers HVM, Plugge CM, Buisman CJN (2011) Biological formation of caproate and caprylate from acetate: fuel and chemical production from low grade biomass. Energ Environ Sci 4(1):216–224
Stephen AJ, Archer SA, Orozco RL, Macaskie LE (2017) Advances and bottlenecks in microbial hydrogen production. J Microbial Biotechnol 10(5):1120–1127
Straub CT, Zeldes BM, Schut GJ, Adams MWW, Kelly RM (2017) Extremely thermophilic energy metabolisms: biotechnological prospects. Curr Opin Biotechnol 45:104–112
Svetlitchnyi VA, Kensch O, Falkenhan DA, Korseska SG, Lippert N, Prinz M, Sassi J, Schickor A, Curvers S (2013) Single-step ethanol production from lignocellulose using novel extremely thermophilic bacteria. Biotechnol Biofuels 6:31
Tapia-Venegas E, Ramirez-Morales J, Silva-Illanes F, Toledo-Alarcón J, Paillet F, Escudie R, Lay C-H, Chu C-Y, Leu H-J, Marone A, Lin C-Y, Kim D-H, Trably E, Ruiz-Filippi G (2015) Biohydrogen production by dark fermentation: scaling-up and technologies integration for a sustainable system. Rev Environ Sci Biotechnol 14(4):761–785
Temudo MF, Kleerebezem R, van Loosdrecht M (2007) Influence of the pH on (open) mixed culture fermentation of glucose: a chemostat study. Biotechnol Bioeng 98(1):69–79
Temudo MF, Mato T, Kleerebezem R, van Loosdrecht MCM (2009) Xylose anaerobic conversion by open-mixed cultures. Appl Microbiol Biotechnol 82(2):231–239
Thauer RK, Jungermann K, Decker K (1977) Energy conservation in chemotrophic anaerobic bacteria. Bacteriol Rev 41(1):100–180

Thauer RK, Kaster AK, Seedorf H, Buckel W, Hedderich R (2008) Methanogenic archaea: ecologically relevant differences in energy conservation. Nat Rev Microbiol 6(8):579–591
Tuck CO, Pérez E, Horváth IT, Sheldon RA, Poliakoff M (2012) Valorization of biomass: deriving more value from waste. Science 337(6095):695–699
Turina P, Petersen J, Gräber P (2016) Thermodynamics of proton transport coupled ATP synthesis. Biochim Biophys Acta Bioenerg 1857(6):653–664
Ueno Y, Sasaki D, Fukui H, Haruta S, Ishii M, Igarashi Y (2006) Changes in bacterial community during fermentative hydrogen and acid production from organic waste by thermophilic anaerobic microflora. J Appl Microbiol 101(2):331–343
Ueno Y, Fukui H, Goto M (2007) Operation of a two-stage fermentation process producing hydrogen and methane from organic waste. Environ Sci Technol 41(4):1413–1419
van der Ha D, Nachtergaele L, Kerckhof F-M, Rameiyanti D, Bossier P, Verstraete W, Boon N (2012) Conversion of biogas to bioproducts by algae and methane oxidizing bacteria. Environ Sci Technol 46(24):13425–13431
Wang X, Wang Y, Zhang X, Xu T (2012) In situ combination of fermentation and electrodialysis with bipolar membranes for the production of lactic acid: operational compatibility and uniformity. Bioresour Technol 125:165–171
Wang X, Wang Y, Zhang X, Feng H, Xu T (2013) In-situ combination of fermentation and electrodialysis with bipolar membranes for the production of lactic acid: continuous operation. Bioresour Technol 147:442–448
Wang H, Zhang Y, Angelidaki I (2016) Ammonia inhibition on hydrogen enriched anaerobic digestion of manure under mesophilic and thermophilic conditions. Water Res 105:314–319
Wang D, Liu Y, Ngo HH, Zhang C, Yang Q, Peng L, He D, Zeng G, Li X, Ni B-J (2017) Approach of describing dynamic production of volatile fatty acids from sludge alkaline fermentation. Bioresour Technol 238:343–351
Wijekoon KC, Visvanathan C, Abeynayaka A (2011) Effect of organic loading rate on VFA production, organic matter removal and microbial activity of a two-stage thermophilic anaerobic membrane bioreactor. Bioresour Technol 102(9):5353–5360
Willquist K, Zeidan A, van Niel E (2010) Physiological characteristics of the extreme thermophile Caldicellulosiruptor saccharolyticus: an efficient hydrogen cell factory. Microb Cell Fact 9(1):89
Wu Y, Wang C, Liu X, Ma H, Wu J, Zuo J, Wang K (2016) A new method of two-phase anaerobic digestion for fruit and vegetable waste treatment. Bioresour Technol 211:16–23
Xu J, Hao J, Guzman JJL, Spirito CM, Harroff LA, Angenent LT (2018) Temperature-phased conversion of acid whey waste into medium-chain carboxylic acids via lactic acid: no external e-donor. Joule 2(2):280–295
Xue C, Liu F, Xu M, Zhao J, Chen L, Ren J, Bai F, Yang S-T (2016) A novel in situ gas stripping-pervaporation process integrated with acetone-butanol-ethanol fermentation for hyper n-butanol production. Biotechnol Bioeng 113(1):120–129
Yentekakis IV, Goula G (2017) Biogas management: advanced utilization for production of renewable energy and added-value chemicals. Front Environ Sci 5:7
Young J, Chung D, Bomble Y, Himmel M, Westpheling J (2014) Deletion of Caldicellulosiruptor bescii CelA reveals its crucial role in the deconstruction of lignocellulosic biomass. Biotechnol Biofuels 7(1):142
Yun Y-M, Kim D-H, Cho S-K, Shin H-S, Jung K-W, Kim H-W (2016) Mitigation of ammonia inhibition by internal dilution in high-rate anaerobic digestion of food waste leachate and evidences of microbial community response. Biotechnol Bioeng 113(9):1892–1901
Zamanzadeh M, Hagen LH, Svensson K, Linjordet R, Horn SJ (2016) Anaerobic digestion of food waste – effect of recirculation and temperature on performance and microbiology. Water Res 96:246–254
Zhang P, Chen Y, Zhou Q (2009) Waste activated sludge hydrolysis and short-chain fatty acids accumulation under mesophilic and thermophilic conditions: effect of pH. Water Res 43(15):3735–3742

Zhang P, Chen Y, Zhou Q, Zheng X, Zhu X, Zhao Y (2010) understanding short-chain fatty acids accumulation enhanced in waste activated sludge alkaline fermentation: kinetics and microbiology. Environ Sci Technol 44(24):9343–9348

Zhang Y, Pinoy L, Meesschaert B, Van der Bruggen B (2011) Separation of small organic ions from salts by ion-exchange membrane in electrodialysis. AIChE J 57(8):2070–2078

Zhang F, Zhang Y, Chen M, Zeng RJ (2012) Hydrogen supersaturation in thermophilic mixed culture fermentation. Int J Hydrogen Energy 37(23):17809–17816

Zhang F, Ding J, Shen N, Zhang Y, Ding Z-W, Dai K, Zeng RJ (2013a) In situ hydrogen utilization for high fraction acetate production in mixed culture hollow-fiber membrane biofilm reactor. Appl Microbiol Biotechnol 97(23):10233–10240

Zhang F, Ding J, Zhang Y, Chen M, Ding Z-W, van Loosdrecht MCM, Zeng RJ (2013b) Fatty acids production from hydrogen and carbon dioxide by mixed culture in the membrane biofilm reactor. Water Res 47(16):6122–6129

Zhang F, Zhang Y, Chen M, van Loosdrecht MCM, Zeng RJ (2013c) A modified metabolic model for mixed culture fermentation with energy conserving electron bifurcation reaction and metabolite transport energy. Biotechnol Bioeng 110(7):1884–1894

Zhang Y, Zhang F, Chen M, Chu P-N, Ding J, Zeng RJ (2013d) Hydrogen supersaturation in extreme-thermophilic (70°C) mixed culture fermentation. Appl Energy 109:213–219

Zhang F, Chen Y, Dai K, Zeng R (2014a) The chemostat study of metabolic distribution in extreme-thermophilic (70 °C) mixed culture fermentation. Appl Microbiol Biotechnol 98(24):10267–10273

Zhang F, Zhang Y, Ding J, Dai K, van Loosdrecht MCM, Zeng RJ (2014b) Stable acetate production in extreme-thermophilic (70°C) mixed culture fermentation by selective enrichment of hydrogenotrophic methanogens. Sci Rep 4:5268

Zhang F, Chen Y, Dai K, Shen N, Zeng RJ (2015a) The glucose metabolic distribution in thermophilic (55 °C) mixed culture fermentation: a chemostat study. Int J Hydrogen Energy 40(2):919–926

Zhang F, Zhang Y, Chen Y, Dai K, van Loosdrecht MCM, Zeng RJ (2015b) Simultaneous production of acetate and methane from glycerol by selective enrichment of hydrogenotrophic methanogens in extreme-thermophilic (70 °C) mixed culture fermentation. Appl Energy 148:326–333

Zhang F, Yang J-H, Dai K, Ding Z-W, Wang L-G, Li Q-R, Gao F-M, Zeng RJ (2016) Microbial dynamics of the extreme-thermophilic (70 °C) mixed culture for hydrogen production in a chemostat. Int J Hydrogen Energy 41(26):11072–11080

Zheng H, O'Sullivan C, Mereddy R, Zeng RJ, Duke M, Clarke WP (2010) Experimental and theoretical investigation of diffusion processes in a membrane anaerobic reactor for bio-hydrogen production. Int J Hydrogen Energy 35(11):5301–5311

Exploring the Selective Lactic Acid Production from Food Waste in Uncontrolled pH Mixed Culture Fermentations Using Different Reactor Configurations

Fabian Bonk, Juan-Rodrigo Bastidas-Oyanedel, Ahmed F. Yousef, and Jens Ejbye Schmidt

1 Introduction

The economic treatment of food waste is a major challenge for the development of sustainable waste management systems. Traditional waste management treatment options include landfilling, composting, or anaerobic digestion for biogas production. In recent years, alternative technologies have been developed to produce value-added products from waste, for example, via the carboxylate platform (Bastidas-Oyanedel et al. 2015). In the carboxylate platform, carboxylic acids are produced via an undefined mixed culture of bacteria. To produce volatile fatty acids (VFAs), e.g., acetic, propionic, or butyric acid (all pK_a ~ 4.8), a pH of around 5 is considered the lower boundary because below that pH, solvent production overtakes acid production due to inhibition (Bastidas-Oyanedel et al. 2015; Aceves-Lara et al.

F. Bonk
Chemical Engineering Department, Khalifa University of Science and Technology, Masdar Institute, Abu Dhabi, United Arab Emirates

Department of Environmental Microbiology, UFZ-Helmholtz Centre for Environmental Research, Leipzig, Germany

J.-R. Bastidas-Oyanedel (✉)
Chemical Engineering Department, Khalifa University of Science and Technology, Masdar Institute, Abu Dhabi, United Arab Emirates
e-mail: juan.bastidas@ku.ac.ae

A. F. Yousef
Chemistry Department, Khalifa University of Science and Technology, Masdar Institute, Abu Dhabi, United Arab Emirates

J. E. Schmidt
SDU-Department of Chemical Engineering, Biotechnology, and Environmental Technology, University of Southern Denmark, Campusvej 55, DK-5230 Odense M, Denmark
e-mail: jesc@kbm.sdu.dk

© Springer Nature Switzerland AG 2019
J.-R. Bastidas-Oyanedel, J. E. Schmidt (eds.), *Biorefinery*,
https://doi.org/10.1007/978-3-030-10961-5_18

2008). A possible explanation is that the protonated VFA form is the more inhibitory form for microorganisms (Penumathsa et al., 2008) and the lower the pH, the more VFAs are present in their protonated form.

Controlling pH can be achieved either by chemical means or co-fermentation with other wastes (Wu et al. 2016). The need to control pH in these systems imposes further economic and logistic constraints and is therefore a drawback for the production of platform chemicals from food waste. Ideally, carboxylic acids are produced using the intrinsic buffer capacity of food waste without pH control or co-fermentation, while retaining a high final product concentration. Another obstacle for the production of carboxylic acids from food waste is that mixtures of acids are often produced, e.g., acetic and butyric acid that have to be separated before utilization (Zhang et al. 2014). Separation and recovery of carboxylic acids from fermentation broths can be achieved by adsorption, either as direct addition of adsorbent in the fermenter or in a separated step (Yousuf et al. 2016). There are other techniques, different to adsorption, that include solvent extraction, membrane-based solvent extraction, electro dialysis, and membrane separation (Bastidas-Oyanedel et al. 2016, Lopez-Garzon and Straathof 2014).

The production of a single acid at a high concentration is highly desirable considering downstream processing (Zhang et al. 2014). Lactic acid (LA) is a promising carboxylic acid that can potentially fulfill all the requirements mentioned above. Its pK_a of 3.9 is lower than the pK_a of VFAs, and therefore, fewer acids are in their protonated form at a low pH. This reduces product inhibition by protonated acids at low pH and allows for higher product concentrations at a low pH compared to VFAs. In addition, previous studies suggest that a high selectivity can be achieved for LA over other products. Kim et al. (2012), for example, achieved a selectivity of LA over total fermentation products from the fermentation of food waste of over 90% using food waste pretreatment at 60 °C. LA can serve as precursor for the production of many useful chemicals and products, such as biodegradable polymers, pyruvic acid, acrylic acid, 1,2-propanediol, and lactate ester (Bastidas-Oyanedel et al. 2015; Gao et al. 2011). Today, it is already produced using bacterial fermentation (López-Garzón and Straathof 2014).

As mentioned before, microbial communities of mixed culture fermentation can be affected/modified by process parameters, e.g., pH, product inhibition. Recent advancements and increased access to novel sequencing technologies have resulted in the generation of a large amount of data regarding microbial populations and their dynamics in environmental and industrial settings (Ju and Zhang 2015). These data could help applied microbiologists and engineers in optimizing bioreactor technologies with an emphasis on process parameters and/or inoculum selection and maintenance. In order to do this, it is necessary to first establish correlations between efficient well-operated reactors and the microbial communities driving them.

The aim of this study was to explore different reactor configurations for LA production by pH-uncontrolled fermentation of food waste. In addition, we were interested in determining the microbial communities that develop under different reactor configurations. Model food waste was fermented mesophilically in three reactor configurations: (a) semi-continuous feeding, (b) batch, and (c) percolation systems. Percolation systems are attractive in the context of waste management because they are easy to handle and allow for subsequent composting (Bonk et al. 2015). In these systems, the hydraulic retention time (HRT) can be separated from food waste retention time and possibly also the retention time of microorganisms if they are able to attach to the food waste. Similar systems already exist on industrial scale for biomethane production from waste, such as the Aikan® system (Bastidas-Oyanedel et al. 2015). Since the percolation system is a practical setting for industrial waste treatment, the economic feasibility of its application to LA production from food waste was also estimated in this study. A major obstacle of all the tested reactor configurations is the recovery of the product. Therefore, in situ product extraction was tested using activated carbon.

2 Materials and Methods

2.1 *Substrate and Inoculum*

In order to have reproducible conditions, a model food waste was used as substrate, consisting of 50% cooked rice (Thai Pathumthani Fragrant Rice) and 50% dry dog food (Purina® Dog Chow Complete and Balanced) based on dry mass (Yousuf et al. 2016). It was prepared fresh for each experiment and feeding. Based on manufacturer's information on the product composition, the model food waste contained 12% protein, 69% carbohydrates, and 5% fat based on dry mass (Yousuf et al. 2016). The theoretical maximum LA yield was estimated to 0.74–1 $gCODLA/gTS_{food\ waste\ fed}$, based on Castillo Martinez et al. (2013) and Grootscholten et al. (2013). Generally, no pretreatment was applied to the model food waste if not indicated.

The semi-continuous fed reactors were originally inoculated with anaerobic sludge from the Mafraq wastewater treatment plant in Abu Dhabi, and fed for 5 months with food waste from the Masdar Institute campus canteen, Abu Dhabi, under varying feeding and pH control strategies before the conditions described below were established. After those 5 months, inoculum for the batch experiments was taken from the semi-continuously fed reactors and kept at 37 °C for 4 days before starting the batch experiments.

2.2 *Reactors Configurations*

2.2.1 Semi-Continuously Fed Reactors

Three fermenters were run in parallel with a working volume of 13 L, 100 rpm stirring, an organic loading rate (OLR) of 5 gTS/L/day, a HRT of 15 days, and a semi-continuous feeding interval of model food waste of 10.5 days. The temperature was controlled with a heating bath at 35 °C. To prevent a possible organic overload, half of the corresponding dog food amount was added with 3–4 days delay. Under these conditions, the reactors were run for 8 months.

2.2.2 Batch Experiments

Batch experiments were conducted in duplicates in glass bottles with a total volume of 327 mL and a working volume of 115 mL. Table 1 shows the experimental conditions of most important batch experiments. As standard conditions, substrate concentrations of 52 gTS/L and 35 mL inoculum from the semi-continuously fed reactors were chosen. Tap water was used to bring the volume up to 115 mL. At the start of the experiment, the headspace was flushed with nitrogen, and the bottles were incubated without any shaking at 37 °C for 11 days.

In addition to these standard conditions, the influence of shaking of the batch bottles at 100 rpm was tested. Furthermore, the model food waste was grinded using pestle and mortar to test the influence of mechanical pretreatment. To test the possibility of nutrient limitation, a mineral solution was added (Angelidaki et al. 2009; Bastidas-Oyanedel et al. 2010). To test the difference between the uncontrolled pH fermentation and controlled pH fermentation, the pH of the inoculum was raised to 7, and sodium bicarbonate was added with final concentrations of 15 and 30 g/L. Substrate concentrations were varied between 10 and 156 gTS/L, and inoculum amounts between 0 and 35 mL. To investigate the impact of an increased retention time of the fermenting bacteria, a sequential batch experiment and a fixed bed batch experiment were run.

Table 1 Experimental conditions of most important batch experiments

Batch no.	Substrate conc. (gTS/L)	Shaking	Inoculum (mL)	pH control	Total initial TS (inoculum + food waste)	Other
First set						
1	0	Yes	35	No	10.27	
2	52.14	Yes	35	No	62.41	
4	52.14	No	35	No	62.41	$T = 50$ °C
5	52.14	Yes	35	No	62.41	Pretreatment (grinding) of food waste
Second set						
6	52.14	No	no	Initial pH = 3	52.15	5 g/L lactic acid added
7	52.14	No	35	pH = 7 with 15 g/L $NaHCO_3$	66.09	
Fourth set						
17	156.42	No	35	No	172.03	
20	156.42	No	35	No	172.03	46 g activated carbon (Dophin®)

2.2.3 Percolation System

The percolation system consisted of a plastic tube with a total volume of 2.7 L and a height of 0.72 m. The percolation tube was completely filled with model food waste mixed with wood chips (size ~5*2*2 cm) as structural material. The fermentation liquid entered at the top of the column, percolated through the mixture of food waste and wood chips and exited at an outlet at the bottom of the tube. To prevent a clogging of tubes by particular matter, this outlet was protected by a sieve with <1 mm diameter and a kitchen sponge. The percolate flowed by gravity into a bioreactor with a working volume of 2 L, a temperature of 35 °C, and stirring of 50 rpm. A peristaltic pump recirculated the fermentation liquid with a dilution rate of 2.76 day^{-1} to the top of the percolation tube. Each experiment was run for 1 week and used 218 gTS of model food waste. After starting the experiment, tap water was added as needed to the reactor to keep a working volume of 2 L. The first experiment was started without the addition of inoculum.

2.2.4 In Situ and Online Extraction with Activated Carbon

Batch experiments with activated carbon were conducted to measure if an increase in lactic acid yield is possible by in situ product removal. Forty-six grams of activated carbon pellets (DoPhin® FM902) were washed before use with warm tap water as recommended by the manufacturer. The activated carbon was put in a fine textile net (~30 mesh) to keep the activate carbon granules in contact with the fermentation broth while preventing granules from mixing with the food waste for simpler separation after the fermentation. The activated carbon was submerged into the fermentation broth at the start of the experiment. After the experiment, the activated carbon was placed without any pretreatment into 115 mL acetone to desorb the fermentation products. Acetone desorption allows recovering the carboxylic acids in an organic solvent, rather than in aqueous solution, offering multiple application in thermochemical/catalytic down streaming of the carboxylic acids (Zanella et al. 2014).

2.3 Analytical Methods

Chemical oxygen demand (COD) was measured using spectrophotometric tests (LCK 014, range 1–10 g/L, Hach Lange GmbH) and a Hach DR 2800 spectrophotometer. Reactor samples were prepared by centrifugation for 5 min at 14,000 rpm, filtration (0.45 μm), and dilution with DI water to fit in the cuvette test linear range.

The concentrations of organic acids and alcohols were measured by high-performance liquid chromatography (HPLC). Samples were prepared by centrifugation for 5 min at 14,000 rpm and filtration (0.45 μm). In addition, samples were diluted five times with 0.01 M H_2SO_4. In the case of the activated carbon experiments, the fermentation products containing acetone samples were diluted 20 times. Samples were run in an Agilent 1260 at 65 °C, using an Agilent Hi-Plex H column, 5 mM H_2SO_4 as mobile phase at 0.6 mL/min, a UV detector at 210 mm for organic acids, and a RID detector at 35 °C for glucose and ethanol. Ethanol could not be quantified for the activated carbon extraction because it eluted at a retention time close to the one of acetone. Error bars in all figures represent the standard error of mean (SEM).

2.4 *Microbial Community Analysis*

2.4.1 DNA Isolation

To extract DNA from the fermentation broth samples, a method for DNA extraction from high-strength wastewater sludge was adopted (Tabatabaei et al. 2010).

2.4.2 16S rRNA Gene Amplicon Sequencing

The extracted DNA samples were sent to Macrogen Inc. (Seoul, South Korea) where 16S rRNA gene libraries were prepared using the universal primers 337F (GAC TCC TAC GGG AGG CWG CAG) and 805R (GAC TAC CAG GGT ATC TAA TC) which amplify the third and fourth variable regions. The barcoded libraries were combined together and sequenced using Illumina MiSeq (Control Software v2.2). The average throughput per sample was approximately 52.5 million bases and 245,000 reads in total (forward and reverse reads combined). The raw sequences were uploaded to the NCBI sequence read archive (SRA), and the accession numbers for the percolation reactor samples are SRX2672533, SRX2672534. The accession numbers for the semi-continuous fed reactors are SRX2672535, SRX2672536, and SRX2672537.

2.4.3 Bioinformatics Pipeline and Analysis

FastQ files were analyzed using the Quantitative Insights Into Microbial Ecology (QIIME V1.9.1) software tool (Kuczynski et al. 2012). The sequences were aligned with the August 2013 version of the Greengenes Database (The Greengenes Database Consortium). Note that the relative abundancies shown were not corrected for differences in 16S rRNA gene copy numbers.

2.5 Techno-Economic Analysis

The techno-economic analysis was based on Bonk et al. (2015), a percolation system followed by a composting stage. The data used for the techno-economic analysis consisted on the product concentrations of the first percolation experiment in this study. All other process parameters and economic assumptions were left identical to the non-pretreatment scenario used by Bonk et al. (2015). No gate fee was applied, i.e., food waste does neither bring cost nor income to the LA producing waste treatment plant. To estimate the separation cost of LA from water, an energy cost of 0.0068 USD/kg_{water} (López-Garzón and Straathof 2014) was used. The income from the by-products (ethanol and carboxylic acid) was neglected.

3 Results and Discussion

3.1 First Process Insights Based on Semi-Continuously Fed Reactor

Figure 1 shows several sampling points of the uncontrolled pH fermentation study, with a semi-continuous fed reactor. LA concentrations reached a maximum of 20.1 $gCOD_{LA}$/L (18.6 g_{LA}/L) after 10.5 days (before the next feeding).

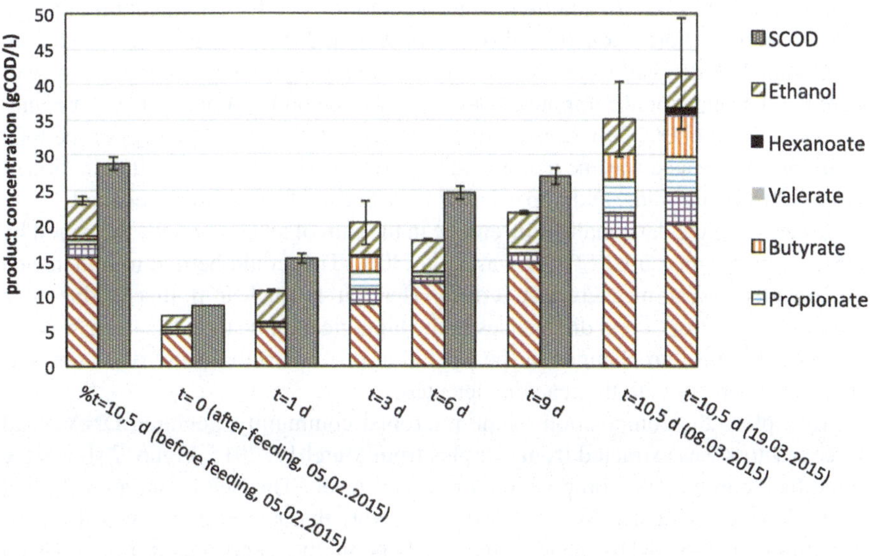

Fig. 1 Results pH-uncontrolled lactic acid production from model food waste: semi-continuous feeding (influent food waste concentration 75 gTS/L; error bars: 1 SEM, representing sum of products

This concentration is much lower than for pH-controlled fermentation of food waste that lead to concentrations of 38 $gCOD_{LA}/L$ (Zhang et al. 2008) or pure culture pH-controlled fermentations leading to 227 $gCOD_{LA}/L$ (Castillo Martinez et al. 2013). A high product concentration is important to decrease the cost of removing water by, for example, evaporation during the product during the product recovery process (López-Garzón and Straathof 2014). The reason for the lower concentration does not lie in a substrate limitation (see Sect. 3.2). Rather, it is probably caused by product inhibition. In comparison to these pH-controlled experiments (minimum pH 5), the pH in our reactors was 2.96. At this low pH, more molecules of LA are protonated and therefore lead to more product inhibition, assuming that protonated carboxylic acids are the more damaging form (Rodríguez et al. 2006). On the other hand, the low pH might have the advantage that a highly stereoisomer-selective L-LA production could occur as previously discovered by Zhang et al. (2008). However, we could not discriminate L and D-lactic isomers with our analytical method. Therefore, further investigations are required in this regard. In conclusion, lactic acid concentration in our pH-uncontrolled fermentation were lower than in pH-controlled experiments and might never reach such high concentrations due to a higher product inhibition at a low pH.

Another important aspect is the selectivity of lactic acid production, i.e., the amount of side products in the fermentation that increase the cost of product separation. In relation to all measured compounds, LA made up 49–67% ($gCOD_{LA}/gCOD_{total\ products}$). Ethanol was found as the major by-product (<4.9 gCOD/L). The contamination by ethanol might impose only minor practical challenges on the recovery of pure LA because its boiling point (78 °C) is quite different to the LA boiling point (122 °C). In addition, ethanol could be used for esterification of LA to yield valuable lactate esters (Castillo Martinez et al. 2013). On the other hand, contamination by other carboxylic acids can result in more problems for the recovery of pure LA, because some compounds have a very close boiling point to LA, for example, acetic acid (118 °C). LA made up 63–88% ($gCOD_{LA}/gCOD_{total\ acids}$) of the total acids, and acetic acid was the major acid by-product. Therefore the contamination of LA by carboxylic acids, and in particular by acetic acid, needs to be reduced.

Concerning the data quality, difference in the sum of all compounds measured by HPLC and the measured SCOD was up to 30%. This could be due to undetected compounds and/or measurement errors. Here, it is unclear if this is due to a measurement error or real differences in the biological replicates. The latter was not expected because no changes in the operation of the reactors were done nor was there any problem with the reactors detected.

To explore the composition of the microbial community, genomic DNA of all three reactors was extracted from samples from March19, 2015. Figure 2 shows the microbial community composition on order level. The semi-continuously fed reactors were dominated by *Lactobacilalles.* Note that no correction was done for the different 16S rRNA gene copy numbers in the genomes of the involved microorganisms. That being said, *Lactobacilalles* represented approximately 53% of aligned sequences in the semi-continuously fed reactors. In this order, species of the genus *Lactobacillus* of the family *Lactobacillaceae* were the most abundant

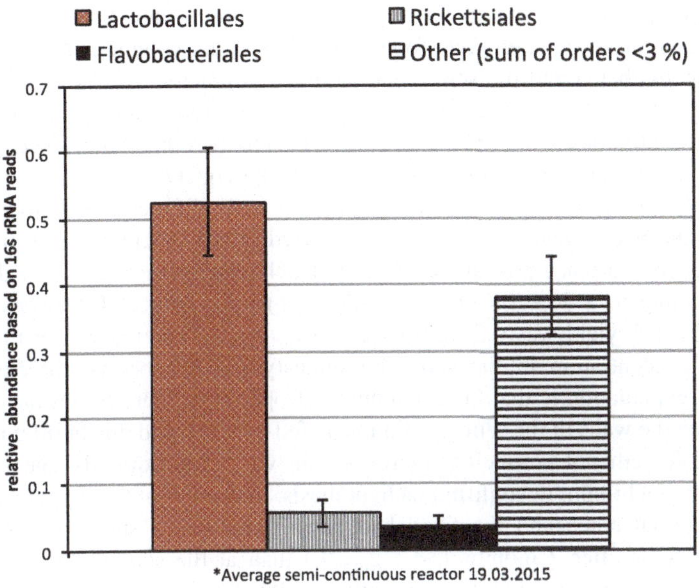

Fig. 2 Bacterial community structure on order level based on 16S rRNA gene amplicon sequencing. All other orders have an abundance of <4%. Biological triplicate, error bars: 1 SEM

species. The dominance of *Lactobacilalles* in the present study is in accordance with the recent discovery of the dominance of *Lactobacillus* in pH-uncontrolled mixed culture fermentation of agricultural peel wastes (Liang et al. 2016). Fermentation of fruit and vegetable waste was also recently found to result in dominance of *Lactobacillus* when controlled at pH = 4 (Wu et al. 2015). The raw FastQ sequencing files were also uploaded to Genbank.

In conclusion, selective lactic acid production was achieved at pH-uncontrolled conditions under semi-continuous feeding. The dominant order was *Lactobacillus*. Yet, higher concentrations, yields, and selectivities for lactic acid would be preferable. Therefore, batch experiments were conducted to evaluate how to improve this performance.

3.2 *Process Understanding and Optimization Based on Batch Experiments*

Batch experiments were conducted to understand better the factors influencing lactic acid production under pH-uncontrolled conditions and finally find conditions to improve concentrations, yields, and selectivities for lactic acid compared to the semi-continuously fed reactor. Only the most important batch results are presented below.

Figure 3 compares the product concentration of the batch experiment No. 3 (no shaking) after 11 days with the concentration of the inoculum (taken from the semi-continuously fed reactor 10.5 days after feeding plus 4 days in incubator). The relative amount of inoculum and food waste to working volume per feeding is the same for this batch and the semi-continuously fed reactor. The batch experiment resulted with 13.95 $gCOD_{LA}/L$ in lower LA concentrations than the inoculum from the semi-continuous reactor (16.53 $gCOD_{LA}/L$). On the other hand, selectivity was higher for the batch experiment with 0.93 $gCOD_{LA}/gCOD_{total\ products}$ compared to 0.8 $gCOD_{LA}/gCOD_{total\ products}$ for the inoculum. A main difference lies in the higher ethanol and acetic acid concentration measured in the inoculum from the semi-continuously fed reactor. Acetic acid in the batch experiment was below detection limit, leading to a 96% ($gCOD_{LA}/gCOD_{total\ acids}$) selectivity of LA over all other acids. The LA yield was 0.28 $gCOD_{LA}/gTS_{food\ waste\ fed.}$ The reason for the higher ethanol concentration in the semi-continuously fed reactors is not clear. One possible explanation is that after 7 months of operation, there might have been a biofilm on the walls of the semi-continuously fed reactors, and this biofilm might be enriched for ethanol producing microbes. But we did not directly check for the existence of a biofilm to confirm this hypothesis. In support of this hypothesis, both the sequential batch and the fixed bed batch experiments show higher ethanol concentrations after 2 months (>4.4 g_{COD}/L) than at the start of the experiment (~0.3 g_{COD}/L). A potential biofilm formation leading to unwanted side products is an important issue and therefore should be followed up in future research. In conclusion, using the same food waste concentration, batch experiment No. 3 resulted in general in similar fermentation products as the semi-continuously fed reactor, and in detail in higher LA selectivity and lower LA concentrations.

The batch experiment at 50 °C (batch No. 4) resulted in a lower LA concentration than the mesophilic control (batch No. 1). Shaking (batch No. 2) and shaking plus grinding (batch No. 5) increased the LA concentration by 27% and 20%,

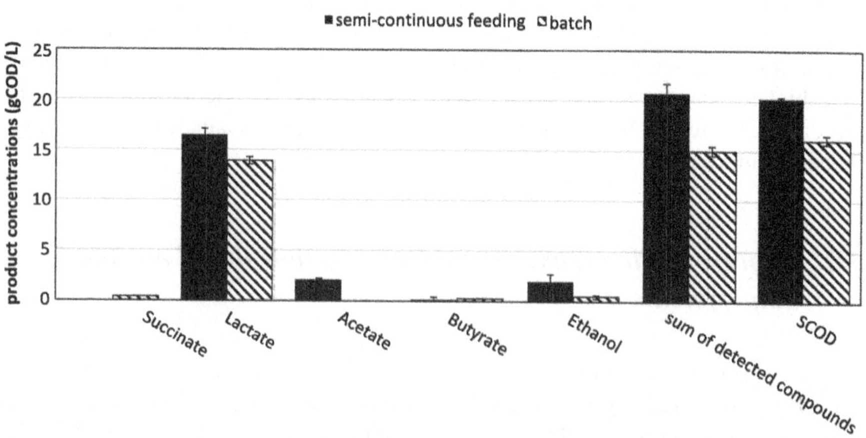

Fig. 3 Comparison of semi-continuously fed reactors with batch No. 3. Semi-continuous reactor sampled 10.5 days after feeding plus 4 days in incubator. Batch sampled at day 11. Error bars: 1 SEM (technical duplicate semi-continuously fed reactor, biological duplicate batch; error bar of first bar for each date is the SEM of the sum of products). Succinate, Formate, Valerate, Hexanoate below detection limit

respectively. The batch experiment with additional mineral medium (batch No. 15) did not show any improvements ruling out nutrient limitation. This is further supported by the pH-controlled batch experiment (batch No. 7) that resulted in much higher product yields (up to 0.79 $gCOD_{total\ products}/gTS_{food\ waste\ fed}$) than the pH-uncontrolled experiments. Therefore, product inhibitions by protonated acids and ethanol as well as the low pH seem to be the limiting factors to higher product concentrations, rather than substrate or nutrient limitations. Increasing the initial food waste concentration to 158 gTS/L (batch No. 17) leads to LA concentration of 31.8 gCOD/L but lowered the yield (to 0.2 $gCOD_{LA}/gTS_{food\ waste\ fed}$). The selectivity over the total fermentation products remained high (0.93 $gCOD_{LA}/gCOD_{total\ products}$, 0.95 $gCOD_{LA}/gCOD_{total\ acids}$). The higher LA concentration might be the result of the higher buffer capacity resulting from the high initial food waste concentration. The final pH of 3.3 was higher the pH in most other batch experiments and the semi-continuously fed reactors. The higher pH lowers product inhibition by protonated acids. Fixed bed and sequential batch experiments (batch No. 11 and No. 18) lead to increased ethanol concentrations but also to higher selectivities of LA over other organic acids. Evaluating the impact of inoculum is challenging because the inoculum itself contains LA and other products and has an impact on the initial pH. Therefore a batch with inoculum (batch No. 10, initial LA concentration 6.1 $gCOD_{LA}/L$, initial pH 3.25) was compared with a batch without inoculum but with LA addition (batch No. 6, initial LA concentration 5.4 $gCOD_{LA}/L$, initial pH 3.23). Both resulted in the same selectivity for LA (93% $gCOD_{LA}/gCOD_{total\ products}$) and a similar LA concentration (15.1 $gCOD_{LA}/L$ for batch 10 and 16.2 $gCOD_{LA}/L$ for batch 6). Therefore, it seems like the microorganisms present in the inoculum did not play a crucial role. The microorganisms present in the model food waste seemed to be sufficient for the fermentation. Nevertheless, the use of inoculum lowers the initial pH resulting in higher LA selectivities.

From a techno-economic point of view, the most interesting result is that a simple batch fermentation system with a high food waste load without shaking resulted in high LA concentrations and selectivities. The high LA concentration combined with a high selectivity over acetic acid is important to reduce product separation costs. The low yield is disadvantageous, especially if food waste becomes a resource in the future that needs to be paid for. In that case, a higher yield strategy will become economically critical.

3.3 *Percolation System as Practicable Large Scale Reactor Configuration*

3.3.1 Experimental Results

Figure 4 shows the fermentation products of the percolation system after 3 and 7 days. After 7 days of percolation, the final LA concentration reached 16.4 gCOD/L with a yield of 0.15 $gCOD_{LA}/gTS_{fed}$. The calculated product COD concentration differed less than 1% from the measured SCOD at day 7. The LA selectivity of 78%

($gCOD_{LA}/gCOD_{total\ acids}$) was much lower compared to the batch experiments. We hypothesize that this is due to the higher buffer capacity and the higher final pH of 3.74, both resulting from the higher food waste amount per fermentation broth. A higher buffer capacity leads to a slower pH drop and therefore results in a longer time period favorable for VFA production. Possible improvements to increase the selectivity are to change the food waste—fermentation liquid ratio, leave some of the percolation liquid for the next batch, or to have a continuous liquid exchange disconnected to the percolation tube. Nevertheless, even with the simple percolation set-up used in this study, lactic acid can be produced as major fermentation product in concentrations similar to batch and semi-continuous experiments.

3.3.2 Techno-Economic Assessment

Section 3.3.1 has shown that LA can be produced in a pH-uncontrolled percolation system but that further process optimization is still required. Nevertheless, this is a practicable system for industrial scale implementation, and therefore a techno-economic evaluation is already interesting at this stage. Excluding the cost of lactic acid separation and purification, the minimum selling price of LA was estimated to 362 USD/t_{LA}. This is low compared to the market price of 1000–1800 USD/t_{LA} (Bastidas-Oyanedel et al. 2015), but the separation and purification of LA from the fermentation broth remains a challenge and uncertainty in the techno-economic assessment. For a first estimate, purification of LA can be neglected assuming that

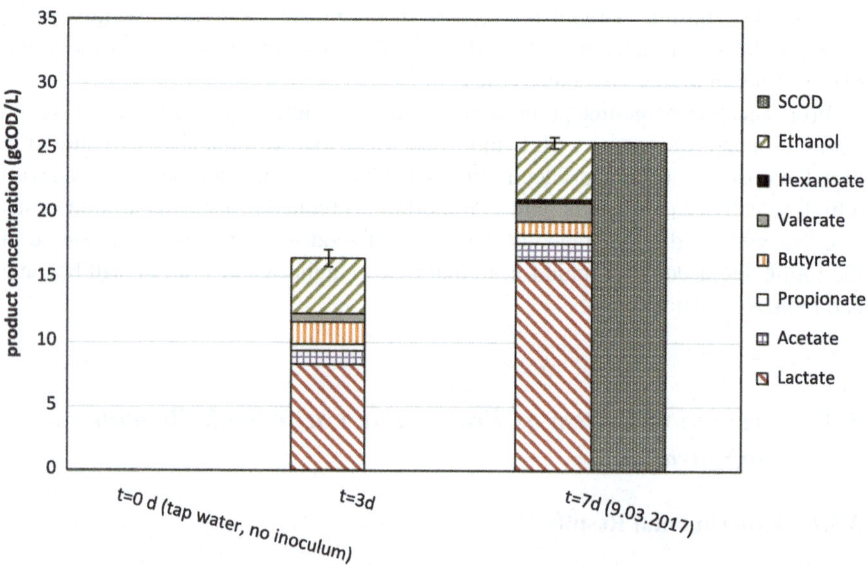

Fig. 4 Results of first percolation system experiment (no inoculum, 109 gTS food waste per liter working volume, technical duplicate; error bars: 1 SEM, error bar of first bar for each date is the SEM of the sum of products)

LA selectivities in the fermentation can be further optimized in the future. In that case, the separation cost of LA from the fermentation broth can be roughly estimated by the cost of removing water. Evaporation is a standard process to achieve this separation, but the costs are high for dilute aqueous solution (López-Garzón and Straathof 2014) such as the fermentation broth from the percolation reactors. Therefore, assuming evaporation as separation technology will result in a conservative separation cost. For the LA concentration of percolation experiment (16.4 $gCOD_{LA}/L$, 15.2 g_{LA}/L), the energy costs to evaporate the necessary 65 t_{water}/t_{LA} were estimated to 442 USD/t_{LA}. This results in a LA production cost of 804 USD/t_{LA}, excluding equipment cost for evaporation. This is still 200–1000 USD/t_{LA} below today's market price. Assuming that the performance of the batch experiments (batch No. 17) can also be achieved in the percolation system, the total LA production cost including separation from water would be 514 USD/t_{LA}.

Evaporation is rather an unpracticable separation technology in our case because of the high energy requirement of evaporation. Therefore, the techno-economic feasibility of other separation processes should be evaluated in the future, for example, chromatography (Thang and Novalin 2008), reverse osmosis (Diltz et al. 2007), electrodialysis (Wang et al. 2013), or adsorption as presented in the experimental results. In particular, technologies are preferable that allow for small-scale decentralized waste management plants. In this context, the idea of a fermentation process that requires no pH control is particularly attractive.

LA can be produced via two different metabolic routes, leading to the isomers L and/or D (Hoelzle et al. 2014). Using pure culture LA fermentation, a higher selectivity for L-LA was found for pH-uncontrolled fermentation (Abdel-Rahman et al. 2013). Unfortunately, the isomeric purity of the LA produced in this study could not be discriminated. But if this is the case, the economic feasibility would again increase for producing a more valuable pure isomer with a price of about 2400–3400 USD/*t* (Taian Health Chemical Co. Ltd. 2016).

This estimate on the techno-economic feasibility showed that the production of LA from food waste has a realistic potential to become a waste treatment option, but before it can be realized, further research and development in the field of separating LA from the fermentation broth in an energy efficient and sustainable way is required.

3.4 In Situ for Product Removal and Yield Increase

The techno-economic analysis showed that the recovery of LA from the fermentation broth remains a major challenge. In addition, the experiments resulted in LA yields as low as 0.15 $gCOD_{LA}/gTS_{food\ waste\ fed}$ for the percolation system and 0.2 $gCOD_{LA}/gTS_{food\ waste\ fed}$ for the batch experiments with the highest LA concentration (batch No. 17). A low yield also has a negative impact on the process economics. In situ extraction using activated carbon could be a solution for both problems. Instead of recovering LA after the fermentation, in situ extraction removes LA from the fermentation broth during the fermentation, thereby decreasing product inhibition and enabling higher product yields. Gao et al. (2011), for example, increased the LA yield in pure culture fermentations by 50% using in situ extraction with activated carbon. But their glucose-based fermentation is not comparable to a "dirty" food waste fermentation because the latter contains particulate matter that potentially clogs the pores of the activated carbon and reduces its performance.

Figure 5 shows the results of the batch fermentations with additional activated carbon after 7 days of fermentation. The LA yield could be increased by 35% ($gCOD/gTS_{food\ waste\ fed}$) by the use of activated carbon (batch 20) compared to the control without activated carbon (batch 17). Note that the adsorption experiment has no duplicates. The LA adsorpted to activated carbon (and desorpted with acetone) was 18.6 $mg_{LA}/g_{activated\ carbon}$. Yousuf et al. (2016) tested the adsorption of carboxylic acids to activated carbon in fermentation broths, containing LA with a concentration of 12.53 gCOD/L and found a very similar adsorption of 18.63 $mg_{LA}/g_{activated\ carbon}$. These numbers are not directly comparable to the present values, since Yousuf et al. (2016) used a different kind of activated carbon, had a different fermentation broth and removed particles from the fermentation broth by centrifugation. Nevertheless, our results suggest that in situ extraction with activated carbon can be achieved even in a "dirty" fermentation broth with particulate matter. This is not an obvious result,

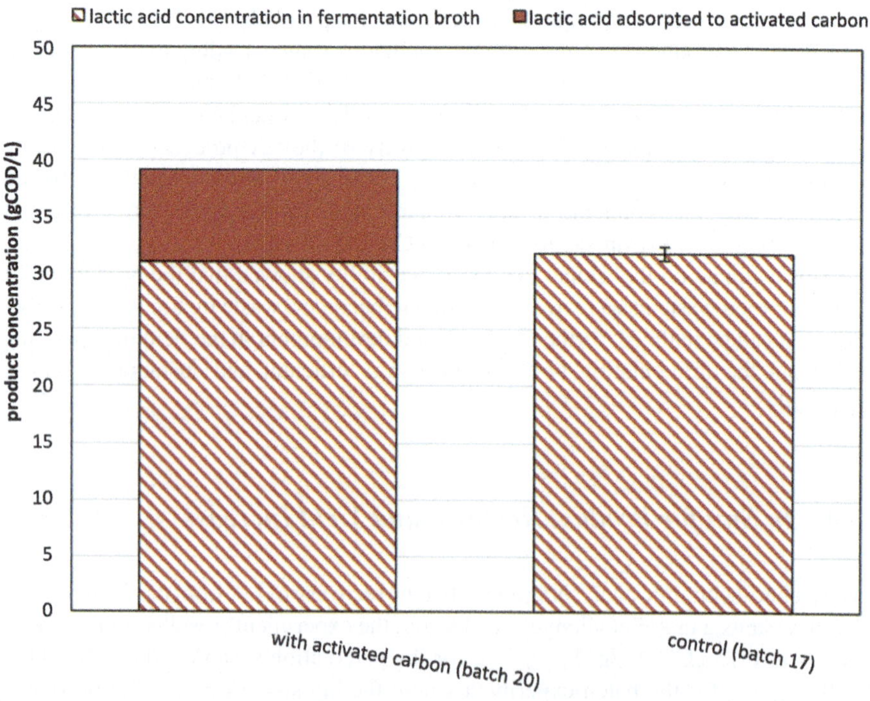

Fig. 5 Increase in lactic acid production by in situ extraction using activated carbon. Initial food waste concentration 158 gTS/L. Amount of lactic acid adsorpted to activated carbon normalized to working volume batch. Error bars: 1 SEM (Biological duplicate batch 17. No duplicate batch 20)

because it could also be that the activated carbon pores get blocked with food waste particles and microorganisms, limiting the mass transfer. It could be that such effects appear after recycling activated carbon several times. It remains uncertain if all lactic acid was desorpted from the activated carbon. In batch experiment No. 19, it was tested a second acetone extraction of LA from activated carbon, and additional 18% LA were desorpted, indicating that most LA was already desorpted in the first extraction.

In the future, the use of in situ (or online) extraction of LA from a pH-uncontrolled fermentation of food waste in a percolation system can be a promising waste management option. In such a scheme, pure, and potentially isomerically pure, LA could be gained via in situ extraction with positive effects on the product yield. The LA remaining in the fermentation broth can be flexibly used depending on the market situation, for example, as a substrate for other bioprocesses, as final product after separation from the fermentation broth, or used for energy production (e.g., by bioelectrochemical systems or via converting LA to biomethane in an anaerobic digester).

In conclusion, in situ extraction has the advantage to remove carboxylic acids already during the fermentation, thereby reducing product inhibition. For a practicable application, activated carbon and acetone need to be recycled several times. This needs to be studied further.

4 Conclusions

Simple and practicable process configurations for lactic acid production in pH-uncontrolled food waste fermentations were successfully tested, in particular batch fermentation without mixing and a percolation system. The percolation system appeared to be techno-economically feasible but product recovery remained a major obstacle. In situ extraction using activated carbon was successfully tested as a simple way of product recovery that additionally increases lactic acid yield. In a future food waste biorefinery scheme, some of the lactic acid could be recovered by in situ extraction, while the remaining lactic acid in the fermentation broth could be fed to another bioreactor, e.g., for bioenergy production.

Acknowledgments The author would like to acknowledge the financial support from Masdar Institute (Project 2BIONRG, 12KAMA4, and BIOREF, 13KAMA1), to help fulfill the vision of the late President Sheikh Zayed Bin Sultan Al Nahyan for sustainable development and empowerment of the United Arab Emirates and humankind.

References

Abdel-Rahman MA, Tashiro Y, Sonomoto K (2013) Recent advances in lactic acid production by microbial fermentation processes. Biotechnol Adv 31:877–902

Aceves-Lara CA, Trably E, Bastidas-Oyanedel JR, Ramirez I, Latrille E, Steyer JP (2008) Bioenergy production from waste: examples of biomethane and biohydrogen. J Soc Biol 202:177–189

Angelidaki I, Alves M, Bolzonella D, Borzacconi L, Campos JL, Guwy AJ, Kalyuzhnyi S, Jenicek P, van Lier JB (2009) Defining the biomethane potential (BMP) of solid organic wastes and energy crops: a proposed protocol for batch assays. Water Sci Technol 59:927–934

Bastidas-Oyanedel JR, Mohd-Zaki Z, Pratt S, Steyer JP, Batstones DJ (2010) Development of membrane inlet mass spectrometry for examination of fermentation processes. Talanta 83:482–492

Bastidas-Oyanedel JR, Bonk F, Thomsen MH, Schmidt JE (2015) Dark fermentation biorefinery in the present and future (bio)chemical industry. Rev Environ Sci Biotechnol 14:473–498

Bastidas-Oyanedel JR, Fang C, Almardeai S, Javid U, Yousuf A, Schmidt JE (2016) Waste biorefinery in arid/semi-arid regions. Bioresour Technol 215:21–28

Bonk F, Bastidas-Oyanedel J, Schmidt J (2015) Converting the organic fraction of solid waste from the city of Abu Dhabi to valuable products via dark fermentation – economic and energy assessment. Waste Manag 40:82–91

Castillo Martinez FA, Balciunas EM, Salgado JM, Domínguez González JM, Converti A, Oliveira RPDS (2013) Lactic acid properties, applications and production: a review. Trends Food Sci Technol 30:70–83

Diltz RA, Marolla TV, Henley MV, Li L (2007) Reverse osmosis processing of organic model compounds and fermentation broths. Bioresour Technol 98:686–695

Gao C, Ma C, Xu P (2011) Biotechnological routes based on lactic acid production from biomass. Biotechnol Adv 29:930–939

Grootscholten TIM, Kinsky dal Borgo F, Hamelers HVM, Buisman CJN (2013) Promoting chain elongation in mixed culture acidification reactors by addition of ethanol. Biomass Bioenergy 48:10–16

Hoelzle RD, Virdis B, Batstone DJ (2014) Regulation mechanisms in mixed and pure culture microbial fermentation. Biotechnol Bioeng 111:2139–2154

Ju F, Zhang T (2015) Experimental design and bioinformatics analysis for the application of metagenomics in environmental sciences and biotechnology. Environ Sci Technol 49(21):12628–12640

Kim DH, Lim WT, Lee MK, Kim MS (2012) Effect of temperature on continuous fermentative lactic acid (LA) production and bacterial community, and development of LA-producing UASB reactor. Bioresour Technol 119:335–361

Kuczynski J, Stombaugh J, Walters WA, Gonzalez A, Caporaso JG, Knight R (2012) Using QIIME to analyze 16 rRNA gene sequences from microbial communities. Curr Protoc Microbiol 27(1):1E.5.1-1E.5.20

Liang S, Gliniewicz K, Gerritsen AT, McDonald AG (2016) Analysis of microbial community variation during the mixed culture fermentation of agricultural peel wastes to produce lactic acid. Bioresour Technol 208:7–12

López-Garzón CS, Straathof AJJ (2014) Recovery of carboxylic acids produced by fermentation. Biotechnol Adv 32:873–904

Rodríguez J, Lema JM, van Loosdrecht MCM, Kleerebezem R (2006) Variable stoichiometry with thermodynamic control in ADM1. Water Sci Technol 54:101–110

Tabatabaei M, Zakaria MR, Rahim RA, Abdullah N, Wright ADG, Shirai Y, Shamsara M, Sakai K, Hassan MA (2010) Comparative study of methods for extraction and purification of environmental DNA from high-strength wastewater sludge. African Journal of Biotechnology 9(31):4926–4937

Taian Health Chemical Co. Ltd (2016) Factory price tech grade 50-21-5 l-lactic acid price [WWW Document]. http://www.alibaba.com/product-detail/food-grade-L-Lactic-acid-CAS_60464833300.html?spm=a2700.7724838.0.0.7BITBV. Accessed 19 June 2016

Thang VH, Novalin S (2008) Green biorefinery: separation of lactic acid from grass silage juice by chromatography using neutral polymeric resin. Bioresour Technol 99:4368–4379

Wang X, Wang Y, Zhang X, Feng H, Xu T (2013) In-situ combination of fermentation and electrodialysis with bipolar membranes for the production of lactic acid: continuous operation. Bioresour Technol 147:442–448

Wu Y, Ma H, Zheng M, Wang K (2015) Lactic acid production from acidogenic fermentation of fruit and vegetable wastes. Bioresour Technol 191:53–58

Wu Q-L, Guo W-Q, Zheng H-S, Luo H-C, Feng X-C, Yin R-L, Ren N-Q (2016) Enhancement of volatile fatty acid production by co-fermentation of food waste and excess sludge without pH control: the mechanism and microbial community analyses. Bioresour Technol 216:653–660

Yousuf A, Bonk F, Bastidas-Oyanedel J-R, Schmidt JE (2016) Recovery of carboxylic acids produced during dark fermentation of food waste by adsorption on Amberlite IRA-67 and activated carbon. Bioresour Technol 217:137–140

Zanella O, Tessaro IC, Feris LA (2014) Desorption- and decomposition-based techniques for the regeneration of activated carbon. Chem Eng Technol 37:1447–1459

Zhang B, He P-J, Ye N-F, Shao L-M (2008) Enhanced isomer purity of lactic acid from the non-sterile fermentation of kitchen wastes. Bioresour Technol 99:855–862

Zhang F, Zhang Y, Ding J, Dai K, van Loosdrecht MCM, Zeng RJ (2014) Stable acetate production in extreme-thermophilic (70°C) mixed culture fermentation by selective enrichment of hydrogenotrophic methanogens. Sci Rep 4:5268

Effect of Total Solid Content and Pretreatment on the Production of Lactic Acid from Mixed Culture Dark Fermentation of Food Waste

Ahasa Yousuf, Juan-Rodrigo Bastidas-Oyanedel, and Jens Ejbye Schmidt

1 Introduction

Excessive generation of the waste is currently a major issue worldwide due to urbanization and human population growth. The food waste composition of the world municipal solid waste is on average 40%, with a range from 23 to 67.5% (Hoornweg and Bhada-Tata 2012). According to United Nations Food and Agriculture Organization (FAO), 1.3 billion tonnes of food were wasted in 2011 (FAO 2011). If used in landfill, the high moisture and volatile solid content of food waste can cause environmental degradation by greenhouse gas emission, odor, and ground water contamination. This situation calls for efficient resource recovery from the food waste to both minimize the environmental issues associated and also create value from waste.

Dark fermentation (DF), a bioprocess, can convert the food waste into carboxylic acids, for example, lactic, propionic, butyric, acetic, and valeric acid and solvents in the liquid phase, and bio-hydrogen and carbon dioxide in the gas phase (Bastidas-Oyanedel et al. 2015). The acids produced, when separated from the broth, can be used as platform chemicals for the production of alcohols, bio-hydrogen, and bioplastics and can also be used as pure acids in various applications including food, pharmaceutical, and chemical industries (Agler et al. 2011; Bastidas-Oyanedel et al. 2015; Dahiya et al. 2015). Among the biological processes available, dark fermentation for carboxylic acid production shows the promise of practical viability

A. Yousuf · J.-R. Bastidas-Oyanedel (✉)
Chemical Engineering Department, Khalifa University of Science and Technology, Masdar Institute, Abu Dhabi, United Arab Emirates
e-mail: juan.bastidas@ku.ac.ae

J. E. Schmidt
SDU-Department of Chemical Engineering, Biotechnology, and Environmental Technology, University of Southern Denmark, Campusvej 55, DK-5230 Odense M, Denmark
e-mail: jesc@kbm.sdu.dk

© Springer Nature Switzerland AG 2019

J.-R. Bastidas-Oyanedel, J. E. Schmidt (eds.), *Biorefinery*,
https://doi.org/10.1007/978-3-030-10961-5_19

due to its high market value in comparison to methane, animal feed, and other products obtained from different bioprocesses (Bastidas-Oyanedel et al. 2016; Vanwonterghem et al. 2015). Moreover, DF is a mixed culture fermentation (MCF) and therefore does not require sterile operating conditions and has the potential to consume a wide spectrum of substrates containing diverse organic compounds (Bastidas-Oyanedel et al. 2015; Jankowska et al. 2015).

Dark fermentation of food waste to carboxylic acids involves a series of biological mediated reactions under anaerobic conditions (Aceves-Lara et al. 2008; Boe et al. 2003; Jankowska et al. 2015). The mechanisms behind the product spectrum shifts of DF are not yet clear (Bastidas-Oyanedel et al. 2008, 2012; Hoelzle et al. 2014; Temudo et al. 2007). A considerable body of research has already been conducted to control and optimize the product spectrum of MCF with the aid of modeling (ADM1 2002). However, these models have been made on assumptions and require experimental support to be applied on a complex substrate such as food waste for optimization (Jankowska et al. 2015). Future interest in research is focused on the maximization of food waste utilization, without losing the specificity of produced carboxylic acid. In this regard, Chen et al. (2017a) have maximized the production of total volatile fatty acids with a maximum concentration of acetic acid of 2.7 g/L, at 55 °C mixed culture fermentation, when stepwise increasing the reaction pH from 7 to 9, using sewage sludge as feedstock with a concentration of 2.7 g/L of total suspended solids (approximately 0.3% TS). Chen et al. (2017b) have also obtained high purity propionic acid (16 g/L, with 90% purity) in a mixed culture fermentation of glucose using different levels of ammonium.

Another important topic to investigate is the purification and esterification of produced carboxylic acids, in order to produce alcohols. Esterification of carboxylic acids has been studied in organic solution using diverse catalysts then (Kaur and Ali 2015; Liu and Wu 2016; Park et al. 2010; Rezende and Pinto 2016). The main challenge is that carboxylic acids produced by fermentation are in aqueous solution. These aqueous solutions then need to be purified and concentrated in an organic solvent, in order to be used in the available esterification technologies.

Total solid (TS) content is one of the parameters that can affect the MCF process. Several studies have been performed to evaluate the effect of TS content on the production of methane, yet to date, there have been no studies conducted for carboxylic acids. The other obstacle of this promising MCF technology is low product yield. Pretreatment methods have been found to have a positive influence on the yield of carboxylic acid and, thus, have become a point of interest for various researchers (Breton-Toral et al. 2016; Liu et al. 2016; Liang et al. 2014, 2015; Yin et al. 2014; Yu et al. 2014).

The objective of the study was to investigate the effect of TS content on the carboxylic acids spectrum, focusing on lactic acid from food waste. The study utilized industrial enzymes for waste degradation and investigated its effect on the product yield. It also investigated the effect of aerobic pretreatment, which involves the supply of air into the waste and is aimed at increasing the growth of microorganisms naturally present in the waste. Thus, in conclusion, the study investigated the effect of TS content and enzymatic and aerobic pretreatment on the product yield and product spectrum of the mixed culture fermentation of food waste.

The experiments were performed in triplicate in a batch operation in a pH-uncontrolled environment. pH-uncontrolled fermentation was conducted as it results in an acidic environment at the end of fermentation and makes the recovery of carboxylic acids feasible (Tang et al. 2016; Yousuf et al. 2016).

2 Materials and Methods

2.1 Food Waste

Synthetic food waste was used in order to reproduce experimental results. The synthetic food waste was prepared mimicking real food waste, based on the work of Nwobi et al. (2015). It contained 11% rice, 5% pasta, 7% potatoes, 2% corn, 4% bread, 1% pineapple, 1% apple, 1% carrot, 2% cucumber, 3% lemon, 1% pawpaw, 6% tomatoes, 3% pickles, 6% meat, 14% fish, 6% dairy, 2% cabbage, 2% lettuce, 2% okra, 2% eggplant, 2% cauliflower, 2% broccoli, 13% vegetable oil, 1% newspaper, 0.6% cardboard, and 0.4% A4 paper in terms of weight. The food was cut into small pieces, cooked, and then mixed thoroughly in the weight ratio mentioned above. Its composition, in g/100g_TS, has carbohydrates, 22.93; protein, 1.99; fats, 18.01; fibers, 2.49; and ash, 4.03, as calculated by Nwobi et al. (2015). A large quantity of food was prepared and then stored at 4 °C, to prevent spoilage.

2.2 Batch Fermentation

Batch fermentation experiments were performed in triplicate and were carried in 500 mL glass bottles. After adding the corresponding food waste, the bottles were flushed with nitrogen gas to obtain anaerobic conditions. Bottles were capped with rubber stoppers and brewing airlocks. The initial pH for the food waste was found to be 4.4 ± 0.1. The bottles were kept at 37 °C, in an orbital shaker at a stirring speed of 120 rpm, without pH control, for 7 days. No external inoculum and nutrients were added.

2.3 Effect of Total Solid Content on Food Waste Fermentation

The experiments were performed in semidry and dry conditions. In general, wet reactors have <10% TS, semidry with 10–20% TS, and dry with >20% TS (Pognani et al. 2009). The experimental conditions to investigate the effect of percentage of total solid were categorized into three sets: (1) semidry fermentation with the total solids of 19.0 ± 1%, (2) dry fermentation with the total solids in a range of 29.0 ± 1%, and (3) extreme dry fermentation with the total solids of 34 ± 1%. This was the maximum possible TS content, as the food waste prepared had an approximate TS content of 34% and the remaining 64% was water present naturally in the waste. The extreme dry fermentation was only investigated for enzymatic pretreated waste. The extreme dry fermentation condition was not possible to be tested for any other condition due to extreme dry environment; sampling was difficult because of absence of free fluids. A fixed amount of food waste, 100 g, was added in all the bottles and then diluted with variable amount of water to achieve TS content of 19 ± 1%, 29 ± 1%, and 34 ± 1%. The total solid concentration of the different TS fermentation was, in g TS/L: 180.0, 243.7, and 284.34 for 19.0 ± 1%, 29.0 ± 1%, and 34.0 ± 1% of TS, respectively. The batch fermentation was performed as described in Sect. 2.2.

2.4 Effect of Enzymatic Pretreatment

The enzymes were obtained from Novozymes A/S Bagsværd, Denmark. The mixture of enzymes contained in weight basis, 34.0% cellulase, 37.0% amylase, 6.8% hemicellulose, 7.1% pectate lyase, 7.3% lipase, and 7.4% protease. The enzymes codes as provided by the manufacturer are cellulase NS81210, amylase NS81217, hemicellulase NS81233, pectate lyase NS81215, lipase NS81022, and protease NS81220 (Kolbl and Stres 2016). For all the experiments performed, 1.631 g enzymes mixture/100 g of food waste was used, as suggested by Nwobi et al. (2015). This ratio was selected based on the manufacturer's (Novozymes A/S, Denmark) recommended loading range.

The enzymatic hydrolysis was carried out in dry conditions with a TS content of 29% and extremely dry conditions with a TS content of 34%. Enzymatic hydrolysis was carried out at 50 °C, as per manufacturer's recommendation, for 24 h in an orbital shaker at 120 rpm. After enzymatic hydrolysis, batch fermentation was conducted as described in Sect. 2.2. The total process time was 7 days, with enzymatic hydrolysis conditions for 1 day and batch fermentation for 6 days.

2.5 Effect of Aeration Pretreatment

The effect of aeration pretreatment was performed using an air pump with a constant flow rate of 1.5 L/min. Air was sparged at the bottom of the glass bottles (as in Sect. 2.2) containing food waste for 24 h. During the aeration, the glass bottles were placed in a water bath at 37 °C. Aeration was performed without any addition of specific inoculum. Aerobic conditions would activate the growth of aerobic and/or facultative microorganisms present naturally in the food waste. Batch fermentation, as described in Sect. 2.2, was performed after the aeration pretreatment. The total process time was 7 days, with aeration pretreatment conditions for 1 day and batch fermentation for 6 days. Aeration pretreatment was performed on a total solid content of 19%.

2.6 Analytical Methods

The TS was determined according to standard methods (APHA, AWWA, WEF 1995). Liquid samples from the different batch fermentation were first centrifuged for 5 min at a speed of 20,238 rcf and filtered through 0.45 μm pore size. The filtered samples were then used for HPLC and total soluble chemical oxygen demand analysis.

The concentrations of lactic acid, acetic acid, propionic acid, butyric acid, isobutyric acid, valeric acid, isovaleric acid, ethanol, and glucose were measured by 1260 HPLC (Agilent Technologies, USA) using an Agilent Hi-Plex H column (65 °C), UV detector at 210 nm for organic acids, and RID detector at 35 °C for ethanol and glucose. The mobile phase was 5 mM H_2SO_4 with a flow rate of 0.6 mL/min, and external standards were used for calibration.

The HPLC equivalent COD results were calculated using their respective COD conversion factors, in g-COD/g compound: lactic acid 1.08, acetic acid 1.08, propionic acid 1.53, butyric and isobutyric acid 1.84, valeric and isovaleric acid 2.06, ethanol 2.02, and glucose 1.067 (Bonk et al. 2015).

Filtered samples were also used to measure the total soluble chemical oxygen demand (SCOD), using LCK 014 (range 1000–10,000 mg/L O_2) cuvettes by Hach Lange GmbH. After incubation in a Hach Lange HT 200S digestor, the soluble chemical oxygen demand (SCOD) was measured using a Hach Lange DR 2800 spectrophotometer.

The HPLC equivalent COD results and the total SCOD results were compared. It is expected that the HPLC equivalent COD is equal or lower than the total SCOD value. When the value is lower, it indicates the presence of undetectable HPLC components.

3 Results and Discussion

3.1 Effect of TS Content

Figure 1 presents the dark fermentation liquid product concentration for the 19% TS and 29% TS conditions. For 19% TS condition, products corresponded to approximately 61% lactic acid (19.56 g/L), 8.3% to acetic acid (2.65 g/L), 6.2% to propionic acid (2.0 g/L), 4.9% isobutyric acid (1.57 g/L), 1.8% of butyric acid (0.59 g/L), 3.1% of valeric acid (1.01 g/L), 7% of isovaleric acid (2.25 g/L), and ethanol (2.19 g/L). This makes a total liquid product concentration of 31.86 g/L for semidry fermentation.

For the dry environment, i.e., 29% TS, the total concentration of liquid products was around 43.32 g/L. Lactic acid accounted for 72% (31.2 g/L) of the total metabolites. Acetic acid is the second dominant metabolite with 2.9 g/L, and the concentration of propionate did not change, 2 g/L, when increasing the TS content.

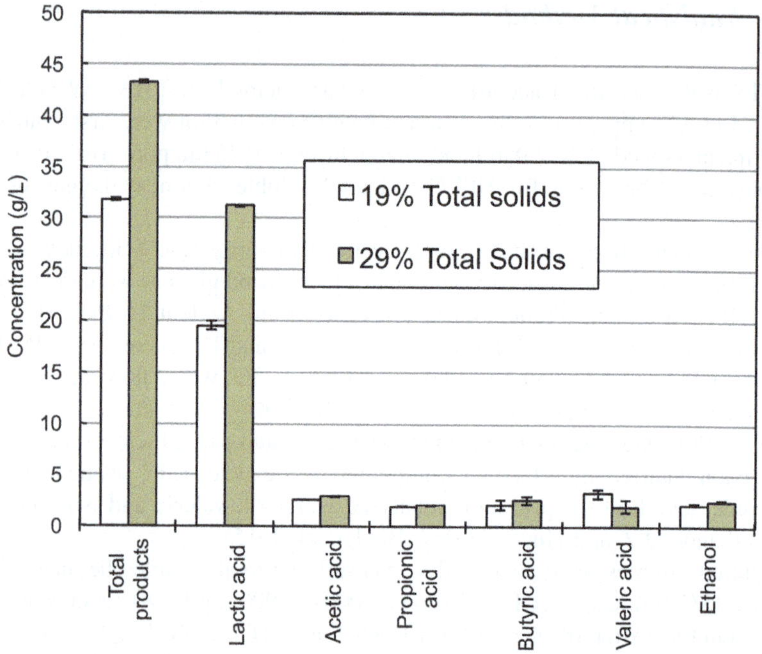

Fig. 1 Fermentative product distribution after 7 days of fermentation at different TS conditions, at temperature 37 °C, without pH control, without inoculum, without pretreatment

The concentration of butyrate increased from 2.1 to 2.6 g/L as TS content increased; again, the concentration of valerate (both valeric and isovaleric acid) increased from 3.27 to 4.0 g/L as TS content of the system increases. The total product yield of liquid products from food waste in the semidry, 19% TS, and dry condition, 29% TS, was 0.18 g_{prod}/g_{TS} and 0.17 g_{prod}/g_{TS}, respectively.

The impact of increasing TS from 19 to 29% resulted in around a 36% increase in the concentration the total liquid fermentation products; however, the product spectrum and the total liquid product yield were similar. Therefore, decreasing water input in the beginning could make downstream processing simple and efficient.

An increase of TS from 19 to 29% has positive effects on the concentration and yields of the total liquid dark fermentation products. However, this fact does not seem to agree with previous literature, where it is believed that increasing water content, i.e., decreasing TS, provides a homogeneous environment and improves the accessibility of microorganisms to substrates, resulting in a higher conversion. Staley et al. (2011) suggested that higher solid content creates a depletion zone at the microscopic level, having no microbial activity. These depletion zones result in lower degradation of the substrate and thus lower conversion rate.

At 19% TS the HPLC equivalent COD was 41.6 g-COD/L, while the measured total SCOD was 42.8 g-COD/L, corresponding to a COD balance of 97.3% and an

HPLC equivalent COD over measured total SCOD. At 29% TS, the HPLC equivalent COD was 53.2 g-COD/L, and the measured total SCOD was 54.4 g-COD/L, with a COD balance of 97.8%. The results were similar confirming the accuracy of the results.

3.2 *Effect of Enzymatic Pretreatment*

Hydrolysis is the rate-limiting step in dark fermentation, and thus an efficient pretreatment method can help to improve food digestibility and the product yield (Aceves-Lara et al. 2008). Figure 2 presents the product spectrum obtained from the enzymatically pretreated food waste at different TS content. For dry conditions, for example, at TS of 29%, there was 31% increment in the total concentration of carboxylic acids in comparison to the non-enzymatically pretreated waste. The total liquid product was 43.3 g/L of non-pretreated food waste and 63 g/L from pretreated waste. The product yield improved from 0.18 g_{prod}/g_{TS} to 0.259 g_{prod}/g_{TS}. The concentration of lactic acid was 44.2 g/L (41% higher) and propionic acid was 9.3 g/L.

The extreme dry fermentation condition, specifically at 34% TS, resulted in a product yield of 0.28 g_{prod}/g_{TS}, the highest observed so far. The total liquid product

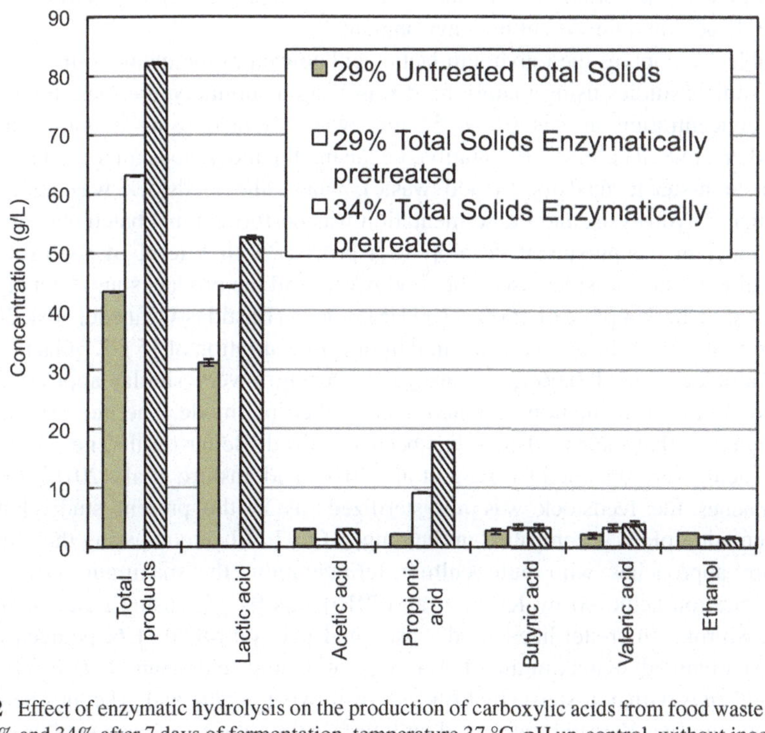

Fig. 2 Effect of enzymatic hydrolysis on the production of carboxylic acids from food waste at TS of 29% and 34% after 7 days of fermentation, temperature 37 °C, pH un-control, without inoculum

concentration was found to be 82.1 g/L, with a lactic acid concentration of 52.5 g/L, followed by propionic acid, 17.7 g/L, acetic acid, 2.8 g/L. The HPLC equivalent COD and total SCOD, for 29% TS, were 89.6 g-COD/L and 113 g-COD/L, respectively, representing a COD balance of 79%. At 34% TS the results were 121 g-COD/L and 167 g-COD/L, respectively, with a COD balance of 72%.

The COD of enzymes was unknown, and thus this COD was not included in the calculation of theoretical COD. The difference in theoretical and measured COD with 29% TS is smaller and could be because of enzymes. However, the difference at a higher TS was slightly higher, and this could be because of the presence of polysaccharides along with the enzymes. While this may be the reason, as yet, experiments have not been conducted to confirm this correlation.

The enzymatic pretreatment of food waste improved the total liquid product concentration and yield, in comparison to the system without enzymatic pretreatment. The system which had the TS of 34% was able to produce a higher concentration of acids, in comparison to all the previous experiments. Secondly, the product yield was also improved. For this system, along with lactic acid, propionic acid was the second dominant component. The important thing that should be noted here is the selectivity of lactic acid. Though the experiments were targeted to produce a mixture of carboxylic acids, the downstream processing of these mixed acids could be costly. The selectivity of lactic acid ranges between 0.4 and 0.7 $g_{lactic}/g_{totalacid}$, lower than the optimum desirable value of 0.9 $g_{lactic}/g_{totalacid}$. Further investigation of the effect of different parameters and their control could help to achieve high selectivity of single acid in a mixed culture environment.

Table 1 compares the maximum lactic acid concentration obtained in this study with similar studies using mainly food waste. As a summary, the maximum lactic acid concentration in this study, 52 g/L, was obtained using a non-externally inoculated mixed culture fermentation, i.e., using the indigenous microbiota present in the non-sterile feedstock (food waste), where the feedstock was treated by enzymatic hydrolysis, and the fermentation was performed in a batch uncontrolled pH mode, in the pursue of developing a process with a level of sophistication according to the feedstock used, the food waste, and the product sought, an organic acid with a market price of 1000–2100 USD/tonne (Bastidas-Oyanedel et al. 2016). Liang et al. (2014) have obtained a maximum concentration of 15 g/L of lactic acid, with a selectivity of 0.66 $g_{lacticacid}/g_{totalproducts}$, using a very similar approach, i.e., mixed culture fermentation, in batch uncontrolled pH mode. The most significant difference is that their feedstock has been sterilized. Better results, i.e., 39 g/L of lactic acid, were obtained by Tang et al. (2016) and Tashiro et al. (2016). In both approaches the feedstock was not sterilized, as in the present study, but the fermentation pH was controlled in the range of 6–7. When comparing these mixed culture approaches with pure culture fermentation, the maximum lactic acid concentration achieved by Kwan et al. (2016) was 94 g/L, using a *Lactobacillus casei* Shirota, filter-sterilized food waste, and pH controlled at 6. Nguyen et al. (2013) obtained a maximum of 186 g/L of lactic acid using a *Lactobacillus coryniformis* with non-sterile hydrolyzed fresh sweet potato, and pH controlled at 6. This encourages the study of mixed culture fermentation to achieve similar results obtained with pure culture fermentation.

Table 1 Maximum lactic acid concentration produced using food waste as feedstock

Feedstock	System	Lactic acid Max. conc. (g/L)	Yield (g_{Lactic}/ g_{TS})	Selectivity (g_{Lactic}/ $g_{totalprod}$)	References
Food waste hydrolyzed by an enzymatic cocktail	Mixed culture fermentation using feedstock indigenous microbiota (non-sterilized feedstock). Uncontrolled pH mode. Temperature controlled at 37 °C	52	0.18	0.64	This study
Potato peel waste	Mixed culture fermentation using activated sludge from a municipal wastewater treatment plant. Batch mode. Sterile-gelatinized feedstock. Uncontrolled pH mode. Temperature controlled at 35 °C	15	0.22	0.66	Liang et al. (2014)
Food waste	Mixed culture fermentation using feedstock indigenous microbiota (non-sterilized feedstock). Continuous mode, OLR 18, using pH and temperature controlled at 6.0 and 37 °C, respectively	39		0.85	Tang et al. (2016)
Food waste	Mixed culture fermentation using compost as fermentation seed. Batch mode. Non-sterilized feedstock, pH and temperature controlled at 7.0 and 50 °C, respectively	39	1.38		Tashiro et al. (2016)
Food waste hydrolyzed by *A. awamori* and *A. oryzae*	*Lactobacillus casei* Shirota. Batch mode. Filter-sterilized food waste hydrolysate, pH and temperature controlled at 6.0 and 37 °C, respectively	94	0.94		Kwan et al. (2016)
Fresh sweet potato hydrolyzed by an enzymatic cocktail	*Lactobacillus coryniformis*. Batch mode. pH and temperature controlled at 6.0 and 37 °C, respectively	186	0.85		Nguyen et al. (2013)

3.3 Effect of Aeration Pretreatment

Figure 3 presents the liquid product concentration obtained from aerobically pretreated food waste. The product spectrum is similar to the experiment without aeration pretreatment, and concentration is higher; around 18% increment is observed with aerated food waste. The total liquid product concentration was around 37 g/L for pretreated waste, in comparison to 32 g/L for non-pretreated waste. While conducting the experiment, the level of water in the fermenter bottle after aeration was lower. This indicates that air has taken some water along with it, which decreases the dilution and could be a reason for the concentration of lactic acid, but it does not explain the unchangeable concentration for acetic and propionic acids. The consistency of the liquid was the same as observed without pretreatment. As in this study, the aeration pretreatment was reported to achieve a higher rate of degradation of organic waste by some researchers (González-González and Cuadros 2015; Peces et al. 2016).

4 Conclusions

The study aimed to provide food waste valorization. The work focused on food waste conversion to carboxylic acids through mixed culture dark fermentation. The process did not require any sterilization or high-tech equipment and was conducted without inoculum and pH control. Experiments were performed at different TS content using enzymatic and aeration as pretreatment methods. Aeration did not produce significant improvement in the process. However, enzymatic pretreatment helped to increase the TS content of the system to 34% and also resulted in the highest liquid product concentration of 82 g/L with a yield of 0.28 $g_{liq\ prod}/g_{TS}$, in comparison to non-pretreated waste. The current platform produced lactic acid as the main product along with other carboxylic acids.

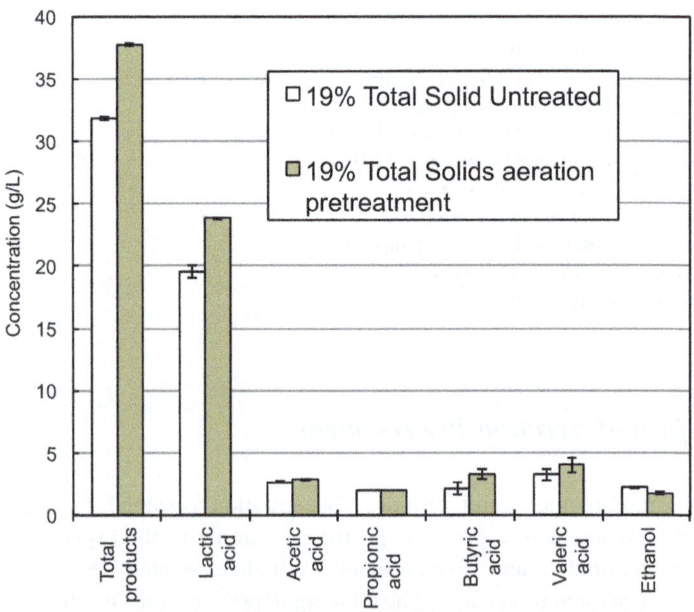

Fig. 3 Comparison of aerobic pretreated and non-pretreated waste on the production of carboxylic acids at TS of 19% after 7 days of fermentation, temperature 37 °C, pH un-control, without inoculum

Acknowledgments The authors would like to acknowledge the financial support of the Masdar Institute of Science and Technology, to help fulfill the vision of the late President Sheikh Zayed Bin Sultan Al Nahyan for sustainable development and empowerment of the United Arab Emirates and humankind, funding project 2GBIONRG (12KAMA4). A. Yousuf would like to acknowledge the Masdar Institute of Science and Technology Student Support Grant: SS2014-000003. The authors would like to acknowledge Dr. Glenda El Gamal for reviewing the manuscript's English.

References

Aceves-Lara CA, Trably E, Bastidas-Oyanedel J-R, Ramirez I, Latrille E, Steyer J-P (2008) Bioenergy production from waste: examples of biomethane and biohydrogen. J Soc Biol 202:177–189

ADM1, Anaerobic Digestion Model No. 1 (2002) IWA Task Group for Mathematical Modeling of Anaerobic Digestion Processes. IWA publishing.

Agler MT, Wrenn BA, Zinder SH, Angenent LT (2011) Waste to bioproduct conversion with undefined mixed cultures: the carboxylate platform. Trends Biotechnol 29:70–78

APHA, AWWA, WEF (1995) Standard methods for the examination of water and wastewater, 19th edn. American Public Health Association, Washington, DC

Bastidas-Oyanedel JR, Aceves-Lara CA, Ruiz-Filippi G, Steyer JP (2008) Thermodynamic analysis of energy transfer in acidogenic cultures. Eng Life Sci 8:487–498

Bastidas-Oyanedel J, Mohd-zaki Z, Zeng RJ, Bernet N (2012) Gas controlled hydrogen fermentation. Bioresour Technol 110:503–509

Bastidas-Oyanedel J-R, Bonk F, Thomsen MH, Schmidt JE (2015) Dark fermentation biorefinery in the present and future (bio)chemical industry. Rev Environ Sci Biotechnol 14:473–498

Bastidas-Oyanedel JR, Fang C, Almardeai S, Javid U, Yousuf A, Schmidt JE (2016) Waste biorefinery in arid/semi-arid regions. Bioresour Technol 215:21–28

Boe K, Batstone DJ, Angelidaki I (2003) Online headspace chromatographic method for measuring VFA in biogas reactor. Water Sci Technol 52:473–478

Bonk F, Bastidas-Oyanedel JR, Schmidt JE (2015) Converting the organic fraction of solid waste from the city of Abu Dhabi to valuable products via dark fermentation – economic and energy assessment. Waste Manag 40:82–91

Breton-Toral A, Trejo-Estrada SR, McDonald AG (2016) Lactic acid production from potato peel waste, spent coffee grounds and almond shells with undefined mixed cultures isolated from coffee mucilage from Coatepec Mexico. Ferment Technol 6:1–6

Chen Y, Jiang X, Xiao K, Shen N, Zeng RJ, Zhou Y (2017a) Enhanced volatile fatty acids (VFAs) production in a thermophilic fermenter with stepwise pH increase – investigation on dissolved organic matter transformation and microbial community shift. Water Res 112:261–268

Chen Y, Shen N, Wang T, Zhang F, Zeng RJ (2017b) Ammonium level induces high purity propionate production in mixed culture glucose fermentation. RSC Adv 7:518–525

Dahiya S, Sarkar O, Swamy YV, Venkata Mohan S (2015) Acidogenic fermentation of food waste for volatile fatty acid production with co-generation of biohydrogen. Bioresour Technol 182:103–113

FAO (2011) Global food losses and food waste – extent, causes and prevention. FAO, Rome

González-González A, Cuadros F (2015) Effect of aerobic pretreatment on anaerobic digestion of olive mill wastewater (OMWW): an ecoefficient treatment. Food Bioprod Process 95:339–345

Hoelzle RD, Virdis B, Batstone DJ (2014) Regulation mechanisms in mixed and pure culture microbial fermentation. Biotechnol Bioeng 111:1–37

Hoornweg D, Bhada-Tata P (2012) What a waste: a global review of solid waste management. World Bank, Washington, DC

Jankowska E, Chwiałkowska J, Stodolny M, Oleskowicz-Popiel P (2015) Effect of pH and retention time on volatile fatty acids production during mixed culture fermentation. Bioresour Technol 190:274–280

Kaur N, Ali A (2015) Preparation and application of Ce/ZrO2-TiO2/SO42- as solid catalyst for the esterification of fatty acids. Renew Energy 81:421–431

Kolbl S, Stres B (2016) Mixture of primary and secondary municipal wastewater sludge as a short-term substrate in 2 MW agricultural biogas plant: site-specific sustainability of enzymatic and ultrasound pretreatments. J Chem Technol Biotechnol 91:2769–2778

Kwan TH, Hu Y, Lin CSK (2016) Valorisation of food waste via fungal hydrolysis and lactic acid fermentation with Lactobacillus casei Shirota. Bioresour Technol 217:129–136

Liang S, McDonald AG, Coats ER (2014) Lactic acid production with undefined mixed culture fermentation of potato peel waste. Waste Manag 34:2022–2027

Liang S, McDonald AG, Coats ER (2015) Lactic acid production from potato peel waste by anaerobic sequencing batch fermentation using undefined mixed culture. Waste Manag 45:51–56

Liu CM, Wu SY (2016) From biomass waste to biofuels and biomaterial building blocks. Renew Energy 96:1056–1062

Liu H, Xiao H, Yin B, Zu Y, Liu H, Fu B, Ma H (2016) Enhanced volatile fatty acid production by a modified biological pretreatment in anaerobic fermentation of waste activated sludge. Chem Eng J 284:194–201

Nguyen CM, Choi GJ, Choi YH, Jang KS, Kim JC (2013) D- and L-lactic acid production from fresh sweet potato through simultaneous saccharification and fermentation. Biochem Eng J 81:40–46

Nwobi A, Cybulska I, Tesfai W, Shatilla Y (2015) Simultaneous saccharification and fermentation of solid household waste following mild pretreatment using a mix of hydrolytic enzymes in combination with *Saccharomyces cerevisiae*. Appl Microbiol Biotechnol 99:929–938

Park JY, Wang ZM, Kim DK, Lee JS (2010) Effects of water on the esterification of free fatty acids by acid catalysts. Renew Energy 35:614–618

Peces M, Astals S, Clarke WP, Jensen PD (2016) Semi-aerobic fermentation as a novel pre-treatment to obtain VFA and increase methane yield from primary sludge. Bioresour Technol 200:631–638

Pognani M, D'Imporzano G, Scaglia B, Adani F (2009) Substituting energy crops with organic fraction of municipal solid waste for biogas production at farm level: a full-scale plant study. Process Biochem 44:817–821

Rezende MJC, Pinto AC (2016) Esterification of fatty acids using acid-activated Brazilian smectite natural clay as a catalyst. Renew Energy 92:171–177

Staley BF, Iii FLDLR, Barlaz MA (2011) Effect of spatial differences in microbial activity, pH, and substrate levels on methanogenesis initiation in refuse. Appl Environ Microbiol 77:2381–2391

Tang J, Wang X, Hu Y, Zhang Y, Li Y (2016) Lactic acid fermentation from food waste with indigenous microbiota: Effects of pH, temperature and high OLR. Waste Manag 52:278–285

Tashiro Y, Inokuchi S, Poudel P, Okugawa Y, Miyamoto H, Miayamoto H, Sakai K (2016) Novel pH control strategy for efficient production of optically active L-lactic acid from kitchen refuse using a mixed culture system. Bioresour Technol 216:52–59

Temudo MF, Kleerebezem R, Van Loosdrecht MCM (2007) Influence of the pH on (open) mixed culture fermentation of glucose: a chemostat study. Biotechnol Bioeng 98:69–79

Vanwonterghem I, Jensen PD, Rabaey K, Tyson GW (2015) Temperature and solids retention time control microbial population dynamics and volatile fatty acid production in replicated anaerobic digesters. Sci Rep 5:8496

Yin J, Wang K, Yang Y, Shen D, Wang M, Mo H (2014) Improving production of volatile fatty acids from food waste fermentation by hydrothermal pretreatment. Bioresour Technol 171:323–329

Yousuf A, Bonk F, Bastidas-Oyanedel J-R, Schmidt JE (2016) Recovery of carboxylic acids produced during dark fermentation of food waste by adsorption on Amberlite IRA-67 and activated carbon. Bioresour Technol 217:137–140

Yu L, Bule M, Ma J, Zhao Q, Frear C, Chen S (2014) Enhancing volatile fatty acid (VFA) and bio-methane production from lawn grass with pretreatment. Bioresour Technol 162:243–249

Use of Syngas for the Production of Organic Molecules by Fermentation

Raymond Jianxiong Zeng and Fang Zhang

1 Introduction

Exploring environment-friendly methods, such as anaerobic fermentation, to convert biodegradable organic matter into biofuels and chemicals have drawn worldwide interest (Henstra et al. 2007; Latif et al. 2014; Miltner et al. 2010). However, direct conversion of recalcitrant organic wastes by biological processes entails difficulty, and a significant amount of non-biodegradable materials remains in effluents. Most biodegradable cellulose (40–50%) and hemicellulose (20–40%) materials in the biomass are packed with lignin (10–40%), which is resistant to microbial degradation (Abubackar et al. 2011; Meng and Ragauskas 2014; Zeng et al. 2014). Gasification, a thermochemical process, can convert mineral fuels or biomass into synthesis gas (syngas) as a mixture of CO, H_2, and minor components CO_2, CH_4, H_2S, and NO_x (Fabbri and Torri 2016; Latif et al. 2014; Shen et al. 2015). Syngas as a type of cleaning chemical feedstock can be further used for production by both chemical methods (e.g., Fischer–Tropsch synthesis) and biotechnological methods (e.g., syngas fermentation) (Latif et al. 2014; Shen et al. 2015).

As an important biotechnological technique, syngas fermentation provides lower operational temperature, lower pressure, as well as higher selectivity and resistance to toxicity that those of Fischer–Tropsch synthesis (Bengelsdorf et al. 2013; Ganigué et al. 2016; Liew et al. 2016; Massaro et al. 2015). Consequently, it provides a potential pathway to use hardly biodegradable organic materials, such as lignocellulose and sludge, for the production of biofuel and volatile fatty acids (VFAs)

R. J. Zeng (✉) · F. Zhang (✉)
Fujian Provincial Key Laboratory of Soil Environmental Health and Regulation, College of Resources and Environment, Fujian Agriculture and Forestry University, Fuzhou, Fujian, China
e-mail: rzeng@ustc.edu.cn; zhfang@ysu.edu.cn

© Springer Nature Switzerland AG 2019
J.-R. Bastidas-Oyanedel, J. E. Schmidt (eds.), *Biorefinery*,
https://doi.org/10.1007/978-3-030-10961-5_20

(Henstra et al. 2007; Jing et al. 2017; Latif et al. 2014; Liew et al. 2016; Zhang et al. 2013b). Thus far, syngas fermentation focuses on pure culture and co-culture under mesophilic conditions and is proposed to convert syngas to VFAs (such as acetate and butyrate), ethanol, butanol, and/or caproate via microbes, such as *Clostridium ljungdahlii*, *Clostridium autoethanogenum*, *Clostridium carboxidivorans*, and *Alkalibaculum bacchi* (Liew et al. 2016; Martin et al. 2016; Ramió-Pujol et al. 2015; Schuchmann and Muller 2014). Compared with mixed culture fermentation (MCF), pure culture or co-culture fermentation is typically challenged by strain degeneration and contamination (Esquivel-Elizondo et al. 2017; Henstra et al. 2007). CO toxicity to some bacteria also impedes CO conversion (Esquivel-Elizondo et al. 2017; Jing et al. 2017). Bertsch and Müller (2015) demonstrated that the hydrogen-dependent CO_2 reductase of *Acetobacterium woodii* is highly sensitive to CO, consequently impeding the growth of *A. woodii* on CO as a sole carbon and energy source. Thus, mixed culture syngas fermentation can potentially facilitate the simultaneous conversion of H_2 and CO by different enriched bacteria.

In syngas MCF, the functional bacteria are acetogenic bacteria, such as *C. ljungdahlii*, *C. autoethanogenum*, and *C. carboxidivorans*. These bacteria can convert CO, H_2, and CO_2 to acetate, ethanol, and other products via the Wood–Ljungdahl pathway (Köpke et al. 2011), as shown in Fig. 1. Other bacteria, such as *Clostridium kluyveri*, can produce longer carbon-chain metabolites, including butyrate, caproate, and caprylate from ethanol and acetate via reverse β-oxidation reaction (Fig. 1) (Seedorf et al. 2008). When the methanogens archaea are enriched in the reactor, the produced metabolites and syngas are also consumed to produce methane. Biochemical reactions in syngas fermentation are also thermodynamically controlled. Thus, the basic bioreactions and thermodynamics are summarized in Sect. 2.

The operating conditions—pH, temperature, CO and H_2 partial pressure, and impurities of tar and NO_x—potentially induce changes in the microbial community or metabolic pathway in mixed culture fermentation. These factors are reviewed in Sect. 3. Meanwhile, the low solubility of H_2 and CO in the water phase also limits syngas utilization (Henstra et al. 2007). The configurations of the reactor, such as the continuous stirred-tank reactor (CSTR), trick biofilm reactor, and hollow fiber membrane biofilm reactor (HfMBR), are summarized in Sect. 3.4.

Lastly, syngas pretreatment was generally disregarded in syngas fermentation, which was demonstrated to reduce bacterial activity; thus, these technologies were indispensable and should be coupled with syngas fermentation (Benalcázar et al. 2017; Sheth and Babu 2010). On the other hand, the inhibition of organic acids, particularly at acidic pH, presents a main challenge for bacteria because the inward diffusion of organic acids over the cytoplasmic membrane leads to the dissipation of the proton-motive force, and bacteria have to transport these metabolites by energy consumption in the form of ATP (Louis et al. 2004; Zhang et al. 2013c). Meanwhile, the accumulation of ethanol causes the hyperpolarization of the bacterial lipid bilayer, which consequently decreases membrane integrity and inhibits bacterial activity (Thammasittirong et al. 2013). Thus, coupling syngas fermentation with other technologies, such as syngas pretreatments and membrane technology, is necessary for its

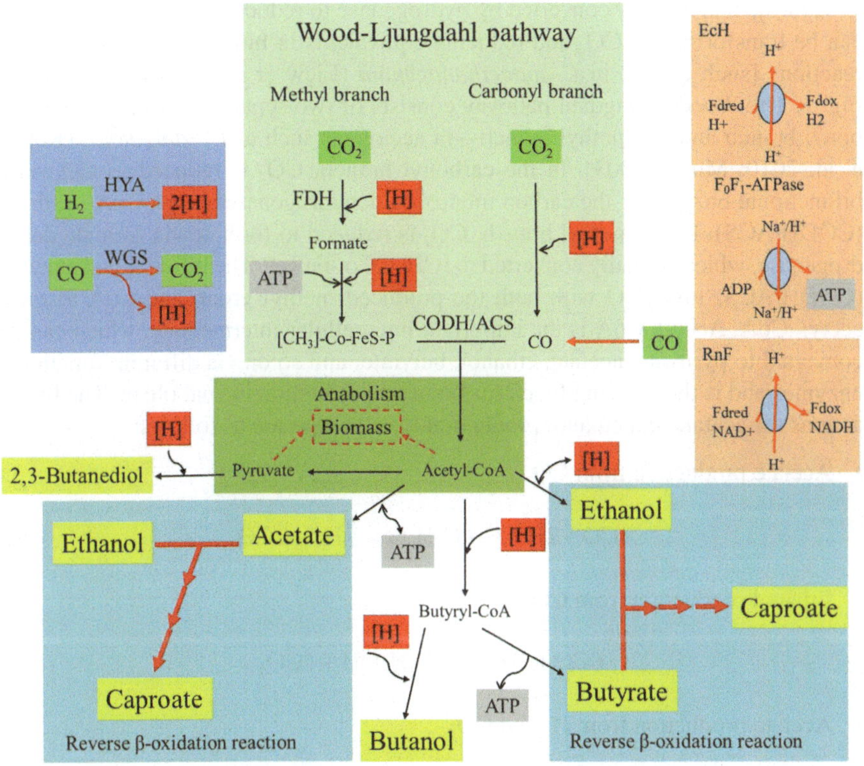

Fig. 1 Metabolic pathways in syngas fermentation (Diender et al. 2015; Schuchmann and Muller 2014; Seedorf et al. 2008)

application (Dai et al. 2017; Liu and Qureshi 2009). Such application is summarized in Sect. 4. Other promising technologies, such as PHA production and microbial fuel cells (MFCs), are also reviewed. Thus, this chapter is expected to promote the development and worldwide application of syngas fermentation in the future.

2 Bioreactions and Thermodynamics in Syngas Fermentation

The metabolic pathways of syngas fermentation are presented in Fig. 1. Energy conservation occurs by substrate-level phosphorylation in a catabolic reaction, ion-motive force, and energy conservation via electron bifurcation reaction, which involves key enzymes such as EcH (e.g., that in *Moorella thermoacetica*), Rnf complex (e.g., that in *C. ljungdahlii*), and ATPase (Angenent et al. 2016; Basen and Müller 2017; Diender et al. 2015; Drake et al. 2008; Schuchmann and Muller 2014; Seedorf et al. 2008). The main metabolites are identified as acetate, butyrate, caproate, and ethanol (Bengelsdorf et al. 2013; Diender et al. 2015; Spirito et al. 2014).

Hydrogen is initially converted by hydrogenase to reducing equivalents, and CO can be transformed to CO_2 and reducing equivalents in biological water–gas shift reactions [such as that in *C. autoethanogenum* (Liew et al. 2016)], as shown in Fig. 1. The Wood–Ljungdahl pathway consists of two separate branches—the carbonyl branch and the methyl branch—in acetogens such as *C. ljungdahlii* (Köpke et al. 2010; Muller 2003). In the carbonyl branch, CO_2 is reduced to CO via a bifunctional enzyme of the carbon monoxide dehydrogenase/acetyl–CoA synthase (CODH/ACS). In the methyl branch, CO_2 is reduced to formate via formate dehydrogenase, which is finally converted to [CH_3]–Co–FeS–P. The bifunctional enzyme (CODH/ACS) fuses CO with both the produced methyl group and CoA to form acetyl–CoA. Acetyl–CoA is the important intercellular intermediate, which can be converted to pyruvate, acetate, ethanol, butyrate, and so on via different functional enzymes and is the building block for biomass production in anabolism. The bioreactions for acetate and ethanol production from syngas are as follows:

- Acetate production from CO

$$4CO + 2H_2O \rightarrow C_2H_3O_2^- + H^+ + 2CO_2 \tag{1}$$

- Ethanol production from CO

$$6CO + 3H_2O \rightarrow C_2H_5OH + 4CO_2 \tag{2}$$

- Acetate production from H_2 and CO_2

$$4H_2 + 2CO_2 \rightarrow C_2H_3O_2^- + H^+ + 2H_2O \tag{3}$$

- Ethanol production from H_2 and CO_2

$$6H_2 + 2CO_2 \rightarrow C_2H_5OH + 3H_2O \tag{4}$$

The produced acetate, butyrate, and ethanol in the Wood–Ljungdahl pathway are chemical building blocks for the production of longer carbon-chain molecules, such as caproate, via reverse β-oxidation reaction in *C. kluyveri* (Seedorf et al. 2008; Spirito et al. 2014) in which electron bifurcation and two membrane-associated, energy-converting enzyme complexes involved in fermentation, ferredoxin:NAD oxidoreductase and ATP synthase, provide the energy source. The bioreactions for caproate production from acetate, butyrate, and ethanol are as follows:

- Caproate production from acetate and ethanol

$$C_2H_3O_2^- + 2C_2H_5OH \rightarrow C_6H_{11}O_2^- + 2H_2O \tag{5}$$

- Caproate production from butyrate and ethanol

$$C_4H_7O_2^- + C_2H_5OH \rightarrow C_6H_{11}O_2^- + H_2O \tag{6}$$

Finally, methanogens can convert acetate and H_2/CO_2 to CH_4, referred to as aceticlastic methanogenesis and hydrogenotrophic methanogenesis, respectively (Dai et al. 2017; Thauer et al. 2008). The former (Eq. 7) is conducted by *Methanosarcinaceae* and *Methanosaetaceae*, whereas the latter (Eq. 8) is performed by *Methanomicrobiales* and *Methanobacteriales* (Karakashev et al. 2006).

$$C_2H_3O_2^- + H^+ \rightarrow CH_4 + CO_2 \tag{7}$$

$$4H_2 + CO_2 \rightarrow CH_4 + 2H_2O \tag{8}$$

The biochemical reactions in Fig. 1 are generally constrained by thermodynamic control (Richter et al. 2016; Schuchmann and Muller 2014). The standard Gibbs free energy of formation and standard enthalpy of formation for the relevant compounds are shown in Table 1 (Speight 2005; Thauer et al. 1977). The detailed calculation of the reaction of Gibbs free energy is provided elsewhere (Bastidas-Oyanedel et al. 2008; Kleerebezem and Van Loosdrecht 2010; Lee et al. 2008).

3 Influencing Factors in Syngas Fermentation

Operating conditions, such as pH, temperature, and CO and H_2 partial pressure, potentially trigger changes in microbial community composition and/or metabolic flow in mixed culture fermentation, consequently affecting the performances of the reactors, as summarized in Table 2.

Table 1 ΔG_f^0 and ΔH_f^0 of syngas fermentation metabolites

Metabolite	State	ΔG_f^0 (kJ/mol)	ΔH_f^0 (kJ/mol)
CO	Gas	−137.16	−110.53
H_2	Gas	0	0
CO_2	Gas	−394.36	−393.50
Acetate	Aqueous	−369.31	−486.01
Ethanol	Aqueous	−181.64	−288.3
Butyrate	Aqueous	−352.63	−535.55
Butanol	Aqueous	−162.5	−327.3
2,3-Butanediol	Aqueous	−322.0	−541.5
Caproate	Aqueous	−336.0	–
Hexanol	Aqueous	−152.3	−377.5
H_2O	Liquid	−237.19	−285.83

3.1 Effect of Temperature

Temperature can shift the dominant bacteria or the main metabolic pathways and play an important role in syngas MCF. Using H_2/CO_2 as the substrates in HfMBR, Zhang et al. found a mixture of acetate (7.4 g/L), butyrate (1.8 g/L), caproate (0.98 g/L), and caprylate (0.42 g/L) that was accumulated at pH 6.0 and in 35 °C syngas MCF, where *Clostridium* spp. (such as *C. ljungdahlii* and *C. kluyveri*) as the dominant bacteria (Zhang et al. 2013b). Wang et al. (2017) demonstrated that with an increase in temperature to 55 °C, *Thermoanaerobacterium* (66%) became the main bacterium and acetate comprised more than 98.5% and 99.1% of total metabolites in batch and continuous modes, respectively. Owing to a decrease in the diffusion coefficients of acidic metabolites at low temperature, metabolite inhibition weakened (Ramió-Pujol et al. 2015; Zhang et al. 2013c). As temperature decreased to 25 °C, acetate, ethanol, butyrate, and caproate were the main metabolites, as determined in the study by Wang et al. (2018b). Caproate concentration (5.7 g/L) was particularly higher than that of pure culture fermentation (*C. carboxidivorans* P7, 1.05 g/L).

Ramió-Pujol et al. (2015) used H_2/CO as the substrate and compared the metabolites in pure culture fermentation of *C. carboxidivorans* P7 at 25 and 35 °C; acetate (1.6 g/L) was found to be the main metabolite, with apparent accumulation of caproate (1.05 g/L) at 25 °C. Meanwhile, as temperature increased to 37 °C, no caproate was produced. Under thermophilic conditions (55 °C), the dominant bacteria in syngas MCF were *Desulfotomaculum* and *Caloribacterium*, and the main product was acetate (0.15 g/L), as determined in the study by Alves et al. (2013). We recently compared metabolite distribution in HfMBR by using CO and H_2 as the substrate; acetate (4.22 g/L), butyrate (1.35 g/L), caproate (0.88 g/L), and caprylate (0.52 g/L) were detected at 35 °C, whereas acetate (24.6 g/L) and minimal butyrate (less than 1 g/L) were detected at 55 °C (unpublished data).

In addition, the changes in Gibbs free energy ($\Delta G'$) of the main bioreactions in syngas (CO/H_2) fermentation under standard conditions, except for pH at 7.0, are shown in Table 3. Except for caproate production, $\Delta G'$ is higher at a low temperature of 25 °C than at 35 °C and 55 °C; thus, more energy can be used for biomass growth or maintenance at low temperature from the viewpoint of thermodynamics. Ramió-Pujol et al. (2015) determined that the maximum OD600 values of *C. carboxidivorans* P7 at 25 °C (OD600, 1.2) was higher than that at 35 °C (OD600, 0.55). On the other hand, all $\Delta G'$ values of acetate and ethanol production from CO (Eqs. 1 and 2) were lower than those from CO_2 and H_2 (Eqs. 3 and 4), allowing bacteria to obtain more energy from CO utilization. Although a high temperature favors caproate production from the viewpoint of thermodynamics, no caproate has been detected in the thermophilic reactor. As a known caproate production bacterium, *C. kluyveri* only lives under mesophilic conditions (Seedorf et al. 2008; Thauer et al. 1968). Thus, the enriched bacteria are also considered a critical factor for the determination of metabolite production.

Table 2 Operating conditions and metabolites in syngas fermentation

Type of fermentation	Reactor configuration	Syngas composition	Temp. (°C)	pH	Main metabolites	References
Pure culture (*Clostridium carboxidivorans*)	Batch	CO (20%), CO_2 (15%), H_2 (5%), N_2 (60%), minor NO (200 ppm)	37	5.7	Ethanol (0.042 g/L)	Ahmed and Lewis (2007)
Pure culture (*Clostridium carboxidivorans*)	External HFM reactor	CO (20%), H_2 (5%), CO_2 (15%), N_2 (60%)	37	4.5–6.0	Acetate (5.0 g/L), ethanol (23.93 g/L)	Shen et al. (2014)
Pure culture (*Clostridium carboxidivorans*)	Batch	CO (32%), H_2 (32%), N_2 (28%), CO_2 (8%)	25 and 37	4.8–5.9	25 °C: acetate (1.6 g/L), caproate (1.05 g/L); 37 °C: acetate (3.5 g/L), butyrate (0.36 g/L)	Ramió-Pujol et al. (2015)
Pure culture (*Clostridium autoethanogenum*)	Batch	CO (100%)	35	4.75	Ethanol (0.87 g/L)	Abubackar et al. (2015)
Pure culture (*Clostridium autoethanogenum*)	Continuous	CO (100%)	35	6.0	Acetate (0.91 g/L), ethanol (0.91 g/L)	Abubackar et al. (2015)
Pure culture (*Clostridium ragsdalei*)	Trickle-bed reactor	CO (38%), CO_2 (28.5%), H_2 (28.5%) N_2 (5%)	37	4.6	Ethanol (5.7 g/L), acetate (12.3 g/L)	Devarapalli et al. (2016)
Pure culture (*Alkalibaculum bacchi*)	Batch	CO (40%), CO_2 (30%), H_2 (30%)	37	8.0	Propanol (0.4 g/L), butanol (0.5 g/L), hexanol (0.8 g/L)	Liu et al. (2014)
Co-culture (*Alkalibaculum bacchi* and *Clostridium propionicum*)	Batch	CO (40%), CO_2 (30%), H_2 (30%)	37	8.0	Propanol (1.0 g/L), butanol (0.8 g/L), hexanol (1.0 g/L)	Liu et al. (2014)
Mixed cultures	Batch	CO (20%), CO_2 (15%), H_2 (20%), CH_4 (3%), N_2 (42%)	37	6.0	Ethanol (2.2 g/L), acetate (0.9 g/L)	Singla et al. (2014)
Mixed culture	Batch	H_2 (1.5 bar)	35	6.0	Ethanol (0.17 g/L), propanol (0.48 g/L), butanol (0.27 g/L)	Steinbusch et al. (2008)
Mixed culture	Batch	CO (100%)	55	7.0	Acetate (0.15 g/L)	Alves et al. (2013)
Mixed culture	Batch	H_2 (32%), CO (32%), CO_2 (8%), N_2(28%)	37	4.8	Ethanol (1.7 g/L), butanol (1.1 g/L), hexanol (0.6 g/L)	Ganigué et al. (2016)

(continued)

Table 2 (continued)

Type of fermentation	Reactor configuration	Syngas composition	Temp. (°C)	pH	Main metabolites	References
Mixed culture	CSTR	CO (55%), H_2 (20%), CO_2 (10%), Ar (15%)	37	7.0	Ethanol (6.50 g/L), acetate (5.43 g/L)	Mohammadi et al. (2012)
Mixed culture	HfMBR	H_2 (60%) and CO_2 (40%)	35	6.0	Acetate (7.4 g/L), butyrate (1.8 g/L), caproate (0.98 g/L), caprylate (0.42 g/L)	Zhang et al. (2013b)
Mixed culture	HfMBR	H_2 (60%) and CO_2 (40%)	35	4.5	Batch: acetate (12.5 g/L) Continuous: acetate (0.4 g/L/d)	Zhang et al. (2013a)
Mixed culture	HfMBR	H_2 (60%) and CO_2 (40%)	25	6.0	Acetate (31.1 g/L), butyrate (4.1 g/L), caproate (5.7 g/L)	Wang et al. (2018b)
Mixed culture	HfMBR	H_2 (60%) and CO_2 (40%)	55	6.0	Batch: acetate (42.4 g/L) Continuous: acetate (10.5 g/L/d)	Wang et al. (2017)
Mixed culture	HfMBR	CO (60%) and H_2 (40%)	35	4.5	Ethanol (16.9 g/L)	Wang et al. (2018a)

3.2 Effect of pH

The inward diffusion of organic acids over the cytoplasmic membrane at acidic pH presents major challenge for bacteria because it leads to energy dissipation (Louis et al. 2004; Zhang et al. 2013c). Wilbanks and Trinh (2017) recently found that higher concentrations and/or hydrophobicity of metabolites cause the increased growth inhibition of *E. coli*. Consequently, acidic pH was generally considered a main factor for the shifting of metabolites to produce alcohol (Datar et al. 2004; Fernández-Naveira et al. 2017; Valgepea et al. 2017). Abubackar et al. (2015) reported that at pH 4.75, no acetate was produced, and ethanol concentration reached a maximum of 0.87 g/L; at pH of 6.0, almost equal amounts of ethanol and acetate were formed from CO, obtaining 0.91 g/L. Ganigué et al. (2016) indicated that at pH of about 4.8 in the batch mode, a mixture of ethanol (1.7 g/L), butanol (1.1 g/L), and hexanol (0.6 g/L) was produced from syngas (32% H_2, 32% CO, 8% CO_2, and 28% N_2). Liu et al. (2014) demonstrated that compared with the *A. bacchi* strain CP15 monoculture (propanol of 0.4 g/L, butanol of 0.5 g/L, and hexanol of 0.8 g/L), the addition of propionic acid, butyric acid, and hexanoic acid to the mixed culture of CP15 and *Clostridium propionicum* resulted in a 50% higher conversion efficiency of these acids to their respective alcohols (propanol of 1.0 g/L, butanol of 0.8 g/L, and hexanol of 1.0 g/L). Singla et al. (2014) enriched several mixed cultures and optimized their growth conditions for ethanol production, obtaining a maximum ethanol concentration of 2.2 g/L.

Using H_2/CO_2 as the substrate, Zhang et al. (2013b) found a mixture of acetate (7.4 g/L), butyrate (1.8 g/L), caproate (0.98 g/L), and caprylate (0.42 g/L). The mixture was accumulated at pH 6.0 and temperature of 35 °C in syngas MCF, with *Clostridium* spp. (such as *C. ljungdahlii* and *C. kluyveri*) as the dominant bacteria; meanwhile, as pH was reduced to 4.0, the metabolite only consisted of acetate (12.5 g/L), and the dominant bacteria were identified as *C. ljungdahlii* and *C. drakei*

Table 3 Change in Gibbs free energy of main bioreactions in syngas (CO/H_2) fermentation under standard conditions

Bioreactions	$\Delta G'$ (kJ/mol) *		
	25 °C	35 °C	55 °C
Acetate production from CO: $4CO + 2H_2O \rightarrow C_2H_3O_2^- + H^+ + 2CO_2$ (1)	−175.0	−172.2	−166.6
Ethanol production from CO: $6CO + 3H_2O \rightarrow C_2H_5OH + 4CO_2$ (2)	−224.5	−220.6	−212.9
Acetate production from CO_2 and H_2: $4H_2 + 2CO_2 \rightarrow C_2H_3O_2^- + H^+ + 2H_2O$ (3)	−95.0	−87.8	−77.4
Ethanol production from CO_2 and H_2: $6H_2 + 2CO_2 \rightarrow C_2H_5OH + 3H_2O$ (4)	−104.5	−96.0	−79.1
Caproate: $C_2H_3O_2^- + 2C_2H_5OH \rightarrow C_6H_{11}O_2^- + 2H_2O$(5)	−77.7	−81.5	−89.1
Caproate: $C_4H_7O_2^- + C_2H_5OH \rightarrow C_6H_{11}O_2^- + H_2O$ (6)	−38.9	−43.0	−51.3

(Zhang et al. 2013a). Thus, apart from acidic pH, CO was also considered a main factor promoting ethanol production. However, the accumulation of ethanol leads to the hyperpolarization of the bacterial lipid bilayer, which consequently decreases membrane integrity and inhibits bacterial activity (Thammasittirong et al. 2013). Thus, removing the accumulation of organic acids and ethanol from the bulk solution could also increase bacterial activity.

The Gibbs free energy ($\Delta G'$) of acetate and ethanol was calculated. The results are listed in Table 4. $\Delta G'$ of acetate (Eq. 1) and ethanol (Eq. 2) production from CO is higher than that from H_2 (Eqs. 3 and 4); thus, CO is the more suitable substrate for syngas fermentation (Diender et al. 2015). On the other hand, at neutral pH, $\Delta G'$ of ethanol production from CO (Eq. 2) is −220.6 kJ/mol and that from H_2 (Eq. 4) is −96.0 kJ/mol. Both values are higher than the values obtained for acetate production (−172.2 and −89.2 kJ/mol); thus, under neutral pH, bacteria can obtain more energy from CO utilization from the viewpoint of thermodynamics.

3.3 *CO and H_2 Partial Pressure*

CO and H_2 can inhibit hydrogenase activity and change the ratio of intercellular redox couplers, such as Fdred/Fdox and NADH/NAD^+, and consequently shift the metabolite distribution (Abubackar et al. 2015; Sancho-Navarro et al. 2016; Zhang et al. 2013c). Several studies demonstrated that hydrogen partial pressure (P_{H_2}) and CO partial pressure (P_{CO}) as factors could shift the dominant bacteria and change the metabolite distribution in syngas MCF (Peintner et al. 2010; Steinbusch et al. 2008; Temudo et al. 2008; Zhang et al. 2013a, b). Steinbusch et al. (2008) indicated that VFAs such as acetic, propionic, and butyric acids were reduced at P_{H_2} of 1.5 bar by MCF: the final alcohol concentrations were ethanol (0.17 g/L), propanol (0.48 g/L), and *n*-butanol (0.27 g/L). Bertsch and Müller (2015) revealed that the hydrogen-dependent CO_2 reductase of *A. woodii* was highly sensitive to CO; consequently, *A. woodii* failed to grow on CO as a sole carbon and energy source.

Table 4 Gibbs free energy ($\Delta G'$) of acetate and ethanol production at acidic pH in syngas (H_2 and CO) fermentation

Bioreactions	$\Delta G'$ (kJ/mol)[a]		
	pH 7.0	pH 6.0	pH 4.5
Acetate production from CO: $4CO + 2H_2O \rightarrow C_2H_3O_2^- + H^+ + 2CO_2$ (1)	−172.2	−166.3	−157.5
Ethanol production from CO: $6CO + 3H_2O \rightarrow C_2H_5OH + 4CO_2$ (2)	−220.6	−220.6	−220.6
Homoacetogenesis: $4H_2 + 2CO_2 \rightarrow C_2H_3O_2^- + H^+ + 2H_2O$ (3)	−89.2	−83.3	−74.4
Ethanol production from H_2: $6H_2 + 2CO_2 \rightarrow C_2H_5OH + 3H_2O$ (4)	−96.0	−96.0	−96.0

[a]All calculated under standard conditions, except for 35 °C and acidic pH

Sancho-Navarro et al. (2016) recently analyzed the methane production pathway from syngas and determined that acetoclastic methanogens were the most sensitive to CO and that high CO concentrations led to a shift in the archaeal population to hydrogen-utilizing methanogens.

3.4 Reactor Configurations

Although syngas fermentation provides a platform for organic waste utilization, the poor aqueous solubility of H_2 and CO is a major limiting factor in syngas fermentation (Esquivel-Elizondo et al. 2017; Lee et al. 2016). Increasing the speed of the impeller (500 rpm in the study by Mohammadi et al. (2012)) in CSTR can provide high gas/liquid mass transfer coefficients with an agitation mechanism that allows the breakdown of large bubbles into smaller ones and improves gas–liquid mass transfer (Fernández-Naveira et al. 2017; Mohammadi et al. 2012). Mohammadi et al. (2012) operated a mesophilic (37 °C) CSTR with an agitation rate of 500 rpm and a working volume of 2 L; the produced metabolites were ethanol (6.50 g/L) and acetate (5.43 g/L). However, high agitation rates can also lead to high-power consumption and may inhibit bacterial activity (Henstra et al. 2007; Yasin et al. 2015; Zhao et al. 2014). The trickle-bed reactor was also proposed to resolve poor solubility; Devarapalli et al. (2016) proposed ethanol production in a semi-continuous trickle-bed reactor and found that the biofilm facilitates syngas utilization; the final ethanol and acetate concentrations were 5.7 and 12.3 g/L, respectively.

Increasing the specific gas–liquid interfacial area by membrane technologies can diminish the poor gas solubility (Henstra et al. 2007; Nerenberg 2016; Zhang et al. 2013b). Shen et al. (2014) found that the volumetric mass transfer coefficients (K_{La}) of the hollow fiber membrane were higher than those of most reactor configurations, such as CSTR and bubble columns. Zhang et al. (2013b) proposed a mesophilic HfMBR for the in situ consumption of H_2 and CO_2 with 100% utilization of H_2, with *Clostridium* spp. (such as *C. ljungdahlii* and *C. kluyveri*) as the dominant bacteria. In addition, the product contained a mixture of acetate (7.4 g/L), butyrate (1.8 g/L), caproate (0.98 g/L), and caprylate (0.42 g/L). In a thermophilic HfMBR (55 °C), acetate comprised more than 99% of total VFAs from H_2 and CO_2 MCF, but no caproate was produced (Wang et al. 2017). HfMBR also provides several advantages, such as low energy consumption and small reactor footprints (Martin and Nerenberg 2012). Moreover, the biofilm formed on the outer surface of the hollow fiber membrane may enhance bacterial resistance to CO toxicity. Jiang et al. (2011) reported that the butyric acid tolerance of *Clostridium tyrobutyricum* increased markedly after being immobilized in a fibrous-bed bioreactor and the final butyric acid concentration reached 86.9 g/L.

However, membrane fouling has been recognized as a key factor for lower running efficiency, higher operating cost, and shorter membrane lifespan (Drews 2010; Meng et al. 2017; Wang et al. 2014). Ayala et al. (2011) assigned a linear trend between membrane permeability loss (due to membrane fouling and cleaning) and operation time, which indicated the recovered membrane permeability to reach a threshold minimum value for virgin membrane after about 7 years of operation. In HfMBR, a sufficient quantity of microorganisms attached to the membrane surface is necessary for efficient and stable operation; however, the smaller the size of the membrane pores, the higher the gas pressure and energy consumption (Munasinghe and Khanal 2010). Consequently, the energy problem still needs to be evaluated in future studies.

3.5 Impurities of Synthesis Gas

Syngas fermentation using artificial syngas formulated only with CO and H_2 remains the focus of research, whereas impurities such as NO_x are rarely studied (Benalcázar et al. 2017; Liew et al. 2016; Xu et al. 2011). Syngas is produced by thermochemical gasification; thus, minor components such as NO_x and ammonia can also potentially affect syngas fermentation (Benalcázar et al. 2017; Sheth and Babu 2010). Datar et al. (2004) found that in *C. carboxidivorans* $P7^T$ fermentation, cell growth stopped (with negligible death) when syngas directly produced from switchgrass was used as feedstock because the components of the original syngas might inhibit the hydrogenase enzyme. Ahmed and Lewis (2007) analyzed NO toxicity on the hydrogenase of *C. carboxidivorans* $P7^T$ and concluded that when NO content was below 40 ppm, inhibition could be tolerated by cells in a syngas fermentation system without compromising hydrogenase activity, cell growth, and product distribution. However, when the NO content was 200 ppm, hydrogenase activity remained completely inhibited, and ethanol concentration was only 0.042 g/L (Ahmed and Lewis 2007). Xu et al. (2011) indicated that the entrained tar particulates (above 0.025 mm), nitric oxide (0.004 mol%), and ammonia (above 0.25 mol/L) negatively affected the syngas fermentation process.

Except for NO_x, other impurities such as cyanide may also lower the performance of syngas fermentation. Benalcázar et al. (2017) recently reported that when lignocellulosic biomass and municipal solid waste were used as feedstock for gasification, ethanol production was rather low, owing to cyanide toxicity; meanwhile, when CO-rich flue gases from the steel industry were used, the project seemed to have successfully developed. Worth 47 ktons per year, this project was the first to produce ethanol by gas fermentation to be built in Europe. Consequently, pretreatment systems that are suitable for raw syngas fermentation need to be urgently developed (Liew et al. 2016).

4 Process Coupling and Perspectives

First, impurities such as NO_x in syngas need to be removed for the use in syngas fermentation. Conventional syngas upgrading includes cyclones (for particulate removal), water quench scrubbers for removal of ammonia and trace impurities, and mixed oxide sorbents for H_2S removal (Torres et al. 2007; Woolcock and Brown

2013; Xu et al. 2011). Shen et al. (2016) recently reviewed syngas cleaning processes and proposed that biochar and bio-oil can be potentially used for gas cleaning in biomass pyrolysis/gasification. Other techniques, such as membrane separation, may also be used to purify syngas (Castro-Dominguez et al. 2017; Parsley et al. 2014). Castro-Dominguez et al. (2017) demonstrated the pilot-scale application of palladium-based membrane technology for the purification of H_2 from coal-derived syngas; the results indicated that the purity of the produced H_2 ranged from 99.87 to 98% and that H_2 production of 2.72 kg/day and recovery of 64% were achieved.

Second, metabolites in syngas fermentation always consist of a mixture; thus, coupling processes are necessary to use the mixed products. As potential substitutes for petroleum fuel, ethanol has attracted more attention for their higher energy density, less corrosiveness, and higher compatibility with gasoline (Xue et al. 2013). For high volatility under high temperature, ethanol can be easily recovered by gas stripping after coupling with syngas fermentation. Löser et al. (2005) showed that more than 30% of produced ethanol in the reactor could be removed. Xue et al. (2016) recently developed two-stage gas stripping and pervaporation integrated with acetone–butanol–ethanol (ABE) fermentation for butanol recovery. The results indicated that considerably more ABE (27.5 g/L acetone, 75.5 g/L butanol, 7.0 g/L ethanol) were produced in fed-batch fermentation.

Third, electrodialysis (ED) is a traditional technology and can be used to separate and concentrate organic acids (Moresi and Sappino 2000; Zhang et al. 2011). Redwood et al. (2012) proposed an integrated hydrogen refinery of food wastes in a synergistic combination of photofermentation, extractive fermentation, and hydrothermal hydrolysis. In this process, ED provided the key link in waste to energy for the selective separation of organic acids. Zhang et al. (2009) proposed the use of a mixture of water and ethanol to be used as a medium for enhancing the solubility of sebacic acid, which can also facilitate the recovery of medium long-chain organic acids, such as caproate and caprylate; this technique requires further study. Except for the bacterial metabolites of organic acid and alcohol, the components of MCF broth normally include inorganic salts, which decrease the real separation factors, such as current efficiency. Zhang et al. (2011) analyzed the ion competition between organic acids (e.g., formate, acetate, propionate, and butyrate) and inorganic salts (e.g., HPO_4^{2-} and Cl^-) and found that membrane selectivity depended on the size, charge, and functional groups of the organic ions. The concentrations of acetate, propionate, and butyrate are decreased more slowly because of the presence of inorganic ions. Current efficiency was even lower than 30%; thus, the development of the selective separation of membranes for specific metabolites is urgently needed. Coupling of syngas fermentation with ED deserves further research.

Fourth, for a longer carbon chain and a lower O/C ratio, the mixture of the produced medium-chain fatty acids could also be upgraded to biofuels by hydrogen reduction (Steinbusch et al. 2011; Zhang et al. 2013b). The produced metabolites in syngas fermentation, such as acetate and ethanol, can be suitable substrates for the production of medium-chain fatty acids (Grootscholten et al. 2014; Kucek et al. 2016). Kucek et al. (2016) achieved high *n*-caprylate productivity (0.33 g/(L•day)) by feeding a high substrate ratio of ethanol to acetate amounting to 15 gCOD/gCOD and extracting the product from the bioreactor broth. Xu et al. (2015) extracted *n*-caproate from the bioreactor broth by a hollow fiber membrane and found that selective phase separation occurred because of the low maximum solubility of this acid, which allowed the separation of simple products into an oily liquid containing 90% *n*-caproic and *n*-caprylic acids. However, the bacterial toxicities of medium-chain carboxylic acids still need to be considered (Zhang et al. 2013b). Khor et al. (2017) recently converted medium-chain fatty acids to decane (0.41$/Kg) via Kolbe electrolysis; the low density and low solubility of decane render it a rather simple product to target in terms of process engineering because the liquid fuel market is extensive and well-entrenched (Khor et al. 2017).

Lastly, apart from the metabolites shown in Fig. 1, syngas can also be converted to biopolymers, such as polyhydroxyalkanoates (PHA) (Revelles et al. 2016). Lagoa-Costa et al. (2017) recently proposed a two-stage syngas utilization system by using *C. autoethanogenum* and a highly enriched PHA-accumulating biomass that could convert syngas to ethanol, 2,3-butanediol, and PHA; the maximum PHA content was 24%. Meanwhile, MFC is a fast-growing environmental biotechnology in which bio-convertible substrates are consumed with simultaneous electron generation (Logan and Regan 2006; Schroder et al. 2015). Syngas can also be converted to electricity in MCF. Hussain et al. (2012) demonstrated electricity generation in a thermophilic MFC operated on syngas (CO and H_2, 50:50 v/v) as the sole electron donor, with volumetric power output ranging from 30 to 35 mW/L and syngas conversion efficiency ranging from 87 to 98%. Foley et al. (2010) showed that MFC provides no significant environmental benefit relative to conventional anaerobic treatment; by contrast, a microbial electrolysis cell provides significant environmental benefits for biochemical production. Consequently, syngas utilization in a microbial electrolysis cell may also need to be evaluated in future research. Thus, syngas fermentation provides a promising platform for biochemical production but requires other related methods, including membrane separation and MFC, to promote its application worldwide.

5 Conclusion

In this chapter, the basic bioreactions of the Wood–Ljungdahl pathway and reverse β-oxidation reaction and thermodynamics are summarized in Sect. 2. The operating conditions—pH, temperature, CO and H_2 partial pressure, and impurities of tar and NO_x—and the reactor configuration are reviewed in Sect. 3. Lastly, syngas fermentation coupled with other technologies, such as syngas pretreatment and membrane technology, was necessary for its application, as summarized in Sect. 4. Similarly, other high-potential technologies such as PHA production and MFC are also reviewed in Sect. 4. Thus, syngas fermentation provides a promising platform for biochemical production, but to promote its application, coupled technologies are still necessary.

Acknowledgments The authors would like to acknowledge the financial support from National Natural Science Foundation of China (51478447 and 51408530), Foundation of Hebei Education Department (BJ2017014), and the Program for Changjiang Scholars and Innovative Research Team in University.

References

Abubackar HN, Veiga MC, Kennes C (2011) Biological conversion of carbon monoxide: rich syngas or waste gases to bioethanol. Biofuels Bioprod Biorefin 5(1):93–114

Abubackar HN, Veiga MC, Kennes C (2015) Carbon monoxide fermentation to ethanol by Clostridium autoethanogenum in a bioreactor with no accumulation of acetic acid. Bioresour Technol 186:122–127

Ahmed A, Lewis RS (2007) Fermentation of biomass-generated synthesis gas: effects of nitric oxide. Biotechnol Bioeng 97(5):1080–1086

Alves JI, Stams AJM, Plugge CM, Madalena Alves M, Sousa DZ (2013) Enrichment of anaerobic syngas-converting bacteria from thermophilic bioreactor sludge. FEMS Microbiol Ecol 86(3):590–597

Angenent LT, Richter H, Buckel W, Spirito CM, Steinbusch KJJ, Plugge CM, Strik DPBTB, Grootscholten TIM, Buisman CJN, Hamelers HVM (2016) Chain elongation with reactor microbiomes: open-culture biotechnology to produce biochemicals. Environ Sci Technol 50(6):2796–2810

Ayala DF, Ferre V, Judd SJ (2011) Membrane life estimation in full-scale immersed membrane bioreactors. J Membr Sci 378(1):95–100

Basen M, Müller V (2017) "Hot" acetogenesis. Extremophiles 21(1):15–26

Bastidas-Oyanedel J-R, Aceves-Lara C-A, Ruiz-Filippi G, Steyer J-P (2008) Thermodynamic analysis of energy transfer in acidogenic cultures. Eng Life Sci 8(5):487–498

Benalcázar EA, Deynoot BG, Noorman H, Osseweijer P, Posada JA (2017) Production of bulk chemicals from lignocellulosic biomass via thermochemical conversion and syngas fermentation: a comparative techno-economic and environmental assessment of different site-specific supply chain configurations. Biofuels Bioprod Biorefin 11(5):861–886

Bengelsdorf FR, Straub M, Dürre P (2013) Bacterial synthesis gas (syngas) fermentation. Environ Technol 34(13–14):1639–1651

Bertsch J, Müller V (2015) CO metabolism in the acetogen Acetobacterium woodii. Appl Environ Microbiol 81(17):5949–5956

Castro-Dominguez B, Mardilovich IP, Ma R, Kazantzis NK, Dixon AG, Ma YH (2017) Performance of a pilot-scale multitube membrane module under coal-derived syngas for hydrogen production and separation. J Membr Sci 523(suppl C):515–523

Dai K, Wen J-L, Zhang F, Zeng RJ (2017) Valuable biochemical production in mixed culture fermentation: fundamentals and process coupling. Appl Microbiol Biotechnol 101(17):6575–6586

Datar RP, Shenkman RM, Cateni BG, Huhnke RL, Lewis RS (2004) Fermentation of biomass-generated producer gas to ethanol. Biotechnol Bioeng 86(5):587–594

Devarapalli M, Atiyeh HK, Phillips JR, Lewis RS, Huhnke RL (2016) Ethanol production during semi-continuous syngas fermentation in a trickle bed reactor using Clostridium ragsdalei. Bioresour Technol 209:56–65

Diender M, Stams AJM, Sousa DZ (2015) Pathways and bioenergetics of anaerobic carbon monoxide fermentation. Front Microbiol 6:1275

Drake HL, Gößner AS, Daniel SL (2008) Old acetogens, new light. Ann N Y Acad Sci 1125(1):100–128

Drews A (2010) Membrane fouling in membrane bioreactors—characterisation, contradictions, cause and cures. J Membr Sci 363(1–2):1–28

Esquivel-Elizondo S, Delgado AG, Rittmann BE, Krajmalnik-Brown R (2017) The effects of CO2 and H2 on CO metabolism by pure and mixed microbial cultures. Biotechnol Biofuels 10(1):220

Fabbri D, Torri C (2016) Linking pyrolysis and anaerobic digestion (Py-AD) for the conversion of lignocellulosic biomass. Curr Opin Biotechnol 38:167–173

Fernández-Naveira Á, Veiga MC, Kennes C (2017) H-B-E (hexanol-butanol-ethanol) fermentation for the production of higher alcohols from syngas/waste gas. J Chem Technol Biotechnol 92(4):712–731

Foley JM, Rozendal RA, Hertle CK, Lant PA, Rabaey K (2010) Life cycle assessment of high-rate anaerobic treatment, microbial fuel cells, and microbial electrolysis cells. Environ Sci Technol 44(9):3629–3637

Ganigué R, Sánchez-Paredes P, Bañeras L, Colprim J (2016) Low fermentation pH is a trigger to alcohol production, but a killer to chain elongation. Front Microbiol 7:702

Grootscholten TIM, Strik DPBTB, Steinbusch KJJ, Buisman CJN, Hamelers HVM (2014) Two-stage medium chain fatty acid (MCFA) production from municipal solid waste and ethanol. Appl Energy 116:223–229

Henstra AM, Sipma J, Rinzema A, Stams AJM (2007) Microbiology of synthesis gas fermentation for biofuel production. Curr Opin Biotechnol 18(3):200–206

Hussain A, Mehta P, Raghavan V, Wang H, Guiot SR, Tartakovsky B (2012) The performance of a thermophilic microbial fuel cell fed with synthesis gas. Enzyme Microb Technol 51(3):163–170

Jiang L, Wang J, Liang S, Cai J, Xu Z, Cen P, Yang S, Li S (2011) Enhanced butyric acid tolerance and bioproduction by Clostridium tyrobutyricum immobilized in a fibrous bed bioreactor. Biotechnol Bioeng 108(1):31–40

Jing Y, Campanaro S, Kougias P, Treu L, Angelidaki I, Zhang S, Luo G (2017) Anaerobic granular sludge for simultaneous biomethanation of synthetic wastewater and CO with focus on the identification of CO-converting microorganisms. Water Res 126(suppl C):19–28

Karakashev D, Batstone DJ, Trably E, Angelidaki I (2006) Acetate oxidation is the dominant methanogenic pathway from acetate in the absence of Methanosaetaceae. Appl Environ Microbiol 72(7):5138–5141

Khor WC, Andersen S, Vervaeren H, Rabaey K (2017) Electricity-assisted production of caproic acid from grass. Biotechnol Biofuels 10(1):180

Kleerebezem R, Van Loosdrecht MCM (2010) A generalized method for thermodynamic state analysis of environmental systems. Crit Rev Environ Sci Technol 40(1):1–54

Köpke M, Held C, Hujer S, Liesegang H, Wiezer A, Wollherr A, Ehrenreich A, Liebl W, Gottschalk G, Dürre P (2010) Clostridium ljungdahlii represents a microbial production platform based on syngas. Proc Natl Acad Sci 107(29):13087–13092

Köpke M, Mihalcea C, Bromley JC, Simpson SD (2011) Fermentative production of ethanol from carbon monoxide. Curr Opin Biotechnol 22(3):320–325

Kucek LA, Spirito CM, Angenent LT (2016) High n-caprylate productivities and specificities from dilute ethanol and acetate: chain elongation with microbiomes to upgrade products from syngas fermentation. Energ Environ Sci 9(11):3482–3494

Lagoa-Costa B, Abubackar HN, Fernández-Romasanta M, Kennes C, Veiga MC (2017) Integrated bioconversion of syngas into bioethanol and biopolymers. Bioresour Technol 239(suppl C):244–249

Latif H, Zeidan AA, Nielsen AT, Zengler K (2014) Trash to treasure: production of biofuels and commodity chemicals via syngas fermenting microorganisms. Curr Opin Biotechnol 27:79–87

Lee HS, Salerno MB, Rittmann BE (2008) Thermodynamic evaluation on H_2 production in glucose fermentation. Environ Sci Technol 42(7):2401–2407

Lee J, Jang N, Yasin M, Lee EY, Chang IS, Kim C (2016) Enhanced mass transfer rate of methane via hollow fiber membrane modules for Methylosinus trichosporium OB3b fermentation. J Ind Eng Chem 39:149–152

Liew F, Martin ME, Tappel RC, Heijstra BD, Mihalcea C, Köpke M (2016) Gas fermentation – a flexible platform for commercial scale production of low carbon fuels and chemicals from waste and renewable feedstocks. Front Microbiol 7:694

Liu S, Qureshi N (2009) How microbes tolerate ethanol and butanol. N Biotechnol 26(3–4):117–121
Liu K, Atiyeh HK, Stevenson BS, Tanner RS, Wilkins MR, Huhnke RL (2014) Mixed culture syngas fermentation and conversion of carboxylic acids into alcohols. Bioresour Technol 152:337–346
Logan B, Regan J (2006) Microbial fuel cells-challenges and applications. Environ Sci Technol 40(17):5172–5180
Löser C, Schröder A, Deponte S, Bley T (2005) Balancing the ethanol formation in continuous bioreactors with ethanol stripping. Eng Life Sci 5(4):325–332
Louis P, Duncan SH, McCrae SI, Millar J, Jackson MS, Flint HJ (2004) Restricted distribution of the butyrate kinase pathway among butyrate-producing bacteria from the human colon. J Bacteriol 186(7):2099–2106
Martin KJ, Nerenberg R (2012) The membrane biofilm reactor (MBfR) for water and wastewater treatment: principles, applications, and recent developments. Bioresour Technol 122:83–94
Martin ME, Richter H, Saha S, Angenent LT (2016) Traits of selected Clostridium strains for syngas fermentation to ethanol. Biotechnol Bioeng 113(3):531–539
Massaro V, Digiesi S, Mossa G, Ranieri L (2015) The sustainability of anaerobic digestion plants: a win–win strategy for public and private bodies. J Clean Prod 104:445–459
Meng X, Ragauskas AJ (2014) Recent advances in understanding the role of cellulose accessibility in enzymatic hydrolysis of lignocellulosic substrates. Curr Opin Biotechnol 27:150–158
Meng F, Zhang S, Oh Y, Zhou Z, Shin H-S, Chae S-R (2017) Fouling in membrane bioreactors: an updated review. Water Res 114:151–180
Miltner A, Wukovits W, Pröll T, Friedl A (2010) Renewable hydrogen production: a technical evaluation based on process simulation. J Clean Prod 18(suppl 1):S51–S62
Mohammadi M, Younesi H, Najafpour G, Mohamed AR (2012) Sustainable ethanol fermentation from synthesis gas by Clostridium ljungdahlii in a continuous stirred tank bioreactor. J Chem Technol Biotechnol 87(6):837–843
Moresi M, Sappino F (2000) Electrodialytic recovery of some fermentation products from model solutions: techno-economic feasibility study. J Membr Sci **164**(1–2):129–140
Muller V (2003) Energy conservation in acetogenic bacteria. Appl Environ Microbiol 69(11):6345–6353
Munasinghe PC, Khanal SK (2010) Syngas fermentation to biofuel: evaluation of carbon monoxide mass transfer coefficient (kLa) in different reactor configurations. Biotechnol Prog **26**(6):1616–1621
Nerenberg R (2016) The membrane-biofilm reactor (MBfR) as a counter-diffusional biofilm process. Curr Opin Biotechnol 38:131–136
Parsley D, Ciora RJ, Flowers DL, Laukaitaus J, Chen A, Liu PKT, Yu J, Sahimi M, Bonsu A, Tsotsis TT (2014) Field evaluation of carbon molecular sieve membranes for the separation and purification of hydrogen from coal- and biomass-derived syngas. J Membr Sci 450(suppl C):81–92
Peintner C, Zeidan AA, Schnitzhofer W (2010) Bioreactor systems for thermophilic fermentative hydrogen production: evaluation and comparison of appropriate systems. J Clean Prod 18:S15–S22
Ramió-Pujol S, Ganigué R, Bañeras L, Colprim J (2015) Incubation at 25 °C prevents acid crash and enhances alcohol production in Clostridium carboxidivorans P7. Bioresour Technol 192:296–303
Redwood MD, Orozco RL, Majewski AJ, Macaskie LE (2012) An integrated biohydrogen refinery: synergy of photofermentation, extractive fermentation and hydrothermal hydrolysis of food wastes. Bioresour Technol 119:384–392
Revelles O, Tarazona N, García JL, Prieto MA (2016) Carbon roadmap from syngas to polyhydroxyalkanoates in Rhodospirillum rubrum. Environ Microbiol 18(2):708–720
Richter H, Molitor B, Wei H, Chen W, Aristilde L, Angenent LT (2016) Ethanol production in syngas-fermenting Clostridium ljungdahlii is controlled by thermodynamics rather than by enzyme expression. Energ Environ Sci 9(7):2392–2399

Sancho-Navarro S, Cimpoia R, Bruant G, Guiot SR (2016) Biomethanation of syngas using anaerobic sludge: shift in the catabolic routes with the CO partial pressure increase. Front Microbiol 7:1188

Schroder U, Harnisch F, Angenent LT (2015) Microbial electrochemistry and technology: terminology and classification. Energ Environ Sci 8(2):513–519

Schuchmann K, Muller V (2014) Autotrophy at the thermodynamic limit of life: a model for energy conservation in acetogenic bacteria. Nat Rev Microbiol **12**(12):809–821

Seedorf H, Fricke WF, Veith B, Brüggemann H, Liesegang H, Strittmatter A, Miethke M, Buckel W, Hinderberger J, Li F, Hagemeier C, Thauer RK, Gottschalk G (2008) The genome of Clostridium kluyveri, a strict anaerobe with unique metabolic features. Proc Natl Acad Sci **105**(6):2128–2133

Shen Y, Brown R, Wen Z (2014) Syngas fermentation of Clostridium carboxidivoran P7 in a hollow fiber membrane biofilm reactor: evaluating the mass transfer coefficient and ethanol production performance. Biochem Eng J **85**:21–29

Shen Y, Jarboe L, Brown R, Wen Z (2015) A thermochemical–biochemical hybrid processing of lignocellulosic biomass for producing fuels and chemicals. Biotechnol Adv **33**(8):1799–1813

Shen Y, Wang J, Ge X, Chen M (2016) By-products recycling for syngas cleanup in biomass pyrolysis – an overview. Renew Sustain Energy Rev **59**(suppl C):1246–1268

Sheth PN, Babu BV (2010) Production of hydrogen energy through biomass (waste wood) gasification. Int J Hydrogen Energy **35**(19):10803–10810

Singla A, Verma D, Lal B, Sarma PM (2014) Enrichment and optimization of anaerobic bacterial mixed culture for conversion of syngas to ethanol. Bioresour Technol **172**:41–49

Speight JG (2005) Lange's handbook of chemistry. McGraw-Hill, New York

Spirito CM, Richter H, Rabaey K, Stams AJM, Angenent LT (2014) Chain elongation in anaerobic reactor microbiomes to recover resources from waste. Curr Opin Biotechnol **27**(0):115–122

Steinbusch KJJ, Hamelers HVM, Buisman CJN (2008) Alcohol production through volatile fatty acids reduction with hydrogen as electron donor by mixed cultures. Water Res **42**(15):4059–4066

Steinbusch KJJ, Hamelers HVM, Plugge CM, Buisman CJN (2011) Biological formation of caproate and caprylate from acetate: fuel and chemical production from low grade biomass. Energ Environ Sci **4**(1):216–224

Temudo MF, Muyzer G, Kleerebezem R, van Loosdrecht MCM (2008) Diversity of microbial communities in open mixed culture fermentations: impact of the pH and carbon source. Appl Microbiol Biotechnol **80**(6):1121–1130

Thammasittirong SN-R, Thirasaktana T, Thammasittirong A, Srisodsuk M (2013) Improvement of ethanol production by ethanol-tolerant Saccharomyces cerevisiae UVNR56. SpringerPlus **2**(1):583

Thauer RK, Jungermann K, Henninger H, Wenning J, Decker K (1968) The energy metabolism of Clostridium kluyveri. Eur J Biochem **4**(2):173–180

Thauer RK, Jungermann K, Decker K (1977) Energy conservation in chemotrophic anaerobic bacteria. Bacteriol Rev **41**(1):100–180

Thauer RK, Kaster AK, Seedorf H, Buckel W, Hedderich R (2008) Methanogenic archaea: ecologically relevant differences in energy conservation. Nat Rev Microbiol **6**(8):579–591

Torres W, Pansare SS, Goodwin JG (2007) Hot gas removal of tars, ammonia, and hydrogen sulfide from biomass gasification gas. Catal Rev **49**(4):407–456

Valgepea K, de Souza Pinto Lemgruber R, Meaghan K, Palfreyman RW, Abdalla T, Heijstra BD, Behrendorff JB, Tappel R, Köpke M, Simpson SD, Nielsen LK, Marcellin E (2017) Maintenance of ATP homeostasis triggers metabolic shifts in gas-fermenting acetogens. Cell Syst **4**(5):505–515.e5

Wang Z, Ma J, Tang CY, Kimura K, Wang Q, Han X (2014) Membrane cleaning in membrane bioreactors: a review. J Membr Sci **468**:276–307

Wang Y-Q, Yu S-J, Zhang F, Xia X-Y, Zeng RJ (2017) Enhancement of acetate productivity in a thermophilic (55 °C) hollow-fiber membrane biofilm reactor with mixed culture syngas (H2/CO2) fermentation. Appl Microbiol Biotechnol **101**(6):2619–2627

Wang H-J, Dai K, Xia X-Y, Wang Y-Q, Zeng RJ, Zhang F (2018a) Tunable production of ethanol and acetate from synthesis gas by mesophilic mixed culture fermentation in a hollow fiber membrane biofilm reactor. J Clean Prod **187**:165–170

Wang Y-Q, Zhang F, Zhang W, Dai K, Wang H-J, Li X, Zeng RJ (2018b) Hydrogen and carbon dioxide mixed culture fermentation in a hollow-fiber membrane biofilm reactor at 25°C. Bioresour Technol **249**:659–665

Wilbanks B, Trinh CT (2017) Comprehensive characterization of toxicity of fermentative metabolites on microbial growth. Biotechnol Biofuels **10**(1):262

Woolcock PJ, Brown RC (2013) A review of cleaning technologies for biomass-derived syngas. Biomass Bioenergy **52**(suppl C):54–84

Xu D, Tree DR, Lewis RS (2011) The effects of syngas impurities on syngas fermentation to liquid fuels. Biomass Bioenergy **35**(7):2690–2696

Xu J, Guzman JJL, Andersen SJ, Rabaey K, Angenent LT (2015) In-line and selective phase separation of medium-chain carboxylic acids using membrane electrolysis. Chem Commun **51**(31):6847–6850

Xue C, Zhao JB, Liu FF, Lu CC, Yang ST, Bai FW (2013) Two-stage in situ gas stripping for enhanced butanol fermentation and energy-saving product recovery. Bioresour Technol **135**:396–402

Xue C, Liu F, Xu M, Zhao J, Chen L, Ren J, Bai F, Yang S-T (2016) A novel in situ gas stripping-pervaporation process integrated with acetone-butanol-ethanol fermentation for hyper n-butanol production. Biotechnol Bioeng **113**(1):120–129

Yasin M, Jeong Y, Park S, Jeong J, Lee EY, Lovitt RW, Kim BH, Lee J, Chang IS (2015) Microbial synthesis gas utilization and ways to resolve kinetic and mass-transfer limitations. Bioresour Technol **177**:361–374

Zeng Y, Zhao S, Yang S, Ding S-Y (2014) Lignin plays a negative role in the biochemical process for producing lignocellulosic biofuels. Curr Opin Biotechnol **27**:38–45

Zhang F, Huang C, Xu T (2009) Production of sebacic acid using two-phase bipolar membrane electrodialysis. Ind Eng Chem Res **48**(16):7482–7488

Zhang Y, Pinoy L, Meesschaert B, Van der Bruggen B (2011) Separation of small organic ions from salts by ion-exchange membrane in electrodialysis. AIChE J **57**(8):2070–2078

Zhang F, Ding J, Shen N, Zhang Y, Ding Z-W, Dai K, Zeng RJ (2013a) In situ hydrogen utilization for high fraction acetate production in mixed culture hollow-fiber membrane biofilm reactor. Appl Microbiol Biotechnol **97**(23):10233–10240

Zhang F, Ding J, Zhang Y, Chen M, Ding Z-W, van Loosdrecht MCM, Zeng RJ (2013b) Fatty acids production from hydrogen and carbon dioxide by mixed culture in the membrane biofilm reactor. Water Res **47**(16):6122–6129

Zhang F, Zhang Y, Chen M, van Loosdrecht MCM, Zeng RJ (2013c) A modified metabolic model for mixed culture fermentation with energy conserving electron bifurcation reaction and metabolite transport energy. Biotechnol Bioeng **110**(7):1884–1894

Zhao H-P, Ontiveros-Valencia A, Tang Y, Kim B-O, VanGinkel S, Friese D, Overstreet R, Smith J, Evans P, Krajmalnik-Brown R, Rittmann B (2014) Removal of multiple electron acceptors by pilot-scale, two-stage membrane biofilm reactors. Water Res **54**:115–122

Bioelectrochemical Systems for the Valorization of Organic Residues

Roman Moscoviz, Elie Desmond-Le Quéméner, Eric Trably, and Nicolas Bernet

1 Principles and Fundamentals of BES

Bioelectrochemical systems (BES) are electrochemical devices in which at least one reaction, oxidation at the anode or reduction at the cathode, is catalyzed by a microorganism (or a microbial community) or an enzyme. Anode and cathode may be separated by a membrane. Enzymatic BES are generally considered as unsustainable and less durable compared to microbial BES (Santoro et al. 2017); this is why this chapter will focus on microbial BES, also called microbial electrochemical technologies (MET).

Microbial fuel cells (MFCs) that directly produce electricity from the oxidation of substrates, and more generally microbial BES that have different other applications, have been the topic of increasing research for nearly 20 years, although the concept of using microorganisms to generate electricity is not new since it has been published more than hundred years ago (Potter 1911). The growing interest in BES can be explained by the fact that they nicely combine the treatment of waste or wastewater through oxidation of organic matter at the anode with the production of energy (electricity, hydrogen, methane) or molecules of interest at the cathode, depending on the type of BES.

Microbial electrochemical technologies are based on the capacity of some microorganisms to exchange electrons with an electrode. In most cases, these bacteria use the anode as an electron acceptor and have been called electroactive bacteria (EAB), anode-respiring bacteria (ARB), electrode-oxidizing bacteria, exoelectrogens, or electricigens (Lovley 2008). Several mechanisms for electron transfer to anodes have been proposed including direct electron transfer via outer-surface c-type cytochromes, long-range electron transfer via microbial nanowires, electron flow through a conductive biofilm matrix containing cytochromes, and soluble electron

R. Moscoviz · E. Desmond-Le Quéméner · E. Trably · N. Bernet (✉)
LBE, Univ Montpellier, INRA, Narbonne, France
e-mail: roman.moscoviz@suez.com; elie.lequemener@inra.fr; eric.trably@inra.fr; nicolas.bernet@inra.fr

© Springer Nature Switzerland AG 2019
J.-R. Bastidas-Oyanedel, J. E. Schmidt (eds.), *Biorefinery*,
https://doi.org/10.1007/978-3-030-10961-5_21

shuttles (Lovley 2008). The major electron transfer mechanisms are presented in Fig. 1. Electromicrobiology has been recently defined as a subdiscipline of microbiology that involves extracellular electron transfer (EET) to (or from) insoluble electron active redox compounds located outside the outer membrane of the cell (Lovley 2012; Nealson and Rowe 2016).

In the case of direct electron transfer, the bacteria are structured as an electroactive biofilm on the electrode. Thus, one strategy to determine which microorganisms contribute to power production is to identify those that selectively colonize anode surfaces. *Geobacteraceae* are generally predominant, from which bacteria of the genus *Geobacter* are the most studied, but *Gammaproteobacteria* (including *Shewanella* another model exoelectrogenic bacteria), *Betaproteobacteria*, *Rhizobiales*, or *Clostridia* may also been identified (Lovley 2008). In many cases, especially in the case of complex substrates, the efficiency of the BES depends on the interactions of the anodic microbial communities (Kokko et al. 2018). An important parameter to characterize the performance of the biofilm to transfer electrons from substrate oxidation to the anode is the coulombic efficiency (CE). It is defined

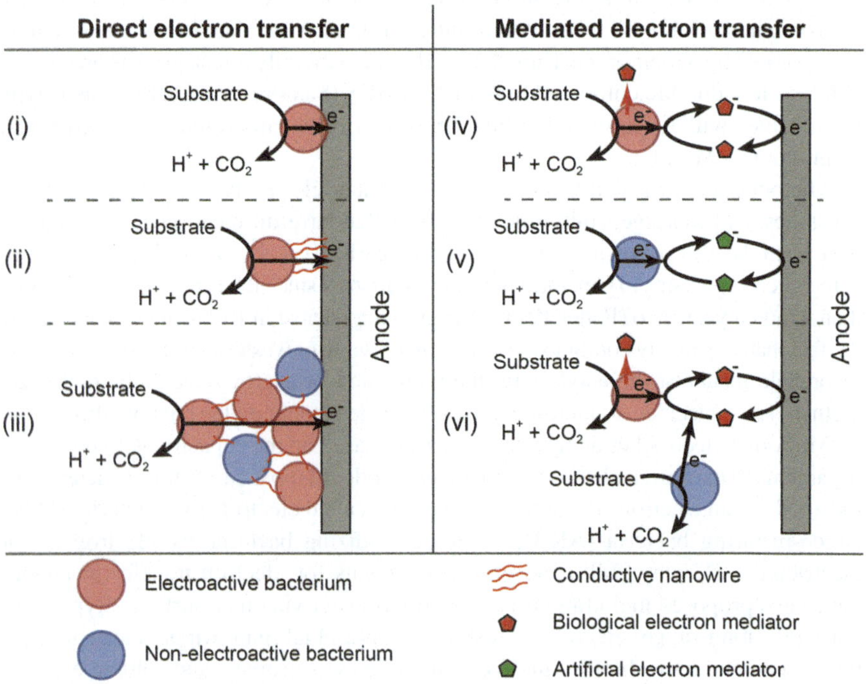

Fig. 1 Extracellular electron transfer mechanisms. Direct electron transfer through (i) membrane-bound cytochromes; (ii) conductive pili called nanowires; (iii) a conductive biofilm containing at least an electroactive species. Mediated electron transfer through (iv) biologically synthesized electron carriers (e.g., flavins); (v) artificially added electron mediators (e.g., neutral red); (vi) biologically synthesized electron carriers used by a non-electroactive species. Adapted from Pham et al. (2009)

as the ratio of total coulombs actually transferred to the anode from the substrate to maximum possible coulombs if all substrate removal produced current (Logan et al. 2006).

More recently, it has been shown that some microorganisms can directly accept electrons from electrodes and therefore be used as biocatalyst at a cathode. Potential electron acceptors for these electrotrophs include carbon dioxide, nitrate, metals, chlorinated compounds, organic acids, protons, and oxygen. These electrotrophs can be used in biocathodes for nitrogen removal (denitrification), bioremediation of chlorinated compounds, or microbial electrosynthesis in which CO_2 and water are converted into organic molecules (Lovley 2011).

Microbial fuel cells (MFCs) were the first and are still the most studied MET. They combine the oxidation of organic substrate at the anode with a thermodynamically favorable reduction at the cathode, in most cases electrochemical oxygen reduction to water (Fig. 2), and will be developed in Sect. 2. Other reactions are possible, such as nitrate reduction involving denitrifying bacteria and making possible to use MFC for nitrogen removal (Virdis et al. 2008). Thus, the main advantage of MFC compared to conventional wastewater treatment processes is the direct conversion of dissolved COD to electricity. However, the development of this technology has to face different challenges such as the low energy produced or the development of efficient cathodes for oxygen reduction (Santoro et al. 2017).

In 2005, it was discovered that augmenting the electrochemical potential achieved by bacteria in a MFC with an additional voltage made it possible to produce

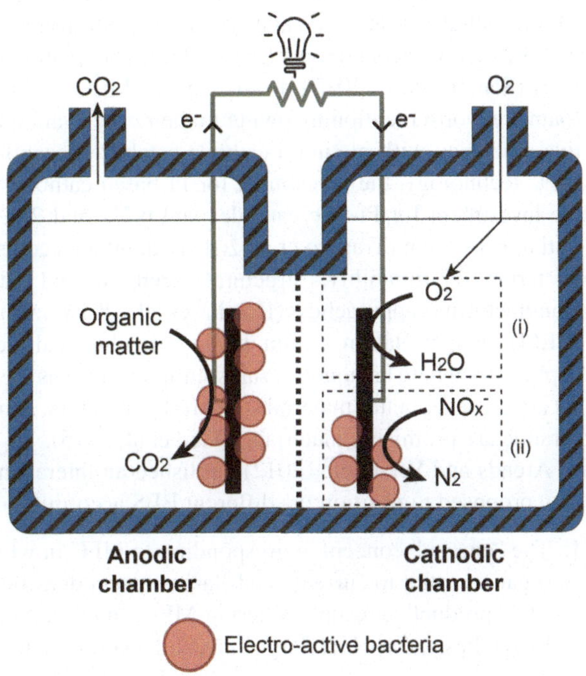

Fig. 2 Microbial fuel cell. Organic matter is fed in the anodic chamber and degraded by electroactive bacteria, while electrons are released in the anode. In the cathodic chamber, electrons are consumed to reduce (i) O_2 either with chemical or biological catalyzers; (ii) nitrite and nitrate into N_2 (i.e., denitrification). The global reactor is thermodynamically favorable and produces electrical energy. The anodic and cathodic chambers can be separated by an anion or cation exchange membrane or left unseparated

hydrogen at the cathode directly from the oxidized organic matter (Liu et al. 2005). A voltage must be applied because hydrogen formation at the cathode using organic substrate at the anode is not spontaneous. However, energy recovered as hydrogen may be theoretically as high as eight times the electrical energy input. This system detailed in Part 3 of this chapter is called microbial electrolysis cell (MEC). As in a MFC, it combines a biocatalyzed oxidation at the anode with an electrochemical reduction at the cathode. In the presence of methanogenic Archaea at the cathode, methane can be directly produced by a process called electromethanogenesis (Cheng et al. 2009). This process could be used to store electrical energy (power to gas), for biogas upgrading or coupled to waste treatment to increase anaerobic digestion performance (Blasco-Gomez et al. 2017).

Many other applications have been proposed and tested, mostly at laboratory scale, including microbial desalination cells and microbial electrosynthesis cells. Microbial desalination cells (MDC) have been developed with the objective of treating wastewater, generating electricity, and desalinating water simultaneously (Cao et al. 2009). Microbial electrosynthesis uses microorganisms to catalyze reactions at the cathode for the production of molecules (Rabaey and Rozendal 2010).

Important elements in BES are the electrodes and membrane materials. Anodes that have to be colonized by the electroactive biofilm should have characteristics favoring interactions between microorganisms and the electrode. They include good electrical conductivity and low resistance, strong biocompatibility, chemical stability and anticorrosion, large surface area, and appropriate mechanical strength and toughness (Zhou et al. 2011). Carbon materials (carbon cloth, brush, felt, rod, etc.) and metals (mainly stainless steel) are the most used anodes. Cathodes used in MFCs include an abiotic catalyst and can be grouped in platinum-based (PGM-based), carbonaceous-based (metal-free), and platinum group metal-free (PGM-free) (Santoro et al. 2017). In the case of MEC, stainless steel, MoS_2, and nickel foam are worth mentioning owing to their easy availability, low cost, and comparative efficiency with platinum cathode catalyst. Recently, in an economic study on MFC technology, the investment for Pt-based cathodes was estimated to 233% of the investment for Pt-free cathodes and 6.2% and 2.8%, respectively, of the total initial investment (Trapero et al. 2017). Stainless steel is probably the best material in terms of durability, structural strength, and easy adaptability to mass-manufacturing approaches (Kundu et al. 2013). Cation exchange membranes (CEM), such as Nafion, commonly used in chemical fuel cells are suitable in MET application where an aqueous salt solution is used as the electrolyte. Porous materials are however better materials for MET separators. Among them, earthenware and ceramic are promising materials (Daud et al. 2015).

Arends and Verstraete (2012) published an interesting review on BES in which they proposed to position the different BES according to three main concepts:

1. The "energy" concept, corresponding to MFC in which the direct conversion of organic matter to current needs large current densities (at least $10\ A\ m^{-2}$).
2. The "product" concept gathering MEC, microbial desalination, microbial electrosynthesis, for which large current densities are needed (at least $10\ A\ m^{-2}$).

3. The "sustainability" concept in which the performance is not directly dependent on the current but small densities of a few mA m^{-2} can be associated with large impacts (removal of soil- or µ-pollutants, control of Eh, heavy metal precipitation, stabilizing anaerobic digestion). Electro-fermentation that will be described later is also included in this concept.

2 Microbial Fuel Cells (MFCs)

Basically, because they involve a biofilm, the applied target of MFC is the treatment of liquid effluents or wastewater. The idea is to couple wastewater treatment at the anode and electricity production with the objective to propose an alternative to conventional wastewater treatment processes such as activated sludge for diluted wastewater or anaerobic digestion for agro-industrial effluents. Many different substrates, including wastewaters, have been tested in laboratory-scale MFC (Pandey et al. 2016; Pant et al. 2010). The principle of a MFC is presented in Fig. 2.

In MFC research, acetate has been the most widely used substrate as it is known to promote the growth of electroactive bacteria, especially *Geobacter sulfurreducens*, and thus generate high current densities. Thus, a power density as high as 4.3 W m^{-2} could be obtained in a single-chamber MFC with a double-cloth electrode assembly fed with acetate (Fan et al. 2012). Moreover, acetate is a major product of fermentation from different complex carbon substrates. However, electroactive biofilms from complex ecosystems grown with acetate have a low microbial diversity.

More complex fermentable substrates such as glucose are considered to generate lower coulombic efficiency compared to simple molecules, such as acetate, because the microbial communities may also contain other microbial species that direct electron flow away from current production. These microbial species include methanogens, aerobic or facultative aerobic bacteria, sulfate- and nitrate-reducers, and homoacetogens (Kokko et al. 2018). However, due to the more diverse metabolic pathways to degrade such substrates, a higher microbial diversity and thus adaptability can be expected (Chae et al. 2009).

Due to the potential application of MFC to wastewater treatment, many organic wastewaters have been successfully tested as substrates. They include food and food processing industry wastewater (composite vegetables and food wastewater, food processing wastewater, protein food industry wastewater, tomato pomace and acidogenic food waste leachate), beverage industry wastewaters (beer brewery wastewater, winery wastewater, fermented apple juice), confectionary industry wastewater, dairy industry wastewater (cheese whey, yoghurt wastewater), agro-processing industry wastewater (cassava mill wastewater, palm oil mill effluent, mustard tuber wastewater), livestock industry wastewater (meat industry wastewater, animal carcass wastewater, swine wastewater), refinery and distillery industry wastewater (biorefinery wastewater, corn stover neutral and acid steam-exploded liquid hydrolysate fraction, ethanol stillage wastewater, molasses wastewater), mining and allied industry wastewater, pharmaceutical industry wastewater (recalcitrant pharmaceutical industrial effluent, steroidal drug production wastewater), paper recycling industry wastewater, textile industry wastewater, petrochemical industry wastewater, and finally domestic and municipal wastewater (Pandey et al. 2016).

Most of the published studies result from experiments carried out at laboratory scale. Upscaling MFCs toward practical application has been the topic of active investigation (Logan 2010), and it still has to face important challenges including low power density and high fabrication and operation costs, even if significant progress has already been made (Butti et al. 2016). Typical maximum power densities in MFCs are ~2–3 W m^{-2} of projected electrode with upper limits of 17–19 W m^{-2} predicted (Logan and Rabaey 2012).

The first large-scale pilot plant (1 m^3) was developed by the Advanced Water Management Centre at the University of Queensland in 2006 for the treatment of brewery wastewater in Australia as reported by Logan (2010). To date, three studies investigated pilot-scale MFC for municipal wastewater treatment under practical conditions. Feng et al. (2014) operated a 250 L stackable horizontal plug flow MFC for 1 year on a WWTP with the aim of using the MFC as a standalone system for wastewater treatment (Feng et al. 2014). Different 2 L tubular MFCs were integrated in another study to the aeration tank of a WWTP to save aeration and infrastructure (Zhang et al. 2013). These articles do not give information for optimal operational strategies and how the removal of COD, suspended solids, and nitrogen will affect the biological treatment. In the most recent work, a 45-L pilot MFC system, consisting of four single-chamber membraneless MFCs, was integrated into a full-scale WWTP and operated with the effluent of the primary clarifier for 9 months in the activated sludge basin (Hiegemann et al. 2016). Energetic calculations of the model WWTP showed that energy savings due to reduced excess sludge production and energy gain of the MFC are significantly higher than the loss of energy due to reduced biogas production. However, these authors also showed that electricity generation and nutrient removal was completely inhibited by impact loads of ammonia rich wastewater (40 mg free ammonia nitrogen L^{-1}) (Hiegemann et al. 2018). Recently, a 1000-L modularized microbial fuel cell (50 modules) was operated for more than 1 year with municipal wastewater (Liang et al. 2018). The COD concentration in the effluent was kept below 50 mg L^{-1} with a removal rate of 70–90%. An unstable maximum power density within the range of 7–60 W m^{-3} (0.42–3.64 W m^{-2}) was obtained.

In 2010, Foley et al. conducted a life cycle assessment (LCA) to compare the environmental impact of anaerobic treatment with biogas generation, a microbial fuel cell treatment with direct electricity generation and a microbial electrolysis cell (MEC) with hydrogen peroxide production (Foley et al. 2010). They concluded that a MFC does not provide a significant environmental benefit compared to anaerobic digestion whereas the MEC does. The authors point out that the assessment is highly dependent on the underlying assumptions, such as the used reactor materials and target performance, which is an issue to evaluate processes that do not exist at full scale. Recently, Trapero et al. (2017) published an economic assessment of a MFC implementation for wastewater treatment in a juice processing plant. Considering different scenarios (optimistic, pessimistic, most likely), they showed that MFC may become an alternative to activated sludge for wastewater treatment.

3 Microbial Electrolysis Cells (MECs) for Hydrogen Production

Microbial electrolysis cells (MECs) are basically based on the same concept of water electrolysers, but biological oxidation of organic substrates is carried out by EAB at the anode instead of water splitting. Oxidation of organic matter by EAB generates electrons that are transferred to the circuit and protons that are released in the bulk. EAB in MECs present similar abilities than EAB in microbial fuel cells (MFCs) since they are also selected on their ability to oxidize the organic matter at the anode. Instead of the reduction of oxygen into water that occurs at the cathode in MFCs, the cathodic compartment of a MEC is anaerobic, and protons are reduced by electrons from the circuit to form H_2 in an abiotic way (Fig. 3).

Direct production of hydrogen was first reported in MECs by Liu et al. (2005) and Rozendal et al. (2006). H_2 conversion yields were ranging from 67 to 100% when operated on various simple organic substrates, such as cellulose, glucose, butyrate, lactate, propionate, ethanol, or acetate (Lee and Rittman 2010). These yields are much higher than those observed during single fermentation process, that is, thermodynamically limited to only 33% (Geelhoed et al. 2010).

The proton reduction into H_2 occurs at a lower redox potential than the chemical reactions at the anode. H_2 generation is therefore not spontaneous but requires a supply of external energy. Nonetheless, this contribution is much lower than in water electrolyser, most of the energy being provided by the oxidation of organic matter (Liu et al. 2010).

In MECs, the reaction can be supported by external supply of electrical energy. Theoretically, the minimum voltage to be applied corresponds to −0.14 V (Rozendal et al. 2006). In comparison, water oxidation in electrolyser occurs at a minimum voltage of −1.234 V (Rozendal et al. 2006). The energy required in MECs is therefore theoretically 8.7 times lower than in water electrolysers, to produce the same amount of hydrogen. However, the potential of the cathode and therefore the equilibrium potential between the electrodes also depend on the protons or cations concentrations in the solution, and therefore on the pH, which itself tends to vary during the reaction. pH values around the neutrality are preferred at the anode to promote the development of electroactive microorganisms but correspond only to very low concentrations of protons. In practice, the anode potential at a neutral pH is generally around −0.2 V (ENH) to favor the growth of EAB or lower depending on the operating conditions. Hydrogen production at the cathode will then be possible from applied potentials higher than 0.22 V. The potentials of each electrode, according to their Nernst equations, depend also on the partial pressure of H_2 (Logan et al. 2008). Under standard conditions (i.e., pH 7, 25 °C), an increase of the H_2 partial

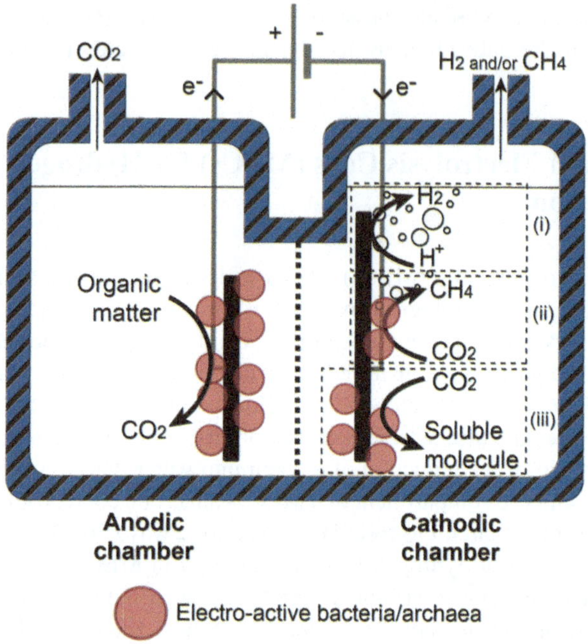

Fig. 3 As in microbial fuel cells, organic matter is fed in the anodic chamber and degraded by electroactive bacteria while electrons are released in the anode. In the cathodic chamber, electrons are consumed to reduce (i) protons into H_2 either with chemical or biological catalyzers; (ii) CO_2 into CH_4 by electroactive methanogens; (iii) CO_2 into value-added soluble compounds (e.g., carboxylic acids, alcohols). The global reactor is not thermodynamically favorable and requires additional electrical energy. The anodic and cathodic chambers can be separated by an anion or cation exchange membrane or left unseparated

pressure from 1 to 10 bar leads to a change in the applied potential from −0.14 to −0.17 V. It appears necessary, in order not to limit hydrogen production and to allow better energy efficiency, to withdraw hydrogen when it is produced. Production yield can thus be increased by decreasing the partial pressure of H_2 (degassing) (Logan et al. 2008). In practice, the potential to produce hydrogen at the cathode is substantially higher than 0.14 V (Rozendal et al. 2006). Indeed, part of the energy is also used by electroactive microorganisms for their own growth and metabolism maintenance. Moreover, the ohmic resistance of the system causes overvoltages at the electrodes leading to energy losses. Among the other possible solutions, removing the membrane between the anode and cathode compartments reduces the ohmic drop but favors the conversion of the hydrogen produced into methane by methanogenic microorganisms (Logan et al. 2008; Liu et al. 2010). One of the major challenges in MECs is thus to reduce these losses to make the MEC energy balance more favorable. Optimally, the applied potential between the electrodes should be less than 0.6 V (Lee et al. 2010). In most studies, MECs are operated at potentials varying between 0.3 and 1.0 V (Liu et al. 2010). Below 0.3 V, the production speed is low and unstable, while above 1 V, the energy consumption is high. Nonetheless, hydrogen production by water electrolysis requires applied voltages above 1.6 V. Overall, the energy recovered as hydrogen in a MEC is about two to four times higher than the electrical energy supplied (Cheng and Logan 2007; Call and Logan 2008; Hu et al. 2008).

Similarly to MFCs, MECs can be operated either in single chamber or with two compartments separated by a membrane, also called dual-chamber systems (Logan et al. 2008). The most commonly used configuration is the two compartments (Liu et al. 2010). In that case, different types of membranes can be used, in particular cation exchange membranes (CEM) such as Nafion, CMA-7000 and Fumasep FKE membranes, and anion exchange membranes (AEM) such as AMI-7001 membrane, bipolar membranes, or mosaic membranes (Liu et al. 2008). The advantage of using a membrane is to avoid interferences between the anodic and cathodic reactions (such as hydrogen consumption at the anode), ensure the purity of the hydrogen produced, and act as a separator to avoid short circuits (Liu et al. 2008). But the use of a membrane has also several disadvantages such as a higher resistance to ion transfer and can generate significant variations in pH and a loss of potential (Rozendal et al. 2007). Tartakovsky et al. (2009) developed a continuously fed MEC consisting of a liquid anode compartment with a carbon felt anode and a gas diffusion cathode. A maximum production of 6.3 $m^3H_2\ m^{-3}\ day^{-1}$ from acetate was obtained under an applied potential of 1 V. If the use of a MEC with only one compartment makes possible to get rid of the resistance generated by the membrane and to reduce the cost linked to the membrane, the direct consumption of hydrogen by certain microbial populations, in particular methanogens which convert hydrogen into methane, makes this configuration difficult to be stabilized in long-term operation (Call and Logan 2008; Hu et al. 2008). Nonetheless, various techniques were tested with varying degrees of success to prevent methane production, such as periodic exposure of cathodes to air (Call and Logan 2008; Hu et al. 2008).

Most published work uses carbon-based anodes because of their good conductivity, biocompatibility, their different forms and high porosity, and their low cost (Logan et al. 2008). The use of materials having high-specific surfaces increases the surface available to the electroactive biofilm and thus the MEC performances. These materials include graphite granules, graphite brushes, graphite felt, carbon fabric, and carbon paper (Liu et al. 2010).

The cathode consists of a metal catalyst and a support material as found in water electrolysers. Platinum is known to be the best catalyst for this reaction and is widely used in the MEC (Liu et al. 2010). However, its high cost (800–1800 USD per once in the last 10 years) has led to develop catalysts excluding precious metals for chemical electrolysis processes. The results obtained in abiotic processes are difficult to be applied in MEC due to the different physicochemical conditions: very acidic or alkaline conditions on one side and neutral in MEC. Nonetheless, cheap catalysts suitable for MECs have been successfully tested such as nickel oxides, nickel- or tungsten-based materials, and stainless steel (Call et al. 2009; Selembo et al. 2009; Hu et al. 2009). The use of these non-precious metals is 10–100 times more cost-effective than platinum. Cheng and Logan (2011) obtained a productivity of 17.8 $m^3H_2\ m^{-3}\ day^{-1}$ with a graphite fiber brush anode and a platinum-containing carbon cathode as catalyst. The highest productivity was obtained by Jeremiasse et al. (2012) with 50 m^3H_2 m^{-3} day^{-1} with a nickel-based cathode at a potential of −0.7 V (vs. standard H_2 electrode).

Several authors reported as well the use of biocathodes in MECs. In that case, the cathodes are colonized with a biofilm that directly catalyzes the reduction of protons to hydrogen. The first biocathode was proposed by Rozendal et al. (2008). While biocathodes have the advantage of being cheap and resistant to corrosion, hydrogen production is generally ten times lower than with a metal catalyst (Rozendal et al. 2008; Jeremiasse et al. 2010, Villano et al. 2011). In terms of hydrogen yield at the cathode (anode electrons recovered as hydrogen at the cathode), biocathodes are lower than metal catalyzed cathodes which can reach up to 93% with nickel (Jeremiasse et al. 2012) and 96% with platinum (Call and Logan 2008). Batlle-Vilanova et al. (2014) compared the operation of two MECs, one with a biotic cathode, the other with an abiotic cathode, in a wide range of potentials applied to the cathode (−0.4 to −1.8 V vs. ESH). In both cases, the rate of hydrogen production increases with decreasing applied potential. The biocathode MEC showed better performances in terms of hydrogen produced per kWh consumed.

The first assays of H_2 production in MEC by treating organic effluents were carried out in very small prototypes. In 2009, Wagner et al. used a laboratory-scale MEC to treat pig manure in batch mode. Their MEC produced 0.9–1 m^3H_2 m^{-3} day^{-1} eliminating up to 72% of the COD. The gas contained up to 77% of hydrogen but also 13% of methane. Lalaurette et al. (2009) coupled a fermentation reactor with a MEC to treat lignocellulosic residues. Hydrogen yields and productivities reached about 1 L_{H2} g^{-1} COD and 1 m^3H_2 m^{-3} day^{-1}. Cusick et al. (2010) evaluated the energy treatment and recovery capabilities of MFCs and MECs treating wine effluents and urban wastewaters. They concluded that hydrogen production could be economically viable. Later on, Cusick et al. (2011) published the first pilot study on a 1000 L MEC treating wine effluents in California. The reactor consisted of 144 pairs of electrodes divided into 24 modules. The development of the electroactive biofilm took 60 days, which is longer than observed in laboratory tests, mainly due to either the absence of VFAs in the feed or a low pH and low and variable temperatures. The control of these different parameters improved the performances in terms of current, with a maximum of 7.4 A m^{-3} after 100 days of operation, at a potential of 0.9 V. Gas production reached a maximum of 0.19 ± 0.04 m^3H_2 m^{-3} day^{-1}. However, in the absence of a membrane, the hydrogen produced was rapidly converted into methane (86 ± 6% of the gas produced) by hydrogenotrophic methanogens. In 2011, a 120 L MEC was installed in an urban wastewater treatment plant in Northern England and fed with raw wastewater without temperature regulation. It was operated for 12 months, facing winter temperatures as low as 6 °C. The applied load was 0.14 kg DCO m^{-3} day^{-1} (Heidrich et al. 2013). The energy cost was about 2.3 KJ g_{DCO}^{-1}, which is lower than the values estimated for activated sludge (about 2.5–7.2 KJ g_{DCO}^{-1}). The reactor produced the equivalent of 0.15 m^3H_2 m^{-3} day^{-1} and converted 70% of the supplied electrical energy. This study is the first "proof of concept" showing the feasibility of treating urban wastewater in MEC at room temperature. Finally, Escapa et al. (2012) conducted an economic analysis on the integration of a MEC in an urban wastewater treatment plant. They estimated an investment cost of 1220€ m^{-3} of anode compartment for a MEC operated at a current density of 5 A m^{-2} with an energy consumption of 0.9 kWh kg_{COD}^{-1} with a return to investment around after 7 years.

4 Using Electrodes for Fermentation Control: Electro-Fermentation

The MFC and MEC for the production of electricity or molecules by (bio)electrosynthesis are currently within the most advanced BES in terms of technological maturity and process performances. However, these systems still suffer from limitations related to the microbes/electrode interface and biological constraints which limit the extracellular electron transfer (EET) rates and ultimately the attainable current densities (Arends and Verstraete 2012). To hinder these limitations while taking advantage of the microbes/electrode interactions, a new type of BES called electro-fermentation (EF) has recently been proposed for a better control of fermentation processes (Moscoviz et al. 2016; Schievano et al. 2016).

Indeed, fermentation is a redox-related process that corresponds to a cascade of oxidation and reduction reactions that must be kept in balance. Such equilibrium can be shifted by providing/removing electrons to/from fermentative bacteria through electrodes, thus permitting to exceed the metabolic limitations of traditional fermentation. To illustrate this concept, a good example is the propionate production during glucose fermentation by *Propionibacterium freudenreichii* (Emde and Schink 1990), according to the following catabolic equations (Fig. 4):

$$\text{glucose} + 2\text{NADH} \rightarrow 2\text{propionate}^- + 2H_2O + 2\text{NAD}^+ + \text{energy} \quad (1)$$

$$\text{glucose} + 4\text{NAD}^+ + 2H_2O \rightarrow 2\text{acetate}^- + 2CO_2 + 4\text{NADH} + 6H^+ + \text{energy} \quad (2)$$

As propionate is more reduced than glucose, its production (Eq. 1) requires reducing equivalents (i.e., NADH) which are generated from the acetate pathway (Eq. 2). Thus, the balanced equation for such fermentation is:

$$3\text{glucose} \rightarrow 2\text{acetate}^- + 4\text{propionate}^- + 2H_2O + 2CO_2 + 6H^+ + \text{energy} \quad (3)$$

This fermentation is not completely satisfactory as a third of the substrate consumed for catabolism is used to ensure the redox balance instead of being converted into propionate, thus limiting the maximum propionate yield to 1.33 $mol_{propionate}$ $mol_{glucose}^{-1}$ (based on the glucose used for catabolism). A way to exceed this limitation is to provide reducing equivalents directly from a cathode. In an early work, Emde and Schink (1990) have shown experimentally that *Propionibacterium freudenreichii* was able to uptake electrons from a cathode when mediated by neutral red

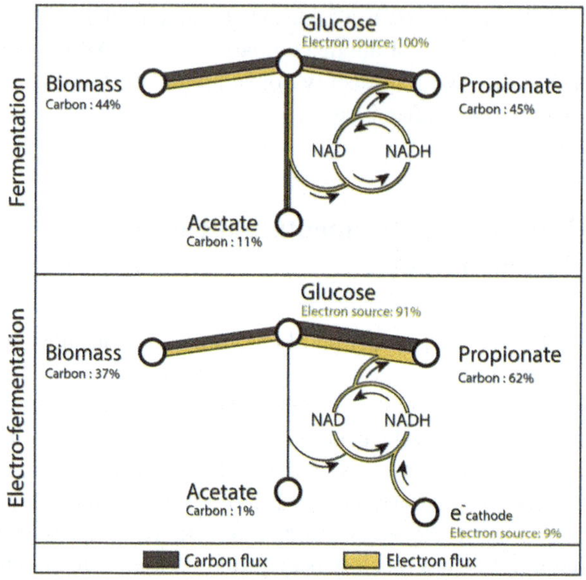

Fig. 4 Example of electro-fermentation. During glucose fermentation by *Propionibacterium freudenreichii*, the metabolic NADH balance (i.e., redox balance) is ensured by producing both acetate and propionate which are more oxidized and reduced than glucose, respectively. When grown in contact with a cathode (cathodic electro-fermentation), *P. freudenreichii* can uptake extracellular electrons which are used to ensure the redox balance instead of the acetate pathway. Experimental data retrieved from Emde and Schink (1990)

(i.e., cathodic EF). As a result, acetate production almost ceased during glucose EF, while the propionate yield reached 1.94 $mol_{propionate}\ mol_{glucose}^{-1}$ (based on the glucose used for catabolism), representing 97% of the theoretical maximum yield.

Similar results were also achieved with *Clostridium tyrobutyricum* (Choi et al. 2012) and *Propionibacterium acidipropionici* (Schuppert et al. 1992) for butyrate and propionate production, respectively. In addition, anodic EF can also be carried out when the product of interest is more oxidized than the substrate. In this case, the excess reducing equivalents (e.g., NADH) can be oxidized at an anode while producing electric current. For instance, the stoichiometric ethanol production from glycerol requires the dissipation of 2 moles of electrons per mole ethanol. Such conversion was experimentally achieved by Flynn et al. (2010) using a genetically engineered *Shewanella oneidensis*.

When compared to bioelectrosynthesis, EF requires lower current densities from the electrochemical system because the main energy source is the fermentation substrate instead of the electrode. In the case of propionate production, the respective equations for bioelectrosynthesis and glucose EF are:

$$3CO_2 + 14e^- + 13H^+ \rightarrow propionate^- + 4H_2O \quad (4)$$

$$\frac{1}{2}\text{glucose} + 2e^- + H^+ \rightarrow \text{propionate}^- + H_2O \quad (5)$$

Thus, seven times less electrons are required for propionate production by glucose EF, underlying that EF is less sensitive to EET rates than bioelectrosynthesis. However, this kind of EF still requires fermentative microorganisms able to uptake significant amount of extracellular electrons, either by direct or mediated EET. A recent study has shown that electro-activity seems to be a widespread capacity within microorganisms although at highly diverse EET rates (Koch and Harnisch 2016).

When EF is carried out with weak electroactive microorganisms (i.e., limited EET rates), the extracellular electron uptake or dissipation is negligible regarding the fermentation electron balance and cannot directly impact metabolite production. Yet it is still possible to use electrodes to redirect and control the fermentation patterns of such microorganisms. Indeed, fermentation reactions are constrained by biological regulations within microorganisms and interspecies interactions in microbial communities. EF is also a way to affect biological regulations related to intracellular or extracellular redox sensing, through electrochemical reactions between electrodes and redox species, either present in the fermentation bulk or bound to bacterial membranes. As a result, metabolic fluxes can be redirected and significantly change the spectrum of final fermentation products. For instance, Choi et al. (2014) have shown that butanol and 1,3-propanediol production by *Clostridium pasteurianum* were significantly increased during glucose and glycerol EF, respectively, while EET rates remained limited. In the case of glucose EF, 0.2% of the total electron input originated from the cathode and could only explain a 1.12-fold increase in butanol production if a direct conversion of electrons and glucose into butanol occurred. In fact, the observed butanol yield increased more than threefold and could be explained by electron fluxes diverted from the H_2 production pathway and biomass synthesis. Similar results showing significant metabolic shift with very low currents were also obtained with pure cultures of *Escherichia coli* (Harrington et al. 2015a, b), *Klebsiella pneumoniae* (Harrington et al. 2015a), *Clostridium autoethanogenum* (Kracke et al. 2016), and *Clostridium acetobutylicum* (Kim and Kim 1988; Gallardo et al. 2016) but also undefined mixed cultures (Dennis et al. 2013; Zhou et al. 2013, 2015; Moscoviz et al. 2017; Villano et al. 2017).

The exact mechanisms of such metabolic shift still remain unknown for most microorganisms and would require further investigation, including, for instance, metatranscriptomics to highlight changes in metabolic regulations and fluxes. The only well-documented EF mechanism so far is the case of neutral red-mediated glucose EF by *E. coli* (Harrington et al. 2015b). In this particular case, the authors identified that neutral red was able to reduce the menaquinone pool that modulates the activity of the *arcB*/*arcA* redox-sensing cascade. In *E. coli*, such regulation cascade controls the expression of numerous operons involved in fermentative and respiratory metabolism, thus allowing the metabolism to adapt to extracellular redox conditions (Malpica et al. 2004). In other words, the electrons provided by neutral red to *E. coli* gave false signals of low extracellular redox potential which ultimately resulted in an altered fermentation pattern favoring efficient electron dissipation pathways. EF mechanisms in other microorganisms are likely to be species-dependent, especially regarding their capacity to adapt to extracellular redox potential (e.g., strict or facultative anaerobes) or their specific electron transport chains.

In addition to interactions between fermentative species and electrodes, either direct or indirect, several benefits can be obtained by carrying out fermentation in BES. For instance, it is possible to remove by-products that build up during fermentation by efficient electroactive species such as *Geobacter sulfurreducens*. Removal of inhibitory products can then enhance fermentation performances such as optimizing final concentration and productivities of the desired product. A good illustration is the glycerol fermentation for ethanol production carried out by Speers et al. (2014). The authors compared co-cultures of *Clostridium cellobioparum* and *G. sulfurreducens* in contact with an anode with pure cultures of *C. cellobioparum*. They observed that the removal of by-products such as acetate and formate by *G. sulfurreducens* was able to stimulate glycerol consumption by 1.6-fold by *C. cellobioparum* in batch experiments, resulting in higher ethanol concentration (1.3-fold increase). With an adequate choice of electroactive species, similar strategies could be implemented to remove all fermentation metabolites but the desired product through oxidation at an anode. As a result, the product of interest would be easily purified as sole soluble compound present in the fermentation broth.

To conclude, EF is still in its early age but is a very promising strategy for the optimization of fermentative metabolites by pure or mixed culture. As for other BES, major challenges concerning the scale-up potential of EF are still to be addressed regarding the use of an electrochemical setup (e.g., power supply) and a specific reactor configuration which, in many cases, includes costly membranes for the separation of the anodic and cathodic compartments. Thus, in the context of bulk chemicals production, future EF processes must be developed as much as possible with low-cost materials. Beyond these structural limitations, EF is a unique way to exceed metabolic limitations and outperform yields of classical fermentation. EF has also the potential to become an effective redox regulation strategy provided that molecular mechanisms are better understood in the future. Additional services such as interactions between fermentative and electroactive species for inhibitor and fermentation by-product removal (i.e., in situ purifying of the fermentation product) are also great opportunities that could reduce the overall operating costs of fermentation processes and be optimized in forthcoming research works.

5 Other Applications of Microbial BES

5.1 Methane Production

MECs can be used for the production of various types of organic molecules using microbial biocathodes (Logan and Rabaey 2012). Among them, methane is currently the most studied organic compound produced in MECs. Methane electrosynthesis indeed provides an additional route for biofuel production in these systems. In comparison with hydrogen, it has a lower specific energy, but its transportation and storage are easier and rely on mature technologies.

A first mechanism for methane production in MEC is the conversion of hydrogen produced on a cathode to methane by hydrogenotrophic methanogens. In that case, the electrochemical processes occurring at the anode and cathode of the MEC are the same as described previously (part 1 of this chapter), but hydrogen is partially converted to methane by methanogenic archaea according to the following catabolic equation:

$$CO_2 + 4H_2 \rightarrow CH_4 + 2H_2O \quad (6)$$

Methane production in a MEC was first documented by Clauwaert et al. (2008), who reported that methane could account for up to 60% of the gas produced by a biocathode. MECs provide highly favorable conditions for hydrogenotrophic methanogens with continuous hydrogen supply and a possible high hydrogen partial pressure in the electrode surroundings. These archaea thus thrive in this type of system as reported in numerous publications with the identification of various genera such as *Methanobacterium*, *Methanobrevibacter*, *Methanococcus*, *Methanocorpusculum*, *Methanoculleus*, *Methanosarcina*, or *Methanothermobacter* from methane-producing biocathodes (Blasco-Gomez et al. 2017).

Alternatively, methane can be produced by methanogenic archaea using electrons directly from the electrode according to the following equation:

$$CO_2 + 8H^+ + 8e^- \rightarrow CH_4 + 2H_2O \quad (7)$$

This pathway was first reported by Cheng et al. (2009) who found that methane production rates on carbon cloth cathodes could not be explained by hydrogen evolution alone. The methane was produced with very high coulombic efficiencies (up to 96%) without the need of a metal catalyst by a microbial consortium dominated by *Methanobacterium palustre*. Since then, several other studies reported methane production with direct electron transfer. Archaeal genera identified as encompassing possible electrotrophic methanogenic species are *Methanobacterium*, *Methanococcus*, *Methanoculleus*, *Methanosaeta*, *Methanosarcina*, and *Methanothermobacter* (Blasco-Gomez et al. 2017). Reaction (7) theoretically allows CO_2 reduction at −0.244 V vs. SHE at pH = 7, a potential higher than the maximal potential for water electrolysis. However the potentials typically used in methane-producing MECs are in the same range as those reported for hydrogen-producing MECs, and in practice both direct and mediated (with hydrogen, formate, acetate, or other mediators) electron transfer mechanisms are probably implicated in methane synthesis on biocathodes (Blasco-Gomez et al. 2017). As for the molecular mechanisms of cathodic electron transfer, they are currently poorly understood. Depending on the electrotrophic species, they could involve conductive pili and redox active proteins or extracellular enzymes catalyzing the synthesis of mediators such as hydrogen or formate (Blasco-Gomez et al. 2017).

Currently, methane-producing MECs are mainly studied in combination with traditional anaerobic digestion (AD) reactors. This coupling allows improving AD performances with increased methane production rates, stabilized performances, and possibly increased methane yields at the expense of electrical power input (Yu et al. 2018). Mechanisms explaining this improvement are however unclear, and De Vrieze et al. attributed those to increased biomass retention on the electrodes (De Vrieze et al. 2014). Several recent studies also indicate that the introduction of conductive material alone improve AD performances without the need of additional electrical power input (Yu et al. 2018).

Another application of this type of technology is to use two-chamber MECs for the production of methane from low-strength effluents such as wastewater. In this context, the technology has several advantages including protection of methanogenic activity against potential inhibitors that are segregated in the anodic compartment, generation of biogas with high methane content and the possibility of running reactors at ambient temperature (Blasco-Gomez et al. 2017). For example, Zeppilli et al. (2015) performed the treatment of artificial wastewater with a two-chamber MEC using graphite granules as material for the electrodes and obtained stable and robust performances with 70% CE at the anode, 70% COD removal, and 80% CE for methane production at the cathode.

5.2 *Production of Soluble Multicarbon Compounds*

The recent discovery of microbial electrosynthesis (MES), i.e., the ability of microbes to synthesize multicarbon organic compounds using electrons from a cathode, greatly broadened the field of application of MEC for environmental biorefinery (Logan and Rabaey 2012). Indeed it introduces the possibility to produce liquid biofuels or platform molecules for green chemistry with higher added value than hydrogen or methane. Nevin et al. demonstrated in a pioneer work the ability of *Sporomusa ovata* to produce acetate using electrons from a graphite electrode (Nevin et al. 2010). The catabolic process proceeded according to the following equation:

$$CO_2 + 8H^+ + 8e^- \rightarrow CH_3COOH + 2H_2O \tag{8}$$

The potential of the cathode used in the experiment was −0.4 V vs. SHE above the potential necessary to produce hydrogen on unpolished graphite thus indicating direct conversion of electrons to acetate. Consistently coulombic efficiency was very high (86% ± 21%). Since then, other acetogenic bacteria with the same ability have been identified such as *Clostridium aceticum*, *Clostridium ljungdahlii*, *Sporomusa sphaeroides*, and *Moorella thermoacetica* (Nevin et al. 2011). Although the metabolic pathways of these electrotrophic bacteria remain largely unclear

(Kracke and Krömer 2014), MES demonstrated the ability of producing different organic molecules such as fatty acids (acetate, butyrate, or caproate) or alcohols (methanol, ethanol) from either defined co-cultures or environmental samples (Marshall et al. 2012; Deutzmann and Spormann 2017; Logan and Rabaey 2012). Recently Jourdin et al. (2018) obtained electrosynthesis of caproate (C6) with mixed microbial consortia on a carbon cloth cathode using carbon dioxide as only carbon source. The synthesis appeared to occur through chain elongation with acetate and butyrate as intermediates. Current density reached up to 175 A m^{-2}, and CEs were comprised between 60 and 100%.

At the moment, however, this type of MES cathode has almost exclusively been used with an abiotic anode and only rarely coupled to a bioanode in a MEC setting. This observation is maybe linked to the difficulty to conciliate microbial kinetics on anode and cathode (Bouchez et al. 2017). However, it appears as a very promising technology for environmental biorefinery that would enable coupling of waste treatment by complex microbial consortia at the bioanode with controlled synthesis of platform molecules in a clean cathodic compartment, thus bridging the gap between gray and white biotechnologies.

5.3 *Bioelectrochemical Systems for the Extraction of Nutrients*

In addition to energy or molecule production, BES cells can be used for the extraction of ions taking advantage of the selectivity of membranes separating compartments and using the electric potential to drive ionic migration. This principle was first used for the implementation of a microbial desalination cell (MDC) by Cao et al. (2009). The cell consisted of three chambers with an anode chamber, a cathode chamber, and a central chamber filled with water to be desalinated (Fig. 5). The membrane separating the anode chamber from the central chamber was an anion exchange membrane (AEM), while the membrane separating the central chamber from the cathode chamber was a cation exchange membrane (CEM). When electrons are transferred by bacteria to the anode and protons are released into the solution, the negatively charged species move from the central chamber to the anode through the AEM. In the cathode chamber, protons are conversely consumed and positively charged species move from the central chamber to the cathode chamber through the CEM.

With this setup Cao et al. (2009) demonstrated the possibility to generate electricity oxidizing acetate on the anode and reducing ferricyanide on the cathode while, at the same time, removing about 90% of the salt (NaCl) in a single desalination cycle. Since then, various technologies where implemented either taking advantage of the electrical field in MFC or MEC to desalinate water or using salinity gradients to provide energy for MFC or MEC with reverse electrodialysis (Sevda et al. 2015).

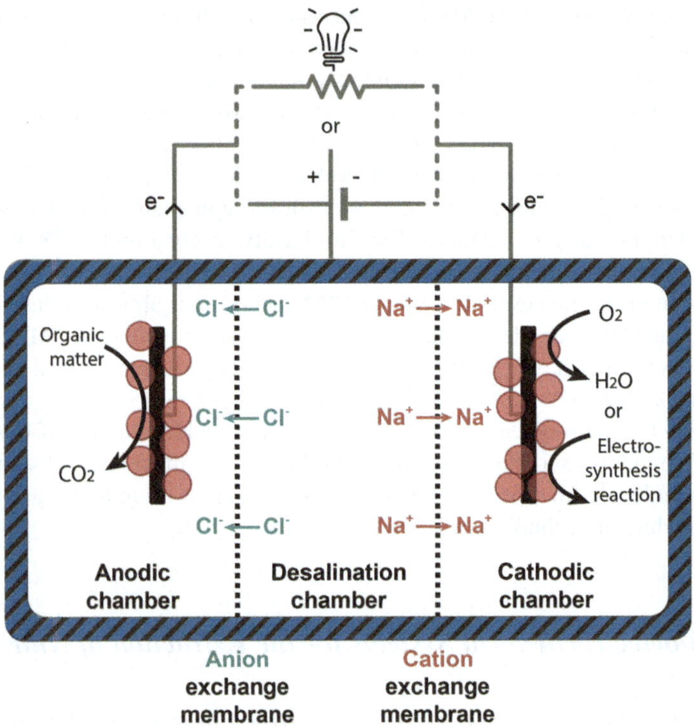

Fig. 5 Ion extraction principle in a microbial desalination cell. Within a microbial fuel cell or a microbial electrolysis cell, a desalination chamber can be added. This chamber is separated by both anion and cation exchange membranes (potentially more than two). When (bio)electrochemical reactions occur at the anode and the cathode, ions from the desalination chamber migrate through the membranes to ensure charge neutrality. As a result, all salts can be removed from the liquid present in the desalination chamber

In terms of nutrient recovery, it was early found that this phenomenon could be used for the extraction of ammonium (NH_4^+) from the anodic compartment in MFCs (Kuntke et al. 2011) or MEC (Villano et al. 2013). Villano et al. (2013) reported efficient methane production in a MEC fed with acetate together with removal of over 50% of the influent nitrogen load due to the migration of ammonium to the cathodic compartment. Interestingly, high pH is typically found in the cathodic compartment because of proton consumption on the cathode thus favoring conversion of ammonium to ammonia (NH_3) and its recovery in the gas phase (Cord-Ruwisch et al. 2011).

In this context urine was identified as a very interesting substrate with high nutrients content and attractive properties as substrate and electrolyte for BES (Ledezma et al. 2015). Indeed, in terms of nutrients, effective urine recovery could provide up to 20% of current macronutrients (nitrogen N, phosphorus P, and potassium K) required for the production fertilizers in the world (Ledezma et al. 2015). As a substrate and electrolyte, it has a high COD content of 2–10 g L^{-1} (Maurer et al. 2006), a good conductivity (around 20 mS cm^{-1}), and a strong buffering capacity that allows bioanodes functioning under optimal neutral pH conditions. Accordingly, interesting performances for ammonium recovery have been demonstrated with both MFC and MEC systems (Rodriguez Arredondo et al. 2015). For example, in the case of a MEC treating diluted urine, current density reached 23.07 ± 1.15 A/m^2 with a CE of 95.9 ± 3.1% and an ammonium removal rate of 173.4 ± 18.1 g N/m^2/day (Kuntke et al. 2014) with competitive energy requirements compared to other technologies such as ammonia stripping, electrodialysis, or struvite crystallization (Rodriguez Arredondo et al. 2015). The fraction of removed ammonium was however limited, reaching only 34.2% due to the low stripping of ammonia in the cathodic chamber. Alternatively, a new reactor design was recently proposed for ammonium recovery through precipitation using a three-chamber reactor with a CEM separating the anode chamber from the central chamber, an AEM separating the central chamber from the cathode chamber, and alimentation of the cathode with anodic effluent after air treatment (Ledezma et al. 2017). In that case, ammonium was extracted from the anode chamber and up concentrated in the central chamber, while bicarbonate (HCO_3^-) was extracted from the cathode chamber and up concentrated in the central chamber allowing precipitation of NH_4HCO_3 crystals using flash cooling. The average current density was 29.3 ± 2.3 A/m^2 with an average coulombic efficiency of 94.43 ± 16.63% and an ammonium removal rate of 431 g N/m^2/day. With this design, the average N recovery efficiency was 49.5 ± 1.8% with 14.30 ± 0.8% of N recovered as solid. Interestingly other nutrients such as phosphorous, potassium, and sodium were also up concentrated in the central chamber allowing recovery efficiencies of 42.8 ± 1.0%, 54.7 ± 1.3%, and 51.9 ± 1.2%, respectively. Altogether current performances obtained for nutrient recovery from urine at Labscale appear promising, but further research is required to improve recovery efficiencies, to lower the costs, and to demonstrate feasibility when up-scaled to real-world conditions.

References

Arends JBA, Verstraete W (2012) 100 years of microbial electricity production: three concepts for the future. Microb Biotechnol 5(3):333–346

Batlle-Vilanova, P, Puig S, Gonzalez-Olmos R, Vilajeliu-Pons A, Baneras L, Balaguer MD, Colprim J (2014) Assessment of biotic and abiotic graphite cathodes for hydrogen production in microbial electrolysis cells, Int J Hydrogen Energ 39:1297–1305.

Blasco-Gomez R, Batlle-Vilanova P, Villano M, Balaguer MD, Colprim J, Puig S (2017) On the edge of research and technological application: a critical review of electromethanogenesis. Int J Mol Sci 18(4):874. https://doi.org/10.3390/ijms18040874

Bouchez T, Bridier A, Le Quéméner E (2017) Method and device for controlling the activity of a bioelectrochemical system comprising both a bioanode and a biocathode. US2017218530A1

Butti SK, Velvizhi G, Sulonen MLK, Haavisto JM, Koroglu EO, Cetinkaya AY, Singh S, Arya D, Modestra JA, Krishna KV, Verma A, Ozkaya B, Lakaniemi AM, Puhakka JA, Mohan SV (2016) Microbial electrochemical technologies with the perspective of harnessing bioenergy: maneuvering towards upscaling. Renew Sust Energ Rev 53:462–476

Call D, Logan BE (2008) Hydrogen production in a single chamber microbial electrolysis cell lacking a membrane. Environ Sci Technol 42(9):3401–3406

Call DF, Merrill MD, Logan BE (2009) High surface area stainless steel brushes as cathodes in microbial electrolysis cells. Environ Sci Technol 43:2179–2183

Cao XX, Huang X, Liang P, Xiao K, Zhou YJ, Zhang XY, Logan BE (2009) A new method for water desalination using microbial desalination cells. Environ Sci Technol 43(18):7148–7152

Chae KJ, Choi MJ, Lee JW, Kim KY, Kim IS (2009) Effect of different substrates on the performance, bacterial diversity, and bacterial viability in microbial fuel cells. Bioresour Technol 100(14):3518–3525

Cheng S, Logan BE (2007) Sustainable and efficient biohydrogen production via electrohydrogenesis. Proc Natl Acad Sci U S A 104(47):18871–18873

Cheng SA, Logan BE (2011) High hydrogen production rate of microbial electrolysis cell (MEC) with reduced electrode spacing. Bioresour Technol 102(3):3571–3574

Cheng SA, Xing DF, Call DF, Logan BE (2009) Direct biological conversion of electrical current into methane by electromethanogenesis. Environ Sci Technol 43(10):3953–3958

Choi O, Um Y, Sang B-I (2012) Butyrate production enhancement by *Clostridium tyrobutyricum* using electron mediators and a cathodic electron donor. Biotechnol Bioeng 109:2494–2502

Choi O, Kim T, Woo HM, Um Y (2014) Electricity-driven metabolic shift through direct electron uptake by electroactive heterotroph *Clostridium pasteurianum*. Sci Rep 4:6961

Clauwaert P, Toledo R, Van der Ha D, Crab R, Verstraete W, Hu H, Udert KM, Rabaey K (2008) Combining biocatalyzed electrolysis with anaerobic digestion. Water Sci Technol 57(4):575–579

Cord-Ruwisch R, Law Y, Cheng KY (2011) Ammonium as a sustainable proton shuttle in bioelectrochemical systems. Bioresour Technol 102(20):9691–9696

Cusick RD, Kiely PD, Logan BE (2010) A monetary comparison of energy recovered from microbial fuel cells and microbial electrolysis cells fed winery or domestic wastewaters. Int J Hydrog Energy 35(17):8855–8861

Cusick RD, Bryan B, Parker DS, Merrill MD, Mehanna M, Kiely PD, Liu GL, Logan BE (2011) Performance of a pilot-scale continuous flow microbial electrolysis cell fed winery wastewater. Appl Microbiol Biotechnol 89(6):2053–2063

Daud SM, Kim BH, Ghasemi M, Daud WRW (2015) Separators used in microbial electrochemical technologies: current status and future prospects. Bioresour Technol 195:170–179

De Vrieze J, Gildemyn S, Arends JBA, Vanwonterghem I, Verbeken K, Boon N, Verstraete W, Tyson GW, Hennebel T, Rabaey K (2014) Biomass retention on electrodes rather than electrical current enhances stability in anaerobic digestion. Water Res 54:211–221

Dennis PG, Harnisch F, Yeoh YK, Tyson GW, Rabaey K (2013) Dynamics of cathode-associated microbial communities and metabolite profiles in a glycerol-fed bioelectrochemical system. Appl Environ Microbiol 79(13):4008–4014

Deutzmann JS, Spormann AM (2017) Enhanced microbial electrosynthesis by using defined co-cultures. ISME J 11:704–714

Emde R, Schink B (1990) Enhanced propionate formation by *Propionibacterium freudenreichii* subsp. *freudenreichii* in a three-electrode amperometric culture system. Appl Environ Microbiol 56:2771–2776

Escapa A, Gomez X, Tartakovsky B, Moran A (2012) Estimating microbial electrolysis cell (MEC) investment costs in wastewater treatment plants: case study. Int J Hydrog Energy 37:18641–18653

Fan YZ, Han SK, Liu H (2012) Improved performance of CEA microbial fuel cells with increased reactor size. Energy Environ Sci 5(8):8273–8280

Feng YJ, He WH, Liu J, Wang X, Qu YP, Ren NQ (2014) A horizontal plug flow and stackable pilot microbial fuel cell for municipal wastewater treatment. Bioresour Technol 156:132–138

Flynn JM, Ross DE, Hunt KA, Bond DR, Gralnick JA (2010) Enabling unbalanced fermentations by using engineered electrode-interfaced bacteria. MBio 1:5 e00190–5 e00110

Foley J, de Haas D, Hartley K, Lant P (2010) Comprehensive life cycle inventories of alternative wastewater treatment systems. Water Res 44(5):1654–1666

Gallardo R, Acevedo A, Quintero J, Paredes I, Conejeros R, Aroca G (2016) In silico analysis of *Clostridium acetobutylicum* ATCC 824 metabolic response to an external electron supply. Bioprocess Biosyst Eng 39(2):295–305

Geelhoed JS, Hamelers HVM, Stams AJM (2010) Electricity-mediated biological hydrogen production. Curr Opin Microbiol 13(3):307–315

Harrington TD, Mohamed A, Tran VN, Biria S, Gargouri M, Park J-J, Gang DR, Beyenal H (2015a) Neutral red-mediated microbial electrosynthesis by *Escherichia coli, Klebsiella pneumoniae*, and *Zymomonas mobilis*. Bioresour Technol 195:57–65

Harrington TD, Tran VN, Mohamed A, Renslow R, Biria S, Orfe L, Call DR, Beyenal H (2015b) The mechanism of neutral red-mediated microbial electrosynthesis in *Escherichia coli*: menaquinone reduction. Bioresour Technol 192:689–695

Heidrich ES, Dolfing J, Scott K, Edwards SR, Jones C, Curtis TP (2013) Production of hydrogen from domestic wastewater in a pilot-scale microbial electrolysis cell. Appl Microbiol Biotechnol 97:6979–6989

Hiegemann H, Herzer D, Nettmann E, Lübken M, Schulte P, Schmelz K-G, Gredigk-Hoffmann S, Wichern M (2016) An integrated 45L pilot microbial fuel cell system at a full-scale wastewater treatment plant. Bioresour Technol 218:115–122

Hiegemann H, Lübken M, Schulte P, Schmelz K-G, Gredigk-Hoffmann S, Wichern M (2018) Inhibition of microbial fuel cell operation for municipal wastewater treatment by impact loads of free ammonia in bench- and 45L-scale. Sci Total Environ 624:34–39

Hu HQ, Fan YZ, Liu H (2008) Hydrogen production using single-chamber membrane-free microbial electrolysis cells. Water Res 42:4172–4178

Hu H, Fan Y, Liu H (2009) Hydrogen production in single-chamber tubular microbial electrolysis cells using non-precious-metal catalysts. Int J Hydrog Energy 34:8535–8542

Jeremiasse AW, Hamelers EVM, Buisman CJN (2010) Microbial electrolysis cell with a microbial biocathode. Bioelectrochemistry 78:39–43

Jeremiasse AW, Hamelers HVM, Croese E, Buisman CJN (2012) Acetate enhances startup of a H_2-producing microbial biocathode. Biotechnol Bioeng 109:657–664

Jourdin L, Raes SMT, Buisman CJN, Strik DPBTB (2018) Critical biofilm growth throughout unmodified carbon felts allows continuous bioelectrochemical chain elongation from CO_2 up to caproate at high current density. Front Energy Res 6:7. https://doi.org/10.3389/fenrg.2018.00007

Kim TS, Kim BH (1988) Electron flow shift in *Clostridium acetobutylicum* fermentation by electrochemically introduced reducing equivalent. Biotechnol Lett 10:123–128

Koch C, Harnisch F (2016) Is there a specific ecological niche for electroactive microorganisms? ChemElectroChem 3:1282–1295

Kokko M, Epple S, Gescher J, Kerzenmacher S (2018) Effects of wastewater constituents and operational conditions on the composition and dynamics of anodic microbial communities in bioelectrochemical systems. Bioresour Technol 258:376–389

Kracke F, Krömer JO (2014) Identifying target processes for microbial electrosynthesis by elementary mode analysis. BMC Bioinformatics 15(1):410

Kracke F, Virdis B, Bernhardt PV, Rabaey K, Krömer JO (2016) Redox dependent metabolic shift in Clostridium autoethanogenum by extracellular electron supply. Biotechnol Biofuels 9(1):249

Kundu A, Sahu JN, Redzwan G, Hashim MA (2013) An overview of cathode material and catalysts suitable for generating hydrogen in microbial electrolysis cell. Int J Hydrog Energy 38(4):1745–1757

Kuntke P, Geleji M, Bruning H, Zeeman G, Hamelers HVM, Buisman CJN (2011) Effects of ammonium concentration and charge exchange on ammonium recovery from high strength wastewater using a microbial fuel cell. Bioresour Technol 102(6):4376–4382

Kuntke P, Sleutels T, Saakes M, Buisman CJN (2014) Hydrogen production and ammonium recovery from urine by a microbial electrolysis cell. Int J Hydrog Energy 39(10):4771–4778
Lalaurette E, Thammannagowda S, Mohagheghi A, Maness PC, Logan BE (2009) Hydrogen production from cellulose in a two-stage process combining fermentation and electrohydrogenesis. Int J Hydrog Energy 34:6201–6210
Ledezma P, Kuntke P, Buisman CJN, Keller J, Freguia S (2015) Source-separated urine opens golden opportunities for microbial electrochemical technologies. Trends Biotechnol 33(4):214–220
Ledezma P, Jermakka J, Keller J, Freguia S (2017) Recovering nitrogen as a solid without chemical dosing: bio-electroconcentration for recovery of nutrients from urine. Environ Sci Technol Lett 4(3):119–124
Lee HS, Rittmann BE (2010) Characterization of energy losses in an upflow single-chamber microbial electrolysis cell. Int J Hydrog Energy 35:920–927
Lee HS, Vermaas WFJ, Rittmann BE (2010) Biological hydrogen production: prospects and challenges. Trends Biotechnol 28:262–271
Liang P, Duan R, Jiang Y, Zhang X, Qiu Y, Huang X (2018) One-year operation of 1000-L modularized microbial fuel cell for municipal wastewater treatment. Water Res 141:1–8
Liu H, Grot S, Logan BE (2005) Electrochemically assisted microbial production of hydrogen from acetate. Environ Sci Technol 39(11):4317–4320
Liu Y, Yu P, Song X, Qu Y (2008) Hydrogen production from cellulose by co-culture of *Clostridium thermocellum* JN4 and *Thermoanaerobacterium thermosaccharolyticum* GD17. Int J Hydrog Energy 33:2927–2933
Liu H, Hu H, Chignell J, YanZhen F, Fan Y (2010) Microbial electrolysis: novel technology for hydrogen production from biomass. Biofuels 1:129–142
Logan BE (2010) Scaling up microbial fuel cells and other bioelectrochemical systems. Appl Microbiol Biotechnol 85(6):1665–1671
Logan BE, Rabaey K (2012) Conversion of wastes into bioelectricity and chemicals by using microbial electrochemical technologies. Science 337(6095):686–690
Logan BE, Hamelers B, Rozendal RA, Schrorder U, Keller J, Freguia S, Aelterman P, Verstraete W, Rabaey K (2006) Microbial fuel cells: methodology and technology. Environ Sci Technol 40(17):5181–5192
Logan BE, Call D, Cheng S, Hamelers HVM, Sleutels T, Jeremiasse AW, Rozendal RA (2008) Microbial electrolysis cells for high yield hydrogen gas production from organic matter. Environ Sci Technol 42:8630–8640
Lovley DR (2008) The microbe electric: conversion of organic matter to electricity. Curr Opin Biotechnol 19(6):564–571
Lovley DR (2011) Powering microbes with electricity: direct electron transfer from electrodes to microbes. Environ Microbiol Rep 3(1):27–35
Lovley DR (2012) Electromicrobiology. In: Gottesman S, Harwood CS, Schneewind O (eds) Annual review of microbiology, vol 66. Annual Reviews, Palo Alto, pp 391–409
Malpica R, Franco B, Rodriguez C, Kwon O, Georgellis D (2004) Identification of a quinone-sensitive redox switch in the ArcB sensor kinase. Proc Natl Acad Sci U S A 101(36):13318–13323
Marshall CW, Ross DE, Fichot EB, Norman RS, May HD (2012) Electrosynthesis of commodity chemicals by an autotrophic microbial community. Appl Environ Microbiol 78(23):8412–8420
Maurer M, Pronk W, Larsen TA (2006) Treatment processes for source-separated urine. Water Res 40(17):3151–3166
Moscoviz R, Toledo-Alarcón J, Trably E, Bernet N (2016) Electro-fermentation: how to drive fermentation using electrochemical systems. Trends Biotechnol 34:856–865
Moscoviz R, Trably E, Bernet N (2017) Electro-fermentation triggering population selection in mixed-culture glycerol fermentation. Microb Biotechnol 11(1):74–83
Nealson KH, Rowe AR (2016) Electromicrobiology: realities, grand challenges, goals and predictions. Microb Biotechnol 9(5):595–600
Nevin KP, Woodard TL, Franks AE, Summers ZM, Lovley DR (2010) Microbial electrosynthesis: feeding microbes electricity to convert carbon dioxide and water to multicarbon extracellular organic compounds. MBio 1(2):e00103–e00110

Nevin KP, Hensley SA, Franks AE, Summers ZM, Ou JH, Woodard TL, Snoeyenbos-West OL, Lovley DR (2011) Electrosynthesis of organic compounds from carbon dioxide is catalyzed by a diversity of acetogenic microorganisms. Appl Environ Microbiol 77(9):2882–2886

Pandey P, Shinde VN, Deopurkar RL, Kale SP, Patil SA, Pant D (2016) Recent advances in the use of different substrates in microbial fuel cells toward wastewater treatment and simultaneous energy recovery. Appl Energy 168:706–723

Pant D, Van Bogaert G, Diels L, Vanbroekhoven K (2010) A review of the substrates used in microbial fuel cells (MFCs) for sustainable energy production. Bioresour Technol 101(6):1533–1543

Pham TH, Aelterman P, Verstraete W (2009) Bioanode performance in bioelectrochemical systems: recent improvements and prospects. Trends Biotechnol 27(3):168–178

Potter MC (1911) Electrical effects accompanying the decomposition of organic compounds. Proc R Soc Lond B 84(571):260–276

Rabaey K, Rozendal RA (2010) Microbial electrosynthesis - revisiting the electrical route for microbial production. Nat Rev Microbiol 8(10):706–716

Rodriguez Arredondo M, Kuntke P, Jeremiasse AW, Sleutels THJA, Buisman CJN, ter Heijne A (2015) Bioelectrochemical systems for nitrogen removal and recovery from wastewater. Environ Sci: Water Res Technol 1(1):22–33

Rozendal RA, Hamelers HVM, Euverink GJW, Metz SJ, Buisman CJN (2006) Principle and perspectives of hydrogen production through biocatalyzed electrolysis. Int J Hydrog Energy 31:1632–1640

Rozendal RA, Hamelers HVM, Molenkmp RJ, Buisman JN (2007) Performance of single chamber biocatalyzed electrolysis with different types of ion exchange membranes. Water Res 41:1984–1994

Rozendal RA, Jeremiasse AW, Hamelers HVM, Buisman CJN (2008) Hydrogen production with a microbial biocathode. Environ Sci Technol 42:629–634

Santoro C, Arbizzani C, Erable B, Ieropoulos I (2017) Microbial fuel cells: from fundamentals to applications. A review. J Power Sources 356:225–244

Schievano A, Pepé Sciarria T, Vanbroekhoven K et al (2016) Electro-fermentation—merging electrochemistry with fermentation in industrial applications. Trends Biotechnol 34:866–878

Schuppert B, Schink B, Trösch W (1992) Batch and continuous production of propionic acid from whey permeate by *Propionibacterium acidi-propionici* in a three-electrode amperometric culture system. Appl Microbiol Biotechnol 37:549–553

Selembo PA, Merrill MD, Logan BE (2009) The use of stainless steel and nickel alloys as low-cost cathodes in microbial electrolysis cells. J Power Sources 190:271–278

Sevda S, Yuan H, He Z, Abu-Reesh IM (2015) Microbial desalination cells as a versatile technology: functions, optimization and prospective. Desalination 371:9–17

Speers AM, Young JM, Reguera G (2014) Fermentation of glycerol into ethanol in a microbial electrolysis cell driven by a customized consortium. Environ Sci Technol 48:6350–6358

Tartakovsky B, Manuel MF, Wang H, Guiot SR (2009) High rate membrane-less microbial electrolysis cell for continuous hydrogen production. Int J Hydrog Energy 34(2):672–677

Trapero JR, Horcajada L, Linares JJ, Lobato J (2017) Is microbial fuel cell technology ready? An economic answer towards industrial commercialization. Appl Energy 185:698–707

Villano M, De Bonis L, Rossetti S, Aulenta F, Majone M (2011) Bioelectrochemical hydrogen production with hydrogenophilic dechlorinating bacteria as electrocatalytic agents. Bioresour Technol 102:3193–3199

Villano M, Scardala S, Aulenta F, Majone M (2013) Carbon and nitrogen removal and enhanced methane production in a microbial electrolysis cell. Bioresour Technol 130:366–371

Villano M, Paiano P, Palma E, Miccheli A, Majone M (2017) Electrochemically driven fermentation of organic substrates with undefined mixed microbial cultures. ChemSusChem 10(15):3091–3097

Virdis B, Rabaey K, Yuan Z, Keller J (2008) Microbial fuel cells for simultaneous carbon and nitrogen removal. Water Res 42(12):3013–3024

Wagner RC, Regan JM, Oh SE, Zuo Y, Logan BE (2009) Hydrogen and methane production from swine wastewater using microbial electrolysis cells. Water Res 43:1480–1488

Yu ZS, Leng XY, Zhao S, Ji J, Zhou TY, Khan A, Kakde A, Liu P, Li XK (2018) A review on the applications of microbial electrolysis cells in anaerobic digestion. Bioresour Technol 255:340–348

Zeppilli M, Villano M, Aulenta F, Lampis S, Vallini G, Majone M (2015) Effect of the anode feeding composition on the performance of a continuous-flow methane-producing microbial electrolysis cell. Environ Sci Pollut Res 22(10):7349–7360

Zhang F, Ge Z, Grimaud J, Hurst J, He Z (2013) In situ investigation of tubular microbial fuel cells deployed in an aeration tank at a municipal wastewater treatment plant. Bioresour Technol 136:316–321

Zhou MH, Chi ML, Luo JM, He HH, Jin T (2011) An overview of electrode materials in microbial fuel cells. J Power Sources 196(10):4427–4435

Zhou M, Chen J, Freguia S et al (2013) Carbon and electron fluxes during the electricity driven 1,3-propanediol biosynthesis from glycerol. Environ Sci Technol 47:11199–11205

Zhou M, Freguia S, Dennis PG et al (2015) Development of bioelectrocatalytic activity stimulates mixed-culture reduction of glycerol in a bioelectrochemical system: bioelectrocatalytic activity in glycerol-fed BESs. Microb Biotechnol 8:483–489

Techno-economic Analysis of Fermentation-Based Biorefinery: Creating Value from Food Residues

Juan-Rodrigo Bastidas-Oyanedel and Jens Ejbye Schmidt

1 Introduction

Unlocking value from organic waste is a feasible idea, in contrast to the disposal of these organic wastes into landfills, that has an associated cost ranging from 40 to 400 USD/t (Diggelman and Ham 2003; Tuck et al. 2012; Pfaltzgraff et al. 2013). Instead, the organic wastes (residues) can be converted into bio-products and/or bioenergy, creating economic value rather than costs, generating value and benefits for the society.

Based on the characteristics of organic wastes/residues, they can be characterized according to their saccharides, lignin, lipids, and protein content (Tuck et al. 2012). The source of the waste is also classified into agricultural waste, food waste, and municipal waste. Here we focus in the creations of economic value from complex organic wastes, e.g., food waste, by anaerobic digestion processes. The creation of value from noncomplex residues, e.g., citrus peels and coffee spends, for the extraction of pigments has been reviewed elsewhere (Tuck et al. 2012; Pfaltzgraff et al. 2013; Arancon et al. 2013).

For the treatment of complex organic waste, anaerobic digestion has been historically the chosen technology. Anaerobic digestion converts the complex wastes into biogas, containing methane (bioenergy) and a digestate that can be valorize as soil improver. However, as noted by Pfaltzgraff et al. (2013), the conversion of biomass to bulk chemicals is 3.5–7.5 times more profitable than its conversion to fuels/energy. This is the main motivation for this techno-economic analysis.

J.-R. Bastidas-Oyanedel (✉)
Chemical Engineering Department, Khalifa University of Science and Technology, Abu Dhabi, United Arab Emirates
e-mail: juan.bastidas@ku.ac.ae

J. E. Schmidt
SDU-Department of Chemical Engineering, Biotechnology and Environmental Technology, University of Southern Denmark, Campusvej 55, DK-5230 Odense M, Denmark
e-mail: jesc@kbm.sdu.dk

© Springer Nature Switzerland AG 2019

J.-R. Bastidas-Oyanedel, J. E. Schmidt (eds.), *Biorefinery*,
https://doi.org/10.1007/978-3-030-10961-5_22

In recent years, several mixed culture anaerobic technologies, different from the conventional anaerobic digestion for biogas production, have emerged. Among these technologies are dark fermentation and mixed culture lactic acid fermentation. The interest in these "new" technologies is their value products, with market prices more attractive than methane and digestate. Their average prices are 400 and 15 USD/t, respectively (Whyte and Perry 2001; Gebrezgabher et al. 2010; Clarke and Alibardi 2010; Kim et al. 2016; https://www.eia.gov/dnav/ng/hist/n3035us3m.htm; http://ec.europa.eu/eurostat/statistics-explained/index.php/Energy_price_statistics#Natural_gas_prices_for_industrial_consumers; http://www.indexmundi.com/commodities/?commodity=natural-gas; Koupaie et al. 2014).

Figure 1 presents the market price range for the products yielded by anaerobic digestion, dark fermentation, and mixed culture lactic acid fermentation. Dark fermentation has been extensively reviewed for the production of value products (Bastidas-Oyanedel et al. 2015; Coma et al. 2017; Kiran et al. 2015; Cope et al. 2014; Queir et al. 2017; Kleerebezem et al. 2015; Ghimire et al. 2015; Aceves-Lara et al. 2008; Birgitte 2017; Agler et al. 2011; Zhou et al. 2018; Jankowska et al. 2017; Chang et al. 2010). Here it is presented the prices of hydrogen, acetic acid, ethanol, propionic acid, butyric acid, and caproic acid as dark fermentation products (Bastidas-Oyanedel and Schmidt 2018).

Hydrogen price (600–1800 USD/t_H_2) is based on the production costs of hydrogen from natural gas. Hence, hydrogen price depends on natural gas market price. Here the hydrogen price was given when the natural price is in the range of 3–6 USD/GJ (Padro and Putsche 1999). As described by Padro and Putsche (1999), hydrogen production costs from natural gas are the cheapest when compared to

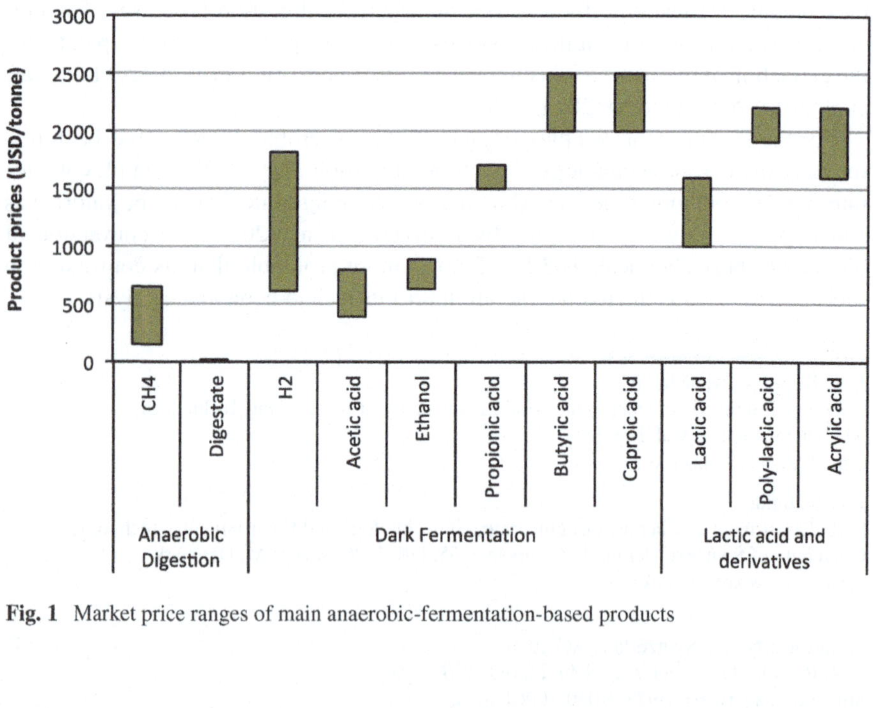

Fig. 1 Market price ranges of main anaerobic-fermentation-based products

other sources, i.e., coal, biomass, and electrodialysis. To illustrate this, Clarke and Alibardi (2010) reported a biomass-based biohydrogen price of 4700 USD/t_H_2.

Market price ranges for acetic acid (Clarke and Alibardi 2010; Bastidas-Oyanedel et al. 2015) and ethanol (Clarke and Alibardi 2010; Bastidas-Oyanedel et al. 2015; Beerthuis et al. 2015) are closer to the methane price range, i.e., 400–900 USD/t, which makes it less attractive to spend effort trying to improve the production of either acetic acid and ethanol from organic wastes by anaerobic technologies. In contrast, price ranges for propionic acid, butyric acid, and caproic acid (Bastidas-Oyanedel et al. 2015), from 1500 to 2500 USD/t, are an incentive to optimize and improve the production of these organic acids by dark fermentation.

Lactic acid and polylactic acid relevance in this assessment is expressed in that that technology has been commercialized and has demonstrated its sustainability and profitability (Tuck et al. 2012). The conversion of food waste into lactic acid by uncontrolled pH mixed culture fermentation has been shown to be feasible (Gao et al. 2011; Bonk et al. 2017; Tang et al. 2016; Li et al. 2015), achieving lactic acid concentrations of 30 g/L with lactic acid selectivity of 93% (in COD base) over other organic acids (Bonk et al. 2017; Yousuf et al. 2018). Lactic acid market price range, 1000–1600 USD/t (Bastidas-Oyanedel et al. 2015; Beerthuis et al. 2015; Datta and Henry 2006; González et al. 2007), and its derivatives, i.e., polylactic acid and acrylic acid, 1600–2200 USD/t (Beerthuis et al. 2015; Datta and Henry 2006; Nampoothiri et al. 2010), make them attractive alternatives to biogas. Below is presented a techno-economic assessment of these three routes, i.e., anaerobic digestion, mixed culture lactic acid fermentation, and dark fermentation, in order to compare their potential revenues, costs, and profit.

2 Methodology

The present techno-economic assessment is based on relevant literature data and using conservative assumptions. The assessment compares three main routes: anaerobic digestion, mixed culture lactic acid fermentation, and dark fermentation. Each route has been divided into two sub-routes: (A1) anaerobic digestion where the produced and upgraded methane is sold to the grid; (A2) methane is used to in-site power generation; (B1) mixed culture lactic acid fermentation, where lactic acid is produced, separated, and purified, and the residues from this process are converted into methane by anaerobic digestion; (B2) as in B1, adding a conversion of lactic acid to polylactic acid step; (C1) dark fermentation producing an upgraded hydrogen stream sold to the grid, where the residues of this process are converted into methane by anaerobic digestion; and (C2) dark fermentation producing an upgraded hydrogen stream and producing, separating, and purifying acetic and butyric acid streams, where the residues are converted into methane by anaerobic digestion. Table 1 presents the capital and operational costs for each route/sub-route, considered in this techno-economic analysis. Capital and operational costs were obtained from literature (Whyte and Perry 2001; Kim et al. 2016;

Table 1 Capital and operational costs, obtained from literature, for the different assessed scenarios

Route	Annualized capital cost (USD/t_VS_fw/year)[a]	Annual operational cost (USD/t_VS_fw/year)	Annualized total investment (USD/t_VS_fw/year)	Capital cost as present value (USD)	Total investment as present value (USD)
(A1) Anaerobic digestion—methane sold to the grid	42[b,c]	11[b,c]	53	1,528,365	1,928,652
(A2) Anaerobic digestion—power generation	167[d]	113[d]	280	6,091,628	10,189,103
(B1) Mixed culture lactic acid fermentation	108[b,c,e]	55[b,c,e]	163	3,930,083	5,931,514
(B2) Polylactic acid production	114[b,c,e]	59[b,c,e]	173	4,148,421	6,295,410
(C1) Dark fermentation—hydrogen and methane sold to the grid	47[b,c,e,f,g]	13[b,c,e,f,g]	60	1,710,314	2,183,379
(C2) Dark fermentation—acetic and butyric acids purified	252[b,c,e,f,g]	148[b,c,e,f,g]	400	9,170,193	14,555,862

All scenarios considered a project time of 20 years, with an annual interest of 5%, and a designed capacity of 50 t/day of food waste, with 16% volatile solid composition

[a]t_VS_fw, tonnes of volatile solids of food waste; in this review it was used volatile solid composition of 16%. (A1) The costs considered an anaerobic digestion reactor, a digestate solid composting facility, and a methane upgrading facility; (A2) as in A1, plus a combined heat and power generator; (B1) considers a mixed culture lactic acid reactor, a lactic acid separator and purification system, an anaerobic digestion reactor, a digestate solid composting facility, and a methane upgrading facility; (B2) as in B1, plus a lactic acid polymerization facility; (C1) considers a dark fermentation reactor, a hydrogen upgrading facility, an anaerobic digestion reactor, a digestate solid composting facility, and a methane upgrading facility; (C2) as in C1, plus an acetic acid separation and purification facility and a butyric acid separation and purification facility

[b]Kim et al. (2016)

[c]Whyte and Perry (2001)

[d]Moriarty (2013)

[e]Nampoothiri et al. (2010)

[f]Bastidas-Oyanedel et al. (2015)

[g]Bonk et al. (2015)

Bastidas-Oyanedel et al. 2015; Nampoothiri et al. 2010; Moriarty 2013; Bonk et al. 2015). For all the scenarios, it was considered a project time of 20 years, with an annual interest of 5%.

Figure 2 presents a scheme of the three main routes and sub-routes. The different assessed scenarios are as follows:

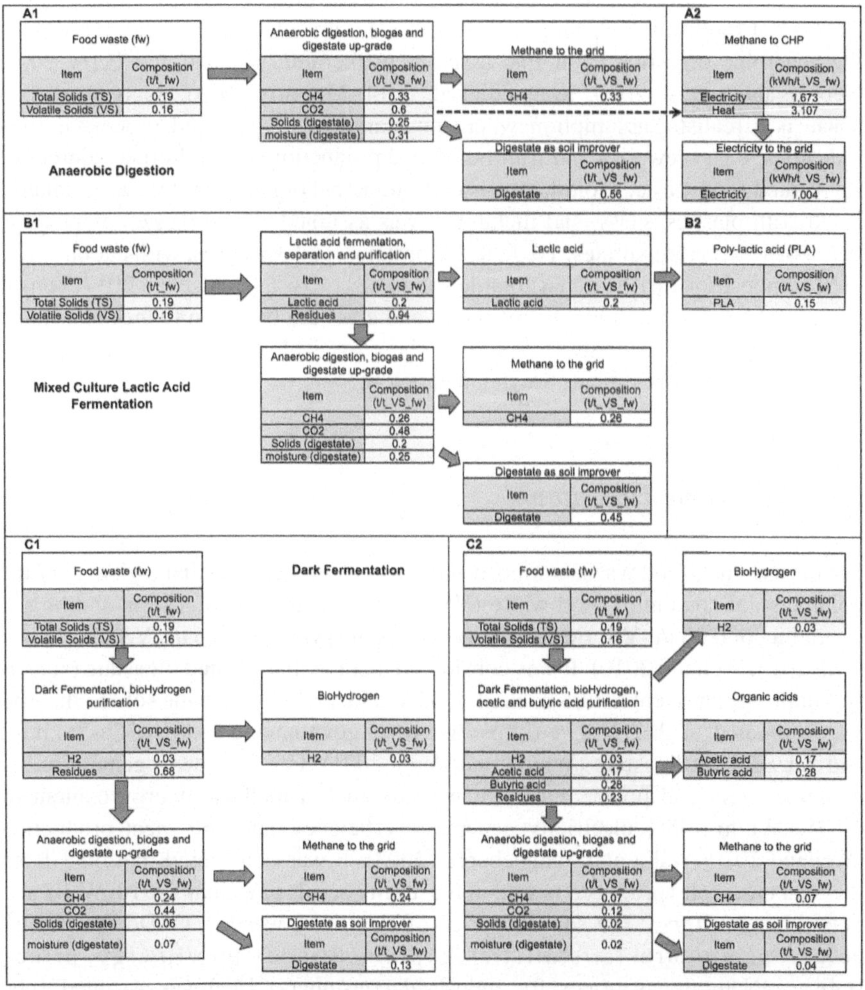

Fig. 2 Mass flow diagram (t/t_VolatileSolid_foodwaste) for the different assessed scenarios. Anaerobic digestion: (**A1**) methane (to the gas grid) and digestate; (**A2**) electricity and digestate. Lactic acid fermentation combined with anaerobic digestion: (**B1**) lactic acid, methane, and digestate; (**B2**) polylactic acid, methane, and digestate. Dark fermentation combined with anaerobic digestion: (**C1**) hydrogen, methane, and digestate; (**C2**) hydrogen, acetic acid, butyric acid, methane, and digestate

Anaerobic digestion: (**A1**) methane (to the gas grid) and digestate; (**A2**) electricity and digestate.

Lactic acid fermentation combined with anaerobic digestion: (**B1**) lactic acid, methane, and digestate; (**B2**) polylactic acid, methane, and digestate.

Dark fermentation combined with anaerobic digestion: (**C1**) hydrogen, methane, and digestate; (**C2**) hydrogen, acetic acid, butyric acid, methane, and digestate.

For the economic assessments, it was assumed a well segregated food waste, from the hospitality and catering sectors, and a biorefinery "valorization" plant of 50 t/day of food waste capacity, which is in the range of what has been reported in the anaerobic digestion economic assessment literature (Whyte and Perry 2001; Gebrezgabher et al. 2010; Koupaie et al. 2014; Moriarty 2013). Segregated food waste is a realistic assumption where postharvest activities and processed food industries, e.g., breweries and fruit pulp/juice production, are additional sources of segregated food waste, avoiding the use of municipal organic solid waste (contaminated with plastics, glass, and metals). It was assumed a food waste composition (*w*/*w*) of 13% carbohydrates, 1% fats, 2% proteins, and 3% ashes, with a total solids (TS) composition of 19% and a volatile solid (VS) of 16% (Bonk et al. 2017; Yousuf et al. 2016). Conversion yields, revenues, and costs are based on tonnes of volatile solid content of food waste (t_VS_fw). The costs include investment and operational costs. The assessments are detailed below.

2.1 Anaerobic Digestion

For the assumed food waste composition, it was estimated a methane yield over the volatile solid content of food waste (VS_fw) after anaerobic digestion and biogas upgrading of 0.33 t/t_VS_fw, and a digestate (solid) yield of 0.25 t/t_VS_fw (Nielfa et al. 2015; Weiland 2010). It was considered that the final treated digestate (sold as soil improver) has a moisture content of 55%. The price of methane sold to the grid was estimated at 207 USD/t (https://www.eia.gov/dnav/ng/hist/n3035us3m.htm) and 5 USD/t of digestate (Whyte and Perry 2001). For the power generation by combined heat and power (CHP), it was assumed a methane energy content of 10.35 kWh/m^3, 35% electricity generation efficiency, and 65% heat production (Weiland 2010; Walla and Schneeberger 2008). It was considered that 60% from the total electricity produced is sold as electricity surplus (Weiland 2010; Walla and Schneeberger 2008). The selected price of electricity was 0.1 USD/kWh (http://ec.europa.eu/eurostat/statistics-explained/index.php/Energy_price_statistics#Natural_gas_prices_for_industrial_consumers). Heat was assumed to be used in the plant, with no market price as discussed by Gebrezgabher (2010). The investment and operational cost for the sub-route A1, production of methane (to the grid) and digestate as soil improver, were set as 53 USD/t_VS_fw (Kim et al. 2016). For A2, power generation and digestate was 280 USD/t_VS_fw (Moriarty 2013).

2.2 *Mixed Culture Lactic Acid Fermentation*

A yield of 0.2 t_lacticacid/t_VS_fw was used for the conversion, separation, and purification of lactic acid (Bonk et al. 2017; Yousuf et al. 2018; Joglekar et al. 2006; Åkerberg and Zacchi 2000; Vaidya et al. 2005). This overall yield takes into account the fermentative production of lactic acid from food waste (Bonk et al. 2017; Yousuf et al. 2018) and the conventional downstream process train, i.e., coarse separation of suspended solids from the broth, acidification of the broth with strong acid (H_2SO_4), removal of gypsum from the lactic acid solution, and distillation (Joglekar et al. 2006; Åkerberg and Zacchi 2000; Vaidya et al. 2005). It was assumed that the residues of the lactic acid process were used to produce methane and digestate. Methane and digestate production and upgrading yield, from the lactic acid residue, were estimated as 0.2 and 0.45 t/t_VS_fw, respectively. Polylactic acid (PLA) yield from the purified lactic acid was 0.75 t_PLA/t_LA (Gruber 2001; Subramanian et al. 2015). In this process, purified lactic acid is first converted to lactide by a chemical process, and this molecule is further polymerized into PLA (Gruber 2001; Subramanian et al. 2015). Market prices for lactic acid and polylactic acid were assumed as 1000 and 1900 USD/t, respectively (Datta and Henry 2006; Nampoothiri et al. 2010). Prices for methane and digestate are as in "Anaerobic Digestion" section. Cost for the production of lactic acid and anaerobic digestion of residues with methane and digestate as products was estimated as 163 USD/t_VS_fw and 173 USD/t_VS_fw for the scenario when lactic acid is converted to polylactic acid, converting the residues to methane and digestate (Kim et al. 2016; Nampoothiri et al. 2010). These values are based on the targeted cost of lactic acid and polylactic acid reported by Nampoothiri et al. (2010) of 0.55 USD/kg_LA and 0.8 USD/kg_PLA, respectively.

2.3 *Dark Fermentation*

Biohydrogen production and purification yield were assumed as 0.03 t/t_VS_fw (Bastidas-Oyanedel et al. 2015; Bonk et al. 2015); this is close to the maximum hydrogen yield. This was chosen, in order to discuss the huge effort in recent years regarding the biohydrogen production optimization from organic wastes by dark fermentation (Ghimire et al. 2015; Ntaikou et al. 2010; Palomo-Briones et al. 2017; Elbeshbishy et al. 2017; Poggi-Varaldo et al. 2014; Bundhoo et al. 2015; Abreu et al. 2016). Purified hydrogen was assumed to be sold to the grid at a price of 1800 USD/t (Padro and Putsche 1999), to make it competitive versus natural gas-based hydrogen.

During the sub-route "hydrogen, methane, and digestate," the residues from the biohydrogen production are converted to methane and digestate with 0.24 and 0.13 t/t_VS_fw yields, respectively. The cost of the dark fermentation and hydrogen upgrade was assumed to be 7.5 USD/t_VS_fw. This cost was based on the hydraulic retention time ratio of dark fermentation (2 days) (Bastidas-Oyanedel et al. 2015) over anaerobic digestion (14 days) (Aceves-Lara et al. 2008), i.e., the cost of dark fermentation was estimated as 1/7; the cost of the anaerobic digestion is as in "Anaerobic Digestion" section, i.e., 53 USD/t_VS_fw (Kim et al. 2016). The total cost for this sub-route was estimated as 60 USD/t_VS_fw.

The second sub-route "hydrogen, acetic acid, butyric acid, methane, and digestate" assumed that acetic acid and butyric acid are separated and purified from the dark fermentation process. The combined production and purification yields of acetic and butyric acids used were 0.17 and 0.28 t/t_VS_fw, respectively (Bastidas-Oyanedel et al. 2015; Bonk et al. 2015; Joglekar et al. 2006; Åkerberg and Zacchi 2000; Vaidya et al. 2005). These overall yields take into account the fermentative production of acetic and butyric acid from food waste (Bastidas-Oyanedel et al. 2015; Bonk et al. 2015) and the conventional downstream processing for other organic acid, i.e., lactic acid (Joglekar et al. 2006; Åkerberg and Zacchi 2000; Vaidya et al. 2005). The residues are converted to methane and digestate by anaerobic digestion. Methane and digestate yields of 0.07 and 0.04 t/t_VS_fw were used. Prices for acetic and butyric acid considered were 400 and 2000 USD/t, respectively (Bastidas-Oyanedel et al. 2015; Bonk et al. 2015). Costs for dark fermentation and anaerobic digestion are as in the previous sub-route. Separation and purification costs of 170 USD/t_VS_fw for acetic acid were assumed. The same cost was assumed for butyric acid. These values are based on the target production cost of polylactic acid (Nampoothiri et al. 2010); these costs can be decreased considerably for organic acids as discussed by Bonk et al. (2015). The total cost for this sub-route was estimated as 400 USD/t_VS_fw.

3 Results and Discussion

3.1 Economic Assessment Results

Figure 3 shows the revenues, cost, and profit of the different assessed scenarios. For the first route, anaerobic digestion only, the food waste valorization plant, producing methane and digestate has profit of 19 USD/t_VS_fw. When considering the conversion into heat and power, the venues from electricity and digestate are close to 100 USD/t_VS_fw, but generating electricity at the assessed plant scale, i.e., 50 tonnes of food waste per day, considerably increases the total costs, 275 USD/t_VS_fw versus 50 USD/t_VS_fw when selling the methane to the grid. The conversion of methane into heat and power at the plant is only economically feasible when considering a minimum tipping fees and/or subsidies of 176 USD/t_VS_fw, i.e., 28 USD/t_foodwaste. These tipping fees/subsidies are in the range of what has been reported in the literature for tipping fees, 40–60 USD/t_foodwaste (Moriarty 2013), and considerably lower than the food waste landfilling costs, 40–400 USD/t_foodwaste (Diggelman and Ham 2003; Tuck et al. 2012; Pfaltzgraff et al. 2013).

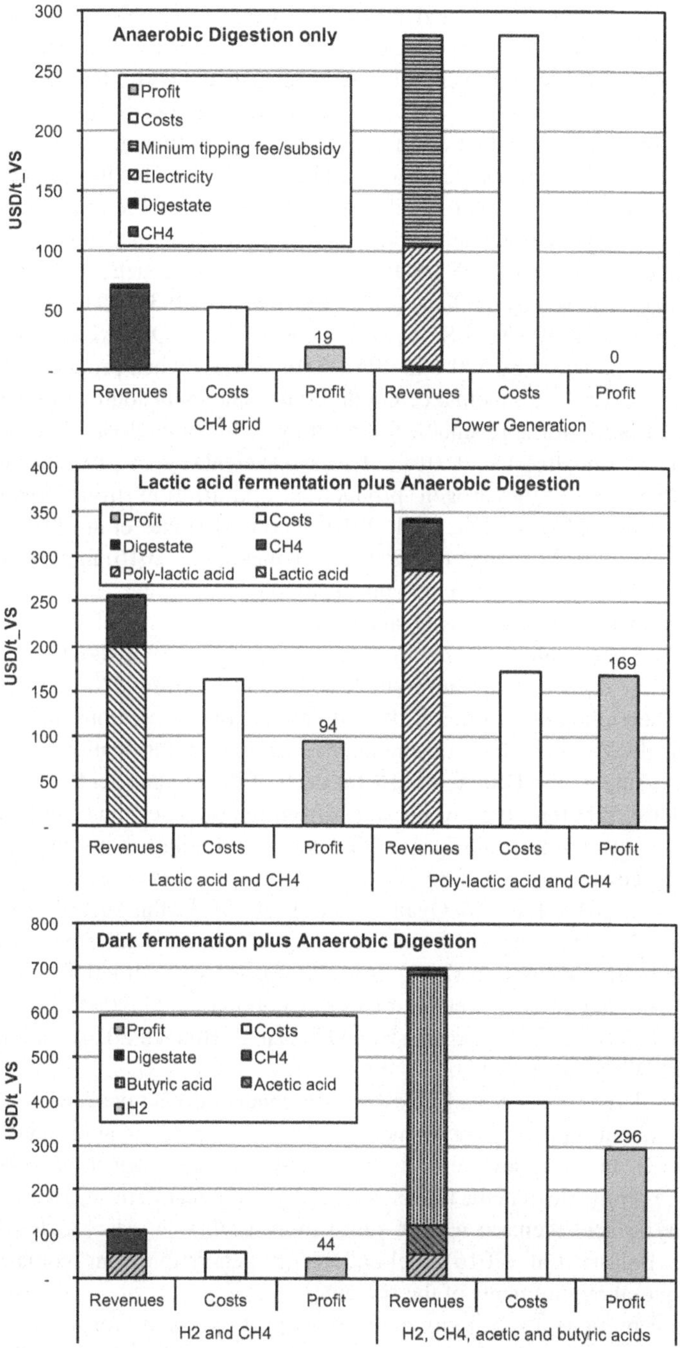

Fig. 3 Economic assessment results for the different assessed scenarios

The mixed culture lactic acid fermentation and anaerobic digestion scenario assessment resulted in higher profit, compared to the anaerobic digestion only scenario. Revenues for the production of lactic acid, methane (to the grid), and digestate were estimated to be 94 USD/t_VS_fw, with total cost of 162 USD/t_VS_fw. The revenues for the production of polylactic acid, methane, and digestate were 169 USD/t_VS_fw, with total cost of 172 USD/t_VS_fw. This implies an 80% revenues increase, with a 6% increase in the costs from lactic acid to polylactic acid.

The dark fermentation with production of hydrogen, methane, and digestate resulted in total revenues of 95 USD/t_VS_fw, where hydrogen represented 57% of the revenues, followed by methane (42%) and digestate (0.3%). This sub-route generated a profit of 44 USD/t_VS_fw, with a cost of 60 USD/t_VS_fw. This profit is considerably increased by 570% to 296 USD/t_VS_fw when producing hydrogen, acetic acid, butyric acid, methane, and digestate. The main source of revenues was butyric acid, 80% of the revenues, followed by acetic acid (10%), hydrogen (7%), methane (2%), and digestate (0.01%). This assessment suggests that the huge scientific effort toward biohydrogen production optimization from organic waste (Ghimire et al. 2015; Ntaikou et al. 2010; Palomo-Briones et al. 2017; Bastidas-Oyanedel et al. 2012; Zhang et al. 2012; Yeshanew et al. 2016) may be re-focused into the production, separation, and purification of the organic acids produced during dark fermentation (Bastidas-Oyanedel et al. 2015; Agler et al. 2011; Jankowska et al. 2017; Chen et al. 2016; 2017a; den Boer et al. 2016). It should be noted that the cost of the organic acids sub-route has also increased by 570% to 400 USD/t_VS_fw, when compared to the 60 USD/t_VS_fw of the previous sub-route. This increase is due to the high acetic and butyric separation-purification costs assumed in this assessment, 170 USD/t_VS_fw for each of the organic acids.

As discussed before, the revenues/cost from the organic acids can be improved as the scientific effort could be (re)-directed into finding more environmentally friendly processes and lowering the cost of organic acid separation-purification (Yousuf et al. 2016; Bastidas-Oyanedel et al. 2016). In this regard, several techniques have been investigated for the recovery of organic acids from fermentation broths, including adsorption (Zhou et al. 2013), solvent extraction (Alkaya et al. 2009), membrane-based solvent extraction (Choudhari et al. 2014), electrodialysis (Lopez and Hestekin 2013; Prochaska and Woźniak-Budych 2014), and membrane separation (Xiong et al. 2015).

In general, the conversion of food waste into valuable chemicals was more profitable than its conversion to fuels (methane and hydrogen), as noted by Pfaltzgraff et al. (2013). In our assessment the lactic and polylactic scenarios were 5 and 9 times more profitable than the methane (to the grid) scenario, respectively, while the acetic-butyric acid scenario was 16 times more profitable. Also, as has been discussed by Belasri et al. (2016), fuel and energy generation from biomass will not match the total requirements of the society.

Figure 4 presents a sensitivity analysis on profit variation for the assessed scenarios. The sensitivity analysis was performed using a 15% decrease/increase in selected parameters. In the case of scenarios (A1) biogas to grid, the parameters that strongly affect the profit variation, i.e., more than 20% variation, are capital cost,

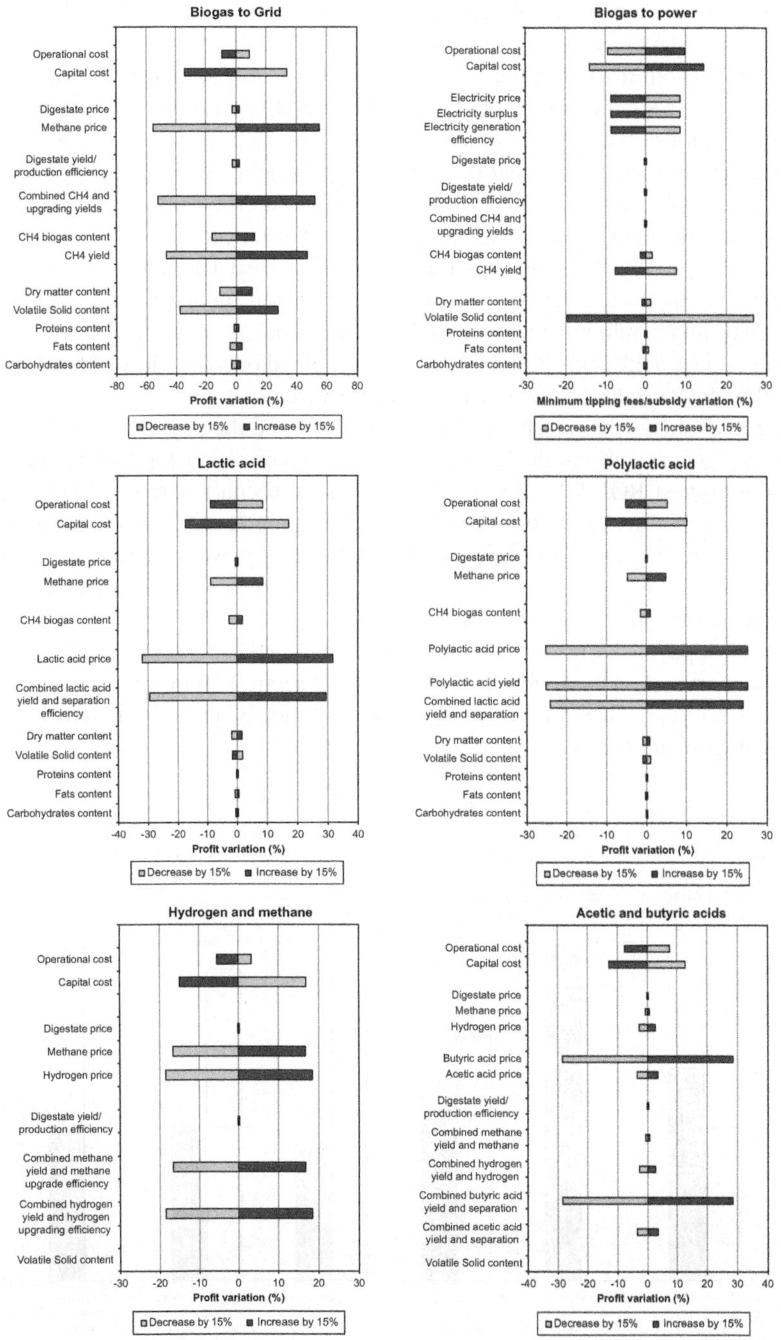

Fig. 4 Economic sensitivity analysis results for the different assessed scenarios. All the scenarios where evaluated in a time frame of 20 years, with an annual interest of 5%. Costs include the capital cost, converted to annual, plus the annual operational cost

methane price, methane yield, methane upgrading yield, and volatile solid content. In the case of (A2) biogas to power, only the volatile solid content produces a variation over 20% in the minimum tipping fees/subsidy. For scenario (B1) lactic acid, lactic acid price and combined lactic acid yield and separation efficiency generate profit variations above 20%. For scenario (B2) polylactic acid, the parameters that produce profit variations higher than 20% are polylactic acid price, polylactic acid yield, and combined lactic acid yield and separation efficiency. For scenario (C1) hydrogen and methane, none of the parameters produced variations higher than 20% on the profit. The parameters that produced at least 10% profit variation were capital cost, methane price, hydrogen price, combined methane yield and upgrade efficiency, and combined hydrogen yield and upgrade efficiency. For case (C2) acetic and butyric acids, butyric acid price and combined butyric acid yield and separation efficiency produced more than 20% profit variation.

Figure 5 shows the return on investment (ROI) and the payback time for all the assessed scenarios. From the ROI perspective, scenario (B2) polylactic acid generates the highest ROI, 98%. From all the assessed scenarios, only (A2) biogas to power does not generate ROI, assuming that tipping fees/subsidies are minimal. This is due to the high cost of the combined heat and power generator and the low prices for electricity and digestate sold as soil improver (Moriarty 2013). All the other scenarios present ROI higher than 30%.

The best payback time was obtained for scenario (B2) polylactic acid, 7.8 years. Both scenarios (C1) hydrogen-methane and (C2) acetic-butyric acids have payback times of 9.1 years. Scenarios (A1) biogas to grid and (B1) lactic acid have payback times of 12.6 and 10.3 years, respectively.

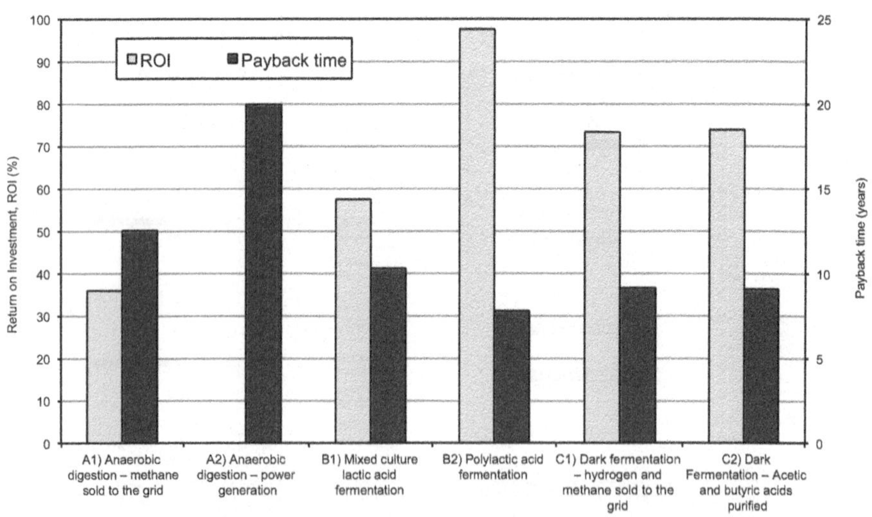

Fig. 5 Return on investment (ROI) and payback time for the different assessed scenarios. All the scenarios were evaluated in a time frame of 20 years, with an annual interest of 5%

3.2 *Emerging Mixed Culture Technologies*

The added benefits of mixed culture fermentation technologies versus pure culture fermentations (where the feedstock has to be sterilized to prevent microbial contamination of the pure culture) have been partially discussed previously in the literature (Bastidas-Oyanedel et al. 2015). Here these benefits are expanded to (1) use of a complex feedstock as food waste and (2) no feedstock sterilization, e.g., autoclaving, which reduces the process investment costs (no autoclave facilities are required) and operational costs (no energy is required for autoclaving). On the other hand, pure cultures are known for their high product selectivity, yielding higher product efficiency, and well controlled by environmental parameters (Bastidas-Oyanedel et al. 2015). However, new technologies in mixed culture dark fermentation are being developed to revert this trend. High product selectivity of one organic acid from the mixed culture dark fermentation has been reported for propionate using glycerol (Chen et al. 2016), propionate by controlling ammonium levels in the culture broth (Chen et al. 2017a), or lactic acid using food waste in an uncontrolled pH mode (Bonk et al. 2017; Yousuf et al. 2018). High selectivity is pursued in order to make separation and purification a less complicated task. In this regard, elongation of carboxylic acids has been explored as an alternative to this issue. As an example, promoting the chain elongation of acetic and butyric acid can enhance caproic acid production (Leng et al. 2017; Liu et al. 2017; Chen et al. 2017b). Caproic acid solubility, 10.8 g/L in water (Liu et al. 2017), is low when compared to the miscibility of acetic and butyric acid (López-Garzón and Straathof 2014). Carboxylic acids esterification has been also reported to improve separation and reduce costs (Cabrera-Rodriguez et al. 2017). Electrodialysis has been explored in order to increase carboxylic acid yield in dark fermentation (Jones et al. 2017; Huang et al. 2007; Villano et al. 2017) and for the separation of lactic acid from the culture broth (Yi et al. 2008). In a complete different perspective, the production of carboxylic acids from syngas using anaerobic biofilms has been reported (Chen and Ni 2016; Gildemyn et al. 2017). This is relevant, since anaerobic technologies can be used in combination to thermochemical processes for the valorization of plastics and lignocellulosic wastes, for the production of value chemicals.

4 Conclusions

The present techno-economic analysis has shown that profitability of food waste conversion to bulk chemicals, e.g., lactic acid or butyric acid, can be increased 5–16 times when compared to the base scenario, i.e., production of methane (sold to the grid). From the discussed scenarios, the highest profit is obtained by dark fermentation with separation and purification of butyric acid, 296 USD/t_VS (47 USD/t_foodwaste). From the return on investment (ROI) and payback time, the best scenario is the production of polylactic acid, with 98% ROI, and 7.8 years payback time. Production of butyric acid ROI and payback time was 74% and 9.1 years. From these profit, ROI, and payback time perspectives, the present techno-economic analysis suggests a change in focus from biogas/biohydrogen into butyric acid and polylactic acid production from food waste. These results suggest that industry may refocus effort on bulk chemicals, e.g., butyric acid and/or polylactic acid, rather than only focusing on biofuels, as H_2 and CH_4.

References

Abreu AA, Tavares F, Alves MM, Pereira MA (2016) Boosting dark fermentation with co-cultures of extreme thermophiles for biohythane production from garden waste. Bioresour Technol 219:132–138. https://doi.org/10.1016/j.biortech.2016.07.096

Aceves-Lara CA, Trably E, Bastidas-Oyanedel JR, Ramirez I, Latrille E, Steyer JP (2008) Bioenergy production from waste: examples of biomethane and biohydrogen. J Soc Biol 202:177–189

Agler MT, Wrenn BA, Zinder SH, Angenent LT (2011) Waste to bioproduct conversion with undefined mixed cultures: the carboxylate platform. Trends Biotechnol 29:70–78. https://doi.org/10.1016/j.tibtech.2010.11.006

Åkerberg C, Zacchi G (2000) An economic evaluation of the fermentative production of lactic acid from wheat flour. Bioresour Technol 75:119–126. https://doi.org/10.1016/S0960-8524(00)00057-2

Alkaya E, Kaptan S, Ozkan L, Uludag-Demirer S, Demirer GN (2009) Recovery of acids from anaerobic acidification broth by liquid-liquid extraction. Chemosphere 77:1137–1142. https://doi.org/10.1016/j.chemosphere.2009.08.027

Arancon RAD, Lin CSK, Chan KM, Kwan TH, Luque R (2013) Advances on waste valorization: new horizons for a more sustainable society. Energy Sci Eng 1:53–71. https://doi.org/10.1002/ese3.9

Bastidas-Oyanedel JR, Schmidt JE (2018) Increasing profits in food waste biorefinery-a techno-economic analysis. Energies 11(6):1551. https://doi.org/10.3390/en11061551

Bastidas-Oyanedel J-R, Mohd-Zaki Z, Zeng RJ, Bernet N, Pratt S, Steyer J-P, Batstone DJ (2012) Gas controlled hydrogen fermentation. Bioresour Technol 110:503–509. https://doi.org/10.1016/j.biortech.2012.01.122

Bastidas-Oyanedel J-R, Bonk F, Thomsen MH, Schmidt JE (2015) Dark fermentation biorefinery in the present and future (bio)chemical industry. Rev Environ Sci Bio/Technology 14:473–498. https://doi.org/10.1007/s11157-015-9369-3

Bastidas-Oyanedel JR, Fang C, Almardeai S, Javid U, Yousuf A, Schmidt JE (2016) Waste biorefinery in arid/semi-arid regions. Bioresour Technol 215:21–28. https://doi.org/10.1016/j.biortech.2016.04.010

Beerthuis R, Rothenberg G, Shiju NR (2015) Catalytic routes towards acrylic acid, adipic acid and ε-caprolactam starting from biorenewables. Green Chem 17:1341–1361. https://doi.org/10.1039/C4GC02076F

Belasri D, Sowunmi A, Bastidas-Oyanedel JR, Amaya C, Schmidt JE (2016) Prospecting of renewable energy technologies for the Emirate of Abu Dhabi: a techno-economic analysis. Prog Ind Ecol 10:301–318

Birgitte K (2017) Ahring biochemical production and separation of carboxylic acids for biorefinery applications biochemical production and separation of carboxylic acids for biorefinery applications. Fermentation 3:1–25. https://doi.org/10.3390/fermentation3020022

Bonk F, Bastidas-Oyanedel J-R, Schmidt JE (2015) Converting the organic fraction of solid waste from the city of Abu Dhabi to valuable products via dark fermentation—economic and energy assessment. Waste Manag 40:82–91. https://doi.org/10.1016/j.wasman.2015.03.008

Bonk F, Bastidas-Oyanedel J-R, Yousef AF, Schmidt JE (2017) Exploring the selective lactic acid production from food waste in uncontrolled pH mixed culture fermentations using different reactor configurations. Bioresour Technol 238:416–424. https://doi.org/10.1016/j.biortech.2017.04.057

Bundhoo MAZ, Mohee R, Hassan MA (2015) Effects of pre-treatment technologies on dark fermentative biohydrogen production: a review. J Environ Manag 157:20–48. https://doi.org/10.1016/j.jenvman.2015.04.006

Cabrera-Rodriguez CI, Moreno-Gonzalez M, de Weerd FA, Viswanathan V, van der Wielen LAM, Straathof AJJ (2017) Esters production via carboxylates from anaerobic paper mill wastewater treatment. Bioresour Technol 237:186–192. https://doi.org/10.1016/j.biortech.2017.02.030

Chang HN, Kim NJ, Kang J, Jeong CM (2010) Biomass-derived volatile fatty acid platform for fuels and chemicals. Biotechnol Bioprocess Eng 15:1–10. https://doi.org/10.1007/s12257-009-3070-8

Chen X, Ni BJ (2016) Anaerobic conversion of hydrogen and carbon dioxide to fatty acids production in a membrane biofilm reactor: a modeling approach. Chem Eng J 306:1092–1098. https://doi.org/10.1016/j.cej.2016.08.049

Chen Y, Wang T, Shen N, Zhang F, Zeng RJ (2016) High-purity propionate production from glycerol in mixed culture fermentation. Bioresour Technol 219:659–667. https://doi.org/10.1016/j.biortech.2016.08.026

Chen Y, Shen N, Wang T, Zhang F, Zeng RJ (2017a) Ammonium level induces high purity propionate production in mixed culture glucose fermentation. RSC Adv 7:518–525. https://doi.org/10.1039/C6RA25926J

Chen WS, Strik DPBTB, Buisman CJN, Kroeze C (2017b) Production of caproic acid from mixed organic waste—an environmental life cycle perspective. Environ Sci Technol 51(12):7159–7168. https://doi.org/10.1021/acs.est/6b06220

Choudhari SK, Cerrone F, Woods T, Joyce K, Flaherty VO, Connor KO, Babu R (2014) Pervaporation separation of butyric acid from aqueous and anaerobic digestion (AD) solutions using PEBA based composite membranes. J Ind Eng Chem 23:163–170. https://doi.org/10.1016/j.jiec.2014.08.010

Clarke WP, Alibardi L (2010) Anaerobic digestion for the treatment of solid organic waste: what's hot and what's not. Waste Manag 30:1761–1762. https://doi.org/10.1016/j.wasman.2010.06.019

Coma M, Martinez Hernandez E, Abeln F, Raikova S, Donnelly J, Arnot TC, Allen M, Hong DD, Chuck CJ (2017) Organic waste as a sustainable feedstock for platform chemicals. Faraday Discuss 202:175–195. https://doi.org/10.1039/C7FD00070G

Cope JL, Hammett AJM, Kolomiets EA, Forrest AK, Golub KW, Hollister EB, DeWitt TJ, Gentry TJ, Holtzapple MT, Wilkinson HH (2014) Evaluating the performance of carboxylate platform fermentations across diverse inocula originating as sediments from extreme environments. Bioresour Technol 155:388–394. https://doi.org/10.1016/j.biortech.2013.12.105

Datta R, Henry M (2006) Lacic acid: recent advances in products processes and technologies—a review. J Chem Technol Biotechnol 81:1119–1129. https://doi.org/10.1002/jctb.1486

den Boer E, Lukaszewska A, Kluczkiewicz W, Lewandowska D, King K, Reijonen T, Kuhmonen T, Suhonen A, Jaaskelainen A, Heitto A, Laatikainen R, Hakalehto E (2016) Volatile fatty acids as an added value from biowaste. Waste Manag 58:62–69. https://doi.org/10.1016/j.wasman.2016.08.006

Diggelman C, Ham RK (2003) Household food waste to wastewater or to solid waste? That is the question. Waste Manag Res 21:501–514. https://doi.org/10.1177/0734242X0302100603

EIA United States natural gas industrial price. https://www.eia.gov/dnav/ng/hist/n3035us3m.htm. Accessed 3 Jul 2017

Elbeshbishy E, Dhar BR, Nakhla G, Lee H-S (2017) A critical review on inhibition of dark biohydrogen fermentation. Renew Sust Energ Rev 79:656–668. https://doi.org/10.1016/j.rser.2017.05.075

Eurostat energy price statistics. http://ec.europa.eu/eurostat/statistics-explained/index.php/Energy_price_statistics#Natural_gas_prices_for_industrial_consumers. Accessed 3 Jul 2017

Gao MT, Shimamura T, Ishida N, Takahashi H (2011) PH-uncontrolled lactic acid fermentation with activated carbon as an adsorbent. Enzym Microb Technol 48:526–530. https://doi.org/10.1016/j.enzmictec.2010.07.015

Gebrezgabher SA, Meuwissen MPM, Prins BAM, Lansink AGJMO (2010) Economic analysis of anaerobic digestion—a case of Green power biogas plant in the Netherlands. NJAS - Wageningen J Life Sci 57:109–115. https://doi.org/10.1016/j.njas.2009.07.006

Ghimire A, Frunzo L, Pirozzi F, Trably E, Escudie R, Lens PNL, Esposito G (2015) A review on dark fermentative biohydrogen production from organic biomass: process parameters and use of by-products. Appl Energy 144:73–95. https://doi.org/10.1016/j.apenergy.2015.01.045

Gildemyn S, Molitor B, Usack JG, Nguyen M, Rabaey K, Angenent LT (2017) Upgrading syngas fermentation effluent using Clostridium kluyveri in a continuous fermentation. Biotechnol Biofuels 10:83. https://doi.org/10.1186/s13068-017-0764-6

González MI, Álvarez S, Riera F, Álvarez R (2007) Economic evaluation of an integrated process for lactic acid production from ultrafiltered whey. J Food Eng 80:553–561. https://doi.org/10.1016/j.jfoodeng.2006.06.021

Gruber PR (2001) Commodity polymers from renewable resources: polylactic acid. In: Carbon management: implications for R&D in the Chemical Sciences and Technology: a workshop report to the chemical sciences roundtable. National Academies Press, Washington, DC, pp 111–127. ISBN 0309503051

Huang C, Xu T, Zhang Y, Xue Y, Chen G (2007) Application of electrodialysis to the production of organic acids: state-of-the-art and recent developments. J Membr Sci 288:1–12. https://doi.org/10.1016/j.memsci.2006.11.026

Indexmundi natural gas price. http://www.indexmundi.com/commodities/?commodity=natural-gas. Accessed 3 Jul 2017

Jankowska E, Chwiałkowska J, Stodolny M, Oleskowicz-Popiel P (2017) Volatile fatty acids production during mixed culture fermentation—the impact of substrate complexity and pH. Chem Eng J 326:901–910. https://doi.org/10.1016/j.cej.2017.06.021

Joglekar HG, Rahman I, Babu S, Kulkarni BD, Joshi A (2006) Comparative assessment of downstream processing options for lactic acid. Sep Purif Technol 52:1–17. https://doi.org/10.1016/j.seppur.2006.03.015

Jones RJ, Massanet-Nicolau J, Mulder MJJ, Premier G, Dinsdale R, Guwy A (2017) Increased biohydrogen yields, volatile fatty acid production and substrate utilisation rates via the electrodialysis of a continually fed sucrose fermenter. Bioresour Technol 229:46–52. https://doi.org/10.1016/j.biortech.2017.01.015

Kim MS, Na JG, Lee MK, Ryu H, Chang YK, Triolo JM, Yun YM, Kim DH (2016) More value from food waste: lactic acid and biogas recovery. Water Res 96:208–216. https://doi.org/10.1016/j.watres.2016.03.064

Kiran EU, Trzcinski AP, Liu Y (2015) Platform chemical production from food wastes using a biorefinery concept. J Chem Technol Biotechnol 90:1364–1379. https://doi.org/10.1002/jctb.4551

Kleerebezem R, Joosse B, Rozendal R, Van Loosdrecht MCM (2015) Anaerobic digestion without biogas? Rev Environ Sci Bio/Technology 14(4):787–801. https://doi.org/10.1007/s11157-015-9374-6

Koupaie EH, Barrantes Leiva M, Eskicioglu C, Dutil C (2014) Mesophilic batch anaerobic co-digestion of fruit-juice industrial waste and municipal waste sludge: process and cost-benefit analysis. Bioresour Technol 152:66–73. https://doi.org/10.1016/j.biortech.2013.10.072

Leng L, Yang P, Mao Y, Wu Z, Zhang T, Lee PH (2017) Thermodynamic and physiological study of caproate and 1,3-propanediol co-production through glycerol fermentation and fatty acids chain elongation, vol 114. Elsevier, New York. ISBN 8522766606

Li X, Chen Y, Zhao S, Chen H, Zheng X, Luo J, Liu Y (2015) Efficient production of optically pure l-lactic acid from food waste at ambient temperature by regulating key enzyme activity. Water Res 70:148–157. https://doi.org/10.1016/j.watres.2014.11.049

Liu Y, He P, Shao L, Zhang H, Lü F (2017) Significant enhancement by biochar of caproate production via chain elongation. Water Res 119:150–159. https://doi.org/10.1016/j.watres.2017.04.050

Lopez AM, Hestekin JA (2013) Separation of organic acids from water using ionic liquid assisted electrodialysis. Sep Purif Technol 116:162–169. https://doi.org/10.1016/j.seppur.2013.05.028

López-Garzón CS, Straathof AJJ (2014) Recovery of carboxylic acids produced by fermentation. Biotechnol Adv 32:873–904. https://doi.org/10.1016/j.biotechadv.2014.04.002

Moriarty K (2013) Feasibility study of anaerobic digestion of food waste in St. Bernard, Louisiana

Nampoothiri KM, Nair NR, John RP (2010) An overview of the recent developments in polylactide (PLA) research. Bioresour Technol 101:8493–8501. https://doi.org/10.1016/j.biortech.2010.05.092

Nielfa A, Cano R, Fdz-Polanco M (2015) Theoretical methane production generated by the co-digestion of organic fraction municipal solid waste and biological sludge. Biotechnol Reports 5:14–21. https://doi.org/10.1016/j.btre.2014.10.005

Ntaikou I, Antonopoulou G, Lyberatos G (2010) Biohydrogen production from biomass and wastes via dark fermentation: a review. Waste and Biomass Valorization 1:21–39. https://doi.org/10.1007/s12649-009-9001-2

Padro CEG, Putsche V (1999) Survey of the economics of hydrogen technologies National Renewable Energy Laboratory; Golden

Palomo-Briones R, Razo-Flores E, Bernet N, Trably E (2017) Dark-fermentative biohydrogen pathways and microbial networks in continuous stirred tank reactors: novel insights on their control. Appl Energy 198:77–87. https://doi.org/10.1016/j.apenergy.2017.04.051

Pfaltzgraff LA, De bruyn M, Cooper EC, Budarin V, Clark JH (2013) Food waste biomass: a resource for high-value chemicals. Green Chem 15:307–314. https://doi.org/10.1039/c2gc36978h

Poggi-Varaldo HM, Munoz-Paez KM, Escamilla-Alvarado C, Robledo-Narvaez PN, Ponce-Noyola MT, Calva-Calva G, Rios-Leal E, Galindez-Mayer J, Estrada-Vazquez C, Ortega-Clemente A, Rinderknecht-Seijas NF (2014) Biohydrogen, biomethane and bioelectricity as crucial components of biorefinery of organic wastes: a review. Waste Manag Res 32:353–365. https://doi.org/10.1177/0734242X14529178

Prochaska K, Woźniak-Budych MJ (2014) Recovery of fumaric acid from fermentation broth using bipolar electrodialysis. J Membr Sci 469:428–435. https://doi.org/10.1016/j.memsci.2014.07.008

Queir D, Sousa R, Pereira S, Serafim S (2017) Valorization of a pulp industry by-product through the production of short-chain organic acids. Fermentation 3:20. https://doi.org/10.3390/fermentation3020020

Subramanian MR, Talluri S, Christopher LP (2015) Production of lactic acid using a new homofermentative Enterococcus faecalis isolate. Microb Biotechnol 8:221–229. https://doi.org/10.1111/1751-7915.12133

Tang J, Wang X, Hu Y, Zhang Y, Li Y (2016) Lactic acid fermentation from food waste with indigenous microbiota: effects of pH, temperature and high OLR. Waste Manag 52:278–285. https://doi.org/10.1016/j.wasman.2016.03.034

Tuck CO, Perez E, Horvath IT, Sheldon RA, Poliakoff M (2012) Valorization of biomass: deriving more value from waste. Science (80–) 337:695–699. https://doi.org/10.1126/science.1218930

Vaidya AN, Pandey RA, Mudliar S, Kumar MS, Chakrabarti T, Devotta S (2005) Production and recovery of lactic acid for polylactide—an overview. Crit Rev Environ Sci Technol 35:429–467. https://doi.org/10.1080/10643380590966181

Villano M, Paiano P, Palma E, Miccheli A, Majone M (2017) Electrochemically-driven fermentation of organic substrates with undefined mixed microbial cultures. ChemSusChem 10(15):3091–3097. https://doi.org/10.1002/cssc.201700360

Walla C, Schneeberger W (2008) The optimal size for biogas plants. Biomass Bioenergy 32:551–557

Weiland P (2010) Biogas production: current state and perspectives. Appl Microbiol Biotechnol 85:849–860

Whyte R, Perry G (2001) A rough guide to anaerobic digestion costs and MSW diversion. Biocycle 42:30–33

Xiong B, Richard TL, Kumar M (2015) Integrated acidogenic digestion and carboxylic acid separation by nanofiltration membranes for the lignocellulosic carboxylate platform. J Membr Sci 489:275–283. https://doi.org/10.1016/j.memsci.2015.04.022

Yeshanew MM, Frunzo L, Pirozzi F, Lens PNL, Esposito G (2016) Production of biohythane from food waste via an integrated system of continuously stirred tank and anaerobic fixed bed reactors. Bioresour Technol 220:312–322. https://doi.org/10.1016/j.biortech.2016.08.078

Yi SS, Lu YC, Luo GS (2008) Separation and concentration of lactic acid by electro-electrodialysis. Sep Purif Technol 60:308–314. https://doi.org/10.1016/j.seppur.2007.09.004

Yousuf A, Bonk F, Bastidas-Oyanedel JR, Schmidt JE (2016) Recovery of carboxylic acids produced during dark fermentation of food waste by adsorption on Amberlite IRA-67 and activated carbon. Bioresour Technol 217:137–140. https://doi.org/10.1016/j.biortech.2016.02.035

Yousuf A, Bastidas-Oyanedel J-R, Schmidt JE (2018) Effect of total solid content and pretreatment on the production of lactic acid from mixed culture dark fermentation of food waste. Waste Manag 77:516–521. https://doi.org/10.1016/j.wasman.2018.04.035

Zhang F, Zhang Y, Chen M, Zeng RJ (2012) Hydrogen supersaturation in thermophilic mixed culture fermentation. Int J Hydrog Energy 37:17809–17816. https://doi.org/10.1016/j.ijhydene.2012.09.019

Zhou J, Wu J, Liu Y, Zou F, Wu J, Li K, Chen Y, Xie J, Ying H (2013) Modeling of breakthrough curves of single and quaternary mixtures of ethanol, glucose, glycerol and acetic acid adsorption onto a microporous hyper-cross-linked resin. Bioresour Technol 143:360–368. https://doi.org/10.1016/j.biortech.2013.06.009

Zhou M, Yan B, Wong JWC, Zhang Y (2018) Enhanced volatile fatty acids production from anaerobic fermentation of food waste: a mini-review focusing on acidogenic metabolic pathways. Bioresour Technol 248(Pt A):68–78. https://doi.org/10.1016/j.biortech.2017.06.121

Part VI
Bio-based Polymers

Recent Advances on Enzymatic Catalysis as a Powerful Tool for the Sustainable Synthesis of Bio-Based Polyesters

Alessandro Pellis, Gibson S. Nyanhongo, and Thomas J. Farmer

Abbreviations

BDO	1,4-Butanediol
CaLB	*Candida antarctica* lipase B
Cut1	Cutinase 1 from *Thermobifida cellulosilytica*
DBTO	Dibutyl tin oxide
DMA	Dimethyl adipate
DMI	Dimethyl itaconate
EGDM	Ethylene glycol dimethacrylate
EHDA	9,10-Epoxy-18-hydroxyoctadecanoic acid
GLC	Glycerol
Glux-diester	2,4:3,5-di-*O*-Methylene-D-glucarate
Glux-diol	2,4:3,5-di-*O*-Methylene-D-glucitol
HEMA	*N*-(2-hydroxyethyl)maleimide
IA	Itaconic acid
iCaLB	CaLB covalently immobilized on epoxy-activated beads
iCut1	iCut1 covalently immobilized on epoxy-activated beads
MPEG	Methoxylated-poly(ethylene glycol)
MWe	Microwave energy
N435	Novozyme 435
ODO	1,8-Octanediol
PBSI	Poly(1,4-butylene succinate-*co*-1,4-butylene itaconate)
PES	Poly(ethylene succinate)
PGA	Poly(glycerol adipate)

A. Pellis (✉) · T. J. Farmer
Department of Chemistry, University of York, Green Chemistry Centre of Excellence, York, UK
e-mail: ale.pellis@york.ac.uk; thomas.farmer@york.ac.uk

G. S. Nyanhongo
Institute of Environmental Biotechnology, University of Natural Resources and Life Sciences, Vienna, Tulln, Austria
e-mail: g.nyanhongo@boku.ac.at

© Springer Nature Switzerland AG 2019

J.-R. Bastidas-Oyanedel, J. E. Schmidt (eds.), *Biorefinery*,
https://doi.org/10.1007/978-3-030-10961-5_23

PHAs	Poly(hydroxy alkanoates)
ROP	Ring-opening polymerization
$Sc(OTf)_3$	Scandium trifluoromethanesulfonate
$scCO_2$	Supercritical CO_2
TBD	1,5,7-Triazabicyclo[4,4,0]dec-5-ene
TMPO	Trimethylolpropane oxetane
YLL	*Yarrowia lipolytica* lipase
ε-CL	ε-Caprolactone

1 Introduction

The enzymatic synthesis of polyesters dates back to the 1980s when various independent groups of scientists discovered the two main reaction pathways, namely, polycondensation and ring-opening polymerization (ROP). The most investigated polycondensation reactions are these between dicarboxylic acid diesters with polyols, despite the fact that condensation polymerization of dicarboxylic acids and oxyacids and their esters were also reported. Regarding ROP reactions instead, the polymerization of lactones (cyclic esters) is predominant over the investigation of other cyclic monomers (Fig. 1) (Kobayashi 2010). For the initial studies on enzymatic polyester synthesis, several halogenated diesters were chosen together with toxic solvents (e.g., pyridine) and the utilization of lyophilized enzymes mainly belonging to the protease and lipase families and used in their powder form (Pellis et al. 2017a). Later on, the interest on enzymatic polymerizations switched to the possibility of recycling the enzyme (commonly the most expensive component of an enzymatic reaction) (Pellis et al. 2017b) that led to the development of several immobilization technologies (adsorption, covalent bindings, cross-linked enzymes)

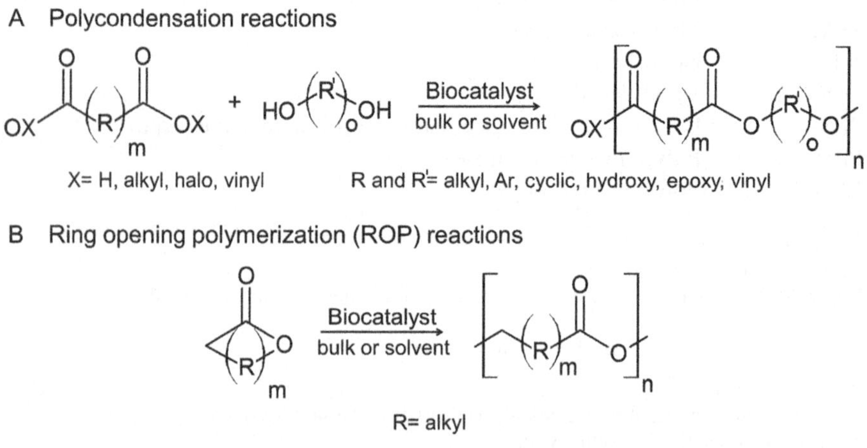

Fig. 1 Most common mechanisms for the enzymatic synthesis of polyesters. (**a**) Polycondensation and (**b**) ring-opening polymerization (ROP) reactions

(Hanefeld et al. 2009; Cantone et al. 2013). In the early 2000s, the research interest shifted to the production of functional aliphatic polyesters carrying lateral hydroxy and epoxy groups that are suitable for further functionalization or cross-linking (Uyama et al. 1999, 2001).

In the last 5 years, the enzymatic polymerization field had a renewed interest by several research groups with a special focus on the production of bio-based polyesters driven by the desire of developing a more sustainable and environmentally friendly polymer industry. This chapter will focus on the most recent advances in the field discussing polymerizations of monomers that could be easily derived from biomass allowing the closure of the carbon cycle (Pellis et al. 2016a) together with traditional and alternative biocatalysts which stability was considerably improved in order to allow their recyclability. Moreover this section will have a special focus on the polymerization of bio-based platform molecules carrying lateral functionalities (Farmer et al. 2015) and therefore representing a key point of applying enzymatic catalysis for polyester synthesis since only reactions conducted in mild conditions allow the selective reactivity of the desired moieties avoiding side reactions.

2 Synthesis of Aliphatic Polyesters

The simplest type of polyesters that is possible to synthetize via enzymatic polycondensation and ROP is indeed represented by aliphatic polyesters that can be divided in copolymeric linear chains composed by a diacid and a diol or homopolymeric chains formed from the ROP of a lactone. The most studied and widespread enzyme for this kind of reactions is without any doubts *Candida antarctica* lipase B (CaLB) in its commercially available immobilized form known as Novozyme 435 (N435) that consist of the abovementioned biocatalyst physically adsorbed within the macroporous resin poly(methyl methacrylate-*co*-butyl methacrylate) known as Lewatit VPOC 1600. This preparation is frequently used as benchmark for comparing other immobilized enzymes due to its remarkable activity and availability on the market. Recently, other enzymes of fungal origin belonging to the serine hydrolases class, namely, cutinases, were widely reported to be active for the degradation of several aliphatic and aromatic polyesters such as poly(1,4-butylene succinate), poly(ethylene terephthalate), and poly(ethylene furanoate) (Gamerith et al. 2017; Pellis et al. 2016b; Weinberger et al. 2017a, b).

Among them, cutinase 1 from *Thermobifida cellulosilytica* (Cut1) was expressed in its recombinant form, immobilized on various supports and tested as alternative to CaLB for polycondensation reactions. Bulk reactions between dimethyl adipate (DMA) (the dimethyl ester of adipic acid) and 1,4-butanediol (BDO) (both derived from the very same fermentative pathway) (Pellis et al. 2016c) were performed using the same conditions (70 °C, 1000 mbar, 10% w/w on the total amount of monomers of the immobilized biocatalyst) for N435 and for CaLB and Cut1 covalently immobilized on epoxy-activated beads (Resindion EC-EP) further called iCaLB and iCut1. Results show how the two CaLB preparations lead to comparable monomer conversions (78% for N435 vs. 76% for iCaLB) while being considerably higher (86%) for the immobilized cutinase. The same trend was also observed for the molecular masses of the polymers with iCut1 reaching almost 2000 Da, while the two CaLB preparations did not lead to M_w higher than around 1000 Da (Pellis et al. 2016d). In the same publication, the reaction selectivity of iCut1 is also discussed comparing bulk and reactions in organic media (toluene) between DMA and diols having linear aliphatic chains of 4, 6, and 8 carbon atoms, respectively. While for bulk reactions the polymerization of the C_4 diol was greatly favorable, in toluene the polymerization efficiency relativity is lower for all different diols. This effect is believed to be due to a combination of viscosity and mass transfer, though some solvent effects on the conformation and accessibility of the biocatalyst's active site cannot be excluded (Pellis et al. 2016d).

A further optimization of the reaction conditions for Cut1 was developed using a factorial design approach where conditions such as temperature, pressure, length of the diol, and water activity were taken into account. Cut1 works efficiently in solvent-free systems and under very mild reaction conditions (50 °C) as compared to conventional polycondensations employing toxic metals ($T > 150$ °C). When compared to CaLB, Cut1 is less sensitive to the presence of water in the reaction system allowing a better chain elongation at pressures closer to 1000 mbar without observing hydrolytic reactions (Pellis et al. 2017c). The very same enzymes, Cut1 and CaLB, were later immobilized on a renewable solid support such as rice husks. Due to the fact that the chosen lignocellulosic biomass contains silica, therefore possessing a clear chemical versatility, physical adsorption, and covalent binding of CaLB using minimal pretreatment processing was achieved, demonstrating rice husks with its worldwide availability and mechanochemical robustness, an inexpensive natural matrix that is a promising candidate for replacing organic fossil-based carriers (such as acrylates and polypropylene beads) for enzyme immobilization (Corici et al. 2016). Regarding the economic aspects of these reactions, when a thin-film system is used, the described iCaLB, iCut1, and rice husk-based preparations can be reused for up to ten cycles without observing any significant reduction of the monomer conversion rates and with an excellent reproducibility (Pellis et al. 2016d; Corici et al. 2016).

Another interesting technology for the immobilization of enzymes is the ionic bonding of His-tagged Cut1 using beads with chelated (nontoxic) Fe(III) ions developed by EnginZyme (Stockholm, Sweden). In this case a > 99% protein binding was observed on all three tested carriers each with differing superficial hydrophilicity (opal-hydrophilic, coral-hydrophobic, and amber-semi-hydrophobic). A selectivity study involving the dimethyl esters of succinic (C_4), DMA (C_6), and sebacic (C_8) acid together with BDO and 1,8-octanediol (ODO) was performed showing a clear preference of the C_6-diester/C_4-diol combination. Recyclability studies showed an excellent stability of the immobilized Cut1 (retained activity >94%) over 24 h cycles when a solvent-free workup was used (Pellis et al. 2017a).

Despite these works describing the potential of alternative biocatalysts, most of the polymerizations are still conducted using N435 as catalyst due to its availability. Kanelli et al., for example, enzymatically polymerized BDO and ODO with diacids or their derivatives (diethyl succinate, sebacic acid, 1,12-dodecanedioic acid, and 1,14-tetradecanedioic acid) and obtained an array of aliphatic polyesters, namely, poly(butylene succinate), poly(octylene sebacate), poly(octylene dodecanate), and poly(octylene tetradecanate) having $3700 < M_n < 8000$ Da and yields up to 97%. Subsequently, in order to increase the molecular weight and improve the physical characteristics and the thermal properties of the prepolymers, chemo-catalytic solid state polymerization was used for the finishing of the polycondensation procedure (Kanelli et al. 2014).

With the aim of obtaining sustainable polymeric nano-carriers and surfactants, a team from the University of Nottingham investigated the polycondensation of azelaic acid with 1,6-hexanediol end-capping the obtained polymers using methoxy poly(ethylene glycol) (MPEG) chains of various molecular masses in a one-pot high pressure (50 bar) reaction at a near-ambient temperatures as low as 35 °C. Bio-based amphiphilic polyesters were successfully produced thanks to the unique properties of supercritical CO_2 ($scCO_2$) in a solvent-free reaction system (Fig. 2a) (Curia and Howdle 2016). Another interesting work reported the possibility of synthetizing disiloxane-containing polyesters. Reactions were conducted in the $35 < T < 150$ °C range in order to test the tolerance of the immobilized enzyme, and long-term thermal stability studies were carried out over ten reaction cycles. A significant decrease in the initial polymerization rate was observed but surprisingly no changes in the final monomer conversion after 24 h was detected (Fig. 2b) (Frampton et al. 2013). In a second study, the authors reported that at all tested temperatures, the rate constants steadily increased until 140 °C when a maximum rate constant was reached. Beyond this temperature the enzyme began to show signs of denaturation, shown by a decline in the rate constants. A comparison between data sets suggests that the increased number of siloxane units led to lowered reaction rates, with the monomer competing with $scCO_2$ as the solvent, thus effectively reducing the frequency of collisions between the reactive end groups of both monomers and the growing polymer chain. It is, however, possible to conclude that incorporation of a single disiloxane unit into a diester or diol can greatly increase the temperature at which CaLB can catalyze reactions (Frampton and Zelisko 2013).

A recent study investigated the CaLB affinity toward the diol chain length. BDO was compared with 1,3-propanediol, with the first leading to higher molecular mass

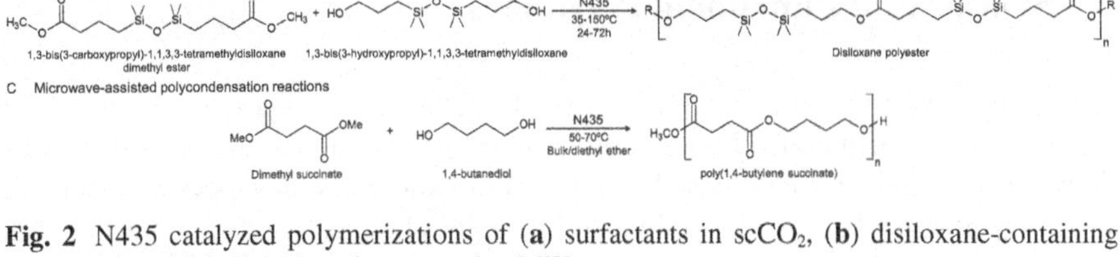

Fig. 2 N435 catalyzed polymerizations of (**a**) surfactants in $scCO_2$, (**b**) disiloxane-containing polyesters, and (**c**) aliphatic polyesters using MWe

polyesters. No preference was observed between diethyl succinate and diethyl adipate. In the same work, two series of random co-polyesters poly(1,4-butylene succinate-ran-1,4-butylene adipate) and poly(1,3-propylene succinate-ran-1,4-butylene succinate) were synthesized at different compositions in order to investigate the effect of the composition on co-polyesters (Debuissy et al. 2017). After biocatalyst, monomer, media selection, and the optimization of the reaction system in order to allow the reuse of the biocatalysts, also various methods of heating the reaction vessel were investigated. Over the years, several reports of microwave energy (MWe) applied to biocatalyzed reactions have been published, but only the ROP of ε-caprolactone (ε-CL) has been successfully carried out. Scientists also investigated the effect of MWe on the polycondensation of dimethyl esters with BDO. The collected data are pretty conclusive since, apart from issues with bulk polymerization, MWe can be efficiently used as a replacement for conventional heating for the enzyme-catalyzed polycondensation of bio-derived polyesters with minimal impact on monomer conversion and polymer chain length (Fig. 2c) (Pellis et al. 2016e). As last example of polycondensation of aliphatic polyesters,

ROP is regularly reported as the most successful biocatalyzed polymerization; indeed the polymerization of ε-CL has been demonstrated using a varied range of conditions including in bulk and organic media systems (Duskunkorur et al. 2014), micro-reactors (Banghale et al. 2012), and ionic liquids (Wu et al. 2013), these always using N435 as biocatalyst. On this regard, an immobilized preparation of the lipase from *Yarrowia lipolytica* (YLL) adsorbed on Lewatit and Amberlite beads was used for the enzymatic ROP of ε-CL in various solvents. The highest number average molecular weight (10,685 Da) was achieved at 90 °C (100% monomer conversion) using Lewatit-immobilized with decane as reaction solvent after 60 h (Barrera-Rivera and Martínez-Richa 2017).

An interesting work by Morales-Huerta et al. (2017) describes the synthesis of poly(ethylene succinate) (PES) with a M_w of over 60 kDa via enzymatic ROP of cyclic oligo(ethylene succinate)s which were prepared initially by enzymatic cyclo-condensation of dimethyl succinate and ethylene glycol, this showing and demonstrating how ROP can achieve higher polymer chain lengths than that achievable through standard polycondensation. The production of several PES-co-polyesters containing butylene succinate, ε-hydroxycaproate, or L-lactate units with a random distribution and high molecular masses was also reported (Morales-Huerta et al. 2017), broadening the scope of this approach to aliphatic bio-based polyesters.

3 Synthesis of Functional Polyesters

It is well known that the enzyme-catalyzed synthesis of polyesters on industrial scale is not yet economically competitive with the traditional chemical processes employing organic or metal-based catalysts. For example, it is suggested that costs of few hundred $/kg are acceptable for specialty chemicals, whereas in the bulk chemical sector, the economic impact must remain below $10/kg but if often close to 0.1$/kg (Pellis et al. 2016a, 2017b). However, the major strength of enzymatic synthesis is the possibility to selectively synthesize polyesters having lateral functional moieties that are otherwise difficult to maintain using traditional chemo-catalysis if not employing complicated protective strategies that usually require additional chemicals, solvents, and workup steps (Chanda and Ramakrishnan 2015).

A recent example where chemical catalysis was compared with enzymatic synthesis was for the synthesis of sorbitol-based hydroxyl-functional, bio-based polyesters. In these works, Gustini and co-workers screened four different catalytic systems: the organo-base catalyst 1,5,7-triazabicyclo[4,4,0]dec-5-ene (TBD) and two metal-based catalysts, dibutyl tin oxide (DBTO) and scandium trifluoromethanesulfonate (also known as scandium triflate, $Sc(OTf)_3$), together with N435 or a covalently immobilized preparation of CaLB (Gustini et al. 2016a). Authors reported that the DBTO-catalyzed polycondensation at 140 °C proved to have limitations in providing a soluble polymer with the targeted molecular weights, whereas no polymerization was observed at 100 °C. Moreover at 140 °C sorbitan was detected as side reaction product. Also when $Sc(OTf)_3$ and TBD were employed as catalyst, despite obtaining polymers above a certain temperature, the degradation of sorbitol to sorbitan was always observed, proving that a traditional chemo-catalyzed route is suitable for maintaining the sorbitol unit fully intact (Gustini et al. 2016a). The CaLB-catalyzed polymerizations of the same polymer instead lead to almost-linear polyesters with side reactions such as degradation to sorbitan or ether formation that were avoided. The obtained polyesters carrying pendant hydroxyl groups were found adequate for coating applications and further characterized and cured via several processes (Gustini et al. 2015, 2016b) (Fig. 3).

Similarly, a family of polyesters, namely, poly(glycerol adipate) (PGA), poly(sorbitol adipate), and PGA-graft-poly(ε-CL) were synthesized using CaLB as biocatalyst and converted in macroinitiators for atom transfer radical polymerization by esterification with α-bromoisobutyryl bromide (Jbeily et al. 2014). Interestingly, in the case of glycerol (GLC), the enzymatic polymerization can also lead to branched polymers since the secondary hydroxyl group can react when the reaction temperature exceeds certain values. This problem can be easily overcome dosing the amount of GLC over the amount of diester (ratio 1.2:1.0) and a $T \leq 50$ °C (Bilal et al. 2017). When the secondary hydroxy group of GLC is maintained, an easy coupling of indomethacin using 1-ethyl-3-(3-dimethylaminopropyl)carbodiimide as catalyst was obtained, opening interesting possibilities for the use of these polymers for drug release nanoformulation systems (Wersig et al. 2017).

Since polymer chemistry is switching from petroleum-derived polyesters toward bio-based alternatives, components derived from biomass are easier to degrade (e.g. the substitution of poly(ethylene terephthalate) with poly(ethylene furanoate) on which several companies are currently working) (Weinberger et al. 2017a), several functional polyesters were enzymatically synthesized using 9,10-epoxy-18-hydroxyoctadecanoic acid (EHDA), an epoxy-functional ω-hydroxy-fatty acid extracted from the outer layer of birch bark. This is the case of polyesters produced from the homopolymerization of EHDA and end capped using ethylene glycol dimethacrylate (EGDM) (Fig. 4a), and subsequently cationic polymerization of the

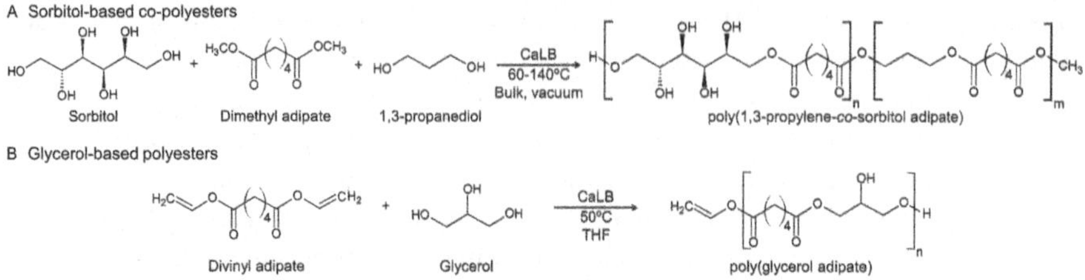

Fig. 3 CaLB-catalyzed synthesis of linear, bio-based aliphatic polyesters containing lateral hydroxy functionalities based on (**a**) sorbitol and (**b**) glycerol

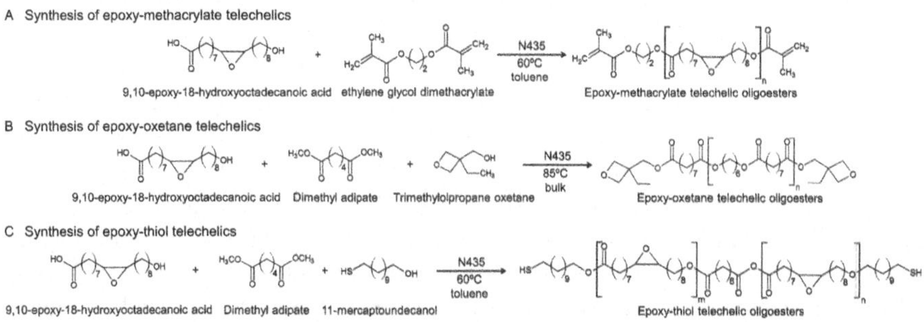

Fig. 4 CaLB-catalyzed synthesis of linear, bio-based telechelic polyesters containing lateral epoxy functionalities and (**a**) methacrylate, (**b**) maleimide, and (**c**) thiol-containing end-cappers

epoxides and free radical polymerization of the methacrylates were performed either as separate or parallel reactions in order to produce cross-linked films with tailored mechanical and thermal properties (Torron et al. 2014).

Later, Semlitsch et al. reported the synthesis of polyesters composed of DMA and EHDA that were end capped using EGDM, *N*-(2-hydroxyethyl)maleimide (HEMA), or trimethylolpropane oxetane (TMPO) (Fig. 4b). Targeted molecular weights and monomer conversions >95% were achieved with both end-functionalities and the epoxy groups that were fully preserved thanks to the mild reaction conditions (60–85 °C, 200 mbar, 17–24 h, 10–20% of N435 as catalyst). The end-cappers enable the application of different cross-linking techniques such as Diels-Alder reactions for the maleimide, radical polymerization for the methacrylate group and cationic polymerization for the oxetane group permitting the use of various chemistries in order to produce thermosets with tuned properties (Semlitsch et al. 2016). Following these ideas, telechelics having 11-mercaptoundecanol as end-capper were synthesized (Fig. 4c), and the production of surface-active hydrophobic films via curing of the thiol moieties was successfully demonstrated (Nameer et al. 2017).

Additional works in the field focused their attention on the synthesis of azelaic-based telechelics having sorbic alcohol, 12-hydroxystearic acid, HEMA, or TMPO as functional end groups in $scCO_2$ (Curia et al. 2015) and on the production of various thermosets using three different hydroxyl cores, namely, neopentyl glycol, trimethylolpropane, and di-trimethylolpropane, and four fatty acids esters, methyl stearate, epoxidized methyl oleate, epoxidized methyl linoleate, and epoxidized methyl linolenate (Torron et al. 2016).

Functional polyesters containing lateral unsaturated moieties in the main chain are also an important class of macromolecules with sites for post-polymerization modification. The model/typical example molecule for such kind of polymers—and also the most investigated one—is indeed itaconic acid (IA) despite other molecules such as *cis*-2-butene-1,4-diol and tall oil fatty acids having also recently been used (Roberts and Friebel 2016).

Like the previously reported examples, also in the case of C=C bonds, enzymatic catalysis is useful in order to obtain linear polyesters that can be further cross-linked or functionalized in a second reaction step. The chemo-catalyzed polymerization of IA and its derivatives (such as dimethyl itaconate (DMI) and itaconic anhydride) are susceptible to the high operational temperatures commonly used for polyester synthesis (150–200 °C) that in most cases cause isomerization to the less reactive mesaconic or citraconic acids or cross-linking of the polymeric chains (Farmer et al. 2015) if protection strategies such as the addition of quinol as a radical quencher are not put in place (Chanda and Ramakrishnan 2015). Enzymatic catalysis on the other hand, performed under mild conditions ($T \leq 80$ °C) allows the synthesis of linear chains maintaining fully the C=C moiety.

The synthesis of poly(butylene succinate-*co*-itaconate) (PBSI) was optimized regarding solvent selection with dodecane, diethylene glycol dimethyl ether, and diphenyl ether, and cyclohexane/toluene mixtures were evaluated together with various synthetic protocols (two-stage method and azeotropic polymerization) with the best results in terms of molecular masses that were obtained using the azeotropic protocol in cyclohexane/toluene mixtures (Fig. 5a) (Jiang et al. 2013, 2014a). When using the inactivated dicarboxylic acid (itaconic acid) vs. its dimethyl ester, only oligomers were obtained since the reactions were heterogeneous due to the non-solubility of the monomer. PBSI polyesters were reported to be semicrystalline with the T_g of the cured polymers increasing significantly as the cross-linking density rises. The cured unsaturated polyesters on the other hand are brittle materials with a Young's modulus, ultimate tensile stress, and rupture strain of 11–66 MPa, 3–12 MPa, and 26–34%, respectively (Jiang et al. 2014a).

The same group investigated the synthesis of DMI-based co-polyesters using various dicarboxylic acid diethyl esters. The synthetic aliphatic polyesters reached high weight average molecular weight (M_w) values up to 94 kDa with the used enzyme, CaLB, showing the highest specificity for diethyl adipate among the tested diacid ethyl esters ($n = 2$–10) (Jiang et al. 2015a).

Other works on DMI focused instead on the possibility of synthesizing oligoesters in benign solventless conditions and optimizing the reaction system in order to render the biocatalyst's preparation recyclable for over ten cycles, thus constituting a considerable leap forward for the scaling up of the process at an industrial level (Pellis et al. 2015). The same group reported a detailed study on the limitations of the polycondensation of DMI with BDO and determined the optimal DMI/BDO ratio to be used for such reactions (initial DMI/BDO ratio = 1.0:0.5 in order to favor the elongation). Moreover 1,4-cyclohexanedimethanol, a cyclic diol, and GLC were also

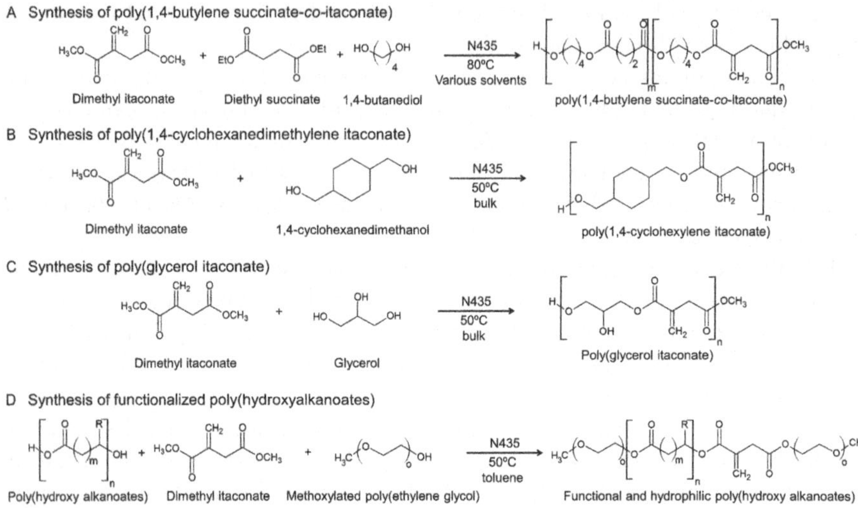

Fig. 5 CaLB-catalyzed synthesis of (**a**) poly(butylene succinate-*co*-itaconate) (PBSI), (**b**) poly(1,4-cyclohexandimethylene itaconate), (**c**) poly(glycerol itaconate), and (**d**) vinyl functionalized and hydrophilic poly(hydroxy alkanoates) (PHAs)

considered in the study as interesting diols for the preparation of aliphatic-cyclic oligoesters, showing chain rigidity and hydroxy-vinyl functional polyesters having several available moieties for further derivatization (Fig. 5b, c) (Corici et al. 2015).

The exciting potential of enzymatic catalysis was also recently shown for the modification of medium chain poly(hydroxy alkanoates) (PHA) produced via microbial fermentation using *Escherichia coli* as host. In this work, Vastano and co-workers present an elegant approach for the enzymatic coupling of PHA with dimethyl itaconate and methoxylated-poly(ethylene glycol) (MPEG) for the production of PHA-based telechelic polyesters having vinyl functionalities and hydrophilic end-cappers that totally reversed the hydrophobic nature of the initial PHAs (Fig. 5d), therefore opening the possibility of using these conjugated for the development of drug delivery systems (Vastano et al. 2017).

Unsaturated oligoesters composed of *cis*-2-butene-1,4-diol, dimethyl succinate and 6-mercapto-1-hexanol have been prepared via a N435-catalyzed polycondensation in a solvent-free reaction system. The obtained oligomers having a degree of polymerization between $2 < DP < 4$ were subsequently UV-cure using the photo initiator Irgacure 651, successfully producing thermoset resins (Finnveden et al. 2016). The production of linear aliphatic polyesters is not always desired, for example, the one-pot synthesis (in bulk) of purposely highly branched polyesters based on a feed of bio-based monomers such as GLC, pentaerythritol, azelaic acid, and tall oil fatty acid were was also reported with the purpose of cross-linking the C=C and obtaining films with good UV stability, water contact angles up to 141°, and a glass transition temperature that could be controlled through the feed composition (Nguyen et al. 2016).

4 Synthesis of Aliphatic-Aromatic Polyesters

The last part of this collection of advancements of the enzymatic synthesis of bio-based polyesters focuses on the possibility to produce via polycondensation a wide array of aliphatic-aromatic polyesters. In fact, most of the polyesters currently used for beverage bottling, food packaging, and automotive parts belong to this specific class due to their good mechanical and barrier properties. Following the boom of furan-based compounds, one of the most reported reactions involves the polytrans-esterification of dimethyl furan-2,5-dicarboxylate and linear α,ω-aliphatic diols (chain length ranging from C_2 to C_{12}) using N435 in dry organic solvents (Fig. 6a). Authors reported 90–95% substrate conversions when C_4–C_{12} diols were used, while in the case of ethylene glycol and 1,3-propanediol, the conversion were instead only moderate (≤75%) (Cruz-Izquierdo et al. 2015). Other authors confirm this observation, stating that the higher molecular weight furanic-aliphatic polyesters were produced using long-chain alkane-α,ω-aliphatic linear diols and, extending the work to other diols, found that it was possible to obtain only oligomers when using 2,3-butanediol, isosorbide, GLC, or sorbitol (Jiang et al. 2015b). These results clearly confirm that CaLB is more active on longer-chain aliphatic linear diols and that the reaction temperature (increased from 80 to 140 °C) is a key point for obtaining high molecular weight furan-based polyesters.

As an alternative 2,5-bis(hydroxymethyl)furan was used in combination with several aliphatic diacid ethyl esters with a backbone chain of C_4 to C_{12} to produce aliphatic-aromatic polyesters where the furan moiety derives from the diol component (Fig. 6b). In this case the synthetic approach was relatively unsuccessful since only low M_n products of around 2000 Da were obtained. The likely cause of these low chain lengths was that etherification occurred during the enzymatic polymerization procedure, hence a different approach based on lower reaction temperature,

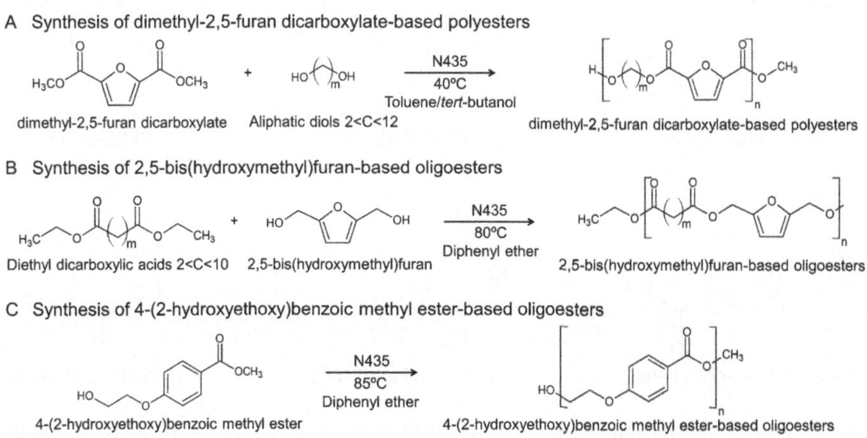

Fig. 6 Enzymatic synthesis of aliphatic-aromatic polyesters based on (**a**) dimethyl-2,5-furan dicarboxylate, (**b**) 2,5-bis(hydroxymethyl)furan, and (**c**) 4-(2-hydroxyethoxy)benzoic methyl ester

shorter reaction time, or using a different organic media is needed (Jiang et al. 2014b).

Another study described for the production of biodegradable oligoesters in a simple one-pot reaction system involved the screening of various aromatic dicarboxylic acids and their dimethyl esters with C_2–C_{12} aliphatic diols. In this case the authors report how (similarly for what reported for the itaconates) the dimethyl esters are the enzyme's preferred substrate leading to higher monomer conversions and molecular weights of the products since methanol is a way easier byproduct to be removed if compared with the water released from the reaction of the dicarboxylic acids (Fig. 7). In this case also the stereochemistry of the diester is very important, in fact dimethyl phthalate did not lead to any detectable esterification products while comparing dimethyl terephthalate with dimethyl isophthalate, the latter gave conversions almost double than the 1,4-disubstituted moiety (e.g., 88% vs. 43% when BDO was used as a diol) (Pellis et al. 2016f). This is a "substitution pattern effect" as the ortho-, meta-, and para-substitution can also greatly affect the electron density on the carboxylic acid groups.

In an attempt to improve the enzymatic synthesis of poly(ethylene terephthalate)-like polyesters, the polycondensation of 4-(2-hydroxyethoxy)benzoic methyl ester to obtain oligo(2-hydroxyethoxy benzoate) was attempted using various polymerization protocols. In this case low number average molecular weights (M_n) of around 2000 Da or lower were obtained, while a lower degree of polymerization or no oligomer formation was observed from the enzymatic reaction of compounds modified by one or more methoxy groups. The molecular weight of the obtained polyesters was, in a second stage, successfully increased to 13,050 Da by using it as a macromolecular monomer in a sequential polymerization (Fodor et al. 2017).

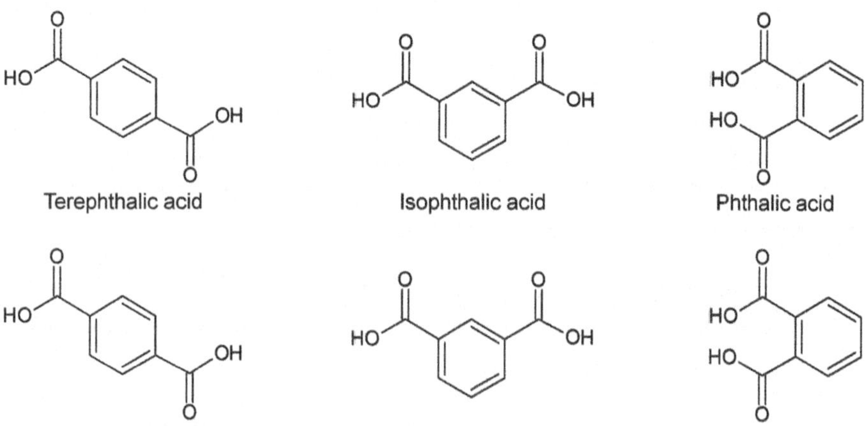

Fig. 7 Aromatic dicarboxylic acids and their dimethyl esters used for the CaLB-catalyzed synthesis of aromatic-aliphatic polyesters in toluene

Despite not belonging to the aromatic-aliphatic class of polyesters, it is worth noting the synthesis of polyesters made from BDO, sebacic acid, and rigid monomer from D-glucose in the melt and in solution via enzymatic polycondensation. In this piece of work, two rigid bicyclic diacetals, namely, 2,4:3,5-di-*O*-methylene-D-glucitol (Glux-diol) and dimethyl 2,4:3,5-di-*O*-methylene-D-glucarate (Glux-diester), were prepared from glucose and reacted in order to obtain bio-based polyesters containing up to 50% of sugar-based units. The enzymatic synthesis has proven to be effective to produce polyesters with $M_w \sim 10{,}000$, whereas $M_w \sim 40{,}000$ were obtained by melt polycondensation. In both cases the molecular weight of the prepared co-polyesters decreased with the content in Glux units, and their microstructure was almost random for whichever used composition (Japu et al. 2015).

5 Outlook

The extraordinary potential of enzymes for the synthesis of aliphatic, functional, and aromatic polyesters is immediately clear to the reader. Potential that till now is unfortunately limited to the lab-scale research since, in such polymerization reactions, the biocatalyst is indeed the most expensive component of the whole reaction system (Pellis et al. 2017b). An attempt to scale up the process was done in the 1990s by the British company Baxenden Chemicals, which aimed to produce highly regular-structured polymers used in coating and adhesive applications via enzymatic polycondensation of polyols and diacids. Evidently, biocatalyzed polyester synthesis requires the use of immobilized enzymes to enable the recovery of the expensive biocatalyst (as we saw before generally lipase B from *Candida antarctica*) and to avoid the contamination of the product with the enzymes. As such, in 2010, a thin-film concept operating under solventless conditions and characterized by high viscosity was experimentally validated at a 10-kg scale using a turbo reactor and successfully patented but to date was never scaled to industrial production (Cerea et al. 2012).

The biocatalyst cost toward polymerization can be surely reduced by using cheap and available renewable biomass (e.g., rice husks) as immobilization supports (Corici et al. 2016) and further enhanced by using reactions systems that allow effective reuse over many reaction cycles. Improvements to the thermal stability of enzymes should also allow for longer release and for higher chain length polyesters, and the condensate is more easily removed (Brogan and Hallett 2016).

Near-endless opportunities are still open in this field since the use of bio-derived monomers and environmentally friendly synthetic strategies that maximize the atom efficiency and reduce byproducts are also needed to aid progression toward sustainable chemistry. New monomers from biomass are being prepared and investigated all the time, and the often high level of functionality, including significant quantities of oxygen, means that the properties of any resulting polymers can alter significantly from their petrochemical rivals. Research on the enzymatic synthesis of aromatic-aromatic and aromatic-functional polyesters, innovative immobilization strategies, and novel biocatalysts together with the development of novel green solvents are only few of the most recent and intriguing challenges that scientists will face during the next decade in the enzymatic polyester synthesis field.

Acknowledgment Alessandro Pellis thanks the FWF Erwin Schrödinger fellowship (grant agreement J 4014-N34 for financial support).

References

Banghale AS et al (2012) Enzyme-catalyzed polymerization of end-functionalized polymers in a microreactor. Macromolecules 45:7000–7008. https://doi.org/10.1021/ma301178k

Barrera-Rivera KA, Martínez-Richa A (2017) Yarrowia lipolytica extracellular lipase Lip2 as biocatalyst for the ring-opening polymerization of ε-caprolactone. Molecules 22:1917. https://doi.org/10.3390/molecules22111917

Bilal MH et al (2017) Synthesis of poly(glycerol adipate)-g-oleate and its ternary phase diagram with glycerol monooleate and water. Eur Polym J 91:162–175. https://doi.org/10.1016/j.eurpolymj.2017.03.057

Brogan APS, Hallett JP (2016) Solubilizing and stabilizing proteins in anhydrous ionic liquids through formation of protein–polymer surfactant nanoconstructs. J Am Chem Soc 138:4494–4501. https://doi.org/10.1021/jacs.5b13425

Cantone S et al (2013) Efficient immobilisation of industrial biocatalysts: criteria and constraints for the selection of organic polymeric carriers and immobilisation methods. Chem Soc Rev 42:6262–6276. https://doi.org/10.1039/C3CS35464D

Cerea G et al (2012) Process for the production of polyesters through synthesis catalyzed by enzyme. WO2013110446A1, World Patent

Chanda S, Ramakrishnan S (2015) Poly(alkylene itaconate)s—an interesting class of polyesters with periodically located exo-chain double bonds susceptible to Michael addition. Polym Chem 6:2108. https://doi.org/10.1039/c4py01613k

Corici L et al (2015) Understanding potentials and restrictions of solvent-free enzymatic polycondensation of itaconic acid: an experimental and computational analysis. Adv Synth Catal 357:1763–1774. https://doi.org/10.1002/adsc.201500182

Corici L et al (2016) Large scale applications of immobilized enzymes call for sustainable and inexpensive solutions: rice husks as renewable alternatives to fossil-based organic resins. RSC Adv 6:63256-63270. https://doi.org/10.1039/C6RA12065B

Cruz-Izquierdo A et al (2015) Lipase-catalyzed synthesis of oligoesters of 2,5-furandicarboxylic acid with aliphatic diols. Pure Appl Chem 87:59–69. https://doi.org/10.1515/pac-2014-1003

Curia S, Howdle S (2016) Towards sustainable polymeric nano-carriers and surfactants: facile low temperature enzymatic synthesis of bio-based amphiphilic copolymers in scCO2. Polym Chem 7:2130–2142. https://doi.org/10.1039/C6PY00066E

Curia S et al (2015) Green process for green materials: viable low-temperature lipase-catalysed synthesis of renewable telechelics in supercritical CO2. Phil Phil Trans R Soc A 373:20150073. https://doi.org/10.1098/rsta.2015.0073

Debuissy T et al (2017) Lipase-catalyzed synthesis of biobased and biodegradable aliphatic copolyesters. Eur Polym J 97:328–337

Duskunkorur HO et al (2014) Lipase catalyzed synthesis of polycaprolactone and clay-based nanohybrids. Polymer 55:1648–1655. https://doi.org/10.1016/j.polymer.2014.02.016

Farmer TJ et al (2015) Synthesis of unsaturated polyester resins from various bio-derived platform molecules. Int J Mol Sci 16:14912–14932. https://doi.org/10.3390/ijms160714912

Finnveden M et al (2016) One-component thiol-alkene functional oligoester resins utilizing lipase catalysis. Macromol Chem Phys 217:1335–1341. https://doi.org/10.1002/macp.201500490

Fodor C et al (2017) Enzymatic approach for the synthesis of biobased aromatic aliphatic oligo-polyesters. Polym Chem 8(44):6795–6805. https://doi.org/10.1039/x0xx00000x

Frampton MB, Zelisko PM (2013) Synthesis of lipase-catalysed silicone-polyesters and silicone-polyamides at elevated temperatures. Chem Commun 49:9269. https://doi.org/10.1039/c3cc45380d

Frampton MB et al (2013) Synthesis of polyesters containing disiloxane subunits: structural characterization, kinetics, and an examination of the thermal tolerance of Novozym-435. J Mol Catal B Enzym 85–86:149–155. https://doi.org/10.1016/j.molcatb.2012.09.010

Gamerith C et al (2017) Enzymatic degradation of aromatic and aliphatic polyesters by P. Pastoris expressed cutinase 1 from thermobifida cellulosilytica. Front Microbiol 8:938. https://doi.org/10.3389/fmicb.2017.00938

Gustini L et al (2015) Enzymatic synthesis and preliminary evaluation as coating of sorbitol-based, hydroxy-functional polyesters with controlled molecular weights. Eur Polym J 67:459–475. https://doi.org/10.1016/j.eurpolymj.2014.12.025

Gustini L et al (2016a) Green and selective polycondensation methods toward linear sorbitol-based polyesters: enzymatic versus organic and metal-based catalysis. ChemSusChem 9:1–12. https://doi.org/10.1002/cssc.20160062

Gustini L et al (2016b) Sustainable coatings from bio-based, enzymatically synthesized polyesters with enhanced functionalities. Polym Chem 7(43):6586. https://doi.org/10.1039/c6py01339b

Hanefeld U et al (2009) Understanding enzyme immobilisation. Chem Soc Rev 38:453–468. https://doi.org/10.1039/B711564B

Japu C et al (2015) Copolyesters made from 1,4-butanediol, sebacic acid and D-glucose by melt and enzymatic polycondensation. Biomacromolecules 16:868–879. https://doi.org/10.1021/bm501771e

Jbeily M et al (2014) Enzymatically synthesized polyesters with pendent OH groups as macroinitiators for the preparation of well-defined graft copolymers by atom transfer radical polymerization. Polym Int 63:894–901. https://doi.org/10.1002/pi.4676

Jiang Y et al (2013) Enzyme-catalyzed synthesis of unsaturated aliphatic polyesters based on green monomers from renewable resources. Biomol Ther 3:461–480. https://doi.org/10.3390/biom3030461

Jiang Y et al (2014a) Fully biobased unsaturated aliphatic polyesters from renewable resources: enzymatic synthesis, characterization, and properties. Macromol Chem Phys 215:2185–2197. https://doi.org/10.1002/macp.201400164

Jiang Y et al (2014b) Enzymatic synthesis of biobased polyesters using 2,5-bis(hydroxymethyl) furan as the building block. Biomacromolecules 15:2482–2493. https://doi.org/10.1021/bm500340w

Jiang Y et al (2015a) Environmentally benign synthesis of saturated and unsaturated aliphatic polyesters via enzymatic polymerization of biobased monomers derived from renewable resources. Polym Chem 6:5451. https://doi.org/10.1039/c5py00660k

Jiang Y et al (2015b) A biocatalytic approach towards sustainable furanic–aliphatic polyesters. Polym Chem 6:5198. https://doi.org/10.1039/c5py00629e

Kanelli M et al (2014) Production of biodegradable polyesters via enzymatic polymerization and solid state finishing. J Appl Polym Sci 131:40820. https://doi.org/10.1002/app.40820

Kobayashi S (2010) Lipase-catalyzed polyester synthesis—a green polymer chemistry. Proc Jpn Acad, Ser B 86:338–365. https://doi.org/10.2183/pjab.86.338

Morales-Huerta JC et al (2017) A green strategy for the synthesis of poly(ethylene succinate) and its copolyesters via enzymatic ring opening polymerization. Eur Polym J 95:514–519. https://doi.org/10.1016/j.eurpolymj.2017.08.043

Nameer S et al (2017) One-pot enzyme-catalyzed synthesis of dual functional polyester macromers towards surface active hydrophobic films. RSC Adv 7:50294. https://doi.org/10.1039/c7ra09828f

Nguyen HD et al (2016) Highly branched bio-based unsaturated polyesters by enzymatic polymerization. Polymers 8:363. https://doi.org/10.3390/polym8100363

Pellis A et al (2015) Towards feasible and scalable solvent-free enzymatic polycondensations: integrating robust biocatalysts with thin film reactions. Green Chem 17:1756–1766. https://doi.org/10.1039/C4GC02289K

Pellis A et al (2016a) The closure of the cycle: enzymatic synthesis and functionalization of bio-based polyesters. Trends Biotechnol 34:316–328. https://doi.org/10.1016/j.tibtech.2015.12.009

Pellis A et al (2016b) Ultrasound-enhanced enzymatic hydrolysis of poly (ethylene terephthalate). Bioresour Technol 218:1298–1302. https://doi.org/10.1016/j.biortech.2016.07.106

Pellis A et al (2016c) Renewable building blocks for sustainable polyesters: new biotechnological routes for greener plastics. Polym Int 65:861–871. https://doi.org/10.1002/pi.5087

Pellis A et al (2016d) Enlarging the tools for efficient enzymatic polycondensation: structural and catalytic features of cutinase 1 from Thermobifida cellulosilytica. Cat Sci Technol 6:3430. https://doi.org/10.1039/c5cy01746g

Pellis A et al (2016e) On the effect of microwave energy on lipase-catalyzed polycondensation reactions. Molecules 21:1245. https://doi.org/10.3390/molecules21091245

Pellis A et al (2016f) Exploring mild enzymatic sustainable routes for the synthesis of bio-degradable aromatic-aliphatic oligoesters. Biotechnol J 11:642–647. https://doi.org/10.1002/biot.201500544

Pellis A et al (2017a) His-tag immobilization of cutinase 1 from Thermobifida cellulosilytica for solvent-free synthesis of polyesters. Biotechnol J 12:1700322. https://doi.org/10.1002/biot.201700322

Pellis A et al (2017b) Evolving biocatalysis to meet bioeconomy challenges and opportunities. New Biotechnol 40(Pt A):154–169. https://doi.org/10.1016/j.nbt.2017.07.005

Pellis A et al (2017c) Fully renewable polyesters via polycondensation catalyzed by Thermobifida cellulosilytica cutinase 1: an integrated approach. Green Chem 19:490. https://doi.org/10.1039/c6gc02142e

Roberts T, Friebel S (2016) Itaconic acid—a versatile building block for renewable polyesters with enhanced functionality. Green Chem 18:2922–2934. https://doi.org/10.1039/C6GC00605A

Semlitsch S et al (2016) Enzymatic catalysis as a versatile tool for the synthesis of multifunctional, bio-based oligoester resins. Green Chem 18(7):1923–1929. https://doi.org/10.1039/c5gc02597d

Torron S et al (2014) Polymer thermosets from multifunctional polyester resins based on renewable monomers. Macromol Chem Phys 215:2198–2206. https://doi.org/10.1002/macp.201400192

Torron S et al (2016) Biocatalytic synthesis of epoxy resins from fatty acids as a versatile route for the formation of polymer thermosets with tunable properties. Biomacromolecules 17(12):4003–4010. https://doi.org/10.1021/acs.biomac.6b01383

Uyama H et al (1999) Regioselective polymerization of divinyl sebacate and triols using lipase catalyst. Macromol Rapid Commun 20:171–174. https://doi.org/10.1002/(SICI)1521-3927(19990401)20:4<171::AID-MARC171>3.0.CO;2-2

Uyama H et al (2001) Regioselectivity control in lipase-catalyzed polymerization of divinyl sebacate and triols. Macromol Biosci 1:40–44. https://doi.org/10.1002/1616-5195(200101)1:1<40::AID-MABI40>3.0.CO;2-T

Vastano M et al (2017) Enzymatic production of clickable and PEGylated recombinant polyhydroxyalkanoates. Green Chem 19:5494–5504. https://doi.org/10.1039/C7GC01872J

Weinberger S et al (2017a) Enzymatic surface hydrolysis of poly(ethylene furanoate) thin films of various crystallinities. Green Chem 19(22):5381–5384. https://doi.org/10.1039/C7GC02905E

Weinberger S et al (2017b) Enzymatic degradation of poly(ethylene 2, 5-furanoate) powders and amorphous films. Catalysts 7:318. https://doi.org/10.3390/catal7110318

Wersig T et al (2017) Poly(glycerol adipate)—indomethacin drug conjugates—synthesis and in vitro characterization. Int J Pharm 531:225–234. https://doi.org/10.1016/j.ijpharm.2017.08.093

Wu C et al (2013) Enzymatic synthesis of poly(e-caprolactone) in monocationic and dicationic ionic liquids. Biotechnol Lett 35:879–885. https://doi.org/10.1007/s10529-013-1160-3

Enzymatic Processing of Technical Lignins into Materials

G. M. Guebitz, Alessandro Pellis, and Gibson S. Nyanhongo

1 Introduction

Fossil-based resources supply approximately 86% of energy and 96% of organic chemicals and polymers in the world economy (Pfaltzgraff and Clark 2014). Unfortunately, the depletion of these natural resources and the negative environmental pollutions' impact during their processing, e.g. emission of greenhouse gases implicated in global warming, have stimulated interest in exploiting bio-based renewable resources using green technologies. Lignin, which represents 30% of all non-fossil-based organic carbon source worldwide, is emerging as a potential raw material that can replace many fossil-based raw materials (Huber et al. 2016). Although lignin, as an aromatic biopolymer, has the potential to replace many fossil-based products including packaging plastics, its exploitation has been limited to only 2% of the 50 million tonnes per year of lignin isolated from pulping process (Humpert et al. 2016), due to the complex and heterogeneous nature of lignin, undesirable modifications introduced during the pulping process, impurities and its strong inter- and intra-hydrogen bonding. Lignin therefore requires processing to make it a suitable industrial raw material. Many physical, chemical and physicochemical strategies aimed at purifying (Gillet et al. 2017; Toledano et al. 2010a, b; Jönsson and Wallberg 2009; Duval et al. 2015; Ouyang et al. 2010) and modifying lignin end groups (hydroxyl, methoxyl, carbonyl and carboxyl groups) (Stewart 2008; Chen et al. 2011; Maldhure et al. 2011; Laurichesse and Avérous 2014) and

G. M. Guebitz · G. S. Nyanhongo (✉)
Institute of Environmental Biotechnology, University of Natural Resources and Life Sciences, Vienna, Tulln, Austria
e-mail: guebitz@boku.ac.at; g.nyanhongo@boku.ac.at

A. Pellis
Department of Chemistry, University of York, Green Chemistry Centre of Excellence, York, UK
e-mail: ale.pellis@york.ac.uk

© Springer Nature Switzerland AG 2019
J.-R. Bastidas-Oyanedel, J. E. Schmidt (eds.), *Biorefinery*,
https://doi.org/10.1007/978-3-030-10961-5_24

depolymerizing lignin into low molar mass chemicals or the use of plasticizers (Bouajila et al. 2006; Sivasankarapillai and McDonald 2011) are actively being investigated. In line with the desire to exploit lignin for industrial applications using green chemistry technologies, enzyme-based technologies are emerging as highly promising catalysts for modifying and synthesizing lignin-based materials. This chapter presents advances in the development of enzyme-based strategies to make lignin suitable for the synthesis of high-value lignin-based materials. As an emerging field, this chapter places special emphasis in giving an overview of the reactivity of different lignins (from different types of lignocellulose plants and different pulping processes) with different enzymes under different reaction conditions, the characteristics of the modified lignins and their potential application.

2 Lignin as a Precious Renewable Resource: Opportunities and Challenges

Lignin is a structural component of plant biomass encrusting hemicellulose and cellulose. Lignin makes 18–30% dry weight of wood, while cellulose and hemicellulose constitute 35–50% and 25–35% dry weight, respectively (Janusz et al. 2017). Lignin is responsible for transporting water and nutrients, providing mechanical strength, preventing damage and protecting plant from pests and pathogens (Gordobil et al. 2016). It is primarily made from three major monolignol precursor molecules, namely, *p*-coumaryl alcohol (H), guaiacyl (G) and sinapyl alcohol (S) moieties (Fig. 1) (Mottiar et al. 2016). These molecules are synthesized in the cytoplasm and deposited on the cell wall where they are oxidized by peroxidases and/or laccases generating phenoxyl radicals (Wang et al. 2013) which cross-link randomly leading to the formation of lignin (Fig. 1).

Laccases (benzenediol: oxygen oxidoreductase, EC1.10.3.2) are multicopper-containing enzymes widely distributed in plants, fungi and bacteria involved in the oxidation of various aromatic substrates (phenols, hydroxyindoles, methoxy-substituted phenols, polyphenols, anilines, aryl diamines, inorganic/organic metal

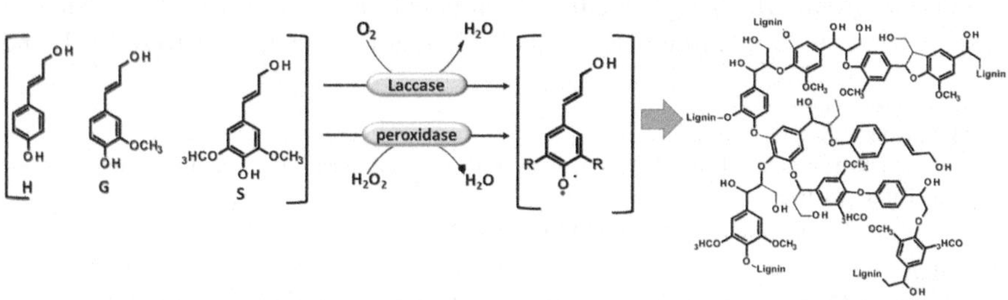

Fig. 1 Basic lignin precursor monomers [*p*-coumaryl alcohol (H), guaiacyl alcohol (G) and sinapyl alcohol(S)] oxidized by laccase or peroxidases during lignin biosynthesis to produce complex lignin polymer

compounds and many others) while simultaneously reducing molecular oxygen to water (Nyanhongo et al. 2010). In plants, laccases are involved in lignin biosynthesis, wound healing, maintenance of cell wall structure and integrity, plant responses to environmental stresses (Wang et al. 2015; Dwivedi et al. 2011). Unlike laccases, peroxidases (EC 1.11.1.7) use hydrogen peroxide as cofactor to oxidize substrates resulting in the generation of a phenoxyl radical. Plants' peroxidases, as summarized in Pandey et al. (2017), are involved in a wide range of physiological processes such as cell wall metabolism, lignin biosynthesis, suberization, auxin metabolism, wound healing, reactive oxygen species and reactive nitrogen species metabolism, fruit growth and ripening defence against pathogens. During cross-linking of laccase- and peroxidase-oxidized lignin monomers, the type of bond formed depends on the position of the electron on the generated phenoxyl radical or the presence/absence of functional group on lignin monomers, e.g. methoxyl groups. Generally the random coupling occurring at different positions produces C–O and C–C linkages at various positions resulting in β-O-4, β-5, β–β, 4-O-5, 5–5 or β-1 linkages (Behling et al. 2016) (Fig. 1). Although the most common linkage between lignin monomers is the β-O-4 ether bond, its prevalence varies between plants. For example, in soft wood and hardwood, it accounts for 50% and 65% of the total linkages, respectively (Behling et al. 2016). This is attributed to the fact that lignin in softwood (gymnosperm) contains predominantly G units that yield abundant β-5, 5–5 and β–β carbon–carbon bonds, while hardwood (angiosperm) is rich in guaiacyl-syringyl (S units rich in ether-type β-O-4 linkages), and grass or annual plant (graminaceous) are rich in H units (Crestini et al. 2011; Derkacheva 2013).

In addition to this natural lignin complexity, industrial lignins are by-products of different wood-pulping processes (kraft and sulphite, soda, organosolv) which modify the lignin. During the pulping process, lignin is solubilized under harsh environmental conditions in the presence of chemical catalysts that randomly degrade and modify lignin. For example, during the kraft pulping process, ether bonds (β-O-4) are cleaved by nucleophilic sulphide or bisulphide ions resulting in the introduction of sulphur in the lignin structure (Fig. 2). Similarly, during the sulphite pulping process, acidic cleavage of ether bonds results in electrophilic carbocation products which react with bisulphite ions (HSO_3^-) to give sulphonates (Fig. 2). Further, since

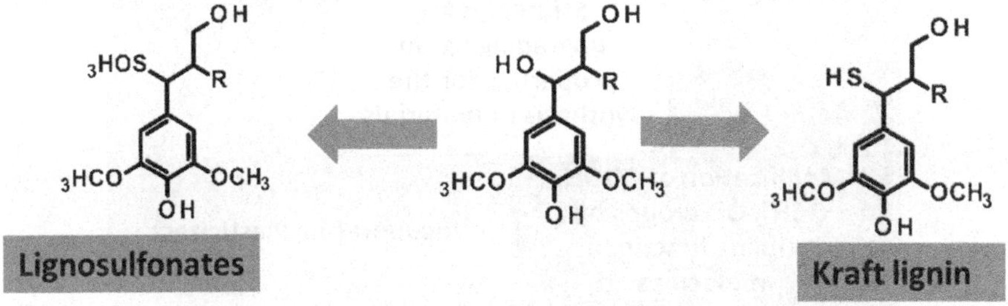

Fig. 2 Simplified modification of lignin showing introduction of sulphur groups into lignin during sulphite and kraft pulping processing

traditionally lignin is produced as a by-product of the pulping process, it is found in waste streams associated with many other impurities such as plant extractives, inorganic and organic impurities, suspended solids and residual chemicals used during the pulping process.

3 Enzymes as Emerging Versatile Catalysts for Lignin Modification

Based on the knowledge that lignin acts as glue which binds hemicellulose and cellulose together in plants, and that peroxidases and laccases are involved its synthesis, pioneering studies explored the possibility of using lignin as an adhesive in situ, incorporating unmodified and modified lignin and treating with enzymes for making wood boards (Kharazipour et al. 1998; Felby et al. 1997) and wood laminates (Jin et al. 1990; Unbehaun et al. 2000; Milstein et al. 1994). It was speculated that the enzymes would oxidize the industrial lignins as well as the lignin in the wood leading to strong cross-linking/cross-bonding. Although some degree of adhesion was observed, they were not as those produced by fossil-based resins. Since these early studies, many have investigated the possibility of upgrading lignin properties using physical, chemical and biological techniques as summarized in Fig 3. Enzymes from different sources under different reaction conditions oxidize different industrial lignins leading to either polymerization or depolymerization. For example, *Trametes versicolor* laccase (redox potential 0.75 V) was able to oxidize and increase the molecular weight of sodium lignosulphonates although no activity was detected with *Rhus vernicifera* plant (redox potential 0.34 V) (Madad et al. 2013). Bae and Kim observed depolymerization of lignosulphonates at pH 4.5 and

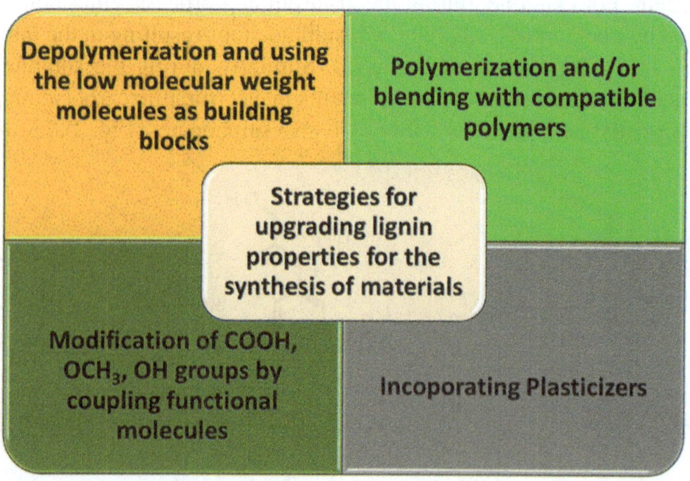

Fig. 3 Different strategies used to upgrade lignin properties during the synthesis of polymeric materials

polymerization at pH 6 (Bae and Kim 1996) suggesting the importance of pH on directing the reaction outcome. Hataaka et al. (1996) noted depolymerization of lignosulphonates with molecular weight <10,000 g/mol and polymerization with larger molecular weight fragments, suggesting the importance of the initial molecular weight of the lignin (Hataaka et al. 1996). Indeed, Arerkogh et al. (2010a, b) also confirmed that the lignosulphonate concentration is an important factor in order to obtain increase in molecular weight and indicated that the low solubility of kraft lignin is an important drawback to scale up the polymerization process. Increased molecular weight and purity of the lignosulphonate enhance the plasticising effect as well as the reduction of viscosity of the concrete (Areskogh et al. 2010a), a market which accounts for approximately 50% of lignosulphonate use worldwide (Ansari and Pawlik 2007). Indeed Nugroho Prasetyo et al. (2010) demonstrated that the increase in molecular weight of laccase-polymerized lignosulphonates led to improved dispersion properties.

In another comparative study, *Trametes villosa* (TVL) laccases (redox potential 0.8 V and pH optima at 5) were able to oxidize nonphenolic end groups, while *Myceliophthora thermophila* laccase (MTL) with lower redox potential of 0.43 V and pH optima at 7.5 was able to oxidize phenolic end groups only (Areskogh et al. 2010b). ^{1}H–^{13}C HSQC 2D-NMR analysis of TVL- and MTL-treated lignosulphonates indicated the formation of new C–O–C and C–C linkages and pronounced lignin condensation (Nugroho Prasetyo et al. 2010), decrease in the organic sulphur content, a slight increase in inorganic sulphur content, lignin demethylation, increase in molecular weight and polydispersity of lignosulphonates than in MTL-treated lignosulphonates (Areskogh et al. 2010b; Nugroho Prasetyo et al. 2010). Alkaline laccases from *Mycelia sterilia* YY-5, *Melanocarpus albomyces* (MAL) and *Streptomyces ipomoea* CECT 3341 (SIL) (Moya et al. 2011) were able to polymerize both kraft lignin and organosolv lignin under alkaline conditions (pH 7–9) (Moya et al. 2011; Weihua and Hongzhang 2008). This is an extremely important development since kraft lignin accounts for more than 85% of technical lignin produced worldwide which is insoluble under acidic conditions or neutral conditions (Tejado et al. 2007). This effectively makes it possible to oxidize lignin using enzymes from pH as low as 3–9, a development which was not possible a decade ago.

4 Laccase-Mediator System

Laccase activity is limited by its redox potential (highest 0.8 V). In order to extend the ability of laccase to oxidize molecules with higher redox potential and/or even oxidize molecules that are not its substrate, mediators are used. As shown in Fig. 4, laccase oxidizes a mediator by abstracting an electron which it uses to reduce molecular oxygen to water. The oxidized mediator attacks and oxidizes lignin which can either to depolymerization or polymerization of lignin. Mediators are small substrate molecules which when oxidized form highly reactive species able to in turn oxidize other molecules in the surrounding milieu even those which are not laccase

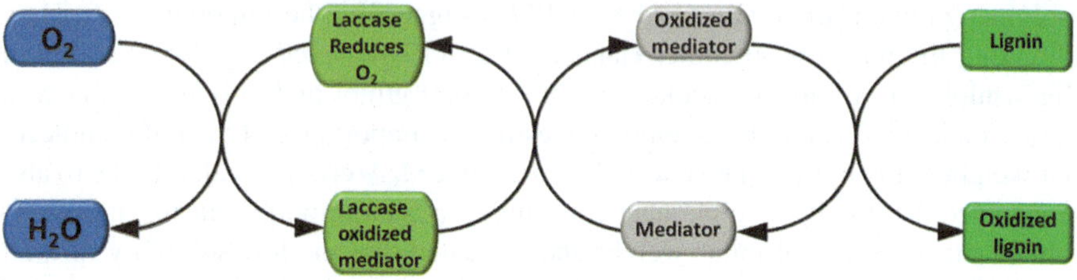

Fig. 4 Schematic presentation of laccase-mediator system. Laccase oxidizes a mediator by abstracting an electron which it uses to reduce molecular oxygen to water. The oxidized mediator attacks and oxidizes lignin

substrates or substrates with high redox potential. Laccase-mediator systems are extensively explored for degrading or modifying chemicals/polymers with higher redox potential than that of laccase and oxidize nonphenolic moieties in lignin (Call and Mücke 1997; Morozova et al. 2007; Elegir et al. 2005; Trovaslet-Leroy et al. 2010).

Different mediators result in different effects during the treatment of lignins. TVL, *Pycnoporus cinnabarinus* laccase (PCL), *Botrytis cinerea* laccase and MTL in the presence of either 1-hydroxybenzotriazole, violuric acid or ABTS [2,2′-azino-bis (3-ethylbenzothiazoline-6-sulphonic acid)] were able to oxidize nonphenolic lignin dimer, 1-(3,4-dimethoxyphenyl)-2-(2-methoxyphenoxy)propan-1,3-diol and a phenolic lignin model compound which is otherwise not oxidized by laccases alone (Li et al. 1999). The TVL and PCL-violuric acid system oxidized nonphenolic lignin dimer much faster than 1-hydroxybenzotriazole. The *Bacillus subtilis* laccase (BCL) in the presence of violuric acid oxidized nonphenolic lignin dimer more slowly than TVL laccase and 1-hydroxybenzotriazole (Li et al. 1999). The violuric acid and 1-hydroxybenzotriazole mediators also inactivated the laccases (Li et al. 1999; Xu et al. 2000). In a screening study, laccases from TVL and *Trametes hirsuta* (THL) were effective in oxidizing lignosulphonates under acidic conditions, while the MTL and BCL laccases were more effective at neutral pH in the presence of 1-hydrobenzotriazole as mediators (Nugroho Prasetyo et al. 2010). Acetosyringone and violuric acid were more effective in polymerizing sodium lignosulphonates in the presence of *Trametes versicolor* laccase (TVRL) than 1-hydroxy-benzotriazole, acetovanillone and 2,2′-azino-bis(3-ethylbenzothiazoline-6-sulphonic acid) and 1-hydroxybenzotriazole (Madad et al. 2013). SIL and MAL were able to extensively polymerize the lignosulphonates in the presence of acetosyringone and 2,2,6,6-tetramethylpiperidine 1-oxyl (TEMPO) as mediators than in the presence of small lignin monomers (*p*-coumaric, ferulic and sinapic acid) as mediators under alkaline conditions (Moya et al. 2011). Although many studies reported increasing molecular during laccase-mediator oxidation, the use of some mediators, e.g. 2,2′-azino-bis(3-ethylbenzothiazoline-6-sulphonic acid), repeatedly resulted in depolymerization of lignin (Hernández Fernaud et al. 2006), even when added after the initiation of oxidation of lignin with laccase (Bourbonnais et al. 1995). Elegir et al. (2005) provided evidence for the ability of laccase to cleave β-5 and a 5–5′

linkages present in lignin in the presence of 2,2′-azino-bis(3-ethylbenzothiazoline-6-sulphonic acid) as mediators (Elegir et al. 2005). Similarly, natural mediators acetovanillone and acetosyringone were shown to depolymerize fractionated lignosulphonates in the presence of laccase from *Cerrena unicolor* (Cho et al. 2004). Some of the mediators have been shown to couple covalently on lignin, thereby blocking chain elongation (Bourbonnais et al. 1995; Cho et al. 2004; Matsumura et al. 1986; Rittstieg et al. 2002). These studies show that laccase-mediator system can either lead to polymerization or depolymerization reactions. This makes it imperative to screen laccase-mediator systems and optimize reaction conditions in order to produce the desired outcome. As a general comment, if mediators are intended to activate lignin, then they should not easily couple onto the lignin, as they should be recycled and facilitate lignin modification.

5 Mediators vs. Sufficient Oxygen Supply

In another recent development which seems to question the need to use mediators, laccase-mediated polymerization of different types of lignins [organosolv, indulin AT, sodium lignosulphonates and magnesium lignosulphonates] was much higher in reactions supplied with oxygen as compared to the reaction supplemented with mediators (TEMPO, vanillin, HBT and 2,6-dimethoxyphenol) (Gouveia et al. 2012; Ortner et al. 2015; Abou-Mansour et al. 2010). Through these studies, it became very clear that oxygen was limiting in most of the previously reported laccase-mediated polymerization studies. Oxygen levels decreased to undetectable levels soon after adding laccase and remained undetectable for a long period in shaking samples without external oxygen supply. This discovery is critical as it seem to discredit the need to use laccase mediators which may have been erroneously popularized as the only way to achieve extensive polymerization of technical lignins. It is speculated that some of the laccase oxidized lignin substrates could act as powerful oxidants or mediators during the supply of oxygen. However, it must also be stressed that prior purification/processing of lignin seems to be a critical step in order to achieve polymerization. Nevertheless, this new process is set to make laccase lignin-based polymerization simpler, cheaper, easier, efficient and more environmentally friendly.

6 The Influence of Lignin H, G and S Ratio and Pulping Process on Enzyme Reactivity

The rates of oxidation and extent of lignin polymerization depend on the H, G and S content of a particular lignin, modifications introduced during the pulping process and enzyme and reaction conditions. Generally, kinetic studies show that laccase reactivity increases with increasing number of hydroxyl groups and/or methoxyl

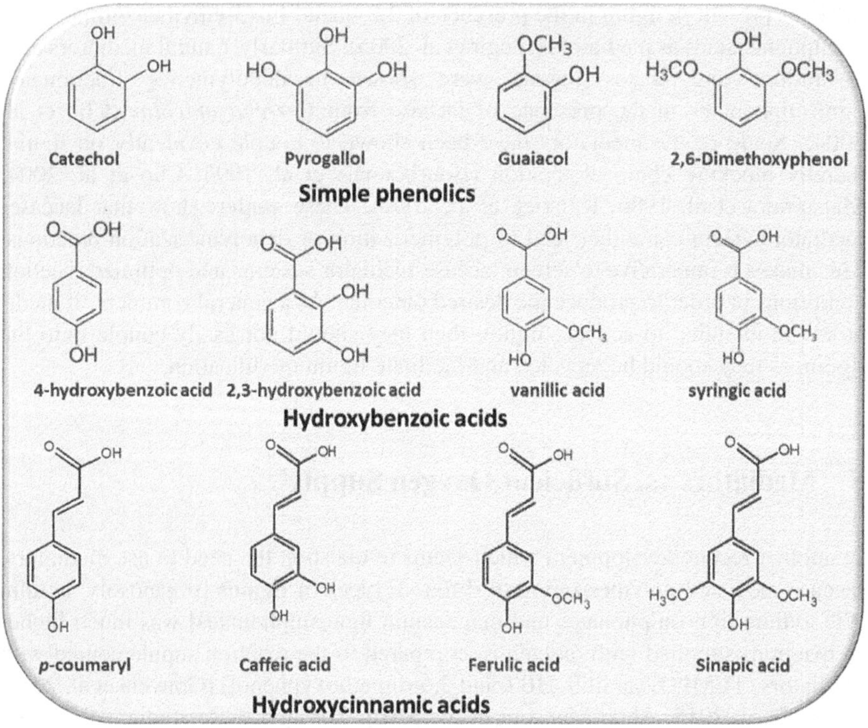

Fig. 5 Shows some of the representative simple phenolics, hydroxybenzoic acids and hydrocinnamic acids

groups on the *meta*- or *para*-positions in lignin building blocks (simple phenolics, hydroxybenzoic acids and hydroxycinnamic acids) (Fig. 5), for example, simple phenolics (pyrogallol > catechol; 2,6-dimethoxyphenol > guaiacol); hydroxybenzoic acids (2,3-dihydroxybenzoic acid > 4-hydroxybenzoic acid; vanillic acid > 2,3-dihydroxybenzoic acid), hydroxycinnamic acids (sinapic acid > caffeic acid > ferulic acid > *p*-coumaric acid) (Abou-Mansour et al. 2010; Si et al. 2013; Fonseca et al. 2015; Nyanhongo et al. 2006; Frasconi et al. 2010). Generally, laccase reactivity with hydroxycinnamic acids is higher compared to corresponding hydroxybenzoic acids attributed to the presence of long aliphatic chain which ensures greater H-donating ability and radical stabilization than the –COOH group acid group directly linked to the aromatic ring in hydroxybenzoic acids. The easy oxidation of phenolic molecules with increasing methoxyl group is attributed to the electron-donating property of methoxy groups that makes it easier for laccases to abstract an electron from the hydroxyl group on the *ortho*-positions. Similarly, degree of polymerization and/or coupling/grafting of functional molecules on lignin fragments depends on the size and substituents on lignin fragments and their reactivity. For example, laccase was able to mediate the coupling of dodecylamine and dihexylamine on catechol. However, the two dodecylamine and dihexylamine molecules were only grafted per dimer of polymerized guaiacol and ferulic acid (Kudanga et al. 2010a), and only one dodecylamine or dihexylamine molecule was coupled onto complex lignin model compounds (syringylglycerol β-guaiacyl ether, guaiacylglycerol β-guaiacyl ether and dibenzodioxocin) possibly due to steric effects and the presence of methoxyl groups at C-5 position which prevent 5–5 linkage, thereby only facilitating 4-O-5 couplings only (Kudanga et al. 2010a).

Again to demonstrate effect of size of lignin, methyl linoleate was successfully coupled onto syringylglycerol β-guaiacyl ether, and no coupling product was detected with sinapic acid (Kudanga et al. 2010a), attributed to the rapid establishment of intramolecular coupling of laccase oxidized sinapic acid. The hydroxyl group on the benzene ring is *ortho* or *para* directing and molecules with free C-5 position usually easily cross-couple through 5–5 linkages due to stability of C–C bonds. As pointed out above, this allows molecules like catechol to easily oligomerize via 5–5 bonds; however, lignin molecules with syringyl or sinapyl units usually couple through 4–O–5 linkages to molecules with a free C-5 position, while the unsaturated side chains can also be utilized for oxidative coupling. Laccase-mediated amine coupling onto lignin occurs establish C–N bonds either through radical coupling or via Michael addition (Kudanga et al. 2010c; Witayakran and Ragauskas 2009a). Although phenolic molecules like sinapic acid are very valuable for mechanistic investigations, the fact is that they easily polymerize and the blocking of C-5 position by the methoxyl group could be limiting. Although *Thielavia arenaria* laccases, TVL, *and* MAL were able to oxidize Alcell, birch organosolv lignins and steam-exploded pine and eucalypt (S units) lignins, THL was more effective in oxidizing all S-rich hardwood Alcell, eucalypt and birch organosolv lignins than the G-rich pine (van de Pas et al. 2011). The easy oxidation of S-rich hardwood lignin is attributed to the large presence of electron-donating methoxy substituents that makes it easy for laccases to abstract electrons from the hydroxyl group on the *ortho*-positions. Even the low redox potential MAL was also able to oxidize all S-rich hardwood Alcell, eucalypt and birch organosolv lignins to the same extent as THL (van de Pas et al. 2011). Moya et al. (Kiiskinen et al. 2004) found out that the oxygen consumption curves of birch and spruce organosolv lignins as well as of hardwood and softwood kraft lignins treated by MAL and SIL at different pH values varied depending on the redox potential of the enzyme. MAL oxidized G-rich softwood kraft lignin faster and at similar rates between pH 7 and 9, although the optimum pH range of the enzyme is 5–7.5 (Kiiskinen et al. 2004). Similarly, MTL produced higher lignin polymers with hard wood kraft lignin than softwood (Gouveia et al. 2013). THL (high redox potential) quickly reacted with both S-and G-rich lignin, whereas MAL (low redox potential) was slightly faster with syringyl models (Moya et al. 2011). Surprisingly, oxidation of hardwood kraft lignin by MAL was slower than that of softwood and birch organosolv lignins (Moya et al. 2011). This may be attributed to the changes introduced during lignin pulping or purification process. For example, it is known that fractionation or purification of lignin using acid or alkali conditions/catalysts induces lignin condensation (formation of inert C–C linkages), which reduces later enzyme oxidation. In an extensive screening study, *Trametes versicolor* laccase was able to polymerize all tested lignins (soda grass/wheat straw, organosolv hardwood, alkali-pretreated wheat straw and indulin AT, a kraft lignin from softwood). Although organosolv hardwood lignin (S) was the most reactive, the highest increase in molecular weight as well as the extensive changes in functional group content and the formation of more condensed structures were observed for alkali-pretreated wheat straw (H) lignin (Fiţigău et al. 2013). Oxidation of soda grass/wheat straw by laccase was somewhat slower than that of the alkali-pretreated wheat straw, showing that the pretreatment method affects the rate of oxidation of wheat straw lignin (Fiţigău et al. 2013). Laccase oxidation rate of G-rich indulin AT kraft lignin from softwood was about 60% slower (Fiţigău et al. 2013) than S-rich hardwood lignin. Among the wheat straw lignins, alkali-pretreated wheat straw was the most reactive, but the initial oxidation rate was lower compared to the soda grass/wheat straw lignin (Fiţigău et al. 2013). Regarding changes in molecular weight, as expected organosolv hardwood lignin (S) had the lowest M_w and is the least condensed (i.e. lowest condensed phenolic and aliphatic-OH content). Grass lignins with high H-units (free *ortho*-positions in the phenolic ring) were more polymerized (Fiţigău et al. 2013). The authors further suggested the obtained polymers had potential applications in bioplastics and adhesives and as polymeric dispersants. It is naturally expected that lignin with high S content is easily oxidized; however, since syringyl units have C-5 position occupied, this prevents formation of 5–5′ and β-5′ linkages which logically leads to low polymerization than in G- and H-rich lignins. Areskogh et al. (2010c) also observed that if position 5 of a lignin end group is blocked like in S units, several undesirable reactions occur including coupling at position 1 leading to unproductive couplings which limit extensive polymerization.

The nature of the pulping process affects enzymatic polymerization of lignin. Electron paramagnetic resonance spectroscopy (EPR) studies reveal that enzymes oxidized industrial lignin and produce different concentrations of phenoxyl radicals with different lignins (Munk et al. 2017a). TVL and MTL oxidation of organosolv lignin from Sigma (CAS No. 8068-03-9), a beech organosolv lignin and wheat straw lignin, resulted in immediate and enzyme dose-dependent increase in intensity in EPR signals. MTL produced phenoxyl radicals faster than the TVL on all three types of lignin substrates. Enzymatic radical formation rate on the wheat straw lignin residue was consistently higher than those of the organosolv lignins (Munk et al. 2017a). Laccase treatment of beech organosolv lignin generated the highest concentration of radicals, while the organosolv lignin from Sigma phenoxy radicals increased but then declined faster. Wheat straw lignin had the lowest radicals attributed to extensive washing. The remarkably higher radical concentration achieved with beech organosolv lignin could be due to a higher concentration of small lignin molecules compared to those in the organosolv lignin from Sigma and wheat straw lignin (Munk et al. 2017a). Laccase-catalysed cross-linking of organosolv lignin and lignosulphonate showed a slow increase in organosolv lignin molecular weight, whereas oxidation of lignosulphonates resulted in a very rapid increase in molecular weight (Gillgren et al. 2017). The author suggested that the differences were due to the fact that lignosulphonates cross-linked through 5–5′ and 4-O-5 bonds, whereas organosolv lignin cross-linked mainly through β-O-4′ coupling (Gillgren et al. 2017). Increased molecular weight and purity of the lignosulphonate enhanced the plasticization effect as well as the reduction of viscosity of the concrete (Areskogh et al. 2010a). Laccase-catalysed reaction resulted in extensive polymerization of lignosulphonates as compared to organosolv and indulin AT (Nugroho Prasetyo et al. 2010; Ortner et al. 2015; Johansson et al. 2014). This is indeed an important information which need consideration when synthesizing lignin-based materials. This information is critical since it will increase lignin application in existing applications (as stabilizers in colloidal suspensions, as dispersing agents and binders for drilling agents, industrial detergents, glues, animal feed, particle board, technical surfactants, oil well drilling and cement additives) (Bouajila et al. 2006; Blinkovsky and Dordick 1993; Kai et al. 2016; Norgren and Edlund 2014) and many other novel products. By far, the largest utilization of lignosulphonates is as a dispersant for cement in concrete, which accounts for approximately 50% of lignosulphonate use worldwide (Ansari and Pawlik 2007). Addition of lignosulphonates to a concrete mix improves the workability of the mixture, reduces the amount of water and thus facilitates a quicker strength development (Ansari and Pawlik 2007). Indeed treatment of lignosulphonates with laccases from TVL, MTL, THL and BCL improved the dispersion properties of calcium lignosulphonates by increasing molecular weight (Nugroho Prasetyo et al. 2010). Polymerization of lignosulphonates with laccase followed by ozonolysis to increase carboxylic groups produced polymers with potential applications as plasticizers (Areskogh et al. 2010a). Enzyme modifications show the possibility of generating lignosulphonates with similar properties as the petroleum-based superplasticizers.

Although fewer studies have explored the possibility of developing peroxidase-based lignin processing technologies, understandably due to the problems associated with supplying H_2O_2, a required cofactor, versatile peroxidase and lignin peroxidase, successfully polymerized softwood and hardwood lignosulphonates (Sáez-Jiménez et al. 2016). Just like laccases, peroxidases oxidize aromatic phenolic groups to produce phenoxyl radicals that undergo random radical couplings. ^{1}H-NMR analysis of lignin incubated with horseradish peroxidase showed increase in β–β′-, β-5′-, β-O-4′-, 4-O-5′- and 5–5′-type bonds although β-O-4′ predominated and decreased in methoxyl protons (Kubo and Kadla 2004; Boerjan et al. 2003). The authors also noted that due to the steric hindrance of the methoxyl groups, the available cross-linking position in the S units are less than those in G and H units. It should be also noted that peroxidases, e.g. lignin peroxidases, uses veratryl alcohol as mediator. Thus Tobimatsu et al. (2010) also observed low polymerization with syringyl lignins. Just like with laccase, peroxidases decreased sulphur content (Ferrari et al. 1999; Muralikrishna and Renganathan 1993) and methoxyl content (Kersten et al. 1990). Several other studies investigating lignin-coupling reactions have also provided valuable information that can be exploited in order to achieve desired products. Through these studies, it is now clear that *p*- and *m*-substituted lignin can be extensively oxidatively polymerized, while no copolymerization reaction occurs with acetylated lignin or methylated lignin (phenolic groups blocked). The reactivity of the different lignins with different S, G and H units provides very important information that can help polymer/material scientists with valuable information for choosing the right type of lignin depending on intended applications.

7 Enzymatic Coupling of Functional Molecules on Lignin

7.1 Converting Kraft Lignin into Lignosulphonates

Owing to the easy of processing and expanding role of lignosulphonates, intense research is focusing on converting kraft lignin into lignosulphonates through sulphonation/sulphomethylation in order to increase its market value (Hu et al. 2018). Chemically, kraft lignin is sulphonated under harsh reaction conditions by reacting it with sodium sulphite or sodium bisulphite and formaldehyde at 100–160 °C and pH ≥ 9 for 4 h (Berlin and Balakshin 2014). However, the sulphonation (or sulphomethylation) reaction introduces sulphonic acid groups into the aromatic ring of the lignin structure, which is different from lignosulphonates that originate from the sulphite pulping process, in which sulphonic acid groups are located on the aliphatic chain (Berlin and Balakshin 2014). Increasing evidence shows the ability of laccases and peroxidases to mediate the sulphonation of kraft lignin. Sulphonation of kraft lignin by horseradish peroxidase was accompanied by increased formation of β-O-4′ and β–β′ linkages, molecular weight and decrease in methoxyl groups (Berlin and Balakshin 2014; Yang et al. 2014a; Dong-jie et al. 2013). In another study, horseradish peroxidase-mediated sulphomethylation of pine and wheat straw alkali lignin improved the doping of polyaniline (Yang et al. 2017). Horseradish peroxidase sulphonation has been shown to be more efficient in improving the molecular weight and sulphonation of kraft lignin than laccase (Grönqvist et al. 2005). Laccase sulphonation of wheat straw kraft lignin showed that sulphonation reactivity decreased under low oxygen pressure (Sun et al. 2013). Co-incubation of laccase and xylanase with kraft lignin enhanced the sulphomethylation by 33% (Zhou et al. 2016). This was attributed to the ability of xylanases to hydrolyse lignin-carbohydrate complexes, thereby exposing lignin for oxidation by laccase (Zhou et al. 2016). Yang et al. (2014a, b) also demonstrated the ability of laccases to increase the sulphomethylation kraft lignin and promoting its polymerization although the change in molecular weight was minimal. Lund and Ragauskas (2001) and Chandra and Ragauskas (2008) use laccase- and peroxidase-incorporated water-soluble phenols with carboxylic and sulphonic acid groups into kraft lignin. These studies show that laccases can be used to introduce a variety of hydrophilic molecules in insoluble lignins that improve their processing into valuable materials. Although in their infancy, these studies are highly promising.

7.2 Enzymatic Coupling of Functional Groups onto Lignin End Groups

Lignin is brittle due to strong inter- and intra-chain hydrogen bonding mainly caused by abundant hydroxyl groups. This is one of the major challenges limiting the industrial exploitation of lignin in materials. To improve lignin mechanical properties, the use of plasticizers or modification of lignin end groups is necessary. The pioneering studies by Mai, who successfully copolymerized acrylamide with softwood organosolv lignin, produced evidence for the possibility of grafting functional molecules onto lignins (Mai et al. 1999, 2000). The produced hybrid polymers proved successful in creating novel plastics, thickeners, fillers and adsorbents with biodegradable properties (Chandra and Ragauskas 2008; Mai et al. 1999). Further studies provided evidence for the possibility to modify a variety of technical lignosulphonates with acrylamide and acrylic acid in the presence of laccase and *t*-butyl hydroperoxide. Blinkovsky and Dordick (1993) successfully coupled *p*-cresol and *p*-phenylphenol onto kraft lignin using horseradish peroxidase. The produced lignin *p*-cresol and lignin *p*-phenylphenol polymeric material showed lower glass transition temperatures and higher curing exotherms with excellent thermoset properties similar to conventional adhesives (Blinkovsky and Dordick 1993). Liu et al. (Duval et al. 2015) demonstrated the ability to control the molecular weight of lignin-cresol polymer material during peroxidase catalysis by adjusting the concentration of surfactant, enzyme, cresol, lignin and the ratio of alcohol to hydrocarbon in the organic phase. Several studies have also provided mechanistic insights into enzyme-mediated coupling of low-molecular-weight functional molecules onto lignin/lignin model compounds, e.g. radical coupling and Michael addition as summarized in the detailed reviews by Witayakran and Ragauskas (2009b) and Mikolasch and Schauer (2009). Laccases were able to mediate the coupling of amine-bearing molecules (tyramine and 3-*O*-methyldopamine and 4-hydroxy-3-methoxybenzylamine) (Kudanga et al. 2009, 2010c), long-chain alkylamines (dodecylamine and dihexylamine) (Kaplan 1979; Kudanga et al. 2010d), fluorophenols (Kudanga et al. 2010e) and oxiranes (Kudanga et al. 2010d). Enzymes are also used to mediate the coupling of reactive molecules. For example, laccase was able to mediate the coupling of tyramine, 3-hydroxytyramine, 3-*O*-methyldopamine and 4-hydroxy-3-methoxybenylamine onto dibenzodioxocin, an important substructure constituting 18–25% of linkages in softwood lignin (Argyropoulos et al. 2002) via a 4-O-5 bond leaving the $-NH_2$ group free for further attachment of functional molecules (Kudanga et al. 2009). The grafted reactive amino groups can be used to further couple other functional molecules into lignin. In another study laccase mediated the coupling of *p*-aminobenzoic acid onto syringic acid; syringaldehyde and eucalyptus kraft lignin showed the presence of identical products (Ibrahim et al. 2013a). Recently, Tzanko and colleagues (Ibrahim et al. 2013b; Aracri et al. 2014) used laccase to synthesize various hardwood kraft lignin-based adhesives grafted with polyethylenimine, chitosan, soy protein, gallic acid, tannic acid or dopamine. The lignin was first oxidized by laccase followed by a phenolation step where natural phenolic compounds were copolymerized with lignin in order to increase its content of reactive quinone, flexibility and bonding strength. The adhesives produced by mixing lignin with soy protein yielded an adhesive with greater than 50% of the strength of commercial polyurethane adhesive and good water-resistance properties and adhesives with antimicrobial properties (Ibrahim et al. 2013a; Aracri et al. 2014). The authors concluded that the synthesized adhesives could be used as binders in paper and cardboard boxes or to replace synthetic latex in the formulation of adhesive used for wool floor coverings (Aracri et al. 2014). Recently, laccase-assisted copolymerization of kraft lignin with hyper-branched copolymers of methyl-hydroquinone resulted in lignin with moderate glass transition temperature and exhibits good thermostability, with potential application as a lignin-based thermoplastic (Cannatelli and Ragauskas 2017). Laccases oxidized both lignin and methyl hydroquinone to generate phenoxy radicals, which can couple with one another via either C–C or C–O bonds. The reaction produced a brown precipitate paste which hardened into a solid glossy material that was insoluble in organic solvents(Cannatelli and Ragauskas 2017). It was speculated that the curing took place during the drying process, possibly due to the presence of methylquinone moieties as well as reactive free thiol end groups, which has been known to occur for hyper-branched polymers synthesized with *tris*(2-mercaptoethyl)amine and ethylene glycol diacrylate(Cannatelli and Ragauskas 2017; Sun et al. 2012). The synthesized material has potential use as a lignin-based adhesive for particleboards and wool floor coverings (Ibrahim et al. 2013a, b). Kraft lignin-treated MTL was grafted on Scots pine and European beech mini-blocks resulting in the formation of a stable coating, and the incorporation of copper improved the decay resistance and copper leaching (Fernández-Costas et al. 2017).

In another study which shows different reactivity of lignin with laccases, five different lignins (organosolv hardwood lignin; soda wheat straw lignin; soda lignin from mixed sarkanda grass/wheat straw; indulin AT, a Kraft lignin from softwood) were oxidized by laccases in the presence of glucosamine and glycil-tyrosyl-glycine(Fiţigău et al. 2015). Only organosolv hardwood lignin and alkali-pretreated wheat straw lignin were successfully modified with glucosamine and glycyl-tyrosyl-glycine (Fiţigău et al. 2015). These results once again, although demonstrating the possibility of enzymes modifying lignin, also show that the reactivity of lignin varies depending on origin, pulping process and properties of the enzyme used. Zhang et al. (2016) synthesized a polyaniline-lignosulphonate complex via laccase catalysis. The lignosulphonate acted as a template for the synthesis of linear polyaniline immobilized onto the surface of cotton with potential use in textile electronic devices. Recently, laccase-mediated synthesis of lignin-carbohydrate and lignin-peptide conjugates (Fiţigău et al. 2015) provide a framework for further functionalization and formulation of materials with distinct properties for various applications, e.g. laccase-synthesized lignin-soy protein adhesive exhibited more than half the strength of commercial polyurethane adhesive and retained 70% of its initial strength after two cycles of 1 h boiling and drying (Ibrahim et al. 2013b). Lipase-transesterified kraft lignin with long acyl chains exhibited interesting thermal and textural properties, different from those of the original kraft lignin (Hulin et al. 2015). Wheat straw and beech organosolv lignin grafted with –N–OH molecules (Munk et al. 2017b) as well as kraft lignin grafted with vanillic acid-PEG ester and ether derivatives (Kalliola et al. 2014) produced hydrophilic derivatives expected to soften lignin and improve its utilization in composite materials. Laccase-iodide treatment of spruce wood produced washout-resistant antimicrobial surface (Ihssen et al. 2014). Laccase converted *o*-vanillin, ethyl vanillin, acetovanillone and methyl vanillate into iodovanillin products. Biocatalytically produced iodovanillin and iodo-ethyl vanillin had significant growth inhibitory effects on several wood degrading fungal species (Kalliola et al. 2014). Elemental iodine (I_2) is a well-known antiseptic and disinfectant that acts against bacteria, fungi and viruses at millimolar concentrations (McDonnell and Russell 1999). The biocidal mechanism is thought to involve the modification of nucleotides, fatty acids, as well as cysteine and methionine groups of proteins (McDonnell and Russell 1999). Horseradish peroxidase and laccase from *TVL* successfully mediated the grafting of poly(*N*-isopropylacrylamide) polymer brushes on lignin nano-surfaces (Gao et al. 2014). Manipulating the type of enzyme, the concentration of reducing reagent and solution pH resulted in poly(*N*-isopropylacrylamide) brushes of various molecular weight, thickness and grafting density and ionic-responsive and temperature-sensitive characteristics comparable to pure poly(*N*-isopropylacrylamide) (Gao et al. 2014). These results indicate that the enzymatic process can be used to build a new platform for controllable surface functional polymer properties. Nugrohojo Prasetyo et al. (2012) copolymerized lignosulphonates with aminosilanes or ethoxytrimethylsilane to form novel polymer blends with potential application multifunctional thin coating films. *Trametes pubescens* laccase cross-linked low-molecular-weight ultra-filtered lignin to improve mechanical properties of kraft liner pulp and chemi-thermo-mechanical pulp leading to the increase in wet strength of kraft liner pulp hand sheets without losing other critical mechanical properties (Elegir et al. 2007). Recently, Andreas Ortner et al. (2018) developed a novel process for enzymatic modification of lignosulphonates to substitute fossil-based styrene-butadiene latex as binders in conventional paper coating formulations. Lignosulphonate purification was found to be a critical step in obtaining lignin that could efficiently be polymerized. Laccase polymerization of ultrafiltrated lignosulphonates resulted in an increase of the molecular weight. Laccase-polymerized lignosulphonates improved printing properties by reduced picking compared to non-polymerized, reducing lignin penetration into the base paper and improving Abo Akademi Gravimetric Water Retention properties of the paper comparable to those obtained with latex. The authors concluded that their results demonstrated the possibility of substituting fossil-based styrene-butadiene latex binders with lignosulphonates although as expected colour remained a challenge that requires further lignin processing. The incorporation of different lignins (alkali lignin, organosolv lignin and lignosulphonates) into starch, clay, latex films and co-immobilization with TVRL produced films with oxygen scavenging capabilities. The oxygen-scavenging capability varied depending on the type of lignin used with lignosulphonates and organosolv lignin being rated as the most suitable (Sáez-Jiménez et al. 2016; Johansson et al. 2012). The authors suggested that coatings based on aqueous dispersions of latex, clay, lignosulphonates, starch and laccase could be applied on packaging boards (Johansson et al. 2012). The water stability of lignosulphonate-containing latex-based coatings and starch-based films was improved after laccase-catalysed oxidation of lignosulphonates, which indicates polymerization to products with lower solubility in water. Graft copolymers of waxy maize starch and sodium lignosulphonate were prepared by TVRL catalysis in aqueous solution (Shogren and Biswas 2013). The synthesized modified starch produced polymers with antioxidant properties with potential application in food, cosmetic and packaging applications (Shogren and Biswas 2013). The strong UV absorption of lignin could help serve as a sunscreen (Elegir et al. 2008; Cruz et al. 2001), wound dressings, coatings for fruits and vegetables or other packaging (Areskogh et al. 2010a), starch-sodium lignosulphonate graft copolymers should be anionic and thus could be useful for cation binding and water absorption when cross-linked (Shogren and Biswas 2013). In summary, the size of the lignin; nature of H, G and S ratio; pulping process; and enzyme reactivity (redox potential, pH) are extremely important issues that must be considered carefully in order to increase the chances of synthesizing value-added materials from lignin. A lot of effort has to be directed at increasing efficiency and predictability of modification of lignins.

8 Conclusions

Laccases and peroxidases are powerful biocatalysts able to reduce the heterogeneity lignins and reduce the strong inter- and intra-hydrogen bonding of lignins by grafting functional molecules which confer new properties to the lignin making it suitable for synthesizing lignin-based materials or composites. The extent of lignin modification by the enzymes depends on the nature and origin of lignin (S, G and H ratio), properties of the enzyme (redox potential, optimum reaction conditions) as well as the modifications introduced by the pulping process. These are critical issues to consider for successful development of lignin-based materials. The continued search for novel enzymes is increasing the pool of enzyme able to react with lignin under different reaction conditions, e.g. the recently discovered *Streptomyces ipomoea* laccase and MAL able to oxidize lignin at pH 9. This discovery increases the likelihood of modifying kraft lignin under alkaline conditions. Advances in reaction engineering (laccase-mediator systems, continuous oxygen supply systems) have also made the lignin modification more efficient. Building on insights gained in understanding the reactivity of enzymes with different lignins (lignin with different S, G and H ratios as well as lignin from different pulping processes) and possibility of coupling functional molecules onto lignin raises the prospect of increasing lignin applications and expanding lignin application is fast becoming a reality. Although significant progress has been made in developing enzyme-based systems for synthesizing lignin-based materials, great effort is needed to increase efficiency of modifications in order to achieve a significant change in lignin properties towards the desired product. This step is necessary and critical as it provides important information on the possibilities of modifying lignin in a much more predictable way.

References

Abou-Mansour E, Polier J, Pezet R, Tabacchi R (2010) Purification and partial characterisation of a 60 KDa laccase from Fomitiporia mediterranea. Phytopathol Mediterr 48:447–453

Ansari A, Pawlik M (2007) Floatability of chalcopyrite and molybdenite in the presence of lignosulfonates. Part I. Adsorption studies. Miner Eng 20:600–608

Aracri E, Díaz Blanco C, Tzanov T (2014) An enzymatic approach to develop a lignin-based adhesive for wool floor coverings. Green Chem 16:2597
Areskogh D, Li J, Gellerstedt G, Henriksson G (2010a) Structural modification of commercial lignosulphonates through laccase catalysis and ozonolysis. Ind Crop Prod 32:458–466
Areskogh D, Li J, Gellerstedt G, Henriksson G (2010b) Investigation of the molecular weight increase of commercial lignosulfonates by laccase catalysis. Biomacromolecules 11:904–910
Areskogh D, Li J, Nousiainen P, Gellerstedt G, Sipilä J, Henriksson G (2010c) Oxidative polymerisation of models for phenolic lignin end-groups by laccase. Holzforschung 64:21–34
Argyropoulos DS, Jurasek L, Krištofová L, Xia Z, Sun Y, Paluš E (2002) Abundance and reactivity of dibenzodioxocins in softwood lignin. J Agric Food Chem 50(4):658–666. https://doi.org/10.1021/JF010909G
Bae HJ, Kim YS (1996) Degradation of lignosulfonates by simultaneous action of laccase and Mn-peroxidase. In: Biotechnology in the pulp and paper industry. ACS, New York
Behling R, Valange S, Chatel G (2016) Heterogeneous catalytic oxidation for lignin valorization into valuable chemicals: what results? What limitations? What trends? Green Chem 18:1839–1854
Berlin A, Balakshin M (2014) Industrial lignins, analysis, properties, and applications. In: Gupta VK (ed) Bioenergy research: advances and applications. Elsevier, New York, pp 315–336
Blinkovsky AM, Dordick JS (1993) Peroxidase-catalyzed synthesis of lignin–phenol copolymers. J Polym Sci Part A Polym Chem 31:1839–1846
Boerjan W, Ralph J, Baucher M (2003) Lignin biosynthesis. Annu Rev Plant Biol 54:519–546
Bouajila J, Dole P, Joly C, Limare A (2006) Some laws of a lignin plasticization. J Appl Polym Sci 102:1445–1451
Bourbonnais R, Paice MG, Reid ID, Lanthier P, Yaguchi M (1995) Lignin oxidation by laccase isozymes from Trametes versicolor and role of the mediator 2,2′-azinobis(3-ethylbenzthiazoline-6-sulfonate) in kraft lignin depolymerization. Appl Environ Microbiol 61:1876–1880
Call HP, Mücke I (1997) History, overview and applications of mediated lignolytic systems, especially laccase-mediator-systems (Lignozym®-process). J Biotechnol 53:163–202
Cannatelli MD, Ragauskas AJ (2017) Laccase-mediated synthesis of lignin-core hyperbranched copolymers. Appl Microbiol Biotechnol 101:6343–6353
Chandra RP, Ragauskas AJ (2008) Modification of high-lignin kraft pulps with laccase. Part 2. Xylanase-enhanced strength benefits. Biotechnol Prog 21:1302–1306
Chen F, Dai H, Dong X, Yang J, Zhong M (2011) Physical properties of lignin-based polypropylene blends. Polym Compos 32:1019–1025
Cho N-S, Shin W-S, Jeong S-W, Leonowicz A (2004) Degradation of lignosulfonate by fungal laccase with low molecular mediators. Bull Korean Chem Soc 25:1551–1554
Crestini C, Melone F, Sette M, Saladino R (2011) Milled wood lignin: a linear oligomer. Biomacromolecules 12:3928–3935
Cruz JM, Domínguez JM, Domínguez H, Parajó JC (2001) Antioxidant and antimicrobial effects of extracts from hydrolysates of lignocellulosic materials. J Agric Food Chem 49(5):2459–2464. https://doi.org/10.1021/JF001237H
Derkacheva OY (2013) Estimation of aromatic structure contents in hardwood lignins from ir absorption spectra. J Appl Spectrosc 80:670–676
Dong-jie Y, Hai-feng Z, Shao-qu X, Xiao-lei W, Xue-qing Q (2013) Sulfomethylation reactivity of alkali lignin with laccase modification. Acta Polym Sin 13:232–240
Duval A, Molina-Boisseau S, Chirat C (2015) Fractionation of lignosulfonates: comparison of ultrafiltration and ethanol solubility to obtain a set of fractions with distinct properties. Holzforschung 69:127–134
Dwivedi UN, Singh P, Pandey VP, Kumar A (2011) Structure–function relationship among bacterial, fungal and plant laccases. J Mol Catal B: Enzym 68:117–128
Elegir G, Daina S, Zoia L, Bestetti G, Orlandi M (2005) Laccase mediator system: oxidation of recalcitrant lignin model structures present in residual kraft lignin. Enzyme Microb Technol 37:340–346

Elegir G, Bussini D, Antonsson S, Lindström ME, Zoia L (2007) Laccase-initiated cross-linking of lignocellulose fibres using a ultra-filtered lignin isolated from kraft black liquor. Appl Microbiol Biotechnol 77:809–817

Elegir G, Kindl A, Sadocco P, Orlandi M (2008) Development of antimicrobial cellulose packaging through laccase-mediated grafting of phenolic compounds. Enzyme Microb Technol 43:84–92

Felby C, Pedersen LS, Nielsen BR (1997) Enhanced auto adhesion of wood fibers using phenol oxidases. Holzforschung 51:281–286

Fernández-Costas C, Palanti S, Sanromán MÁ, Moldes D (2017) Enzymatic grafting of kraft lignin as a wood bio-protection strategy. Part 2: effectiveness against wood destroying basidiomycetes. Effect of copper entrapment. Holzforschung 71:689–695

Ferrari RP, Laurenti E, Trotta F (1999) Oxidative 4-dechlorination of 2,4,6-trichlorophenol catalyzed by horseradish peroxidase. J Biol Inorg Chem 4:232–237

Fiţigău IF, Peter F, Boeriu CG (2013) Oxidative polymerization of lignins by laccase in water-acetone mixture. Acta Biochim Pol 60:817–822

Fiţigău IF, Boeriu CG, Peter F (2015) Enzymatic modification of different lignins through oxidative coupling with hydrophilic compounds. Macromol Symp 352:78–86

Fonseca MI, Fariña JI, Sadañoski MA, D'Errico R, Villalba LL, Zapata PD (2015) Decolorization of Kraft liquor effluents and biochemical characterization of laccases from Phlebia brevispora BAFC 633. Int Biodeter Biodegr 104:443–451

Frasconi M, Favero G, Boer H, Koivula A, Mazzei F (2010) Kinetic and biochemical properties of high and low redox potential laccases from fungal and plant origin. Biochim Biophys Acta Proteins Proteomics 1804:899–908

Gao G, Karaaslan MA, Kadla JF, Ko F (2014) Enzymatic synthesis of ionic responsive lignin nanofibres through surface poly(N-isopropylacrylamide) immobilization. Green Chem 16:3890–3898

Gillet S, Aguedo M, Petitjean L, Morais ARC, da Costa Lopes AM, Łukasik RM, Anastas PT (2017) Lignin transformations for high value applications: towards targeted modifications using green chemistry. Green Chem 19:4200–4233

Gillgren T, Hedenström M, Jönsson LJ (2017) Comparison of laccase-catalyzed cross-linking of organosolv lignin and lignosulfonates. Int J Biol Macromol 105:438–446

Gordobil O, Moriana R, Zhang L, Labidi J, Sevastyanova O (2016) Assessment of technical lignins for uses in biofuels and biomaterials: structure-related properties, proximate analysis and chemical modification. Ind Crop Prod 83:155–165

Gouveia S, Fernández-Costas C, Sanromán MA, Moldes D (2012) Enzymatic polymerisation and effect of fractionation of dissolved lignin from Eucalyptus globulus Kraft liquor. Bioresour Technol 121:131–138

Gouveia S, Fernández-Costas C, Sanromán MA, Moldes D (2013) Polymerisation of Kraft lignin from black liquors by laccase from Myceliophthora thermophila: effect of operational conditions and black liquor origin. Bioresour Technol 131:288–294

Grönqvist S, Viikari L, Niku ML (2005) Oxidation of milled wood lignin with laccase, tyrosinase and horseradish peroxidase. Appl Microbiol Biotechnol 67(4):489–494

Hataaka A, Mettala A, Toikka B, Hortling B, Brunow G (1996) Modification of lignin by laccase and manganese peroxidase. In: Biotechnology in the pulp and paper industry

Hernández Fernaud JR, Carnicero A, Perestelo F, Hernández Cutuli M, Arias E, Falcón MA (2006) Upgrading of an industrial lignin by using laccase produced by Fusarium proliferatum and different laccase-mediator systems. Enzyme Microb Technol 38:40–48

Hu J, Zhang Q, Lee D-J (2018) Kraft lignin biorefinery: a perspective. Bioresour Technol 247:1181–1183

Huber D, Ortner A, Daxbacher A, Nyanhongo GS, Bauer W, Guebitz GM (2016) Influence of oxygen and mediators on laccase-catalyzed polymerization of lignosulfonate. ACS Sustain Chem Eng 4:5303–5310

Hulin L, Husson E, Bonnet J-P, Stevanovic T, Sarazin C (2015) Enzymatic transesterification of kraft lignin with long acyl chains in ionic liquids. Molecules 20:16334–16353

Humpert D, Ebrahimi M, Czermak P (2016) Membrane technology for the recovery of lignin: a review. Membranes (Basel) 6:42

Ibrahim V, Volkova N, Pyo S-H, Mamo G, Hatti-Kaul R (2013a) Laccase catalysed modification of lignin subunits and coupling to p-aminobenzoic acid. J Mol Catal B: Enzym 97:45–53

Ibrahim V, Mamo G, Gustafsson P-J, Hatti-Kaul R (2013b) Production and properties of adhesives formulated from laccase modified Kraft lignin. Ind Crop Prod 45:343–348

Ihssen J, Schubert M, Thöny-Meyer L, Richter M (2014) Laccase catalyzed synthesis of iodinated phenolic compounds with antifungal activity. PLoS One 9:e89924

Janusz G, Pawlik A, Sulej J, Świderska-Burek U, Jarosz-Wilkołazka A, Paszczyński A (2017) Lignin degradation: microorganisms, enzymes involved, genomes analysis and evolution. FEMS Microbiol Rev 41:941–962

Jin L, Sellers T, Schultz TP, Nicholas DD (1990) Utilization of lignin modified by brown-rot fungi. I. properties of flakeboard produced with a brown-rotted lignin modified phenolic adhesive. Holzforschung 44:207–210

Johansson K, Winestrand S, Johansson C, Järnström L, Jönsson LJ (2012) Oxygen-scavenging coatings and films based on lignosulfonates and laccase. J Biotechnol 161:14–18

Johansson K, Gillgren T, Winestrand S, Järnström L, Jönsson LJ (2014) Comparison of lignin derivatives as substrates for laccase-catalyzed scavenging of oxygen in coatings and films. J Biol Eng 8(1):1

Jönsson A-S, Wallberg O (2009) Cost estimates of kraft lignin recovery by ultrafiltration. Desalination 237:254–267

Kai D, Tan MJ, Chee PL, Chua YK, Yap YL, Loh XJ (2016) Towards lignin-based functional materials in a sustainable world. Green Chem 18:1175–1200

Kalliola A, Asikainen M, Talja R, Tamminen T (2014) Experiences of kraft lignin functionalization by enzymatic and chemical oxidation. BioResources 9:7336–7351

Kaplan DL (1979) Reactivity of different oxidases with lignins and lignin model compounds. Phytochemistry 18:1917–1919

Kersten PJ, Kalyanaraman B, Hammel KE, Reinhammar B, Kirk TK (1990) Comparison of lignin peroxidase, horseradish peroxidase and laccase in the oxidation of methoxybenzenes. Biochem J 268:475–480

Kharazipour A, Mai C, Hüttermann A (1998) Polyphenols for compounded materials. Polym Degrad Stab 59:237–243

Kiiskinen L-L, Kruus K, Bailey M, Ylösmäki E, Siika-aho M, Saloheimo M (2004) Expression of Melanocarpus albomyces laccase in Trichoderma reesei and characterization of the purified enzyme. Microbiology 150:3065–3074

Kubo S, Kadla JF (2004) Macromolecules 37:6904–6911

Kudanga T, Nugroho Prasetyo E, Sipilä J, Eberl A, Nyanhongo G, Guebitz G (2009) Coupling of aromatic amines onto syringylglycerol β-guaiacylether using Bacillus SF spore laccase: a model for functionalization of lignin-based materials. J Mol Catal B: Enzym 61:143–149

Kudanga T, Prasetyo EN, Sipilä J, Guebitz GM, Nyanhongo GS (2010a) Reactivity of long chain alkylamines to lignin moieties: Implications on hydrophobicity of lignocellulose materials. J Biotechnol 149(1–2):81–87

Kudanga T, Prasetyo EN, Sipilä J, Nyanhongo GS, Guebitz GM (2010c) Enzymatic grafting of functional molecules to the lignin model dibenzodioxocin and lignocellulose material. Enzyme Microb Technol 46:272–280

Kudanga T, Prasetyo EN, Sipilä J, Nyanhongo GS, Guebitz GM (2010d) Chemo-enzymatic functionalisation of lignocellulose materials using oxiranes. Process Biochem 45(9):1557–1562

Kudanga T, Prasetyo EN, Widsten P, Kandelbauer A, Jury S, Heathcote C, Sipilä J, Weber H, Nyanhongo GS, Guebitz GM (2010e) Laccase catalyzed covalent coupling of fluorophenols increases lignocellulose surface hydrophobicity. Bioresour Technol 101(8):2793–2799

Laurichesse S, Avérous L (2014) Chemical modification of lignins: towards biobased polymers. Prog Polym Sci 39:1266–1290

Li K, Xu F, Eriksson KE (1999) Comparison of fungal laccases and redox mediators in oxidation of a nonphenolic lignin model compound. Appl Environ Microbiol 65:2654–2660

Lund M, Ragauskas AJ (2001) Enzymatic modification of kraft lignin through oxidative coupling with water-soluble phenols. Appl Microbiol Biotechnol 55:699–703

Madad N, Chebil L, Charbonnel C, Ioannou I, Ghoul M (2013) Enzymatic polymerization of sodium lignosulfonates: effect of catalysts, initial molecular weight, and mediators. Can J Chem 91:220–225

Mai C, Milstein O, Hüttermann A (1999) Fungal laccase grafts acrylamide onto lignin in presence of peroxides. Appl Microbiol Biotechnol 51:527–531

Mai C, Milstein O, Hüttermann A (2000) Chemoenzymatical grafting of acrylamide onto lignin. J Biotechnol 79:173–183

Maldhure AV, Chaudhari AR, Ekhe JD (2011) Thermal and structural studies of polypropylene blended with esterified industrial waste lignin. J Therm Anal Calorim 103:625–632

Matsumura E, Yamamoto E, Numata A, Kawano T, Shin T, Murao S (1986) Structures of the laccase-catalyzed oxidation products of hydroxybenzoic acids in the presence of ABTS (2,2′-azino-di-(3-ethylbenzothiazoline-6-sulfonic acid)). Agric Biol Chem 50:1355–1357

McDonnell G, Russell AD (1999) Antiseptics and disinfectants: activity, action, and resistance. Clin Microbiol Rev 12:147–179

Mikolasch A, Schauer F (2009) Fungal laccases as tools for the synthesis of new hybrid molecules and biomaterials. Appl Microbiol Biotechnol 82:605–624

Milstein O, Hüttermann A, Fründ R, Lüdemann H-D (1994) Enzymatic co-polymerization of lignin with low-molecular mass compounds. Appl Microbiol Biotechnol 40:760–767

Morozova OV, Shumakovich GP, Shleev SV, Yaropolov YI (2007) Laccase-mediator systems and their applications: a review. Appl Biochem Microbiol 43:523–535

Mottiar Y, Vanholme R, Boerjan W, Ralph J, Mansfield SD (2016) Designer lignins: harnessing the plasticity of lignification. Curr Opin Biotechnol 37:190–200

Moya R, Saastamoinen P, Hernández M, Suurnäkki A, Arias E, Mattinen M-L (2011) Reactivity of bacterial and fungal laccases with lignin under alkaline conditions. Bioresour Technol 102:10006–10012

Munk L, Andersen ML, Meyer AS (2017a) Direct rate assessment of laccase catalysed radical formation in lignin by electron paramagnetic resonance spectroscopy. Enzyme Microb Technol 106:88–96

Munk L, Punt AM, Kabel MA, Meyer AS (2017b) Laccase catalyzed grafting of –N–OH type mediators to lignin via radical–radical coupling. RSC Adv 7:3358–3368

Muralikrishna C, Renganathan V (1993) Peroxidase-catalyzed desulfonation of 3,5-dimethyl-4-hydroxy and 3,5-dimethyl-4-aminobenzenesulfonic acids. Biochem Biophys Res Commun 197:798–804

Norgren M, Edlund H (2014) Lignin: recent advances and emerging applications. Curr Opin Colloid Interface Sci 19:409–416

Nugroho Prasetyo E, Kudanga T, Østergaard L, Rencoret J, Gutiérrez A, del Río JCJC, Ignacio Santos J, Nieto L, Jiménez-Barbero J, Martínez ATAT et al (2010) Polymerization of lignosulfonates by the laccase-HBT (1-hydroxybenzotriazole) system improves dispersibility. Bioresour Technol 101:5054–5062

Nyanhongo GS, Couto SR, Guebitz GM (2006) Coupling of 2,4,6-trinitrotoluene (TNT) metabolites onto humic monomers by a new laccase from Trametes modesta. Chemosphere 64(3):359–370

Nyanhongo GS, Kudanga T, Prasetyo EN, Guebitz GM (2010) Mechanistic insights into laccase-mediated functionalisation of lignocellulose material. Biotechnol Genet Eng Rev 27:305–330

Ortner A, Huber D, Haske-Cornelius O, Weber HKHK, Hofer K, Bauer W, Nyanhongo GS, Guebitz GM (2015) Laccase mediated oxidation of industrial lignins: Is oxygen limiting? Process Biochem 50:1277–1283

Ortner A, Hofer K, Bauer W, Nyanhongo GS, Guebitz GM (2018) Laccase modified lignosulfonates as novel binder in pigment based paper coating formulations. React Funct Polym 123:20–25

Ouyang XP, Zhang P, Tan CM, Deng YH, Yang DJ, Qiu XQ (2010) Isolation of lignosulfonate with low polydispersity index. Chin Chem Lett 21:1479–1481

Pandey VP, Awasthi M, Singh S, Tiwari S, Dwivedi UN, Dwivedi UN (2017) A comprehensive review on function and application of plant peroxidases. Biochem Anal Biochem 6:1
Pfaltzgraff LA, Clark JH (2014) Green chemistry, biorefineries and second generation strategies for re-use of waste: an overview. In: Waldron KW (ed) Advances in biorefineries. Woodhead Publishing, Sawston, pp 3–33
Prasetyo EN, Kudanga T, Fischer R, Eichinger R, Nyanhongo GS, Guebitz GM (2012) Enzymatic synthesis of lignin-siloxane hybrid functional polymers. Biotechnol J 7(2):284–292
Rittstieg K, Suurnakki A, Suortti T, Kruus K, Guebitz G, Buchert J (2002) Investigations on the laccase-catalyzed polymerization of lignin model compounds using size-exclusion HPLC. Enzyme Microb Technol 31:403–410
Sáez-Jiménez V, Rencoret J, Rodríguez-Carvajal MA, Gutiérrez A, Ruiz-Dueñas FJ, Martínez AT (2016) Role of surface tryptophan for peroxidase oxidation of nonphenolic lignin. Biotechnol Biofuels 9:198
Shogren RL, Biswas A (2013) Preparation of starch–sodium lignosulfonate graft copolymers via laccase catalysis and characterization of antioxidant activity. Carbohydr Polym 91:581–585
Si J, Peng F, Cui B (2013) Purification, biochemical characterization and dye decolorization capacity of an alkali-resistant and metal-tolerant laccase from Trametes pubescens. Bioresour Technol 128:49–57
Sivasankarapillai G, McDonald AG (2011) Synthesis and properties of lignin-highly branched poly (ester-amine) polymeric systems. Biomass Bioenergy 35:919–931
Stewart D (2008) Lignin as a base material for materials applications: chemistry, application and economics. Ind Crop Prod 27:202–207
Sun M, Hong C-Y, Pan C-Y (2012) A unique aliphatic tertiary amine chromophore: fluorescence, polymer structure, and application in cell imaging. J Am Chem Soc 134:20581–20584
Sun Y, Qiu X, Liu Y (2013) Chemical reactivity of alkali lignin modified with laccase. Biomass Bioenergy 55:198–204
Tejado A, Peña C, Labidi J, Echeverria JM, Mondragon I (2007) Physico-chemical characterization of lignins from different sources for use in phenol–formaldehyde resin synthesis. Bioresour Technol 98:1655–1663
Tobimatsu Y, Takano T, Kamitakahara H, Nakatsubo F (2010) Reactivity of syringyl quinone methide intermediates in dehydrogenative polymerization I: high-yield production of synthetic lignins (DHPs) in horseradish peroxidase-catalyzed polymerization of sinapyl alcohol in the presence of nucleophilic reagents. J Wood Sci 56:233–241
Toledano A, Serrano L, Garcia A, Mondragon I, Labidi J (2010a) Comparative study of lignin fractionation by ultrafiltration and selective precipitation. Chem Eng J 157:93–99
Toledano A, García A, Mondragon I, Labidi J (2010b) Lignin separation and fractionation by ultrafiltration. Sep Purif Technol 71:38–43
Trovaslet-Leroy M, Jolivalt C, Froment M-T, Brasme B, Lefebvre B, Daveloose D, Nachon F, Masson P (2010) Application of laccase-mediator system (LMS) for the degradation of organophosphorus compounds. Chem Biol Interact 187:393–396
Unbehaun H, Dittler B, Kühne G, Wagenführ A (2000) Investigation into the biotechnological modification of wood and its application in the wood-based material industry. Acta Biotechnol 20:305–312
van de Pas D, Hickson A, Donaldson L, Lloyd-Jones G, Tamminen T, Fernyhough A, Mattinen ML (2011) Characterization of fractionated lignins polymerized by fungal laccases. BioResources 6:1105–1121
Wang Y, Chantreau M, Sibout R, Hawkins S (2013) Plant cell wall lignification and monolignol metabolism. Front Plant Sci 4:220
Wang J, Feng J, Jia W, Chang S, Li S, Li Y (2015) Lignin engineering through laccase modification: a promising field for energy plant improvement. Biotechnol Biofuels 8:145
Weihua Q, Hongzhang C (2008) An alkali-stable enzyme with laccase activity from entophytic fungus and the enzymatic modification of alkali lignin. Bioresour Technol 99:5480–5484
Witayakran S, Ragauskas AJ (2009a) Modification of high-lignin softwood kraft pulp with laccase and amino acids. Enzyme Microb Technol 44:176–181

Witayakran S, Ragauskas AJ (2009b) Synthetic applications of laccase in green chemistry. Adv Synth Catal 351:1187–1209
Xu F, Kulys JJ, Duke K, Li K, Krikstopaitis K, Deussen HJ, Abbate E, Galinyte V, Schneider P (2000) Redox chemistry in laccase-catalyzed oxidation of N-hydroxy compounds. Appl Environ Microbiol 66:2052–2056
Yang D, Chang Y, Wu X, Qiu X, Lou H (2014a) Modification of sulfomethylated alkali lignin catalyzed by horseradish peroxidase. RSC Adv 4:53855–53863
Yang D, Wu X, Qiu X, Chang Y, Lou H (2014b) Polymerization reactivity of sulfomethylated alkali lignin modified with horseradish peroxidase. Bioresour Technol 155:418–421
Yang D, Huang W, Qiu X, Lou H, Qian Y (2017) Modifying sulfomethylated alkali lignin by horseradish peroxidase to improve the dispersibility and conductivity of polyaniline. Appl Surf Sci 426:287–293
Zhang Y, Dong A, Fan X, Wang Q, Zhang Y, Yu Y, Cavaco-Paulo A (2016) Laccase-catalyzed synthesis of conducting polyaniline-lignosulfonate composite. J Appl Polym Sci 133:42941
Zhou H, Qiu X, Yang D, Xie S (2016) Laccase and xylanase incubation enhanced the sulfomethylation reactivity of alkali lignin. ACS Sustain Chem Eng 4:1248–1254

GINEXTRA®: A Small-Scale Multipurpose Modular and Integrated Biorefinery Technology

L. Infelise, J. Kazimierczak, J. Wietecha, and E. Kopania

1 Introduction to the GINEXTRA®: Case History

As widely recognised, biorefineries are the physical embodiment of bioeconomy. The challenge is to develop flexible biorefining processes capable of handling feedstock of varying origins, composition and quantities and ensure an optimal valorisation of the side streams generated during the process (Dupont and Borg 2018, p. 141).

The current practice is to divert these streams to low-value applications such as energy and fuels.

The present scenario features a concentration in technologies dedicated to the production of biofuel and a lack of attention paid to both the optimisation of the sustainable use of biomass and the optimised valorisation of the side streams to produce biomaterials (Krüger et al. 2018, p. 145).

So far, the prevailing business model has been inspired by the classical petrochemical refinery scheme and based on intensive financial investments in large plants, which require a huge quantity of biomass available at a very low price, cultivated in wide dedicated lands in the proximity of the biorefinery (Krüger et al. 2018; Dupont and Borg 2018).

Infelise, L. is the coordinator of the whole chapter and author of the following Sects. 1, 2, 3, 4, 5, and 7.

Kazimierczak J., Wietecha J., and Kopania E.—Institute of Biopolymers and Chemical Fibres—are the authors of Sect. 6.

L. Infelise (✉)
ARTES_Innovation for Sustainable Growth, Bologna, Italy
e-mail: artes@artes-research.com

J. Kazimierczak · J. Wietecha · E. Kopania
Institute of Biopolymers and Chemical Fibres, Łódź, Poland

© Springer Nature Switzerland AG 2019

J.-R. Bastidas-Oyanedel, J. E. Schmidt (eds.), *Biorefinery*,
https://doi.org/10.1007/978-3-030-10961-5_25

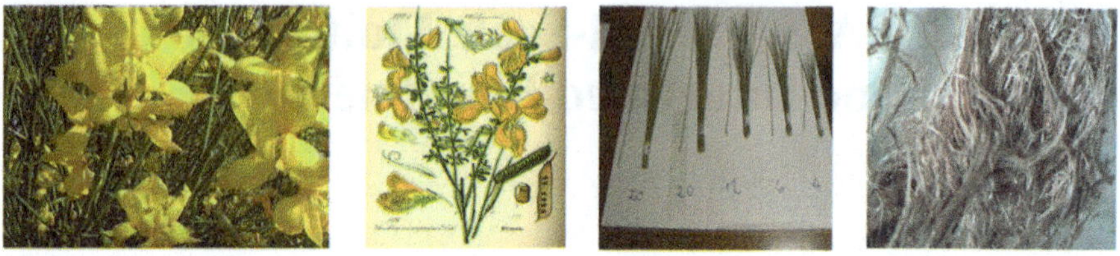

Fig. 1 Ginestra from flowers to yarns and biomaterials—15 years of research and experimental work (2001–2016). Internal research report—PO FESR 2007/2013

While this model is suitable for suburban rural areas or degraded deindustrialised spaces, it has a devastating impact on landscape resilience and biodiversity preservation and valorisation and is not suitable for rural areas that are characterised by a strong commitment to ecotourism or located near precious nature reserves. This is often the case for beautiful ecosystems situated close to small cities and areas rich in archaeological patrimony and natural monuments, which can be found all over our planet. Many of these unique areas have been recognised by the UNESCO as World Heritage Sites, while others are almost unknown, such as the 14 nature reserves and protected areas in the small country of Burundi or villages located in remote zones in the Aspromonte National Park, in Italy.

At the second Global Bioeconomy Summit, held on April 18–20, 2018, in Berlin, experts and stakeholders from all hemispheres met to review the state of the bioeconomy in various parts of the world and called attention to the fact that countries with a rich biodiversity are challenged more than others; many of these countries are less-developed countries on the African continent (Global Bioeconomy Summit 2018).

This chapter presents the case history of GINEXTRA®, a tireless attempt to create a sustainable value chain enabling the economic regeneration and improvement of civil and cultural life in unique rural ecosystems. This biorefinery technological model combines biodiversity preservation, landscape valorisation and competitive economic development through green businesses and their competitive entrance as full protagonists in the fast-growing niche markets of branded bio-products (from yarns, fabrics and garments to high value-added components for food, feed, nutraceuticals or pharmaceuticals with new functionalities, along with bio-composites and bio-based lubricants).

GINEXTRA® is already a European registered brand (Registration Number 7055312, classes 01, 07, 16, 22, 23 and 40) which identifies a patented and proprietary multipurpose modular biorefinery technology. This technology, through enzymatic catalysis[1] (Bommarius and Bommarius 2004), enables the extraction of high-quality fibres and biochemicals from the stems of the Spartium junceum (Fig. 1), known in Italy as Ginestra, an extremely resistant, completely unused and widely diffused plant, which grows spontaneously in the Mediterranean basin.

[1] Biocatalysis is the general term for the transformation of natural and non-natural compounds by enzymes.

The aim of the authors is to propose the GINEXTRA® case as a reference model for those territories which need protecting biodiversity while enhancing bio-based industrial developments.

2 *Spartium junceum*: An Unexploited Almost Unknown Biomass

A robust bio-based economy will require access to renewable feedstocks in sufficient quantities, of guaranteed quality and at a competitive price (Bell et al. 2018). This must of course also be achieved without disrupting food supply chains.

In this perspective, agriculture as a whole will need to find new ways to increase yields in the face of climate change, reduced soil quality and unpredictable growing conditions.

If the market for bio-based products is to develop, it is needed the establishment of efficient and cost-effective supply chains, which increasingly involve new unexploited areas and not yet industrially used biomasses.

A good example is represented by *Spartium junceum*, a perennial shrub growing in poor, semiarid soils.

Spartium junceum is a classified Euro-Mediterranean bast fibre (Code: GI), very suitable for textile, suitable to be complementary to flax which is undergoing an extraordinary crisis due to climatic changes and represents a good substitute of cotton.

It can be cultivated and mechanically harvested with similar yield per hectare as hemp and flax in a variety of climate and soil conditions (Angelini et al. 1998, 1999, 2000; Angelini and Ceccarini 2003).

Spartium junceum is the most resistant variant that gives tissue fibre of the best quality, amongst the botanical species found in the genus *Genista* (*sagittalis, sibirica, lincloria, scoparia* and *juncea*). This shrub spontaneously grows throughout the whole Mediterranean region and represents a shared know-how of the art and craft heritage in rural communities, especially in the Mediterranean Basin.

These plants are often found growing together as dense thickets, in waste areas, abandoned pastures and roadsides, preferring poor, infertile soils. Its penetrating root structure indicates that *Spartium junceum* is an important pioneer species. It is indigenous (Fig. 2) to temperate Europe, in Northern as well as in South Asia, but also it spontaneously grows in Latin America.

Spartium junceum is also suitable to be cultivated in a large variety of pedoclimatic environments.

For botanists, *Spartium junceum* presents some interesting challenges by its tendency to interbreed so much that many plants found growing ‘wild’ are actually hybrids. Vegetable matrices contain a wide range of substances that are available in different parts of the plants and that are interesting from a biological standpoint.

Since ancient times it has been used as a raw material for the manufacture of ropes, nets, bags, sails and even high-quality yarns, fabric and garments in all

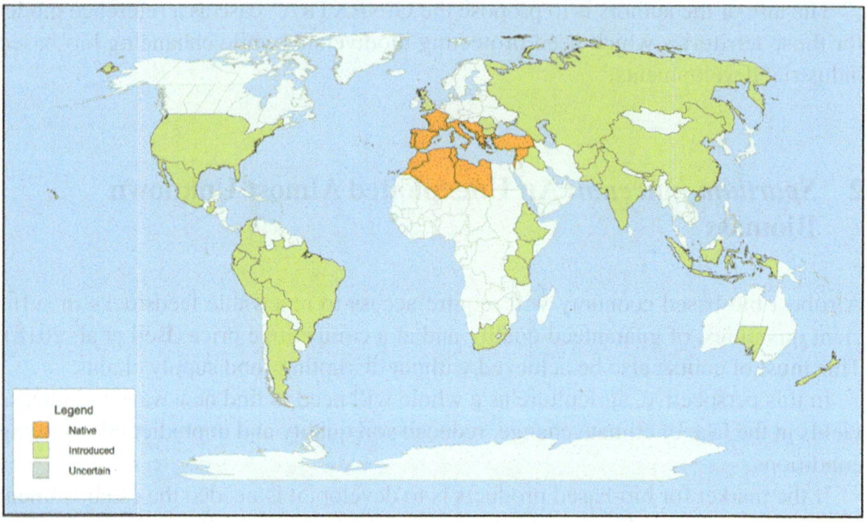

Fig. 2 Spartium junceum geographical distribution (www.ildis.org)

Mediterranean countries (mainly Albania, France, Greece, Spain and Italy). It is still used in rural remote zones of the Mediterranean Basin communities for the extraction of aromatic essences, natural dyeing pigments and other components with various medicinal properties.

Carlo Berti Pichat (1858), in the 'Institutions of Agricultural Sciences and Technology' collection in Book XX, Chapter 725, places *Spartium junceum* amongst the best plants for producing textiles, recalling the robust canvases that were made from its fibres and hence recommended the care in the choice of stems. Gerolamo Boccardo (1880) reports that in 1856 England had obtained 50,000 tonnes of *Spartium junceum* from Spain to make paper. This number increased to 191,000 tonnes in the year 1871. This initiative was then repeated by France but later halted due to problems with the harvesting of annual stems.

Savorgnan D'Osoppo (1891), in his 'Manual on the Cultivation and Industrial Exploitation of Textile Plants', reports the tests made to assess the strength of *Spartium junceum* fibre, slightly less strong than linen, and mentions the production of 'very fine fabrics and valued laces'.

A huge industrial research effort was deployed in between the two world wars, documented by numerous patents by German and Italian authors, and it was industrially processed. After that period the *Spartium junceum* was fully abandoned, and any trace disappeared in both the technical education and research laboratories.

The natural fibre industry in Europe is facing severe problems, mainly related to the shortage of traditionally used raw materials such as flax and hemp and consequent increase of prices, which greatly affect the competitiveness of several industrial sectors, starting from the textile one (both fashion and technical, woven and non-woven) and including the construction, the marine and the automotive industries.

Through a holistic approach that involves cultivation, biorefinery, end-product design and manufacturing and selected market tests, the **GINEXTRA®**biorefinery technology could contribute to the diffusion of bioeconomy models in valuable ecosystems by unlocking the great potential of *Spartium junceum.*

3 GINEXTRA® Biorefinery Technology as the Driver of Precious Rural Ecosystem Development

In 1998, ARTES, a 20-year-old research institute operating with a distinctive multidisciplinary approach in the field of innovation and competitive development of fragile regions, started a community regeneration programme, entitled *Alliance—An alliance for enterprise development and employment in rural areas*. The project was part of the NOW—Employment Programme 1997–2000, funded by the European Community and the Italian Ministry of Labour and Social Policies.

The programme was implemented in rural areas characterised by a unique biodiversity ecosystem, beautiful landscape and cultural heritage in two regions of the South of Italy (Calabria and Basilicata regions) and in the Alba Julia region (Romania). The project aimed at recovering the most authentic and positive roots of the rural culture with a view to fostering the growth of the productive capacities and developing a strong presence in the international market.

The theoretical foundations of the Alliance model were illustrated in a monograph published at the beginning of 2000 (Infelise 2000); it was then presented and discussed during three international conferences, respectively, held in Le Mans (Infelise 1998), in Dortmund (Infelise 2002) and in Pisa (Infelise 2003). The operational model was also evaluated by the Scientific Committee and included in 30 international exemplary experiences at the first World Cluster Conference, organised by OCSE and DATAR (Paris, January 2001).

An extensive discussion took place at the international seminar held on 21–22 September 2006 in Lugano, entitled 'The crisis of the boundaries. Towards an engineering of regional development' (Infelise 2007).

More than 100 articles were published in magazines and newspapers, amongst which, for two times, is the International Herald Tribune.

Within the project developed by ARTES, three ecotourism clusters were created (the Alliance routes) with dozens of small green enterprises operating in the textile, arts and crafts and tourism services sectors. A consortium, named ATENA, composed by microbusinesses (Infelise 2006), was formed by unemployed artisans which started selling worldwide hand-made garments produced using ancient rare textile arts and manually processing the stems of *Spartium junceum* to extract the textile fibre and produce yarns.

The consortium of artisans had a very successful start, with an invitation at the 2005 World Expo, held in Nagoya—Aichi Prefecture—Japan, and at the fair *Carrousel du Louvre*, in Paris, December 2005, which resulted in promising contacts with well-known brands, in the fashion industry.

The items most appreciated by the market were those produced using *Spartium junceum* fibre, which was available at a very limited quantity as it was produced according to ancient rural craft techniques. To satisfy the market demand, a regular production of a much greater quantity was necessary, but a suitable technology enabling the environmentally sustainable production at acceptable market conditions was not available. This technology gap brought to the market failure of the ATENA consortium and to its closure in February 2008. The closure of the consortium had a severe impact on the microbusinesses created thanks to the European project Alliance. Many craft microbusinesses, approaching national and international markets, were forced to close. The black economy revamped.

4 GINEXTRA®: A Small-Scale Multipurpose Modular and Integrated Biorefinery Technology

In order to enhance the competitiveness of the textile cluster, at beginning of 2005, ARTES started a cooperation with a Finnish biotechnological laboratory to identify a cost-effective enzymatic retting process and extract high-quality intact fibres from the *Spartium junceum* shrub. In 2006, after 1-year research, thanks to a deliberation of the whole Government Assembly of the Calabria Region, the small-scale biorefinery (2.5 kg biomass processed in 22 h) built in Oulu was transferred to Italy, in a laboratory managed by the Rural Development Agency of the Calabria Region.

The joint effort between ARTES and the Calabrian Regional Laboratory led to the isolation of strains and to the development of an enzymatic retting process which uses a proprietary enzymatic cocktail branded as GINEXTRA® (European Registration Number 7055312, classes 01, 07, 16, 22, 23 and 40) with high degradative capacities on lignin and hemicellulose, but does not affect cellulose and allows the extraction of intact high-quality fibres.

In 2008 the project ended with a new scaled-up biorefinery plant with 900 L macerating liquid and 75 kg *Spartium junceum* stems processed in a cycle of 48 h (Figs. 3, 4, and 5). It was patented under the title 'Machine, procedure and combined plant for the separation of fibres for textile by macerated stems of fibre plant' (Patent No. 0001396855).

The biorefinery technology which was fully autonomous from the market of enzymes used a close cycle of water circulation, with almost zero waste of enzymatic liquor, which is regenerated, thanks to a combined interaction amongst process parameters. The plant patented in 2008 was a dedicated pilot biorefinery, only applicable to extracting, through the use of enzymatic retting and mechanical extraction and singularisation, high-quality *Spartium junceum* fibre for the fashion industry.

Furthermore, it was based on the current practice to divert the side streams (90% of the wild harvested stems) to low-value applications such as producing compost for energy.

An experimental cultivation with mechanical harvesting was also accomplished with the aim to reduce the costs along the whole value chain.

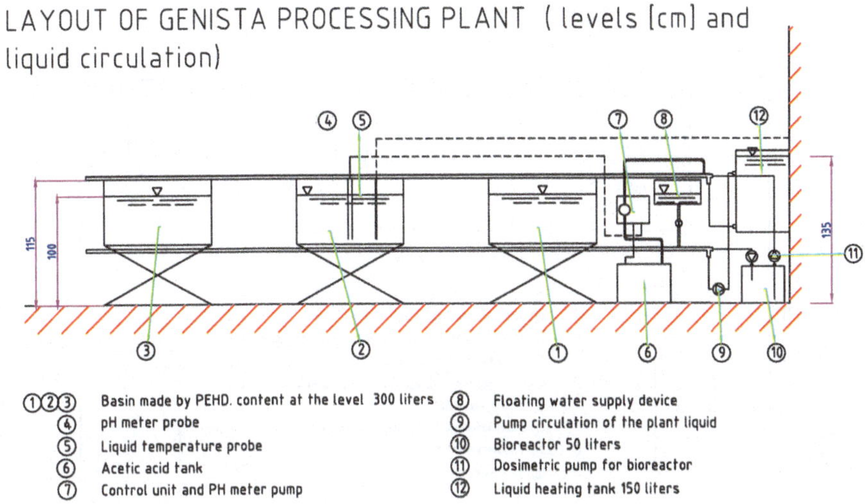

Fig. 3 Pilot biorefinery (Patent No. 0001396855)

A further study envisioned a more advanced automation, and a business plan, for the whole value chain, was devised: from cultivation to harvesting, bio-retting and defibering down to spinning and weaving.

The business plan demonstrated that the key economic performance indicators (Tables 1 and 2) were too low to attract industrial investments, in order to build a scalable pilot plant and assess industrial viability till market uptake.

Side stream valorisation emerged to be the fundamental condition to ensure competitiveness and sustainability and raised interest in private investors in the upscaling and full automation of the pilot small plant, patented in 2008.

In 2014, in order to maximise the cost-efficiency of the valuable fibre extracted and enhance competitiveness of the whole value chain, a new research programme was started by ARTES specifically focused on the enhancement of valuable components contained in processing residue of Spartium junceum fibre with contribution of the Calabria Regional Structural Funds (L.I. 1131 PO FESR 2007/2013–D.D n.15820 del 22/11/2013). The research programme ended in July 2016.

The research programme was organised in a set of components: the first one was dedicated to the study of the isolated organisms and engineering of more efficient enzymatic cocktail, the second component was dedicated to study a more advanced automation of the enzymatic retting and defibering (primary biorefinery technology) and the third component was dedicated to the side streams valorisation. Details are given in paragraph 1.6.

Furthermore, a component of the research was dedicated to the study of the genotypes and to a comparative analysis of the fibre contents.

The research programme ended in 2016 with the improved efficiency of the proprietary mix of enzymes and the development of radical changes to the biorefinery technology (Figs. 6 and 7) which included a preparatory phase to be accomplished before starting the enzymatic retting.

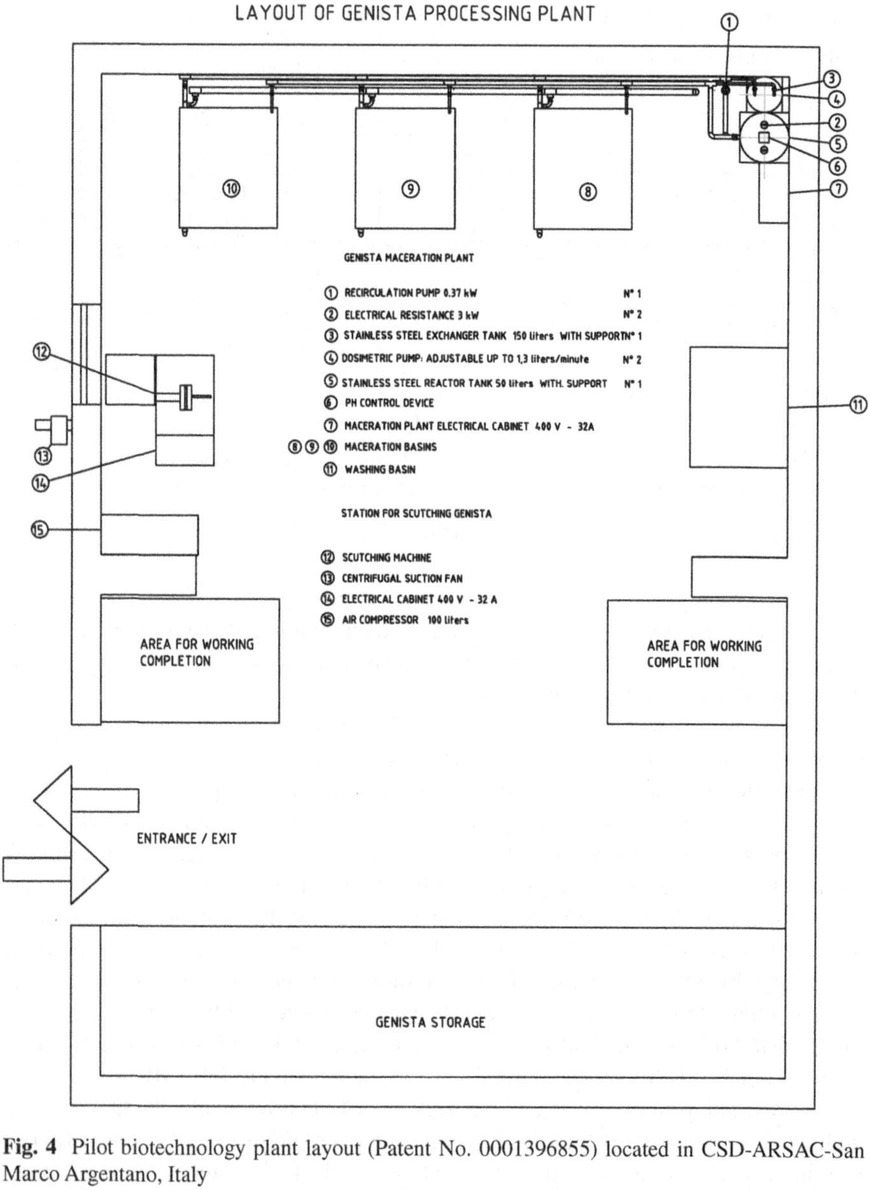

Fig. 4 Pilot biotechnology plant layout (Patent No. 0001396855) located in CSD-ARSAC-San Marco Argentano, Italy

The new biorefinery technology, which uses the proprietary enzymatic mix, combined with a pretreatment with saturated steam, not only reduces enormously the time length of the process (from 48 to 8 h) it can process a variety of lignocellulosic biomasses, extract high-quality singularised fibres and, changing the process parameters, operate as traditional biofuels biorefinery. Furthermore, it has also a higher yield, in terms of intermediates obtainable per unit of biomass waste compared with the technology patented in 2008 (Table 3).

Fig. 5 Pilot biotechnology plant installation (Patent No. 0001396855) located in CSD-ARSAC-San Marco Argentano, Italy

Table 1 Small-scale industrial plant-forecasted yield

Industrial plant	Unit/year
Dry stems	714.28 tonnes
Wet stems	1285.71 tonnes
Raw sellable fibre	67.14 tonnes
Dry sellable biomass	647.14 tonnes
Liquid sellable biomass	0
Stems production for Ha	20.00 tonnes
Land requirement	35.71 ha
Water requirement	1,020,000.00 L
Energy requirement	1512.00 kWh

Table 2 Economic indicators

Investments			
Agricultural investments			294,107.14 €
Investments for the transformation plant			1,925,428.57 €
Laboratory investments			–
Total investments			2,219,535.71 €
Target price	kg	€/kg	€
Output 1: raw fibre	67,143	4.50	302,142.86
Output 2: dry biomass	647,143	0.62	399,262.19
Total	714,286	0.98	701,405.05
Financial indexes			
ROI	7.80%		
ROS	51.20%		
NPV	1,000,000.00 €		
Payback (months)	74		

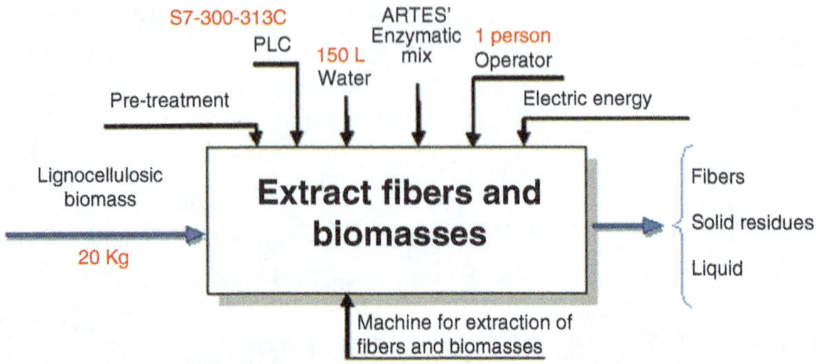

Fig. 6 Schematic structure of the multipurpose modular biorefinery (Patent Demand No. 102017000097949)

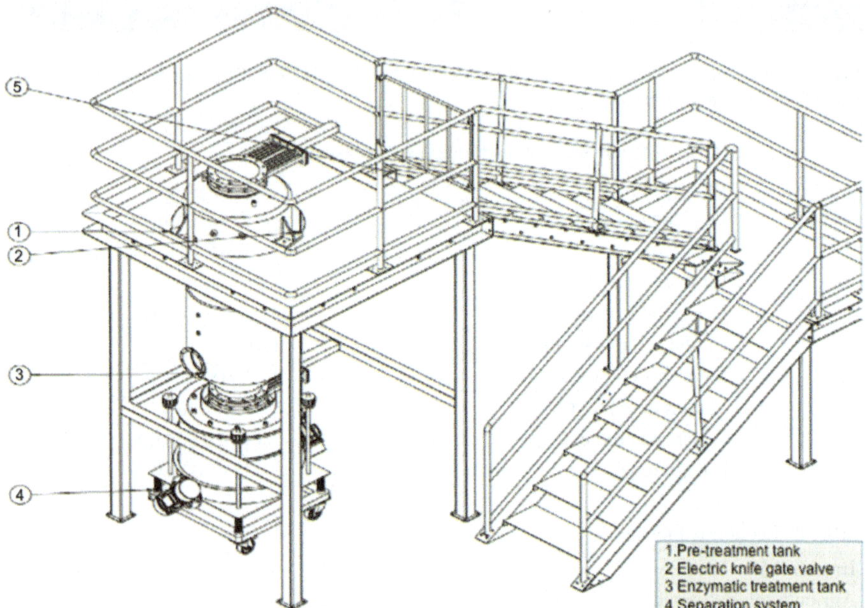

Fig. 7 Layout multipurpose modular biorefinery (Patent Demand No. 102017000097949)

In order to start approaching the market, an important investment is yet needed to scale up the GINEXTRA® biorefinery technology, optimising the integration amongst enzyme production, fibre extraction and waste exploitation in view of obtaining a fully flexible adaptable modular multipurpose biorefinery system, with zero environmental impact and zero waste.

The next planned research programme envisions the engineering and realisation of two pilot plants: **108 tonnes dry biomass per year (research laboratory market) and a scalable version of 714 tonnes dry biomass per year (commercial markets).**

Table 3 Valorisation of solid and liquid waste—Ginestra: from a flower to fine yarns biomaterials and biochemicals for industrial exploitation. (L.I. 1131 PO FESR 2007/2013–D.D n.15820 del 22/11/2013)

Chemicals	Solid waste obtained from the 2008 patented process (Patent No. 0001396855)	Solid waste obtained after the process developed in 2016 (Patent Demand No. 102017000097949)
Cellulose %	37	49
Hemicellulose %	26	11
Total Lignin %	17	31
Pectin %	3.0	0.00
Mineral substances %	1.8	0.6
Ethanol extractable substances %	9.4	7.7

5 The *Spartium junceum* Fibre Characterisation

The experimentation has given excellent results in terms of extraction of the fibre. The fibres extracted from *Spartium junceum* using the two GINEXTRA® process technologies have mechanical properties (Table 4) similar to flax and superior water absorption capabilities (Tables 5 and 6, Fig. 8).

Laboratory assessment run by the Spanish laboratory AITEX demonstrated that the fibres could be spun 100% or even mixed with other fibres like cotton or flax.

Spartium junceum fibres could be also processed in a non-woven manufacturing device if the length of the fibres is reduced. Currently, the average length is 5–10 cm (more or less), and for feasible processability, the length should be maximum 1.5 cm. These non-wovens could find end applications in the building industry, as isolation materials or for textile reinforcement in textile-based composites (Kovačević et al. 2016).

The fibre characterisation showed the following key fibre properties:

- The value of the surface resistivity is the same as that of cotton, viscose and other man-made fibres: <1010 while flax is <1011 and wool is <1012.
- The resistance to acids is weak (it bears only weak acids), while it has a good resistance to alkalis.

With the aim of measuring the water absorption in the fibre, the ASTM D570 method has been run. The analysis revealed that *Spartium junceum* fibre is characterised by eight-time absorption capacities compared with cotton and viscose.

The fibre was spun and woven in industrial plants.[2] Two samples of blended yarns were obtained (70% cotton and 30% *Spartium junceum* and 70% cotton, 15% *Spartium junceum*, 15% wool) for high-quality garments (Fig. 9).

[2] Spinning: Best Yarns (Brescia, Italy); weaving: Grandi & Rubinelli (Novara, Italy); manufacturing: ARCA Textile Lab and Sailor Uomo (Palermo, Italy)

Table 4 Mechanical properties of *Spartium junceum* fibres extracted with the GINEXTRA® biorefinery technology

Average breaking strength (BS)	19
Breaking strength (BS) min	10
Breaking strength (BS) max	29
Average elongation %	3.3
Min. elongation %	2
Max. elongation %	4.9
Tenacity (g/den)	8.5

Table 5 Rate of absorption of the *Spartium junceum* fibre—ASTM D570 method

Fibre	m_{dry} (g)	m_{humid} (g)	Absorption (%)	Imbibement (s)	Abs intrinsic (mL/g)	Abs extrinsic (mL/m^2)
Non-mercerised *S. junceum*	0.127	0.47	270	20	2.7	ND
Mercerised *S. junceum*	0.176	1	466	3	4.66	ND

Table 6 Rate of absorption of cotton and viscose—ASTM D570 method

Fibre	m_{dry} (g)	m_{humid} (g)	Absorption (%)	Imbibement (s)	Abs intrinsic (mL/g)	Abs extrinsic (mL/m^2)
Non-mercerised cotton	1.32	1.95	47.5	15	0.47	63
Mercerised cotton	1.58	2.49	57.6	10	0.57	91
Viscose	1.12	2.14	91.1	7	0.91	102

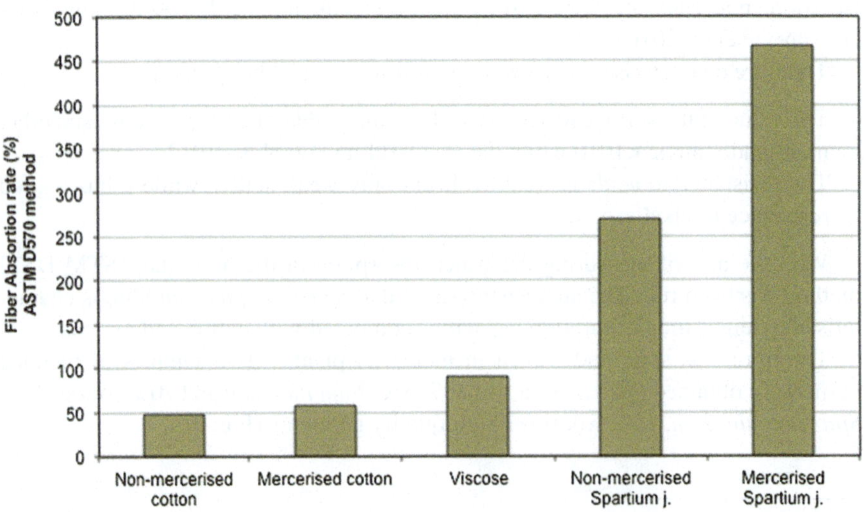

Fig. 8 Comparative analysis of the absorption rate

6 The GINEXTRA® Biorefinery Side Stream Valorisation Programme for Industrial Uses[3]

During processing of *Spartium junceum* stems, after extraction of high-quality fibres, two side streams are generated: post-maceration liquid (PML) and solid waste (shives and tows).

The core aim of the research programme funded by ARTES and which involved the Institute of Biopolymers and Chemical Fibres (IBWCH) was to investigate the possibility of valorisation of liquid and solid wastes, generated during *Spartium junceum* processing dedicated to the extraction of the high-quality fibre, with intact properties in accordance with the concept of the **GINEXTRA®** biorefinery.

The biomass utilisation generally goes in two directions:

- Attainments of building block chemicals, e.g. furfural, xylitol and organic acids (Werpy and Petersen 2004), which can further be used to synthesise other compounds
- Attainments of energy and liquid/gas biofuels

The GINEXTRA® philosophy was instead to preserve the intact fibre and process only the waste of the primary biotech process.

Besides the production of high-quality yarns, fabrics and garments, lignin, hemicelluloses and cellulose pulp were extracted from the solid waste and purified by the Institute of Biopolymers and Chemical Fibres. Preparation and fractionation of biomass for cellulose, lignin and hemicelluloses were based mainly on thermomechanical, chemical or biochemical processing (Kopania et al. 2014). The lab research results also showed that it is possible to use the PML as a component of culture medium for cellulose-producing bacteria in biosynthesis of bacterial nanocellulose (BNC). It was also possible to reduce the cost of biosynthesis by 25% by elimination or reduction of the amount of individual medium components and to obtain

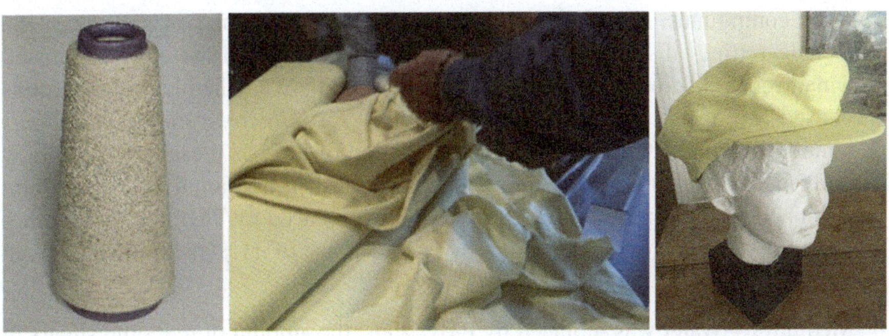

Fig. 9 Yarns, fabrics and garments obtained from processing GINEXTRA® fibres in industrial plants

[3]Written by Kazimierczak J., Wietecha J., Kopania E.—Institute of Biopolymers and Chemical Fibres.

BNC with a comparable or even up to 29% higher yield than in the standard culture medium. It was demonstrated that besides the extracted low-value intermediates such as lignin, residual cellulose and hemicellulose, also bacterial nanocellulose (BNC) could be obtained from which various end products could be produced by a wide range of industries such as automotive, paints and coating, cleaning textile, agro-textile, biomedicine, furniture and interior design and bio-building and bio-fuel, with measurably higher market value compared with the compost for energy uses. The estimated production cost at the lab scale was 0.35 euro/kg for lignin, 0.60 euro/kg for cellulose and hemicellulose and 22 euro/m^2 for BNC.

Bacterial nanocellulose (BNC) is secreted extracellularly by some bacterial genera, e.g. *Gluconacetobacter*, *Sarcina* and *Agrobacterium* (Cannon and Anderson 1991). In contrast to plant cellulose, BNC is devoid of hemicelluloses, lignins and pectins and possesses unique physical and mechanical properties resulting from its three-dimensional nanosized fibrous network. BNC has larger specific surface area, higher water retention value, moldability and high tensile strength. Its diameter is generally 0.1 μm, which is 300 times smaller than wood fibrils (Yoshino et al. 1996). Recently, BNC is receiving increased attention from various industries, as it may be used as a component of dietary foods (Shi et al. 2014), as membranes for electro-acoustic transducers (Ciechanska et al. 2002), as a filtration material or as a raw ingredient for extremely strong paper (Wanichapichart et al. 2002). A number of applications of bacterial cellulose in medicine and veterinary medicine have also been reported (Ciechanska 2004; Piasecka-Zelga et al. 2018). Biofill®, Bioprocess® and Gengiflex® are products of BNC that find applications in surgery and dental implants (Czaja et al. 2006). BNC can be produced in a liquid medium using different natural sources, such as surplus fruits or coconut water (Kim et al. 2017). *Gluconacetobacter* is one of the most frequently characterised acetic acid bacteria for BNC production (Yeo et al. 2004); nonetheless, new bacterial strains have been genetically designed for a better control of BNC production (Florea et al. 2016).

Post-maceration liquid (PML) from enzymatic retting of *Spartium junceum* may supplement the liquid culture medium for bacteria which can synthesise BNC with nitrogen compounds, sugars and microelements. PML obtained as a waste of the enzymatic retting of *Spartium junceum* stems by using a proprietary optimised enzymatic cocktail containing enzymes from the cellulase and hemicellulase groups was used as a part of the culture medium for *Gluconacetobacter xylinus* (Kazimierczak et al. 2016). Chemical composition of PML is presented in Table 7. Culture media were modifications of standard Hestrin-Schramm (H-S) medium (Hestrin and Schramm 1954) in which water was replaced by PML and some other ingredients were removed or limited. Modifications of H-S medium are given below:

- Medium I—Water was replaced by PML; thus, the amount of reducing sugars in the medium increased from 20 to 24.5 g/L.

Table 7 Chemical composition of PML (Kazimierczak et al. 2016)

Component	Amount [g/L]
Hemicellulose	3.5
Lignin	2.1
Protein	3.7
Reducing sugars (expressed as glucose)	3.8
Glucose	0.3
Aliphatic hydrocarbons (C10–C18)	Total: 0.03
Butylene acetate	
Dimethylbenzene	
Other benzene derivatives including vanillin	
Lupanine	
N-Methylcytisine	
Cyclo(leucylprolyl)	

- Medium II—Water was replaced by PML; no peptone was added so PML was the only source of protein for bacterial growth.
- Medium III—Water was replaced by PML; again, no peptone was added, and glucose amount was reduced so that the concentration of protein and reducing sugars in the medium was similar to the control medium.

Biosynthesis of BNC with the use of *Gluconacetobacter xylinus* ATTC 700178 was carried out in sterile flat-bottomed glass vessels with glass covers over 7 days at 30 °C in an incubator (Fig. 10). The standard (H-S) culture medium was used as a control. In static culture bacterial cellulose was produced on the top of liquid culture medium (Fig. 10a, b). The obtained BNC pellicles were washed with distilled water until the complete removal of culture medium components (Fig. 10c). BNC displays unique properties, including high water absorption capacity (Fig. 10c, d). BNC pellicles in wet form typically contain 99.5% water (Table 8). Figure 10e, f present SEM micrographs of BNC. All the BNC samples obtained are characterised by a fine, entangled fibrillar network with an average microfibril width of 100 nm.

The yield of bacterial cellulose biosynthesis and some properties of BNC pellicles are presented in Table 8.

The results presented show that the PML can be a good substitute for sugars and nutrients in the culture media, thus reducing the consumption of water for their preparation. In case of a medium enriched with an additional amount of sugars and proteins (Medium I) by using a PML instead of water, the biosynthesis yield was approximately 30% higher than that recovered with H-S medium. The use of Medium III allowed obtaining of BNC with yields improved about 17%. Medium II composition was the closest to the H-S medium—in this case BNC biosynthesis yields were almost the same. The content of α-cellulose in pellicles obtained from cultures in growth medium containing PML is similar to or a little higher than those obtained from cultures in H-S medium. It may indicate that bacterial cellulose obtained from cultures in growth medium containing PML is of such high chemical purity as BNC obtained in standard culture. Cellulose content in wet pellicles obtained in modified media and in control medium was similar. Only in Medium III the α-cellulose content in wet pellicles was lower—the structure of bacterial cellulose was looser, so larger water amount was included in BNC structure.

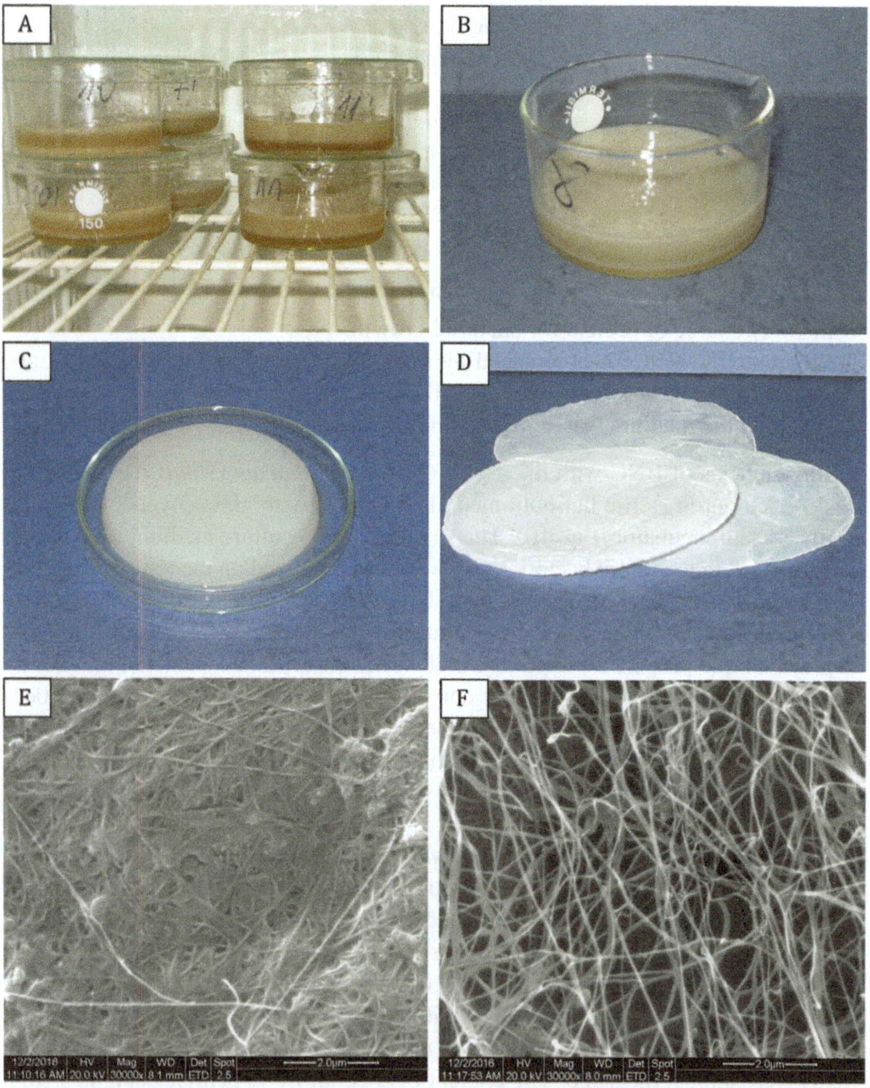

Fig. 10 BNC pellicles: (**a**) during biosynthesis in an incubator; (**b**) after biosynthesis; (**c**) purified, in wet form; (**d**) purified, in dry form; (**e**) SEM micrograph, surface; (**f**) SEM micrograph, cross section (Kazimierczak et al. 2016)

Table 8 Yield of BNC biosynthesis using *G. xylinus* ATCC 700178 and selected properties of BNC pellicles (Kazimierczak et al. 2016)

Culture medium (see above for the medium description)	BNC yield (g/L culture medium)	Cellulose content in wet pellicles (%)	Surface density of dry pellicles (g/m²)	α-cellulose content (%)	Degree of polymerisation DP (–)
H-S (control)	4.68	0.477	52.97	87.85	2243
Medium I	6.07	0.483	68.73	89.67	2414
Medium II	4.72	0.491	61.94	90.79	1564
Medium III	5.47	0.446	53.46	89.58	2277

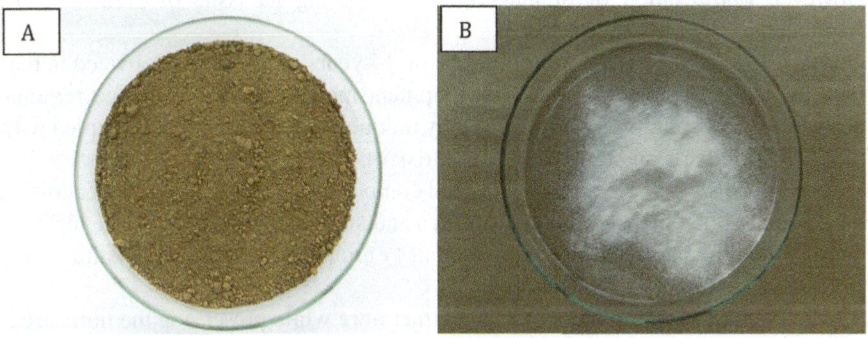

Fig. 11 Biopolymers obtained from black liquor: (**a**) lignin, (**b**) hemicellulose (Kopania, own unpublished work)

The results of the experiments show that it is possible to use PML as a medium component for the biosynthesis of BNC. It is possible to reduce the cost of biosynthesis by at least 25% due to elimination or reduction of the addition of some components and to obtain cellulose with a comparable or even over a dozen percent higher yield than in the standard H-S medium.

The solid waste after textile-grade fibre removal was subjected to chemical treatment by using cooking liquor. As a result of this process, lignin and hemicelluloses dissolved in black liquor, whereas residual cellulose was obtained in the fibrous form. Cellulose fibres after separation from black liquor may be used directly for various applications or be subjected to mechanical disintegration to obtain cellulose nanofibrils (CNF). By acid treatment, lignin precipitated from the black liquor and could be separated, while hemicelluloses after separation of the lignin fraction were precipitated from the solution by addition of ethanol (Fig. 11). Hemicellulose as galactoglucomannan (GGM) with molar ratio of galactose, glucose and mannose 1:2:2 and molecular mass of about 10 kDa were extracted with a yield of 5–15% with respect to the feedstock. The yield of lignin extraction was up to 30%. Both hemicellulose and lignin fractions may be subjected to further chemical modifications. Alternatively, dissolved hemicelluloses (DP below 300) can be subjected to enzymatic treatment to obtain simple sugars and oligomers. Most of oligomers have DP of less than 10, and monomeric xylose comprises about 10% of the total xylo-oligomer content (Puls et al.1985).

7 Final Considerations

The current pattern of production, consumption and materials used in industrialised economies endangers the availability of the natural resources on which our well-being is based. Natural resource used in these economies exceeds availability.

Worldwide population growth (from 7 billion to more than 9 billion by 2050) and economic development will lead to an increased demand for natural resources, many of which are finite.

Biorefineries were initially developed according to traditional oil-based refineries and typically operated by converting biomass to biofuels (Krüger et al. 2018). However, optimisation of biomass use is becoming an issue of great concern (Dupont and Borg 2018, p. 141).

The GINEXTRA® case history reports on 14 years of research, motivated in part by a desire to provide a potential development model for less developed regions, especially small countries characterised by unique natural ecosystems, coupled with landscapes and arts and crafts heritage at risk of disappearing.

The GINEXTRA® technology is a vital component of a community regeneration model, embedded in a sustainable logistics and supply chain strategy. The technology is rooted in an overall operational strategy aimed at preserving the cultural and natural environment of rural areas.

The GINEXTRA® philosophy is to extract fibre while preserving the fibre structure and mechanical properties to produce high-quality yarns, fabrics and garments for the fashion industry while also processing the side streams to obtain intermediates suitable for high value-added end uses (components for food, feed, nutraceuticals or pharmaceuticals with new functionalities, along with bio-composites and bio-based lubricants).

This technology was primarily developed to contribute to the sustainability and competitiveness of the Euro-Mediterranean natural fibre industry which is facing severe problems, mainly related to a shortage and consequent price increase of biomass, particularly cotton and flax. These factors greatly affect the competitiveness of several industrial sectors, from textiles (both fashion and technical, woven and non-woven) to the construction, marine and automotive industries.

Moreover, the GINEXTRA® technology is also meant to give feasible responses to the severe problems faced by the poorest countries with unique natural ecosystems and landscapes at risk of disappearing where agro-food wastes could be processed to produce biomaterials which will hasten job and green business creation and promote landscape resilience.

Fourteen years of intense research activity, led by ARTES in partnership with the European R&D bodies specialised in white and environmental biotechnologies and their application to bast fibres, has produced the following tangible results:

1. Two proprietary non-commercially released microbial strains selected and used to produce an enzyme cocktail with a high capacity to degrade lignin and hemi-

cellulose, but not capable of attacking the cellulose component, which is separated but not modified in its physical and mechanical characteristics
2. Realisation and patenting of a small-scale pilot biotechnology plant (extraction of 5 kg per day of crude fibre) with low power consumption which uses only enzyme retting and mechanical processes to extract fibre (Patent No. 0001396855 entitled 'Machine, procedure and combined plant for the separation of fibres for textile by macerated stems of fibre plant')
3. Conception, engineering and lab testing (300 g, 2.5 kg) of a multipurpose new fast biorefinery plant, which reduces the fibre extraction time eightfold compared with the already-patented plant and produces liquid and solid wastes of particular potential interest for the extraction of hemicelluloses, lignin and other biochemicals (Patent Request Number 102017000009794)
4. Cultivation and mechanised harvesting of *Spartium junceum* successfully tested with parameter definitions of profitability in comparison with hemp and flax
5. Industrial spinning and weaving of the resulting fibres
6. Production of samples of yarn, fabric, garments, paper, bacterial nanocellulose, lignin and hemicellulose
7. Conception and registration of the brand GINEXTRA® (Registration Number 7055312, classes 01, 07, 16, 22, 23 and 40)

For the full exploitation of these potentialities, it is vital to invest in the joint upscaling and integration of fibre extraction and side stream valorisation fully exploiting suitable integration between enzyme applications and engineering processes.

In order to develop these efforts at an industrial marketable scale, further research is required.

The new research will concentrate on the evolution of the current GINEXTRA® biorefinery technology towards a single-step multipurpose zero-waste biorefinery by enhancing the role of enzyme processes and minimising the use of thermochemical and thermomechanical processes.

The planned new research activities aim to:

1. Integrate upstream biorefinery process with downstream technologies to reduce costs (30%) compared to the present ones obtained at the lab scale and reduce costs compared to competing products obtained by petrochemical processes and more traditional materials (minimum 10%).
2. Extend the use of different feedstocks from diverse sources such as agriculture, food industry, forests as well as residues or other biowastes from forest maintenance (current experiments focus on artichoke and cardoon).
3. Tailor and optimise the enzyme mix with the final aim to combine mild thermochemical and thermomechanical technologies with enzyme applications throughout the entire process, from pretreatment until extraction of primitive fibre and side stream processing.
4. Develop a whole value chain business model with demonstrated economic, cultural and environmental sustainability.

The research is now supported by committed industries especially interested in investigating, developing and validating end products obtained from the bio-based fibres and from purified lignin, hemicellulose, cellulose and bacterial nanocellulose in pre-industrial conditions.

This new research effort is feasible thanks to the scientific advances in the biocatalysis technologies (Pellis et al. 2018) which now have the potential to translate into effective innovation.

The major international organisations focusing on environmental policies in developing countries may want to invest in building similar programmes to enhance the bioeconomy, even in the poorest countries. This work, in the field of environmental and white biotechnologies, would aim to foster micro- and small enterprises and to bridge the gap between laboratory research achievements, pilot pre-industrial plant implementation and market uptake of both process and product technologies.

References

Angelini L, Ceccarini L (2003) Improving cultivation of Spartium junceum as new crop for natural fibres for automotive industry. In: Proceeding international South Europe symposium IENICA nonfoodcrops: from agriculture to industry, Bologna

Angelini L, Macchia M, Ceccarini L, Bonari E (1998) Screening of kenaf (Hibiscus cannabinus L.) genotypes for low temperature requirements during germination and evaluation of feasibility of seed production in Italy. Field Crop Res 59:7379

Angelini L, Lazzeri A, Levita G, Fontanelli D, Bozzi C (1999) Physical and mechanical properties of Ramie (Boehmeria nivea Gaud.) and Spanish Broom (Spartium junceum) fibres for fibrereinforced composite materials. In: Proceeding fourth European symposium on industrial crops and products, 2325 March, Bonn, Germany, Edited by FNR Band 14, Gülzow, 787788

Angelini L, Lazzeri A, Levita G, Fontanelli D, Bozzi C (2000) Ramie (Boehmeria nivea Gaud. L.) and Spanish Broom (Spartium junceum L.) fibres for composite materials: agronomical aspects, morphology and mechanical properties. Ind Crop Prod 11(2–3):145–161

Bell J et al (2018) EU ambition to build the world's leading bioeconomy—uncertain times demand innovative and sustainable solutions. New Biotechnol 40(Part A):25–30

Bommarius AS, Bommarius BR (2004) Biocatalysis: fundamentals and applications. Wiley-Blackwell, Hoboken

Cannon RE, Anderson SM (1991) Biogenesis of bacterial cellulose. Crit Rev Microbiol 17:435–447

Ciechanska D (2004) Multifunctional bacterial cellulose/chitosan composite materials for medical applications. Fibres Text East Eur 12(4):69–72

Ciechanska D, Struszczyk H, Kazimierczak J, Guzinska K, Pawlak M, Kozlowska E, Matusiak G, Dutkiewicz M (2002) New electro-acoustic transducers based on modified bacterial cellulose. Fibres Text East Eur 1(36):27–30

Czaja W, Krystynowicz A, Bielecki S, Brown RM Jr (2006) Microbial cellulose—the natural power to heal wounds. Biomaterials 27(2):145–151

Dupont JI, Borg A (2018) Destination bioeconomy—the path towards a smarter, more sustainable future. New Biotechnol 40(Part A):140–143

Florea M, Hagemann H, Santosa G, Abbott J, Micklem CN, Spencer-Milnes X, de Arroyo GL, Paschou D, Lazenbatt C, Kong D, Chughtai H, Jensen K, Freemont PS, Kitney R, Reeve B, Ellis T (2016) Engineering control of bacterial cellulose production using a genetic toolkit and a new cellulose producing strain. PNAS 113(24):E3431–E3440. https://doi.org/10.1073/pnas.1522985113

Bocardo Gerolamo (1880) Trattato teoretico-pratico di economia politica, vol 2. Economia pratica: produzione territoriale, industriale, commerciale; agricoltura, boschi, ... crisi, credito e banche (Classic Reprint)
Global Bioeconomy Summit (2018) Global bioeconomy summit communiqué. In: Innovation in the global bioeconomy for sustainable and inclusive transformation and wellbeing, 19–20 April 2018, Berlin, Germany
Hestrin S, Schramm M (1954) Synthesis of cellulose by *Acetobacter xylinum*. Biochem J 58:345–352
Infelise L. (1998) Le Mans. Diagnosis/development of innovation potential in rural areas throughout innovative networking strategies. A critical appraisal of theoretical works and innovative experiences
Infelise L (2000) "Apprendere e cambiare in una comunità locale" (Learning and change in a local community). In: Amietta P (ed) I luoghi dell'apprendimento—metodi, strumenti e casi di eccellenza delle nuove formazioni. Franco Angeli, Milan
Infelise L (2002) ALLIANCE—an alliance for the development of business and employment in rural areas: a community learning model. In: 42nd congress of the European regional science association, Dortmund
Infelise L (2003) ALLIANCE—an alliance for the development of business and employment in rural areas: a community learning model. In: Regional studies association—reinventing regions in the global economy gateway, 4—rural development and the new rural economy, Pisa, Italy
Infelise L (2006) Territoire anticipateur: «un être vivant » en apprentissage (An in-advance territory: a learning living being) in POUR, Paris, 192
Infelise L (2007) Nuove relazioni tra approcci cognitivi ed economici, nuovi spazi e forme di democrazia partecipativa. Per una pedagogia del cambiamento in aree fragili (New relations between cognitive and economic approaches, new spaces and forms of partecipative democracy. for a pedagogy of changing in fragile areas). In: Del Don C (ed) La crisi dei confini. Verso un'ingegneria dello sviluppo regionale. Franco Angeli, Milano
Kazimierczak J, Wietecha J, Kopania E, Szadkowski M, Infelise L (2016) Utilization of post-maceration liquid from *spartium junceum* enzymatic retting in biosynthesis of bacterial nanocellulose. Fibres Text East Eur 24(6(120)):45–50
Kim SS, Lee SY, Park KJ, Park SM, An HJ, Hyun JM, Choi YH (2017) *Gluconacetobacter sp.* gel_SEA623-2, bacterial cellulose producing bacterium isolated from citrus fruit juice. Saudi J Biol Sci 24:314–319
Kopania E, Wiśniewska-Wrona M, Wietecha J (2014) Galactoglucomannans (GGMs) extracted from spruce sawdust for medical applications. Fibres Text East Eur 22(2(104)):29–34
Kovačević Z, Bischof S, Vujasinović, Fan M (2016) The influence of pre-treatment of Spartium junceum L. fibres on the structure and mechanical properties of PLA biocomposites. Arab J Chem
Krüger A, Schäfers C, Schröder C, Antranikian G (2018) Towards a sustainable bio-based industry—highlighting the impact of extremophiles. New Biotechnol 40:144–156
Pellis A, Cantone S, Ebert C, Gardossi L (2018) Evolving biocatalysis to meet bioeconomy challenges and opportunities. New Biotechnol 40:154–169
Piasecka-Zelga J, Zelga P, Szulc J, Wietecha J, Ciechańska D (2018) An in vivo biocompatibility study of surgical meshes made from bacterial cellulose modified with chitosan. Int J Biol Macromol 116:1119–1127. https://doi.org/10.1016/j.ijbiomac.2018.05.123
Pichat CA (1858) Chapter 725: Institutions of agricultural sciences and technology, collection in Book XX
Puls J , Poutanen K , Körner HU, Viikari L (1985) Biotechnical utilization of wood carbohydrates after steaming pretreatment. Appl Microbiol Biotechnol 22:416–423
Savorgnan D'Osoppo MA (1891) Coltivazione ed industria delle piante tessili. (Manual on the cultivation and industrial exploitation of textile plants). Hoepli, Milano
Shi Z, Zhang Y, Phillips GO, Yang G (2014) Utilization of bacterial cellulose in food. Food Hydrocoll 35:539–545

Wanichapichart P, Kaewnopparat S, Buaking K, Puthai W (2002) Characterization of cellulose membranes produced by Acetobacter xylinum. Songklanakarin J Sci Technol 24, 855(Suppl):–862. Membrane Sci. &Tech
Werpy T, Petersen G (2004) Top value added chemicals from biomass, vol I: results of screening for potential candidates from sugars and synthesis gas. U.S. Department of Energy Report
Yeo SH, Lee OS, Lee IS, Kim HS, Yu TS, Jeong YJ (2004) *Gluconacetobacterpersimmonis* sp. nov., isolated from Korean traditional persimmon vinegar. J Microbiol Biotechnol 14:276–293
Yoshino T, Asakura T, Toda K (1996) Cellulose production by *Acetobacter pasteurianus* on silicone membrane. J Ferment Bioeng 91:32–36

Polyhydroxyalkanoate (PHA) Bioplastics from Organic Waste

Steven Pratt, Luigi-Jules Vandi, Daniel Gapes, Alan Werker, Adrian Oehmen, and Bronwyn Laycock

1 Introduction

The sheer scale of global plastic production (in excess of 300 Mt/year (Plastic Europe (PEMRG) 2016)) raises serious concerns, particularly regarding end-of-life disposal and contamination of waterways: landfill capacity is being challenged, while 10–20 Mt/year of plastic accumulates in the oceans (Gourmelon 2015). These concerns motivate ongoing initiatives towards the development of regional circular economies, with heavy regulation on landfill waste, efficient recycling and recapture/reuse and mandates for the use of bioplastics for products, from shopping bags to automobiles (Kaeb et al. 2016; Nastu 2011). Bioplastics are bio-based (i.e. with formulation ingredients sourced from renewables) and/or biodegradable. Amongst the bioplastics, formulations with polyhydroxyalkanoates (PHAs) are particularly interesting due to their combined reduced environmental impact (since PHA biopolymers are truly biodegradable in soil and marine environments, which is not the case for alternative biopolymers like polylactic acid (PLA)) and potential for high functionality (their mechanical and physical properties are tuneable).

However, despite much anticipation for many years, PHAs are still not making a significant impact in the global market. Broadly speaking, commercial uptake has been restricted by uncertainty of supply, performance and cost. Early experiences with varying batch-to-batch polymer properties alongside limited knowledge base to manage readily solvable challenges for processing and application related to

S. Pratt (✉) · L.-J. Vandi · A. Oehmen · B. Laycock
School of Chemical Engineering, The University of Queensland, St Lucia, QLD, Australia
e-mail: s.pratt@uq.du.au

D. Gapes
Scion, Te Papa Tipu Innovation Park, Rotorua, New Zealand

A. Werker
Promiko AB, Lomma, Sweden

© Springer Nature Switzerland AG 2019

J.-R. Bastidas-Oyanedel, J. E. Schmidt (eds.), *Biorefinery*,
https://doi.org/10.1007/978-3-030-10961-5_26

brittleness and secondary crystallisation require investment developments in the downstream. Many consider PHAs as likely candidates as drop in materials for products and services where traditional fossil polymers are used today. Current high production costs for PHAs generate polymer prices that do not compete against the non-bio-based fossil-derived polymers that feed the plastic industry today. The cost of polypropylene (PP) and polyethylene PE-HD (high density) is reported to be around 1.6 US$/kg and 1.7 US$/kg, respectively (Granta Design Ltd 2017), whereas the cost of PHA as found in various reports and claims, summarised by Vandi et al. (2018), suggests a price in the range of 2–8 US$/kg. Examples include PHA pellets being sold by TianAn Biopolymers at approximately 6.8 US$/kg (TianAn Biopolymer 2012) and Newlight Technologies who claim a production cost in the range of 2–3 US$/kg (Dietrich et al. 2016; Williams 2016). However, for a comparison of the true cost incurred by plastics, one must consider the impact to the environment and associated financial losses. It has been reported that the 10–20 Mt/year of plastic accumulating in the oceans result in an estimated US$13 billion a year in losses from damage to marine ecosystems, including financial losses to fisheries and tourism as well as time spent cleaning beaches (Gourmelon 2015).

Microbial pure culture production typically requires expensive refined substrates and sterilisation of process units, limiting widespread commercialisation as a commodity material (Chanprateep 2010). Currently, commercial-scale pure culture biotechnology utilising refined feedstocks only becomes viable for the production of high-value materials, for example, medical implants. Techno-economic studies have shown that a major cost of pure culture production is the carbon feedstock, estimated to be up to 40% of the product cost (Rehm 2003; Chen 2010; Verlinden et al. 2007). Also, the use of those plant-derived carbon feedstocks upon which they rely can mean both competition with food supply and indirect adverse impacts on natural environments. Consequently many research groups are investigating the potential for using waste organics for PHA production, with the motivation being sustainable waste management in addition to sourcing cheap and sustainable feedstock. The potential to use such waste streams for PHA production along with production strategies has been widely documented and already well reviewed (Chee et al. 2010; Solaiman et al. 2006; Nikodinovic-Runic et al. 2013; Nielsen 2017); details on the properties of the PHAs that are produced are less well documented. The review by Nickodinovic-Runic et al. in 2013 (Nikodinovic-Runic et al. 2013), which considers organic residuals from, for example, food, forest and biodiesel manufacturing (glycerol), is particularly comprehensive. But, in short, the vast majority of studies documented on waste to PHA were at laboratory scale—and 5 years on, little has changed. Nielsen reviewed PHA production from wastes from food production (including whey, plant oils and sugars) in 2017 (Nielsen 2017) and highlighted the many opportunities but presented no examples of commercial operation. Rodriguez-Perez et al. (2018) reviewed the challenges of scaling-up PHA production from waste streams and made a strong case for the potential while quantifying the exponentially growing interest but made no reference to any progress at industrial scale, and only three of the reports were at pilot scale (>100 L). That said, there are grounds for optimism, with now concrete interest in commercialisation in

Europe, the USA and Asia (Williams 2016; Bio-on S.p.A 2013, 2016; Lyons Hardcastle 2016; PlastEurope 2016).

This chapter builds on existing literature, with details of material properties of PHAs from waste organics along with coverage of this recent interest in commercialisation of PHA bioplastics from such wastes—tempered with consideration of the capacity for further advances. Therefore, the specific objectives of this chapter are to:

1. Outline material performance as a function of feedstock and production strategy;
2. Consider the opportunities and challenges for commercial-scale PHA production from organic waste.

The typical strategies for producing PHA from organic wastes are presented for context. Specific details on production strategies and PHA productivity, albeit at lab scale, can be readily found in the literature (Serafim et al. 2008; Lee et al. 1999).

2 Production Strategies

Polyhydroxyalkanoates are produced by many heterotrophic bacteria as intracellular carbon and energy storage molecules. Aerobic environments are the most productive, with oxygen being the electron acceptor for carbon transformations. For a generic organic feedstock (CH_aO_b), and in aerobic conditions, PHA (CH_yO_z) accumulation can be represented as follows (Pratt et al. 2004):

$$\frac{1}{Y_p}CH_aO_b + \left(\frac{\theta_s}{4Y_p} - \frac{\theta_p}{4}\right)O_2 \rightarrow CH_yO_z + \left(1 - \frac{1}{Y_p}\right)CO_2$$

where Y_p is the yield of storage polymer (PHA), θ_s is the degree of reduction of the substrate (the organic feedstock) and θ_p is the degree of reduction of the polymer (PHA). The polymers can be accumulated also under anoxic conditions (Beun et al. 2002), and other studies have shown feasibility for production of PHA with glycogen accumulating organisms which store PHA anaerobically (Bengtsson 2009). Significantly, the natural substrate for PHA accumulation is organic carbon, which is, by definition, the basis of organic waste streams.

The application of open mixed culture biotechnology for PHA production can lower production costs relative to pure culture biotechnology (Serafim et al. 2008; Dias et al. 2006). This is because (1) sterilisation is not necessary and (2) various complex feedstocks, including (cheap or even negative-cost) organic wastes, can be used (Serafim et al. 2008; Dias et al. 2006; Albuquerque et al. 2007; Bengtsson et al. 2008; Dai et al. 2007; Kleerebezem and van Loosdrecht 2007; Reis et al. 2003). Gurieff and Lant (2007) assessed the economics and carbon footprint of mixed culture PHA production using an industrial wastewater as the feedstock and found that

PHA production was preferable to converting the organics to biogas and was financially attractive in comparison to pure culture PHA production, while both PHA production processes had similar carbon footprints that were significantly lower than high-density polyethylene (HDPE) production.

2.1 Feedstock Preparation

Organic waste streams are rarely used directly. Volatile fatty acids (VFAs) are the preferred substrate for PHA accumulation, so most studies on PHA production in open mixed cultures have been carried out using organic acids (such as acetic, propionic, butyric and valeric acids) as the feedstock (Dionisi et al. 2004; Beccari et al. 2002; Serafim et al. 2004; Lemos et al. 2006). Fatty acids are energetically advantageous substrates from a metabolic perspective since their complete β-oxidation generates more chemical energy than the complete oxidation of a molar equivalent of glucose (Solaiman et al. 2006).

These VFAs are generated by fermenting organic wastes, and, as such, feedstock preparation by fermentation is key to utilising organic wastes as a resource for PHA bioplastic production (and for utilisation of organic wastes more broadly). This is discussed further in Sect. 2.2. A wide (and widening) range of wastes have been fermented to generate streams rich in VFA, which have subsequently been used as feedstocks for PHA production as well as for bioenergy generation and as readily degradable carbon to support biological nutrient removal. The production and applications of waste-derived volatile fatty acids for these purposes have been reviewed by Lee et al. (2014). With regard to PHA production, Lee et al. (2014) reported that VFAs generated from palm oil mill wastewater, pulp and paper mill effluent, sugar cane molasses, waste activated sludge and food wastes have been used to produce PHA in open mixed cultures—but this list is certainly not exhaustive. Nielsen (2017) reviewed production on PHA by both open mixed cultures and pure cultures using fermented wastes. Pure culture technology utilising organic waste as the feedstock can be made possible by employment of a membrane for transport of VFA to the PHA production system (Du and Yu 2002a; Hafuka et al. 2011). Table 1 presents some highlights, with commentary additional to the literature included in the reviews by Lee et al. (2014) and Nielsen (2017).

2.2 PHA Production

PHA production on fermented waste is typically a two-step process. Step 1 involves generation of biomass with capacity to accumulate PHA. Step 2 involves accumulation of PHA in the biomass, whereby significant accumulation is trigged by nutrient deficiency. Most often, these steps are carried out in sequence (in some cases, as per pure culture biotechnology, this takes place in a single vessel—though the typical

Table 1 Generation of feedstocks for PHA production from waste streams

Organic waste	VFAs produced from waste	Utility as a feedstock	Reference
Food waste (canteen scraps)	Butyric (27 g/L) and lactic (33 g/L)	Transport of VFAs from mixed culture fermentation through a membrane allowed for pure culture PHA accumulation	Du and Yu (2002b) **Generation of VFA, and subsequently PHA, from food waste is being commercialised by Full Cycle Bioplastics (Full Cycle Bioplastics* 2018*)*
Pulp and paper mill wastewater	Acetic and propionic (total VFA 5–6 g/L)	High fraction of propionic (and valeric) acids, making up about half of the VFAs (on a chemical oxygen demand [COD] basis)—useful for generating HV monomers	Bengtsson et al. (2008)
Waste activated sludge (biosolids)	Acetic and butyric (total VFA 15–20 g/L)	Low degradability of biosolids means pretreatment is required. High-pressure thermal hydrolysis leads to a 2–5 × increase in VFA yield	Morgan-Sagastume et al. (2011)
Sugar cane molasses	(Total VFA 1.2 gC/L)	pH could be used to shift the product spectrum: high pH resulted in acetic and propionic acids, but lower pH resulted in butyric and valeric acids	Albuquerque et al. (2007)

Table 2 Growth and storage yield

Feedstock	Step 1: growth	Step 2: accumulation	Max PHA content
	Growth yield	Storage yield	
Generic (Gurieff 2007)	0.3 ($gCOD_{Biomass}$/ $gCOD_{VFA}$)	0.5 (gPHA/ gCOD)	0.7 (gPHA/ gVSS)
Pulp and paper mill (Bengtsson et al. 2008)	0.3 ($gTSS_{Biomass}$/ $gCOD_{consumed}$)	0.3 (gPHA/ gCOD)	0.48 (gPHA/ gTSS)

approach is to harvest biomass from a growth reactor and then accumulate PHA in a separate vessel). Table 2 shows typical yields for the two steps and highlights that the loss of carbon (as CO_2) is significant, particularly for Step 1: the generation of biomass with capacity to accumulate PHA.

For open mixed culture biotechnology, Step 1 relies on selection pressures to enrich for organisms with capacity to store and later use PHA (Kleerebezem and van Loosdrecht 2007; Daigger and Grady 1982; Salehizadeh and Van Loosdrecht 2004). The main strategy to enrich for PHA-accumulating organisms is to cycle the culture through periods of feast (carbon feedstock is present) and famine (carbon feedstock is absent). This feast-famine approach is also known as aerobic dynamic feeding, or ADF (Beun et al. 2002; Beccari et al. 1998). During feast periods, organisms rapidly take up the external carbon substrates and store them as intracellular storage products. The stored carbon is used for further growth and maintenance during famine periods (Daigger and Grady 1982). Microorganisms that are capable of storing substrates have a strong competitive advantage over those that simply grow when substrate is present and decaying in the absence of external substrates (Van Loosdrecht et al. 1997). Other selection pressures have been applied or implied for mixed cultures, such as the strategic application of selected nutrient limitation such as nitrogen and/or phosphorus (Cavaillé et al. 2016; Valentino et al. 2015). Selective concurrent polymer storage, and active biomass growth, can be sustained, and in this way the PHA-storing phenotype can become dominant in the biomass. Combination of ADF with nutrient restriction has also been applied with the strategy to increase the selection stringency (Silva 2017).

As outlined above, Step 2 then involves accumulation of PHA in the biomass, whereby significant accumulation can be trigged by nutrient deficiency but some presence of nutrients has been suggested to be beneficial (Valentino et al. 2015). When the feedstock is a waste stream, then nutrient (typically nitrogen) deficiency generally cannot be assured. This means it can become difficult to only promote for or sustain exclusive PHA storage activity over active biomass growth as all necessary growth nutrients would always be available (Valentino et al. 2015). This can allow, as indicated, for deliberate simultaneous growth and accumulation (Morgan-Sagastume et al. 2015). Alternatively, to decouple storage from growth, one option is to limit oxygen (Pratt et al. 2012); Satoh et al. (1998) concluded that PHA accumulation in microaerophilic (low DO) environments results in elevated PHA content. Oxygen limitation has been shown to result in slower accumulation kinetics (Pratt et al. 2012). The rates of both growth and PHA generation increase with increased oxygen supply, but the fraction of substrate converted to PHA relative to that which is used for growth is reduced (Third et al. 2003), because PHA generation is less sensitive than biomass growth to dissolved oxygen concentration.

Most of the currently operated mixed microbial cultures (MMC) comprise aerobic organisms, and, therefore, they require sufficient aeration, which can represent a significant fraction of the overall production costs. To mitigate aeration costs, a new MMC PHA producing system has been proposed, using a photosynthetic mixed culture (PMC) composed of a consortium of bacteria and algae that, unlike aerobic MMCs, does not require aeration (Fradinho et al. 2013). Alternatively, it uses light energy as a means of producing ATP and oxygen produced by algae through photosynthesis. The PHA-accumulating PMC can be selected under illuminated conditions in a feast-famine regime, with photosynthetic bacteria accumulating PHA during the feast phase and consuming it during the famine phase using the oxygen produced by algae.

An alternative approach to overcome competition from non-PHA-accumulating organisms and thereby maintain a culture with high capacity to accumulate PHA is to employ PHA-accumulating extremophiles (Koller 2017) and operate in thermophilic or halophilic environments. The application of extremophilic conditions offers significant benefits beyond the potential to maintain a stable population of PHA-accumulating organisms. For example, (1) thermophilic operation could allow the use of cooling water instead of refrigerant for temperature control of exothermic aerobic cultures (Levett et al. 2016), and (2) halophilic operation could allow for hypo-osmotic shock for cell disruption for PHA recovery, thereby reducing the costs of downstream processing (Quillaguamán et al. 2010).

2.3 PHA Composition

A broad range of PHA compositions can be easily produced using mixed cultures based on differing feedstocks, with other monomer units such as 3-hydroxyvalerate (3HV), 4-hydroxybutyrate (4HB), 3-hydroxy-2-methylbutyrate (3H2MB), 3-hydroxy-2-methylvalerate (3H2MV), and 3-hydroxyhexanoate (3-HHx) being common components in P(3HB)-based copolymers (Lemos et al. 2006; Bengtsson et al. 2010; Dai et al. 2008; Takabatake et al. 2000; Lee et al. 2008). For example, P(3HB-*co*-3HV) copolymers with mol% HV ranging from 17 to 85% have been recorded based on the propionate fraction in acetate/propionate feed (Lemos et al. 2006; Takabatake et al. 2000). Pure cultures by contrast often need large amounts of co-substrates to produce polymers with relatively small fractions of monomers other than P(3HB) (see for example (Lee et al. 2008) where P(3HB-*co*-3HV) with only 2–8 mol% HV was produced from mixtures containing equal amounts of vegetable oils and propionic acid (used to produce HV)). This may be due to the fact that mixed cultures contain a diversity of organisms that are likely to employ a range of PHA production pathways. Cellular PHA contents and production rates comparable to or superior to those of pure culture have now been achieved by Johnson et al. (2009, 2010) using mixed cultures, producing P(3HB) at 89 wt% of the dried biomass in less than 8 h.

Given the multiple organisms (and possibly also PHA production pathways) present in a mixed culture environment, one of the key questions will be the compositional distribution of the polymer produced and effect of these blend compositions on chemo-mechanical properties. The outlook is optimistic since a recent pilot production campaign described below has shown that mixed culture processes can be robust in meeting demands of production and product quality.

3 Material Performance

Research and development for PHA from organic waste has long been heavily focused on the enrichment of the PHA producing phenotype and achieving high biomass accumulation capacities. Thus much attention has been paid towards increasing the quantity of PHA in the biomass; systematic investigations towards a fundamental understanding of the resulting polymer quality and processability have been lacking in the literature, yet these are often critical factors in selecting polymeric materials for specific applications (Steinbüchel and Valentin 1995).

3.1 Mechanical Properties

The mechanical properties of PHA from wastes have often been inferred from data on chemical composition, thermal properties and molecular weight (Kim and Lenz 2001; Lenz and Marchessault 2005; Hazer and Steinbüchel 2007; Sudesh 2010; Rehm 2010). But the literature shows that the relationship between mechanical properties and copolymer composition, blend composition, thermal properties, molecular weight, type of polymer processing (e.g. solvent cast versus extrusion) and other characteristics such as microstructure and crystallisation kinetics is complex (Orts et al. 2008; Castilho et al. 2009; Laycock et al. 2014). In addition, micro-block copolymers could be forming when using mixed waste feed based on the preference of the bacterial culture for different components of the mix—for example, Bengtsson et al. (2010) found that in the early stages of each pulse of a mixed VFA feed, propionate was present and PHA rich in 3HV was produced. At the end of the pulse, propionate had all been consumed, and PHA dominated by 3HB was produced. While thermal properties have been characterised, there is little on the mechanical properties of PHA from waste streams. Ultimately, mechanical properties should be directly measured and not inferred. Table 3 below summarises properties for different types of PHAs obtained from wastes. The mechanical properties of PHA produced at pilot scale are considered in Sect. 4.3.

In addition, the possibility exists to manipulate the material properties through adjusting the compositional distribution by alternating or varying the feed composition, resulting in materials with monomer unit distributions that are non-random. Overall, while materials obtained as block copolymeric compounds do exhibit unusual properties, there is inconsistency in results to date, and some detailed structure-property relationships that correlate comonomer sequence distributions with end-use properties need to be developed and detailed characterisation undertaken.

Other aspects that need to be considered include the effect of contaminants and residues in the extracted polymer that may have a large influence on the material properties, such as having a plasticising effect, promoting or inhibiting crystallisation, affecting the extent and rate of secondary crystallisation and decreasing thermal stability.

3.2 Crystallisation

One of the key governing characteristics of PHAs is their degree of crystallinity as well as the rate of crystallisation (which can be extremely slow, depending on the copolymer composition as well as the additives present) and the rate and extent of secondary crystallisation. Pure PHB, for example, has a very high crystallinity (60–70%) and can have very large spherulites if not processed in ways to avoid this occurring. With extended ageing, this crystallinity can further increase due to secondary crystallisation. This means that, in the absence of correct processing and the

Table 3 Mechanical properties of PHBV copolymers obtained from wastes (Laycock et al. 2014)

PHA type	mol% monomer unit	Waste stream	Processing method	Ageing time	Testing	Tensile strength (MPa)	Tensile modulus (GPa)	Strain at break (%)	References
	18 mol% HV 82 mol% HB	Fermented whey permeate waste from cheese production	Solvent cast film	2 weeks	ASTM-D882-02 Film strip (135 mm × 22 mm)	–	1.48	4.3	
	23 mol% HV 35 mol% HB					–	1.76	3.0	
	35 mol% HV 65 mol% HB					–	2.89	3.0	
P(3HB-*co*-3 HV)	43 mol% HV 57 mol% HB					–	0.78	58	Arcos-Hernández et al. (2013)
	54 mol% HV 46 mol% HB					–	0.91	6.0	
	62 mol% HV 38 mol% HB					–	0.87	5.0	
	72 mol% HV 28 mol% HB					–	0.87	5.0	
	0.05 mol% HV 99.95 mol% HB					5.91	–	–	

(continued)

Table 3 (continued)

PHA type	mol% monomer unit	Waste stream	Processing method	Ageing time	Testing	Tensile strength (MPa)	Tensile modulus (GPa)	Strain at break (%)	References
	22.6 mol% HV 77.4 mol% HB	Activated sludge from municipal wastewater	Film making (unspecified)	Unspecified	Film strip (unspecified standard)	4.05	–	–	Tsz-Chun et al. (2005)
P(3HB-*co*-3 HV)	39.2 mol% HV 60.8 mol% HB					3.21	–	–	
	54.1 mol% HV 45.9 mol% HB					2.68	–	–	
	69.1 mol% HV 30.9 mol% HB					1.80	–	–	

PHA type	mol% monomer unit	Waste stream	Processing method	Ageing time	Testing	Tensile strength (MPa)	Tensile modulus (GPa)	Strain at break (%)	References
P(3HB-*co*-3HV)	15.4 mol% HV 84.6 mol% HB	Waste glycerol-based substrates	Solvent casting of PHA pellets in chloroform		Film strips (90 × 20 mm)	12.2 ± 1.1	0.47 ± 0.99	14.2 ± 3.1	Ntaikou et al. (2018)
P(3HB-*co*-3 HV)	14.2 mol% HV 85.8 mol% HB					8.5 ± 1.0	0.46 ± 0.78	9.3 ± 1.2	
P(3HB-*co*-3HHx)	97.5 mol% HB 2.5 mol% HHx					10.3 ± 1.5	2.35 ± 0.18	1.1 ± 0.3	
P(3HB-*co*-3HHx)	94.1 mol% HB 5.9 mol% HHx					6.0 ± 1.0	0.89 ± 0.97	2.2 ± 0.5	

use of appropriate additives, an already stiff and brittle material becomes even more brittle, leading to product failure. The crystallisation rate for PHB is fastest between 80 and 100 °C but slow below 60 °C, whereas for other PHA homo- and copolymers, it is faster below 40 °C.

The crystallisation characteristics of PHAs from waste have been explored in a number of studies, based in the main on their thermal properties as assessed using differential scanning calorimetry, with limited studies to date on X-ray diffraction and/or SAXS/WAXS as a tool to analyse these materials. Of particular note is that there is still a lack of appreciation of the fact that any slow-crystallising components in an as-produced blended product will not crystallise during the first cooling/second reheating cycle during such thermal analyses and, as such, their contribution to the material has often been ignored. Since these components will slowly crystallise over time (another form of secondary crystallisation), this can lead to a distorted understanding of the polymer properties. Of note also is that the assumption that the degree of crystallinity can be approximated from the enthalpy of melt for a PHA copolymer based on the $\varnothing H_{\mathrm{f}}^{\mathrm{o}}$ (the bulk enthalpy of fusion, i.e. the enthalpy of melting of a theoretical 100% crystalline PHB crystal, at 146 J/g) is also problematic, since other 100% PHA homopolymer or pure copolymer crystals will have a different bulk enthalpy of fusion.

3.3 Processing Window

There is a significant amount of information available now in the literature on the thermal (melt) characteristics of PHAs from waste, although this information in the main has not been translated into an understanding of the thermal stability or the rheological properties of these materials. Along with melt temperature and crystallisation temperature and rate, this information is needed in order to optimise the thermal processing of these polymers and to identify the processing window. In extrusion, for example, it is necessary to understand the combination of range of temperatures along the barrel, screw speeds and screw configurations as well as the throughput rate/residence times and initial moisture content in the polymer that can deliver final material properties appropriate for the envisaged applications. PHB has a very narrow processing window, with its melting peak being very close to the temperature at which it thermally degrades, while the addition of HV monomer units, for example, lowers the melt temperature and thus broadens the processing window, since there is improved melt stability at lower processing temperatures. The use of a reverse temperature profile during extrusion of PHBV has demonstrated to minimise thermal degradation (Zhang et al. 2004) and further assist in retaining the full mechanical properties of the polymer. There are an increasing number of studies being reported for extrusion of PHBV polymers, indicating that adequate processability and comparable mechanical properties to polyolefins can be achieved through optimisation of processing parameters.

4 Opportunities and Challenges for PHA Bioplastics from Waste Organics

There is no doubt that organic wastes are potential resources for PHA production. The list of wastes successfully trialled for PHA production will continue to expand. Rodriguez-Perez et al. (2018) recently reviewed the volume of literature on PHA from wastes and found a steady increase since the mid-2000s (now >40 publications per year since 2012).

And the reviews on utilising organic wastes for PHA production are far from exhaustive. After all, organic wastes are simply carbohydrates, fatty acids and proteins, which are the natural (direct and/or indirect) substrates for PHA accumulation.

But to be viable resources, the organic waste must be relatively abundant, and readily degradable. If the raw materials need transportation to the production site, then they should be as concentrated (with as low moisture content) as possible. Solaiman et al. (2006) concluded that acquiring sufficient amounts of actual fermentable substrates is a chief hurdle for industrial PHA production from organic wastes. Ten years on, Rodriguez-Perez et al. (2018) reviewed the challenges to the scale-up of PHA production from waste and concluded that adequate and constant supply will be critical.

Further, for organic wastes to be utilised as feedstocks, the application of some pretreatment technologies will become widespread, to generate the required VFAs but to also manage the presence of non-fermentable components (Solaiman et al. 2006).

In the following sections, the availability of organic wastes and the options for preparing those materials for utilisation as a feedstock for PHA production are considered. Recent interest in commercialising PHA from organic wastes is then outlined.

4.1 *Mass Flows*

Currently, world PHA production capacity is in the order of 100,000 t/a. This is up an order of magnitude over the past 10 years (Chen 2010) and is projected to be up a further order of magnitude in the coming decade. The central players produce in the order of 10–20,000 t/a each. Essentially this is all produced from refined/pure feedstocks such as sugars. It is worth considering the mass of organic waste that would be required to generate 10–20,000 tPHA/a.

A hypothetical mass balance is presented below (Fig. 1). The product stream is set to 10,000 tPHA/a. Assuming biomass PHA content of 70% (see Table 2) and typical PHA accumulation and biomass growth yields (both 0.3 gBiomass/gCOD and 0.5 gPHA/gCOD; see Table 2), this would require in the order of 67,000 t/a of VFA. The yield for feedstock preparation, for readily fermentable wastes, could be assumed to be 0.75 $gCOD_{VFA}$/gCOD (Bengtsson et al. 2008) achieved 0.75 $gCOD_{VFA}$/g

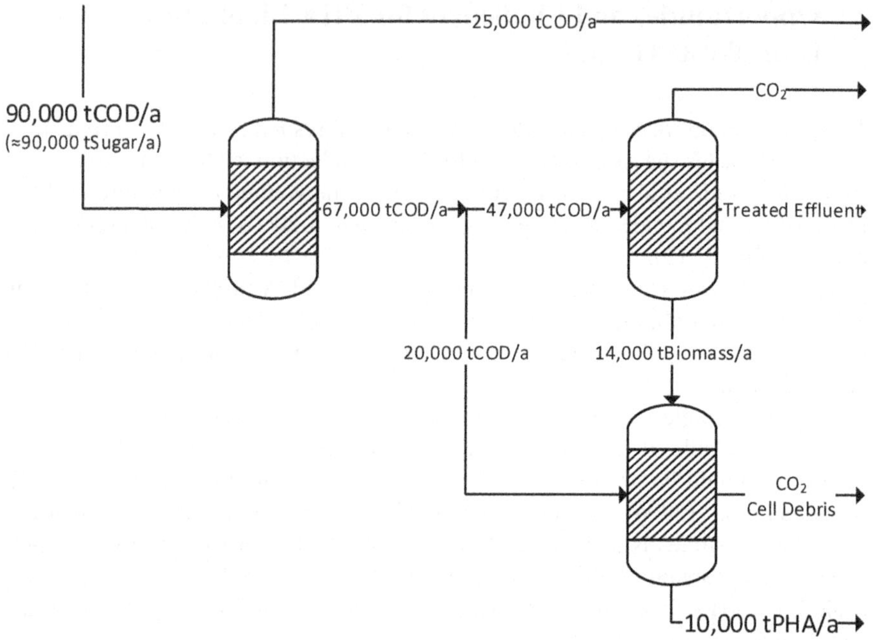

Fig. 1 Material flows to produce an industrially relevant 10,000 tPHA/a (assumed fermentation yield of 0.75; growth yield of 0.3 gVSS/gCOD; accumulation yield of 0.5 gPHA/gCOD)

Table 4 Summary of wastewater streams from Australian industries—estimates from surveys (unpublished data)

	Wine and beer	Dairy	Sugar	Pulp and paper	Biosolids
Volume (GL/y)	13	12	30	16.4	4.5
COD (g/L)	10.5	1.2	70	10	100
COD (t/a)	136,500	144,000	2,100,000	164,000	450,000

COD for pulp and paper wastewater and 0.93 gCOD/gCOD(soluble) for whey). Therefore, 90,000 t/a of organic waste (as carbohydrates) would be required. Such holistic estimations of PHA yield in terms of PHA produced per unit of resource are rarely considered (Rodriguez-Perez et al. 2018).

This is a substantial requirement. For some perspective, mass flows of organic wastewater streams from some relevant Australian industries are presented in Table 4. Only if all the wastes from any given industry were collected would there be sufficient feedstocks for production in the order of 10,000 tPHA/a.

4.2 Pretreatment

Organic waste biomass is not usually present in a form that will maximise its PHA production potential. In particular, the complex macromolecules (cellulose, hemicellulose, lignin, starch, protein) that make up the majority of biomass are often poorly fermentable and benefit significantly from some form of pretreatment, involving the *deconstruction* of the incoming biomass feedstock. In a deconstruction process, biomacromolecules are hydrolysed from their polymeric state and transformed into smaller aqueous phase molecules that are significantly more amenable to subsequent biological or enzymatic processing.

The importance of pretreatment as a unit operation is clear. A review of PHA process developments found that between 2014 and 2016, just over 50% of 36 reviewed laboratory studies and all six pilot studies employed some form of pretreatment prior to PHA fermentation (Rodriguez-Perez et al. 2018). Many of the pretreatment options described for a bioenergy biorefinery have potential application for waste biomass pretreatment for PHA production. Relevant pretreatments include physical, chemical and biochemical approaches. The pool of technologies that target sugars (oligomer, monomer) production from starchy or lignocellulosics range from communition (combined with enzymes or fermentative hydrolysis), sonication, steaming, wet explosion, acid or alkaline hydrolysis, ammonia treatment (AFEX) and ionic liquids. Coverage in the literature is substantial. For example, two complete issues of the journal *Bioresource Technology* (in 2005 (Wyman et al. 2005) and 2016 (Lee et al. 2016)) have been dedicated to biomass pretreatment, with a majority focusing on sugars production (Carrère et al. 2010). In the 2005 Issue, the reported work had been coordinated, allowing comparative analysis of outcomes for different pretreatments on the same feedstock (lignocellulosics in the form of corn stover). For those pretreatments where deconstruction targets sugars production, acidogenic fermentation to VFAs is the next step in a PHA production process (Albuquerque et al. 2011).

4.2.1 Hydrothermal Processing

Amongst the various pretreatment technologies, hydrothermal processing is a group that has emerging utility for PHA production. The hydrothermal processes are characterised by wet processing of organics through the application of heat and pressure, under varying redox, catalyst and pH environments (Möller et al. 2011; Yang et al. 2018; Hii et al. 2014). Hydrothermal processing has potential for transformation of wastes into biochemical feedstocks, including for VFA production (Shanableh 2000; Jin et al. 2005). A wide range of materials can be accepted by these technologies—ranging from wastewater sludges, pulp and paper solid wastes, to municipal organic wastes. Thus, unlike the majority of the sugar platform processes, mixed substrates including proteinaceous and lipid-rich wastes can be included as feedstocks for conversion.

Thermal hydrolysis involves application of heat to an aqueous biomass mixture, to temperatures below 200 °C, for a fixed time. Conventionally applied to improvement of biogas production from biological wastewater treatment plant solids, conditions of 140–175 °C for 30–60 min appear to be the norm (Carrère et al. 2010), with higher temperatures being associated with production of inhibitory components. The mode of action in thermal hydrolysis is deconstruction of biomacromolecules, into a soluble substrate that is more amenable for subsequent fermentations (Morgan-Sagastume et al. 2011; Wilson and Novak 2009). Thermal hydrolysis has been applied as a pretreatment technology within a mixed culture PHA production process (Morgan-Sagastume et al. 2011; Tao et al. 2016).

Subcritical wet oxidation involves processing at temperatures up to the critical point of water (374 °C), under an oxidising environment, usually through the addition of oxygen (Amstel and Rietema 1973; Mishra et al. 1995; Yousefifar et al. 2017). Organic compounds undergo high rates of conversion, with free radical-driven oxidation to CO_2 and water. In addition to complete oxidation, soluble organics are also formed, with acetic acid being the major identified product, this latter reaching yields of 10–20% of the incoming organic load (Jin et al. 2005; Baroutian et al. 2016; Strong and Gapes 2012). In contrast to thermal hydrolysis, wet oxidation has a higher processing rate, results in greater biomass solids destruction and yields greater levels of VFA as a direct product (Strong et al. 2011). Thus, wet oxidation liquor would appear ideally suited as a candidate feedstock for mixed culture PHA production. Unfortunately, while the opportunity has been alluded to (Morgan-Sagastume et al. 2014), currently no published literature exists that provides technical description of PHA production from wet oxidation liquors.

One (significant) disadvantage of conventional wet oxidation is the cost of oxidant addition (usually as oxygen or air) to a high temperature/pressure system. Some recent developments demonstrate promise in obviating this problem, including direct hydrothermal production of lactic acid under non-oxidative conditions (Zhang et al. 2011; Yan et al. 2010) and an innovative application of copper oxide as an alternative oxidant (Wang et al. 2014).

Concluding remarks on pretreatment Biomass pretreatments clearly hold significant promise in benefitting yield and productivity of PHA production from mixed culture fermentations. As is pointed out by Yang and Wyman (2008), pretreatment is pervasive—the implications of its adoption ripple through all prior and subsequent process stages. The challenges that can be anticipated for the application of pretreatments will ultimately revolve around cost—the ongoing cost of pretreatment normalised to the overall process improvement that is accrued via the treatment. Sources of this cost will include (but are not limited to) capital and operating cost, maintainability, reliability at full scale, process scalability, health and safety and complexity. Many of these challenges will only be properly addressed when operating at (or near) commercially relevant scales—this being the ultimate challenge for the technology development as a whole.

4.3 Coupling with Water and Waste Management Services

There is interest for producing PHA from waste streams, motivated by both a desire for circular economies for renewable materials from sustainable waste management practices and for utilising a cheap organic feedstock that does not compete with food crops. It has been demonstrated that biomass involved in biological wastewater treatment can accumulate commercial-grade PHA (Arcos-Hernández et al. 2013). Thus, a service of water quality management can concurrently provide a valued and justifiable function of biomass production (Morgan-Sagastume et al. 2016). Surplus activated sludge may be considered a raw material for a PHA production process. Different methods to enrich for PHA accumulation potential (PAP) in biomass treating municipal and industrial wastewater up to pilot scale (Morgan-Sagastume et al. 2015; Bengtsson et al. 2017a; Tamis et al. 2014; Arcos-Hernández et al. 2015). By exploiting the influent readily biodegradable organic matter (RBCOD), process biomass disposed to feast under anoxic or aerobic conditions become enriched with PAP in the range from 40 to 70% gPHA/gVSS. Pre-fermentation for dominant VFA content of the RBCOD is not a requirement as it was shown that the nature of the process that generates the quality of the feast conditions is perhaps more important than the RBCOD composition (Bengtsson et al. 2017a).

Notwithstanding the breadth of literature examples demonstrating the feasibility of producing biomass from wastewater treatment as a raw material for PHA production (Valentino et al. 2017), there have been no studies until recently that have clearly tackled the question of if a mixed culture process could be reliable with goals of quality of product supply to the downstream. Bio-based value chains of biopolymer production require security of polymer supply both in terms of quantity *and* quality. The Dutch water authorities have been developing insight into value chains for producing polyhydroxyalkanoates (PHAs) as an integral component of regional municipal water quality management. Towards this end, a pilot study (PHARIO) was conducted to establish potential for commercial quality biopolymer production using full-scale Dutch municipal activated sludge from 2015 to 2016 (Bengtsson et al. 2017b). Full-scale municipal activated sludge can be enriched with PHA producing potential by applying and optimising selection principles in the bioprocess. Biological nutrient removal (BNR) treatment plants with predenitrification exert an anoxic feast on the biomass, and this promotes such development of a PHA-storing biomass. To this end a number of Dutch WWTPs were found to already produce surplus biomass with significant enough potential (up to 50% gPHA/gVSS) to be used for demonstrating a PHA production chain. Methods to enhance enrichment for PAP were recognised as being principally related to the nature of the process feast environment. The bacteria that comprise the biomass in the process need to be disposed to at least some well-defined feast periodically. Feast environments can be established and tuned without onerous changes to existing infrastructure.

PHA production potential was measured in the surplus activated sludge from 15 WWTPs, and these in total suggested a potential source of supply of 25,000 tPHA/year. Surplus activated sludge was used from 1 of the 15 wastewater treatment plants (Bath, Brabantse Delta) in pilot-scale production of PHAs over the 10 months of operations. Copolymers of poly-(3 hydroxybutyrate-co-3 hydroxyvalerate) or PHBV were produced in kilogramme batches from 59 production runs (2 per week) with mean 3 HV contents in the range from 0 to 50% by weight. The type of PHBV polymer could be predicted by the type of feedstock and was independent of the activated sludge biomass supplied over the four seasons of pilot operations. The copolymer blends were miscible (single T_g value) and exhibited extraction, melt and mechanical properties that were consistent functions of the mean 3 HV polymer content. A consistently high thermal stability was also observed, and the polymer melt degradation properties were deterministic and independent of polymer type. The polymers were compounded with existing and prototype commercial bioplastic formulations. These bioplastic formulations demonstrated expected modulation of mechanical properties as found for existing commercial products, and even suggested attributes that were superior to the commercial materials that were used for comparison. Polymer average molecular mass was influenced by the methods of materials handling, but PHAs with Mw of 1500 kDa were found to be produced in the activated sludge. Even in the cases found for lower resulting molecular weights due to specific methods of material handling at the time, mechanical properties were not influenced by molecular weight in the range from 200 to 600 kDa. The PHA in the dried biomass (about 3% moisture content) was stable for at least 3 years when stored at room temperature. In summation, the PHARIO project clearly demonstrated for the first time that full-scale surplus municipal activated sludge can be used as a robust and reliable raw material towards regional PHA supply value chains given available sources of fermented organic residual feedstocks.

Since feedstock fermentation product composition determines the polymer type, given an available surplus biomass supply, feasibility for security of supply quantity and quality of polymers depends on securing feedstocks for the PHA accumulation process. Feedstocks can include regional municipal, industrial and agricultural organic residuals which may be also readily available, but the logistics coupled to socio-economic relationships to facilitate supply of VFAs to biomass or biomass supply to VFAs demands a level of coordinated regional commitment. The vision for larger-scale production therefore demands process and stakeholder integration, whereby volatile fatty acids—the feedstocks for PHA production—are generated by fermentation of organic wastes, while functional biomass is generated in and harvested from activated sludge plants that manage municipal and industrial effluent, and recovery of PHA, which requires scales of economy, is to be strategically located to serve a network of PHA production facilities (Arcos-Hernández et al. 2015).

In California, Full Cycle Bioplastics is commercialising PHA production from organic wastes, including food waste and agricultural byproducts, for biomaterial production in conjunction with waste management. Again, key to the technology is fermentation of the waste to fatty acids. Full Cycle Bioplastics then controls the PHA composition by controlling the mix of acids that are fed for PHA production (US20160145659A1).

4.4 *Organic Waste as a 'Conventional' Feedstock*

While additional revenue, through waste to biomaterials and bioenergy, is attractive for waste managers, and is offering potential entry-to-market for waste-to-PHA, there is also commercial interest in considering wastes, particularly sugars, alongside conventional PHA feedstocks. Two notable examples are Bio-on and BluePHA (Kourmentza et al. 2017). In Italy, Bio-on is aiming to utilise sugar beets and wastes from beet processing for production at a scale of 10,000 t/a, and in China, BluePHA is applying halophilic biotechnology so to be able to utilise wastes from corn starch. It is significant to finish with these last two examples as they highlight that organic wastes can refer simply to carbohydrate streams that are consistent with conventional feedstocks and therefore compatible with conventional production systems. That said, in the case of BluePHA, the use of extremophiles avoids the necessity to sterilise those streams.

5 Conclusion

There is no doubt that organic wastes are potential resources for PHA production, and there is now interest in commercialisation of PHA bioplastics from such wastes. After all, organic wastes are simply carbohydrates, fatty acids and proteins, which are the natural (direct and/or indirect) substrates for PHA accumulation. But be viable resources, the organic waste must be relatively abundant, concentrated and readily degradable. As such, it is concluded that (1) wastes should be collected and consolidated in order to ensure sufficient feedstock supply for industrial-scale production and (2) the application of some pretreatment technologies should be applied so to generate the required VFAs but to also manage the presence of non-fermentable components.

Furthermore, while it is known that polyhydroxyalkanoate (PHA) bioplastics can be produced in pure and mixed microbial cultures, knowledge of how material properties are controlled—particularly when using organic wastes as the feedstock—is still lacking. Further research into the relationships between, for example, material performance and copolymer composition, blend composition, molecular weight, microstructure and crystallisation kinetics all govern the mechanical properties, is needed.

References

Albuquerque MGE et al (2007) Strategies for the development of a side stream process for polyhydroxyalkanoate (PHA) production from sugar cane molasses. J Biotechnol 130(4):411–421

Albuquerque MGE et al (2011) Mixed culture polyhydroxyalkanoate (PHA) production from volatile fatty acid (VFA)-rich streams: effect of substrate composition and feeding regime on PHA productivity, composition and properties. J Biotechnol 151(1):66–76

Amstel JJAPV, Rietema K (1973) Wet-air oxidation of sewage sludge. Part II: the oxidation of real sludges. Chem Ing Tech 45(20):1205–1211

Arcos-Hernández MV et al (2013) Physicochemical and mechanical properties of mixed culture polyhydroxyalkanoate (PHBV). Eur Polym J 49(4):904–913

Arcos-Hernández M et al (2015) Value-added bioplastics from services of wastewater treatment. Water Pract Technol 10(3):546–555

Baroutian S et al (2016) Formation and degradation of valuable intermediate products during wet oxidation of municipal sludge. Bioresour Technol 205:280–285

Beccari M et al (1998) A bulking sludge with high storage response selected under intermittent feeding. Water Res 32(11):3403–3413

Beccari M et al (2002) Effect of different carbon sources on aerobic storage by activated sludge. Water Sci Technol 45(6):157–168

Bengtsson S (2009) The utilization of glycogen accumulating organisms for mixed culture production of polyhydroxyalkanoates. Biotechnol Bioeng 104(4):698–708

Bengtsson S et al (2008) Production of polyhydroxyalkanoates by activated sludge treating a paper mill wastewater. Bioresour Technol 99(3):509–516

Bengtsson S et al (2010) Molecular weight and thermal properties of polyhydroxyalkanoates produced from fermented sugar molasses by open mixed cultures. J Biotechnol 147(3):172–179

Bengtsson S et al (2017a) A process for polyhydroxyalkanoate (PHA) production from municipal wastewater treatment with biological carbon and nitrogen removal demonstrated at pilot-scale. New Biotechnol 35:42–53

Bengtsson S et al (2017b) Stepping stone to a value chain for PHA bioplastic using municipal activated sludge. STOWA, The Netherlands

Beun JJ et al (2002) Poly-β-hydroxybutyrate metabolism in dynamically fed mixed microbial cultures. Water Res 36(5):1167–1180

Bio-on S.p.A (2013) FLOS to be world's first company to use revolutionary bioplastic designed by Bio-on. http://www.bio-on.it/project.php?lin=portoghese

Bio-on S.p.A (2016) Bio-on announces new multi-license maxi agreement for revolutionary PHAs bioplastic. http://www.bio-on.it/news.php. Accessed 22 Dec 2016

Carrère H et al (2010) Pretreatment methods to improve sludge anaerobic degradability: a review. J Hazard Mater 183(1):1–15

Castilho LR, Mitchell DA, Freire DMG (2009) Production of polyhydroxyalkanoates (PHAs) from waste materials and by-products by submerged and solid-state fermentation. Bioresour Technol 100(23):5996–6009

Cavaillé L et al (2016) Understanding of polyhydroxybutyrate production under carbon and phosphorus-limited growth conditions in non-axenic continuous culture. Bioresour Technol 201:65–73

Chanprateep S (2010) Current trends in biodegradable polyhydroxyalkanoates. J Biosci Bioeng 110(6):621–632

Chee J-Y et al (2010) Bacterially produced polyhydroxyalkanoate (PHA): converting renewable resources into bioplastics. In: Current research, technology and education topics in applied microbiology and microbial biotechnology, pp 1395–1404

Chen G-Q (2010) Plastics completely synthesized by bacteria: polyhydroxyalkanoates. In: Chen GG-Q (ed) Plastics from bacteria: natural functions and applications. Springer, Berlin, pp 17–37

Dai Y et al (2007) Production of targeted poly(3-hydroxyalkanoates) copolymers by glycogen accumulating organisms using acetate as sole carbon source. J Biotechnol 129(3):489–497

Dai Y et al (2008) Microstructure of copolymers of polyhydroxyalkanoates produced by glycogen accumulating organisms with acetate as the sole carbon source. Process Biochem 43(9):968–977

Daigger GT, Grady CPL (1982) The dynamics of microbial growth on soluble substrates: a unifying theory. Water Res 16(4):365–382
Dias JML et al (2006) Recent advances in polyhydroxyalkanoate production by mixed aerobic cultures: from the substrate to the final product. Macromol Biosci 6(11):885–906
Dietrich K et al (2016) Producing PHAs in the bioeconomy—towards a sustainable bioplastic. Sustain Prod Consum
Dionisi D et al (2004) Biodegradable polymers from organic acids by using activated sludge enriched by aerobic periodic feeding. Biotechnol Bioeng 85(6):569–579
Du G, Yu J (2002a) Metabolic analysis on fatty acid utilization by Pseudomonas oleovorans: mcl-poly(3-hydroxyalkanoates) synthesis versus β-oxidation. Process Biochem 38(3):325–332
Du G, Yu J (2002b) Green technology for conversion of food scraps to biodegradable thermoplastic polyhydroxyalkanoates. Environ Sci Technol 36(24):5511–5516
Fradinho JC et al (2013) Polyhydroxyalkanoates production by a mixed photosynthetic consortium of bacteria and algae. Bioresour Technol 132:146–153
Full Cycle Bioplastics (2018) Full cycle bioplastics. http://fullcyclebioplastics.com/
Gourmelon G (2015) Global plastic production rises, recycling lags—new worldwatch Institute analysis explores trends in global plastic consumption and recycling. http://www.worldwatch.org/global-plastic-production-rises-recycling-lags-0
Granta Design Ltd (2017) Cambridge Engineering Selector CES EduPack
Gurieff N (2007) Production of biodegradable polyhydroxyalkanoate polymers using advanced biological wastewater treatment process technology. School of Chemical Engineering, The University of Queensland, Brisbane, p 160
Gurieff N, Lant P (2007) Comparative life cycle assessment and financial analysis of mixed culture polyhydroxyalkanoate production. Bioresour Technol 98(17):3393–3403
Hafuka A et al (2011) Effect of feeding regimens on polyhydroxybutyrate production from food wastes by Cupriavidus necator. Bioresour Technol 102(3):3551–3553
Hazer B, Steinbüchel A (2007) Increased diversification of polyhydroxyalkanoates by modification reactions for industrial and medical applications. Appl Microbiol Biotechnol 74(1):1–12
Hii K et al (2014) A review of wet air oxidation and thermal hydrolysis technologies in sludge treatment. Bioresour Technol 155:289–299
Jin F et al (2005) Controlling hydrothermal reaction pathways to improve acetic acid production from carbohydrate biomass. Environ Sci Technol 39(6):1893–1902
Johnson K et al (2009) Enrichment of a mixed bacterial culture with a high polyhydroxyalkanoate storage capacity. Biomacromolecules 10(4):670–676
Johnson K, Kleerebezem R, van Loosdrecht MCM (2010) Influence of ammonium on the accumulation of polyhydroxybutyrate (PHB) in aerobic open mixed cultures. J Biotechnol 147(2):73–79
Kaeb H et al (2016) Market study on the consumption of biodegradable and compostable plastic products in Europe 2015 and 2020. Nova-Institute, Hürth, p 6
Kim YB, Lenz RW (2001) Polyesters from microorganisms. In: Babel W, Steinbüchel A (eds) Biopolyesters. Springer, Berlin, pp 51–79
Kleerebezem R, van Loosdrecht MCM (2007) Mixed culture biotechnology for bioenergy production. Curr Opin Biotechnol 18(3):207–212
Koller M (2017) Production of polyhydroxyalkanoate (PHA) biopolyesters by extremophiles. MOJ Polym Sci 1(2)
Kourmentza C et al (2017) Recent advances and challenges towards sustainable polyhydroxyalkanoate (PHA) production. Bioengineering 4(2):55
Laycock B et al (2014) The chemomechanical properties of microbial polyhydroxyalkanoates. Prog Polym Sci 39(2):397–442
Lee SY, Choi J-I, Wong HH (1999) Recent advances in polyhydroxyalkanoate production by bacterial fermentation: mini-review. Int J Biol Macromol 25(1):31–36
Lee W-H et al (2008) Biosynthesis of polyhydroxyalkanoate copolymers from mixtures of plant oils and 3-hydroxyvalerate precursors. Bioresour Technol 99(15):6844–6851

Lee WS et al (2014) A review of the production and applications of waste-derived volatile fatty acids. Chem Eng J 235:83–99

Lee J-S, Saddler J, Binod P (2016) Pretreatment of biomass. Bioresour Technol 199:1

Lemos PC, Serafim LS, Reis MAM (2006) Synthesis of polyhydroxyalkanoates from different short-chain fatty acids by mixed cultures submitted to aerobic dynamic feeding. J Biotechnol 122(2):226–238

Lenz RW, Marchessault RH (2005) Bacterial polyesters: biosynthesis, biodegradable plastics and biotechnology. Biomacromolecules 6(1):1–8

Levett I et al (2016) Techno-economic assessment of poly-3-hydroxybutyrate (PHB) production from methane—the case for thermophilic bioprocessing. J Environ Chem Eng 4(4, Part A):3724–3733

Lyons Hardcastle J (2016) Will Lego's $150 million sustainable plastics challenge make biobased the norm? https://www.environmentalleader.com/2016/06/will-legos-150-million-sustainable-plastics-challenge-make-biobased-the-norm/. Accessed 20 June 2016

Mishra VS, Mahajani VV, Joshi JB (1995) Wet air oxidation. Ind Eng Chem Res 34(1):2–48

Möller M et al (2011) Subcritical water as reaction environment: fundamentals of hydrothermal biomass transformation. ChemSusChem 4(5):566–579

Morgan-Sagastume F et al (2011) Production of volatile fatty acids by fermentation of waste activated sludge pre-treated in full-scale thermal hydrolysis plants. Bioresour Technol 102(3):3089–3097

Morgan-Sagastume F et al (2014) Polyhydroxyalkanoate (PHA) production from sludge and municipal wastewater treatment. Water Sci Technol 69(1):177–184

Morgan-Sagastume F et al (2015) Integrated production of polyhydroxyalkanoates (PHAs) with municipal wastewater and sludge treatment at pilot scale. Bioresour Technol 181:78–89

Morgan-Sagastume F et al (2016) Techno-environmental assessment of integrating polyhydroxyalkanoate (PHA) production with services of municipal wastewater treatment. J Clean Prod 137:1368–1381

Nastu P (2011) Italy carries out plastic bag ban. In: Environmental Leader

Nielsen PH (2017) Microbial biotechnology and circular economy in wastewater treatment. Microb Biotechnol 10(5):1102–1105

Nikodinovic-Runic J et al (2013) Carbon-rich wastes as feedstocks for biodegradable polymer (polyhydroxyalkanoate) production using bacteria. Adv Appl Microbiol 84:139–200

Ntaikou I et al (2018) Comparison of yields and properties of microbial polyhydroxyalkanoates generated from waste glycerol based substrates. Int J Biol Macromol 112:273–283

Orts WJ et al (2008) Poly(hydroxyalkanoates): biorefinery polymers with a whole range of applications. The work of Robert H. Marchessault. Can J Chem 86(6):628–640

PlastEurope (2016) Collaboration with newlight technologies on "AirCarbon" methane-based biopackaging. https://www.plasteurope.com/news/detail.asp?id=233308. Accessed 11 Feb 2016

Plastic Europe (PEMRG) (2016) Plastics—the Facts 2016. An analysis of European latest plastics production, demand and waste data. Plastics Europe

Pratt S, Yuan Z, Keller J (2004) Modeling aerobic carbon oxidation and storage by integrating respirometric, titrimetric, and off-gas CO_2 measurements. Biotechnol Bioeng 88(2):135–147

Pratt S et al (2012) Microaerophilic conditions support elevated mixed culture polyhydroxyalkanoate (PHA) yields, but result in decreased PHA production rates. Water Sci Technol 65(2):243–246

Quillaguamán J et al (2010) Synthesis and production of polyhydroxyalkanoates by halophiles: current potential and future prospects. Appl Microbiol Biotechnol 85(6):1687–1696

Rehm BHA (2003) Polyester synthases: natural catalysts for plastics. Biochem J 376(1):15–33

Rehm BHA (2010) Bacterial polymers: biosynthesis, modifications and applications. Nat Rev Microbiol 8:578

Reis MAM et al (2003) Production of polyhydroxyalkanoates by mixed microbial cultures. Bioprocess Biosyst Eng 25(6):377–385

Rodriguez-Perez S et al (2018) Challenges of scaling-up PHA production from waste streams. A review. J Environ Manag 205:215–230
Salehizadeh H, Van Loosdrecht MCM (2004) Production of polyhydroxyalkanoates by mixed culture: recent trends and biotechnological importance. Biotechnol Adv 22(3):261–279
Satoh H et al (1998) Activated sludge as a possible source of biodegradable plastic. Water Sci Technol 38(2 pt 2):103–109
Serafim LS et al (2004) Optimization of polyhydroxybutyrate production by mixed cultures submitted to aerobic dynamic feeding conditions. Biotechnol Bioeng 87(2):145–160
Serafim LS et al (2008) Strategies for PHA production by mixed cultures and renewable waste materials. Appl Microbiol Biotechnol 81(4):615–628
Shanableh A (2000) Production of useful organic matter from sludge using hydrothermal treatment. Water Res 34(3):945–951
Solaiman DKY et al (2006) Conversion of agricultural feedstock and coproducts into poly(hydroxyalkanoates). Appl Microbiol Biotechnol 71(6):783–789
Steinbüchel A, Valentin HE (1995) Diversity of bacterial polyhydroxyalkanoic acids. FEMS Microbiol Lett 128(3):219–228
Strong PJ, Gapes DJ (2012) Thermal and thermo-chemical pre-treatment of four waste residues and the effect on acetic acid production and methane synthesis. Waste Manag 32(9):1669–1677
Strong PJ, McDonald B, Gapes DJ (2011) Combined thermochemical and fermentative destruction of municipal biosolids: a comparison between thermal hydrolysis and wet oxidative pretreatment. Bioresour Technol 102(9):5520–5527
Sudesh K (2010) Practical guide to microbial polyhydroxyalkanoates. S.R. Technology, Shrewsbury
Takabatake H et al (2000) Recovery of biodegradable plastics from activated sludge process. Water Sci Technol 42(3):351–356
Tamis J et al (2014) Enrichment of Plasticicumulans acidivorans at pilot-scale for PHA production on industrial wastewater. J Biotechnol 192:161–169
Tao B et al (2016) Recovery and concentration of thermally hydrolysed waste activated sludge derived volatile fatty acids and nutrients by microfiltration, electrodialysis and struvite precipitation for polyhydroxyalkanoates production. Chem Eng J 295:11–19
Third KA, Newland M, Cord-Ruwisch R (2003) The effect of dissolved oxygen on PHB accumulation in activated sludge cultures. Biotechnol Bioeng 82(2):238–250
TianAn Biopolymer (2012) TanAn Biopolmer ENMAT company details
Tsz-Chun M et al (2005) Microbial synthesis and characterization of physiochemical properties of polyhydroxyalkanoates (PHAs) produced by bacteria Isolated from activated sludge obtained from the municipal wastewater works in Hong Kong. In: Davison BH et al (eds) Twenty-sixth symposium on biotechnology for fuels and chemicals. Humana Press, Totowa, NJ, pp 731–739
Valentino F et al (2015) Polyhydroxyalkanoate (PHA) storage within a mixed-culture biomass with simultaneous growth as a function of accumulation substrate nitrogen and phosphorus levels. Water Res 77:49–63
Valentino F et al (2017) Carbon recovery from wastewater through bioconversion into biodegradable polymers. New Biotechnol 37:9–23
Van Loosdrecht MCM, Pot MA, Heijnen JJ (1997) Importance of bacterial storage polymers in bioprocesses. Water Sci Technol 35(1):41–47
Vandi LJ et al (2018) Wood-PHA composites: a market analysis. Materials. In: Submission (Special Issue "Green composites: preparation, properties, and applications")
Verlinden RAJ et al (2007) Bacterial synthesis of biodegradable polyhydroxyalkanoates. J Appl Microbiol 102(6):1437–1449
Wang F et al (2014) One-pot hydrothermal conversion of cellulose into organic acids with CuO as an oxidant. Ind Eng Chem Res 53(19):7939–7946
Williams H (2016) IKEA and newlight sign major partnership agreement. https://www.biobasedworldnews.com/huge-investment-as-ikea-newlight-sign-10-billion-pounds-biotechnology-production-licence. Accessed 7 July 2016

Wilson CA, Novak JT (2009) Hydrolysis of macromolecular components of primary and secondary wastewater sludge by thermal hydrolytic pretreatment. Water Res 43(18):4489–4498

Wyman CE et al (2005) Coordinated development of leading biomass pretreatment technologies. Bioresour Technol 96(18):1959–1966

Yan X et al (2010) Hydrothermal conversion of carbohydrate biomass to lactic acid. AICHE J 56(10):2727–2733

Yang B, Wyman CE (2008) Pretreatment: the key to unlocking low-cost cellulosic ethanol. Biofuels Bioprod Biorefin 2(1):26–40

Yang B, Tao L, Wyman CE (2018) Strengths, challenges, and opportunities for hydrothermal pretreatment in lignocellulosic biorefineries. Biofuels Bioprod Biorefin 12(1):125–138

Yousefifar A et al (2017) Fundamental mechanisms and reactions in non-catalytic subcritical hydrothermal processes: a review. Water Res 123:607–622

Zhang J, McCarthy S, Whitehouse R (2004) Reverse temperature injection molding of Biopol™ and effect on its properties. J Appl Polym Sci 94(2):483–491

Zhang S et al (2011) Improvement of lactic acid production from cellulose with the addition of Zn/Ni/C under alkaline hydrothermal conditions. Bioresour Technol 102(2):1998–2003

Techno-economic Analysis for the Production of Novel Bio-derived Elastomers with Modified Algal Proteins as a Reinforcing Agent

Tomasz Bochenski, Wui Yarn Chan, Bradley D. Olsen, and Jens Ejbye Schmidt

1 Introduction

Depletion of fossil fuels (Hook and Tang 2013), climate change (Thornton et al. 2014), uncertainty in energy markets (Ang et al. 2014), and waste disposal issues (Tabata et al. 2016) force us to reconsider our approach to how we utilize the Earth's resources. A traditional petroleum biorefinery separates the fractions of crude oil, which after upgrading form a variety of products such as fuels, plastics, and chemicals. Therefore, sustainable development conducted through shifting the economy from fossil fuels to bio-based resources should consider all of these groups of commodities. Bio-derived plastics (bioplastics) are defined as plastics containing all carbon atoms derived from renewable, natural feedstocks; hence they have increased biodegradability and/or recyclability (Soroudi and Jakubowicz 2013). Currently, the global market is highly dependent on fossil fuel-derived plastics. In 2016, only slightly over four million tons of both biodegradable and bio-based plastics were produced worldwide, which only makes up to about 1% of total global production (300 million tons). Among different types of biomaterials, polyurethanes are the most predominant group of bioplastics, contributing 41.2% of the global market for biomaterials (Bioplastics Market Data 2016). In general, polyurethanes have

T. Bochenski
Department of Chemistry, Khalifa University of Science and Technology, Masdar Campus, Abu Dhabi, United Arab Emirates

W. Y. Chan · B. D. Olsen
Department of Chemical Engineering, Massachusetts Institute of Technology, Cambridge, MA, USA
e-mail: bdolsen@mit.edu

J. E. Schmidt (✉)
SDU-Department of Chemical Engineering, Biotechnology, and Environmental Technology, University of Southern Denmark, Campusvej 55, DK-5230 Odense M, Denmark
e-mail: jschmidt@masdar.ac.ae; jesc@kbm.sdu.dk

© Springer Nature Switzerland AG 2019

J.-R. Bastidas-Oyanedel, J. E. Schmidt (eds.), *Biorefinery*,
https://doi.org/10.1007/978-3-030-10961-5_27

numerous forms and applications, ranging from insulation foams to hard, thermostable plastics used in a variety of commodities. Polyurethanes are synthesized by the condensation of polyols with isocyanates to form urethane bonds between prepolymer segments (Howard 2002). A number of approaches to replace polyols by utilizing plant-derived oils have been demonstrated (Alagi and Hong 2015), including a study reporting on algal oils successfully incorporated in the process (Kumar et al. 2014). Due to the hazardous nature of isocyanates (Baur et al. 1994), a second component of polyurethanes, there has been significant interest in developing non-isocyanate polyurethanes. Some of the examples of alternatives include polyurethanes based on polymerization of cyclic carbonates (Figovsky et al. 2013; Guan et al. 2011) or carbonated oils (Javni et al. 2008; Bahr and Mulhaupt 2012) and amines, which contribute to the hardness of the materials. Recently, hybrid materials comprised of proteins and synthetic rubbery polymers have been proposed as alternatives to these engineering plastics, where bio-derived proteins play the role of reinforcing hard domains (Chan et al. 2017).

In the last decades, large investments were made in the research and development of renewable biofuels. First-generation feedstocks (starch-rich) for producing biofuels are often quoted as unsustainable, whereas second-generation ones (waste and lignocellulose) are often economically and energetically infeasible (Naik et al. 2010). Remarkable interest has been paid into third-generation biofuels—microalgae, which are aquatic organisms that convert sunlight energy and carbon dioxide into lipids, which further can be used as substrates for biodiesel production. This concept has numerous benefits: algae have much higher lipid productivity per area than terrestrial plants (Mata et al. 2010), they can be used in carbon dioxide sequestration and waste water treatment, they have relatively low nutrient requirements and can be grown on variety of nonarable lands, and they are not directly competitive with food (Gouveia 2011). However, microalgal fuels have not been commercialized yet, mainly due to high costs associated with producing algal biodiesel (Bahadar and Bilal Khan 2013). One suggested approach is to incorporate the biorefinery concept into microalgae processing, utilizing other higher-value compounds that are present in the biomass (Brennan and Owende 2010). It has previously been concluded that algal biodiesel derived from *Chlorella vulgaris* lipids can be sold at a market price if the residual soluble proteins' stream can find its customer and be sold for more than $800/ton (Bochenski et al. 2016).

In order to solve both of the issues explained in the previous paragraphs—a need for sustainable and nontoxic replacements for isocyanate-based elastomers and a need for perspective customer for residual algal proteins—we have proposed a novel way to produce strong and tough thermoset materials. The process has been tested out in the laboratory with whey protein (Chan et al. 2017) and residual algal proteins (Bochenski 2017) as feedstock. A two-stage reaction is conducted to obtain the protein copolymers (Fig. 1). In the first step, proteins undergo reaction with methacrylic anhydride at basic conditions to yield methacrylated proteins and methacrylic acid as a by-product. In this step, primary amine groups located in lysine side chain and in N-terminal amino acids are substituted by methacrylamide groups, making the proteins reactive in copolymerization. The degree of methacrylation is

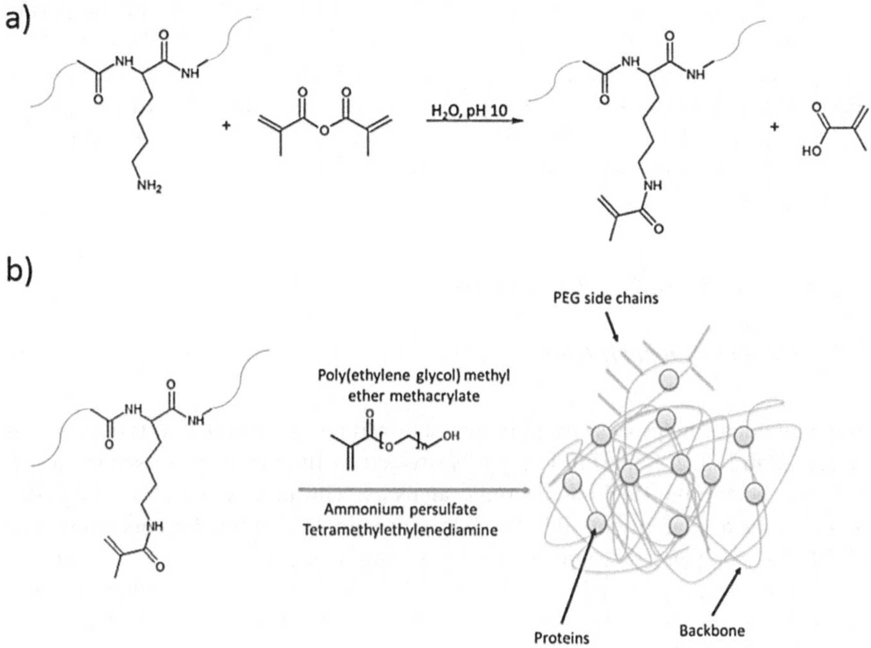

Fig. 1 Two-step reaction for synthesizing polyurethane-like elastomers: (**a**) methacrylation of proteins, (**b**) copolymerization with PEGMA in the presence of AMPS and TEMED

dependent on the ratio of methacrylic anhydride to crude protein. Variation in this step influences the final properties of the product as it determines cross-linking within the polymer. In principle, proteins are responsible for the stiffness and rigidity of the material, whereas poly(ethylene glycol) methyl ether methacrylate (PEGMA) contributes to its elasticity.

In the second reaction, methacrylated proteins copolymerize with comonomer in the presence of ammonium persulfate (APS) as initiator and tetramethylethylenediamine (TEMED) as catalyst to create a cross-linked material consisting of stiff proteins and flexible synthetic polymer chains. The likely random copolymerization leads to randomly distributed hard and soft domains. In the course of this process, a hydrogel is formed, which requires molding and further drying in order to obtain the final product. Specific data on processing parameters were listed in the relevant studies for whey (Chan et al. 2017) and algae (Bochenski 2017) proteins as feedstocks. The first manifestation of this concept resulted in hydrophilic thermoset materials; however, projected further developments are expected to yield products with variety of characteristics. These can be obtained by changing processing parameters, comonomer, or the protein modifications.

Based on the currently available data on the first demonstration of this technology, the objective was to perform technical and economic assessment (TEA) whether the suggested method for production of novel protein-based, polyurethane-inspired elastomers is economically feasible at its current technology readiness level (TRL). The process was modelled using appropriate software (SuperPro Designer by Intelligen©) with adjustment of necessary variables. As the technology is expected to undergo future improvements, this model serves as a framework for these kinds of processes and allows quick economic assessment. In order to investigate model boundaries and influence of specific parameters on the final model performance, sensitivity analysis was also performed.

2 Approach and Assumptions

2.1 Process Description

The manufacturing process for polyurethane-inspired biomaterials was designed as presented in Fig. 2. SuperPro Designer provided by Intelligen© was used in modelling and performing techno-economic analysis. The process has been modelled based on the availability of 40 kg/h of algae protein. With a standard operating time of 7920 h per year, the annual protein requirement for this process is 316.8 t. Assuming 50 wt% protein in dry algal biomass, this input value is within the range of currently operating large-scale open pond systems (Carr 2015). However, with possible improvements in yields and aerial productivity in the future, scaling up of the process was investigated to analyze the extent to which the process feasibility

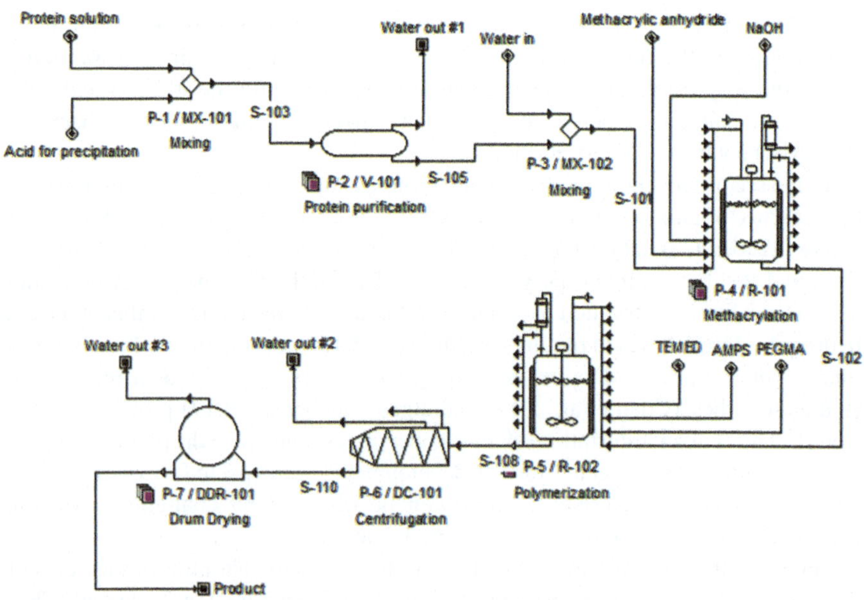

Fig. 2 Process scheme for production of polyurethane-like elastomers derived from algal proteins

can be enhanced. It is assumed that the biomaterial producer has its facility located next to the microalgae processing plant and the feedstock is transferred directly to the facility by piping installation. In other words, the biomaterial producer is considered a customer who is buying the entire liquid fraction that outflows from cell disruption using, for example, high-pressure homogenization. It has been assumed that the input stream consists of 2% proteins, 1% nonprotein water-soluble compounds, and 97% water. In a previous study, it has been calculated that in order for biofuel producers to make a profit, the protein stream should be sold at the price of \$800/ton of crude protein (feasibility of producing biodiesel is then achieved) or \$1000/ton of crude protein (production of both biodiesel and renewable diesel is feasible) (Bahadar and Bilal Khan 2013). In this model, the second price has been used. In the first unit operation, proteins are precipitated using concentrated hydrochloric acid at pH 3. The aim of this step is to remove impurities; however, additional effects might include alterations in accessibility of primary amine groups due to protein denaturation. It is assumed that 50% of the impurities are removed at this stage. This process is carried out in the decanter, and water loss is replaced with the same amount of fresh water. It is assumed that no loss in solubility has occurred, and denatured proteins can still be easily solubilized as a 2% solution. Moreover, only the water that is used for re-solubilizing proteins has been assigned a price. In the next step, methacrylic anhydride is added to the protein solution at 0.5% (v/v)—a proportion which was experimentally found to be optimal. Sodium hydroxide is added to shift the pH to 11, taking into account the effect of hydrochloric acid that has been added in protein precipitation. For simplicity of stoichiometry, it is assumed that methacrylation increases the total weight of proteins by no more than 1%. Methacrylic acid is a side product of the reaction. The retention time of this reaction was set to 4 h to ensure maximum extent of the reaction based on the experimental results. The stream is further transferred into another reactor, where polymerization takes place. Comonomer PEGMA is added at the mass ratio of 4:1 to the methacrylated protein in order to produce a 20% protein copolymer. Ammonium persulfate is added as polymerization initiator in the amount of 2% of the PEGMA mass, and TEMED is added as catalyst at the proportion of 0.2% of the mass of PEGMA. Retention time for this reaction was set to 2 h. In laboratory conditions, both methacrylic anhydride and PEGMA were purchased with inhibitors; here, it is assumed that no pretreatment of either of the two chemicals is taking place. This assumption was made as suppliers of bulk amounts of these compounds did not mention any information about the addition of inhibitors. On the downstream side, water is removed in two stages: first through centrifugation (water content is reduced by 50%) and further through drum drying at 70 °C (water content reduction down to 2%). All other soluble compounds are assumed to be 80% removed through these two procedures. Drum drying has been selected as possibly the most universal unit procedure for this stage of the process. Wet biomass is being dried and subsequently cut off by the knife from internal surface. The process was modelled for manufacturing flat layers of elastomer. Overall, the choice of drying procedure would be dictated, for instance, by the form of the input material and the desired final application of the specimen. Thus, a general scenario was applied for the purposes of this model.

2.2 Assumptions: Substrates, Utility Pricing, and Cost Factors

Prices used in the process were estimated using data from various suppliers of chemicals. For each compound, a mean of at least five prices was used, and these numbers were further compared to their relative prices from Sigma Aldrich to check for consistency. The prices used in this study are as follows: \$3000/ton for PEGMA, \$4200/ton for TEMED, \$8000/ton for methacrylic anhydride, \$800/ton for APS, \$200/m^3 of 32% HCl, and \$300/ton for NaOH. Default values provided by SuperPro were used for prices of electricity (\$0.1/kW h), steam (\$12/MT), and water (\$2.13/MT).

Techno-economic assessment was performed using the Economic Evaluation tool provided by SuperPro. The method uses discounted cash flow analysis for the entire life of the project. Equipment prices were calculated using in-built models; other components of investment costs were estimated based on multiplication factors taken from SuperPro and well-established values found in the literature (Peters and Timmerhaus 1991). The values were as follows: insulation 3%, electrical facilities 10%, auxiliary facilities 40%, engineering/supervision 25% (all from SuperPro Designer database), piping 31%, instrumentation 28%, buildings 22%, yard improvements 10%, and construction 34% (all from Peters and Timmerhaus 1991). Prices of the equipment were calculated using SuperPro's built-in model. Rate of depreciation is calculated as an individual contribution from all undepreciated equipment costs. The inflation rate was estimated as 2%, and the interest rate as 7%. Relevant miscellaneous costs such as insurance, local taxes, and factory expenses were included in the economic evaluation using default SuperPro values.

3 Results and Discussion

3.1 Techno-economic Evaluation for the Base Case Scenario

In this study, a novel method for the production of sustainable biomaterials derived from natural sources has been explored from the perspective of techno-economic investigation of the process. Results have been calculated using the SuperPro Designer economic evaluation tool, and the major items are presented in Table 1. In the base case scenario, the cost of manufacturing 1 ton of material is \$6490. Assigning a profit margin of 7.8% (product price of \$7000/ton) results in an 11.9% return on investment with a payback time of 8.4 years. Total capital investment was estimated to be \$16.2 million, with \$8 million in plant direct costs and \$4.8 million in plant indirect costs. There is a surplus of \$882,000 between revenues and operating costs.

Table 1 Main economic parameters calculated for the process

Executive summary	
Total capital investment	$16,205,000
Total plant direct cost	$8,058,000
Total plant indirect cost	$4,754,000
Direct fixed capital cost	$14,735,000
Operating cost	$11,283,000
Revenues	$12,165,000
Unit production cost per ton	$6490
Unit production revenue per ton	$7000
Return on investment	11.9%
Payback time	8.4 years

Table 2 Distribution of equipment costs

Equipment unit	Size	Total cost
Decanter tank	7.82 m^3	$136,000
Reactor (R-1)	8927.05 L	$874,000
Reactor (R-2)	4837.20 L	$803,000
Centrifuge	2176.74 L/h	$319,000
Drum dryer	9.60 m^2	$152,000
Unlisted equip.		$571,000
Total equipment		$2,855,000

Total capital investment is proportionally related to costs of the equipment. Details about specific unit procedures, sizing, and costs are presented in Table 2. The costs of equipment are relatively equally distributed between unit operations with the highest costs associated with reactors for methacrylation, followed by sedimentation tanks for protein precipitation and reactors for polymerization. Total equipment costs are $2,855,000, a reasonable value, taking into account that the technology itself is relatively simple and does not require advanced or sophisticated apparatus. In terms of volume, the continuously stirred tank reactor used for methacrylation uses the largest volume of almost 9 m^3.

The profitability of this investment is driven by the fact that there is a substantial gap between the revenues and operating costs under the given conditions. Operating costs consist of materials, utilities, labor, facility-dependent costs, and laboratory/QC/QA. The first two—materials and utilities—are calculated based on the quantity and price of a given chemical/utility, labor costs are dependent on complexity of the process and number of unit operations, and the last two—facility and laboratory costs—are calculated by multiplying their respective factors and total equipment costs. Breakdown of operating costs is presented in Table 3.

Almost half of the operating costs are associated with materials costs. In contrast, utilities (steam and electricity) constitute less than 1% of the total operating costs. This can be explained by the fact that the processing pathway does not contain operations which would require adding significant amounts of heat. Therefore, the main costs are associated with mixing in two reactors and drying in the drum dryer.

Materials contribute to a large majority of all operating costs; therefore, fluctuations in prices and required amounts will play the most significant role in determining process feasibility. Figure 3 represents a breakdown of material costs. PEGMA is just one of the possible copolymers to be used in this process. Since the cost contribution related to the price of comonomer is so significant, it would be of great interest in the future to consider lower-cost alternative acrylate and vinyl monomers, as this would possibly decrease processing costs to the greatest extent.

In general, the process of producing elastomers by copolymerization of PEGMA with methacrylated proteins derived from algae biomass is economically feasible. Acceptably short payback time is a result of relatively low investment costs as compared to revenues and operating costs. The surplus of over $882,000 between plant incomes and annual costs allows a reasonably quick return on investment. The scale of the plant is also within a reasonable range with regard to both inlet concentrations and outlet product. Forty kilograms of algal proteins per hour is an amount that can be produced by a large open pond facility. Annual production of 1735 tons of biomaterial is also a reasonable amount as compared to 11.7 million tons of all polyurethanes produced worldwide. It can be claimed that such productivity will not influence the market prices in a significant manner, and as much as the polyurethane market is growing, it is assumed that the lack of demand is not a threat to this project.

Table 3 Detailed description of the process operating costs

Cost unit	Value	Percentage
Materials	$5,233,000	46.38%
Utilities	$101,000	0.90%
Labor	$2,732,000	24.22%
Facility dependent	$2,807,000	24.88%
Laboratory/QC/QA	$410,000	3.63%

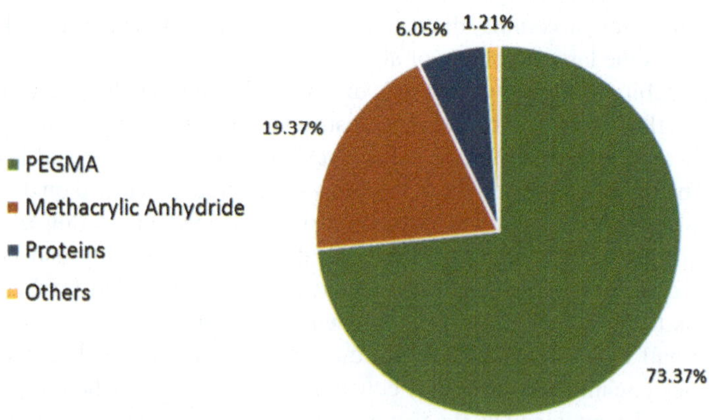

Fig. 3 Contribution of materials costs

3.2 Sensitivity Analysis: Substrate Pricing

Market prices for bulk amounts of substrates used in the process are based on estimates calculated by comparing prices of these chemicals at a laboratory scale with mean bulk prices found through online distributers (May 2017). Sensitivity analysis in this field is crucial in order to account for possible price fluctuations. As the technology develops, a range of other copolymers can be considered instead of PEGMA. Cheaper acrylate monomers such as hydroxypropyl acrylate could possibly be implemented in the same process, leading to a decrease in costs and consequently improving the economic balance. Similarly, if future progress in the field allows incorporation of hydrophobic comonomers, cheaper monomers such as butyl acrylate can be considered. This part of the study was carried out by varying the price of a relevant chemical (with price of other compounds remaining constant) and plotting the values against unit manufacturing cost and payback time (Fig. 4). Purchase costs of PEGMA were previously observed to constitute to over 73% of the costs of raw materials. Several values of prices for PEGMA, methacrylic anhydride, and protein solutions were tested in this study, and results are presented in Fig. 4.

The purchase price of PEGMA is the major cost driver in the production of these materials (Fig. 4a). The manufacturing cost excluding the cost of PEGMA is $4.28/kg, and for each $1/kg increase in price of PEGMA, the manufacturing costs increase linearly by $0.73/kg. A possible decrease in monomer costs to below $3/kg can accommodate much larger fluctuations in price of a final product (Fig. 4b). The range of prices of polyurethane elastomers is relatively wide and depends on many factors, such as the physical characteristics, application, and position in the value chain. Therefore, in this study the emphasis is put on the production costs and minimum price that can be assigned to the product so that the process is economically feasible and payback time is no greater than 10 years. Meanwhile, fluctuations in pricing for proteins (Fig. 4c) and methacrylic anhydride (Fig. 4e) have minor influence on unit manufacturing cost and hence on the process economics. These relationships are also reflected in the investment payback time (Fig. 4d, f) under conditions established in the base case scenario. It is crucial to emphasize that final product price can be lowered by blending the biomaterial with fillers or clays, which is a common practice in the production of polyurethanes (Correa Altafim et al. 2003).

3.3 Sensitivity Analysis: Processing Techniques

The second part of sensitivity analysis involves modifications to the process in order to establish the extent to which this manufacturing technology is economically feasible. Results of economic evaluation are shown in Table 4. Three scenarios have been applied and compared to the base case scenario. In the first one, polymerization is performed with heating (70 °C) as catalyst instead of TEMED. It has been

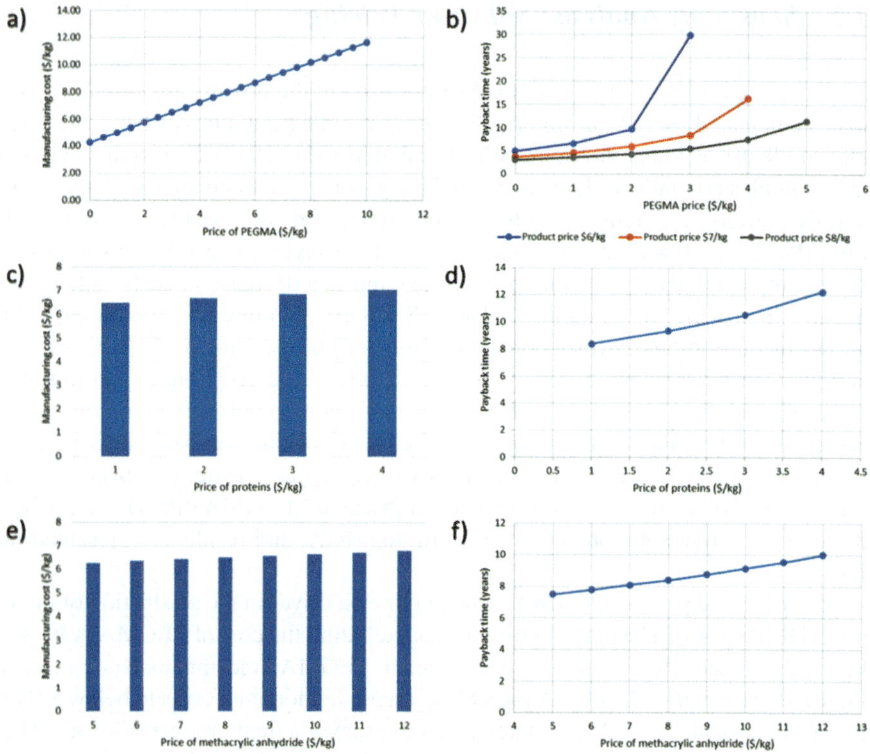

Fig. 4 Sensitivity analysis with regard to prices of substrates. Manufacturing costs and payback time for PEGMA (**a**, **b**), proteins (**c**, **d**), and methacrylic anhydride (**e**, **f**)

shown in laboratory experiments that heat allows even faster polymerization. Therefore, retention time in this scenario has been changed from 2 to 1 h to capture the competitive advantage of the more efficient method. Other process parameters remain the same as in base case operations. In the second scenario, retention time of the polymerization is extended from 2 to 10 h to evaluate the extent to which increasing this parameter influences the final process economics. In the third scenario, both reactions—methacrylation and polymerization—have the retention time increased to 10 h. To assure fair comparison, all scenarios are based on the same input and output flow rates.

Substitution of TEMED with heat catalysis saves almost $0.5M of investment cost as compared to the base case project. Shorter retention time for polymerization allows downsizing of the reactors, decreasing equipment costs and related costs that are calculated as a percentage of these. However, the unit costs for the reactor are higher due to extra heating coatings that have to be purchased in order to generate heat. Increasing retention times of the two reactions significantly increases project investment costs; however, the return on investment is still reasonably high and allows payback times of 9.3 and 9.8 years, respectively.

Table 4 Executive summary of techno-economic evaluation for different processing scenarios

	Base case scenario	Heating and no TEMED	Polymerization RT (10 h)	Both reactions RT (10 h)
Total capital investment	\$16,205,000	\$15,718,000	\$17,639,000	\$18,460,000
Total plant direct cost	\$8,058,000	\$7,805,000	\$8,805,000	\$9,233,000
Total plant indirect cost	\$4,754,000	\$4,605,000	\$5,195,000	\$5,447,000
Direct fixed capital cost	\$14,735,000	\$14,271,000	\$16,101,000	\$16,882,000
Operating cost	\$11,283,000	\$11,199,000	\$11,544,000	\$11,693,000
Revenues	\$12,165,000	\$12,165,000	\$12,165,000	\$12,165,000
Unit production cost per kg	\$6.49	\$6.44	\$6.64	\$6.73
Return on inv.	11.90%	12.31%	10.78%	10.22%
Payback time	8.4 years	8.12 years	9.27 years	9.78 years

3.4 *Sensitivity Analysis: Product Formulation*

Laboratory work on this type of material has shown a high capacity for variation in terms of product formulation. While the relationships between product formulation and performance which will ultimately govern sales price are still being discovered, it is clear that this formulation latitude will be critical to maximizing commercial value of the resulting products. Therefore, four new scenarios have been tested to investigate how changing stoichiometry of reactions and concentrations influences the overall economics of the process. The relevant economic indicators are presented in Table 5. The first scenario considers the weight of the methacrylated protein stream to be equal to the weight of unmodified proteins. In other words, it is assumed that the mass gain by protein modification is so small that it can be omitted in calculations. This scenario is included in the study to test the influence of minor weight fluctuations on the process economics. In order to keep the 20:80 protein-to-monomer ratio, there is a slight decrease in the amount of PEGMA. In the second scenario, the protein-to-monomer ratio is increased to 30:70. It was previously observed that such a formulation also yields materials with relevant properties. This scenario was performed by adjusting the final product stream flow to the base case scenario—to keep the amount of final product constant. The advantage of such an approach is a reduction of the PEGMA input flow rate (which was previously found to be significant factor due to price). On the other hand, equipment costs are increased due to higher inflow of proteins in water solution. In the third scenario, protein concentration in the input stream was changed from 2 to 5%. The major outcome of this modification is the decrease in equipment costs due to lower water input. Finally, in the fourth scenario, the inflow rate of methacrylic acid is decreased by 25%, as the previous experimental work (Bochenski 2017) has indicated that both formulations can be applied to produce polymers with similar properties.

Table 5 Executive summary for process of manufacturing biomaterials using different formulations of substrates and final product compositions

	Base case scenario	No weight gain by M.A.	Protein to monomer ratio	Protein concentration	M.A. levels
Total capital investment	$16,205,000	$16,205,000	$17,362,000	$13,914,000	$16,182,000
Total plant direct cost	$8,058,000	$8,058,000	$8,650,000	$6,868,000	$8,058,000
Total plant indirect cost	$4,754,000	$4,754,000	$5,103,000	$4,052,000	$4,754,000
Direct fixed capital cost	$14,735,000	$14,735,000	$15,816,000	$12,559,000	$14,735,000
Operating cost	$11,283,000	$11,283,000	$11,730,000	$10,798,000	$11,030,000
Revenues	$12,165,000	$12,143,000	$12,648,000	$11,965,000	$12,054,000
Unit product cost per kg	$6.49	$6.50	$6.49	$6.32	$6.41
Return on investment	11.90%	11.82%	11.83%	13.61%	12.45%
Payback time	8.40 years	8.46 years	8.46 years	7.35 years	8.03 years

This last part of the sensitivity analysis again shows that significant changes in process parameters have relatively limited influence on the economic balance of the process. Modifications in stoichiometry and mass gain related to methacrylic anhydride have the lowest (almost negligible) impact on investment payback time, which again is a result of lesser contribution of this chemical in the course of the process. Changes in the ratio of protein to monomer from 20:80 to 30:70 result in an increased payback time (8.46 years), which clearly indicates that costs associated with upstream processing of proteins, as well as those resulting from increased reactor sizing due to dilution of protein solution, significantly surpass those related to the purchase costs of PEGMA. The highest savings on the process costs are achieved by increasing protein concentration in the feed from 2 to 5%. Protein solubility depends on specific composition and sizes of peptides in solution and hence varies between different protein sources. Consequently, improving protein solubility is highly beneficial from the perspective of this process and can lead to decreasing payback time to below 7.5 years.

These advancements, however, require additional investments in research and development (R&D). Chemical companies generally spend 5% of their sales on R&D as of 2017 (Percentage of global research 2017), with the major 17 players on the market using no more than 3.5% per year in the period 2006–2016 (Reisch 2016). Assuming 5% of the sales revenues to be reinvested in R&D, the additional cost to this project is equal to $0.6M, which still allows the economic balance to be positive. Based on this knowledge, it is apparent that future work on decreasing production costs should be primarily focused on developing materials made of the lowest-cost monomers and proteins with high water solubility.

3.5 Process Upscaling Analysis

The analyzed process is relatively small, as the given input (40 kg/h) is close to the maximum of what currently operating large-scale open pond algae plants can supply. However, with the advancement of technologies, such as implementation of photo-bioreactors, which allow higher yields per area, the availability of proteins is expected to increase in the near future. Moreover, manufacturing of these polyurethane-inspired biomaterials can accommodate a variety of proteins from different sources. At the current stage of development, this might be a viable option to increase the process feasibility.

Scaling up the process can allow the prices of the final product to decrease without compromising the return on investment. The payback rate of this project falls below 10 years if the protein availability reaches 100 kg/h at a final product price of at least $5/kg and falls below 5 years at a protein inflow rate of 150 kg/h (Fig. 5a). Moreover, the unit manufacturing cost is significantly lower when the protein availability exceeds 100 kg/h (Fig. 5b). These findings show clearly that one of the most crucial challenges of the technology is to assure a higher protein input than is currently available using existing open pond algae manufacturing.

Diverse sensitivity analysis shows that there is a high potential of this process to be economically feasible. This technology is neither very complex in terms of number of unit operations nor does it consume large amounts of energy. Two major cost drivers of this process are the price of copolymer (which is understandable as this chemical constitutes to 80% weight of the final product) and the protein solubility. It is concluded that the feasibility of this manufacturing technology will be highly dependent on the market price of comonomer and the ability to work with lower-cost monomers could have a significant positive impact on the process. In fact, future developments and a rise in interest in this type of sustainable polymer might lead to a significant decrease in the price of components, as their market demand will be higher. Higher protein concentration in the field is desired in order to decrease costs associated with drying.

Production of biomaterials using this technique displays advantages also with regard to energy consumption. Traditional processes for polyurethane manufacturing are carried out through polymerization reaction between polyols, short diols, and diisocyanates. Both this traditional method and the method suggested here follow a similar production pathway. Components are first mixed in the reactor and then transferred to the reaction chamber where high temperatures (over 100 °C) are applied (Heintz et al. 2003). The resulting material then undergoes similar drying procedure and molding. The major advantage of the new technology is the fact that polymerization occurs in the room temperature, which allows significant savings on energy—more than 25% less steam is used as calculated using SuperPro Designer tool for energy assessment. Additionally, the process does not involve toxic chemicals (such as isocyanates); therefore the costs of process apparatus can also be further decreased. The energy consumption is significantly higher in the downstream processing, with the entire steam being used to fuel the drum drying process and

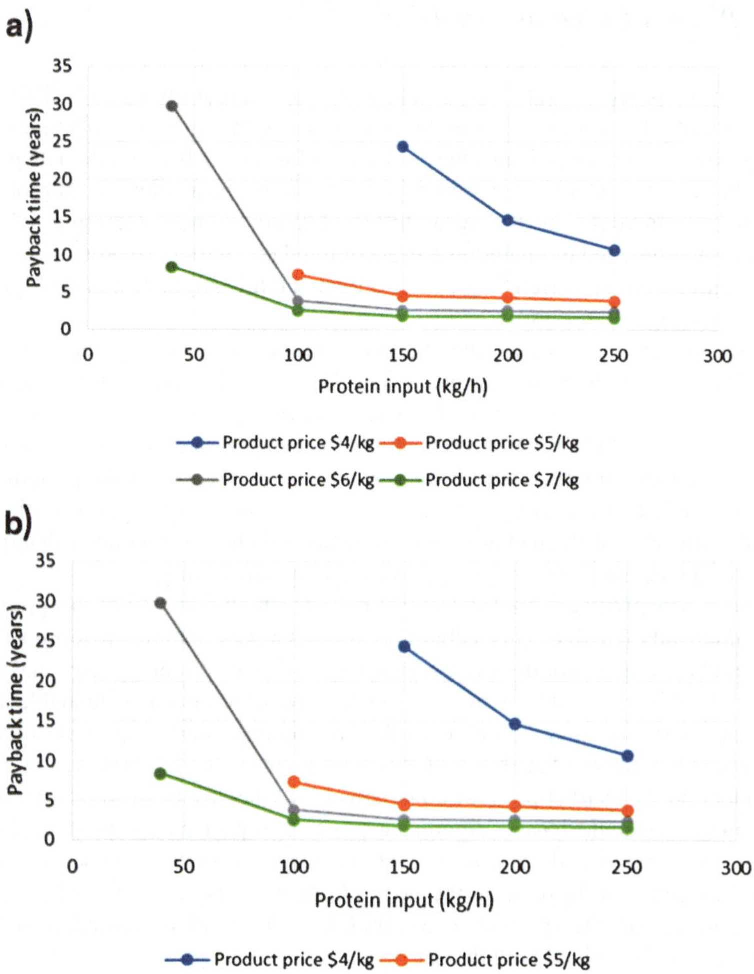

Fig. 5 Influence of protein input on payback time with regard to product price (**a**) and on unit manufacturing cost (**b**)

78% of electricity being utilized in centrifugation. Two percent of electricity is used in drying, and remaining 20% is divided between reactors and piping installation.

Beyond providing positive feedback on process economics for producing novel protein-based elastomers using the suggested method, this study also shows that the method allows valorization of waste algal proteins and hence contributes to economic biodiesel production. If the given process is incorporated in the algae processing facility, it would allow selling biodiesel at a competitive price (with regard to fossil fuels), as the intermediate price of proteins does not influence the economics of producing elastomers using the method explored in this study. Results of this study show an alternative approach on how to apply the biorefinery concept to algal biomass—instead of searching for high-value compounds, which are usually costly to purify and delicate in handling, a mix of bulk proteins is utilized as smaller-volume yet crucial substrate for another chemical process. Variability of acceptable transfer prices for proteins is a result of lesser weight contribution of proteins in the final product—a polyurethane-like biomaterial. That is particularly important if the microalgae biorefinery is examined not through the decomposition approach, but as a factory that is performing all operations on-site, from algae cultivation to final product development. Surplus revenues achieved from selling biomaterials can successfully be used to cover expenses of producing biofuels and therefore justify the concept of the microalgae biorefinery.

4 Conclusions

In this study, a techno-economic analysis of producing novel elastomers from algal protein biomass was conducted. Sensitivity analysis shows that there is a high potential of this process to be economically feasible. Protein input rate, price of comonomer, and protein solubility were found to be major cost drivers of the process, and future developments should ideally target the two issues. Additional value is created through utilization of waste proteins, as well as possible replacement of toxic isocyanates. Utilization of algal proteins for manufacturing these materials is also beneficial from the perspective of justification of producing algal biofuels. The crucial factor driving the profitability of this project is its scale. Unit manufacturing cost decreases significantly with increasing protein input rate; however, additional improvements in algal productivity yields have to be achieved in order to satisfy this goal.

Acknowledgments This work was funded by the Cooperative Agreement between the Masdar Institute of Science and Technology (Masdar Institute), Abu Dhabi, UAE, and the Massachusetts Institute of Technology (MIT), Cambridge, MA, USA—Reference BIOREFINERY 02/MI/MI/CP/11/07633/GEN/G/00 for work under the Second Five-Year Agreement.

References

Alagi P, Hong SC (2015) Vegetable oil-based polyols for sustainable polyurethanes. Macromol Res 23(12):1079–1086

Ang BW, Choong WL, Ng TS (2014) Energy security: definitions, dimensions and indexes. Renew Sust Energ Rev 42:1077–1093

Bahadar A, Bilal Khan M (2013) Progress in energy from microalgae: a review. Renew Sust Energ Rev 27:128–148

Bahr M, Mulhaupt R (2012) Linseed and soybean oil-based polyurethanes prepared via the non-isocyanate route and catalytic carbon dioxide conversion. Green Chem 14(2):483–489

Baur X, Marek W, Ammon J, Czuppon AB, Marczynski B, Raulfheimsoth M, Roemmelt H, Fruhmann G (1994) Respiratory and other hazards of isocyanates. Int Arch Occup Environ Health 66(3):141–152

Bioplastics Market Data (2016) European bioplastics, Nova-Institute. http://www.european-bioplastics.org/market/. Accessed 30 June 2017
Bochenski T (2017) Techno-economic analysis of microalgae biorefinery in arid environments for co-production of biofuels and value-added compounds. Unpublished doctoral thesis, Masdar Institute of Science and Technology, Abu Dhabi, United Arab Emirates
Bochenski T, Torres AI, Ashraf MT, Schmidt JE, Stephanopoulos G (2016) Evaluation of the production of lipids for fuels and proteins from microalgae using a two-level Lagrangian decomposition. Comput Aided Chem Eng 38:1635–1640
Brennan L, Owende P (2010) Biofuels from microalgae—a review of technologies for production, processing, and extractions of biofuels and co-products. Renew Sust Energ Rev 14:557–577
Carr M (2015) Algae industry project book, Algae Biomass Organization. http://algaebiomass.org/wp-content/gallery/2012-algae-biomass-summit/2015/07/ABO_project_book_July2015.pdf
Chan WY, Bochenski T, Schmidt JE, Olsen BD (2017) Peptide domains as reinforcement in protein-based elastomers. ACS Sustain Chem Eng 5(10):8568–8578
Correa Altafim RA, Murakami CR, Neto SC, Ribeiro Araujo LC, Chierice GO (2003) The effects of fillers on polyurethane resin-based electrical insulators. Mater Res 6(2):187–191
Figovsky O, Shapovalov L, Leykin A, Birukova O, Potashnikova R (2013) Recent advances in the development of non-isocyanate polyurethanes based on cyclic carbonates. PU Mag Int 10(4):256–263
Gouveia L (2011) Microalgae as a feedstock for biofuels. In: Springer briefs in microbiology
Guan J, Song Y, Lin Y, Yin X, Zuo M, Zhao Y, Tao X, Zheng Q (2011) Progress in study of non-isocyanate polyurethane. Ind Eng Chem Res 50(11):6517–6527
Heintz AM, Dffy DJ, Hsu SL (2003) Effects of reaction temperature on the formation of polyurethane prepolymer structures. Macromolecules 36:2695–2704
Hook M, Tang X (2013) Depletion of fossil fuels and anthropogenic climate change—a review. Energy Policy 52:797–809
Howard GT (2002) Biodegradation of polyurethane: a review. Int Biodeterior Biodegrad 49(4):245–252
Javni I, Hong DP, Petrovic ZS (2008) Soy-based polyurethanes by nonisocyanate route. J Appl Polym Sci 108(6):3867–3875
Kumar S, Hablot E, Garcia Moscoso JL, Obeid W, Hatcher PG, DuQuette BM, Graiver D, Narayan R, Balan V (2014) Polyurethanes preparation using proteins obtained from microalgae. J Mater Sci 49(22):7824–7833
Mata TM, Martins AA, Caetano NS (2010) Microalgae for biodiesel production and other applications: a review. Renew Sust Energ Rev 14:217–232
Naik SN, Goud VV, Rout PK, Dalai AK (2010) Production of first and second generation biofuels: a comprehensive review. Renew Sust Energ Rev 14(2):578–597
Percentage of global research and development spending in 2017, by industry, Statista—the statistics portal. https://www.statista.com/statistics/270233/percentage-of-global-rundd-spending-by-industry/. Accessed 25 Feb 2018
Peters MS, Timmerhaus KD (1991) Plant design and economics for chemical engineers, 4th edn. McGraw-Hill, New York
Reisch MS (2016) Chemical makers hold research spending steady, but capital spending will stall in 2016. Chem Eng News 94(16):18–20
Soroudi A, Jakubowicz I (2013) Recycling of bioplastics, their blends and biocomposites: a review. Eur Polym J 49:2839–2858
Tabata T, Zhang O, Yamanaka Y, Tsai P (2016) Estimating potential disaster waste generation for pre-disaster waste management. Clean Techn Environ Policy 18:1735–1744
Thornton PK, Ericksen PJ, Herrero M, Challinor AJ (2014) Climate variability and vulnerability to climate change: a review. Glob Chang Biol 20:3313–3328

Part VII
Seawater/Saltwater-based Biorefineries

Considerations for Producing Bioenergy from Halophyte Feedstocks

J. Jed Brown

1 Background

One of the perennial criticisms of producing biofuels as renewable energy sources from conventional crops is that it displaces or competes with food production by using arable land and high-quality freshwater that should or could be used to produce food for humans or for animal production (e.g., Chakravorty et al. 2009). This criticism usually applies to first-generation biofuels, which are produced from widely grown crops such as corn, sugarcane, or soybeans. The food versus fuel debate has been reviewed extensively in both the scientific and popular literature, but the criticisms of these first-generation biofuels have led to the development of alternative schemes for producing biofuel.

One potential way around the food versus fuel conundrum would be to grow biofuels feedstocks on land or using irrigation water that is not suitable for conventional crops. These are the so-called second-generation biofuels—nonfood crops produced on nonarable land (e.g., Carriquiry et al. 2011). More specifically, one way forward would be to consider the production of feedstocks using water or land that is too saline to support the growth of conventional crops/feedstocks. If species could be identified to produce biomass under such conditions, then perhaps biofuel could be produced in areas where conventional feedstocks would not grow and thereby not compete with food production.

Halophytes, or salt-tolerant plants, are species that have evolved to grow and complete their life cycles under high-salinity environments that would kill or severely retard the growth of conventional plants (glycophytes). One commonly used definition of a halophyte is a plant that is able "to complete the life cycle in a salt concentration of at least 200 mM NaCI under conditions similar to those that might be encountered in the natural environment" (Flowers et al. 1986). Two

J. J. Brown (✉)
Center for Sustainable Development, College of Arts and Sciences, Qatar University, Doha, Qatar

© Springer Nature Switzerland AG 2019

J.-R. Bastidas-Oyanedel, J. E. Schmidt (eds.), *Biorefinery*,
https://doi.org/10.1007/978-3-030-10961-5_28

hundred millimolar is roughly one-third of the salinity of typical full-strength seawater. (Note: Salinity units in this paper are expressed in the same format that they were reported in, e.g., 1 g L^{-1} TDS, the weight of total dissolved salts to volume of water equals 1 part per thousand or ppt, which is approximately equal to 1.8 dS m^{-1} by electrical conductivity or a 17 mM NaCl solution.) By contrast, the most salt-tolerant conventional crops like barley can only tolerate less than 10% of the salinity of seawater without yield reduction (NSW Department of Primary Industries 2017), and more typical conventional crops are very sensitive to salt in the irrigation water, and yields will generally begin to decrease with irrigation water conductivities of 1–3 dsS/m (Shannon and Grieve 1998) or approximately 1/35 the salinity of seawater. Figure 1 presents potential halophyte candidates for biofuel production.

2 Saline Soils and Water

There certainly is adequate land and water to grow halophytes. The amount of salt-affected land has been estimated as high as 1128 Mha (Wicke et al. 2011) or slightly greater than the total land area of Canada, the world's second largest country by area. Of this land area, Wicke et al. (2011) estimated that the majority of the soils were slightly affected by salinity (65%), followed by 20% moderately, 10% extremely, and 5% highly salt-affected. They also estimated that approximately 56 exajoules of energy could be produced annually from biosaline biomass grown on this land, which is about half of the energy consumed yearly by the United States. Looking more specifically at salinized irrigated farm lands, FAO estimated that worldwide, 34 Mha (11% of the irrigated area) is affected by some level of salinity, with Pakistan, China, the United States, and India representing more than 60% of the total, and that an additional 60–80 Mha is affected to some extent by waterlogging and related salinity issues (FAO 2011). With respect to the salinity of the world's water, 97.5% is salt water and 2.5% is fresh water, but of this freshwater, only 0.3% is liquid surface water. So whether you consider the total salinized

Fig. 1 Potential halophyte biofuel candidate species. (**a**) mangrove species *Avicennia marina* growing in natural habitat. (**b**) *Salicornia bigelovii* growing in a greenhouse

land area or only salinized irrigated land, there are huge areas of land where halophytes could be cultivated, as well as huge volumes of seawater or saline ground water that could be utilized.

3 Halophytes

Halophytes are found worldwide in various climates and in both coastal (e.g., salt marshes) and inland areas (e.g., saline seeps). Halophytes are members of a variety of plant families and can vary in their salt tolerance from slightly salt-tolerant to highly salt-tolerant and capable of growing in salinities of full-strength seawater or greater. If any halophyte species could be domesticated, then perhaps there is a possibility of producing a biofuel crop on saline water or in saline soils.

Broadly, halophytes can be divided into two groups—the grasses (monocots) and the dicots. The dicots tend to incorporate salt into their tissues to maintain osmotic potential and therefore tend to be succulent and can be very salt-tolerant. The grasses tend to exclude salt from their tissue and are generally not as salt-tolerant as the dicots. Flowers et al. (1986) noted that the dicot halophytes tend to grow optimally or have their growth stimulated by salinities in the range of 50–250 mM NaCl. By contrast, the monocots tend to be inhibited by salinity or might be stimulated by low salinities around 50 mM NaCl (Flowers and Colmer 2008).

Using the 200 mM definition for halophyte, it is estimated that there are approximately 350 halophyte species (Flowers et al. 2010). Alternatively, as Cheeseman (2015) notes, the eHALOPH Halophyte Database (Flowers 2014) currently identifies many more species as salt-tolerant (more than 1400), perhaps due to a more relaxed definition of the salt tolerance of a halophyte. These species are comprised of members of several plant orders, with the most species found in the Caryophalles, which contains the Amaranthaceae family, which now includes the members of the former goosefoot family Chenopodiaceae, where many of the most salt-tolerant species were classified.

A variety of useful products can be obtained from halophytes. Many of the halophytes produce useful oilseeds (Glenn et al. 1991, 2013; Weber et al. 2007) or grains (Glenn et al. 2013) that can be used for food or biofuel. Biomass can be used as firewood or timber, e.g., mangroves (Fig. 1a) (Bandaranayake 1998), or as fresh vegetables (Ventura and Sagi 2013; Ventura et al. 2013). They are used as medicine (Qasim et al. 2010; Xi et al. 2003); as forages for livestock (Swingle et al. 1996; Masters et al. 2007; El Shaer 2010; Glenn et al. 2013); for phytoremediation (Manousaki and Kalogerakis 2010); and as biofilters for saline aquaculture effluent (Brown and Glenn 1999; Shpigel et al. 2013) and ornamentals (Ali et al. 2012).

Additionally, since halophytes are exposed to extreme environmental stress, namely, high salinity, they face potential risk of damage to their cells from the production of reactive oxygen species (ROS) (Jithesh et al. 2006). One of the biochemical pathways that plants have evolved to combat damage from the ROS is the production of antioxidants such as tocopherol, carotenoids, ascorbate, phenolic compounds, alkaloids, glutathione, and nonprotein amino acids that scavenge free radicals (Gill and Tuteja 2010). It is likely that many of these compounds may have potential economic/commercial value as medicines, vitamins, and nutraceuticals (e.g., Cybulska et al. 2014a). Such bioproducts could typically be isolated from the extractives fraction during biomass processing.

3.1 Halophytes for Bioenergy Production

Recently there has been increased interest in exploring the potential of using halophytes to produce bioenergy or biofuels in particular. Interest in this topic may have arisen from the criticism of first-generation biofuels, by the high price of crude oil from the mid-2000s to 2014 or by the increased salinization of irrigated land from unsustainable farming practices. Reflecting this interest, there have been several recent overviews and reviews published on using halophytes as sources for bioenergy (e.g., Abideen et al. 2011, 2012, 2015; Brown et al. 2014a, ; Sharma et al. 2016; Debez et al. 2017). Most of these reviews have focused on producing ethanol from lignocellulosic biomass via fermentation and biodiesel from oilseeds via transesterification.

Halophytes have two products that can be used for bioenergy production: (1) oil produced from oilseed-producing halophytes and (2) the lignocellulosic biomass or straw after the oilseed is removed. The oil from oilseeds can be converted into biodiesel via transesterification for use in diesel engines as transport fuel or to use as a heating oil or converted to Bio-derived Synthetic Paraffinic Kerosene (Bio-SPK) for use as a partial replacement for jet fuel in aircraft jet engines (Rahmes et al. 2009). In the Bio-SPK process, vegetable oils are converted to shorter-chain diesel-range paraffins using a proprietary process, which removes oxygen molecules from the oil and converts olefins to paraffins by reaction with hydrogen. A subsequent reaction cracks and isomerizes the diesel-range paraffins to paraffins and isoparaffins with carbon numbers in the jet range (Kinder and Rahmes 2009). This process is not dependent on the composition of the oil, so ostensibly a wide range of halophyte oils could be used to produce Bio-SPK.

Halophyte oil has been found to be potentially useful for biodiesel production. Abideen et al. (2015) studied the oil yield and chemical composition of oilseeds from 20 species and found that of these, nine species yielded more than 25% oil. They found that the oil of seven species had ideal combinations of several different engine performance parameters and was comparable in composition to other glycophyte oil crops. Similarly Moser et al. (2013) found that oil from seashore mallow (*Kosteletzkya pentacarpos*) seeds was able to be processed to biodiesel that generally met international and national specification standards.

The lignocellulosic biomass of halophytes can be used as a feedstock to produce cellulosic ethanol via biochemical pathways (i.e., fermentation) (e.g., Bañuelos et al. 2018) for use as a gasoline additive for ground transport fuels. Alternatively, thermochemical pathways could be employed. For example, furfural, a chemical building block in biofuel production could be produced (Almardeai et al. 2017), by catalytic conversion at mild operational temperatures, 160–190 °C (Mariscal et al. 2016). Alternatively, biomass can be gasified to produce syngas for heat, electricity, or transport fuel and can undergo pyrolysis to produce syngas, pyrolysis oil, and biochar, which can be used as soil amendment to increase its productivity; or the Fischer-Tropsch process which converts gasses from gasification into synthetic fuels could be utilized (e.g., Warshay et al. 2017).

In a recent review, Sikarwar et al. (2017) presented a table from data from Speight (2011) showing that the ranges for energy yields for bioethanol production from biomass via enzymatic hydrolysis were roughly equal to those produced by thermochemical production of syngas to Fischer-Tropsch diesel and that both yielded more energy at the high end of the range than syngas to ethanol, thus indicating that the thermochemical pathways are feasible so long as they are economically viable.

Several studies have confirmed that the biomass of various species of halophytes is suitable/ideal for bioethanol production, with the straw having a beneficial composition of cellulose, hemicellulose, and low lignin content (Abideen et al. 2011; Bañuelos et al. 2018; Cybulska et al. 2014b, c), with the caveat that the straw will usually have a high ash content (Cybulska et al. 2014b), but not always (Bañuelos et al. 2018). High ash content in the biomass can negatively impact the pretreatment efficiency as well as enzymatic hydrolysis (He et al. 2014). However the majority of the ash can be removed by a freshwater wash (Cybulska et al. 2014b). Similarly, for pyrolysis even low concentrations of metal salts negatively impact the thermal degradation of cellulose and the production of residual char (Williams and Horne 1994). As above, a freshwater wash can remove much of the salts from the biomass.

Given that the characteristics of the biomass are adequate for bioenergy production, the next question is whether halophytes can produce yields similar to conventional bioenergy crops. Despite all of the uses or potential uses, no halophyte crops have been domesticated to date. The one dicot species that has probably received the most attention from researchers is *Salicornia bigelovii* (Fig. 1b), commonly known as glasswort of samphire (hereafter referred to as salicornia), a highly salt-tolerant annual oilseed-producing member of the Amaranthaceae family that is found in North American salt marshes (Lonard et al. 2011). Salicornia was cultivated directly on seawater of about 40 ppt salt in a long-term research project in plots in a coastal desert from 1982 to 1988 in Northwestern Mexico (Glenn et al. 1991). The yield of oilseeds averaged about 200 g m^{-1} $year^{-1}$, similar to soybeans and other oilseed crops (Glenn et al. 1991). Similarly, the yield of dry biomass was also high, ranging from 1.39 to 2.46 kg m^{-2} $year^{-1}$ (Glenn et al. 2013; Anwar et al. 2002). Oilseeds contained 26–33% oil and had 30–33% protein (Glenn et al. 1991). The protein can be used as a component in animal feed (Glenn et al. 1991; Ríos-Durán et al. 2013). In general, the oilseeds do not have a high ash content, unlike the straw.

In any case, to make the bioenergy production process economically viable, it is more advantageous to look at a complete biorefinery concept (Berntsson et al. 2012). One useful definition of a biorefinery comes from the National Renewable Energy Laboratory "A biorefinery is a facility that integrates biomass conversion processes and equipment to produce fuels, power, and chemicals from biomass."

Although ethanol is a high-volume, low-price commodity, if other high-value, low-volume compounds could be captured during the bioconversion process, then perhaps a halophyte bioenergy operation could be economically viable. In particular, if any high-value compounds such as medicinals, vitamins, or nutraceuticals could be removed from the extractive fraction during biomass processing, then this would certainly increase the viability of the operation.

So, for example, in the case of salicornia, at harvest, the oilseeds could be removed from the dry straw. The seeds could be pressed and the oil removed for biodiesel or bio-SPK production. The seed meal that remains after oil extraction has a high protein content and could be used as a component of an animal feed diet. The straw could be processed into bioethanol, and any high-value compounds could be removed from the extractive component. In this fashion, multiple products are produced from a single bioenergy feedstock.

3.2 Considerations for Producing Halophytes for Biofuel

Since currently there are no commercial scale farms producing biofuel from halophytes, the remainder of this article will focus on how to conduct high-salinity agriculture to make such a halophyte bioenergy project sustainable. In no sense does this constitute an endorsement on my part of the concept of turning large tracts of land (coastal or otherwise) into biofuel feedstock farms but rather represent some considerations for those who might want to undertake such an effort.

3.2.1 Site Selection

The first decision that must be made is where to locate a facility. The siting will depend upon the climate, the soil, and the water quality of the area under consideration, as well as the distance to markets, access to labor, and access to facilities/electricity/utilities that will be needed to process the biomass.

3.2.2 Climate

The area under consideration should have a climate with a long growing season to maximize biomass or oilseed production. Very high temperatures can negatively impact plant growth and development, in particular in the reproductive stages, especially if you are not using germplasm from species or populations from high heat environments. For example, maize seed yield has been shown to decline by over 80% when grown at high temperatures (Hatfield and Prueger 2015). Similarly very high temperatures will increase the irrigation deficit and require more frequent irrigation, which will be more costly and require more energy. Obviously lower temperatures will negatively impact growth and freezes or frosts will kill many frost-sensitive plant species.

3.2.3 Water Source

Choice of the water source used to irrigate halophytes will be an important consideration to increase the sustainability of an operation. For halophytes, the water sources might include seawater, brackish ground water, treated sewage effluent, or produced water—water that is derived from the process of extracting oil and gas from the ground (Brown et al. 2018).

For seawater, the actual salinity of seawater can vary greatly among different areas of the world. Generally seawater is 35 ppt salt, but in the Persian/Arabian Gulf, it is typically 40 ppt and can approach 60 ppt in certain areas/seasons (John et al. 1990). By contrast, salinity of seawater in the Baltic Sea can be less than 10 ppt (Omstedt and Axell 1998). The salinity of saline groundwater, produced water, and treated sewage effluent also varies; however treated sewage effluent will generally possess relatively lower salinity as it derived from domestic wastewater (Brown et al. 2018).

The salinity of the irrigation water will have a large effect on the growth of the plant biomass. Typically true halophytes will have an optimum salinity for growth that is somewhere in the range of one third to one half the salinity of seawater (Ayala and O'Leary 1995; Katschnig et al. 2013). Plant growth (and biomass production) will generally be reduced at higher salinities.

Similarly, since dicot halophyte tends to take up salts to maintain osmotic balance, plants grown at higher salinities will accumulate more ash in their tissue (Cybulska et al. 2014b). However, most of the ash is extractable (up to 90%). Therefore if the biomass is washed in freshwater, the resulting ash content is similar to the ash content found in conventional glycophyte biomass.

3.2.4 Land Area

Firstly, sites that do not have the potential to produce conventional crops should be selected—e.g., do not locate an operation on arable land. Such siting will remove any potentially contentious issues surrounding the food versus fuel debate. Secondly, there should be adequate drainage to allow water to drain below the plant root zone. If the drainage is not adequate or if the roots of halophyte are too close to the water table, then there is a potential for the soil to become waterlogged. Waterlogged soils result in oxygen deprivation to the roots, which will stunt or kill plants (Bailey-Serres and Colmer 2014). A medium like sand has the advantage of being fast draining, thereby minimizing the risk of waterlogging as long as plant roots are above the water table. However, on the other hand, the fast-draining nature of sand may also require more frequent irrigation as the soil salinity will increase more rapidly as the water moves more quickly below the plant root zone.

If a facility can be located at low elevation adjacent to the ocean or sea, then the pumping costs can be minimized, thus increasing sustainability. Similarly, for a farm using saline groundwater, the economics are more favorable if water is pumped from shallower depths rather than deeper. For example, at medium pressure, the pumping energy consumption would double for water pumped from 30 m versus 8 m (Martin et al. 2011).

3.2.5 Species Selection

The consideration of which species to use will depend on what the target product is. For biodiesel, for ground transport applications, one should select an oilseed halophyte with high oil content, high yield, and good combustion properties (Abideen et al. 2015). If the goal is to produce biofuel for the aviation sector and to therefore produce Bio-SPK, the combustion properties aren't critical, but yield and oil content are important. So ostensibly, more species might be suitable for Bio-SPK production than for biodiesel production. For lignocellulosic ethanol production, the criteria are high yield and low ash. Additionally, because lignin is difficult to process, beneficial characteristics of a feedstock will include low lignin content and high cellulose and hemicellulose content (Jung et al. 2015). Abideen et al. (2011) recommend halophyte grasses for bioethanol production due to their high productivity and low ash content. But as mentioned above, ash can generally be removed via washing in dicot halophytes, so the dicot halophyte should also be considered as potential feedstocks. Similarly, as mentioned previously, to maximize returns in a biorefinery operation, an ideal feedstock should produce multiple economically valuable products.

In all cases, high productivity (yield) is needed, along with other traits of typical domesticated crops such as the lack of seed shattering and synchronized ripening. (Brown et al. 2014b). Basically, wild candidate species will need to be domesticated. As the domestication process can be long and costly, there will need to be an economic need or incentive to domesticate and develop a halophyte crop.

3.3 Halophyte Bioenergy Cultivation Operations

As mentioned above, to my knowledge, there are currently no large-scale halophyte cultivation projects for producing biofuels. There were two that existed previously, one in Mexico and one in Eritrea. Bailis and Yu (2012) describe these operations in detail. The first farm was located in Eritrea and was about 340 ha at the maximum size and operated from 1998 to 2003. The second farm was located in Bahia Kino, Sonora, Mexico, on the west coast of Mexico. This farm operated from 2005 to 2010 and may have never exceeded 40 ha in size. Both of these farms were integrated seawater aquaculture/agriculture farms. This means that seawater was pumped into ponds or tanks to grow shrimp. The water from the shrimp ponds, which is enriched in nutrients from the uneaten feed plus the wastes from the shrimp, was subsequently used to irrigate fields of salicornia. The salicornia, which would remove the nutrients from the aquaculture wastewater (and thereby potentially sparing the need to use additional fertilizers), could be used for bioenergy production from the oilseeds and/or from the straw. The water that drained from the salicornia fields would flow into a mangrove wetland, where the water would be additionally polished, as the mangroves remove additional nutrients. Trimmings from the mangroves could be used as biofuel feedstocks, and the mangroves also serve to enhance the environment by providing habitat for fish and wildlife. It is efficient to combine lower-value halophyte production with higher-value seafood production from aquaculture, because costs are shared among the whole system (e.g., seawater only needs to be pumped once from the sea), and the wastewater from the aquaculture ponds becomes an asset in the form a fertilizer for the halophytes.

Both of these operations failed due to "political instability, mismanagement and community opposition" (Bailis and Yu 2012). Although these two projects ceased to operate due to financial or political issues, these types of integrated agriculture/aquaculture projects are likely sustainable with respect to greenhouse gas emissions. A life cycle assessment was conducted for a theoretical integrated seawater facility located in the Persian/Arabian Gulf that aimed to produce aviation biofuel, electricity, and seafood. Results suggested that such an operation would release 38–68% less greenhouse gasses than conventional jet fuel production and yield an overall positive net energy balance (Warshay et al. 2017).

In a recent review of halophyte biofuel projects, Marriott and Pourazadi (2017) examined several halophyte bioenergy projects, including the two mentioned above, and they concluded that "Saline cultivation needs to be bought into the realm of industrial engineering if it is expected to provide enough energy supply to make a difference to the global energy situation." However, they didn't address any potential economic benefits associated with the development of new saline biofuel crops. If progress is to be made, there needs to be a strong economic incentive to commercialize saline agriculture to produce biofuels from halophytes. For example, is there currently a shortage of biofuel feedstocks that necessitates the development of saline bioenergy farms? Just because it may be technically feasible to produce bioenergy from halophyte feedstocks using saline water, it does not mean that there is a compelling economic need to do so.

References

Abideen Z, Ansari R, Khan MA (2011) Halophytes: potential source of ligno-cellulosic biomass for ethanol production. Biomass Bioenergy 35(5):1818–1822

Abideen Z, Ansari R, Gul B, Khan MA (2012) The place of halophytes in Pakistan's biofuel industry. Biofuels 3(2):211–220

Abideen Z, Qasim M, Rizvi RF, Gul B, Ansari R, Khan MA (2015) Oilseed halophytes: a potential source of biodiesel using saline degraded lands. Biofuels 6(5–6):241–248

Ali A, Iqbal N, Ali F, Afzal B (2012) *Alternanthera bettzickiana* (Regel) G. Nicholson, a potential halophytic ornamental plant: growth and physiological adaptations. Flora Morphol Distrib Funct Ecol Plants 207(4):318–321

Almardeai S, Bastidas-Oyanedel JR, Schmidt JE (2017) Avicennia marina biomass characterization towards bioproducts. Emirates J Food Agric 29(9):710–715

Anwar F, Bhanger MI, Nasir MKA, Ismail S (2002) Analytical characterization of *Salicornia bigelovii* seed oil cultivated in Pakistan. J Agric Food Chem 50(15):4210–4214

Ayala F, O'Leary JW (1995) Growth and physiology of *Salicornia bigelovii* Torr. at suboptimal salinity. Int J Plant Sci 156(2):197–205

Bailey-Serres J, Colmer TD (2014) Plant tolerance of flooding stress–recent advances. Plant Cell Environ 37(10):2211–2215

Bailis R, Yu E (2012) Environmental and social implications of integrated seawater agriculture systems producing *Salicornia bigelovii* for biofuel. Biofuels 3(5):555–574

Bandaranayake WM (1998) Traditional and medicinal uses of mangroves. Mangrove Salt Marshes 2(3):133–148

Bañuelos JA, Velázquez-Hernández I, Guerra-Balcázar M, Arjona N (2018) Production, characterization and evaluation of the energetic capability of bioethanol from Salicornia Bigelovii as a renewable energy source. Renew Energy 123:125–134

Berntsson T, Sanden BA, Olsson L, Åsblad A (2012) What is a biorefinery?

Brown JJ, Glenn EP (1999) Reuse of highly saline aquaculture effluent to irrigate a potential forage halophyte, Suaeda esteroa. Aquac Eng 20:91–111. https://doi.org/10.1016/S0144-8609(99)00009-6

Brown JJ, Cybulska I, Chaturvedi T, Thomsen MH (2014a) Halophytes for the production of liquid biofuels. In: Sabkha ecosystems. Springer, Dordrecht, pp 67–72

Brown JJ, Glenn EP, Smith SE (2014b) Feasibility of halophyte domestication for high-salinity agriculture. In: Sabkha ecosystems. Springer, Dordrecht, pp 73–80

Brown JJ, Das P, Al-Saidi M (2018) Sustainable agriculture in the Arabian/Persian Gulf region utilizing marginal water resources: making the best of a bad situation. Sustainability 10(5):1–16

Carriquiry MA, Du X, Timilsina GR (2011) Second generation biofuels: economics and policies. Energy Policy 39(7):4222–4234

Chakravorty U, Hubert MH, Nøstbakken L (2009) Fuel versus food. Ann Rev Resour Econ 1(1):645–663

Cheeseman JM (2015) The evolution of halophytes, glycophytes and crops, and its implications for food security under saline conditions. New Phytol 206(2):557–570

Cybulska I, Brudecki G, Alassali A, Thomsen M, Brown JJ (2014a) Phytochemical composition of some common coastal halophytes of the United Arab Emirates. Emirates J Food Agric 26(12):1046

Cybulska I, Chaturvedi T, Alassali A, Brudecki GP, Brown JJ, Sgouridis S, Thomsen MH (2014b) Characterization of the chemical composition of the halophyte Salicornia bigelovii under cultivation. Energy Fuel 28(6):3873–3883

Cybulska I, Chaturvedi T, Brudecki GP, Kádár Z, Meyer AS, Baldwin RM, Thomsen MH (2014c) Chemical characterization and hydrothermal pretreatment of Salicornia bigelovii straw for enhanced enzymatic hydrolysis and bioethanol potential. Bioresour Technol 153:165–172

Debez A, Belghith I, Friesen J, Montzka C, Elleuche S (2017) Facing the challenge of sustainable bioenergy production: could halophytes be part of the solution? J Biol Eng 11(1):27

El Shaer HM (2010) Halophytes and salt-tolerant plants as potential forage for ruminants in the Near East region. Small Rumin Res 91(1):3–12

FAO (2011) The state of the world's land and water resources for food and agriculture (SOLAW)—managing systems at risk. Food and Agriculture Organization of the United Nations, Rome

Flowers TJ (2014) eHALOPH halophytes database. [WWW document]. http://www.sussex.ac.uk/affiliates/halophytes/. Accessed 1 Apr 2018

Flowers TJ, Colmer TD (2008) Salinity tolerance in halophytes. New Phytol 179(4):945–963

Flowers TJ, Hajibagheri MA, Clipson NJW (1986) Halophytes. Q Rev Biol 61(3):313–337

Flowers TJ, Galal HK, Bromham L (2010) Evolution of halophytes: multiple origins of salt tolerance in land plants. Funct Plant Biol 37(7):604–612

Gill SS, Tuteja N (2010) Reactive oxygen species and antioxidant machinery in abiotic stress tolerance in crop plants. Plant Physiol Biochem 48(12):909–930

Glenn EP, O'Leary JW, Watson MC, Thompson TL, Kuehl RO (1991) Salicornia bigelovii Torr.: an oilseed halophyte for seawater irrigation. Science 251(4997):1065–1067

Glenn EP, Anday T, Chaturvedi R, Martinez-Garcia R, Pearlstein S, Soliz D, Nelson SG, Felger RS (2013) Three halophytes for saline-water agriculture: an oilseed, a forage and a grain crop. Environ Exp Bot 92:110–121. https://doi.org/10.1016/j.envexpbot.2012.05.002

Hatfield JL, Prueger JH (2015) Temperature extremes: effect on plant growth and development. Weather Clim Extremes 10:4–10

He Y, Fang Z, Zhang J, Li X, Bao J (2014) De-ashing treatment of corn stover improves the efficiencies of enzymatic hydrolysis and consequent ethanol fermentation. Bioresour Technol 169:552–558

Jithesh MN, Prashanth SR, Sivaprakash KR, Parida AK (2006) Antioxidative response mechanisms in halophytes: their role in stress defence. J Genet 85(3):237

John VC, Coles SL, Abozed AI (1990) Seasonal cycles of temperature, salinity and water masses of the western Arabian Gulf. Oceanol Acta 13(3):273–281

Jung SJ, Kim SH, Chung IM (2015) Comparison of lignin, cellulose, and hemicellulose contents for biofuels utilization among 4 types of lignocellulosic crops. Biomass Bioenergy 83:322–327

Katschnig D, Broekman R, Rozema J (2013) Salt tolerance in the halophyte Salicornia dolichostachya Moss: growth, morphology and physiology. Environ Exp Bot 92:32–42

Kinder JD, Rahmes T (2009) Evaluation of bio-derived synthetic paraffinic kerosene (bio-SPK). The Boeing Company Sustainable Biofuels Research & Technology Program. http://www.safug.org/assets/docs/biofuel-testing-summary.pdf

Lonard RI, Judd FW, Stalter R (2011) The biological flora of coastal dunes and wetlands: Salicornia bigelovii J. Torrey. J Coast Res 28(3):719–725

Manousaki E, Kalogerakis N (2010) Halophytes present new opportunities in phytoremediation of heavy metals and saline soils. Ind Eng Chem Res 50(2):656–660

Mariscal R, Maireles-Torres P, Ojeda M, Sadaba I, Lopez Granados M (2016) Furfural: a renewable and versatile platform molecule for the synthesis of chemicals and fuels. Energy Environ Sci 9:1144–1189

Marriott L, Pourazadi E (2017) Industrialisation of saline cultivation for second-generation biofuels: progress and challenges. Environ Tech Rev 6(1):15–25

Martin DL, Dorn TW, Melvin SR, Corr AJ, Kranz WL (2011) Evaluating energy use for pumping irrigation water. In: Proceedings of the 23rd annual central plains irrigation conference, pp 22–23

Masters DG, Benes SE, Norman HC (2007) Biosaline agriculture for forage and livestock production. Agric Ecosyst Environ 119(3–4):234–248

Moser BR, Dien BS, Seliskar DM, Gallagher JL (2013) Seashore mallow (*Kosteletzkya pentacarpos*) as a salt-tolerant feedstock for production of biodiesel and ethanol. Renew Energy 50:833–839

NSW Department of Primary Industries (2017) Salinity tolerance in irrigated crops Primefact 1345, 2nd edn. Agriculture NSW Water Unit. https://www.dpi.nsw.gov.au/__data/assets/pdf_file/0005/523643/Salinity-tolerance-in-irrigated-crops.pdf

Omstedt A, Axell LB (1998) Modeling the seasonal, interannual, and long-term variations of salinity and temperature in the Baltic proper. Tellus A Dyn Meteorol Oceanogr 50(5):637–652

Qasim M, Gulzar S, Shinwari ZK, Aziz I, Khan MA (2010) Traditional ethnobotanical uses of halophytes from Hub, Balochistan. Pak J Bot 42(3):1543–1551

Rahmes T, Kinder J, Crenfeldt G, LeDuc G, Abe Y, McCall M, Andac M (2009) Sustainable bio-derived synthetic paraffinic kerosene (Bio-SPK) jet fuel flights and engine tests program results. In: 9th AIAA aviation technology, integration, and operations conference (ATIO) and aircraft noise and emissions reduction symposium (ANERS)

Ríos-Durán MG, Valencia IR, Ross LG, Martinez-Palacios CA (2013) Nutritional evaluation of autoclaved Salicornia bigelovii Torr. seed meal supplemented with varying levels of cholesterol on growth, nutrient utilization and survival of the Nile tilapia (Oreochromis niloticus). Aquac Int 21(6):1355–1371

Shannon MC, Grieve CM (1998) Tolerance of vegetable crops to salinity. Sci Hortic (Amsterdam) 78:5–38

Sharma R, Wungrampha S, Singh V, Pareek A, Sharma MK (2016) Halophytes as bioenergy crops. Front Plant Sci 7:1372

Shpigel M, Ben-Ezra D, Shauli L, Sagi M, Ventura Y, Samocha T, Lee JJ (2013) Constructed wetland with Salicornia as a biofilter for mariculture effluents. Aquaculture 412–413:52–63. https://doi.org/10.1016/j.aquaculture.2013.06.038

Sikarwar VS, Zhao M, Fennell PS, Shah N, Anthony EJ (2017) Progress in biofuel production from gasification. Prog Energy Combust Sci 61:189–248

Speight JG (ed) (2011) The biofuels handbook. Royal Society of Chemistry, London

Swingle RS, Glenn EP, Squires V (1996) Growth performance of lambs fed mixed diets containing halophyte ingredients. Anim Feed Sci Technol 63(1–4):137–148

Ventura Y, Sagi M (2013) Halophyte crop cultivation: the case for salicornia and sarcocornia. Environ Exp Bot 92:144–153. https://doi.org/10.1016/j.envexpbot.2012.07.010

Ventura Y, Myrzabayeva M, Alikulov Z, Cohen S, Shemer Z, Sagi M (2013) The importance of iron supply during repetitive harvesting of Aster tripolium. Funct Plant Biol 40(9):968–976

Warshay B, Brown JJ, Sgouridis S (2017) Life cycle assessment of integrated seawater agriculture in the Arabian (Persian) Gulf as a potential food and aviation biofuel resource. Int J Life Cycle Assess 22(7):1017–1032

Weber DJ, Ansari R, Gul B, Khan MA (2007) Potential of halophytes as source of edible oil. J Arid Environ 68(2):315–321

Wicke B, Smeets E, Dornburg V, Vashev B, Gaiser T, Turkenburg W, Faaij A (2011) The global technical and economic potential of bioenergy from salt-affected soils. Energy Environ Sci 4(8):2669–2681

Williams PT, Horne PA (1994) The role of metal salts in the pyrolysis of biomass. Renew Energy 4(1):1–13

Xi JB, Zhang FS, Mao DR, Yan P (2003) The utilization of halophytes for traditional medicine in Xin jiang. Rev China Agric Sci Technol 5:43–48

Characterization of Avicennia marina: An Arid-Coastal Biomass—Toward Biorefinery Products

Saleha Almardeai, Juan-Rodrigo Bastidas-Oyanedel, Sabeera Haris, and Jens Ejbye Schmidt

1 Introduction

Mangroves are found in 112 countries around the world covering approximately 160,000 km^2 (Kathiresan and Bingham 2001; Mijan Uddin et al. 2014). In the United Arab Emirates (UAE), the only naturally occurring mangrove species is *Avicennia marina* (EAD 2016). The area occupied by these single-species mangrove forests has increased from 40 to 155 km^2 between 2005 and 2014 (EAD 2016) due to cultivation, public awareness, and conservation efforts. Mangroves in the gulf coast of the UAE grow in high salinity conditions of over 40 g/kg (Smith et al. 2007) as compared to the world seawater reference salinity of 35 g/kg (Millero et al. 2008).

In arid regions such as the UAE, where average annual rain precipitation is scarce (120 mm) (Böer 1997), water scarcity limits biomass availability and constrains forestry and agricultural activities (Bastidas-Oyanedel et al. 2016). Therefore, the well-managed use of *A. marina* as a lignocellulosic feedstock may provide usable biomass that would otherwise be unavailable due to water scarcity (Almardeai et al. 2017). Lignocellulose biorefinery for the production of high-value chemicals has seen a boost in interest in recent years (Zhang 2008; FitzPatrick et al. 2010), where 10^9 million tonnes of lignocellulose is estimated to be utilized worldwide (Alvira et al. 2010). In addition to its potential as a lignocellulosic bioresource, conservation and cultivation of *A. marina* has environmental benefits, such as increase in

S. Almardeai · J.-R. Bastidas-Oyanedel (✉) · S. Haris
Chemical Engineering Department, Khalifa University of Science and Technology, Masdar Institute, Masdar City, Abu Dhabi, United Arab Emirates
e-mail: juan.bastidas@ku.ac.ae

J. E. Schmidt
SDU-Department of Chemical Engineering, Biotechnology and Environmental Technology, University of Southern Denmark, Campusvej 55, DK-5230 Odense M, Denmark
e-mail: jesc@kbm.sdu.dk

© Springer Nature Switzerland AG 2019
J.-R. Bastidas-Oyanedel, J. E. Schmidt (eds.), *Biorefinery*,
https://doi.org/10.1007/978-3-030-10961-5_29

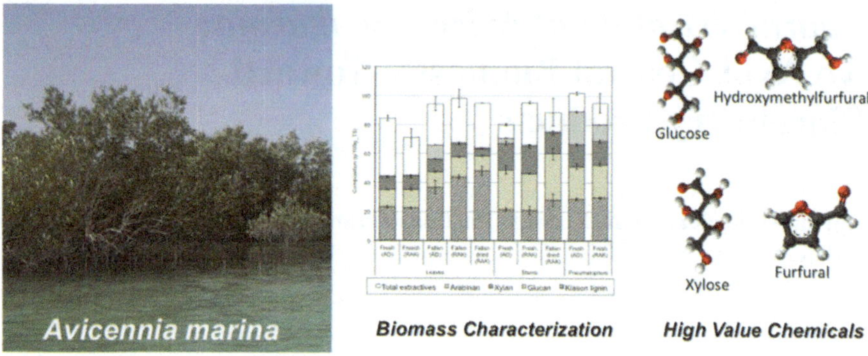

Fig. 1 Scheme of the manuscript objective. Left: *Avicennia marina* forest in the coast of Abu Dhabi, United Arab Emirates. Center: lignocellulosic characterization. Right: furfural and hydroxymethylfurfural, high-value molecules, molecular model

fishery, biodiversity, ecotourism, protection against erosion, and feedstock (Kairo et al. 2001; Han 2003; Sato et al. 2005; Shaifullah et al. 2009).

The objective of the present study (Fig. 1) is the characterization of lignocellulosic components of *A. marina* leaves, stem, and pneumatophore in the UAE. This characterization is the first in this region and represents a first attempt at highlighting the importance of this plant in a future biorefinery-based economy in this arid region. The study also contributes to the creation of a database of local endemic plants.

2 Materials and Method

2.1 *Biomass Collection and Preparation*

Avicennia marina biomass samples were collected from two different locations, namely, Ras Al Khaimah and Abu Dhabi, United Arab Emirates. The coordinates of these locations are 25°46′31.33″N 55°57′47.98″E and 24°27′20.58″N 54°24′18.19″E, respectively. *A. marina* biomass samples consisted of fresh leaves, fallen leaves, fresh stems, fallen stems, and fresh pneumatophores. All the samples were washed with milliQ water, air-dried at 25 °C, and milled to 2 mm.

2.2 *Material Characterization*

All the *A. marina* samples were quantified for total solids (TS), volatile solids (VS), total ashes, glucan, xylan, Klason lignin, structural ashes, and extractives content as previously described (Fang et al. 2015). Structural carbohydrates and lignin composition were determined by a two-step acid hydrolysis of extractive-free biomass. The hydrolyzates were analyzed by HPLC (Agilent 1260 Infinity Bio-Inert Binary LC), using a Hi-Plex H column (Agilent) and refractive index detector, for determination of glucose, xylose, and arabinose concentration at 65 °C with 0.005 M H_2SO_4 as mobile phase with 0.6 mL/min flow rate.

2.3 *Biomethane Potential Analysis*

For the biomethane potential analysis, *A. marina* collected from Ras Al Khaimah was used as substrate, seawater from Ras Al Khaimah as media, and sludge from Al Wathba Wastewater treatment plant in Abu Dhabi as inoculum. Seawater pH was measured using commercially available probes. Seawater salinity was determined by drying a known mass of sample at 150 and 570 °C. Seawater salinity was 43.2 ± 0.1 g/kg and a pH was 7.29. Biomethane potential experiments were conducted at 37 °C in serum bottles with 115 mL liquid volume and 212 mL headspace. The experiment was conducted using 1.09 gVSmangrove/gVSsediment and an inoculum concentration of 11.4 g/L. Biomethane was measured by gas chromatography.

3 Results and Discussion

3.1 *Determination of Dry Matter (Total Solids) and Ash in* A. marina *Samples*

Figure 2 shows the total solids as the sum of volatile solids and ashes. All samples tested contained an average of 70% volatile solids. Leaf samples contained the highest ash values (20% average) and least moisture content (10%). The volatile solid content indicates the organic fraction of the biomass, which can be used for the production of high-value chemicals (Zhang 2008; FitzPatrick et al. 2010) and bioenergy as biogas (Bastidas-Oyanedel et al. 2016).

3.2 *Determination of Klason Lignin and Oligosaccharides in Biomass*

Figure 3 presents *A. marina* biomass composition as total extractives, arabinan, xylan, glucan, and Klason lignin. In general, we did not detect any significant differences between samples collected from Ras Al Khaimah and Abu Dhabi. Stem and pneumatophore samples contained the highest sugar content (arabinan, xylan,

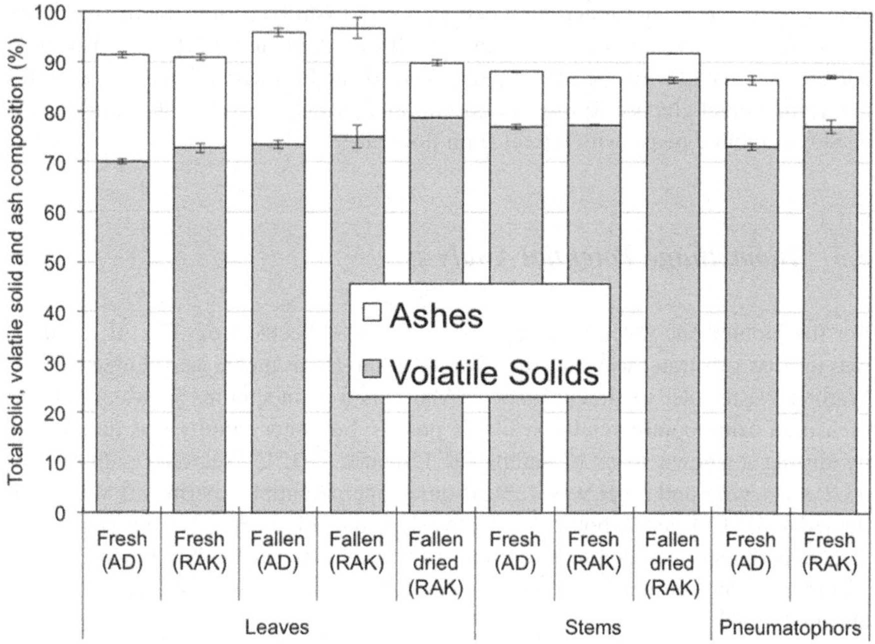

Fig. 2 *A. marina* biomass composition as total solids (TS) expressed as volatile solids (VS) and ashes

and glucan) with an average lignin composition of 22 g/100 g_TS. This makes them convenient for the production of high-value chemicals such as furfural, hydroxymethylfurfural (HMF), and levulinic acid using thermochemical processes (Alvira et al. 2010) or glucaric acid using fermentation processes (Gupta et al. 2016a). The total extractives, which are the main constituent of the leaves, can be of high-value compounds such as naphthoquinones (anticarcinogenic activity), betaine (dietary supplement), iridoid glycosides, triacontane, trimethylglycine, and bioactive lipids (Popp 1984; Konig and Rimpler 1985; Wu et al. 2008; Ramadan et al. 2009; Liebezeit 2012).

Several studies have investigated the potential of different feedstocks including mangrove species for biofuel production in arid regions. One study showed that lignocellulosic biomasses, such as date palm, *A. marina*, and *Moringa peregrina*, had promising result for bioenergy production, with the highest sugar yields observed in the date palm fruit stalks (Ashraf et al. 2016). Another study demonstrated that seedless *S. bigelovii* showed high ash content (43.08 g/100 gDM) with low sugar composition (9.1 ± 1.5 g/100 gDM glucan, 7.7 ± 0.4 g/100 gDM xylan) (Cybulska et al. 2014).

Figure 4 shows the concentration of acetic acid, furfural, and HMF produced during the thermochemical hydrolysis. These compounds are fermentation inhibitors formed during pentose and hexose degradation (Jönsson and Martín 2015), e.g., in the downstream processing of thermochemical-treated biomass by yeast

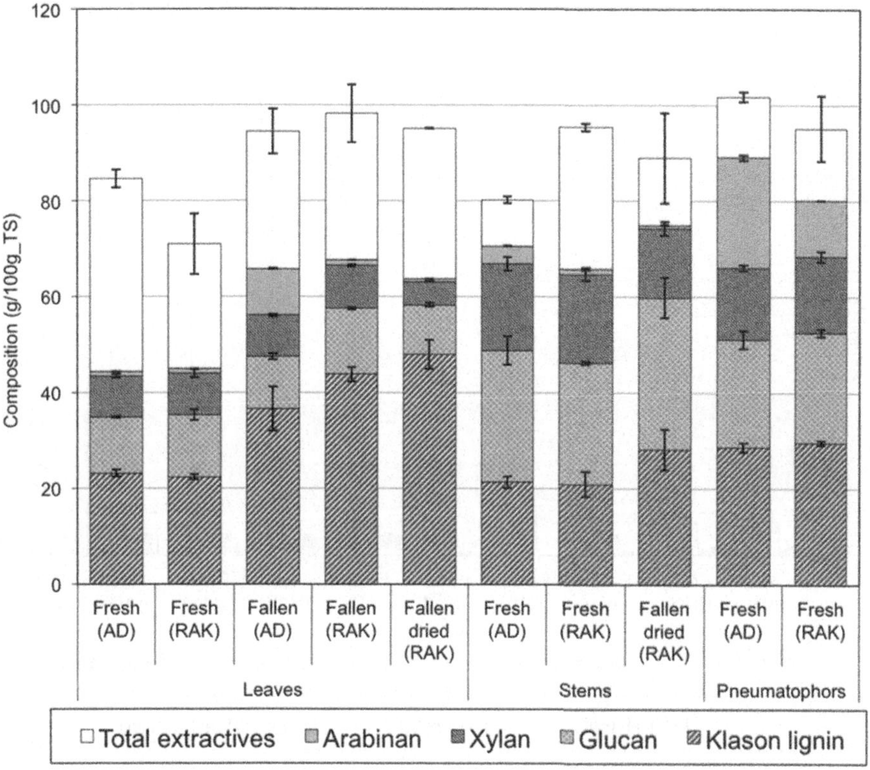

Fig. 3 *A. marina* biomass composition as total extractives, arabinan, xylan, glucan, and Klason lignin

fermentation. Furfural and HMF are among the ten most valuable bioproducts listed by the US Department of Energy, and they are a highly potential renewable chemical feedstock for the production of valuable chemicals and biofuels (Agustina et al. 2013; Cai et al. 2013). In fact, production of furfural from cellulose can be commercially more attractive than bioethanol by yeast fermentation. Moreover, HMF can be converted to 2,5-dimethylfuran (DMF), which has 40% higher energy density than ethanol (Cai et al. 2013).

Furthermore, anaerobic digestion of *A. marina* biomass (and/or residues of its processing) is possible, using UAE Arabian Gulf seawater instead of freshwater (Bastidas-Oyanedel et al. 2016), which would contribute to the production of biogas. Figure 5 shows the cumulative biomethane produced from *A. marina* using Arabian Gulf seawater (43.2 ± 0.1 g/kg salinity).

Demonstrating that seawater can be used in the production of biomethane in this study opens new research doors related to biofuel and high-value chemical production based on seawater lignocellulosic biomass. The substitution of freshwater by seawater is economically and environmentally important since the consumption of water here in the UAE is approximated to be 550 L per individual per day, which is

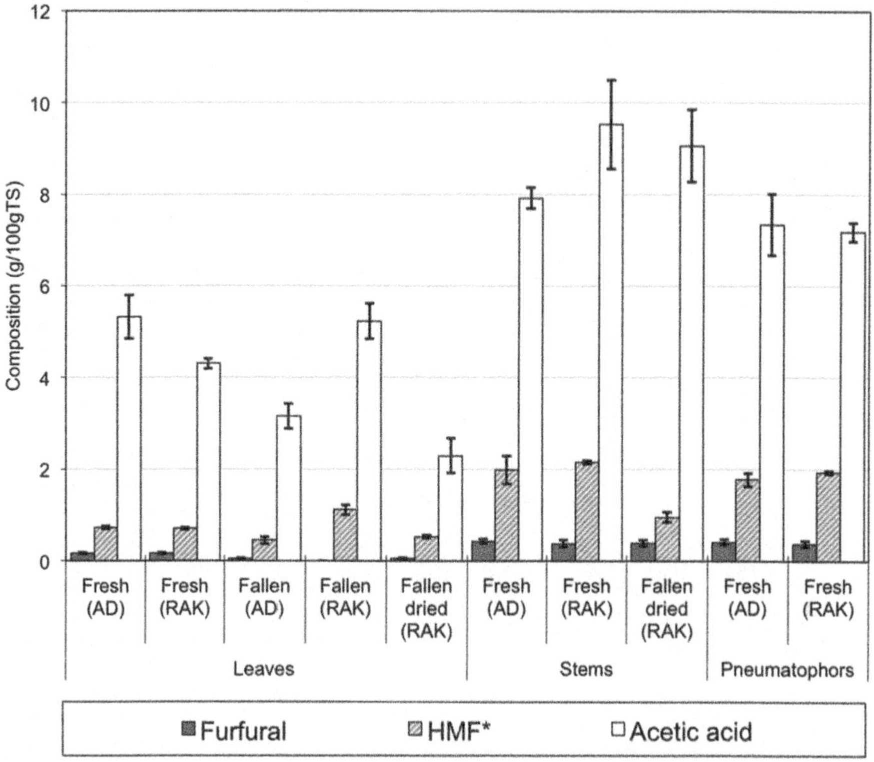

Fig. 4 *A. marina* biomass acid hydrolysis products at 121 °C with 72% sulfuric acid (* hydroxymethylfurfural)

twice the global average ("Conserving every drop," 2014). Seawater can thus be a promising alternative to freshwater for biorefinery purpose as it helps in lowering down the production and energy costs as compared to a process involving freshwater.

The sustainable utilization of mangrove forest in coastal arid regions has not only the benefits of producing lignocellulosic biomass without freshwater but also aims at the conservation and plantation of *A. marina* in the UAE because of the biodiversity it hosts. In addition to being a crucial habitat for fish and crustaceans (Kathiresan and Bingham 2001; Han 2003), mangroves are a habitat to other living organisms with biotechnological potential (Kathiresan and Bingham 2001). Indeed, polyalkanoates and terpenoids producing bacteria (Moorkoth and Nampoothiri 2016; Mitra et al. 2008) have been isolated from mangrove soil. Furthermore, enzymes, antimicrobials, and biopesticides have been isolated from mangrove fungi (Cheng et al. 2009). Long-chain omega-3 fatty acids producing microalgae have also been found in mangrove forests (Gupta et al. 2016b). Taken together, this study demon-

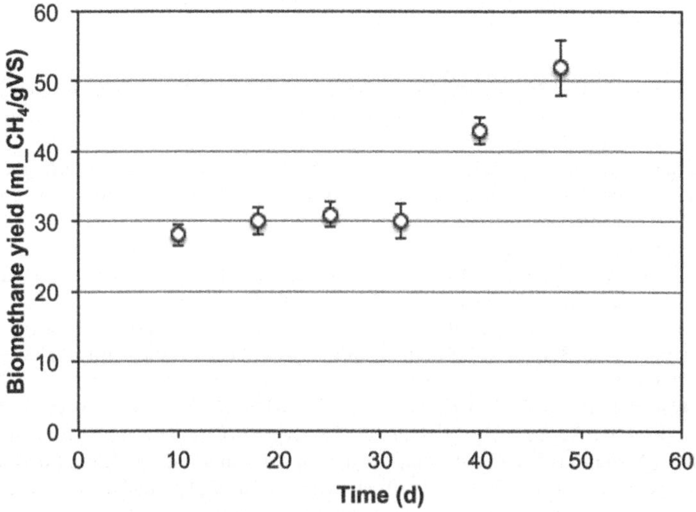

Fig. 5 Biomethane potential of *Avicennia marina* at seawater conditions over an incubation period of 48 days. Seawater, 43.2 g/kg salinity. The substrate over inoculum ratio is 1.09 gVS_*Avicennia marina*/gVS_inoculum. *VS* volatile solids

strates that *A. marina* represents a promising biomass for seawater biorefinery, as it is adapted to UAE seawater salinity conditions and would contribute to the development of alternative energy and wealth-generating activities in the UAE and the world.

4 Conclusions

The chemical composition of *Avicennia marina* biomass, leaves, stem, and pneumatophores, has been quantified. Samples were collected from two locations in the United Arab Emirates (UAE), Ras Al Khaima and Abu Dhabi. The results showed no significant variation between these two locations. Average volatile solids for all the biomasses was 70%. Higher content of sugars was found in stems and pneumatophores. The highest extractive content was found in leaves. The present study highlights the importance of *A. marina* in the UAE as a lignocellulose resource for biorefinery processes while overcoming the water scarcity issue of arid regions.

Acknowledgments The authors would like to acknowledge the financial support from Masdar Institute of Science and Technology, to help fulfill the vision of the late President Sheikh Zayed Bin Sultan Al Nahyan for sustainable development and empowerment of the United Arab Emirates and humankind, funding project 2GBIONRG (12KAMA4).

References

Agustina D, Kumagai S, Nonaka M, Sasaki K (2013) Production of 5-hydroxymethyl furfural from sugarcane Bagasse under hot compressed water. Procedia Earth Planet Sci 6:441–447. https://doi.org/10.1016/j.proeps.2013.01.058

Almardeai S, Bastidas-Oyanedel JR, Haris S, Schmidt JE (2017) Avicennia marina biomass characterization towards bioproducts. Emirates J Food Agric 29(9):710–715. https://doi.org/10.9755/ejfa.2017.v29.i9.109

Alvira P, Tomás-Pejó E, Ballesteros M, Negro MJ (2010) Pretreatment technologies for an efficient bioethanol production process based on enzymatic hydrolysis: a review. Bioresour Technol 101:4851–4861. https://doi.org/10.1016/j.biortech.2009.11.093

Ashraf M, Fang C, Bochenski T et al (2016) Estimation of bioenergy potential for local biomass in the United Arab Emirates. Emirates J Food Agric 28(2):99–106. https://doi.org/10.9755/ejfa.2015-04-060

Bastidas-Oyanedel JR, Fang C, Almardeai S et al (2016) Waste biorefinery in arid/semi-arid regions. Bioresour Technol 215:21–28. https://doi.org/10.1016/j.biortech.2016.04.010

Böer B (1997) An introduction to the climate of the United Arab Emirates. J Arid Environ 35:3–16

Cai CM, Zhang T, Kumar R, Wyman CE (2013) Integrated furfural production as a renewable fuel and chemical platform from lignocellulosic biomass. J Chem Technol Biotechnol 89(1):2–10. https://doi.org/10.1002/jctb.4168

Cheng ZS, Pan JH, Tang WC et al (2009) Biodiversity and biotechnological potential of mangrove-associated fungi. J For Res 20:63–72. https://doi.org/10.1007/s11676-009-0012-4

Cybulska I, Chaturvedi T, Brudecki GP et al (2014) Chemical characterization and hydrothermal pretreatment of Salicornia bigelovii straw for enhanced enzymatic hydrolysis and bioethanol potential. Bioresour Technol 153:165–172. https://doi.org/10.1016/j.biortech.2013.11.071

EAD (2016) Status of mangroves in Abu Dhabi—biodiversity annual report 2014. Abu Dhabi

Fang C, Thomsen MH, Brudecki GP et al (2015) Seawater as alternative to freshwater in pretreatment of date palm residues for bioethanol production in coastal and/or arid areas. ChemSusChem 8:3823–3831. https://doi.org/10.1002/cssc.201501116

FitzPatrick M, Champagne P, Cunningham MF, Whitney RA (2010) A biorefinery processing perspective: treatment of lignocellulosic materials for the production of value-added products. Bioresour Technol 101:8915–8922. https://doi.org/10.1016/j.biortech.2010.06.125

Gupta A, Hicks MA, Manchester SP, Prather KLJ (2016a) Porting the synthetic D-glucaric acid pathway from Escherichia coli to Saccharomyces cerevisiae. Biotechnol J 11(9):1201–1208. https://doi.org/10.1002/biot.201500563

Gupta A, Singh D, Byreddy AR et al (2016b) Exploring omega-3 fatty acids, enzymes and biodiesel producing thraustochytrids from Australian and Indian marine biodiversity. Biotechnol J 11:345–355. https://doi.org/10.1002/biot.201500279

Han W-D (2003) Present status and conservation strategies of mangrove resource in Guangdong, P.R. China. J For Res 14:151–154

Jönsson LJ, Martín C (2015) Pretreatment of lignocellulose: formation of inhibitory by-products and strategies for minimizing their effects. Bioresour Technol 199:103–112. https://doi.org/10.1016/j.biortech.2015.10.009

Kairo JG, Dahdouh-Guebas F, Bosire J, Koedam N (2001) Restoration and management of mangrove systems—a lesson for and from the east African region. S Afr J Bot 67:383–389. https://doi.org/10.1016/S0254-6299(15)31153-4

Kathiresan K, Bingham BL (2001) Biology of mangroves and mangrove ecosystems. Adv Mar Biol 40:81–251. https://doi.org/10.1016/S0065-2881(01)40003-4

Konig G, Rimpler H (1985) Iridoid glucosides in Avicennia marina. Phytochemistry 24:1245–1248

Liebezeit G (2012) Natural products chemistry of the mangrove species Avicennia spp.—review and new data. Nat Prod Indian J 8:10–20. https://doi.org/10.1007/978-3-642-22144-6

Mijan Uddin SM, Rafiqul Hoque ATM, Abdullah SA (2014) The changing landscape of mangroves in Bangladesh compared to four other countries in tropical regions. J For Res 25:605–611. https://doi.org/10.1007/s11676-014-0448-z

Millero FJ, Feistel R, Wright DG, McDougall TJ (2008) The composition of standard seawater and the definition of the reference-composition salinity scale. Deep Sea Res I Oceanogr Res Pap 55:50–72. https://doi.org/10.1016/j.dsr.2007.10.001

Mitra A, Santra SC, Mukherjee J (2008) Distribution of actinomycetes, their antagonistic behaviour and the physico-chemical characteristics of the world's largest tidal mangrove forest. Appl Microbiol Biotechnol 80:685–695. https://doi.org/10.1007/s00253-008-1626-8

Moorkoth D, Nampoothiri KM (2016) Production and characterization of poly(3-hydroxy butyrate-co-3 hydroxyvalerate) (PHBV) by a novel halotolerant mangrove isolate. Bioresour Technol 201:253–260. https://doi.org/10.1016/j.biortech.2015.11.046

Popp M (1984) Chemical composition of Australian mangroves II. Low molecular weight carbohydrates. Z Pflanzenphysiol 113:411–421. https://doi.org/10.1016/S0044-328X(84)80096-3

Ramadan MF, Amer MMA, Mansour HT et al (2009) Bioactive lipids and antioxidant properties of wild Egyptian Pulicaria incise, Diplotaxis harra, and Avicennia marina. J Consum Prot Food Saf 4:239–245. https://doi.org/10.1007/s00003-009-0300-0

Sato G, Fisseha A, Gebrekiros S et al (2005) A novel approach to growing mangroves on the coastal mud flats of Eritrea with the potential for relieving regional poverty and hunger: reply. Wetlands 25:776–779. https://doi.org/10.1672/0277-5212(2006)26[638:RANATG]2.0.CO;2

Shaifullah KM, Haque SMS, Sujauddin M, Karmakar S (2009) Coastal afforestation effects on soil properties at Hatiya in Bangladesh. J For Res 20:243–248. https://doi.org/10.1007/s11676-009-0035-x

Smith R, Purnama A, Al-Barwani HH (2007) Sensitivity of hypersaline Arabian gulf to seawater desalination plants. Appl Math Model 31:2347–2354. https://doi.org/10.1016/j.apm.2006.09.010

Wu J, Xiao Q, Xu J et al (2008) Natural products from true mangrove flora: source, chemistry and bioactivities. Nat Prod Rep 25:955–981. https://doi.org/10.1039/b807365a

Zhang YHP (2008) Reviving the carbohydrate economy via multi-product lignocellulose biorefineries. J Ind Microbiol Biotechnol 35:367–375. https://doi.org/10.1007/s10295-007-0293-6

Techno-economic Assessment of Microalgae Biorefinery as a Source of Proteins, Pigments, and Fatty acids: A Case Study for the United Arab Emirates

Mariam Ali AlMahri, Kichul Jung, Mashael Alshehhi, Juan-Rodrigo Bastidas-Oyanedel, and Jens Ejbye Schmidt

1 Introduction

Microalgae can grow 20–30 times faster than food crops (some species can double their mass several times a day) (Ullah et al. 2014), and they can be cultivated on nonproductive land (deserts included) or industrially in photo-bioreactors or open ponds. Microalgae are microscopic photosynthetic organisms and sunlight-driven cell factories that convert CO_2 to potential biofuels, foods, feeds, and high-value bioactive molecules (Fu et al. 2016). Depending on the type of microalgae and cultivation environment, certain systems are chosen from multiple pathways to produce chemical products. However, given the diverse ways in cultivating microalgae, it is not simple to predict whether microalgal cultivation in open (e.g., open ponds), closed (e.g., photobioreactor), or hybrid systems will prevail in certain region (Thomas and Herbert 2005).

Microalgae have been found to gain decent production levels of oil per acre compared to other oil seed crops such as soybeans, but the economic viability of algae-

M. A. AlMahri
Department of Applied Mathematics and Sciences, College of Science,
Khalifa University of Science and Technology, Abu Dhabi, United Arab Emirates

K. Jung (✉)
Department of Civil and Environmental Engineering, Konkuk University, Seoul, South Korea
e-mail: jkichul11@konkuk.ac.kr

M. Alshehhi · J.-R. Bastidas-Oyanedel
Chemical Engineering Department, Khalifa University of Science and Technology,
Petroleum Institute, Abu Dhabi, United Arab Emirates

J. E. Schmidt
SDC-Department of Chemical Engineering, Biotechnology, and Environmental Technology,
University of Southern Denmark, Campusvej 55, DK-5230 Odense M, Denmark
e-mail: jesc@kbm.sdu.dk

© Springer Nature Switzerland AG 2019

J.-R. Bastidas-Oyanedel, J. E. Schmidt (eds.), *Biorefinery*,
https://doi.org/10.1007/978-3-030-10961-5_30

based biofuels is still unfeasible (Dasgupta 2015). However, the high metabolic versatility of microalgae allows the production of nonfuel products such as pigments (chlorophyll), biopolymers, proteins, and polyunsaturated fatty acids (PUFAs), which have a very high value and could play a major role in turning economic and energy balances more favorable (Gouveia 2015). This versatility and huge potential of microalgae could support a microalgae-based biorefinery and microalgae-based bio-economy opening up vast opportunities in the global algae business (Gouveia 2015). Although there are clear differences in the compositions of the microalgal classes and species, the protein is the major organic constituent, followed usually by fatty acid (lipid) and then by carbohydrate (Becker 2007).

A biorefinery is a facility that processes the biomass into a variety of marketable products and energy (Cherubini 2010). The concept of biorefinery is parallel to today's petroleum refinery, which makes various fuels and products from petroleum (Aresta et al. 2012). Producing various products in a biorefinery helps in maximizing the value acquired from the biomass feedstock. There are three main processes that can be involved in the algae biorefinery scheme. The first process is the growth of the algae under special optimal conditions in order to derive maximum returns. The second process will then involve harvesting and dewatering of the algae. The last process will involve processing the algae into the desired high-value chemicals (Seth and Wangikar 2015). The UAE has an arid climate with scarce rainfall summers, stretching from April through the month of September, with temperatures rising up to 50 °C in the southern desert regions (Böer 1997; Hasanean 2013). Therefore, for a microalgae biorefinery to thrive in the UAE, there should be a type that has a high temperature and high salinity tolerance as this will help in reducing the overall cost (Bastidas-Oyanedel et al. 2016).

This paper explores the potential use of microalgae as a source of biomass in the UAE. This is done in two steps, first by studying several strains available in land and the second by designing a biorefinery based on experimental results. For the first step, three strains such as *Chlorella*, *Euglena*, and *Scenedesmus* were isolated and tested under different experimental conditions including temperature and salinity. The experiment was carried out with three different salinities (30, 40, and 50 g/L) of BMM medium (Agrawal and Sarma 1982) and three temperatures (25, 30, and 40 °C). For the second step, a microalgae biorefinery was designed to produce proteins, chlorophyll, and fatty acids relaying on the experimental data using the SuperPro Designer software. The software produces the evaluation for the mass and economic impacts in the process.

2 Materials and Methods

The experiments study the productivity of biomass under different environmental conditions including temperature and salinity. The experiments were carried out with three different salinities (30, 40, and 50 ppt) of BMM medium (Agrawal and Sarma 1982); these conditions are chosen to simulate seawater salinity in the UAE

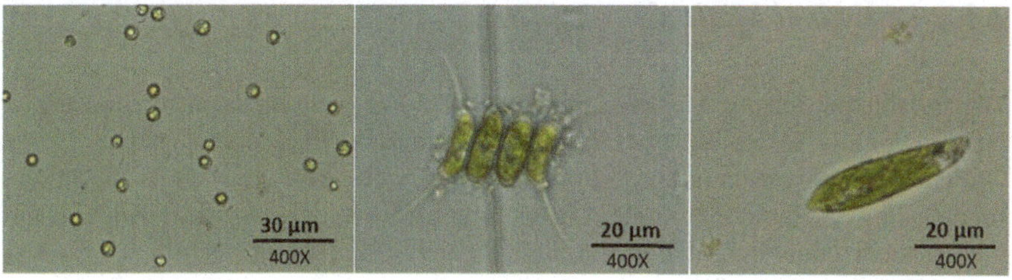

Fig. 1 Algae strains isolated from United Arab Emirates. From left to right: *Chlorella*, *Euglena*, and *Scenedesmus*

Arabian Gulf coast. Moreover, the effect of three temperatures (25, 30, and 40 °C) is investigated on the growth of algae strains.

2.1 Microorganism

The collection of different strains began during the months of November and December 2015, in some main locations in the UAE, which included Abu Dhabi, Al Ain, Ras Al Khaimah, and Fujairah. The samples were stored as batches for later examination. Filtration and cleaning were carried out to remove the unwanted specimens and other impurities from the samples as they are collected from the wild. Each sample contained a mixture of different microalgae strains.

Three strains were recognized to be the most abundant in all samples. The three strains were *Chlorella, Euglena*, and *Scenedesmus*; see Fig. 1. Other strains were also identified including *Anabaena, Diatoms Microcystis, Oscillatoria, Spirulina, Spirogyra*, and *Phormidium*, but they were not abundant.

2.2 Experimental Setup

The microalgae were cultured in 200 mL BMM medium (Agrawal and Sarma 1982), in 500-mL Erlenmeyer flasks through which atmospheric gases are continuously bubbled. Waste gases, such as O_2, are blown out of the tubes by the incoming gases, thus preventing changes in the atmospheric composition due to cellular metabolism (Thomas and Herbert 2005). The air is supplied through 220–250 V pumps at a follow rate of 2.8×10^{-3} m^3/s. The selected algae species are cultured up to 20–30 days under illumination of white fluorescent lamps [lumen: 2400 ± 100] for 8 h daily. Water baths were used to control the culturing temperature, and thermometers were connected to each water bath to monitor any deviation in temperatures. For the salinity tests, artificial seawater media were prepared using Instant Ocean salt (Aquarium Systems, Inc.).

2.3 Analytical Methods

The microalgal density was determined using a hemocytometer, a method developed originally for counting cells in blood samples (Phelan and Lawler 2001). Growth kinetics were monitored by counting the cells every day under the microscope using the hemocytometer. After counting the cells, several calculations were made to count the cell concentration in 1 mL of the sample as follows (Lobban et al. 1988):

1. Percent of viable cells:

$$\%\text{viable cells} = \frac{\text{number of viable cells}}{\text{total number of cells}} \times 100$$

2. Average number of cells per square of the hemocytometer:

$$\text{Average number of cells per square} = \frac{\text{number of viable cells}}{\text{number of squares}}$$

3. Dilution factor:

$$\text{Dilution factor} = \frac{\text{Final volume}}{\text{Cell volume}}$$

4. Concentration (viable cells/mL):

$$\text{Concentration} = \text{average number of cells per square} \times \text{Dilution factor} \times 10^4$$

Plots of number of cells against time (in days) are made (available in the results section); from these curves specific growth rate (u) and generation time (t_g) are calculated.

In the stationary phase, growth can be limited by resources such as light or nutrients. Growth rate (u) is calculated with the following equation (Lobban et al. 1988; Andersen 2005):

$$u = \frac{\text{In } X_2 - \text{In } X_1}{t_2 - t_1} \quad (1)$$

where X_1 and X_2 are densities at times t_1 and t_2.

The division time (t_g) is calculated with the following equation:

$$t_g = \frac{0.6931}{u} \quad (2)$$

The generation or doubling time (G_t) can also be calculated once the division time is known as follows:

$$G_t = \frac{1}{t_g} \tag{3}$$

2.4 Process Description for Biorefinery

To evaluate the potential use of microalgae as source of protein, chlorophyll, and fatty acids in the UAE, a full-scale biorefinery was designed. The processing pathways are illustrated in Fig. 2 with descriptions of the inputs, outputs, and processes for each station in producing protein, stabilized chlorophyll, and fatty acids. The processes chosen for each module were selected to meet three feasibility criteria which are scalability, low energy demand, and low cost.

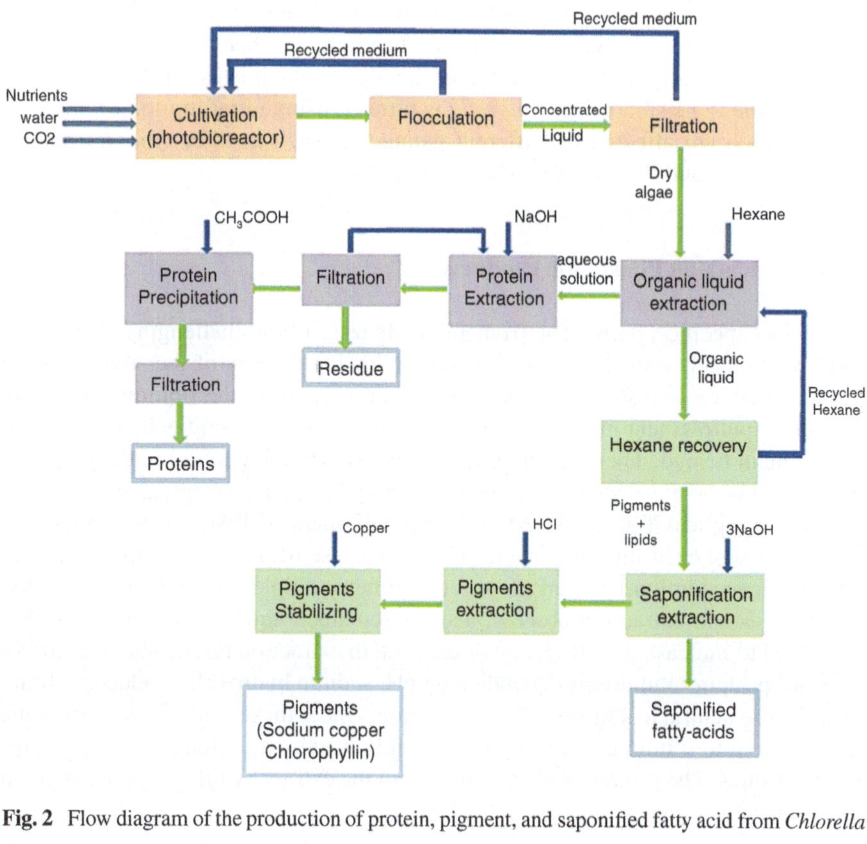

Fig. 2 Flow diagram of the production of protein, pigment, and saponified fatty acid from *Chlorella*

Large-scale production of microalgae has several limitations; these include maintaining mass culture stability, developing low cost harvesting methods, and optimizing strains for energy yield (Lundquist et al. 2010). The experiment was carried out to obtain optimum conditions for biomass cultivation and harvesting of local strain as an attempt to overcome these barriers.

2.4.1 Cultivation and Harvesting Phases for Microalgae

The cultivation process is considered to take place in tubular photobioreactor, in the presence of phosphorus, nitrogen, and carbon dioxide. These elements are essential during the process of photosynthesis where carbon dioxide is ingested by the algae to make the by-products which are useful in determination of the end products. The photobioreactor will therefore be given optimal conditions for the growth of the algae. This can greatly improve the quantity and quality of the harvested algae (Pulz 2001).

The harvesting of the algae can be accomplished by combining two techniques to maximize recovery of the biomass, flocculation followed by filtration. Flocculation allows the treatment of large quantities of microalgal culture and can be applied to a wide range of species (Safi et al. 2014b). The product is then further dewatered by filtration to increase the concentration of the algae, which is essential for the next process. Due to its small size, *Chlorella* is filtrated using ultrafiltration or microfiltration which are affected by different parameters such as filter type, transmembrane pressure, and flow velocity (Safi et al. 2014b).

2.4.2 Extraction Phases for Products

Extracting specific component from microalgae is often challenging due to the rigidity of its cell wall (Safi et al. 2014a). Following the aim of the study, organic liquid extraction unit is used to separate the proteins from the lipid/pigment. The nonpolar lipid/pigment molecules are soluble in nonpolar solvent, whereas the proteins remain behind. Therefore, hexane is used as extracting liquid in the proposed design as it is the most widely accepted extracting liquid in conceptual algal process designs (Gong and You 2014). After the lipid/pigment is dissolved in hexane, the solution is separated into two layers. This is because hexane is a nonpolar solvent and the algae solution is based on water. The hexane lipid/pigment system is then sent to a second extraction process, and the remained aqueous solution is further processed to increase the efficiency of the protein extraction (Gong and You 2014).

Since protein solubility is dependent on pH, sodium hydroxide is added, to bring the pH from neutral to slightly alkaline reaction. Under these conditions, most of the proteins go into solution, and this process is followed by filtration to remove microalgae residues. The protein is precipitated from the extract by bringing the pH down to the isoelectric region using acetic acid. The type of acid used or the temperature of precipitation does not affect the yield or purity of precipitated protein. The precipitated protein is then separated by filtration (Piorreck et al. 1984).

Pigment extraction is conducted by a saponification process. Several studies have been performed to extract pigments from photosynthetic organisms including algae (Wall 1951; Han et al. 2013; Soares et al. 2016). For instance, Wall (1951) provided the process to sodium copper chlorophyllin for pigment production from chlorophyll using alkaline hydrolysis and copper. Han et al. (2013) used simultaneous aqueous two-phase extraction and saponification reaction of chlorophyll based on silkworm excrement to obtain sodium chlorophyllin. Soares et al. (2016) performed the analysis for the extraction of pigments from microalgae by identifying the effect of solvent types on productions. Soares et al. (2016) evaluated potassium hydroxide alkaline saponification with the use of ethanol and methanol to extract pigments.

Similarly, in this work, saponification reaction is used to obtain the sodium copper chlorophyllin for pigment industries from chlorophyll. The nature of the chlorophyll is unstable, and the chlorophyll can be easily decomposed from the light, heat, acid, and alkali conditions (Soares et al. 2016). To solve this problem, the chlorophyll is made stable by adding copper acetate solution in chlorophyll, which results in sodium copper chlorophyllin. This solution is used for food, cosmetic, and pharmaceutical coloring (Tumolo and Lanfer-Marquez 2012).

Pigment extraction was achieved following the study of Wall (1951), which utilizes saponification reaction. For the alkaline saponification reaction, chlorophyll is saponified with NaOH, and the resulting mixture is acidified with HCI to produce chlorine. Copper acetate solution is added to obtain copper chlorine. The water-soluble sodium copper chlorophyllin obtained can be used for pigment industries. Besides the pigments, saponified fatty acids are also produced which are commercialized for the production of soap. Chlorophyll-a ($C_{55}H_{72}O_5N_4Mg$) and chlorophyll-b ($C_{55}H_{70}O_5N_6Mg$) are present in photosynthetic organisms. Generally, the chlorophyll-a form is the predominant. Thus, we focus on the chlorophyll derivative with the chlorophyll-a in the present work, and the chlorophyll-a produces sodium copper chlorophyllin. The chlorophyll derivative can be as follows:

1. Alkaline hydrolysis for saponification reaction shows

$$C_{55}H_{72}O_5N_4Mg\left(\text{Chlorophyll a,water insoluble}\right)+3NaOH\rightarrow C_{34}H_{31}O_6N_4MgNa_3$$
$$\left(\text{Sodium magnesium chlorophyllin,water soluble}\right)$$

2. Add hydrogen for replacement of magnesium and sodium:

$$C_{34}H_{31}O_6N_4MgNa_3\left(\text{Sodium magnesium chlorophyllin,water soluble}\right)$$
$$+5HCI\rightarrow C_{34}H_{36}O_6N_4\left(\text{Chlorine,water insoluble}\right)$$

3. Add copper for replacement of hydrogen:

$$
\begin{aligned}
&C_{34}H_{36}O_6N_4\,(\text{Chlorin e,water insoluble})\\
&+CuC_4H6O_4 \rightarrow C_{34}H_{34}N_4CuO_6\\
&(\text{Copperchlorin e,water insoluble / copperchlorophyllin})
\end{aligned}
$$

4. Sodium copper chlorophyllin:

$$
\begin{aligned}
&C_{34}H_{34}N_4CuO_6\,(\text{Copperchlorin e,water insoluble})\\
&+3NaOH \rightarrow C_{34}H_{31}N_4CuNa_3O_6\\
&(\text{Sodiumcopperchlorin e,water soluble / sodium copper chlorophyllin})
\end{aligned}
$$

The final production of $C_{34}H_{31}N_4CuNa_3O_6$ (sodium copper chlorophyllin) is used for the industry of pigments.

3 Results and Discussion

3.1 Experimental Results

3.1.1 Temperature Test

Figure 3 illustrates the effect of the temperature on the biomass growth of the three isolated strains. The initial growth of *Chlorella*, *Euglena*, and *Scenedesmus* biomass at an early lag phase is similar at the three tested temperatures. However, differences in the exponential phase are observed. A sharp increase in concentration is

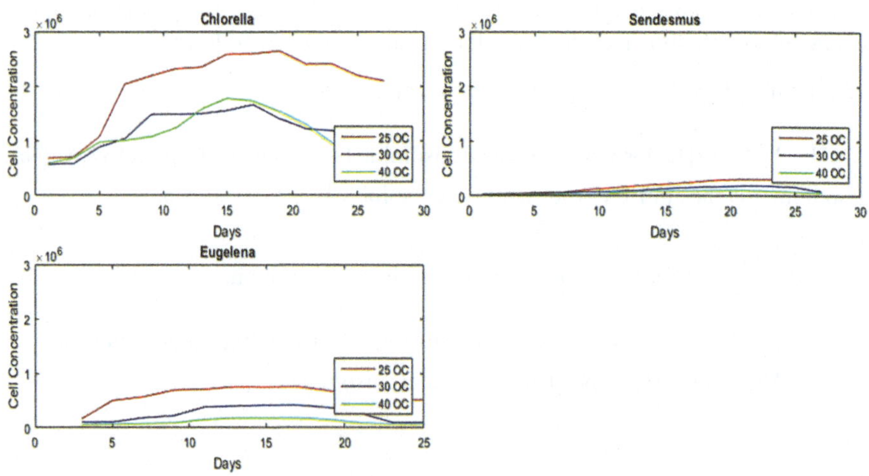

Fig. 3 Effect of temperature on the growth of *Chlorella*, *Scenedesmus*, and *Euglena*

characteristic of the growth medium at 25 °C during the exponential phase of the cultivation. Algae strains survive under the investigated temperatures, but the biomass growth decreases with the temperature increase. Cultivation at 25 °C shows the highest growth for all species with doubling time every 48 h for *Chlorella* and *Scenedesmus*. *Euglena*, on the other hand, shows a lower growth rate with doubling time every 72 h.

Table 1 illustrates the calculated growth rate, division time, and doubling time of each strain at the three tested temperatures using Eqs. 1, 2, and 3. The critical day of the biomass growth for *Chlorella* at the three tested temperatures is day 20, when the biomass concentration starts to decrease rapidly until it reaches a constant concentration. The critical days of the biomass growth for *Euglena* and *Scenedesmus* are days 17 and 23, respectively.

3.1.2 Salinity Test

The experiment is carried out with three different salinities (30, 40, 50 g/L) at 25 °C of BMM medium. These conditions are chosen to simulate seawater salinity. The growth rate, division time, and doubling time are calculated at the tested salinities based on each strain. In this section a comparison of the different test results is shown for each strain. Figure 4 presents the effect of the salinity on the biomass growth of the three different algae strains. It can be noticed from the figures that the initial growth of *Chlorella*, *Euglena*, and *Scenedesmus* biomass at early lag phase is similar on the three tested salinities; however, some variations in the exponential phase are observed. The highest algae growth is found out to be at 30 g/L followed by 40 and 50 g/L. Algae strains survive under the tested salinities, but the biomass growth decreases as the salinity increases.

From Table 2, it can be observed that *Scenedesmus* can stand high salinities and it has the highest growth compared to the other species with doubling time every

Table 1 Algae growth results (temperature test)

Strain	Condition (°C)	X_1[a]	X_2	Growth rate [%]	Division time/day	Generation time[b]
Chlorella	25	1,088,000	2,038,947	0.31	0.45	2.21
	30	596,000	1,048,421	0.14	0.20	4.91
	40	604,000	991,579	0.12	0.18	5.59
Euglena	25	168,421	505,263	0.22	0.32	3.15
	30	113,684	227,368	0.17	0.25	4.00
	40	33,684	100,000	0.14	0.20	5.10
Scenedesmus	25	36,000	260,000	0.33	0.48	2.10
	30	23,000	145,000	0.15	0.22	4.52
	40	20,000	93,000	0.13	0.18	5.41

[a]X_1 and X_2 are defined in Eq. 1
[b]Doubling time

48 h. *Chlorella* and *Euglena*, on the other hand, show a lower growth rate with doubling time every 72 h.

Table 2 illustrates the calculated growth rate, division time, and doubling time of each strain. The critical days of the biomass growth for *Chlorella* at the three tested salinities are identified as day 21, when the biomass concentration starts to decrease rapidly until it reaches a constant concentration. The critical days of the biomass growth for *Euglena* and *Scenedesmus* are days 17 and 19, respectively.

3.1.3 Selection of Microalgae for Biorefinery

Based on the experimental results of the three isolated microalgae, *Chlorella* containing 45% protein, 20% fat, 20% carbohydrate, 5% fiber, and 10% minerals and vitamins (5% chlorophyll) shows the highest potential to be used as a biomass feedstock in the UAE. Considering the environment of the UAE, *Chlorella* could survive

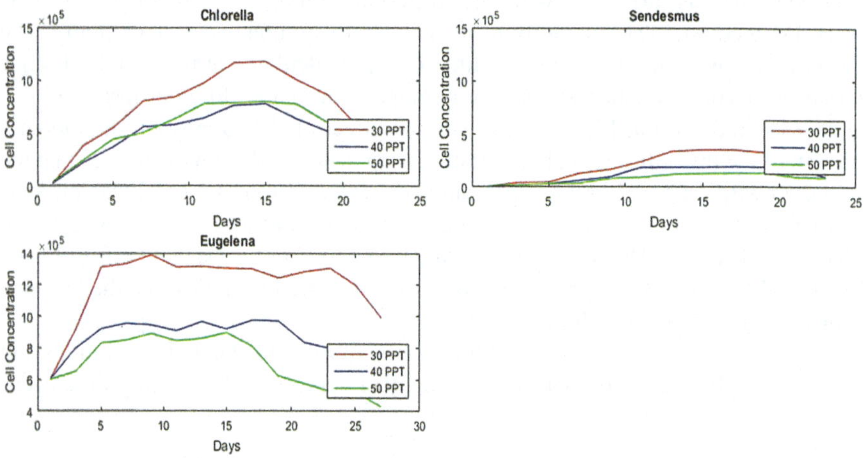

Fig. 4 Effect of salinity on *Chlorella, Scenedesmus, and Euglena* growth

Table 2 Algae growth results (salinity test)

Strain	Condition (ppt)	X_1	X_2	Growth rate	Division time/day	Generation time
Chlorella	30	604,000	1,312,632	0.19	0.28	3.57
	40	604,000	800,000	0.14	0.20	4.93
	50	652,632	831,579	0.12	0.17	5.72
Euglena	30	20,000	70,000	0.18	0.26	3.87
	40	29,000	42,000	0.19	0.27	3.74
	50	40,000	60,000	0.20	0.29	3.42
Scenedesmus	30	6316	101,000	0.28	0.40	2.50
	40	2105	7000	0.20	0.29	3.46
	50	10,526	24,211	0.17	0.24	4.16

in high temperatures and high salinities, indicating the possible use of seawater as a medium for growth. Moreover, although *Euglena* survived with a higher growth rate than *Chlorella* and *Scenedesmus* in a very high salinity, the lipid content of *Euglena* was very low compared to *Chlorella*. Lab scale experiments help to maintain the specified conditions for each strain of algae, and this provides a better understanding of the whole process to design a large-scale production.

All the experiments that are performed have shown that microalgae can be used as a source of biomass in the UAE.

3.2 Simulation for Economic Evaluation

We used the software SuperPro Designer to simulate the proposed process; this is done by providing input parameters for the biorefinery process. The process is assumed to be continuous. In the simulation for the economic evaluation, direct cost segments are chosen based on typical percentage of fixed-capital investment values for multipurpose plants (Peters and Timmerhaus 1991). The input parameters for each stream in the SuperPro Designer are shown in Table 3.

The purchasing prices of materials and the selling prices of the productions are assigned from market websites for chemical and industrial markets and from default price provided by the SuperPro Designer. The purchasing prices of materials used in each reactor are 8$/kg for nitrogen, 2.756$/kg for phosphorus, 10$/m^3 for water, 0.304$/kg for hexane, 5.250$/kg for sodium hydroxide, 0.085$/kg for hydrochloric acid, and 9.440$/kg for copper. The prices for nitrogen, phosphorus, hexane, and hydrochloric acid are obtained from ICIS which is the petrochemical market information provider. Also, the prices of sodium hydroxide and copper solution are gained from the price list of SD Fine-Chem Limited which is one of the largest lab chemical companies. The price of industrial water in Abu Dhabi is obtained from Abu Dhabi Water and Electricity Authority (Regulation and Supervision Bureau).

As proven by the experiments, *Chlorella* has high resistance against harsh conditions and invaders, which make it ideal as a source of proteins, enzyme, lipids, carbohydrates, pigments, vitamins, and minerals. Because it consists mostly of proteins, it is reasonable to design the biorefinery based on the protein production first.

Table 3 Input parameters in SuperPro Designer for streams

Streams	Input parameters
S-101	Carbon dioxide, nitrogen, phosphorus, water
S-107	Hexane
S-119	Sodium hydroxide
S-120	Copper
S-121	Hydrochloric acid

A conventional protein biorefinery from soy can produce up to 30 tons of protein per day (Miner 1976). Accordingly, we set our biorefinery protein production based on that but a bit lower as microalgae are not as mature industry as soy. The production of protein is targeted to be 10 tons of protein per day, and based on that the inputs are calculated.

Based on the input parameters and the biorefinery process as shown in Fig. 2, a simple design for the microalgae biorefinery process was made (see Fig. 5).

Using the SuperPro Designer, we obtain the productions such as proteins including water (S-109), saponified fatty acid (S-117), and pigment (S-118). Other streams produce residuals including ash and water. Based on the general market prices for the production, we provide the estimated revenue as 5$/kg for proteins, 1.3$/kg for saponified fatty acids, and 26$/kg for sodium copper chlorophyllin (pigments). The prices of the products are obtained from an online marketplace company (Alibaba.com). Then, we evaluate the economic impact for the productions with the software.

From the SuperPro Designer, we have Table 4 representing executive summary for the economic evaluation.

Based on Table 4, we can identify that the protein provides the main revenue and the pigment creates a large amount of the revenue. The total revenues with the proteins, pigments, and saponified fatty acids are 173,906,000$/year, and the payback time is 2.62 years. If we produce only the protein from the *Chlorella* biorefinery process, the total revenues are 107,977,000$/year with the payback time of 6.38 years. Thus, by generating the pigment and the saponified fatty acid from the *Chlorella* biorefinery process, we can improve the total revenues and reduce the pay time. Figure 6 illustrates the revenues for the three productions.

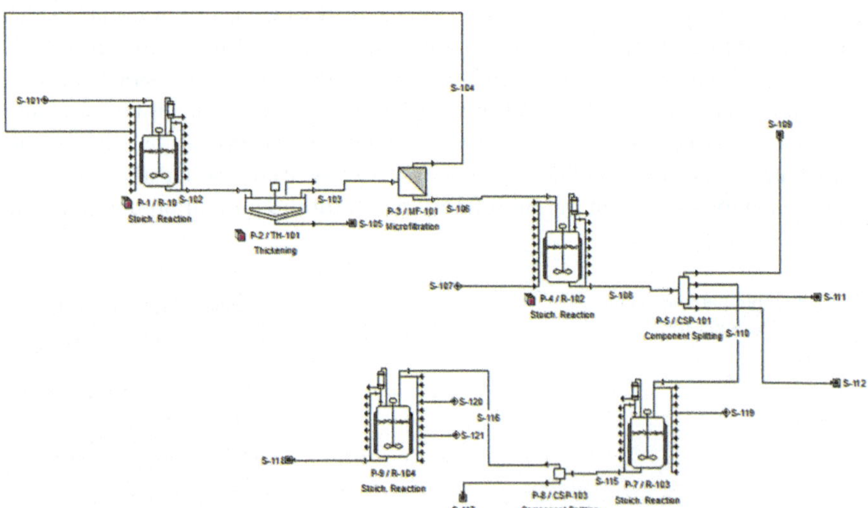

Fig. 5 Chlorella biorefinery process scheme based on the SuperPro Designer

Table 4 Executive summary for the economic evaluation derived from the SuperPro Designer based on the *Chlorella* biorefinery process

Total capital investment	138,127,000$
Operating cost	105,613,000$/year
Protein revenue	107,977,000$/year
Pigment revenue	62,514,000$/year
Saponified fatty acid revenue	3,415,000$/year
Total revenues	173,906,000$/year
Gross margin	39.27%
Return on investment	38.22%
Payback time	2.62 years

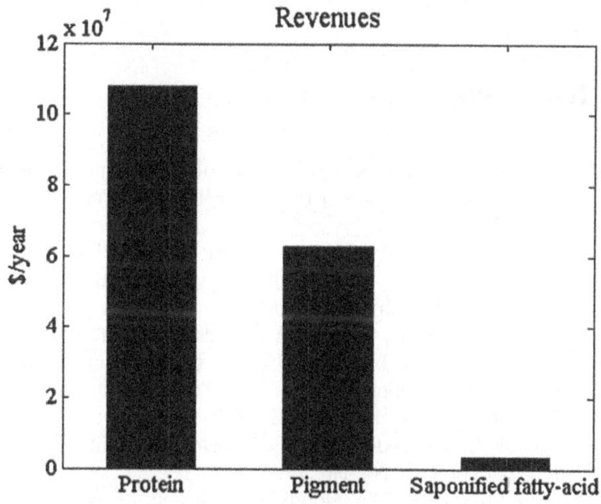

Fig. 6 The three revenues for the protein of 107,977,000$/year, the pigment of 62,514,000$/year, and the saponified fatty acid of 3,415,000$/year

4 Conclusions

The paper provides a detailed design of the microalgae biorefinery production process of proteins, pigments, and lipids. For the analysis of the microalgae biorefinery, three strains such as *Chlorella*, *Euglena*, and *Scenedesmus* are investigated to determine the appropriate alga in different environments and to generate precious products for industries. Based on the study, we select *Chlorella* which has the strong resistance against harsh conditions and invaders as arid and semiarid conditions. This alga contains several essential nutrients including proteins, fats, and chlorophylls which provide benefits for the society and economy.

From the composition of *Chlorella*, the microalgae biorefinery is designed based on the SuperPro Designer to produce 10 tons of proteins per day with pigments defined as the sodium copper chlorophyllin and lipids in the form of saponified fatty acids. All the design aspects are incorporated in the paper. We provide a discussion on individual steps that are involved in the algae biorefinery. The processes are algae growth, harvesting, lipid extraction, conversion of lipids into proteins, and generation of pigments with fatty acids through saponification reaction. A detailed design of the biorefinery is illustrated by indicating the various parameters relevant for actual implementation of the whole process.

In the SuperPro Designer process, an economical evaluation is performed for the three products such as proteins, pigments, and saponified fatty acids. The three products generate the total revenues of 173,906,000$/year, and the payback time shows 2.62 years. When only one product (proteins) is created from the biorefinery, the total revenues present 107,977,000$/year with the payback time of 6.38 years. Thus, the products for pigments and fatty acids improve the total revenues. The microalgae biorefinery using *Chlorella* has the potential revenues by providing valuable products to industries.

References

Agrawal SC, Sarma YSRK (1982) Effects of nutrients present in Bold's basal medium on the green alga Stigeoclonium pascheri. Folia Microbiol (Praha) 27:131–137. https://doi.org/10.1007/BF02879772

Andersen RA (ed) (2005) Algal culturing techniques, 1st edn. Academic, Cambridge

Aresta M, Dibenedetto A, Dumeignil F (eds) (2012) Biorefinery: from biomass to chemicals and fuels. [Walter] de Gruyter, Berlin

Bastidas-Oyanedel JR, Fang C, Almardeai S et al (2016) Waste biorefinery in arid/semi-arid regions. Bioresour Technol 215:21–28. https://doi.org/10.1016/j.biortech.2016.04.010

Becker EW (2007) Micro-algae as a source of protein. Biotechnol Adv 25:207–210. https://doi.org/10.1016/j.biotechadv.2006.11.002

Böer B (1997) An introduction to the climate of the United Arab Emirates. J Arid Environ 35:3–16. https://doi.org/10.1006/jare.1996.0162

Cherubini F (2010) The biorefinery concept: using biomass instead of oil for producing energy and chemicals. Energy Convers Manag 51:1412–1421. https://doi.org/10.1016/j.enconman.2010.01.015

Dasgupta CN (2015) Algae as a source of phycocyanin and other industrially important pigments. In: Das D (ed) Algal biorefinery: an integrated approach. Springer, Berlin, pp 253–276

Fu W, Chaiboonchoe A, Khraiwesh B et al (2016) Algal cell factories: approaches, applications, and potentials. Mar Drugs 14:1–19. https://doi.org/10.3390/md14120225

Gong J, You F (2014) Global optimization for sustainable design and synthesis of algae processing network for CO2 mitigation and biofuel production using life cycle optimization. AICHE J 60:3195–3210. https://doi.org/10.1002/aic.14504

Gouveia L (2015) From tiny microalgae to huge biorefineries. In: Serrano-Ruiz JC (ed) New microbial technologies for advances biofuels: toward more sustainable production methods. Apple Academic Press, Ontario, pp 55–76

Han J, Wang Y, Ma J et al (2013) Simultaneous aqueous two-phase extraction and saponification reaction of chlorophyll from silkworm excrement. Sep Purif Technol 115:51–56. https://doi.org/10.1016/j.seppur.2013.04.047

Hasanean H (2013) Middle east meteorology. Int Pacific Res Cent. http://iprc.soest.hawaii.edu/users/yqwang/Middle-eastmeteorology.doc. Accessed 19 Feb 2013

Lobban CS, Chapman DJ, Kremer BP, Phycological Society of America (1988) Experimental phycology: a laboratory manual. Cambridge University Press, Cambridge

Lundquist TJ, Woertz IC, Quinn NWT, Benemann JR (2010) A realistic technology and engineering assessment of algae biofuel production. Energy Biosciences Institute, California

Miner BD (1976) Edible soy proteins: operational aspects of production and marketing. Farmers Coop Service, US Department of Agri, FCS Res, Report 33:1–11
Peters MS, Timmerhaus KD (1991) Plant design and economics for chemical engineers. McGraw-Hill, New York
Phelan MC, Lawler G (2001) Cell counting. Curr Protoc Cytom Appendix 3:Appendix 3A. https://doi.org/10.1002/0471142956.cya03as00
Piorreck M, Baasch KH, Pohl P (1984) Biomass production, total protein, chlorophylls, lipids and fatty acids of freshwater green and blue-green algae under different nitrogen regimes. Phytochemistry 23:207–216. https://doi.org/10.1016/S0031-9422(00)80304-0
Pulz O (2001) Photobioreactors: production systems for phototrophic microorganisms. Appl Microbiol Biotechnol 57:287–293. https://doi.org/10.1007/s002530100702
Safi C, Ursu AV, Laroche C et al (2014a) Aqueous extraction of proteins from microalgae: effect of different cell disruption methods. Algal Res 3:61–65. https://doi.org/10.1016/j.algal.2013.12.004
Safi C, Zebib B, Merah O et al (2014b) Morphology, composition, production, processing and applications of Chlorella vulgaris: a review. Renew Sust Energ Rev 35:265–278. https://doi.org/10.1016/j.rser.2014.04.007
Seth JR, Wangikar PP (2015) Challenges and opportunities for microalgae-mediated CO2 capture and biorefinery. Biotechnol Bioeng 112:1281–1296. https://doi.org/10.1002/bit.25619
Soares AT, Júnior JGM, Lopes RG et al (2016) Improvement of the extraction process for high commercial value pigments from Desmodesmus sp. microalgae. J Braz Chem Soc 27:1083–1093. https://doi.org/10.5935/0103-5053.20160004
Thomas DJ, Herbert SK (2005) An inexpensive apparatus for growing photosynthetic microorganisms in exotic atmospheres. Astrobiology 5:75–82. https://doi.org/10.1089/ast.2005.5.75
Tumolo T, Lanfer-Marquez UM (2012) Copper chlorophyllin: a food colorant with bioactive properties? Food Res Int 46:451–459. https://doi.org/10.1016/j.foodres.2011.10.031
Ullah K, Ahmad MS, Sharma VK et al (2014) Algal biomass as a global source of transport fuels: overview and development perspectives. Prog Nat Sci Mater Int 24:329–339. https://doi.org/10.1016/j.pnsc.2014.06.008
Wall ME (1951) Preparation of chlorophyll derivatives for industrial and pharmaceutical use. US Dept of Agriculture, Bureau of Agricultural and Industrial Chemistry, 299

Factors Affecting Seawater-Based Pretreatment of Lignocellulosic Date Palm Residues

Chuanji Fang, Mette Hedegaard Thomsen, Christian Grundahl Frankær, Juan-Rodrigo Bastidas-Oyanedel, Grzegorz Przemyslaw Brudecki, and Jens Ejbye Schmidt

1 Introduction

Biorefinery in semiarid and arid regions is challenging due to the limitation of biomass availability and scarcity of freshwater (Bastidas-Oyanedel et al. 2016). The innovative concept of using seawater as reaction media for large-scale applications in biorefineries, to our knowledge, was firstly proposed in 2013 (Domínguez de María 2013). Seawater, a renewable and widely available water resource containing various inorganic salts, was speculated to be an interesting option for biorefineries. A number of proof-of-concept seawater-based applications for chemoenzymatic, chemocatalytic, and fermentative processes have been developed in recent years (Grande and Domínguez de María 2012; Hongsiri et al. 2015, 2014; Lin et al. 2011; Mao et al. 2013; Ren et al. 2016).

Date palm (*Phoenix dactylifera* L.) is a major fruit crop in most Middle East countries, with an estimated annual yield of over six million tonnes of lignocellulosic residues (Bastidas-Oyanedel et al. 2016). Despite of that, valorization of date palm residues through a biorefinery concept is still in its infancy. A recent study of bioenergy potential toward local biomass in the United Arab Emirates

C. Fang · J.-R. Bastidas-Oyanedel (✉) · G. P. Brudecki
Chemical Engineering Department, Khalifa University of Science and Technology, Masdar Institute, Masdar City, Abu Dhabi, United Arab Emirates
e-mail: juan.bastidas@ku.ac.ae

M. H. Thomsen
Department of Energy Technology, Aalborg University, Esbjerg, Denmark

C. G. Frankær
Department of Chemistry, University of Copenhagen, København Ø, Denmark

J. E. Schmidt
SDU-Department of Chemical Engineering, Biotechnology, and Environmental Technology, University of Southern Denmark, Campusvej 55, DK-5230 Odense M, Denmark
e-mail: jesc@kbm.sdu.dk

© Springer Nature Switzerland AG 2019

J.-R. Bastidas-Oyanedel, J. E. Schmidt (eds.), *Biorefinery*,
https://doi.org/10.1007/978-3-030-10961-5_31

(UAE) showed that lignocellulosic date palm residues possessed the most promising potential for bioethanol production compared with other local biomass resources including mangroves, garden trimmings, halophytes, seagrass, macroalgae, and municipal solid waste (MSW) (Ashraf et al. 2016). Fang et al. (2015a) recently studied the valorization of date palm residues through bioethanol production, showing that hydrothermal pretreatment could be a promising approach for the sake of producing high digestible and fermentable solids and low inhibitory liquids. However, the conflict of large water consumption of biorefineries, 1.9–5.9 m^3 of water per 1 m^3 of biofuel produced (Wu and Chiu 2011), with the scarcity of freshwater, limits the application of hydrothermal pretreatment in the arid/semiarid regions. Seawater as a potential alternative to freshwater in the pretreatment of lignocellulosic date palm residues was first investigated by Fang et al. (2015b). It was encouraging to observe that using seawater-based pretreatment produced comparably digestible and fermentable solids to those produced by using freshwater (Fang et al. 2015b).

Many factors are hypothesized to affect the efficiency of seawater-based pretreatment. Temperature is one of the most important parameters affecting pretreatment severity (Ko et al. 2014; Nitsos et al. 2013). Increasing pretreatment severity by raising temperature tends to facilitate the destruction of cell wall and the degradation of cellulose, hemicellulose, and lignin, which accordingly would increase the enzymatic digestibility of biomass (Nitsos et al. 2013). However, the inhibitors generated at high temperature deactivate the enzymes and consequently decrease the digestibility (Ko et al. 2014; Nitsos et al. 2013). The combined inorganic chloride salts in seawater is another influencing factor. Many previous reports have presented the degradation capacities of inorganic salts (e.g., $FeCl_3$, NaCl, and $MgCl_2$) toward cellulose, hemicellulose, and lignin (Liu and Wyman 2006; Liu et al. 2009a, b; Loow et al. 2017, 2015). Catalysts tend to bring significant enhancement of pretreatment performance (Chen et al. 2011; da Costa Sousa et al. 2016; Harun et al. 2011). Acids and alkalis, as two major catalysts, have been frequently used in the leading pretreatment technologies such as dilute acid pretreatment (Chen et al. 2011), alkali pretreatment (Harun et al. 2011), and ammonia fiber expansion (AFEX) (da Costa Sousa et al. 2016). Dilute acid favors hydrolysis of the hemicellulose, whereas alkaline hydrolysis targets the lignin fraction (Loow et al. 2016).

The aim of this work is to study the effects of factors influencing seawater-based pretreatment of lignocellulosic date palm residues, accordingly paving the way of further process optimization. Three factors including pretreatment temperature (180–210 °C), salinity of seawater (20–50 parts per thousand, ppt), and catalysts (H_2SO_4, Na_2CO_3, and NaOH) were studied. Cellulose crystallinity analysis was used for physical characterization of pretreated solids. The enzymatic digestibility and fermentability of pretreated solids were used to compare the impacts of the three factors upon the performance of seawater-based pretreatment by statistical analysis.

2 Materials and Methods

2.1 *Raw Materials*

Leaflets were collected from date palm trees (*Phoenix dactylifera* L.) grown in Abu Dhabi in 2014. They were air-dried (98% dry matter) and stored at room temperature before use. The dried material was milled using a knife mill (IKA, 10 MF Basic) to pass through a 1 mm screen. Artificial seawater with different salinity (ppt) was made by dissolving sea salts (Aqua Medic, Bissendorf, Germany) into deionized water.

2.2 *Pretreatment*

Hydrothermal pretreatment with seawater was performed at different temperatures 180, 190, 200, and 210 °C with constant salt concentration of 35 ppt; at different salt concentrations 0, 20, 35, and 50 ppt with constant temperature of 190 °C; and using three different catalysts H_2SO_4, Na_2CO_3, and NaOH with constant temperature and salinity of 190 °C and 35 ppt, respectively. The catalyst was used at a ratio of 0.025 g/g dry biomass for each catalyst. The pretreatment was performed for 20 min, at the respective process temperatures. The biomass was used at 10 wt% dry matter loading in a Parr reactor (Parr Instrument Company, Moline, Illinois) with a working volume of 1 L. After the treatment, the reactor was cooled to 40 °C, and the pretreated materials were separated by vacuum filtration into solids (fibers) and liquids (filtrations) fraction. Water with around ten times weight of pretreated solids was used to wash out residue compounds in solid fraction. Both fractions were kept at 4 °C until analysis and further processing.

2.3 *X-Ray Diffraction (XRD)*

X-ray diffraction (XRD) was performed as described previously (Fang et al. 2015a). Date palm leaflets before and after pretreatment were dewaxed before being analyzed by Empyrean X-ray diffractometer (PANalytical, Eindhoven, Netherlands) equipped with a PIXcel3D detector and operated at 45 kV and 40 kA using Cu Kα radiation ($\lambda = 1.5418$ Å). Powder diffraction data were collected in reflection geometry in the 2θ range of 10–40° with a step size of 0.008° and a counting time of 10 s per step.

With respect to the overlapping and widely broadened diffraction peaks of cellulose, Rietveld method was employed to estimate the percent crystallinity in the biomass samples (Nishiyama et al. 2003). Refinements were performed by using the crystal structure of cellulose Iβ as input (Thygesen et al. 2005). The peak shapes were modelled by the Voigt function, and a total of 11 parameters were refined to find the best fit for all samples. The best parameter set included refinement of one overall scale factor, eight background parameters, one parameter accounting for the sample transparency effect, and one Voigt peak profile parameter accounting for the crystallite size perpendicular to the fiber direction. The crystallite size parallel to the fiber direction was fixed to 350 Å. This value was determined from another XRD experiment in which the sample was aligned according to the fiber direction. The background was modelled using an experimentally recorded pattern of amorphous cellulose, which was later smoothed and scaled. In addition to a scaling parameter for the amorphous pattern, seven Chebyshev background parameters were refined. Preferred orientation along the fiber direction was observed to some extent and included in the model by a fixed preferred orientation parameter along (001). Unit cell parameters were fixed as reported previously (De Figueiredo and Ferreira 2014), except for the untreated samples in which the unit cell parameter was set to 8.08 Å.

2.4 *Chemical Composition Analysis of Pretreated Solids*

Structural carbohydrates and lignin content in extractive-free biomass before and after pretreatment were determined according to the analytical procedure of the National Renewable Energy Laboratory (NREL) by two-step acid hydrolysis (Sluiter et al. 2011). The sugar analyses of hydrolyzates were performed using High Performance Liquid Chromatography (HPLC, Agilent 1260 Infinity Bio-inert Binary LC). A Hi Plex-H column (Agilent) and refractive index detector (RID) were used to determine the concentrations of glucose, xylose, and arabinose at 65 °C using 0.005 M H_2SO_4 as the mobile phase with a flow rate of 0.6 mL/min.

2.5 *Enzymatic Hydrolysis of Pretreated Solids*

Enzymatic hydrolysis was performed according to National Renewable Energy Laboratory protocol (Selig et al. 2008) using 100 g/L dry biomass loading, 15 FPU cellulase/g dry matter of biomass. Cellic® CTec2 (117 FPU/mL) as cellulase was used in this assay. The process was performed in the presence of 50 mM citrate buffer (pH 5.0) at 50 °C, and samples were shaken at 150 rpm for 72 h. Glucose released during the enzymatic hydrolysis were quantified by HPLC at the same operating conditions as applied in the acid hydrolyzed samples (described in Sect. 2.4). Glucan-to-glucose conversion was calculated by Eq. 1:

$$\begin{aligned}&\text{Glucan to glucose conversion}(\%)\\&=\left(C_{\text{glucose}}/\left(L*C_{\text{fibers}}/100\text{g TS}\right)\right)*100\%\end{aligned} \tag{1}$$

where $C_{glucose}$ = concentration of glucose measured in the enzymatic hydrolyzate [g/L], L = fibers loading [g/L], and C_{fibers} = content of glucan in the fiber fraction after the pretreatment [g/100 g TS].

2.6 *Simultaneous Saccharification and Fermentation (SSF) of Pretreated Solids.*

Simultaneous saccharification and fermentation followed the same procedures as described previously (Fang et al. 2015a). Pre-hydrolysis was conducted for 24 h at 50 °C and 150 rpm in a shaking bed incubator with 10% dry matter (DM) biomass loading. After the pre-hydrolysis, fermentation was performed by loading 2 g/L dry Baker's yeast (*S. cerevisiae*) at 32 °C, with constant shaking (90 rpm) for 72 h. Weight measurements of the flasks were recorded throughout the duration of the process. The weight loss (representing release of carbon dioxide) was translated into ethanol yield using Eq. 2:

$$W_{ethanol} = \left((1 \text{ mol EtOH})/(1 \text{ mol CO}_2)\right) * \left(M_{ethanol} / M_{CO_2}\right) * W_{CO_2} \quad (2)$$

where $W_{ethanol}$ = weight of ethanol (EtOH) produced [g], $M_{ethanol}$ = molar mass of ethanol [46 g/mol], M_{CO_2} = molar mass of CO_2 [44 g/mol], and W_{CO_2} = weight of CO_2 produced = weight loss of the fermentation flask [g].

Theoretical ethanol yield was determined based on the glucan content in the raw material following Eq. 3. Ethanol yield (%, $Y_{ethanol}$) was calculated as a percent ratio of the actual ethanol amount produced to the theoretical ethanol yield (Eq. 4):

$$\text{TY}_{ethanol} = 0.511 * C_{glucan} * L * 1.111 \quad (3)$$

where $\text{TY}_{ethanol}$ = theoretical ethanol yield [g], C_{glucan} = glucan content in fibers, L = fibers loading [g/L], 0.511 = glucose-to-ethanol factor, and 1.111 = glucan-to-glucose factor.

$$Y_{ethanol} = W_{ethanol} / \text{TY}_{ethanol} * 100\% \quad (4)$$

2.7 Analysis of Pretreatment By-products in Pretreatment Liquids

Liquid fractions were analyzed for the pretreatment by-products (mostly sugar degradation products) including acetic acid, furfural, and HMF. The analysis was performed using HPLC with the same operating conditions as described above in Sect. 2.4.

2.8 Chemical Composition Analysis of Pretreatment Liquids

To determine the quantity of the oligomers of xylose, arabinose, and glucose in the pretreated liquid fraction, dilute acid (8% w/w H_2SO_4 solution) hydrolysis was performed as reported previously (Cybulska et al. 2014). The ratio of the sample to acid was 1:1, producing final acid concentration of 4% w/w. The solution was autoclaved at 121 °C for 10 min. Sugars released in hydrolysis were analyzed by HPLC using the same method described in Sect. 2.4.

2.9 Inhibition Test of Pretreatment Liquids

Inhibition test of pretreatment liquids was determined using fermentation studies as reported previously (Cybulska et al. 2014). The liquid fractions obtained from the pretreatment were fermented in order to evaluate their inhibition to ethanol production by *S. cerevisiae*. Since the liquid fraction contains fewer hexoses (as compared to solid fractions) and more inhibitors, glucose and additional external nutrients were added to the fermentation solution to avoid the inhibition from lacking of nutrients. The fermentation solution contained 35 g/L of glucose and 4% of the stock solution (consisting of 62.5 g/L ammonium sulfate, 2.50 g of mono potassium phosphate, 5.00 g of yeast extract, and 0.75 g of hydrous magnesium sulfate). Sodium hydroxide was used to adjust the pH to 4.8. Flasks containing the fermentation solution with 2 g/L of *S. cerevisiae* were incubated at 32 °C and 100 rpm. Weight loss measurements were taken over the following 72 h, and the weight loss was converted to ethanol yield using Eq. 4.

2.10 Effect Screening of Factors for Pretreatment Using Seawater

Data of glucan-to-glucose conversions in enzymatic hydrolysis and ethanol yields in SSF were fitted by the variable combinations of temperature (180–210 °C), salinity (0–50 ppt), and catalysts (H_2SO_4, Na_2CO_3, and NaOH). The statistical analysis software, JMP (V12.0.1), was used to perform the fitting of models and the sequentially factor effects screening. As for the regression of the model, standard least squares were selected. The Analysis of Variance (ANVOA) of the model and the scaled estimates were performed by JMP software afterward.

3 Results and Discussion

3.1 Carbohydrates and Lignin Recoveries

The glucan, xylan, and lignin recoveries of solids treated at different temperature, salinity, and catalysts are shown in Fig. 1. Slight decreases of glucan recovery were observed when increasing temperature (from 180 to 210 °C) and salinity (from 0 to 50 ppt). The glucan recovery decreased gradually from 100.3% at 180 °C to 94.1% at 210 °C. Similarly, there was a 7.4% decrease of glucan recovery when raising the salinity from 0 ppt (102.1%) to 50 ppt (94.7%).

No changes of glucan recovery were observed for pretreatment of date palm residues with addition of different catalysts. Unlike kglucan recovery, obvious varia-

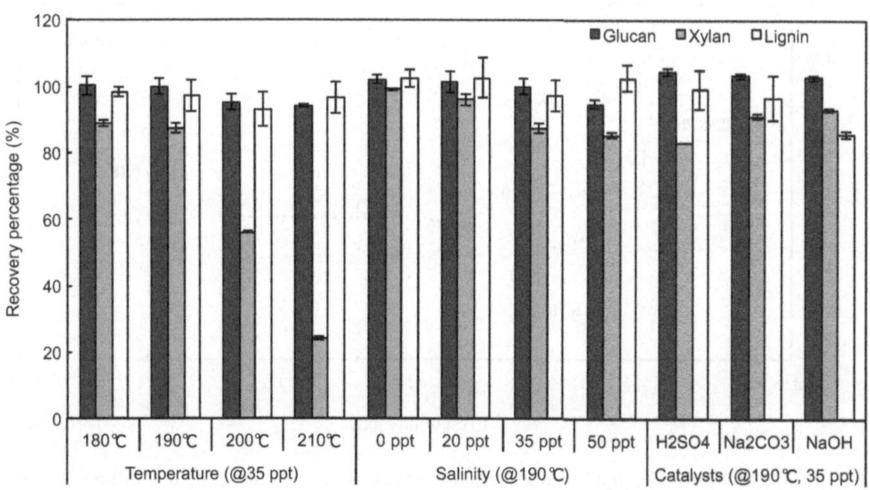

Fig. 1 Glucan, xylan and lignin recoveries of solids treated at different conditions by varying temperature, salinity, and catalysts

tions of xylan recovery were obtained by changing pretreatment conditions. Xylan recovery reduced drastically from 89.0% at 180 °C to only 24.1% at 210 °C. Salinity increase also facilitated the degradation of xylan, ranging from 99.1% at 0 ppt to 85.4% at 50 ppt. Addition of NaOH showed the highest xylan recovery (93.0%), while the lowest xylan recovery (83.1%) was obtained by addition of H_2SO_4. The variations of lignin recovery were small when changing temperature (less than 5.4%) and salinity (less than 0.3%). However, changing catalysts from H_2SO_4 to NaOH brought significant reduction of lignin recovery from 99.2 to 85.8%.

Cellulose is the most recalcitrant component in the cell wall. This can explain why only slight variations of glucan recovery are observed when changing temperature, salinity, and catalysts. By contrast, the significant changes of xylan recovery reflect the sensitivity of xylan. Compared with the variations of salinity and catalysts, changing temperature causes the largest variations of xylan recovery as shown

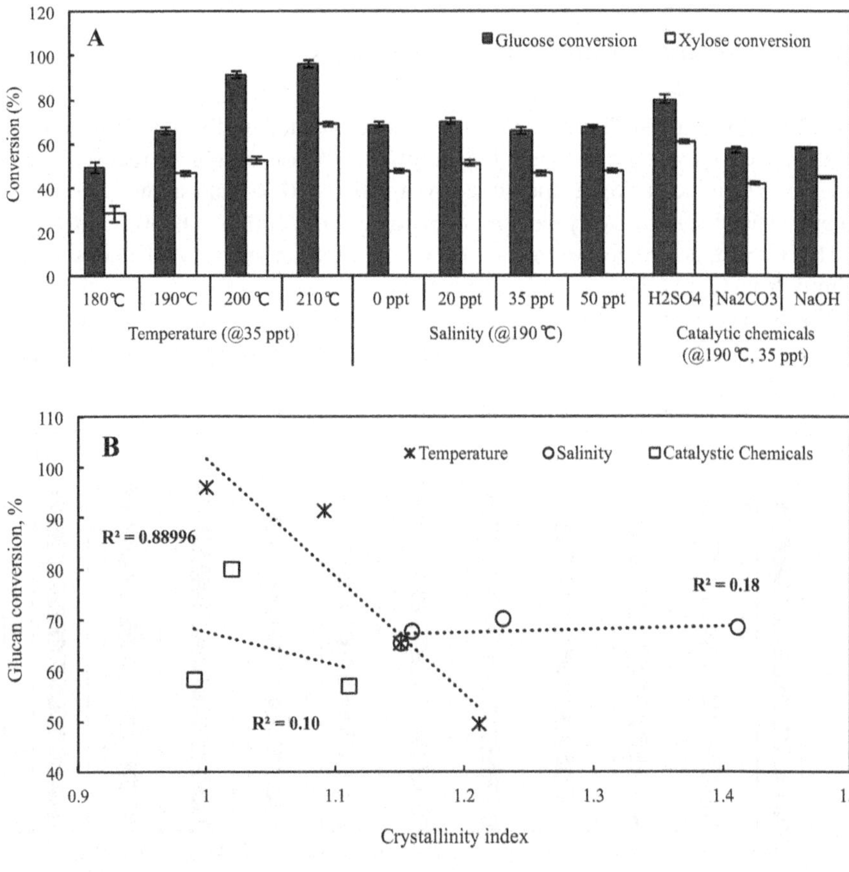

Fig. 2 Glucan-to-glucose and xylan-to-xylose conversion (**a**) of solids treated at different conditions and the corresponding (glucan-to-glucose conversion only) correlations with crystallinity index (**b**)

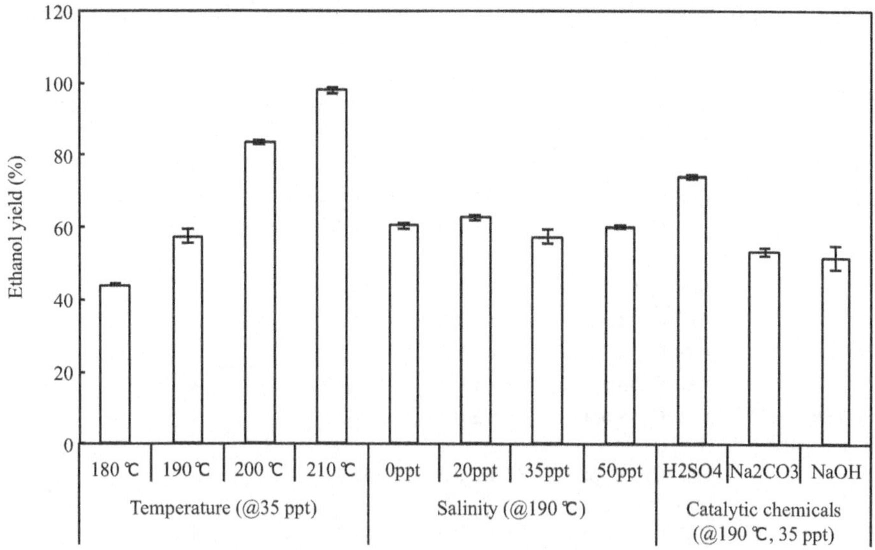

Fig. 3 Ethanol yield of pretreated solids obtained at different temperature, salinity, and catalysts in SSF

in Fig. 1. Addition of H_2SO_4 removes the largest portion of xylan compared with that done by pretreatment with addition of Na_2CO_3 and NaOH. It is possibly because acids are effective catalysts governing the degradation of xylan in lignocellulosic biomass (Chen et al. 2011; Marcotullio et al. 2011; Pu et al. 2013). Significant changes of lignin content were observed when using different catalysts. Higher removal of lignin was obtained using NaOH than using H_2SO_4. It is because alkalis are more competent for extraction of lignin than acids are (Chen et al. 2013; Hussin et al. 2013).

It should be noted that although high temperatures (200 and 210 °C) were observed to facilitate enzymatic digestibility and fermentability of pretreated solids (Figs. 2 and 3) in Sects. 3.3 and 3.4, they were not used in this study when investigating the effects of salinity and catalysts on pretreatment due to the low xylan recoveries (56.2% at 200 °C and 24.1 at 210 °C) as shown in Fig. 1. Instead, 190 °C was selected because of the higher xylan recovery (87.3%).

3.2 *Crystallinity Changes*

Crystallinity is one of the most crucial parameters to represent the recalcitrance of biomass, which is widely used for biomass characterization (Kumar et al. 2009). Table 1 shows the crystallinity variations of biomass treated at different conditions as well as the corresponding crystallinity index taking consideration of glucan content (Singh et al. 2015). The variations of crystallinity of biomass were small when

Table 1 Glucan content, crystallinity of biomass, and crystallinity index with consideration of glucan content of date palm residues treated at different conditions

Factors	Conditions	Glucan content (g/100 g dry biomass)	Crystallinity of biomass (%)	R_{wp}[a] (%)	Crystallinity index[a]
Temperature	180 °C, 35 ppt	31.0 ± 0.9	37.38	3.73	1.21
	190 °C, 35 ppt	32.1 ± 0.8	37.03	4.01	1.15
	200 °C, 35 ppt	33.2 ± 0.8	36.20	4.20	1.09
	210 °C, 35 ppt	33.1 ± 0.1	33.02	4.10	1.00
Salinity	190 °C, 0 ppt	27.9 ± 0.4	39.40	4.13	1.41
	190 °C, 20 ppt	32.8 ± 1.0	40.23	4.08	1.23
	190 °C, 35 ppt	32.1 ± 0.8	37.03	4.01	1.15
	190 °C, 50 ppt	33.3 ± 0.5	38.54	4.07	1.16
Catalytic chemicals[b]	190 °C, 35 ppt, H_2SO_4	33.5 ± 0.3	34.22	4.18	1.02
	190 °C, 35 ppt, Na_2CO_3	32.4 ± 0.3	35.97	3.93	1.11
	190 °C, 35 ppt, NaOH	34.5 ± 0.3	34.23	3.82	0.99

[a]Crystallinity index = Crystallinity of biomass/glucan content
[b]The loading concentration of catalytic chemicals is 0.025 g/g DM of biomass
[c]The weighted profile residual (Young 1995) given as $R_{wp} = [\sum_i w_i(I_{obsi} - I_{calci})^2/\sum_i w_i(I_{obsi})^2]^{1/2}$, where I_{obs} and I_{calc} are the observed and calculated intensity, respectively, and w is the weighting function, $1/I_{obs}$

changing temperature (33.02–37.38%), salinity (37.0–40.2%), and catalysts (34.2–36.0%). The weighted profile residual (R_{wp}) is varying from 3.73 to 4.2% for the fitting of diffraction data obtained from biomass treated at different conditions. In order to truly reflect the effects of pretreatment on cellulose crystallinity, cellulose content should be taken into account (Singh et al. 2015). The calculated crystallinity index decreased from 1.21 to 1.00 as the temperature increased from 180 to 210 °C. Similar trend was observed when the salinity increased from 0 to 50 ppt with crystallinity index reducing from 1.41 to 1.16 accordingly. No differences of crystallinity index were found when adding H_2SO_4 (1.02), Na_2CO_3 (1.11), and NaOH (0.99).

The set of refinement parameter used resulted in good fits ($R_{wp} \leq 4.20\%$) of all samples, as well as realistic backgrounds, which is a prerequisite for accurate determination of crystallinity. The reduction of crystallinity index caused by increased temperature from 180 to 210 °C is mainly because of pretreatment severity raised at high temperature, which subsequently favors the destruction of lignocellulosic cell wall. Nevertheless, to date, the reaction mechanism of inorganic salts on lignocellulosic biomass pretreatment has yet to be understood. Two mechanisms have been proposed to explain the behavior of inorganic salts. In the first mechanism, metal salts act as Lewis acid by dissolving in aqueous solvents to produce complex cations. The formation of the metal cations would aid in the cleavage of the glycosidic linkages (Peng et al. 2010; Román-Leshkov and Davis 2011). In the study of degradation of cellulose by zinc chloride, the zinc was found to coordinate to the glycosidic oxygen to help break down the glycosidic linkage, leading to the reduction of reaction energy. Then the coordinated water molecules of the metal cation complex participated as nucleophiles to yield D-glucose (Amarasekara and Ebede 2009). In the second mechanism, metal ions can undergo hydrolysis when they are mixed with water to produce the H_3O^+ ion. The aqueous metal solutions, therefore, present a Brønsted acid character similar to that of HCl (Zhang et al. 2014). Based on these mechanisms, it can be hypothesized that increasing the concentration of metal solution would expedite the breakdown of glycosidic linkage of cellulose and consequently reduce crystallinity of cellulose as shown in the present study.

3.3 *Effects of Different Pretreatment Conditions on Enzymatic Hydrolysis of Pretreated Solids*

Glucan-to-glucose and xylan-to-xylose conversions were used to evaluate the digestibility of solids treated at different conditions (Fig. 2a). Glucan-to-glucose conversion increased significantly as the temperature increased from 180 °C (49.3%) to 210 °C (96.0%). However, there were almost no changes of glucan-to-glucose conversion when increasing the salinity. The glucan-to-glucose conversions ranged from 67.3 to 69.8%. Compared with other two catalysts (Na_2CO_3 and NaOH), only the addition of H_2SO_4 showed boosting effects on glucan-to-glucose conversion. It raised glucan-to-glucose conversion from 65.6 to 80.0%. The xylan-to-xylose conversions for all pretreatment conditions were between 28.2 and 68.5%. The pattern of variations of xylan-to-xylose conversion was the same as that of glucose conversion. The correlations of glucan-to-glucose conversion with crystallinity index (Fig. 2b) were high when modifying the temperature ($R^2 = 0.89$). But there were no correlations between glucan-to-glucose conversion and crystallinity when changing salinity ($R^2 = 0.18$) and using different catalysts ($R^2 = 0.10$).

Correlations of physical (e.g., crystallinity, specific surface area (SSA), and degree of polymerization (DP)) and chemical (e.g., lignin, xylan, and acetyl group) features of biomass with enzymatic digestibility have been well studied (Jiang et al. 2016; Trajano et al. 2013; Zhang et al. 2012; Zhao et al. 2012; Zhu et al. 2008). Lignin, xylan, and acetyl group tend to bring negative effects to digestibility. As for crystallinity, there have not been clear correlations with digestibility (Zhao et al. 2012). Crystallinity was found to have negative effects on digestibility in the hydrothermal pretreatment of date palm residues (Fang et al. 2015a). The high correlations between crystallinity index and enzymatic digestibility when changing temperature (Fig. 2b) are consistent with Fang et al. (Fang et al. 2015a). However, no correlations between crystallinity index and enzymatic digestibility in the case of varying salinity and using different catalysts reflect the indeterminate effect of crystallinity on enzymatic digestibility as discussed in previous review (Zhao et al. 2012).

Increasing pretreatment temperature facilitates not only the removal of xylan and lignin (Fig. 1) but the reduction of crystallinity index; this can explain the positive effects of temperature on the digestibility. However, the decrease of xylan content and crystallinity observed when increasing salinity did not contribute to the digestibility enhancement, which indicates that other negative factors might influence the overall digestibility. It is worth noting that increasing salinity did not facilitate the removal of lignin (Fig. 2) that is inhibitory to enzymatic hydrolysis (Li et al. 2014; Zeng et al. 2014). Moreover, most of the ions in seawater (Ca^{2+}, Mg^{2+}, Na^{+}, K^{+}, Br^{-}, and Cl^{-}) could impose destabilizing effects on enzymes according to the Hofmeister series (Zhao 2005). In the present study, the enzymatic digestibility was reduced by 22% (data not shown) when the hydrolysis of untreated date palm leaflets was performed in artificial seawater (35 ppt) than in freshwater. The above two negative effects, therefore, are speculated to offset the positive effects derived from xylan removal and crystallinity reduction. Co-catalysis, using different salts together with the acidic catalysts (e.g., hydrochloric acid, sulfuric acid, and oxalic acid), has been shown to improve the dehydration of lignocellulosic derives (Hongsiri et al. 2015, 2014; Nguyen and Tucker 2002), which is also observed in our study when adding H_2SO_4 in seawater-based pretreatment. Addition of alkali (NaOH) or alkali salts (Na_2CO_3) shows inferior performance to pretreatment when using seawater alone, indicating that alkali catalysts may not be suitable for pretreatment with seawater.

3.4 *Effects of Different Pretreatment Conditions on Simultaneous Saccharification and Fermentation (SSF) of Pretreated Solids*

Figure 3 shows the ethanol yield of solids pretreated at the different temperature, salinity, and catalysts in SSF. The highest ethanol yield was obtained at 210 °C with 35 ppt salinity (98.0%), while the least was obtained at 180 °C with 35 ppt salinity (44.0%) within all the pretreatment conditions. It is noted that the trends in the profiles of ethanol yield against temperature, salinity, and catalysts were the same as that in the profiles of glucan-to-glucose conversion against above three variables (Fig. 2a).

The variations of glucan-to-glucose conversion and ethanol yield against different pretreatment conditions match very well. This phenomenon was also observed by Fang et al. (2015a) using hydrothermal pretreatment for processing date palm residues. The plausible explanation is that the enzymatic saccharification process throughout SSF determines the yield of glucose. The yield of glucose determines the yield of ethanol in fermentation. Also, the matching trends of ethanol yield and glucan-to-glucose conversion indicate that the remaining compounds (e.g., degrading compounds existing in pretreatment liquids) in pretreatment solids do not exert inhibitions to *S. cerevisiae* in fermentation under the experimental conditions tested. This was proved in the inhibition test of pretreatment liquids to *S. cerevisiae* (see

Sect. 3.5). Furthermore, the good match between glucan-to-glucose conversion and ethanol yield facilitates the further optimization of pretreatment conditions. In other words, the optimization objective can be enzymatic digestibility only due to the good match of glucan-to-glucose conversion with ethanol yield.

3.5 Effects of Different Pretreatment Liquids on the Inhibition of Yeast

In order to evaluate the inhibitory effects of pretreatment liquids on ethanol fermentation by *S. cerevisiae*, fermentability assays of pretreatment liquids at different conditions were performed (Fig. 4). The ethanol yield in fermentation with addition of liquids obtained from pretreatment using freshwater (Seawater_0ppt_190°C) was treated as the control. The liquid obtained at 210 °C with the salinity of 35 ppt (Seawater_35ppt_210°C) showed the most inhibitory effect on ethanol production. There was a drop of ethanol yield from 92.51% for liquids obtained at condition "Seawater_0ppt_190°C" to 66.98% for liquids obtained at condition "Seawater_35ppt_210°C" after 24 h' fermentation, while the difference of ethanol yield between them decreased in the fermentation at 48 and 72 h. As for the fermentation of all the other pretreatment liquids, slight or no reductions of ethanol yield were observed at 24, 48, and 72 h compared with that in the fermentation with addition of liquids obtained from pretreatment using freshwater.

The lag phase was only found in the fermentation with addition of pretreatment liquids obtained at 210 °C with the salinity of 35 ppt. While there is no lag phase in

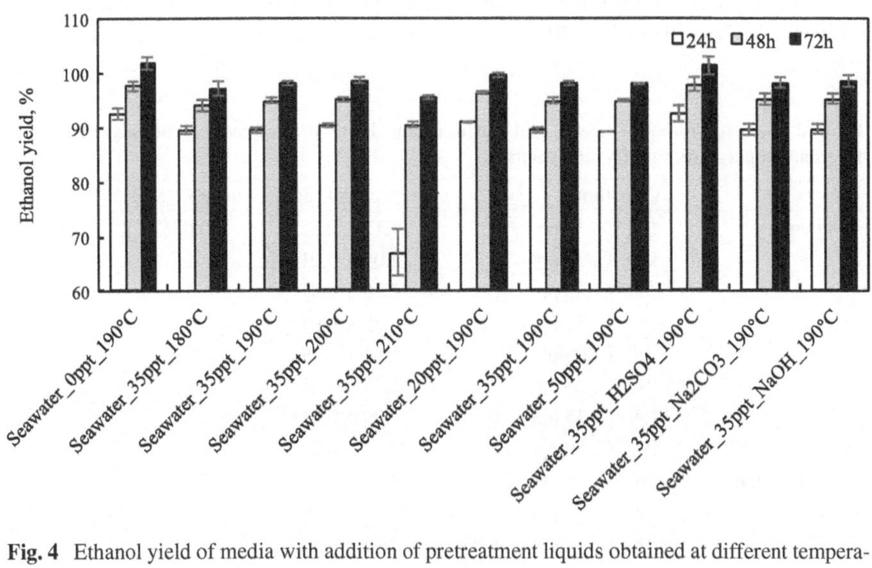

Fig. 4 Ethanol yield of media with addition of pretreatment liquids obtained at different temperature, salinity, and catalysts

the fermentation with addition of pretreatment liquids obtained at 200 °C with the same salinity, indicating the threshold temperature is between 200 and 210 °C. It was reported that the main toxic degradation compounds generated in the pretreatment of lignocellulosic biomass includes acetic acid, furfural, 5-(hydroxymethyl) furfural (HMF), and phenolic compounds (Huang et al. 2011; Loow et al. 2015). The production of inhibitors is heavily dependent on inorganic salt characteristics and pretreatment harshness (Loow et al. 2015). When the concentration of $FeCl_3$ increased from 100 to 200 mmol/L, Marcotullio et al. (Marcotullio et al. 2011) observed an increase of acetic acid formation from 1.6 to 2.6 wt% during the pretreatment of wheat straw at 100 °C. Nonetheless, the significant growth of acetic acid formation disappeared at 120 °C (Marcotullio et al. 2011). The results indicate that temperature determines the extent of acetic acid production rather than the inorganic salt does in the pretreatment, which was also observed in the present study. There was only a slight increase of the concentration of acetic acid in pretreatment liquids when increasing salinity from 0 ppt (1.96 g/L) to 50 ppt (2.25 g/L) at 190 °C (Table 2), while the concentration of acetic acid increased significantly from 1.15 g/L at 180 °C to 3.29 g/L at 210 °C (Table 2).

3.6 Screening of Factors for Pretreatment Using Seawater

Screening of temperature, salinity, and catalyst effect on enzymatic digestibility and fermentability of pretreated solids was performed by the statistic software, JMP (V12.0.1), to determine the most influential parameter of pretreatment using seawater. Two models were fitted for two different responses, i.e., enzymatic digestibility and ethanol yields of pretreated solids. The high R^2 vales (0.96 for enzymatic digestibility, 0.98 for ethanol yield) and low P values (<0.0001 for both enzymatic

Table 2 Concentrations of acetic acid, furfural, and HMF in pretreatment liquids obtained at different temperature, salinity, and catalysts

Factors	Conditions	Acetic acid (g/L)	Furfural (g/L)	HMF (g/L)
Temperature	180 °C, 35 ppt	1.15 ± 0.02	<0.005	<0.005
	190 °C, 35 ppt	2.09 ± 0.01		
	200 °C, 35 ppt	2.90 ± 0.01		
	210 °C, 35 ppt	3.29 ± 0.01		
Salinity	190 °C, 0 ppt	1.96 ± 0.19		
	190 °C, 20 ppt	2.25 ± 0.01		
	190 °C, 35 ppt	2.09 ± 0.01		
	190 °C, 50 ppt	2.25 ± 0.02		
Catalytic chemicals[a]	190 °C, 35 ppt, H_2SO_4	2.44 ± 0.02		
	190 °C, 35 ppt, Na_2CO_3	2.33 ± 0.00		
	190 °C, 35 ppt, NaOH	2.45 ± 0.03		

[a]The loading concentration of catalytic chemicals is 0.025 g/g DM of biomass

a

Term	Scaled Estimate	Std Error	t Ratio	Prob>|t|
Intercept	0.689	0.008	81.336	<.0001*
Temperature, °C	0.248	0.012	20.398	<.0001*
Catalytic chemicals[H2SO4]	0.143	0.016	9.099	<.0001*
Catalytic chemicals[Na2CO3]	−0.09	0.016	−5.550	<.0001*
Catalytic chemicals[NaOH]	−0.08	0.016	−4.913	<.0001*
Catalytic chemicals[0]	0.021	0.01	2.127	0.0439*
Salinity, ppt	0.005	0.012	0.449	0.6575

b

Term	Scaled Estimate	Std Error	t Ratio	Prob>|t|
Intercept	0.639	0.005	117.559	<.0001*
Temperature, °C	0.288	0.008	36.861	<.0001*
Catalytic chemicals[H2SO4]	0.139	0.010	13.829	<.0001*
Catalytic chemicals[NaOH]	−0.077	0.010	−7.677	<.0001*
Catalytic chemicals[Na2CO3]	−0.071	0.010	−7.015	<.0001*
Catalytic chemicals[0]	0.0087	0.006	1.345	0.1911
Salinity, ppt	−0.003	0.008	−0.359	0.7226

Fig. 5 Scaled estimates of temperature, salinity, and catalysts in effects screening in terms of enzymatic digestibility (**a**) and fermentability (**b**) of pretreated solids. The R^2 and significance of the fitting model are 0.96 and $P < 0.0001$ in (**a**), and 0.98 and $P < 0.0001$ in (**b**), respectively

digestibility and ethanol yield) obtained from Analysis of Variance (ANOVA) indicate the models adequately fit the experimental data. Figure 5 shows the scaled estimates of temperature, salinity, and catalysts in effect screening. The absolute values of "*t* Ratio" determine the influences of parameters affecting the response, i.e., either enzymatic digestibility or fermentability. In other words, the most influential parameter is the one with the maximal absolute value of "*t* Ratio." As shown in Fig. 5, the maximal absolute value of "*t* Ratio" comes from temperature in terms of both enzymatic digestibility (20.398) and fermentability (36.861). The orders of other parameters in the effect test in terms of enzymatic digestibility and fermentability are the same, which is catalysts [H_2SO_4] > catalysts [Na_2CO_3] > catalysts [NaOH] > salinity.

The statistical analyses show that temperature plays the most influential role in determining the performance of pretreatment using seawater considering enzymatic digestibility and fermentability of pretreated solids. The next comes to catalysts, while the least influential factor is salinity. The future optimization direction, accordingly, will focus on pretreatment temperature and catalysts (e.g., H_2SO_4) as well. However, it is worth noting that the independence of pretreatment performance on salinity widens the application spectrum of saline water sources including not only seawater but also brines discharged after desalination.

4 Conclusions

This work presents the effects of temperature (180–210 °C), salinity (0–50 ppt), and catalysts (H_2SO_4, Na_2CO_3, and NaOH) on the pretreatment of lignocellulosic date palm residues using seawater. Temperature has the highest impact on carbohydrate recovery, enzymatic digestibility, and fermentability, while salinity showed the lowest impact on enzymatic digestibility and fermentability. At 210 °C with 35 ppt, salinity was achieved the highest digestible (96.0%) and fermentable (98.0%) solid but lowest xylan recovery (24.1%). These results are expected to be relevant for process optimization and economics, stimulating academic and industrial interest of seawater biorefinery. Still, corrosion related issues/costs challenges are considered for perspectives.

Acknowledgments This work was supported by Cooperative Agreement between the Masdar Institute of Science and Technology (Masdar Institute), Abu Dhabi, UAE and the Massachusetts Institute of Technology (MIT), Cambridge, MA, USA—Reference 02/MI/MI/CP/11/07633/GEN/G/00.

References

Amarasekara AS, Ebede CC (2009) Zinc chloride mediated degradation of cellulose at 200 °C and identification of the products. Bioresour Technol 100:5301–5304. https://doi.org/10.1016/j.biortech.2008.12.066

Ashraf MT, Fang C, Bochenski T, Cybulska I, Alassali A, Sowunmi A, Farzanah R, Brudecki GP, Chaturvedi T, Haris S, Schmidt JE, Thomsen MH (2016) Estimation of bioenergy potential for local biomass in the United Arab Emirates. Emirates J Food Agric 28:99–106. https://doi.org/10.9755/ejfa.2015-04-060

Bastidas-Oyanedel J-R, Fang C, Almardeai S, Javid U, Yousuf A, Schmidt JE (2016) Waste biorefinery in arid/semi-arid regions. Bioresour Technol 215:21–28. https://doi.org/10.1016/j.biortech.2016.04.010

Chen W-H, Tu Y-J, Sheen H-K (2011) Disruption of sugarcane bagasse lignocellulosic structure by means of dilute sulfuric acid pretreatment with microwave-assisted heating. Appl Energy 88:2726–2734. https://doi.org/10.1016/j.apenergy.2011.02.027

Chen Y, Stevens MA, Zhu Y, Holmes J, Xu H (2013) Understanding of alkaline pretreatment parameters for corn stover enzymatic saccharification. Biotechnol Biofuels 6:8. https://doi.org/10.1186/1754-6834-6-8

Cybulska I, Chaturvedi T, Brudecki GP, Kádár Z, Meyer AS, Baldwin RM, Thomsen MH (2014) Chemical characterization and hydrothermal pretreatment of Salicornia bigelovii straw for enhanced enzymatic hydrolysis and bioethanol potential. Bioresour Technol 153:165–172. https://doi.org/10.1016/j.biortech.2013.11.071

da Costa Sousa L, Jin M, Chundawat SPS, Bokade V, Tang X, Azarpira A, Lu F, Avci U, Humpula J, Uppugundla N, Gunawan C, Pattathil S, Cheh AM, Kothari N, Kumar R, Ralph J, Hahn MG, Wyman CE, Singh S, Simmons BA, Dale BE, Balan V (2016) Next-generation ammonia pretreatment enhances cellulosic biofuel production. Energy Environ Sci 9:1215–1223. https://doi.org/10.1039/C5EE03051J

De Figueiredo LP, Ferreira FF (2014) The Rietveld method as a tool to quantify the amorphous amount of microcrystalline cellulose. J Pharm Sci 103:1394–1399. https://doi.org/10.1002/jps.23909

Domínguez de María P (2013) On the use of seawater as reaction media for large-scale applications in biorefineries. ChemCatChem 5:1643–1648. https://doi.org/10.1002/cctc.201200877

Fang C, Schmidt JE, Cybulska I, Brudecki GP, Frankær CG, Thomsen MH (2015a) Hydrothermal pretreatment of date palm (*Phoenix dactylifera* L.) leaflets and rachis to enhance enzymatic digestibility and bioethanol potential. Biomed Res Int 2015:1–13. https://doi.org/10.1155/2015/216454

Fang C, Thomsen MH, Brudecki GP, Cybulska I, Frankaer CG, Bastidas-Oyanedel J-R, Schmidt JE (2015b) Seawater as alternative to freshwater in pretreatment of date palm residues for bioethanol production in coastal and/or arid areas. ChemSusChem 8:3823–3831. https://doi.org/10.1002/cssc.201501116

Grande PM, Domínguez de María P (2012) Enzymatic hydrolysis of microcrystalline cellulose in concentrated seawater. Bioresour Technol 104:799–802. https://doi.org/10.1016/j.biortech.2011.10.071

Harun R, Jason WSY, Cherrington T, Danquah MK (2011) Exploring alkaline pre-treatment of microalgal biomass for bioethanol production. Appl Energy 88:3464–3467. https://doi.org/10.1016/j.apenergy.2010.10.048

Hongsiri W, Danon B, de Jong W (2015) The effects of combined catalysis of oxalic acid and seawater on the kinetics of xylose and arabinose dehydration to furfural. Int J Energy Environ Eng 6:21–30. https://doi.org/10.1007/s40095-014-0146-9

Hongsiri W, Danon B, De Jong W (2014) Kinetic study on the dilute acidic dehydration of pentoses toward furfural in seawater. Ind Eng Chem Res 53:5455–5463. https://doi.org/10.1021/ie404374y

Huang H, Guo X, Li D, Liu M, Wu J, Ren H (2011) Identification of crucial yeast inhibitors in bioethanol and improvement of fermentation at high pH and high total solids. Bioresour Technol 102:7486–7493. https://doi.org/10.1016/j.biortech.2011.05.008

Hussin MH, Rahim AA, Mohamad Ibrahim MN, Brosse N (2013) Physicochemical characterization of alkaline and ethanol organosolv lignins from oil palm (Elaeis guineensis) fronds as phenol substitutes for green material applications. Ind Crop Prod 49:23–32. https://doi.org/10.1016/j.indcrop.2013.04.030

Jiang W, Chang S, Qu Y, Zhang Z, Xu J (2016) Changes on structural properties of biomass pretreated by combined deacetylation with liquid hot water and its effect on enzymatic hydrolysis. Bioresour Technol 220:448–456. https://doi.org/10.1016/j.biortech.2016.08.087

Ko JK, Kim Y, Ximenes E, Ladisch MR (2014) Effect of liquid hot water pretreatment severity on properties of hardwood lignin and enzymatic hydrolysis of cellulose. Biotechnol Bioeng 112:252–262. https://doi.org/10.1002/bit.25349

Kumar R, Mago G, Balan V, Wyman CE (2009) Physical and chemical characterizations of corn stover and poplar solids resulting from leading pretreatment technologies. Bioresour Technol 100:3948–3962. https://doi.org/10.1016/j.biortech.2009.01.075

Li H, Pu Y, Kumar R, Ragauskas AJ, Wyman CE (2014) Investigation of lignin deposition on cellulose during hydrothermal pretreatment, its effect on cellulose hydrolysis, and underlying mechanisms. Biotechnol Bioeng 111:485–492. https://doi.org/10.1002/bit.25108

Lin CSK, Luque R, Clark JH, Webb C, Du C (2011) A seawater-based biorefining strategy for fermentative production and chemical transformations of succinic acid. Energy Environ Sci 4:1471–1479. https://doi.org/10.1039/c0ee00666a

Liu C, Wyman CE (2006) The enhancement of xylose monomer and xylotriose degradation by inorganic salts in aqueous solutions at 180 °C. Carbohydr Res 341:2550–2556. https://doi.org/10.1016/j.carres.2006.07.017

Liu L, Sun J, Cai C, Wang S, Pei H, Zhang J (2009a) Corn stover pretreatment by inorganic salts and its effects on hemicellulose and cellulose degradation. Bioresour Technol 100:5865–5871. https://doi.org/10.1016/j.biortech.2009.06.048

Liu L, Sun J, Li M, Wang S, Pei H, Zhang J (2009b) Enhanced enzymatic hydrolysis and structural features of corn stover by FeCl3 pretreatment. Bioresour Technol 100:5853–5858. https://doi.org/10.1016/j.biortech.2009.06.040

Loow Y-L, Wu TY, Lim YS, Tan KA, Siow LF, Jahim JM, Mohammad AW (2017) Improvement of xylose recovery from the stalks of oil palm fronds using inorganic salt and oxidative agent. Energy Convers Manag 138:248–260. https://doi.org/10.1016/j.enconman.2016.12.015

Loow Y-L, Wu TY, Jamaliah Md. J, Mohammad AW, Teoh WH (2016) Typical conversion of lignocellulosic biomass into reducing sugars using dilute acid hydrolysis and alkaline pretreatment. Cellulose 23:1491–1520. https://doi.org/10.1007/s10570-016-0936-8

Loow YL, Wu TY, Tan KA, Lim YS, Siow LF, Md. Jahim J, Mohammad AW, Teoh WH (2015) Recent advances in the application of inorganic salt pretreatment for transforming lignocellulosic biomass into reducing sugars. J Agric Food Chem 63:8349–8363. https://doi.org/10.1021/acs.jafc.5b01813

Mao L, Zhang L, Gao N, Li A (2013) Seawater-based furfural production via corncob hydrolysis catalyzed by FeCl3 in acetic acid steam. Green Chem 15:727. https://doi.org/10.1039/c2gc36346a

Marcotullio G, Krisanti E, Giuntoli J, de Jong W (2011) Selective production of hemicellulose-derived carbohydrates from wheat straw using dilute HCl or FeCl3 solutions under mild conditions. X-ray and thermo-gravimetric analysis of the solid residues. Bioresour Technol 102:5917–5923. https://doi.org/10.1016/j.biortech.2011.02.092

Nguyen Q, Tucker M (2002) Dilute acid/metal salt hydrolysis of lignocellulosics. US Patent 6,423,145

Nishiyama Y, Sugiyama J, Chanzy H, Langan P (2003) Crystal structure and hydrogen bonding system in cellulose Iα from synchrotron X-ray and neutron fiber diffraction. J Am Chem Soc 125:14300–14306

Nitsos CK, Matis KA, Triantafyllidis KS (2013) Optimization of hydrothermal pretreatment of lignocellulosic biomass in the bioethanol production process. ChemSusChem 6:110–122. https://doi.org/10.1002/cssc.201200546

Peng L, Lin L, Zhang J, Zhuang J, Zhang B, Gong Y (2010) Catalytic conversion of cellulose to levulinic acid by metal chlorides. Molecules 15:5258–5272. https://doi.org/10.3390/molecules15085258

Pu Y, Hu F, Huang F, Davison BH, Ragauskas AJ (2013) Assessing the molecular structure basis for biomass recalcitrance during dilute acid and hydrothermal pretreatments. Biotechnol Biofuels 6:15. https://doi.org/10.1186/1754-6834-6-15

Young RA (1995) Introduction to the Rietveld method. In: Young RA (ed) The Rietveld method. Oxford University Press, Oxford, pp 1–39

Ren H, Zong M-H, Wu H, Li N (2016) Utilization of seawater for the biorefinery of lignocellulosic biomass: ionic liquid pretreatment, enzymatic hydrolysis, and microbial lipid production. ACS Sustain Chem Eng 4:5659–5666. https://doi.org/10.1021/acssuschemeng.6b01562

Román-Leshkov Y, Davis ME (2011) Activation of carbonyl-containing molecules with solid Lewis acids in aqueous media. ACS Catal 1:1566–1580. https://doi.org/10.1021/cs200411d

Selig MJ, Weiss N, Ji Y (2008) Enzymatic saccharification of lignocellulosic biomass. National Renewable Energy Laboratory, Golden. NREL/TP-510-42629

Singh S, Cheng G, Sathitsuksanoh N, Wu D, Varanasi P, George A, Balan V, Gao X, Kumar R, Dale BE, Wyman CE, Simmons BA (2015) Comparison of different biomass pretreatment techniques and their impact on chemistry and structure. Front Energy Res 2:1–12. https://doi.org/10.3389/fenrg.2014.00062

Sluiter A, Hames B, Ruiz RO, Scarlata C, Sluiter J, Templeton D, Crocker D (2011) Determination of structural carbohydrates and lignin in biomass. National Renewable Energy Laboratory, Golden. NREL/TP-510-42618

Thygesen A, Oddershede J, Lilholt H, Thomsen AB, Ståhl K (2005) On the determination of crystallinity and cellulose content in plant fibres. Cellulose 12:563–576. https://doi.org/10.1007/s10570-005-9001-8

Trajano HL, Engle NL, Foston M, Ragauskas AJ, Tschaplinski TJ, Wyman CE (2013) The fate of lignin during hydrothermal pretreatment. Biotechnol Biofuels 6:110. https://doi.org/10.1186/1754-6834-6-110

Wu M, Chiu Y (2011) Consumptive water use in the production of ethanol and petroleum gasoline-2011 updated. Argonne National Laboratory, Lemont

Zeng Y, Zhao S, Yang S, Ding S-Y (2014) Lignin plays a negative role in the biochemical process for producing lignocellulosic biofuels. Curr Opin Biotechnol 27:38–45. https://doi.org/10.1016/j.copbio.2013.09.008

Zhang J, Tang M, Viikari L (2012) Xylans inhibit enzymatic hydrolysis of lignocellulosic materials by cellulases. Bioresour Technol 121:8–12. https://doi.org/10.1016/j.biortech.2012.07.010

Zhang L, Yu H, Wang P, Li Y (2014) Production of furfural from xylose, xylan and corncob in gamma-valerolactone using FeCl3·6H2O as catalyst. Bioresour Technol 151:355–360. https://doi.org/10.1016/j.biortech.2013.10.099

Zhao H (2005) Effect of ions and other compatible solutes on enzyme activity, and its implication for biocatalysis using ionic liquids. J Mol Catal B Enzym 37:16–25. https://doi.org/10.1016/j.molcatb.2005.08.007

Zhao X, Zhang L, Liu D (2012) Biomass recalcitrance. Part I: the chemical compositions and physical structures affecting the enzymatic hydrolysis of lignocellulose. Biofuels Bioprod Biorefin 6:465–482. https://doi.org/10.1002/bbb.1331

Zhu L, O'Dwyer JP, Chang VS, Granda CB, Holtzapple MT (2008) Structural features affecting biomass enzymatic digestibility. Bioresour Technol 99:3817–3828. https://doi.org/10.1016/j.biortech.2007.07.033

Pyrolysis Kinetics of Arid-Land Biomasses

Prosper Dzidzienyo, Juan-Rodrigo Bastidas-Oyanedel, and Jens Ejbye Schmidt

1 Introduction

Any arid region biomass-based renewable energy form has to confront a number of challenges, such as the unavailability of enough arable land and lack of freshwater (Bastidas-Oyanedel et al. 2016). The two biomasses considered in this study were halophyte *Salicornia bigelovii*, which can be grown in arid lands using saltwater (Dzidzienyo et al. 2018), and date palms (*Phoenix dactylifera*), a native arid land biomass and one of the most abundant agricultural residues in arid regions. Pyrolysis of biomass was identified as an effective way of producing precursors for jet-fuel production. Pyrolysis of biomass yields pyro-oils, pyro-char, and gases in varying proportions depending on process parameters (Isahak et al. 2012; Czernik et al. 1995; Bridgwater 2012). Pyro-oil yields of up to 35% have been obtained by various authors (Demirbas 2004), while fast pyrolysis yields have reached up to 75% (Bridgwater 2012). Other authors have studied the effects of temperature on pyrolysis oil yields (Uçar and Karagöz 2009; Solar et al. 2016).

Arid-land lignocellulose biomass pyrolysis literature is scarce (Sait et al. 2012; Conti et al. 1994; Putun et al. 1996), and co-pyrolysis of these types of biomass have not been reported before. Both *S. bigelovii* and date palm have attracted attention in the biorefinery of arid-land biomass (Bastidas-Oyanedel et al. 2016; Warshay et al. 2017; Abideen et al. 2012). From the sustainability perspective, both plants are adapted to the harsh conditions of arid land. *S. bigelovii* can grow in arid-land

P. Dzidzienyo · J.-R. Bastidas-Oyanedel (✉)
Chemical Engineering Department, Khalifa University of Science and Technology, Masdar City Campus, Abu Dhabi, United Arab Emirates
e-mail: juan.bastidas@ku.ac.ae

J. E. Schmidt
SDU-Department of Chemical Engineering, Biotechnology, and Environmental Technology, University of Southern Denmark, Campusvej 55, DK-5230 Odense M, Denmark
e-mail: jschmidt@masdar.ac.ae; jesc@kbm.sdu.dk

© Springer Nature Switzerland AG 2019

J.-R. Bastidas-Oyanedel, J. E. Schmidt (eds.), *Biorefinery*,
https://doi.org/10.1007/978-3-030-10961-5_32

coastal areas using seawater, thereby reducing the stress on freshwater demand (Warshay et al. 2017; Abideen et al. 2012). Date palm leaf residues', including leaflets and rachis, annual global production is estimated to be over six million tonnes (Bastidas-Oyanedel et al. 2016). Hence, the pyrolysis of this biomass is an opportunity to divert these residues from landfilling.

Understanding the kinetic parameters and effect of heating rates on reaction rate at different stages in the pyrolysis process is important in the design of pyrolysis reactors specifically for the investigated biomasses and in scaling up the process. It is also necessary in predicting the extent of reaction under different experimental conditions at different times. This work studies the pyrolysis reactions using thermogravimetric analysis for *S. bigelovii* and date palm and co-pyrolysis of both at different mass ratios. The effects of the heating rate on the reaction rates and key Arrhenius kinetic parameters, i.e., activation energy and pre-exponential factor, were determined. By obtaining thermal loss data from thermogravimetric analysis (Buratti et al. 2015), these kinetic parameters were determined by using predetermined non-isothermal models such as the Kissinger model, the Flynn-Wall-Ozawa model (FWO), and the improved Kissinger-Akahira-Sunose (KAS) model derived from the generalized Arrhenius model. Results from these methods are compared.

2 Results and Discussion

2.1 Effect of Heating Rate on Conversion and Reaction Rate

Heating rates were varied for date palm biomass at 5, 10, and 15 K/min to observe their effects on peak temperatures and also to obtain isoconversional data points for kinetic analysis. The mass loss curve for the two biomasses and co-pyrolysis are shown in Fig. 1. Each mass loss curve (α) in Fig. 1 is the resulting average of triplicates. The mass loss (α) is explained in Sect. 3.

The mass loss curves for both biomasses show pyrolysis took place through an identical pathway. Co-pyrolysis biomass followed a pathway between the two pure biomasses as might be predicted; theoretical and kinetic parameters lie between those of the two pure biomasses. The pathway can be categorized into three main phases: evaporation of water, passive pyrolysis, and active pyrolysis which corresponds to the first two phases having characteristic peaks associated with them on the DTG diagram in Fig. 2. The evaporation of water occurs around 373 K, corresponding to the first peak on the curve. Active pyrolysis was observed to take place between 473 and 633 K as seen with the two peaks in this region. Passive pyrolysis began after active pyrolysis and continued till the end of mass loss.

Gasparovic et al. (2010) studied the decomposition of hemicellulose, cellulose, and lignin and concluded that the decomposition of these three components typically occurred at temperature ranges between 473–653, 523–653, and 453 up to 1075 K, respectively. This observation reveals that the decomposition of all hemicellulose and cellulose took place during active pyrolysis but decomposition of lignin took place in both active and passive pyrolysis phases.

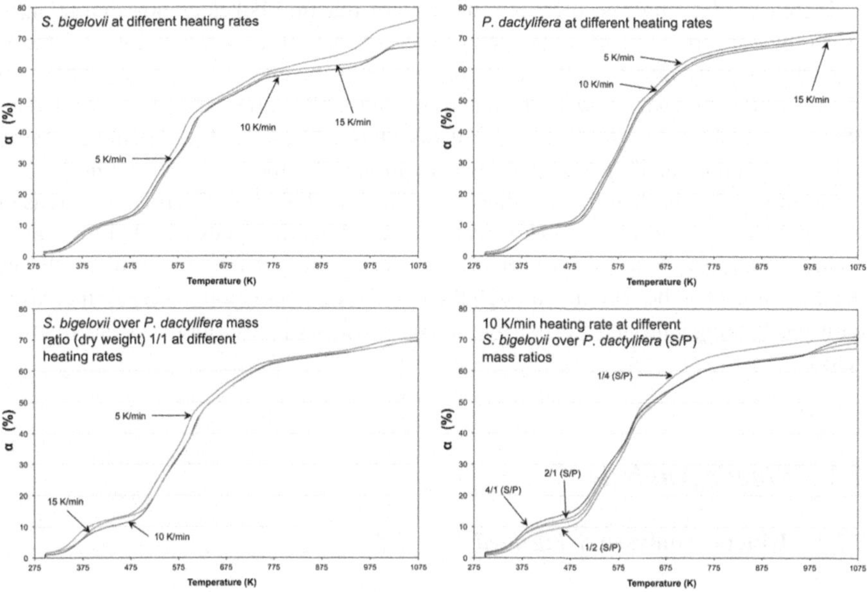

Fig. 1 Comparison of the mass loss (α) curves of *S. bigelovii*, *P. dactylifera*, and mixtures of them, at different heating rates (K/min)

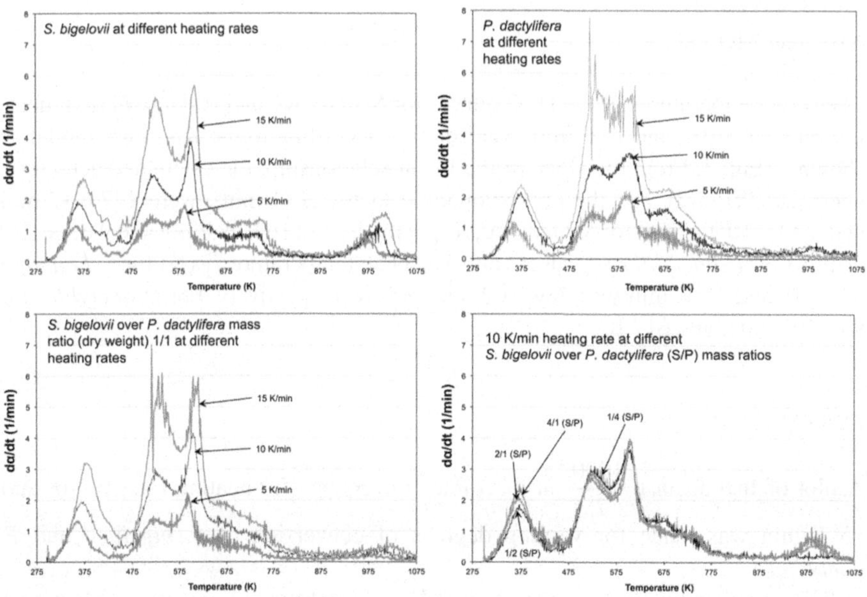

Fig. 2 Decomposition rates ($d\alpha/dt$) of *S. bigelovii*, *P. dactylifera*, and mixtures of them, at different heating rates (K/min) showing peak temperatures

Increasing heating rate increases the rate of reaction but does not significantly affect the conversion yields at the end of the experiment. Lower heating rates for *S. bigelovii* actually increased the total conversion at the end of the process although the rate of reaction was slower. The reaction rate for all heating rates peaked between 588 and 602 K during the active pyrolysis phase. These peak temperatures were employed in the kinetic parameter determination using the Kissinger method.

From Fig. 2, it is clear that the rate of reaction at all phases of pyrolysis increased with increasing heating rates. However, increased heating rates also led to a more nonuniform reaction decomposition process during the active pyrolysis phase. From the diagram, it can be seen that though the rate of decomposition generally increases with the heating rate and during the active pyrolysis phase, there are no distinct peaks.

2.2 *Kinetic Analysis*

2.2.1 Kinetic Analysis: *S. bigelovii*

Kinetic parameters were determined for different conversion rates corresponding to different segments of the pyrolysis reaction, and the activation energy and pre-exponential factors were determined. The three utilized methods, i.e., Kissinger, FWO, and KAS, are explained in Sect. 3.

Kissinger Method

The peak temperatures of the DTG curves for *S. bigelovii* and *P. dactylifera* at heating rates of 5, 10, and 15 K/min were plotted according to the Kissinger model as shown in Fig. 3. From the slope of the linear relationship, the calculated activation energy, E (kJ/mol), and the pre-exponential factor, A (1/min), were 147.6 kJ/mol and 3.13×10^9 (1/min) for *S. bigelovii*, respectively, and 164.7 kJ/mol and 9.55×10^{10} (1/min) for *P. dactylifera*, respectively. The peak temperature T_p (K) for *S. bigelovii* at 5, 10, and 15 K/min was 589, 602, and 609 K, respectively. For *P. dactylifera* it was 592, 603, and 612 K.

FWO Method

A plot of $\ln b$ against $\frac{1000}{T_i}$ at the same conversions for heating rates 5, 10, and 15 K/min was made for varying degrees of conversion of *S. bigelovii* and *P. dactylifera*.

The calculated activation energy, E (kJ/mol), and the pre-exponential factor, A (1/min), were 146.7 kJ/mol and 2.92×10^{20} (1/min) for *S. bigelovii*, respectively,

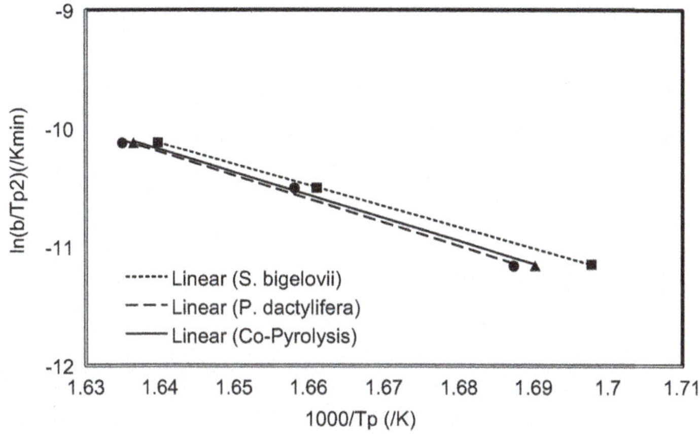

Fig. 3 Kissinger curve for *S. bigelovii*, *P. dactylifera* and co-pyrolysis of them at a mass ratio of 1/1, at different isoconversional points

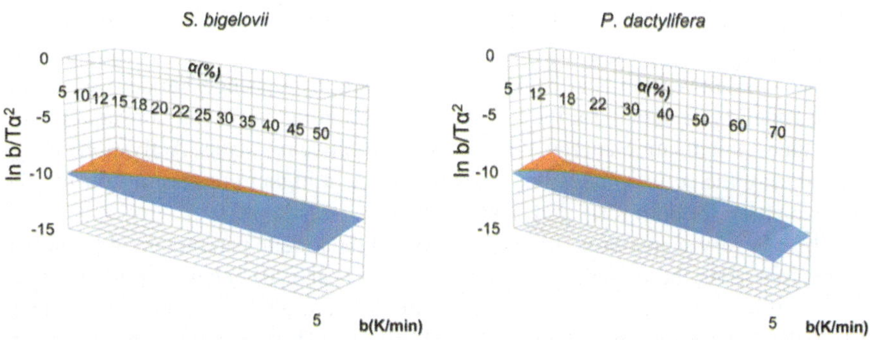

Fig. 4 FWO plot for *S. bigelovii* and *P. dactylifera*

and 204.3 kJ/mol and 1.93 × 10^{37} (1/min) for *P. dactylifera*, respectively. The FWO plot is shown in Fig. 4.

KAS Method

Figure 5 shows a plot of $\ln \frac{b}{T_\alpha^2}$ against $\frac{1000}{T_\alpha}$ at different heating rates 5 K/min, 10 K/min, and 15 K/min at the same conversion for varying conversions from 5 to 50%, during which active pyrolysis takes in *S. bigelovii* and *P. dactylifera*. The calculated activation energy, E (kJ/mol), and the pre-exponential factor, A (1/min), were 147.3 kJ/mol and 3.57 × 10^{14} (1/min) for *S. bigelovii*, respectively, and 201.4 kJ/mol and 8.54 × 10^{31} (1/min) for *P. dactylifera*, respectively.

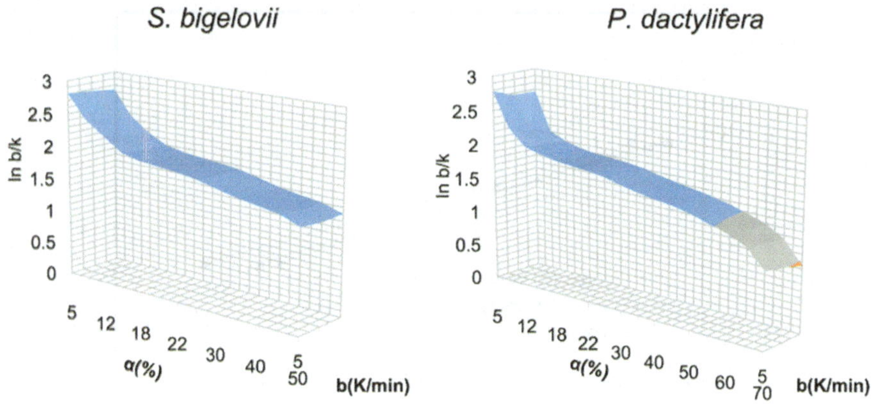

Fig. 5 KAS plot for *S. bigelovii* and *P. dactylifera*

2.2.2 Summary of Parameters

The kinetic parameters obtained for *S. bigelovii* and *P. dactylifera* leaves from all three methods are summarized in Table 1. The Kissinger method gives one value for the activation energy for the whole process. This value tends to correspond more closely with conversion near peak values at which the Kissinger model was developed. Kinetic parameters for the FWO and KAS method vary with conversion with low values of activation energy at stages prior to active pyrolysis. Activation energy rises during the active phases but tends to reduce at higher conversions during the passive phase for both KAS and FWO methods.

There is very scarce or no literature study on the kinetic parameters of these two biomasses using these methods to compare. Sait et al. (2012) have reported activation energies below 44 kJ/mol for date palm biomass, which significantly deviates from the values reported here, 146–204 kJ/mol. The activation energy values reported here are in the same order of what has been reported for other non-arid-land lignocellulosic biomass (Gasparovic et al. 2010; Bartocci et al. 2017; Kongkaew et al. 2015). Gasparovic et al. (2010) determined the kinetic parameters of wood chip using the generalized isoconversional method and activation energy values of between 131.56 and 215.94 kJ/mol depending on the conversion. Kongkaew et al. (2015) also determined the kinetic parameters for pyrolysis of rice straw using Kissinger, FWO, and KAS methods and obtained 172.62 kJ/mol for the Kissinger method and activation energy values between 180.54–220.27 and 181.95–221.72 kJ/mol for FWO and KAS, respectively (Kongkaew et al. 2015). The same author also obtained 1.46×10^{11} min^{-1} as the pre-exponential factor using the Kissinger method as well. Bartocci et al. (2017) have reported activation energies for the three main components of lignocellulosic biomass, 154.1, 224.7, and 190.5 kJ/mol for hemicellulose, cellulose, and lignin, respectively.

This shows the activation energy values obtained for both *P. dactylifera* leaves and *S. bigelovii* are within the range of values obtained by other authors using other

Table 1 Summary of kinetic parameters *E* (kJ/mol) and *A* (1/min) for *S. bigelovii* and *P. dactylifera* by different methodologies

	S. bigelovii		*P. dactylifera*	
Method	*E* (kJ/mol)	*A* (min^{-1})	*E* (kJ/mol)	*A* (min^{-1})
Kissinger	147.6	3.13×10^{9}	164.7	9.55×10^{10}
FWO	146.7	2.92×10^{20}	204.3	1.93×10^{37}
KAS	147.3	3.57×10^{14}	201.4	8.54×10^{31}

non-arid-land lignocellulosic biomasses as reactants. Since the Kissinger method adopts the same method to obtain an average value of activation during active pyrolysis, these values are more comparable with values obtained from other authors using different biomasses. FWO and KAS methods, though generally more accurate than the Kissinger method (Starink 2003), tend to cite averages which also depend on data points. These estimates, however, help to understand the pyrolysis reaction and calculate reaction constants at various stages of the pyrolysis. Also, the fact that the present results are similar to those obtained for non-arid-land lignocellulosic biomass has interesting implications, e.g., the scale-up of the studied arid-land biomass can benefit from the non-arid-land technology.

3 Materials and Methods

Dried *S. bigelovii* (whole plant) were obtained from ISEAS farms in Abu Dhabi. Dried *P. dactylifera* leaves were also obtained from a farm in Abu Dhabi. All feedstock were shredded, milled, and filtered through a sieve to obtain particle sizes below 0.5 mm. A sample of *S. bigelovii* and *P. dactylifera* were milled together in the ratio 1:1 for co-pyrolysis. Thermogravimetric analysis was conducted with a Netzsch STA 449F3 STA449F3A-0625-M instrument and an aluminum crucible (Al_2O_3) under a nitrogen atmosphere. The mass loss of biomass (α) was observed over a temperature program from 283 to 1073 K. For each test, the mass loss (α) was performed in triplicates. The heating rate was kept constant at 10 K/min. Isothermal conditions were created at the start and end of the temperature program to eliminate noise. The heating rate was then varied at 15 and 5 K/min, respectively, for both biomasses.

3.1 Isoconversional Methods for Kinetic Parameter Estimation Theory

The pyrolytic reaction of biomass converts biomass to char, oil, and gases. The kinetics of the reaction can be described by defining a degree of conversion and using the Arrhenius equation. The isoconversional method is based on the fact that activation energy and pre-exponential factors are not constant throughout the decomposition but depend on the degree of conversion. Data points at the same conversion are gathered for different heating rates, and each isoconversional curve is used to estimate the kinetic parameters at that conversion.

If m_i is the initial mass of sample placed in the crucible, and m_f is the mass of sample left after pyrolysis, a degree of conversion (α) for a given sample mass m at any temperature T during the process can be defined as:

$$\alpha = \frac{m_i - m}{m_i - m_f} \tag{1}$$

From the Arrhenius equation,

$$k(T) = Ae^{\frac{-E}{RT}} \tag{2}$$

where k is the reaction rate constant (varies with temperature), E is the activation energy, A is the pre-exponential factor, R is the gas constant, and T is temperature in Kelvin.

The rate of decomposition can then be defined as a function of conversion and temperature:

$$\frac{d\alpha}{dt} = f(\alpha)k(T)$$

$f(\alpha)$ can be expressed as:

$f(\alpha) = (1 - \alpha)^n$, where n is the reaction order.

Since the temperature program was run with a constant heating rate from 298 to 1073 K, the temperature at any time t can be written as:

$$T = T_i + b_t$$

where T = initial temperature = 298 K and b is the constant heating rate.

The rate of decomposition can be written as:

$$\frac{d\alpha}{dt} = (1-\alpha)^n Ae^{\frac{-E}{RT}} \tag{3}$$

3.1.1 Kissinger Method

For a predetermined reaction order, the rate of decomposition can be plotted at the same conversion for different heating rates. Also the decomposition rate is maximum at peak temperatures (T_p).

$$\frac{d}{dt}\frac{d\alpha}{dt} = \frac{d}{dt}\left((1-\alpha)^n Ae^{\frac{-E}{RT}}\right) = 0 \tag{4}$$

$$\text{In}\frac{b}{T_p^2} = -\frac{E}{RT_p} + \text{In}\frac{AR}{E} \tag{5}$$

A plot of $\text{In}\frac{b}{T_p^2}$ against $\frac{1000}{T_p}$ gives a straight line with slope $-\frac{E}{R}$.

3.1.2 FWO Method

Flynn-Wall-Ozawa developed a method where the activation energy is found by plotting the heating rates against the temperature at which a given conversion is obtained at that heating rate.

$$\text{In}b = -1.052\frac{E_\alpha}{RT_\alpha} + \text{In}\frac{A_\alpha E_\alpha}{Rf(\alpha)} - 5.331 \tag{6}$$

A plot of $\text{In}b$ against $\frac{1000}{T_i}$ gives a straight line with slope $-1.052\frac{E_\alpha}{R}$. This is used to find E_α.

3.1.3 KAS Method

The Kissinger-Akahira-Sunose Method

$$\text{In}\frac{b}{T_\alpha^2} = -\frac{E_\alpha}{RT_\alpha} + \text{In}\frac{A_\alpha R}{E_\alpha f(\alpha)} \tag{7}$$

A plot of $\text{In}\frac{b}{T_\alpha^2}$ against $\frac{1000}{T_\alpha}$ at different heating rates at the same conversion gives a straight line with slope $-\frac{E_\alpha}{R}$. The activation energy at that conversion E_α is found from the slope.

4 Conclusions

Activation energy obtained using the Kissinger method for *S. bigelovii* and *P. dactylifera* leaves was 147.6 and 164.7 kJ/mol, respectively, while pre-exponential factors of 3.13×10^9/min and 9.55×10^{10}/min for *S. bigelovii* and *P. dactylifera* leaves were obtained from DTG data. Other values at different stages during the pyrolysis process were also obtained using other isoconversional methods like the FWO, KAS, and the generalized isoconversional theories. The mass loss data also revealed the peak temperatures at which reaction is fastest during active pyrolysis. Results of DTG also showed the reaction proceeded in similar phases for the two sampled biomass types and co-pyrolysis of them at different mass ratios.

Acknowledgments Special thanks goes to ISEAS for supplying the biomasses needed for the project.

Funding This research was funded by Masdar Institute, Project 2BIONRG (12KAMA4) and BIOREF (13KAMA1), to help fulfil the vision of the late President Sheikh Zayed Bin Sultan Al Nahyan for sustainable development and empowerment of the United Arab Emirates and humankind.

References

Abideen Z, Ansari R, Gul B, Khan MA (2012) The place of halophytes in Pakistan's biofuel industry. Biofuels 3:211–220. https://doi.org/10.4155/bfs.11.158

Bartocci P, Anca-Couce A, Slopiecka K, Nefkens S, Evic N, Retschitzegger S, Barbanera M, Buratti C, Cotana F, Bidini G et al (2017) Pyrolysis of pellets made with biomass and glycerol: kinetic analysis and evolved gas analysis. Biomass Bioenergy 97:11–19. https://doi.org/10.1016/j.biombioe.2016.12.004

Bastidas-Oyanedel JR, Fang C, Almardeai S, Javid U, Yousuf A, Schmidt JE (2016) Waste biorefinery in arid/semi-arid regions. Bioresour Technol 215:21–28. https://doi.org/10.1016/j.biortech.2016.04.010

Bridgwater AV (2012) Review of fast pyrolysis of biomass and product upgrading. Biomass Bioenergy 38:68–94. https://doi.org/10.1016/j.biombioe.2011.01.048

Buratti C, Barbanera M, Bartocci P, Fantozzi F (2015) Thermogravimetric analysis of the behavior of sub-bituminous coal and cellulosic ethanol residue during co-combustion. Bioresour Technol 186:154–162. https://doi.org/10.1016/j.biortech.2015.03.041

Conti L, Scano G, Boufala J (1994) Bio-oils from arid land plants: flash pyrolysis of Euphorbia characias bagasse. Biomass Bioenergy 7:291–296. https://doi.org/10.1016/0961-9534(94)00071-Z

Czernik S, Scahill J, Diebold J (1995) The production of liquid fuel by fast pyrolysis of biomass. J Sol Energy Eng 117:2. https://doi.org/10.1115/1.2847714

Demirbas A (2004) Determination of calorific values of bio-chars and pyro-oils from pyrolysis of beech trunkbarks. J Anal Appl Pyrolysis 72:215–219. https://doi.org/10.1016/j.jaap.2004.06.005

Dzidzienyo P, Bastidas-Oyanedel JR, Schmidt JE (2018) Pyrolysis kinetics of the arid land biomass halophyte Salicornia bigelovii and Phoenix dactylifera using thermogravimetric analysis. Energies 11(9):1–8. https://doi.org/10.3390/en11092283

Gasparovic L, Korevnova Z, Jelemensky L (2010) Kinetic study of wood chips decomposition by TGA. Chem Pap 64:174–181

Isahak WNRW, Hisham MWM, Yarmo MA, Hin T-YY (2012) A review on bio-oil production from biomass by using pyrolysis method. Renew Sust Energ Rev 16:5910. https://doi.org/10.1016/j.rser.2012.05.039

Kongkaew N, Pruksakit W, Patumsawad S (2015) Thermogravimetric kinetic analysis of the pyrolysis of rice straw. Energy Procedia 79:663–670. https://doi.org/10.1016/j.egypro.2015.11.552

Putun A, Gercel H, Kockar O, Ege O, Snape C, Putun E (1996) Oil production from an arid-land plant fixed bed pyrolysis and hydropyrolysis. Fuel 75:1307–1312. https://doi.org/10.1016/0016-2361(96)00098-1

Sait HH, Hussain A, Salema AA, Ani FN (2012) Pyrolysis and combustion kinetics of date palm biomass using thermogravimetric analysis. Bioresour Technol 118:382–389. https://doi.org/10.1016/j.biortech.2012.04.081

Solar J, de Marco I, Caballero BM, Lopez-Urionabarrenechea A, Rodriguez N, Agirre I, Adrados A (2016) Influence of temperature and residence time in the pyrolysis of woody biomass waste in a continuous screw reactor. Biomass Bioenergy 95:416–423

Starink MJ (2003) The determination of activation energy from linear heating rate experiments: a comparison of the accuracy of isoconversion methods. Thermochim Acta 404:163–176. https://doi.org/10.1016/S0040-6031(03)00144-8

Uçar S, Karagöz S (2009) The slow pyrolysis of pomegranate seeds: the effect of temperature on the product yields and bio-oil properties. J Anal Appl Pyrolysis 84:151–156. https://doi.org/10.1016/j.jaap.2009.01.005

Warshay B, Brown JJ, Sgouridis S (2017) Life cycle assessment of integrated seawater agriculture in the Arabian (Persian) Gulf as a potential food and aviation biofuel resource. Int J Life Cycle Assess 22:1033. https://doi.org/10.1007/s11367-016-1240-4

Screening and Production of Biogas from Macro Algae Biomass of *Padina boergesenii*, *Colpomenia sinuosa*, and *Ulva* sp.

Rashed H. Farzanah, Grzegorz Przemyslaw Brudecki, Iwona Cybulska, Juan-Rodrigo Bastidas-Oyanedel, Jens Ejbye Schmidt, and Mette Hedegaard Thomsen

1 Introduction

One of the most urgent global problems is to provide enough water and land for meeting the world's future food and bioenergy needs (Foley et al. 2011). Alternative sources of water and land are needed, and using seawater for crop production along coastal deserts could potentially provide 1.3×10^8 ha of new cropland (Lu et al. 2010). Marine and freshwater habitats have immense biodiversity and potential as feedstock for production of bioenergy and other bio-products. Algae and sea grass are considered to be one of the most rapid growing sources of biomass compared to terrestrial plants, and their natural habitat make them ideal biomasses for biofuel production as they do not compete with land-based crops in terms of land use and food production (Rosenberg et al. 2008).

R. H. Farzanah · G. P. Brudecki (✉) · J.-R. Bastidas-Oyanedel
Chemical Engineering Department, Khalifa University of Science and Technology, Masdar Institute, Masdar City, Abu Dhabi, United Arab Emirates
e-mail: grzegorz.brudecki@ku.ac.ae; juan.bastidas@ku.ac.ae

I. Cybulska
Chemical Engineering Department, Khalifa University of Science and Technology, Masdar Institute, Masdar City, Abu Dhabi, United Arab Emirates

Catholic University of Louvain, Louvain-la-Neuve, Belgium
e-mail: iwona.cybulska@uclouvain.be

J. E. Schmidt
SDC-Department of Chemical Engineering, Biotechnology and Environmental Technology, University of Southern Denmark, Campusvej 55, DK-5230 Odense M, Denmark
e-mail: jschmidt@masdar.ac.ae; jesc@kbm.sdu.dk

M. H. Thomsen
Department of Energy Technology, Aalborg University, Esbjerg, Denmark
e-mail: mht@et.aau.dk

© Springer Nature Switzerland AG 2019
J.-R. Bastidas-Oyanedel, J. E. Schmidt (eds.), *Biorefinery*,
https://doi.org/10.1007/978-3-030-10961-5_33

Over 100 different species of macro algae are utilized for food, fertilizer, cosmetics, pharmaceuticals, and recently energy production all around the world. Macro algae should be distinguished from sea grass as the latter has roots, as terrestrial plants, while macro algae do not have roots and they drive their food (nutrition) through their tissue by diffusion. Macro algae are categorized based on their pigments or chlorophylls such as blue-green algae, green algae, brown algae, and red algae (Kim 2015).

Macro algae biomass contains no lignin, which makes it less recalcitrant compared to terrestrial lignocellulosic biomass. Generally, macro algae are rich in carbohydrates, proteins, lipids, and vitamins; thus this aquatic biomass has high potential for co-production of fuel and biochemicals (Rosenberg et al. 2008). Macro algae also have high growth rates such as *Ulva lactuca*, which has daily growth rate up to 20% of its weight (Fortes and Luning 1980; Waite et al. 1972). Macro algae mainly grow in wide bands in the intertidal zone during the colder months, while in the warmer months, they grow in a narrower band in the lower part of intertidal zone (Lee 2008).

Compared to micro algae, macro algae have been less investigated in terms of their chemical composition, and only very few papers exist on chemical characterization of macro algae biomass except those used for food consumption in East Asian countries such as Japan. However, increasing effort is observed in the area of conversion of macro algae biomass into biofuels. Several papers on *Ulva* sp. bioenergy feasibility have agreed that the high carbohydrate contents of *Ulva* sp., which in some cases reached 67%, and protein contents of 27% could be a significant feedstock for biorefinery processing into platform chemicals (Reich 2011; Fleurence 1999). Seaweed-based bioenergy literature has indicated that especially green seaweeds have high potential for production of glucan and ethanol (van der Wal et al. 2013; Lahaye et al. 1994). Several efforts were also made to convert seaweeds into biodiesel (Chen et al. 2015; Demirbas and Demirbas 2011). Some species of macro algae have been tested for production of biogas (Montingelli et al. 2015; Hughes et al. 2012), and some showed high potential in this field. However, due to complex structure and relatively high content of sulfur and nitrogen, the biochemical path for conversion is complex and can result in low bioconversion efficiency or total inhibition of the process (Montingelli et al. 2015).

This study focused on bioprospecting of macro algae species in the Arabian Gulf out the coast of Abu Dhabi. Three species of macro algae (*Padina boergesenii, Colpomenia sinuosa,* and *Ulva* sp.) were selected for detailed characterization and biogas potential quantification.

2 Materials and Methods

2.1 Raw Materials

The samples of biomass of *Padina boergesenii, Colpomenia sinuosa,* and *Ulva* sp. were collected in several locations at coastal areas of Abu Dhabi. *Ulva* sp. was found primarily onshore while *Padina boergesenii* and *Colpomenia sinuosa* were found primarily offshore. *Ulva* sp. was by far the most abundant biomass found at the coast of Abu Dhabi. All macro algae samples were washed using freshwater at room temperature to remove sand, sea snails, and foreign matter. After cleaning, the biomass was packed in zip-type plastic bags and stored at −18 °C until further analysis and processing.

2.2 *Chemical Characterization*

In order to obtain uniform and representative material for chemical characterization and further processing, the clean algal biomass was dried at 105 °C and grounded using 1 mm screen. The ground biomass was used for composition analysis of biomass including ash, total solids, structural carbohydrates, and acid insoluble lignin.

2.2.1 Total Solids and Ash Determination

The total solids and ash determination was done using the NREL protocols (Sluiter et al. 2008a, b). The amount of inorganic material in the biomass, either structural or extractable, was measured as part of the total composition. Structural ash in inorganic material is bound in the physical structure of the biomass, while extractable ash is in the inorganic material that can be removed by washing or extracting the material with water. Crucibles were prepared by pre-drying at 575 °C until constant weight. The biomass sample was placed in the crucibles and dried at 105 °C overnight to remove moisture and volatile components. The difference between the initial and final weight after drying was used for calculating the total solids (Eqs. 1 and 2). In order to measure the ash, the crucibles were heated at 575 °C in the muffle furnace. The total ash content was calculated using Eq. 3.

$$\text{Total Solids of Biomass }(\text{TS})\ \% = \frac{W_\text{d}}{W_\text{i}} * 100 \tag{1}$$

$$\text{Total Moisture Content }\% = \left(1 - \frac{W_\text{d}}{W_\text{i}}\right) * 100 \tag{2}$$

$$W_\text{d} = \text{Weight of the biomass after drying at } 105^\circ\text{C}\,(\text{g})$$
$$W_\text{i} = \text{Weight of the initial biomass}\,(\text{g})$$

$$\text{Total Ash }\% = \frac{W_\text{r}}{W_\text{d}} * 100 \tag{3}$$

$$W_\text{r} = \text{Weight of the residue after drying at } 575^\circ\text{C}\,(\text{g})$$
$$W_\text{d} = \text{Weight of the biomass after drying at } 105^\circ\text{C}\,(\text{g})$$

2.2.2 Determination of Extractives

First steps of characterization were water and ethanol extraction using a Soxhlet apparatus to determine total weight of the extractives. Dried macro algae sample (approx. 10 g) was loaded to a pre-weighed, tarred, dry cellulose thimble. Two hundred milliliters of deionized water was added to the Soxhlet boiling flask, and the flask was heated in heating mantle (Electrothermal, EMV). When the solvent started refluxing, the timer was set to 10 h. The heating mantle setting provided four to five siphon cycles per hour (siphon cycle = reflux of the solvent). Upon completion, the extracts were collected, weighed, and stored in the freezer. The same procedure was carried out using water and ethanol as the solvent. Small aliquot of extracts were used for the determination of total solids in the extracts.

Water- and ethanol-soluble extractive (both volatile and nonvolatile) contents in the raw biomass were calculated using Eqs. 4–7.

$$\text{Oven Dry Weight}\left(\text{ODW}\right)=\frac{W_{\text{air dried sample}} * \text{TS\%}}{100} \tag{4}$$

$W_{\text{air dried sample}}$ = Sample is dried at 60 °C overnight and then weighed (g)

$$\text{Nonvolatile extractives}\left(\text{NE}\right)\%=\frac{W_{\text{dried water or ethanol extract}}}{\text{ODW}} * 100 \tag{5}$$

$W_{\text{dried water or ethanol extract}}$ = The extract is dried overnight at 105 °C for 1 h, then weighed (g).

$$\text{Total extractives}\left(\text{TE}\right)\%=\frac{\text{ODW}-W_{\text{dried extracted biomass}}}{\text{ODW}} * 100 \tag{6}$$

$W_{\text{dried extracted biomass}}$ = The extracted biomass from the thimble is dried overnight at 60 °C and subsequently dried at 105 °C for 1 h, then weighed (g).

$$\text{Volatile extractives\%}=\text{TE\%}-\text{NE\%} \tag{7}$$

2.2.3 Carbohydrates Content Determination by Strong Acid Hydrolysis

The extractives-free material was subjected to strong acid hydrolysis following NREL protocol (Sluiter et al. 2011). Dried samples were treated with 72% (w/w) H_2SO_4 at 30 °C for 1 h. The solutions were then diluted with water to produce 4% H_2SO_4 and autoclaved at 121 °C for 1 h. The hydrolysates were filtered, and the Klason lignin content was determined as the difference between the weight of the acid insoluble residue and the ash content. The sugar content in the filtrate was determined using high-performance liquid chromatography (HPLC) system (Agilent 1260 Infinity Bio-inert Binary LC).

Equations 8–13 summarize the calculations made for the carbohydrates and Klason lignin content in the dry biomass.

The recovery factor (R_f) of the individual sugars was calculated according to Eq. 8 in order to account for any sugar degradation arising during the weak acid hydrolysis (after the strong acid step), sample work-up, and influences from the matrix. The measurement was performed by spiking two replicates of the weak acid hydrolysis samples with a solution of the three sugars (glucose, xylose, and arabinose) at concentration of 30 g/L of each component.

$$R_f = \frac{C_{h+s(\text{measured})}}{C_{s(\text{added})} + C_{h(\text{measured})}} \tag{8}$$

C_{h+s}: Sugar in acid hydrolysate with standard addition (g/100 g DM)
C_s: Sugar standard added (g/100 g DM)
C_h: Sugar in acid hydrolysate without standard addition (g/100 g DM)

The corrected amount of sugar (C_{corr}) was then calculated by Eq. 9:

$$C_{corr} = \frac{C_{h(\text{measured})}}{R_f} \tag{9}$$

The level of standard addition does not influence the recovery factor; hence one level of standard addition was applied in the standard procedure.

The corrected amount of sugar (C_{corr}) was then calculated by Eq. 10:

$$C_{corr} = \frac{C_{h(\text{measured})}}{R_f} \tag{10}$$

Concentration of the polymeric sugars from the concentration of the corresponding monomeric sugars (e.g., glucose to glucan or xylose to xylan conversion) was calculated using an anhydro correction of 0.88 (or 132/150) for C-5 sugars (xylose and arabinose) and a correction of 0.90 (or 162/180) for C-6 sugars (glucose, galactose, and mannose).

$$C_{anhydro} = C_{corr} * \text{Anhydro correction} \tag{11}$$

The percentage of each sugar on an extractives-free basis was calculated using Eq. 12:

$$\%\text{Sugar}_{\text{extractives free}} = \frac{C_{anhydro} * V_{hydrolysate} * \frac{1\text{g}}{1000\text{mg}}}{\text{ODW}(\text{oven dry weight})} * 100 \tag{12}$$

The percentage of each sugar on an as-received basis (including extractives) was calculated to demonstrate the true content of carbohydrates in the raw samples (Eq. 13).

$$\%\text{Sugar}_{\text{as received}} = \%\text{Sugar}_{\text{extractives free}} * \frac{(100 - \%\text{Extractives})}{100} \quad (13)$$

Acid insoluble lignin (Klason lignin) content in the extractives-free material was calculated using Eq. 14, while the lignin content in the "as-received" sample was calculated using Eq. 15.

$$\%\text{AIL}_{\text{extractives free}} = \frac{(W_b - W_a)}{W_s} * 100\% \quad (14)$$

where

%AIL = acid insoluble lignin (%)
W_b = weight of residue ***before*** drying at 105 °C (g)
W_a = weight of residue ***after*** drying at 105 °C (g)
W_s = weight of a sample (g)

$$\%A_{\text{as received}} = \%\text{AIL}_{\text{extractives free}} * \frac{(100 - \%\text{Extractives})}{100} \quad (15)$$

2.3 Enzymatic Digestibility

Enzymatic hydrolysis was carried out using 3% dry matter loading. Commercial enzymes (CTec2 and HTec2, Novozymes A/S, Denmark) were used to measure enzymatically available cellulose and hemicellulose. The hydrolysis was run in the presence of 50 mM citrate buffer (pH 4.8) at 50 °C and 170 rpm, for 72 h in shaking bed incubator (New Brunswick, Innova 42R) according to NREL protocol (Selig et al. 2008). The enzymes dosage was 15 FPU cellulase enzyme loading (with cellulase to hemicellulase ratio of 1:9). The enzymatic hydrolysis was performed in centrifuge tubes (15 mL) with total enzymatic hydrolysis mixture volume of 10 mL. The enzymatic hydrolysates were analyzed for sugars using HPLC system (Agilent 1260 Infinity Bio-inert Binary LC) with the conditions as described in below section, Sect. 2.5.

2.4 Biogas Potential

Inoculum for the biogas batch test was obtained from a wastewater treatment plant company (Al Wathba Water Treatment Plant 2). The total solids (TS), volatile solids (VS), and ash content were measured to be 4.25%, 2.73%, and 1.52%, respectively (all presented per wet matter basis, as received). The pH of the sludge was 7.56.

The experiments were performed in glass bottles with a thick rubber septum at 37 °C in a shaking bed incubator at 120 rpm under anaerobic conditions. Macro algae biomass was mixed with 50 g wet weight of inoculum and filled with nutrient medium to reach 100 g in total (Angelidaki et al. 2009; Bastidas-Oyanedel et al. 2010). The headspace was flushed with nitrogen to achieve anaerobic conditions. The final pH was 7.0–7.1. The experiment was run for 27 days. Gas samples from the headspace were collected using a pressure-lock syringe every second day for the first 2 weeks and then once a week and analyzed on gas chromatogram as described below. Blank samples containing only sludge and nutrient medium were run in parallel to determine the biogas production from the inoculum only. All the experiments were performed in triplicates.

In the initial stage of experiments, two different solid loadings (0.2 and 0.5% of volatile solids) were tested for evaluation of biogas potential of the macro algae.

Another batch was set up with three different loading concentrations for *Ulva* sp. algae (1, 3, and 5% of volatile solids (VS)) since *Ulva* sp. showed higher methane production compared to the other two macro algae species.

2.5 *Chemical Analysis Methods*

2.5.1 High-Performance Liquid Chromatography (HPLC)

Samples produced in the strong and weak acid hydrolysis, enzymatic hydrolysis, and pretreatment liquid fractions were analyzed using high-performance liquid chromatography (HPLC) technique. Samples were analyzed by HPLC through Agilent 1260 Infinity Bio-inert Quaternary LC which determine concentration of components (cellobiose, glucose, xylose, lactic acid, acetic acid, ethanol, and arabinose) in the medium sampled. The Hi Plex-H column and refractive index (RI) detector were used to determine the concentrations of analytes at 65 °C using 5 mM H_2SO_4 as mobile phase (eluent) with flow rate of 0.6 mL/min.

2.5.2 Gas Chromatography (GC)

Gas samples from the biogas experiments were analyzed on SRI GC (SRI 8610C) using silica gel packed column (3″ Silica Gel column) and equipped with flame ionization detector (FID) and thermal conductivity detector (TCD). The carrier gas was helium with inlet pressure set to 20 psi. The column compartment was set to 80 °C, the FID temperature was set to 150 °C, and TCD set to 100 °C. The analysis time for elution of methane and carbon dioxide was 5 min.

2.5.3 Elemental Analysis

Organic elemental analysis (OEA) was performed using automated elemental analyzer (ThermoFisher Scientific, Flash 2000). Five basic elements including carbon, nitrogen, hydrogen, sulfur (CHNS), and oxygen (O) were measured. For the CHNS analysis, each sample was weighed to be within the range of 2–3 mg into the tin capsule and approx. 10 mg of vanadium pentoxide was added to capsule as an oxidation catalyst. For the oxygen analysis, each sample was weighed to be within the range of 1–2 mg into the silver capsule, and no catalyst was added. The instrument calibration was done with the use of 2,5-(Bis(5-*tert*-butyl-2-benzo-oxazol-2-yl) thiophene (BBOT) with the approximate element percentages C = 72.52%, H = 6.09%, N = 6.51%, O = 7.43%, and S = 7.44%.

Protein content of the algae was determined from the nitrogen content using a conversion factor which depends on the particular amino acid sequence of the measured protein. Protein content through nitrogen-to-protein conversion was calculated as:

$$\%\text{Protein} = \%\text{N} \times \text{N factor}$$

where %N is elemental nitrogen content determined by combustion or Kjeldahl methods and N factor is the specific conversion factor determined for algae (4.78) (Laurens 2013).

3 Results and Discussion

3.1 Chemical Characterization

Table 1 shows the chemical compositions of *Padina boergesenii, Colpomenia sinuosa,* and *Ulva* sp. biomass.

The collected local *Ulva* sp. had the lowest ash content of around 40% ash (w/w %). Reported values for the ash contents in this type of algae vary from 14 to 50% depending on place of collection (Golberg et al. 2012). The amount of extractable

Table 1 Chemical compositional analysis of *Padina boergesenii, Colpomenia sinuosa,* and *Ulva sp.* (w/w %, all per dry matter basis)

Components [% DM]	*Padina boergesenii*	*Colpomenia sinuosa*	*Ulva* sp.
Total ash (raw material)	60.31 ± 1.90	51.71 ± 1.83	40.36 ± 0.43
Structural ash	65.03 ± 0.79	62.52 ± 1.96	19.42 ± 3.81
Extractable ash [%]	34.97 ± 0.79	37.48 ± 1.96	80.58 ± 3.81
Water soluble extractives	33.82 ± 0.33	23.69 ± 0.97	56.53 ± 0.81
Ethanol soluble extractives	1.50 ± 0.12	1.31 ± 0.11	1.54 ± 0.46
Glucan	8.05 ± 0.01	12.56 ± 0.72	9.47 ± 0.09
Xylan	3.66 ± 0.01	1.67 ± 0.05	3.31 ± 0.06
Arabinan	0.47 ± 0.02	1.17 ± 0.03	0.25 ± 0.02
Acid insoluble residue	13.05 ± 0.54	21.13 ± 0.44	18.94 ± 1.73
Acid insoluble residue corrected for ash	13.05 ± 0.54	8.92 ± 0.24	16.85 ± 1.87
Sum (%)	**83.26**	**70.73**	**73.82**

Table 2 Structural carbohydrates and acid insoluble residue content in extractives-free materials of three types of macro algae (w/w %, all per dry matter basis)

Component	*Padina boergesenii*	*Colpomenia sinuosa*	*Ulva* sp.
Glucan [%]	9.78 ± 0.34	11.28 ± 0.22	12.15 ± 0.08
Xylan [%]	3.39 ± 0.02	2.94 ± 0.09	2.56 ± 0.06
Arabinan [%]	0.48 ± 0.04	0.31 ± 0.04	2.65 ± 0.001
Acid insoluble residue [%]	16.33 ± 0.72	23.05 ± 1.83	35.14 ± 0.58
Acid insoluble residue corrected for ash [%]	13.85 ± 0.11	18.61 ± 2.10	27.76 ± 0.08

ash from *Ulva* sp. was around 81% with structural ash only representing 20%. *Padina boergesenii* had an ash content of around 60% distributed as 35% extractable ash and 65% structural ash. *Colpomenia sinuosa* contained 51% of ash of which 38% was extractable and 63% structural ash.

Soxhlet water extraction resulted in high fraction of water-soluble extractives out of which inorganic matter is dominant as a result of high ash content in the raw material.

The total sugar content was between 12 and 16%.

The sugar content was also measured in extractives-free materials after sequential water and ethanol extraction (Table 2) and resulted in between 10 and 12%.

3.2 Elemental Analysis

The results of elemental analysis including carbon, hydrogen, nitrogen, oxygen, and sulfur are presented in Table 3. The sulfur is a typical component of marine algal polysaccharides, related to high salt concentration in the environment and with specific functions in ionic regulation. Such sulfated mucilages are not found in land plants. Different amounts of sulfated heteropolysaccharides can be found in green macro algae, while other sulfated polysaccharides such as laminaran, alginate, and fucan are present mostly in brown macro algae, and sulfated galactans such as agar and carrageenan appear most often in red seaweeds (Jiao et al. 2011).

The nitrogen content and the calculated protein content indicate that the *Ulva* species can be a good source of protein and its application for utilization can be broaden to, e.g., animal feedstock or chemicals (Fleurence 1999; Bikker et al. 2016).

The elemental analysis results were used for the calculation of theoretical biogas potential according to Buswell's formula (Achinas and Euverink 2016). The theoretical biogas potential was calculated to be 411 mL/gVS for *Ulva* sp., 368 mL/gVS for *Padina boergesenii*, and 388 mL/gVS for *Colpomenia sinuosa*.

Table 3 Elemental compositions of macro algae (w/w %, all per dry matter basis)

Element	*Padina boergesenii*	*Colpomenia sinuosa*	*Ulva* sp.
Nitrogen [%]	1.13 ± 0.01	1.62 ± 0.09	2.95 ± 0.20
Carbon [%]	22.82 ± 0.69	24.33 ± 0.91	25.58 ± 0.46
Hydrogen [%]	3.10 ± 0.09	3.24 ± 0.21	4.61 ± 0.31
Sulfur [%]	0.98 ± 0.01	0.99 ± 0.05	3.88 ± 0.11
Oxygen [%]	27.14 ± 0.41	27.22 ± 0.31	30.13 ± 2.52
Protein based on nitrogen[%]	5.40 ± 0.03	7.75 ± 0.45	14.08 ± 0.94

Table 4 Enzymatic production of glucose (calculated with reference to compositional analysis on EFM determined by strong acid hydrolysis)

Material	Enzymatic convertibility [%]
Ulva sp. EFM	101.87 ± 5.35
Padina boergesenii EFM	10.50 ± 1.35
Colpomenia sinuosa EFM	5.25 ± 4.03

3.2.1 Enzymatic Hydrolysis

The enzymatic hydrolysis was carried out on the extractives-free biomass (EFB) to evaluate the sugar release without interference from extractives and ash (salt). Table 4 shows the results in terms of enzymatic hydrolysis efficiency. The enzymatic hydrolysis of *Ulva* sp. resulted in very high release of glucose which suggests testing this material for ethanol fermentation. The convertibility above 100% shows that more glucose was released by enzymatic hydrolysis then by strong acid hydrolysis or sugars were degraded during the high-temperature autoclaving, which is part of the strong acid hydrolysis protocol (see Sect. 2). The other two other types of algae *Padina boergesenii* and *Colpomenia sinuosa* resulted in very low glucose release. Pretreatment studies or testing the different type of enzymes are recommended for these two types of algae to examine the sugar polymer fraction for ethanol production.

3.3 Methane Production from Macro Algae Samples

The result from anaerobic digestion of macro algae samples showed high potential of these biomasses for biogas production. The results of cumulative methane production are presented in Fig. 1. The results for *Ulva* sp. showed that cumulative methane production was highest for samples with 5% VS resulting in 501 mL of methane after 27 days. Cumulative methane production for samples with 3% VS and 1% VS loading produced 454 and 337 mL, respectively, after 27 days. The blank sample produced 114 mL of methane after 27 days.

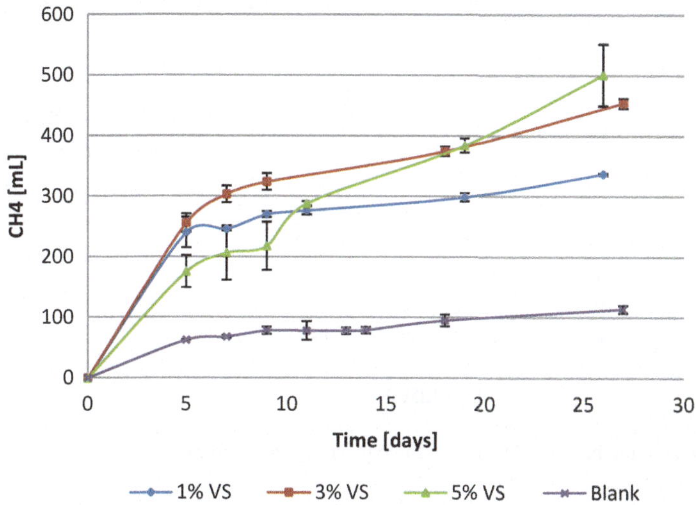

Fig. 1 Cumulative methane production for three different concentrations of *Ulva* sp.

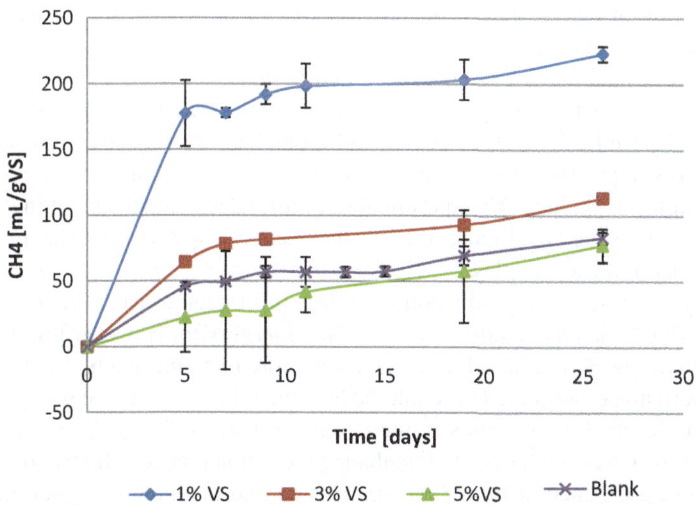

Fig. 2 Specific methane production for three different concentrations of *Ulva* sp.

Specific methane production was calculated as volume of CH4 per gram of VS of algae samples added. The results of specific methane production are presented in Fig. 2. The highest specific methane production for *Ulva* sp. was observed for sample with 1% VS reaching 223 mL g VS-1 after 27 days. Sample with 3% VS reached 113 mL g VS-1 after 27 days, while sample with 5% VS produced 83 mL g VS-1 after 27 days. When comparing the specific methane production to the theoretical methane production, the conversion of 54%, 27%, and 20% was achieved for 1%, 3%, and 5% VS loading rates of *Ulva* sp., respectively.

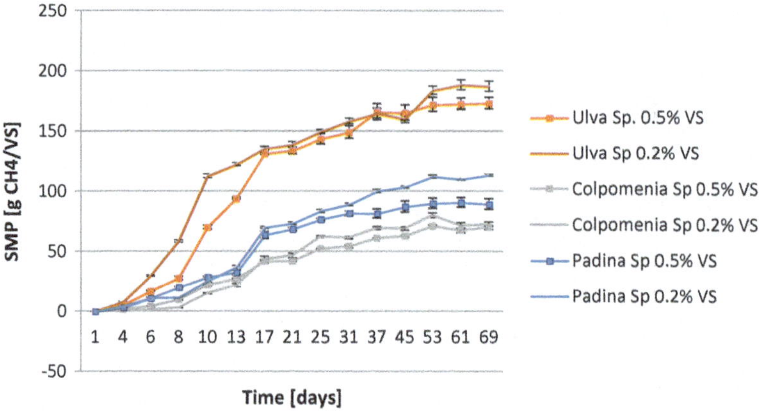

Fig. 3 Specific methane production for three macro algae samples

The results showed that cumulative methane production increased with higher VS loading for both materials while specific methane production decreased with higher VS loading indicating an inhibition of the anaerobic process with increasing concentration of macro algae. This biogas experiment of *Ulva* sp. was preliminary result and was used for the next batch of the experiment (explained below) for all three types of macro algae samples used in this study.

The biogas experiment of *Ulva* sp., *Padina boergesenii,* and *Colpomenia sinuosa* was done in triplicate with respect to different loading capacity of macro algae samples to sludge. The loading capacity was now set to 5:5 and 2:5 ratio of macro algae samples to sludge. The experiment ran for 69 days, and the resulting methane was measured using GC. Figure 3 shows the results of specific methane production from all macro algae samples.

The specific methane production for *Ulva* sp. in general gave the highest yield compared to *Colpomenia sinuosa* and *Padina boergesenii.* However lower biomass loading in terms of volatile solid (VS) yielded more methane that higher VS loading in all macro algae samples. This could be explained by macro algae being known to have antimicrobial components in the extractive part of the biomass (Pérez et al. 2016). As for *Ulva* sp., the yield of methane production reached almost 50% of theoretical biogas production from *Ulva* sp. This can be justified by higher enzymatic convertibility of *Ulva* sp. compared to *Padina boergesenii* and *Colpomenia sinuosa* which also had much lower enzymatic convertibility. As for the cumulative methane production, the highest methane yield was for *Ulva* sp. at 0.5% VS followed by *Padina boergesenii* at 0.5% VS and *Colpomenia sinuosa* as presented in Fig. 3.

4 Conclusions

The macro algae biomass can be valuable feedstock for production of alternative source of energy production. This study showed that tested macro algae have some potential to be utilized as substrate for biogas production. The *Ulva* sp., which was also the most dominant species found around Abu Dhabi's islands, showed the most promising results in terms of bioenergy production. The high yields in enzymatic conversion of sugars suggest it could be a good substrate to bioethanol production. The anaerobic digestion resulted in high value of methane production (411 mL/g VS), which is approximately 50% of theoretical methane production. Furthermore, *Ulva* sp. showed the highest protein content which could be refined and explored in animal feed production. The results found in this study show that *Ulva* sp. could be potential feedstock for third-generation biofuels. Further studies on chemical characterization and bioconversion processes are recommended.

References

Achinas S, Euverink GJW (2016) Theoretical analysis of biogas potential prediction from agricultural waste. Resour Technol 2:143–147. https://doi.org/10.1016/j.reffit.2016.08.001

Angelidaki I, Alves M, Bolzonella D, Borzacconi L, Campos JL, Guwy AJ, Kalyuzhnyi S, Jenicek P, van Lier JB (2009) Defining the biomethane potential (BMP) of solid organic wastes and energy crops: a proposed protocol for batch assays. Water Sci Technol 59:927–934

Bastidas-Oyanedel JR, Mohd-Zaki Z, Pratt S, Steyer JP, Batstone DJ (2010) Development of membrane inlet mass spectrometry for examination of fermentation processes. Talanta 83:482–492

Bikker P, van Krimpen MM, van Wikselaar P, Houweling-Tan B, Scaccia N, van Hal JW, Huijgen WJJ, Cone JW, López-Contreras AM (2016) Biorefinery of the green seaweed Ulva lactuca to produce animal feed, chemicals and biofuels. J Appl Phycol 28:3511–3525. https://doi.org/10.1007/s10811-016-0842-3

Chen H, Zhou D, Luo G, Zhang S, Chen J (2015) Macroalgae for biofuels production: progress and perspectives. Renew Sust Energ Rev 47:427–437. https://doi.org/10.1016/j.rser.2015.03.086

Demirbas A, Demirbas F (2011) Importance of algae oil as a source of biodiesel. Energy Convers Manag 52:163–170. https://doi.org/10.1016/j.enconman.2010.06.055

Fleurence J (1999) Seaweed proteins: biochemical, nutritional aspects and potential uses. Food Sci Technol 10:25–28. https://doi.org/10.1002/ppap.201100070

Foley JA, Ramankutty N, Brauman KA, Cassidy ES, Gerber JS, Johnston M, Mueller ND, O'Connell C, Ray DK, West PC, Balzer C, Bennett EM, Carpenter SR, Hill J, Monfreda C, Polasky S, Rockström J, Sheehan J, Siebert S, Tilman D, Zaks DPM (2011) Solutions for a cultivated planet. Nature 478:337–342. https://doi.org/10.1038/nature10452

Fortes MD, Luning K (1980) Growth rates of North Sea macroalgae in relation to temperature, irradiance and photoperiod. Helgolä Meeresun 34:15–29

Golberg A, Vitkin E, Linshiz G, Khan SA, Hillson NJ, Yakhini Z, Yarmush ML (2012) Proposed design of distributed macroalgal biorefineries: thermodynamics, bioconversion technology, and sustainability implications for developing economies. Biofuels Bioprod Biorefin 8:67–82. https://doi.org/10.1002/bbb

Hughes AD, Kelly MS, Black KD, Stanley MS (2012) Biogas from macroalgae: is it time to revisit the idea? Biotechnol Biofuels 5:1–7. https://doi.org/10.1186/1754-6834-5-86

Jiao G, Yu G, Zhang J, Ewart HS (2011) Chemical structures and bioactivities of sulfated polysaccharides from marine algae. Mar Drugs 9:196–233. https://doi.org/10.3390/md9020196

Kim S-K (2015) Handbook of marine biotechnology. Springer, Berlin. ISBN 9783642539701

Lahaye M, Jegou D, Buleon A (1994) Chemical characteristics of insoluble glucans from the cell wall of the marine green alga Ulva lactuca (L.) Thuret. Carbohydr Res 262:115–125. https://doi.org/10.1016/0008-6215(94)84008-3

Laurens LML (2013) Summative mass analysis of algal biomass—integration of analytical procedures. National Renewable Energy Laboratory; 14. https://doi.org/10.2172/1118072

Lee RE (2008) Phycology, vol 91, 4th edn. Cambridge University Press, Cambridge. ISBN 9780521864084

Lu D, Zhang M, Wang S, Cai J, Zhou X, Zhu C (2010) Nutritional characterization and changes in quality of Salicornia bigelovii Torr. during storage. LWT Food Sci Technol 43:519–524. https://doi.org/10.1016/j.lwt.2009.09.021

Montingelli ME, Tedesco S, Olabi AG (2015) Biogas production from algal biomass: a review. Renew Sust Energ Rev 43:961–972. https://doi.org/10.1016/j.rser.2014.11.052

Pérez MJ, Falqué E, Domínguez H (2016) Antimicrobial action of compounds from marine seaweed. Mar Drugs 14:1–38. https://doi.org/10.3390/md14030052

Reich B (2011) Bioenergy potential of Ulva lactuca: biomass yield, methane production and combustion. Bioresour Technol 102:2595–2604. https://doi.org/10.1016/j.biortech.2010.10.010

Rosenberg JN, Oyler GA, Wilkinson L, Betenbaugh MJ (2008) A green light for engineered algae: redirecting metabolism to fuel a biotechnology revolution. Curr Opin Biotechnol 19:430–436. https://doi.org/10.1016/j.copbio.2008.07.008

Selig MJ, Weiss N, Ji Y (2008) Enzymatic saccharification of lignocellulosic biomass. Renew Energy. NREL/TP-51

Sluiter A, Hames B, Hyman D, Payne C, Ruiz R, Scarlata C, Sluiter J, Templeton D, Nrel JW (2008a) Determination of total solids in biomass and total dissolved solids in liquid process samples. National Renewable Energy Laboratory, Golden. NREL/TP-510-42621

Sluiter A, Hames B, Ruiz R, Scarlata C, Sluiter J, Templeton D (2008b) Determination of ash in biomass: Laboratory Analytical Procedure (LAP). National Renewable Energy Laboratory, Golden. NREL/TP-510-42622

Sluiter A, Hames B, Ruiz RO, Scarlata C, Sluiter J, Templeton D (2011) Energy D. of determination of structural carbohydrates and lignin in biomass. Biomass Anal Technol Team Lab Anal Proced. NREL/TP-51

van der Wal H, Sperber BLHM, Houweling-Tan B, Bakker RRC, Brandenburg W, López-Contreras AM (2013) Production of acetone, butanol, and ethanol from biomass of the green seaweed Ulva lactuca. Bioresour Technol 128:431–437. https://doi.org/10.1016/j.biortech.2012.10.094

Waite TD, Spielman LA, Mitchell R (1972) Growth rate determination of the macrophyte Ulva in continuous culture. Environ Sci Technol 6:1096–1100

Index

© Springer Nature Switzerland AG 2019

J.-R. Bastidas-Oyanedel, J. E. Schmidt (eds.), *Biorefinery*,
https://doi.org/10.1007/978-3-030-10961-5

M

T

U

V

The manufacturer's authorised representative in the EU is Springer Nature Customer Service Centre GmbH, Europaplatz 3, 69115 Heidelberg, Germany. If you have any concerns regarding our products, please contact ProductSafety@springernature.com

Printed and bound by CPI Group (UK) Ltd, Croydon, CR0 4YY
22/09/2026
02221052-0006